American Institute of Chemical Engineers

38th Loss Prevention Symposium 2004

AA003009

Held at the 2004 AIChE Spring National Meeting

April 25-29, 2004
New Orleans, Louisiana, USA

Volume 1 of 2

Printed from e-media with permission by:

Curran Associates, Inc.
57 Morehouse Lane
Red Hook, NY 12571
www.proceedings.com

ISBN: 978-1-60423-457-2

Some format issues inherent in the e-media version may also appear in this print version.

American Institute of Chemical Engineers

38th Loss Prevention Symposium
2004

TABLE OF CONTENTS

Volume 1

A Methodology for Managing Explosion Risks in Refineries and Petrochemical Plants

Authored and presented by G.A.Chamberlain, Shell Global Solutions UK, Cheshire Innovation Park, P.O.Box 1, Chester, CH1 3SH, U.K (Shell Global Solutions is a network of independent technology companies in the Royal Dutch/Shell Group).

Prepared for presentation at the AIChE 38th Loss Prevention Symposium, New Orleans, April 25-29, 2004, Advances in Consequence Modeling I, Unpublished.

ABSTRACT

The accidental release and ignition of flammable vapours or runaway reactions in refineries and petrochemical plants present potential explosion hazards that can impact the safety of plant personnel and plant assets. This paper discusses a methodology to determine the explosion risk to people in occupied buildings on the site. From knowledge of the process layout, the analysis calculates the frequency and distribution of potential explosion overpressures and impulses over the area of the plant. The methodology, known as "explosion exceedance", draws on latest developments in vapour cloud dispersion and explosion science.

The results are expressed as the risk of fatality for the individual who is most exposed to building damage following an explosion in the plant. Also the group risk is calculated in terms of Potential Loss of Life (PLL/yr) and F(N) for the site employees who may also be exposed to the building explosion risk. The calculations are designed to satisfy the Guidance given in API RP752 [1995] and CCPS [1996] on explosion risks.

1. INTRODUCTION

The Health, Safety and Environment Consultancy Group in Shell Global Solutions has developed a methodology to assess the risks posed by vapour cloud explosions in process plant. The methodology has been implemented in a suite of risk tools known as SHEPERD* and is based on a full probabilistic approach to assess explosion hazards [Puttock 2000]. Considerable effort has gone into validation and comparison with historical experience. The methodology is being used for many Shell sites to ensure consistency of approach.

Originally, a full explosion exceedance study was carried out for the processing units at a UK refinery. This involved a complete count of equipment, flanges, fittings and associated pipe-work with the potential to leak flammable fluids. The methodology then takes into account the developments in vapour cloud dispersion and explosion science to calculate the consequences of all foreseeable explosions at occupied buildings on the site. To repeat this exercise for other sites would not be cost effective so a generic version has been developed and implemented into SHEPERD and is reported here. In the generic exceedance methodology, the input and processing of data is considerably reduced over the full exceedance and significant effort has been applied to minimise loss of accuracy.

2. THE RISK ASSESSMENT METHODOLOGY

The full methodology for assessing the tolerability of site personnel in occupied buildings to the risk of process plant explosions is shown in the flow chart given in Figure 2.1. The left and right hand parts of the chart refer to data collection activities whereas the middle column is the core methodology. A hierarchy of analytical tools, ranging from simple screening tools to full state-of-the-art computation, could be applied at each box in the methodology depending on the approach which is best suited to the purpose intended. Using the SHEPHERD exceedance methodology is a simpler, but more pragmatic, version of this full exceedance methodology and is described in Section 3.

The flow chart is designed to start with a single scenario, a flammable release, and to work through its consequences and frequency. Then, on returning to the start, the methodology works the next scenario, finally summing all foreseeable contributions to the risk from explosions to personnel in occupied buildings.

Each box is numbered and explained as follows:

1. The gas cloud size must be calculated, only counting that part of the gas cloud which is both flammable and engulfs a congested area. The gas can be released in several possible directions and in several wind conditions, so boxes 1c and 1d must be linked to each release. The release rate is determined by the fluid properties and hole size. A release frequency must also be assigned using, for example, EP Forum [1992], but then split by the probabilities of release directions and wind conditions.
2. The volume of flammable gas in the congested area follows from box 1, and is required for the explosion source term(DICE (Dispersion in Congested Environments [Chynoweth 2001] was used).
3. An ignition probability must be assigned(The data recommended by Cox et al. [1990]was used).
4. An explosion calculation is performed to determine the source size, pulse duration and overpressure as a function of cloud stoichiometry (assumed constant within each cloud and systematically varied across the flammable range), and the blast decay as the pressure pulse moves away from the source into the surroundings (The Congestion Assessment Method, CAM, [Puttock 1995, 1999, 2001] for vapour cloud explosions (using data from box 4b) and a modified BLEVE (Boiling Liquid Expanding Vapour Explosion) model, [Shield 1993, Baker et al. 1983, Lees 1996] for vessel runaway explosions (using data from Box 4a) was used).
5. The blast loading on buildings depends on the angle of incidence, the incident blast pressure and impulse, reflections from nearby surfaces and shielding effects of the blast wave.
6. The structural response of the building is coupled to the blast loading calculated in Box 5. The dynamic behaviour, ductility of components, strain rate effects, the strength of edge supports, and membrane action, all have a bearing on the structural response and could be analysed explicitly if desired.

7. The result from Boxes 5 and 6 gives the overall building damage. The use of SHEPHERD avoids the detailed calculations of building damage required in Boxes 5 and 6 by using the generic data contained in the 1995 Technology Co-Operative report prepared by Barker et al. [1996].
8. The vulnerability of occupants to building damage has been derived from studies by Oswald et al. [2000], and by Jeffries et al. [1997]. There are 3 vulnerability models programmed into SHEPHERD. The model published in API RP752 [1995], a pressure-only method based on 100ms pulse durations and a Pressure-Impulse (PI) method. The last two methods are based on the generic building types by Barker et al. [1996].
9. The fatality rate is derived from the time each individual spends in that building and the total amount of time that all occupants spend in the building.
10. A return to the start is necessary to complete the full spectrum of foreseeable explosion scenarios.
11. All fatality rates are summed and can be expressed in terms of risk markers such as Individual Risk, risk contours, Potential Loss of Life per annum, or F(N) (group risk) plots.
12. A decision has to be made regarding the tolerability to the explosion risk to occupied buildings. Then the calculated risk values can be compared to the tolerability criteria.
13. At low frequencies, the risk can be regarded as tolerable.
14. At higher frequencies, protection, control and mitigation measures must be assessed to reduce the risks to tolerable or to as low as reasonably practicable (ALARP).

3. SHEPHERD – overview

SHEPHERD forms a family of graphical risk integrators, containing *inter alia* the explosion exceedance methodology. The tool has been developed to carry out fit-for-purpose Quantified Risk Assessment (QRA) for a broad range of onshore industrial sites such as refineries, gas plants, chemicals plants, LPG distribution sites, pipeline systems, etc.

The SHEPHERD exceedance methodology is summarized in the flow chart in Figure 3.1. It is a simplified version of the full methodology described earlier and is designed to make use of generic exceedance curves. The gas dispersion scenarios and their associated explosion overpressures and impulses have been collapsed onto two generic exceedance curves, one for flammable releases originating within the process unit comprising the congestion area and one for releases originating in adjacent areas. The derivation of these curves is described subsequently. Note that SHEPHERD exceedance ends with a prediction of the overall explosion risk per building or site. Application of risk tolerance and effectiveness of safety measures are decisions for the site.

SHEPHERD is not intended to replace the full risk methodology, but it is a considerably more efficient way of carrying out the calculations. It is recommended that the full methodology should be carried out when making critical decisions, such as ones that involve severe consequences with borderline risks.

Figure 2.1 Full Methodology for explosion hazard management.

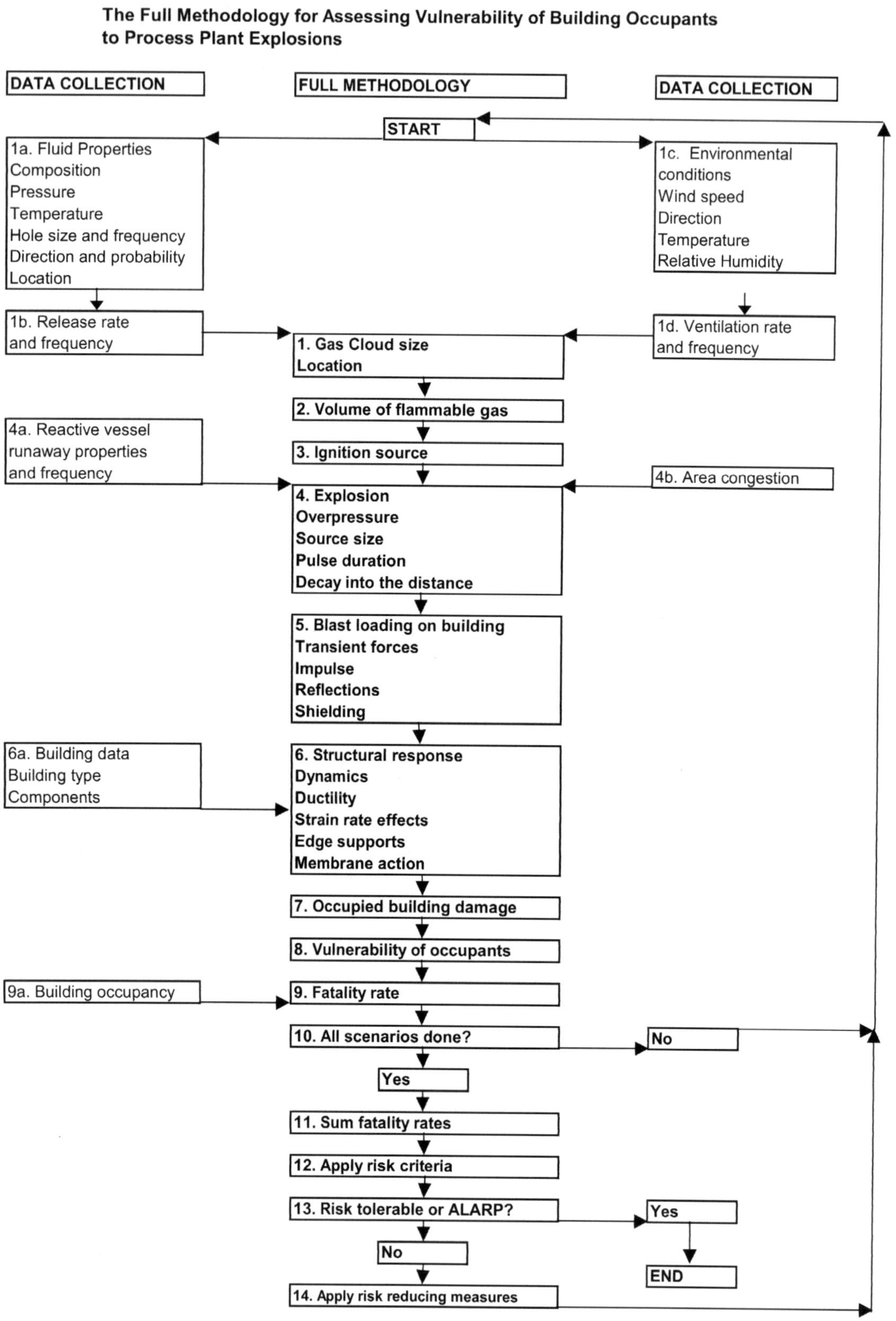

Figure 3.1 The simplified explosion exceedance methodology implemented in SHEPHERD

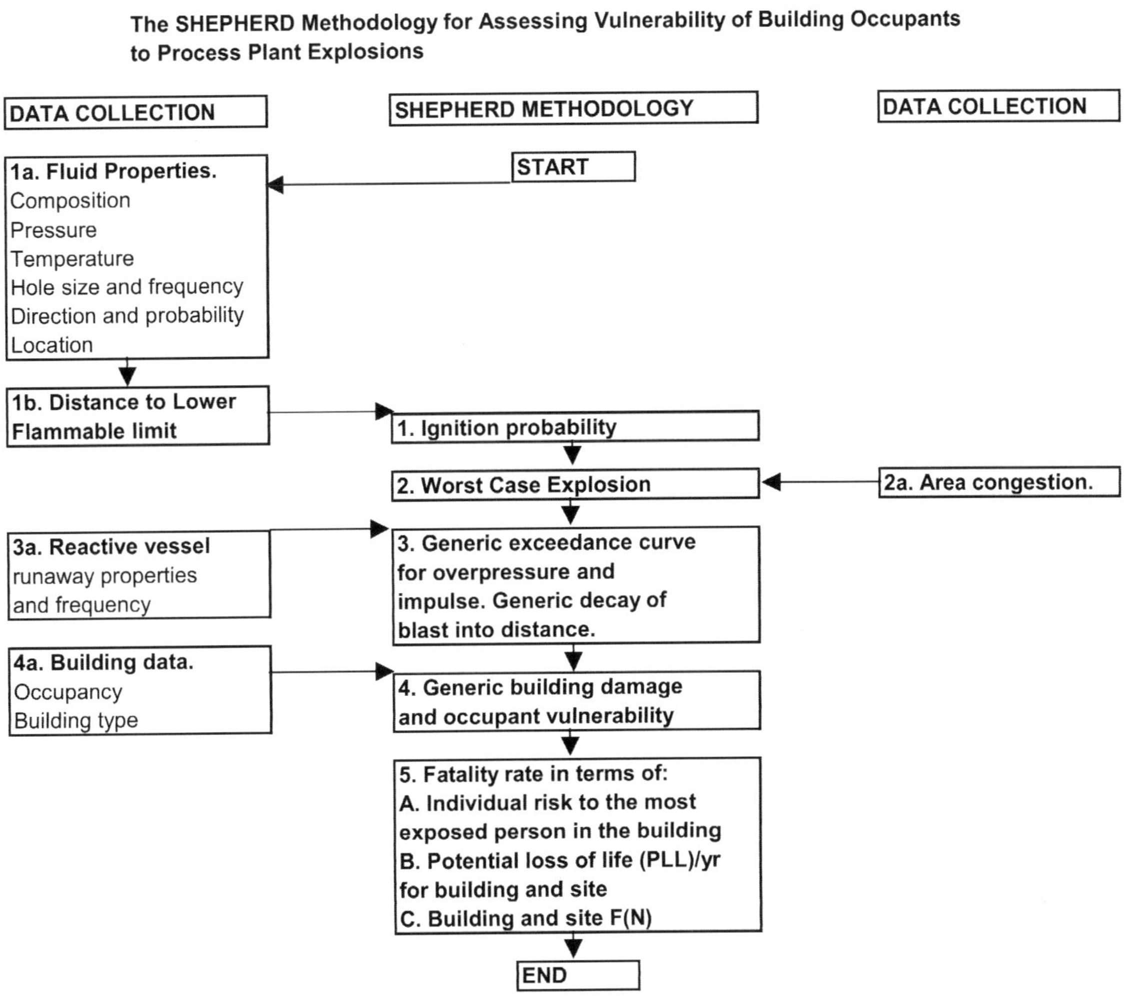

4. DERIVATION OF GENERIC EXCEEDANCE CURVES

For the purposes of the simplified exceedance in SHEPHERD generic explosion overpressure exceedance curves need to be fitted for the internal overpressure in a congested region for a specific fuel. For all the vapour release scenarios (>100,000) in the full exceedance, the following procedure was adopted to prepare the data for subsequent analysis:

1. all releases were sorted into "internal" or "external" according to whether they were within 5m of the boundary of a congestion area;
2. for all internal hits for a given congestion area the gas cloud sizes were calculated using DICE [Chynoweth 2001] taking into account the flammable quantity and direction of release, coupled with the wind flow through the area for each wind direction and speed derived from the wind rose for the site;
3. from the stream composition, the releases were grouped into common "equivalent" fuel types. Seven equivalent fuels were considered representative of the full range of explosive compositions. These are the fuels for which an explosion calculation can be carried out in CAM. For single component fuels the following equivalence rule set was used.

Equivalent fuel	Actual stream fuel
Methane	Methane, natural gas, ethyl chloride, hydrogen sulphide
Ethane	Ethane, toluene, vinyl chloride monomer
Propane	Paraffins C3+, cyclo-paraffins, styrene, ketones, ethanol, aromatics (excluding toluene), PEB, MEG, alcohols
Propylene	Propylene
Butadiene	Butadiene
Ethylene	Ethylene
Hydrogen	Hydrogen, EO, PO, acetylene

Table 4.1 Relationship between actual stream fuel and the CAM equivalent fuel.

For mixtures of fuels, the equivalent laminar burning velocity and expansion ratio were derived and the equivalent fuel nearest to these properties was used;

4. for each gas cloud the overlap with the congestion (% fill) was calculated;
5. the explosion model CAM was run for each % fill and 100% fill taking into account the equivalent fuel type. The ratio of partial fill to full fill overpressure was recorded;
6. each overpressure and impulse result was linked to the frequency of each leak [EP Forum 1992] and probabilities associated with release direction, wind speed and direction, and ignition probability derived from Cox et al. [1990];
7. the procedure was repeated for "external" releases;and
8. no variation was allowed for model uncertainty.

It can be expected that the shape of the exceedance curves to be different when considering releases occurring within the congested area as contrasted to releases occurring elsewhere, with the gas clouds drifting into the area. The latter would have a higher probability of generating clouds that fill a large proportion of the congested volume, as they have some distance to disperse and spread. Plumes from the internal releases would typically still be quite narrow as they leave the area. For this reason, two generic exceedance curves, one for internal releases, and one for external releases should be fitted.

For each congested area, exceedance curves for external releases and internal were obtained, plotting *probability* of exceedance against *relative* overpressure. In this form, with both variables normalized to a maximum value of one, some collapse of the data could be expected.

It was expected that the size of the congested regions could be shown to influence the position of the curves. For example, if the distribution of release sizes and thus gas-cloud sizes is fairly similar for different areas, then a large area will be filled with gas less often than a small area. This would lead to less frequent overpressures near the full-fill overpressure, tending to lower the curve. Similarly, it is difficult to fully fill a long narrow area with gas, because this can only occur when the release direction and wind are closely aligned with the long axis.

To improve the collapse of the data, transformations were applied to the curves depending on the congested area volume, length/width ratio and height/width ratio. It was important that the transformations should not change the asymptotic behaviour of the curves at zero and one, so the form used was:

$$(1-p_s')=(1-p')^n$$

where n is a function of the congestion area volume and aspect ratios, p' is the relative overpressure $P/P_{full-fill}$. This has the effect of shifting a curve left and down (except at the ends) if $n<1$, or right and up if $n>1$. The effect of this transformation was to reduce the variation at, say, $p'=0.3$ by a factor of three (Figure 4.1).

In fitting a generic curve to these results, the aim was to obtain a good representation of the data, with a tendency to conservatism, i.e. a curve towards the upper end of the main group of curves. The result is also shown in Figure 4.1 as a heavy curve.

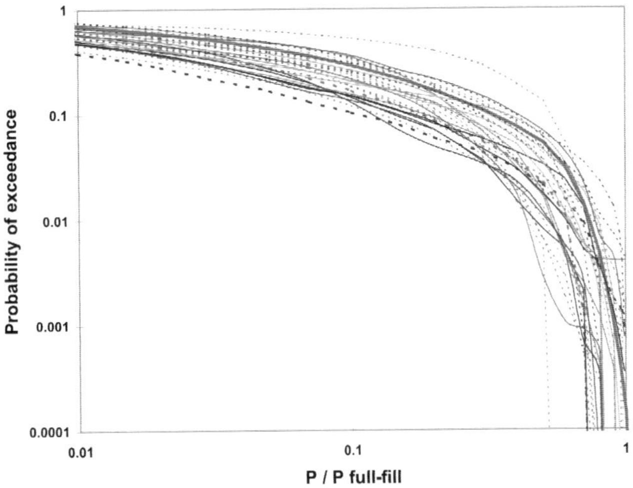

Figure 4.1 Transformed exceedance plot of UK refinery data for external releases. The heavy curve is the generic curve.

The fit for releases inside the congested area followed the same process.

The results give expressions for the probability of exceeding various levels of source overpressure in a plant. If this is to be used to determine the effect on distant receptors, e.g. buildings, further assumptions have to be made about the pressure decay away from the source. The simplest way to do this is to assign an "effective radius" R_0 to each pressure level (or probability level) on the exceedance curve. Then the standard CAM pressure decay relation can be used to give pressure at any distance for that probability level. The issue is what to take for the effective radius.

There are two reasons why the calculated overpressure in a congested region can be lower than the maximum overpressure. One is that the gas cloud might be small; the other is that the gas cloud might not be at stoichiometric concentration. If all the gas clouds are stoichiometric, but of varying sizes, then we can relate the overpressure to the cloud size, and hence effective source size, by running CAM for a number of different gas cloud sizes. At the other extreme, if all the gas clouds are larger than the congested volume, but of varying stoichiometry, then the effective source size is always determined by the congested volume (full-fill effective radius R_{max}), and R_0/R_{max} is always 1.

The reality is that the low overpressures are caused by a combination of smaller cloud sizes and variations in stoichiometry. Thus it can be assumed that there is an effective source radius associated with every point on the source overpressure exceedance curve. Exceedance curves were produced both for the source overpressure and for a number of receptors at distances from 10 m to 400 m from the congested area. The fit was performed by taking successive points on the source curves and the points with the same probability on the receptor curves. Each time, a value for source overpressure and a series of overpressures at various distances was obtained. If a source radius is chosen, the overpressures at the receptors predicted by the CAM correlation can be calculated. These can be compared with the series of values obtained from the exceedance runs. This was done, and the radius was varied to determine a best fit.

For the full range of overpressures, the resulting best fits are shown in Figure 4.2. The plotted values are normalised by the full-fill source radius, and the full-fill stoichiometric overpressure. The exercise was performed for three different congested areas. The fitted line is also shown in Figure 4.2(a).

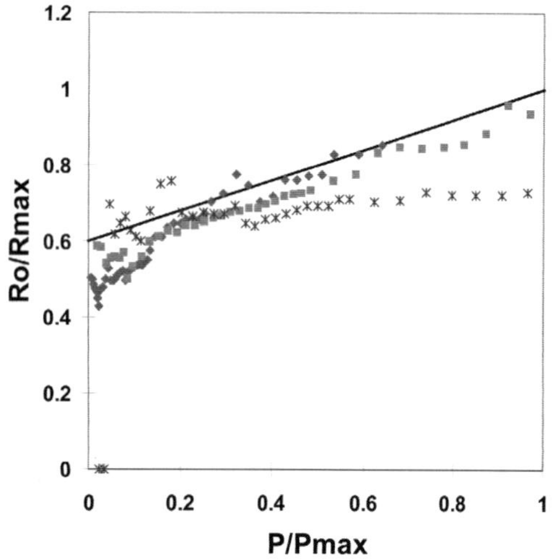

(a)

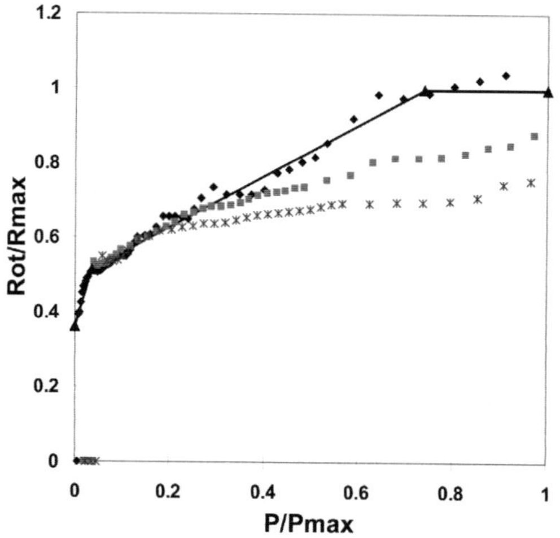

(b)

Figure 4.2 Best fit of the effective source size (effective radius R_o and R_{ot}) for use in explosion (a) overpressure exceedance and (b) impulse exceedance, derived from the UK refinery data.

To determine the effects of an explosion on a structure, it is often necessary to know the impulse received, as well as the overpressure. The Congestion Assessment Method includes correlations for pulse duration as a function of distance from the source, so for any source overpressure and effective radius, a pulse duration at any receptor can be calculated. The pulse shape is assumed to be triangular, so the impulse is half the product of the peak overpressure and the duration. However for calculating the duration at the receptors, we typically found that a different effective radius R_{0t} gave a better fit (Figure 4.2(b)).

To obtain the impulse at a receptor for any given probability level on the exceedance curve, the overpressure at the receptor should be calculated using R_0, the duration using R_{0t}, and the results combined to give the corresponding impulse.

5. COMPARISON WITH API RP752 GENERIC FREQUENCIES OF MAJOR EXPLOSIONS

API RP752 Table C.1 gives a list of generic frequencies of major explosions for different refinery units derived from comprehensive historical databases, and is reproduced in Table 5.1. Overpressure and impulse exceedance curves for the API refinery units were calculated from a number of different refineries using the methodology described in this paper. If an overpressure of 350 mbar is the threshold for major damage and escalation is assumed, the exceedance curves to derive the frequency can be used at which this overpressure is exceeded and then compare directly with the API generic values. The results of this analysis are shown in Table 5.1.

Process Unit	API RP752 frequency of major explosion/yr	Average SHEPHERD result. Frequency of overpressure greater than 350 mbar	Number of refineries in sample	Average pulse duration, ms
Alkylation	5.1×10^{-4}	4.5×10^{-4}	4	60
Cat cracking	6.5×10^{-4}	5.7×10^{-4}	3	73
Cat reforming	2.6×10^{-4}	8.1×10^{-4}	4	55
Crude	4.9×10^{-4}	5.3×10^{-4}	4	69
Hydrotreating	2.0×10^{-4}	7.0×10^{-4}	3	45
Hydrocracking	5.6×10^{-4}	11.0×10^{-4}	2	59
Average of above units	4.5×10^{-4}	6.9×10^{-4}		60

Table 5.1 Comparison of API RP752 generic frequencies of major explosions with the results using the present methodology.

The predicted overall average frequency is about 1.5 times higher with individual comparisons being mostly less than a factor of two. Note that the predicted average pulse durations all fall within the range 45-73 ms, overall average 60ms.

This result is remarkable given the uncertainties in the database, the assumption about "major explosion" overpressures, and in the methodology itself. Although this comparison only tests the explosion source prediction, the result is encouraging and is well within the uncertainty bounds normally associated with risk analysis.

* **THE SHELL GLOBAL SOLUTIONS' RISK TOOL KNOWN AS SHEPERD IS NOT OFFERED AND IS NOT AVAILABLE TO THE PUBLIC IN THE UNITED STATES OF AMERICA.**

REFERENCES

API Recommended Practice 752, "Management of Hazards Associated with Locations of Process Plant Buildings." May 1995.

Baker, W.E., Cox, P.A., Westine, P.S., Kulesz, J.J. and Strehlow, R.A., "Explosion Hazards and Evaluation", Elsevier, 1983.

Barker, D.D., Lowak, M.J., Oswald, C.J., Peterson, J.P., Stahl, M.W. and Waclawczyk, J.H., "Final Report: Conventional Building Blast Performance Capabilities", 1995 Technology Cooperative, Wilfred Baker Engineering Inc., December 1996 (Confidential).

Center for Chemical Process Safety (CCPS) of the American Institute of Chemical Engineers, "Guidelines for Evaluating Process Plant Buildings for External Explosions and Fires." 1996.

Chynoweth, S. "Dispersion in Congested Environments." Major Hazards Offshore 2001, 27-28 Nov, The Paragon Hotel, London.

Cox, A.W., Lees, F.P., and Ang, M.L., "Classification of hazardous locations." I.Chem.E. 1990.

EP Forum Hydrocarbon Leak and Ignition Database, EP Forum Report No. 11.4/180, E&P Forum, London, 1992.

Jeffries, R. M., Gould, L., Anastasiou, D. and Pottrill, R., "Derivation of fatality probability functions for occupants of buildings subject to blast loads", Phase 4. WS Atkins Science and Technology Contract Research Report 147/1997.

Lees, F.P., "Loss Prevention in the Process Industries", Reed Educational and Professional Publishing, 1996, Chapter 17.

Oswald, C. J. and Baker, Q. A., "Vulnerability Model for Occupants of Blast Damaged Buildings", 34th Annual Loss Prevention Symposium, March 6-8, 2000.

Puttock, J. S. "The use of a combination of explosion models in explicit overpressure exceedance calculations for an offshore platform." Proc. Int. Conf. "Major Hazards Offshore - practical implications", London, November 2000.

Puttock, J.S. "Fuel gas explosion guidelines - The Congestion Assessment Method", 2nd European Conf. on Major Hazards Onshore and Offshore, I.Chem.E., Manchester, 1995.

Puttock, J.S., "Improvements in guidelines for prediction of vapour cloud explosions", Intl. Conference and Workshop on Modelling and Consequences of Accidental Releases of Hazardous Materials." San Francisco, Sep-Oct 1999.

Puttock, J.S., "Developments in the congestion assessment method for prediction of vapour cloud explosions", 10th International Symposium, Loss Prevention and Safety Promotion in the Process Industries, Stockholm, 19-21 June 2001, p1107-1133.

SHEPHERD User guide, Shell Global Solutions UK, OGCH/3, Cheshire Innovation Park, 2002.

Shield, S.R., "A model to predict the radiant heat transfer and blast hazards from LPG BLEVEs." AIChE Symp. Series, Vol. 89, 1993.

Consequence Analysis – Using a CFD Model for Industrial Sites

S. Dharmavaram
E.I. du Pont de Nemours & Co.
Wilmington, DE 19898

S. R. Hanna
Hanna Consultants
Kennebunkport, ME 04046

O.R. Hansen
GexCon AS
Bergen, Norway

Prepared for presentation at the 38th Loss Prevention Symposium at the 2004 Spring National Meeting of the American Institute of Chemical Engineers

April 2004

Unpublished

Consequence Analysis – Using a CFD Model for Industrial Sites

S. Dharmavaram, DuPont
S. R. Hanna, Hanna Consultants
O. R. Hansen, GexCon AS

Abstract:

Several models are currently available to model the discharge and dispersion of toxic or flammable materials to the environment. A few of the Gaussian dispersion modeling tools allow the representation of the complex environment within a manufacturing plant or urban area in determining the impact of continuous releases from a plant. For atmospheric dispersion of dense gases, a correction is made for the presence of the buildings and other complexity by using a surface roughness parameter, which is only a crude approximation. A need exists to obtain realistic estimates of plume dispersion in a complex environment, particularly accounting for buildings/obstructions at a plant and the associated turbulence. With the advance of computational technology, and greater availability of computing power, computational fluid dynamics (CFD) tools are becoming more available for solving a wide range of problems. A CFD model, called FLACS, developed originally for explosion modeling, has been upgraded for atmospheric dispersion modeling. CFD tools such as FLACS can now be confidently used to understand the impact of releases in a plant environment consisting of buildings, structures, pipes, and accounting for all complex fluid flow behavior in the atmosphere and predicting toxicity and fire/explosion impacts. With its porosity concept representing geometry details smaller than the grid, FLACS can represent geometry well even when using a coarse grid resolution to limit the simulation time. The performance of FLACS has recently been evaluated using a wide range of field data sets for sulfur dioxide (Prairie Grass), carbon dioxide (Kit Fox), ethylene (EMU), etc. In this paper, details about the improvements made to FLACS, model validation exercises, and results from the modeling of releases from an industrial facility are presented.

Introduction

Many models and methods are currently available to model the dispersion of chemical releases from point, area, and line sources [1]. Gaussian dispersion modeling methods are used for obtaining estimates of long-term averages of concentration resulting from continuous releases of neutral to buoyant gases. Some of these methods allow for inclusion of complex topography. Similarly several models and methods are available for accidental releases of chemicals, including dense gases and two-phase releases, which could potentially result in injuries or fatalities [2]. The complexity of an area in the neighborhood of a release is represented by a surface roughness parameter, which in most cases is only a crude approximation [3]. However, all of these methods are not adequate in properly representing the wind flow-fields in a complex plant environment that includes several buildings, obstructions, and complex topography. In most cases the predictions made by such models overestimate the impacts in the far-field and tend to underestimate the impacts in the near-field.

On a typical industrial facility, there are many obstructions that affect the turbulence and therefore the dispersion behavior of releases. If the facility is located in an urban area there are likely to be a lot of tall buildings that can also influence the turbulence. Considerable research is being conducted in the field of computational fluid dynamics (CFD) and finite element/volume modeling to represent atmospheric flow patterns on the appropriate scales [4, 5]. In this paper a DuPont sponsored development and application of a commercially available CFD tool, i.e. FLACS, for modeling atmospheric dispersion is discussed.

Computational Fluid Dynamics (CFD)

Computational fluid dynamics (CFD) is increasingly being recognized as a powerful and useful tool for predictive analysis of flow, mass, momentum, and energy. Such a predictive analysis methodology can provide results in a faster, cheaper and safer way than experiments. With the availability of increasingly powerful and faster personal computers, the ability to solve complex problems using CFD is getting easier and cheaper to a wide range of users. The application of CFD to atmospheric release & dispersion problems is relatively new. However, because the atmosphere is relatively turbulent, with typical turbulence intensities of 0.1 to 0.5, it is important for a CFD model to be able to properly simulate these large turbulence intensities and then maintain them as the flow moves downwind.

CFD has been applied to a wide range of problems from material design, reaction engineering, solid mechanics, wind models, ocean and weather analysis, and liquid flows in channels and pipelines. Most of these applications deal with problems that have a simple geometry, like airplanes or simple mechanical elements, while others may deal with complex geometry but for a small-scale problem, such as engines, turbo-machinery, and fluidized bed reactors. For these problems grid generation, which is an important aspect of CFD analysis, is easier than when compared to the scale and geometry in a large industrial facility.

In an industrial facility, there are a lot of open and closed structures, piping, buildings, etc. that can in some cases be very confined and congested. Such complex geometry can lead to generation of turbulence that the CFD models need to represent adequately. Also the scale of the problems dealing with atmospheric dispersion of chemicals is much larger than the examples mentioned above, and the proper representation of the atmospheric winds and the boundary layer becomes critical.

There are several CFD models in the process of being developed for atmospheric dispersion simulations. Leading among them is the FEM-MP model that runs on a massively parallel machine at the U.S. Department of Energy's Lawrence Livermore National Laboratory. FEM-MP is a simulation model based on finite element techniques, and is able to take into consideration complex geometry and non-flat terrain effects. It has been used recently for major urban studies to understand the impact of chemical releases in such environments [6, 7]. The general purpose FLUENT CFD model [8] is also being used by several research groups to study the dispersion of chemicals in urban environments such as New York City. Other commercial software that are being applied and tested include STAR-CD [9], CFX [10], and EXSIM [11].

FLACS [12] is a CFD tool developed since the early 1980's by the Christian Michelsen Research Institute (with the FLACS development group now incorporated into GexCon AS), in Norway, primarily for simulating explosions in offshore oil platforms. However, it is capable of modeling ventilation and gas dispersion in complex geometries. After a detailed literature review [13] and discussions with several vendors, DuPont selected FLACS as the CFD tool for solving atmospheric dispersion problems. It is commercially available, from GexCon AS, and has recently been improved for such an application based on DuPont sponsorship. It has been found to be very efficient (short simulation times) and reasonably accurate in predicting concentrations in large domains. Detailed validation studies have also been conducted recently that demonstrates that its performance is reasonable and acceptable.

FLACS

In FLACS, a basic Cartesian numerical grid is used to solve the Navier Stokes equations. The governing equations for mass, momentum, enthalpy, turbulent kinetic energy, and dissipation of turbulence are all solved. The turbulence field effect is described using the k-ε model as well as other terms to analyze the generation of turbulence around the structures and walls present. Arntzen [14] has discussed the mathematical formulations and algorithms used in FLACS for modeling turbulence in extensive detail. To adequately represent geometry in the sub-grid scale, a distributed porosity concept is applied. This allows utilization of coarse grids where necessary to limit the simulation time.

The new developments in FLACS are related to the ability to input the atmospheric flow field parameters as boundary and initial conditions. Either a uniform or a logarithmic velocity profile can be specified, and the winds can be made to fluctuate. The parameters used to characterize turbulence in FLACS are Relative Turbulence Intensity and the Turbulence Length Scale. The Monin-Obukhov similarity theory is used to characterize turbulence in the atmosphere.

According to the Monin- Obukhov theory the stability of the atmospheric layer above the earth's surface can be determined by the ratio of turbulence generated by the temperature gradient and the turbulence generated mechanically by wind shear at the surface. This ratio can be expressed by a characteristic length-scale, the Monin-Obukhov length,

$$L = - \rho_a c_p T_a u^{*3} / \kappa g H_o \qquad (1)$$

The friction velocity u^* is by definition the square root of the shear stress divided by the density of air at the surface. The friction velocity is determined from the mean velocity at a specific reference height. H_o is the sensible heat flux. In the formula, ρ_a , c_p and T_a are the density, specific heat, and near surface absolute temperature of the air, respectively; κ is the von Karman constant (κ= 4.0) and g the gravitational acceleration. The Monin-Obukhov length provides a measure of stability of the surface layer. L can be interpreted, as the height above the ground where turbulence generated by wind shear equals the turbulence dissipated by the heat flux. In unstable conditions there is no such equilibrium,

$L>0 \longrightarrow$ stable

$L<0 \longrightarrow$ unstable

$L=0 \longrightarrow$ neutral

Qualitative schemes to characterize stability are also often used, e.g. the Pasquill scheme, which ranges from class A (unstable) through D (neutral) to F (stable).

L can be directly measured, but it is also possible to calculate L from Pasquill-stability categories (Section 4.3.2 of the TNO Yellow Book [15]). Thus, L is calculated as follows:

$$1/L = (1/L_s) \log_{10} (z_o/z_s) \qquad (2)$$

For 0.001 $m < z_o < 0.5\ m$, L and z are constants, depending on the Pasquill stability class according to Table 1 below. For $z_o > 0.5\ m$, the L calculated for $z_o = 0.5$ m should be used. For Pasquill stability class D, the formula above leads to 1/L = 0.

Pasquill Stability	**L_s [m]**	**Z_s [m]**
A	33.162	1117
B	32.258	11.46
C	51.787	1.324
E	-48.330	1.262
F	-32.325	19.36

Table 1: Calculation of the Monin-Obukhov Length from Pasquill stability categories.

Based on the stability of the atmosphere as determined by the Monin-Obukhov length, the turbulent kinetic energy and the turbulent energy dissipation rate can be determined. In FLACS, several equations relating atmospheric stability to turbulence parameters have been programmed based on the algorithms by Han *et. al.* [16]. With such upgrades to represent atmospheric conditions the performance of FLACS has been improved for dispersion calculations. The logarithmic wind profile and the Pasquill atmospheric stability class can be specified as an initial condition for any simulation.

In addition to representing the atmospheric flow-field, enhancements have also been made to input topography into FLACS. Standard United States Geological Survey digital map files can be incorporated into the program along with geometry information on buildings and structures in any given location. Digital drawing files used to represent buildings and structures can also be imported into FLACS.

Validation Studies

FLACS has been validated for gas explosion calculations in several studies conducted over the years [17, 18, 19]. Included in such studies was an evaluation of the concentration predictions based on the dispersion calculations. However, these studies focused on oil platforms and confined/congested geometries.

In this study, the focus was on evaluating the capabilities of FLACS in predicting atmospheric dispersion in the near and far field. To be sure that FLACS predictions were believable, it was used to simulate releases for evaluation using field measurements in flat terrain i.e. Prairie Grass, and also in terrain with obstacles, i.e. Kit Fox, EMU and MUST. Detailed description of these datasets is provided in other publications [20, 21, 22 & 23]. The EMU L-shaped building was a wind tunnel experiment. The Kit Fox field experiment at the Nevada Test Site consisted of billboard-shaped obstacles (75 obstacles 2.4 m square, and 5500 obstacles 0.2 m by 0.8 m), and the MUST field experiment at the U.S. Army Dugway Proving Ground had 120 obstacles with the size of a shipping container. Detailed validation of FLACS has been completed for the MUST field experiment, which involved tracer releases within an array of the 120 boxes of dimension 2.5 m by 2.5 m by 12 m, but the Department of Defense sponsors of the field experiment have not yet approved publication of the detailed results. Nevertheless, we can give the general conclusion that the FLACS predictions of concentrations were usually within a factor of two of the observations for MUST. Validation results are presented here for only the Prairie Grass and one of the EMU datasets.

In the Prairie Grass field experiment, sulfur dioxide was released at various rates, ranging from 0.03 to 0.01 kg/s, in the middle of a field in Nebraska, U.S.A. No obstacles were present. The wind speeds ranged from 2 to 10 m/s and atmospheric stabilities covered the entire spectrum from A (unstable) to F (stable). Forty-two separate simulations were conducted with FLACS representing each experiment for which the data was available (Figure 1). The simulation times ranged from 9 to 39 hours. The performance of FLACS was found to be very reasonable as shown in Figure 2. Predictions are within a factor of two of the measured concentrations for about 70% of the data points, with little mean bias, and with little trend over distance.

The EMU dataset consisted of measurements made in a wind tunnel for an L-Shaped building. A gas mixture (MW = 28 g/gmol) was continuously released from a door and the wind speed at the boundary was 5 m/s. Figure 3 shows the wind flow patterns around the L-shaped building, with a logarithmic wind profile imposed at the boundary. Figure 4 shows the dispersion of the plume predicted by FLACS. Figure 5 shows the results predicted versus the measured dimensionless concentrations for receptors located at different y and z locations at the x = 10 m distance. Again most of the predictions are within a factor of two. All the predicted concentrations are within a factor of four.

Details of the validation for the other datasets will be provided in other publications at a later date.

Application

FLACS is being utilized within DuPont where realistic prediction of dispersion within the facility and neighboring communities is important. A critical application is to predict the behavior of plumes near buildings & structures to understand the impact of building downwash.

One example where FLACS has been utilized is to understand the flow patterns around a large open topped tank (diameter = 31 m; height = 8.5 m) that under some circumstances may release a toxic chemical. Standard two-dimensional dispersion models are not capable of predicting the downwash characteristics and tend to underestimate the impacts. Figures 6 and 7 show the results obtained from FLACS for two different wind conditions: F stability, 1.5 m/s wind speed (F1.5); and D stability, 5.0 m/s wind speed (D5.0). A chemical that has a molecular weight of 46 g/gmol is released at a rate of 3 g/s. The concentrations predicted are shown on a logarithmic scale where 10 ppm corresponds to –5. The distance to 10 ppm under F1.5, predicted by FLACS, is similar to what might be predicted by other models. However, under D5.0 conditions, FLACS accounts for the downwash in the lee of the large tank and predicts a significant ground-level impact distance. Other models, where downwash is not considered, might suggest that there is no ground-level impact under D5.0. As a result of this analysis, a more realistic estimate of impacts has been obtained leading to actions that minimize risks from such exposure.

Another example of FLACS use was to determine the height of a vent stack that had the potential to emit 0.4 kg/s of chemical with a molecular weight of 58 g/gmol. The concentration of concern was 100 ppm (-4 on the logarithmic scale). Neutral stability (D) and a 5 m/s wind speed were used in the simulations. Figures 8 and 9 show the FLACS predictions that were used to determine the right vent height for a dense gas release, so that it is outside the building downwash cavity.

Summary

CFD modeling is becoming practical for estimating flow around obstacles and dispersion of releases from industrial facilities. DuPont selected the FLACS CFD model, sponsored its improvement for atmospheric dispersion modeling, and sponsored its validation using a variety of field and experimental data. FLACS was originally developed to estimate the effects of explosions from leaks on oil platforms, and has been extended to standard atmospheric flow and dispersion problems, such as release of a toxic chemical from a vent on a building in the middle of a chemical processing plant. FLACS uses a distributed porosity concept, at the sub-grid scale, to allow it to run efficiently.

This paper has summarized the characteristics of FLACS, and has focused on demonstration of the model's performance using a series of field and laboratory databases. These evaluations of atmospheric dispersion are more extensive than any completed previously for other CFD models. In order to demonstrate the model's ability to maintain turbulence in an atmospheric boundary layer that does not contain building obstacles, the evaluation exercise included the flat-field Prairie Grass database. Because of the interest in chemical plants, where there are extensive buildings, tanks and other obstacles, the evaluation exercise has also considered three databases where there are one or more obstacles present. These obstacle databases included the EMU L-shaped building (a wind tunnel experiment), the Kit Fox field experiment at the Nevada Test Site with many billboard-shaped obstacles, and the MUST field experiment at the Dugway Proving Ground with 120 obstacles the size of a shipping container. Some figures have been presented that show the experimental set-ups and the results for the Prairie Grass and EMU experiments. For all experiments, FLACS was able to satisfactorily simulate the flow patterns around the obstacles, and the concentration predictions by FLACS were within a factor of two of the observations about 70 % of the time, with little mean bias and with little trend with distance. This performance is within the acceptable range for atmospheric dispersion models.

Two examples of FLACS applications to gas releases from typical scenarios at a chemical plant have been presented, and the resulting flow and dispersion patterns shown in figures. One scenario involved an open-topped tank and the other scenario involved a vent stack on a building. Both scenarios included consideration of aerodynamic downwash behind the obstacle. It was concluded that FLACS provided realistic results for these scenarios that aided in the estimation of environmental and health impacts and allowed for improved design of the vent stack and other process changes.

References:

1. Turner, D.B., *Workbook of Atmospheric Dispersion Estimates: An Introduction to Dispersion Modeling*, Second Edition, Lewis Publishers, Raleigh, NC, 1994

2. Hanna, S.R., P.J. Drivas, and J.C. Chang, *Guidelines for Use of Vapor Cloud Dispersion Models*, Second Edition, AIChE Center for Chemical Process Safety, New York, NY, 1996

3. Hanna, S.R., and R.E. Britter, *Wind Flow and Vapor Cloud Dispersion at Industrial and Urban Sites*, AIChE Center for Chemical Process Safety, New York, NY, 2002

4. *Tracking and Predicting the Atmospheric Dispersion of Hazardous Material Releases – Implications for Homeland Security*, The National Academies Press, Washington, DC, 2003

5. Seventh Annual George Mason University Conference on Transport and Dispersion Modeling, Meeting Agenda, Fairfax, VA, June 17-19, 2003

6. Chan S.T., D.E. Stevens, and R.L. Lee, "A model for flow and dispersion around buildings and its validation using laboratory measurements," Proc. 3rd Symposium on the Urban Environment, Aug. 14-18, Davis, CA, 2000.

7. Brown, M., M. Leach, R. Calhoun, W.S. Smith, D. Stevens, J. Reisner, R. Lee, N.-H. Chin, and D. DeCroix, "Multiscale modeling of air flow in Salt Lake City and the surrounding region," ASCE Structures Congress 2001, Wash. DC, LA-UR-01-509 (2001).

8. FLUENT6.1, CFD Code, Fluent Inc., Lebanon, NH, 2003

9. STAR-CD, CD Adapco Group, New York, NY, 2003

10. CFX-5, ANSYS Inc., Canonsburg, PA, 2003

11. EXSIM, Telemark Technological R&D Centre, Porsgrunn, Norway, 2003

12. FLACS, GexCon AS, Bergen, Norway, 2003

13. Gentile, M., "Review of CFD Software for Gas Dispersion Modeling in Complex Geometry and Non-Flat Terrain," DuPont Internal Report, 2002

14. Arntzen, B.J., *Modelling of turbulence and combustion for simulation of gas explosions in complex geometries*, Dr. ing. Thesis, Norwegian University of Science and Technology (NTNU), Trondheim, Norway, 1998.

15. *Yellow Book (CPR 14E)*, 3rd Edition, TNO Environment, Energy, and Process Innovation, Apeldoorn, The Netherlands, 1997

16. Han, J., J.P. Arya, S. Shen, and Y-L Lin, "An estimation of Turbulent Kinetic Energy and Energy Dissipation Rate Based on Atmospheric Boundary Layer Similarity Theory," U.S. National Atmospheric and Space Administration, NASA/CR-2000-210298, 2000.

17. Popat, N.R., C.A. Catlin, B.J. Arntzen, R.P. Lindstedt, B.H. Hjertager, T. Solberg, O. Saeter, and A.C. van der Berg, "Investigations to Improve and Assess the Accuracy of Computational Fluid Dynamic Based Explosion Models," J. Haz. Mat., Vo. 45, pp 1-25, 1996.

18. Savvides, C., V. Tam, J.E. Os, O. R. Hansen, K. van Wingerden, and J. Renoult, "Dispersion of Fuel in Offshore Modules: Comparison of Predictions using FLACS and Fullscale Experiments," Major Hazards Offshore, ERA, ISBN 07008 07489, 2001a.

19. Johnson, D.M., R.P. Cleaver, J.S. Puttock, C.J.M. Van Wingerden, "Investigation of Gas Dispersion and Explosions in Off-Shore Modules," Off-Shore Technology Conference, Houston, TX, May 6-9,2002

20. Hanna, S.R., D.G. Strimaitis, and J.C. Chang, *Hazardous Response Modeling Uncertainity (A Quantitative Method), Vol. II Evaluation of Commonly Used Hazardous Gas Dispersion Models,* Sigma Research Corporation, Westford, MA, 1991.

21. Hanna, S.R., and J.C. Chang, "Use of the Kit Fox Field Data to Analyze Dense Gas Dispersion Modeling Issues," Atmos. Envir., Vo. 35, pp 2231-2242, 2001.

22. Cowan, I.R., I.P. Castro, and A.G. Robins, *Project EMU Experimental Data, Case A, Release 1*, ME-FD/96.52, University of Surrey, Fluids Research Center, 1996.

23. Biltoft, C.A., "Customer Report for Mock Urban Setting Test," DTC Project No. 8-CO-160-000-052, West Desert Test Center, U.S. Army Dugway Proving Ground, Dugway, UT, 2001.

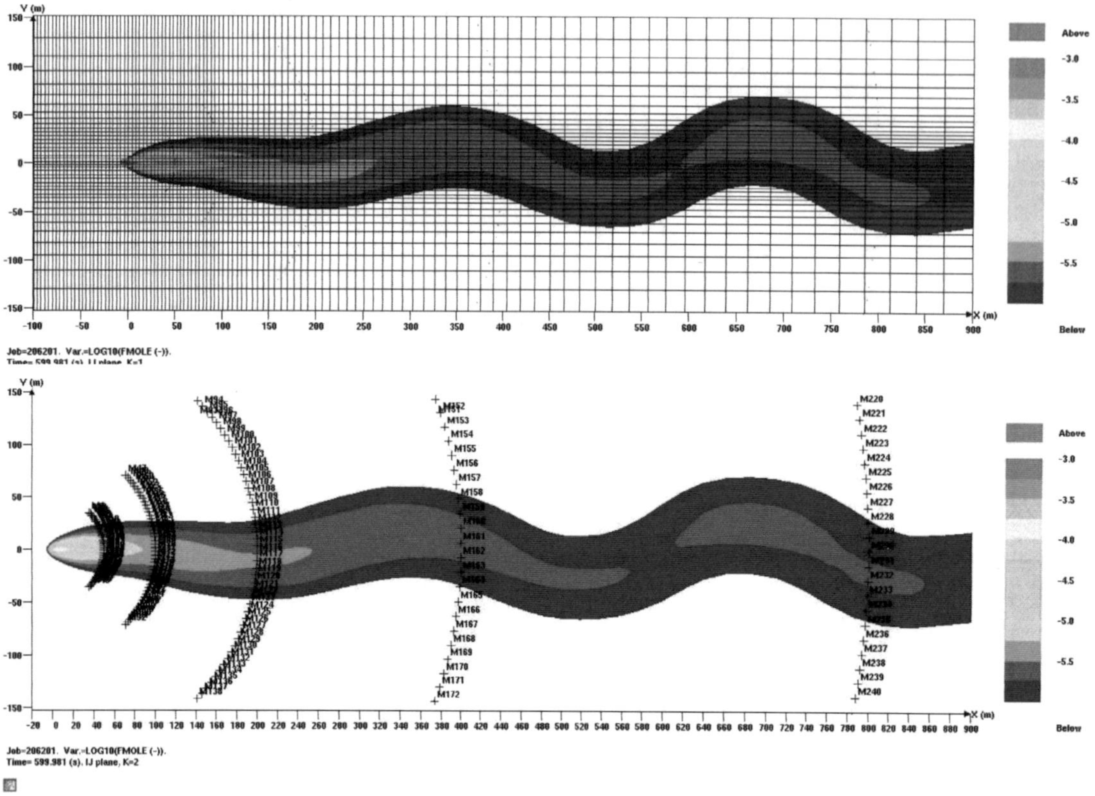

Figure 1: Prairie Grass experiments simulation

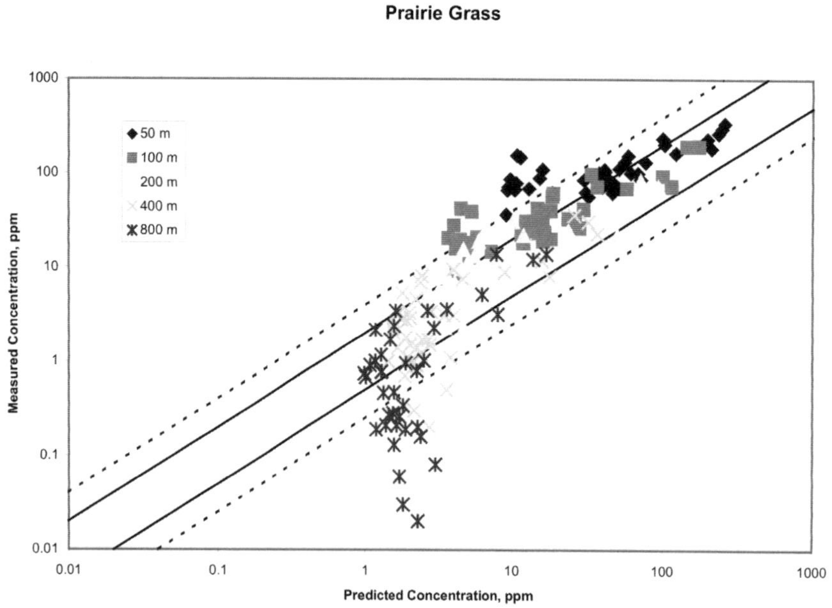

Figure 2: Prairie Grass --- Predicted vs Measured Concentrations at 50, 100, 200, 400, and 800 meters from the release location

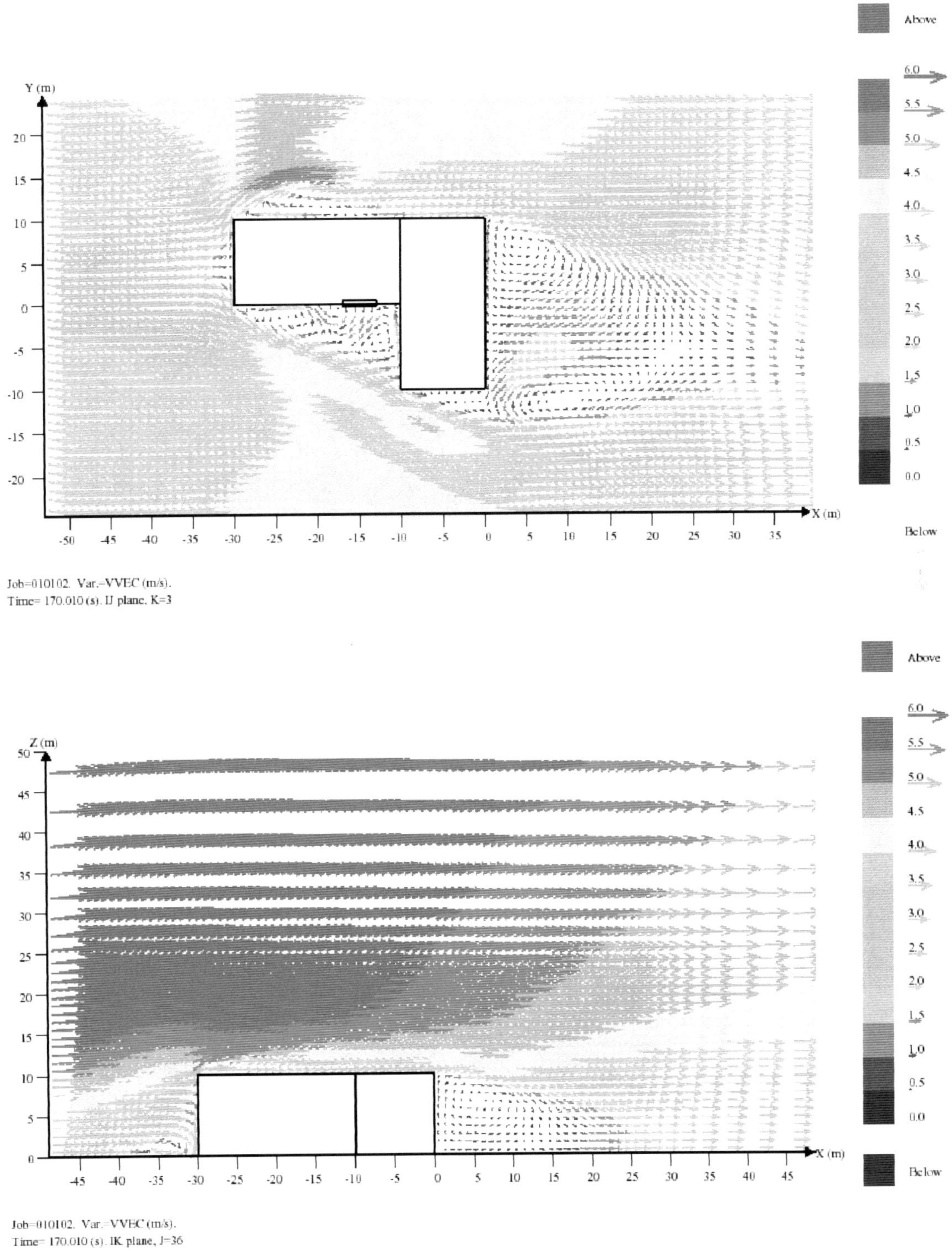

Figure 3: Wind flow patterns around EMU L-Shaped Building with a 5 m/s windspeed at the boundary at a reference height of 10 m.

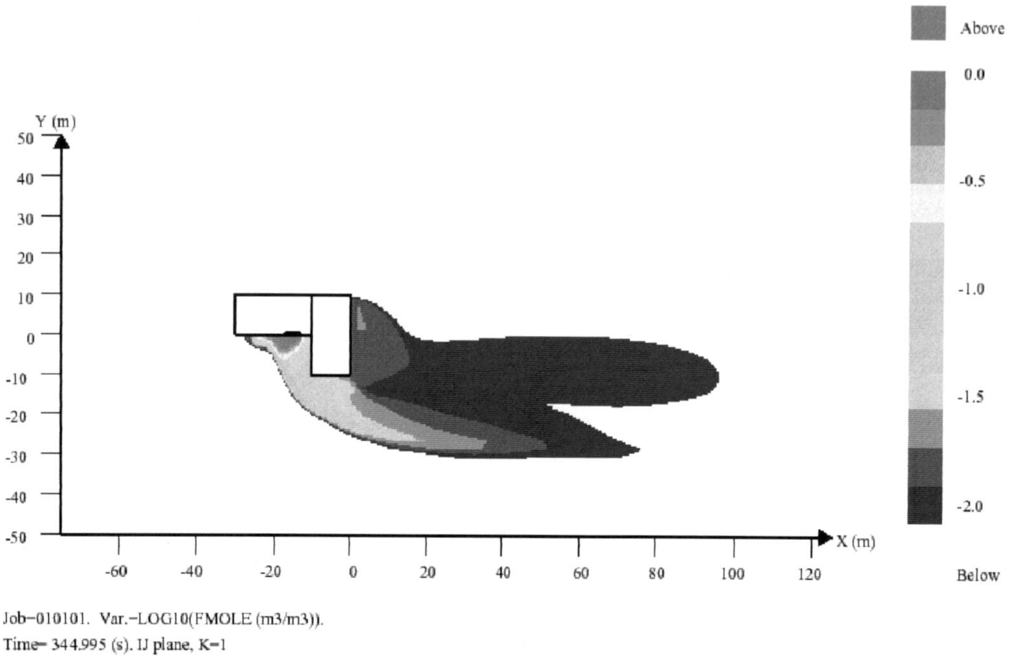

Figure 4: FLACS Simulation for the EMU L-Shaped building

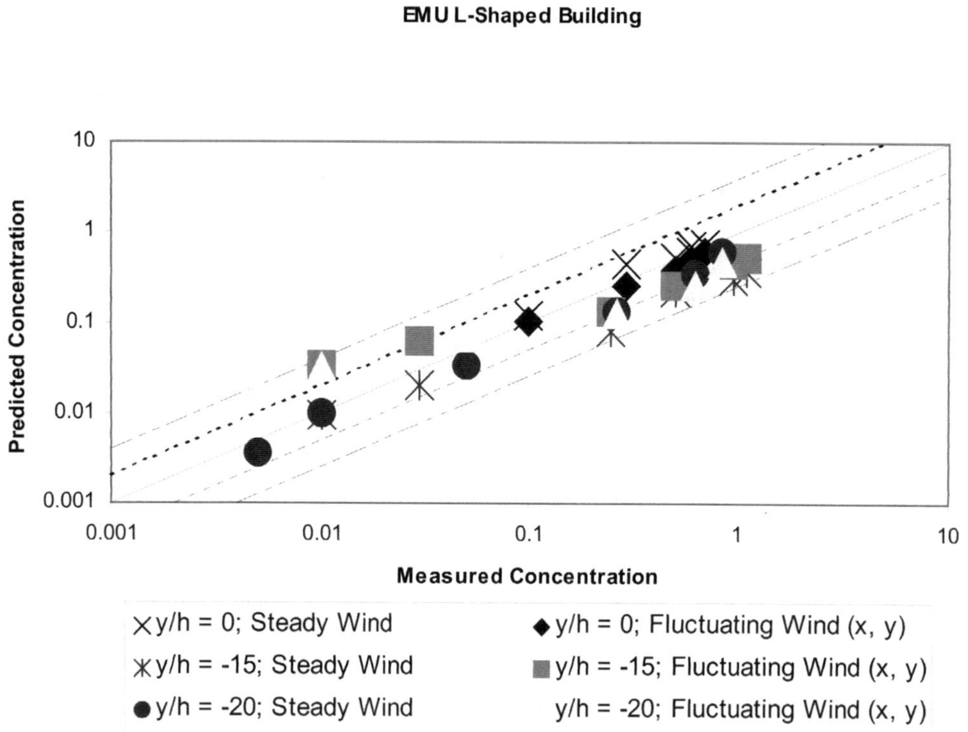

Figure 5: Measured versus predicted concentrations for the EMU L-Shaped building.

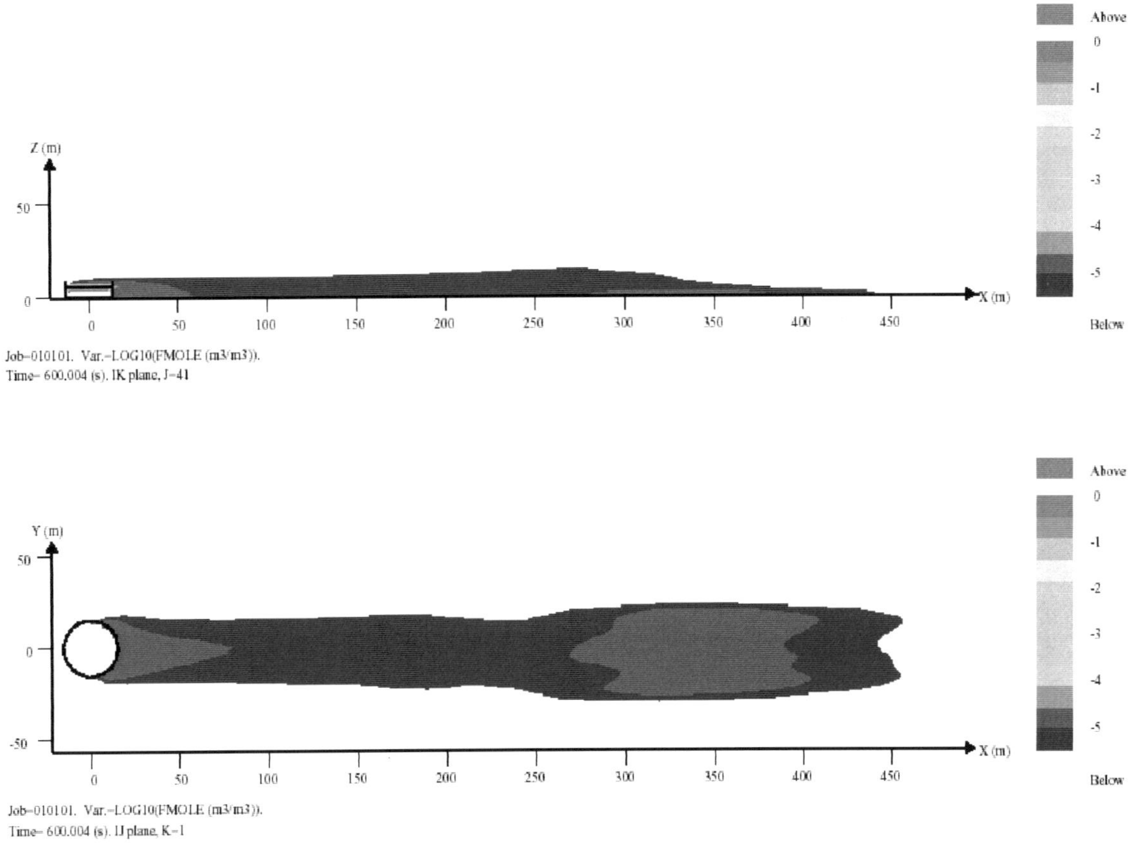

Figure 7: Dispersion under F stability and 1.5 m/s wind speed of a 3 g/s release of chemical (MW = 46 g/gmol) vapors from large (diameter = 31 m) open top tank that is 8.5 m height.

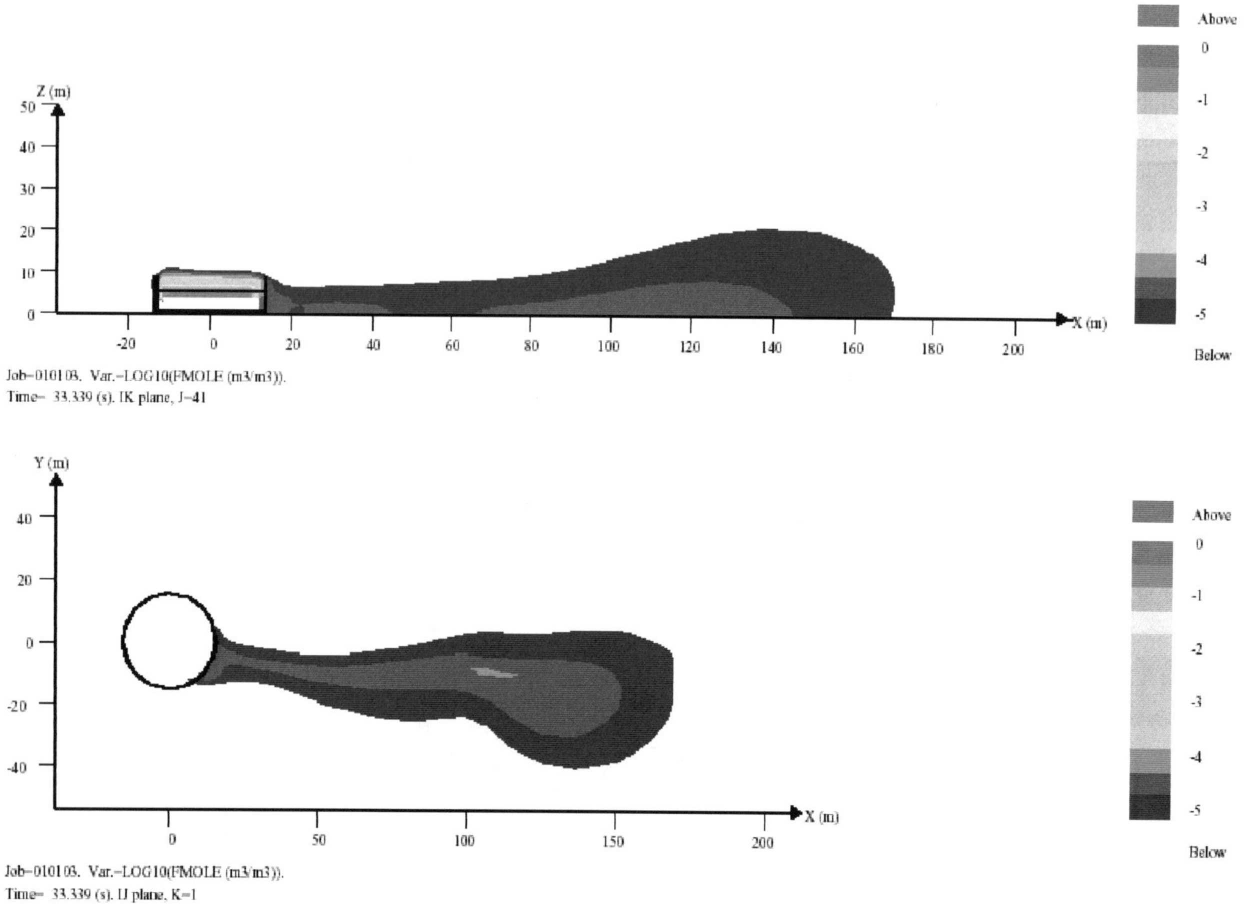

Figure 8: Dispersion under D stability and 5.0 m/s wind speed of a 3 g/s release of chemical (MW = 46 g/gmol) vapors from large (diameter = 31 m) open top tank that is 8.5 m height.

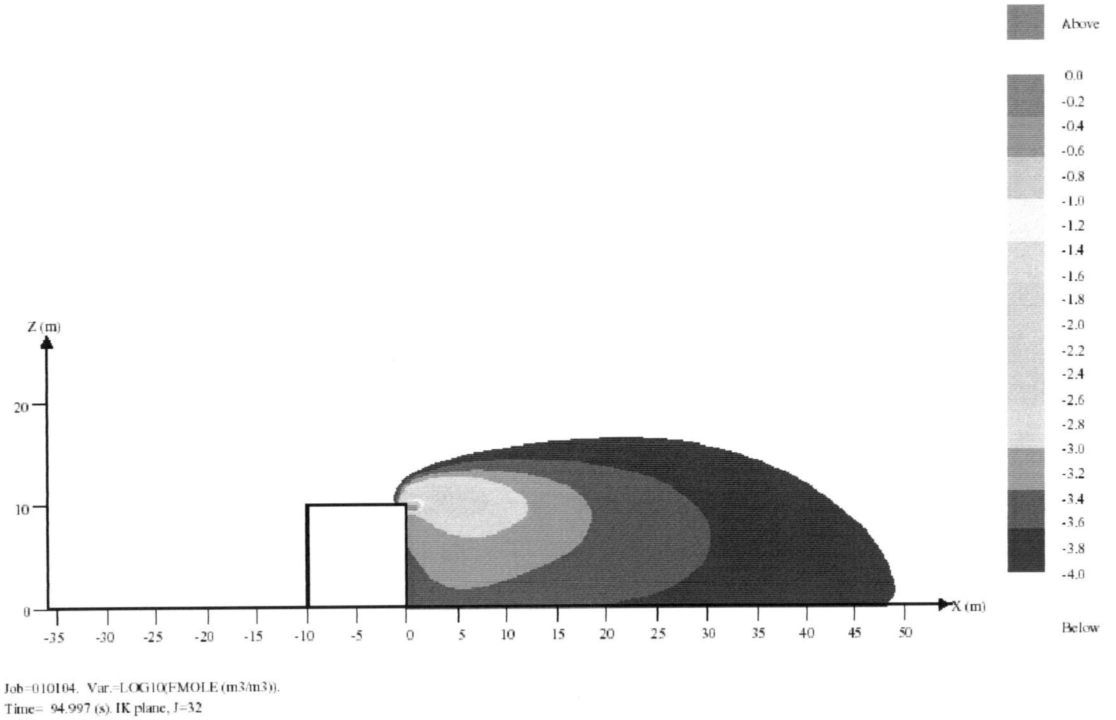

Figure 8: Downwash effect on a 0.4 kg/s chemical (MW=58 g/gmol) release venting close to the building height, under D stability and 5.0 m/s wind speed.

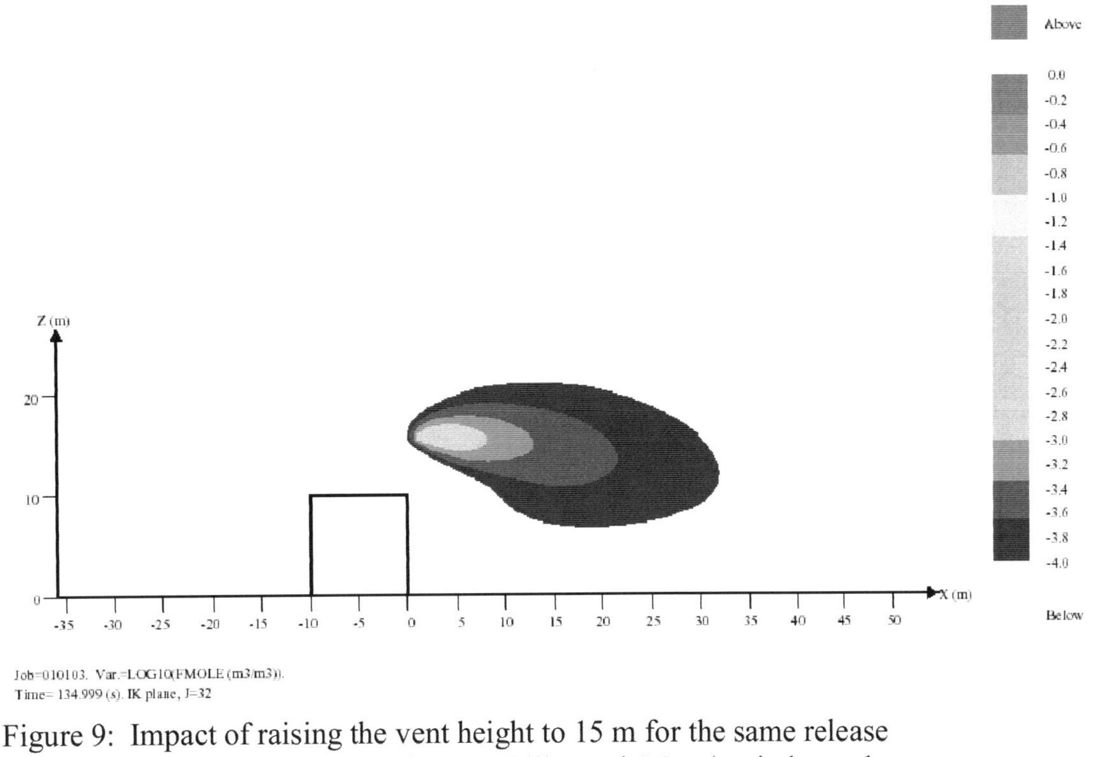

Figure 9: Impact of raising the vent height to 15 m for the same release shown in Figure 8, under D stability and 5.0 m/s wind speed.

An Integrated Approach for Gas Dispersion, Gas Explosion and Structural Impact Analysis for the Offshore Wintershall Q4-C Gas Production Platform on the Dutch Continental Shelf

A.M.A van der Heijden
D. Schaap
ORBITAL Technologies BV
Schiedam, The Netherlands
info@orbin.nl

W. Korndörffer
KCI BV
Schiedam, The Netherlands
info@kcibv.nl

N.H.A. Versloot
TNO Prins Maurits Laboratory
Rijswijk, The Netherlands
versloot@pml.tno.nl

ABSTRACT

The design of the offshore Wintershall gas production platform Q4-C, carried out by KCI, has been subjected to an extensive quantitative risk analysis in particular with regard to its resistance to gas explosions loads. In a joint effort, KCI, ORBITAL Technologies and TNO Prins Maurits Laboratory (TNO-PML) carried out a unique integrated gas dispersion, gas explosion and structural impact analysis to assess the impact and risks of a gas explosion.
Preventive and mitigating measures were taken to lower the probability and effect of a gas explosion to a level as reasonably practicably to safeguard the operators and assets. State of the art computer models (AutoReaGas™ and SACS) have been utilized to analyse the effects of a gas explosion and the overall structural behaviour.
In addition, the probability and effect of gas leakage and dispersion have been modelled utilizing actual environmental wind data for ventilation and generic data for leak probabilities. These analytical and statistical methods were developed to enable rapid assessments to be made.
Finally, numerical physically non-linear time domain computer programs have been developed to properly analyse the local structural components and design blast walls. It was demonstrated that integration of the physical and structural effects of a gas explosion in an early stage of the design results in a safe and economical design.

1. INTRODUCTION

In October 2003 gas was produced for the first time from the Wintershall Q4-C gas production platform located in the North Sea, 30 kilometres offshore the Dutch Coast in a water depth of 24 metres (see Figures 1 and 2).
The recoverable reserves of the gas reservoir at a depth of 2 km are estimated to be 6 BCM. The reservoir is depleted by two wells and presently producing 3,5 million Nm^3 gas per day which is evacuated by a submarine pipeline via Unocal's Hoorn platform. The potential production capacity is 7.5 MNm^3/day. The platform production facilities are based on the gas dewatering process. The platform is normally not manned.
The design of the platform has been carried out by Korndörffer Contracting International (KCI BV), constructed by the construction yard HBG and installed by Seaway Heavy Lifting. The design and construction of the platform have been certified by Bureau Veritas. The Dutch State Supervision of Mines is the supervisory authority.
As part of the safety case, in a joint effort, KCI, ORBITAL Technologies and TNO-PML carried out an integrated gas dispersion, gas explosion and structural impact analysis to assess the impact and risks of a gas explosion.

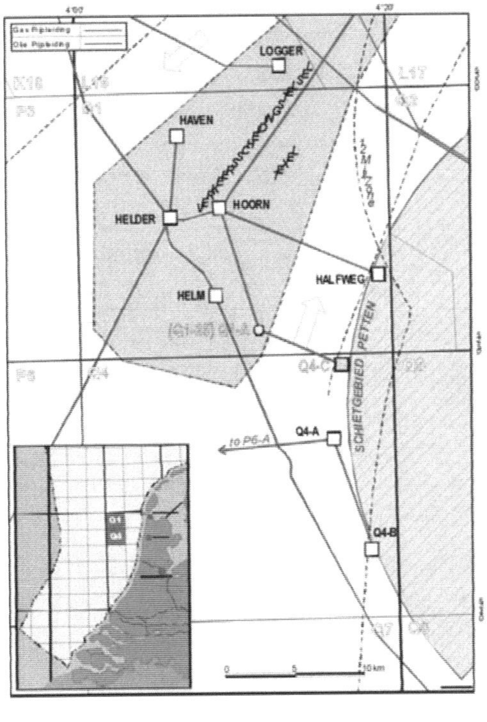

Figure 1: Field location map

Figure 2: Q4-C Gas production platform

2. PLATFORM DESCRIPTION

The Q4-C platform is a fixed structure consisting of a topside deck structure mounted on a jacket type substructure founded with skirt piles to the sea bottom.
The composing components are depicted and described below.

2.1 Platform Modules

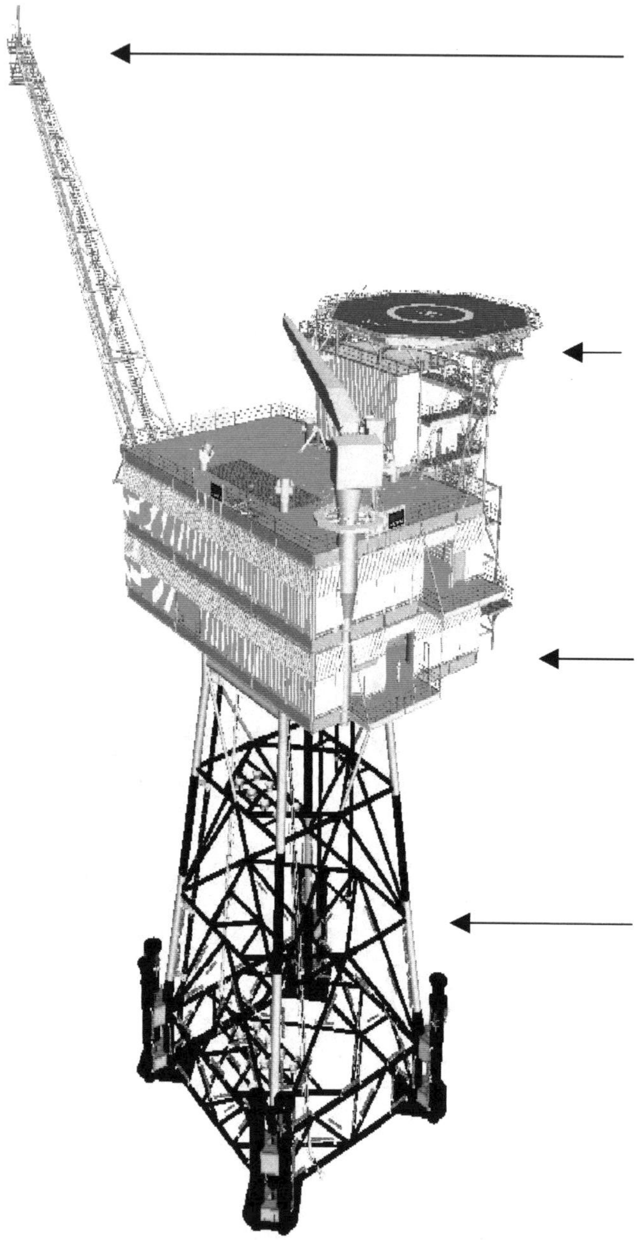

Figure 3: Q4-C Platform

The Vent stack

This 45 meter long structure supports a pipe through which the gas can be vented safely to the atmosphere in case of an over pressure in the process system. The vent stack weighs about 52 tons.

Living Quarter & Helicopter deck

This module consists of a living quarter and a helicopter deck. The living quarter also contains, apart from the kitchen and living rooms, the control room for the platform. The second floor houses the sleeping quarters. This entire module weighs about 400 tons.

Superstructure

The topsides consist of three decks, which house all process equipment that is required to treat the gas for transport to shore via a pipeline. The topsides weigh about 1800 tons.

Substructure

The jacket is fixed to the seabed by means of eight skirt piles driven 35 meters into the seabed and supporting the topside structures. The weight is about 1200 tons.

2.2 Superstructure

The superstructure is composed of three decks, being: the main deck, production deck and cellar deck.

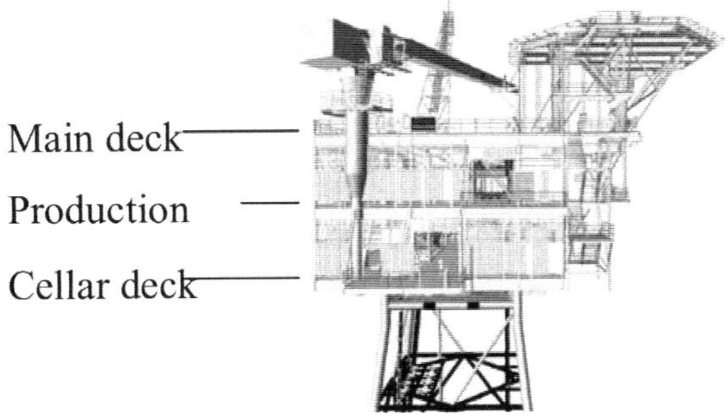

Figure 4: The Q4-C superstructure

Main deck

The main deck supports the deck crane, helicopter deck and living quarters.
This deck will be further used as well work over and maintenance operations and equipment. The remaining open area is reserved for future expansions like compression facilities.

Figure 5: Main deck

Production deck

The production deck houses production facilities and utilities as described below and depicted in Figure 6.

1. ***X-mas tree***; The gas in the well is at a pressure of about 250 bar. The X-mas tree contains a (choke) valve, which reduces the pressure to about 105 bar. Then the gas flows through a pipe below the production deck to the manifold.

2. ***The manifold and HIPPS*** is the location where the gas from the wells is collected. The manifold is located below the main deck. From this manifold the gas flows to the separators. At the end of the manifold two pairs of special valves (HIPPS) are installed. These valves protect the platform equipment against overpressure. As soon as the pressure downstream of the choke valve reaches 125 bar and all other safeguarding systems would fail, these valves will close within 2 seconds. HIPPS is an abbreviation for High Integrity Pressure Protection System.

3. ***Gas/Liquid Separators (2x)***; The gas contains liquids such as condensate (light oil) and water. The separators will separate the free liquids from the gas. At the upper side (left, seen from the stairs) the gas enters the separator. By allowing the gas to flow with a low velocity, the liquids will drop out. The gas then leaves the separator on the top right side. The liquids leave the separator on the bottom and flow to the lower deck, where the condensate will be separated from the water. The maximum capacity of one separator is 3,3 MNm3 of gas per day.

4. ***Microturbines***; four micro turbines generate the required power. They have a capacity of 60 kW each. The produced gas from the platform will be used as fuel.

5. ***Diesel generator;*** In case no gas is available for the micro turbines, a diesel generator is present to provide the required electricity for a so called "black-start".

6. ***Battery room;*** Batteries are present to supply electricity to the main equipment during power failure of the micro turbines.

7. ***Electrical Equipment room***; Almost all electrical equipment will be connected to the cabinets in this room. This room is connected to the control room in the living quarter (at the main deck).

8. ***Sphere Launcher***; See cellar deck (12)

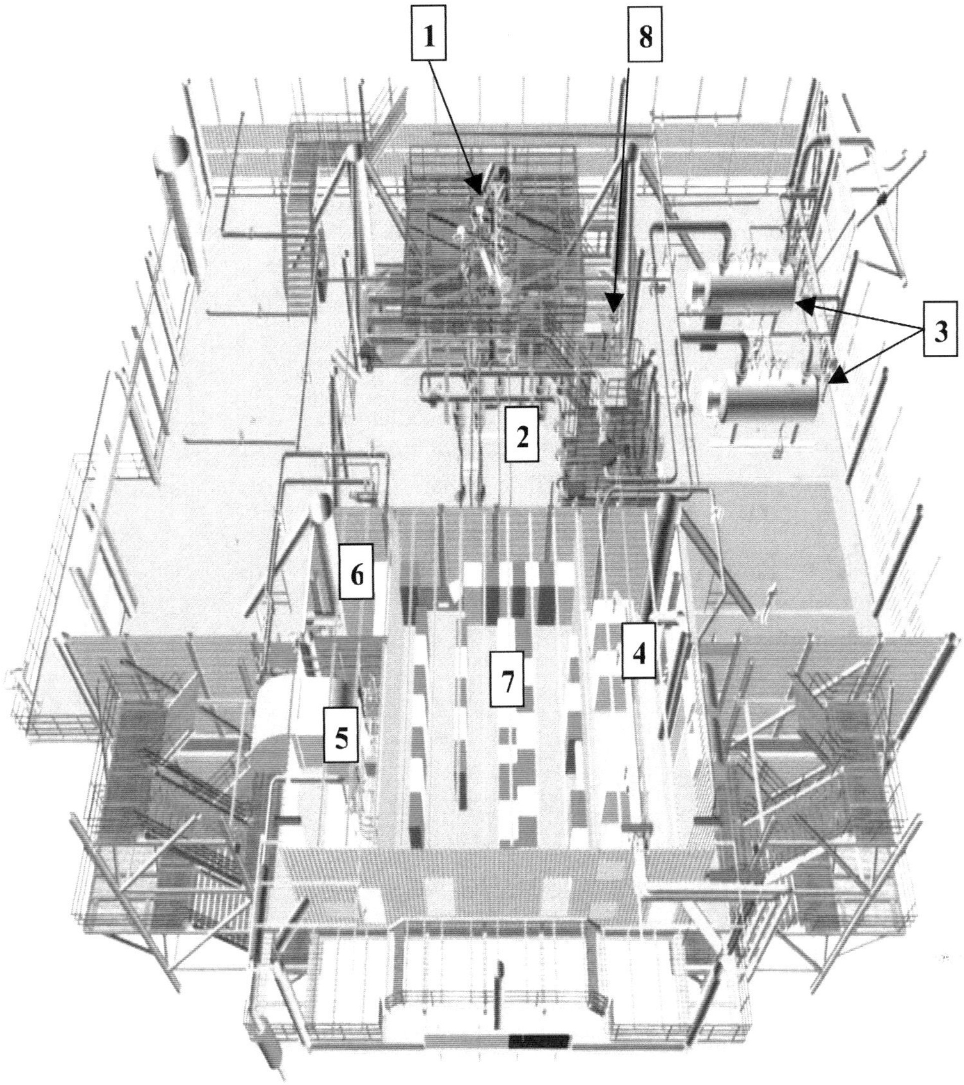

Figure 6: Production deck

Cellar deck

The produced gas is evacuated from this deck. This deck also contains a "tank farm" for storage of chemicals, diesel and drinking water.
The cellar deck houses production facilities and utilities as described below and depicted in Figure 7.

9. ***Liquid/Liquid separators (2x)***; At the production deck the liquids are being separated from the gas (3 in Figure 6). The liquids flow to the liquid/liquid separators located at the cellar deck. Water and condensate will be separated from each other by their difference in density. The condensate floats on top of the water and leaves the separator at the top. The condensate is then being injected again

into the gas stream and transported to shore. The water leaves the separator at the bottom. It will be depressurised and flows to the skimmer tank (14).

10. ***Venturi tube (2x)***; A venturi tube is installed to ensure the condensate is being spiked into the gas stream.

11. ***Export manifold***; The produced gas will be collected in this 16-inch pipe, which is connected to the riser in the jacket. Some liquids are injected before the gas leaves the platform to avoid hydrates. Hydrate is ice forming at high pressure, which can block the pipeline. The chemicals, which are used for injection, are described further below.

12. ***Sphere launcher***; This pipeline functions as a large liquid/gas separator, as the gas cools down over a large distance. A disadvantage of the liquids is that they stay in the pipeline. Therefore a sphere is launched through the pipeline every several days to remove the liquids. The sphere exactly fits into the pipeline. The top of the sphere launcher is at production deck level, where 7 spheres can be loaded into the launcher.

13. ***Vent Knockout vessel***; The gas which is released from the process during a blow down will leave the platform through the vent-stack. Before entering the vent-stack all free liquids are removed from the gas in the knockout vessel. This way a dangerous liquid spray is avoided at the vent tip during blow down.

14. ***Skimmer***; The oil in the liquid from the liquid/liquid separators is further separated from the water in the skimmers. The skimmer contains a filter package, which forces the oily substance to float on top of the water due to the density difference. The clean and heavier water will be at the bottom of the skimmer and leaves the platform from here into the sea. A similar skimmer can be found at the sub-cellar deck for skimming the oil from the rainwater that comes from the decks.

15. ***Methanol tank (10 m^3)***; Methanol will be used to prevent the formation of hydrates in gas (see 11). It will be used in the pipeline but also at start-up of the well. Because the temperatures of the gas are then very low (-50 °C).

16. ***Glycol tank (20 m^3)***; Glycol will be used for opening the safety valve in the well. This valve is located under the ground. The pressure in the gas field, at the bottom side of the valve, is about 250 bars. At the other side of the valve the pressure is atmospheric. The high-pressure difference prevents the valve from opening. To enable opening of the valve glycol will be pumped on top of the valve until it can be opened. The glycol will be taken into the process by the gas and also prevent hydrates.

17. ***Kinetics tank (20 m^3)***; Kinetics is also a liquid to prevent hydrates in the pipeline. Kinetics is more expensive than methanol but less is required.

18. ***Water tank (20 m³);*** This tank contains water for the living quarters. It is mainly used for showers and flushing the toilets.

19. ***Diesel tank (20 m³);*** Diesel is required as fuel for the crane, the boats and the diesel generator.

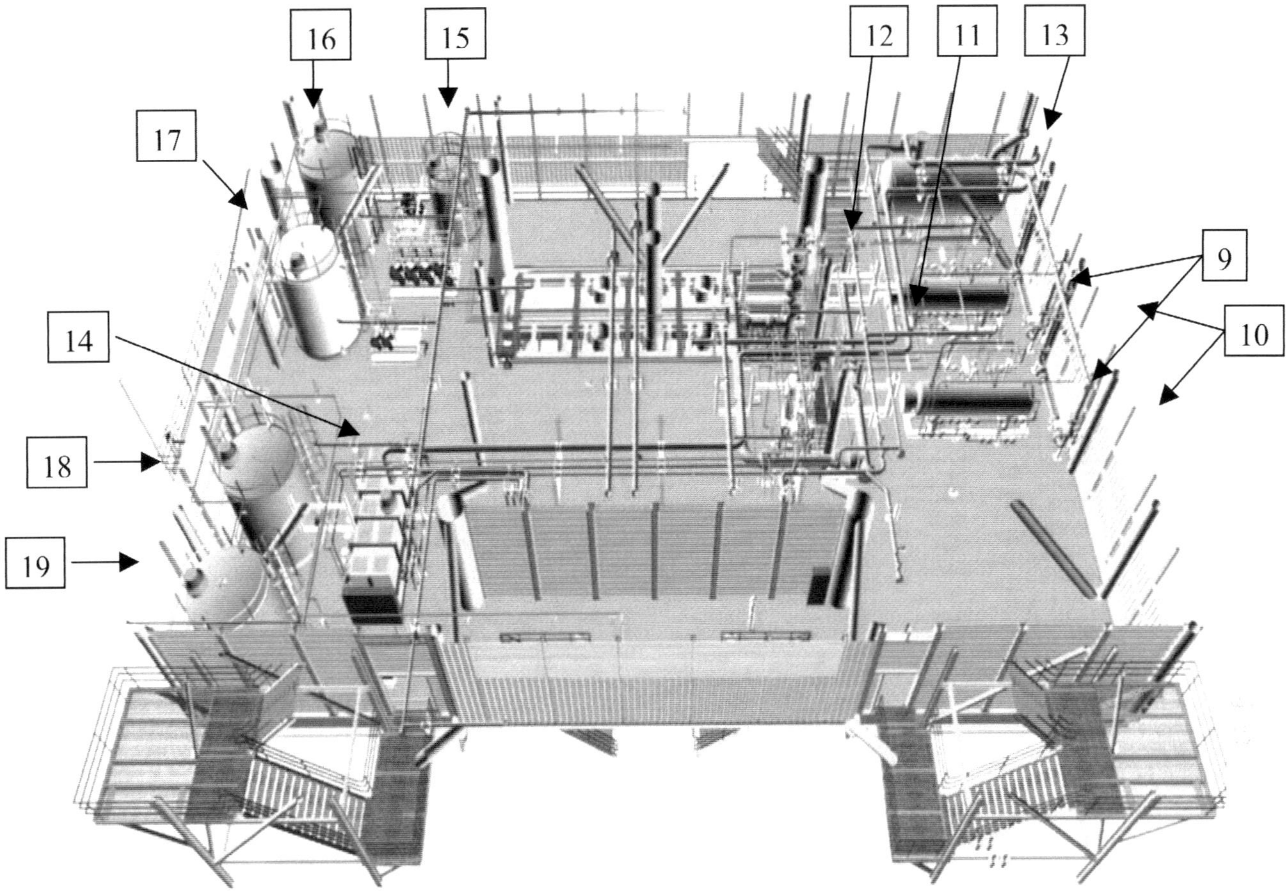

Figure 7: Cellar deck

The production and cellar decks are laid out and provided with facilities to install a future third process train to increase the gas production to 7,5 MNm³/day.

3. EXPLOSION PREVENTIVE AND MITIGATING MEASURES

In order to reduce the risks of a gas explosion on the platform, various preventive and mitigating measures have been taken into account in the design and operational condition.

Prevention of gas explosions and effects on offshore platforms is a major issue to be considered in the quantitative risk assessment as part of the safety case, but also compulsory within the framework of the European ATEX directive, 94/9/EC [1].

Risk reduction of a gas explosion on the Q4-C platform is effectuated by the following preventive and mitigating provisions:

3.1 Preventive Measures

- Optimization of process equipment lay out and sectioning to reduce hazardous area zones and sizes
- Limitation of number of flanges to reduce the probability and leakage rates
- Selection of ATEX classified equipment to reduce the probability of ignition in case of gas leakage
- Fire and gas detection systems
- Venting system to release gas in safe area in case of hazardous conditions
- Process cause and effect control systems

3.2 Mitigating Measures

- Remote operations (normally not manned platform)
- Location of helicopter deck and living quarters on main deck
- Shielding location of survival boat
- Reduction of the wind shielding and creation of maximized natural ventilation to reduce the gas volume in case of gas leakage
- Application of blast type wind shielding to prevent pressure built up due to explosion (maximum overpressure of 50 mbar)
- Reduction of dead ends to prevent gas explosion pressure built up by orientation and lay out of equipment
- Application of blast walls between living quarters and control room and process area
- Blast proof fire/wind screen between stairs and process area
- Reinforced equipment and piping supports to prevent escalation of gas explosion
- Clamping of hatches to prevent blow out
- Structural integrity explosion impact design

4. GAS DISPERSION, GAS EXPLOSION AND STRUCTURAL IMPACT ANALYSES

4.1 Objective

The objective is to achieve a condition where explosion risks at the installation are reduced to as low as reasonably practicable (ALARP). Reference is made to the UKOOA/HSE report "Updated Guidance for Fire and Explosion Hazards" [2].
Apart from the preventing and mitigating measures as discussed in section 3, the following goals define what in practise is necessary to achieve an "ALARP" design with respect to explosion hazard:

- To provide the best practicable protection to personnel in the event of a blast
- To ensure that in the event of a major blast incident, structural integrity will be maintained such as to allow safe evacuation of the platform
- To ensure that the platform structure will, after a major blast, survive a one-year wave loading
- To ensure, by using state of the art modelling, an economically attractive design

The explosion analysis consists of the following steps:

- ❑ Determination of the most likely gas cloud volume on the various decks
- ❑ Determination of the release frequency
- ❑ Determination of the ignition points
- ❑ Simulation of the gas cloud explosions
- ❑ Structural impact analysis

4.2 Gas Release and Frequencies

The gas release frequencies, **P (gas leak)**, have been determined for small (7 mm), medium (22 mm) and large (70 mm) holes. This calculation is based on generic failure rates for pipe work, valves, flanges, instrument connections/small bore connections and pressure vessels. Reference is made to E&P forum report "Hydrocarbon Leak And Ignition Database" [3]. The cumulative frequencies of gas releases are based on the total number of pipe work, valves, flanges, instrument connections/small bore connections and pressure vessels. The gas release rates for various hole diameters have been calculated using the program "Rocalc". Using statistical data for the wind directions and velocities, the relative cumulative probability of the number of air changes, **P (air changes)**, has been calculated for various decks. Combining the results, the most likely gas cloud volumes and probability of occurrence have been calculated for the decks.

The joint probability of explosion is:

P (explosion) = P (air changes) x P (gas leak) x P (ignition)

Event	Release frequency per year			Release probability (%)		
	Small	Medium	Large	Small	Medium	Large
	[7 mm]	[22 mm]	[70 mm]	[7 mm]	[22 mm]	[70 mm]
Release from isolatable	8.31E-02	3.00E-02	1.09E-02	67%	24%	9%
section		Sum	1.24E-01		Sum	100%

Table 1: Gas Release Frequencies and Probabilities

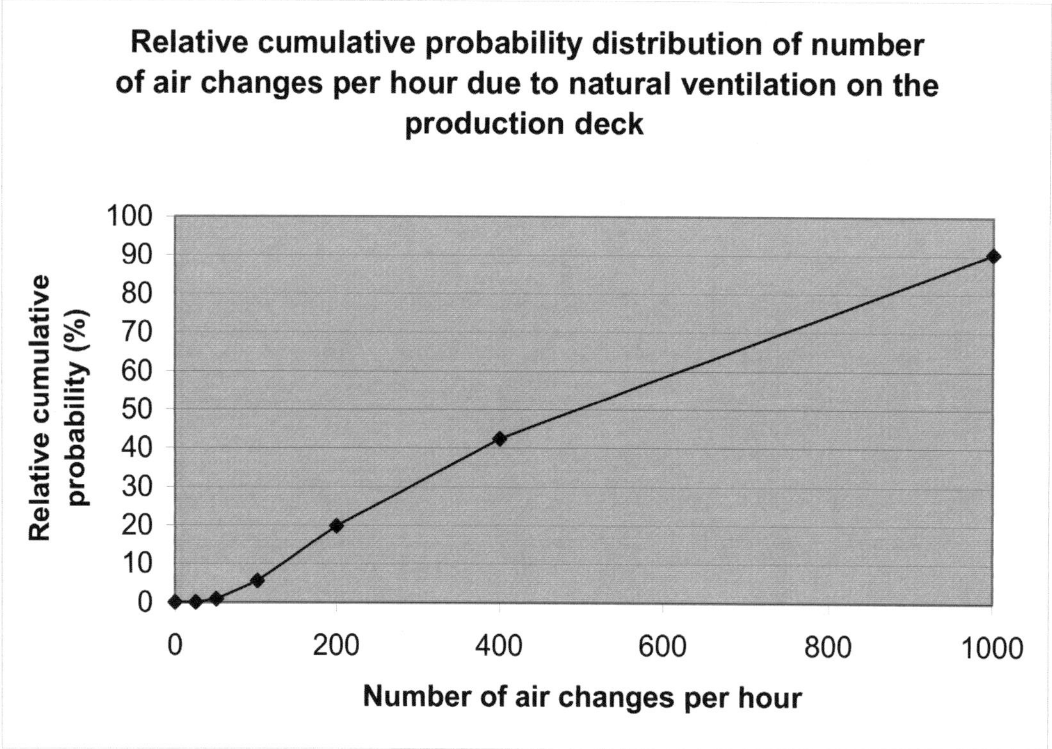

Figure 8: Relative cumulative probability distribution of number of air changes due to natural ventilation on the production deck

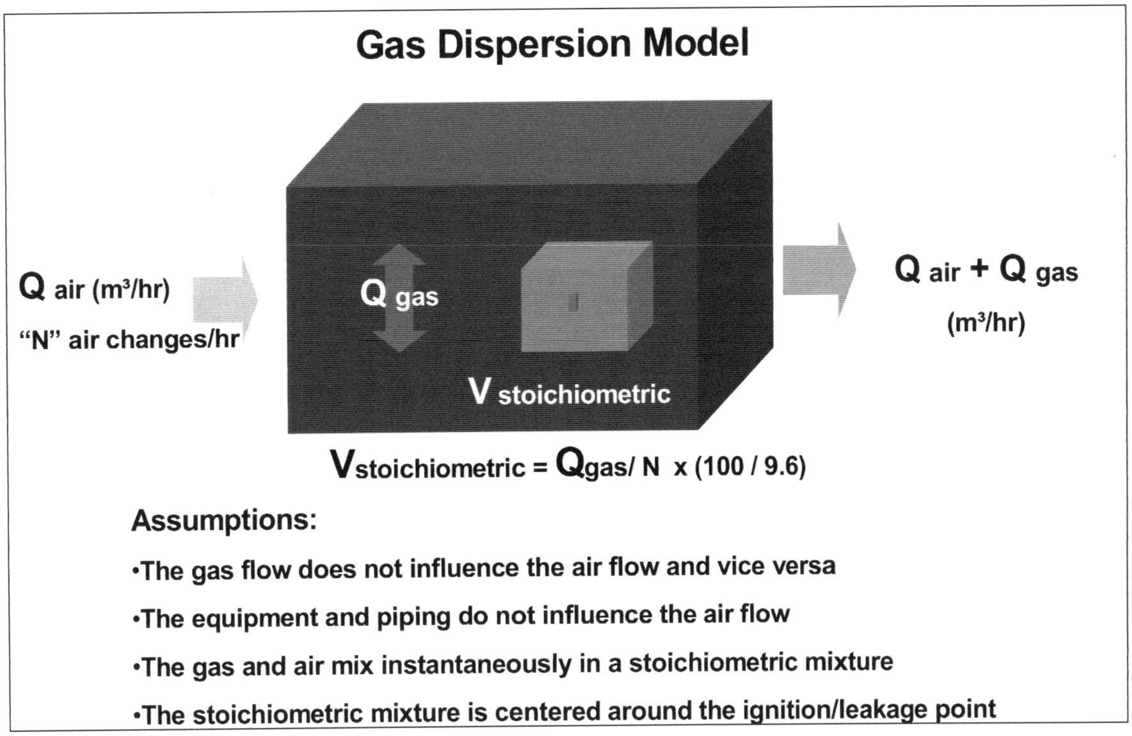

Figure 9: The Gas Dispersion Model

4.3 Ignition Points and Probability

In case of a gas leak and cloud, ignition leading to a gas explosion might occur at various locations within the boundary of the gas cloud. The relative location of the ignition point within the gas cloud determines the origin and effect of the explosion. Sensitivity explosion analyses have been conducted to investigate the influence of the ignition point for a stoichiometric gas cloud of 600 m^3 near the manifold or gas liquid separators on the production deck and the launcher or the condensate water separators on the cellar deck. The ignition point for all cases was selected at the manifold. The results of an explosion (in bar overpressure) for the various explosion scenarios on the decks are presented in Figure 10. It clearly demonstrates that the maximum overpressures on the various decks are dictated by a gas explosion and ignition point at the manifold on the production deck.

The ignition probability of a gas cloud depends on its size. The bigger the cloud, the more ignition sources could be in the cloud and the more likely ignition becomes.
The ignition probability would increase the more ignition sources are engulfed by the gas. In reference [2] a model of the probability of ignition is presented based on historical data. The overall ignition probability of North Sea platforms for small gas leaks is 0.005/year.

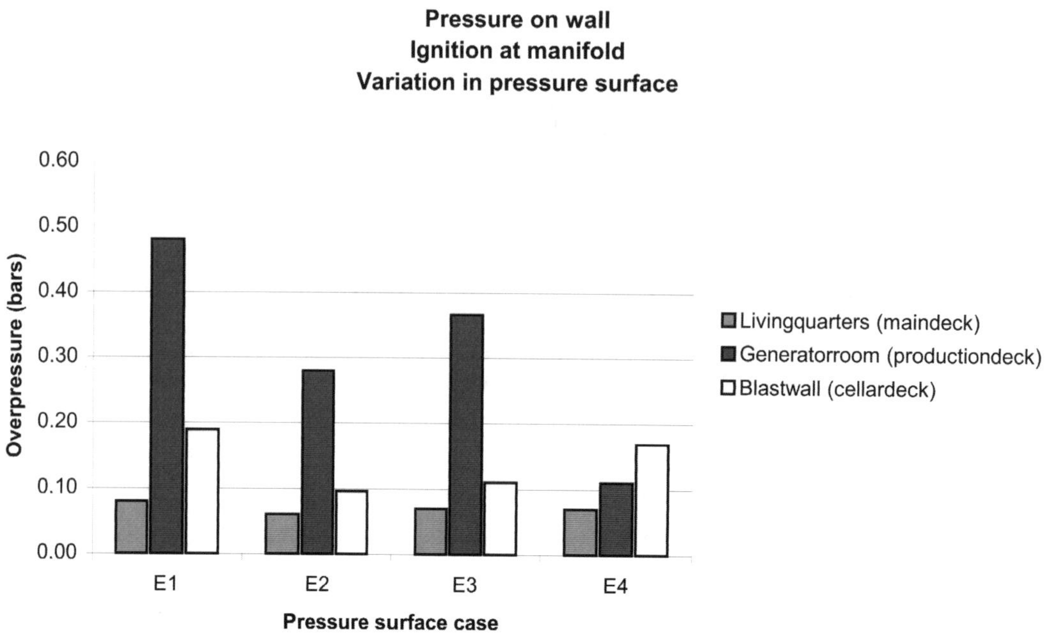

Figure 10: Results of various explosion scenarios

4.4 Gas Explosion Analysis

Modeling

Gas explosion analyses have been carried out using the computer program AutoReaGas™ developed by TNO-PML and Century Dynamics Ltd. AutoReaGas™ is a state of the art 3D computational fluids dynamics computer program for the analysis of combustion in flammable gas mixtures and the subsequent blast effects. It has been extensively validated against various experimental tests, including the large scale tests undertaken as part of the Joint Industry Project on Fire and Blast Engineering for Topside Structures.

Gauge points have been modeled equally distributed on the exposed platform and accommodation module including supports and bulkheads in order to determine the over and under pressures as function of time and location.
In order to be able to transform geometrical data from AutoCad into a format that can be read by AutoReaGas™ an Excel program has been developed.

In the model pipes of 8" and above have been modeled. Both actual and future equipment have been modeled. All the windshields have been modeled as pressure surfaces that disappear at an overpressure of 50 mbar.

Detailed modeling is essential since the explosion escalates when the flame front passes objects and thus increases the explosion velocity and (over)pressures.

Figure 11: Modeling of different objects in AutoReaGas™

Explosion analyses

Simulations have been carried out for ignition in various ignition points and various gas cloud volumes. These simulations yield input data for the structural calculations; specifically overpressure histories and dynamic pressures. The overpressure data are used for the loading on decks, walls and large equipment, whereas dynamic pressures are used for loading on piping and small equipment.

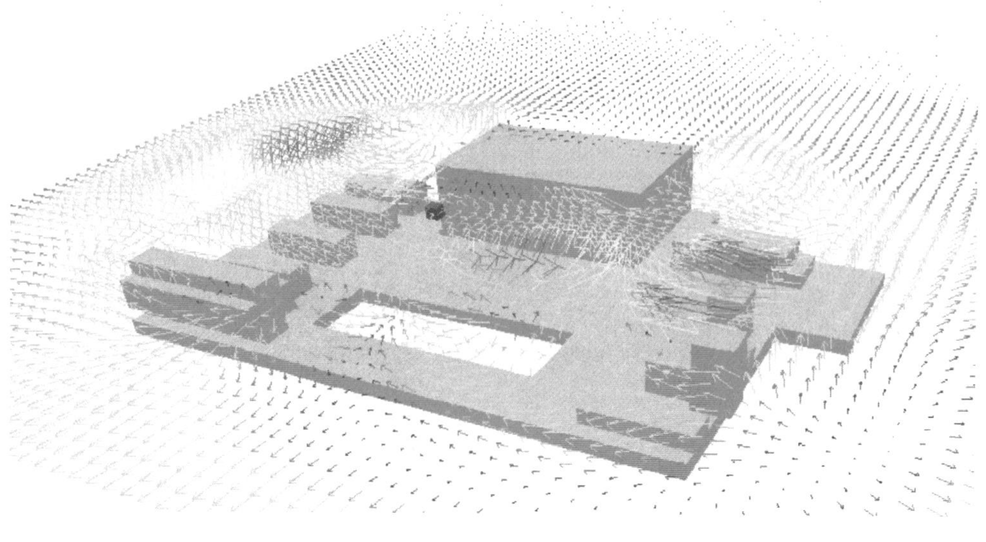

Figure 12: Simulation of an explosion on the main deck

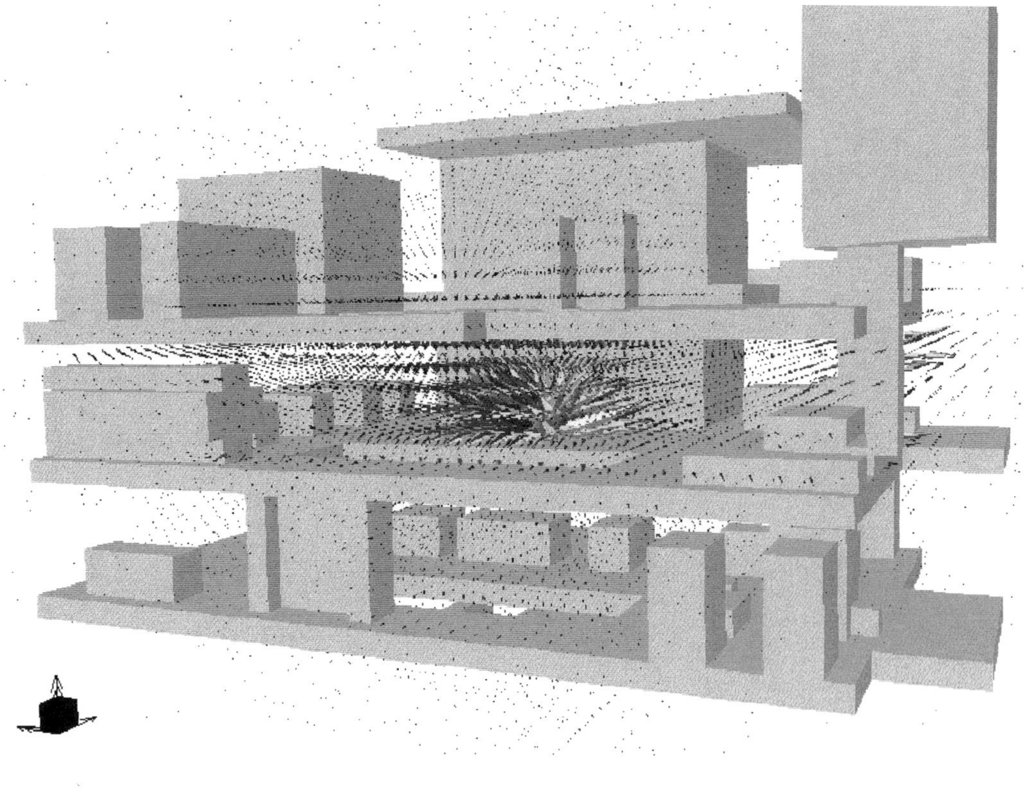

Figure 13: Simulation of an explosion on the production deck

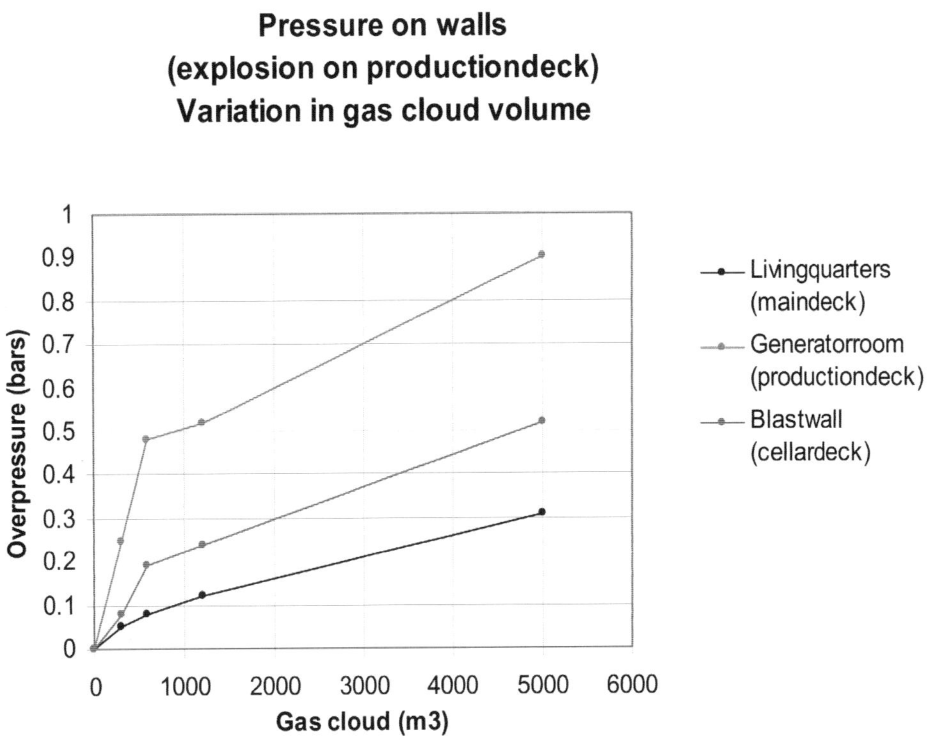

Figure 14: Pressure on walls in relation to gas cloud volume

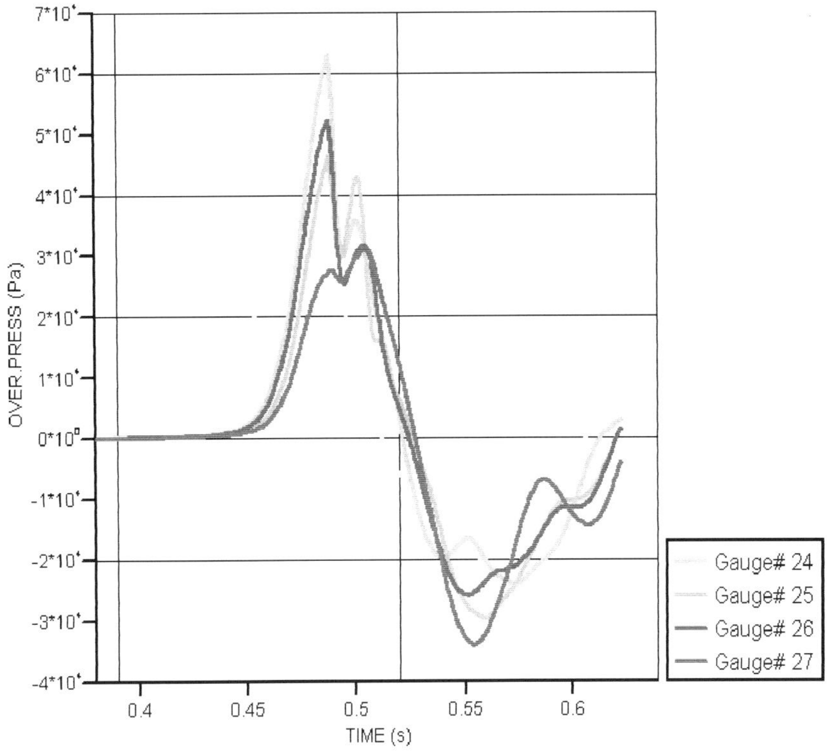

Figure 15: Pressure history plot

Explosion analyses results

The explosion on the production deck generates the maximum pressures, also for the main and cellar deck. The maximum pressures are a bilinear function of the gas cloud volume. The theoretical net volume of the production and cellar deck is 5000 m^3.
The most probable ALARP gas cloud volume is equal to 1250 m^3 with associated pressures of approximately 150 mbar for the main deck, 550 mbar for the production deck and 250 mbar for the cellar deck.
The average pressure pulse duration is in the order of 100 msec, the under pressure following the overpressure is of the same order of magnitude. It is further noticable that the maximum pressures are often elsewhere than near the ignition point, caused by the earlier mentioned turbulence effects when travelling around piping and other objects.

5.0 STRUCTURAL IMPACT ANALYSES

5.1 Structural Impact Resistance Analysis Method

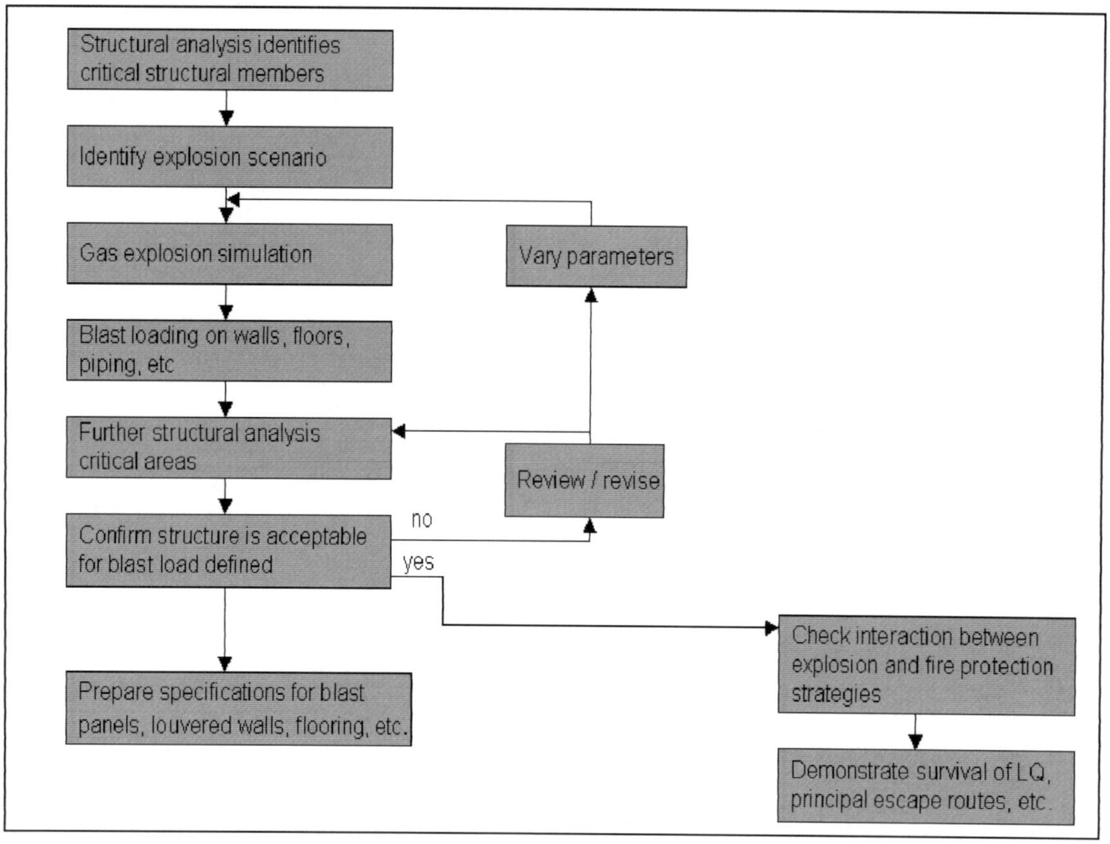

Figure 16: Structural Impact Resistance Analysis Method

The scope of the structural impact assessment encompasses the overall and local effect of explosion loading. The overall effect of the explosion loading is checked with regard to the integrity and stability of the platform. In order to meet the design objectives the following checks have been carried out to safeguard protection and evacuation of the platform in an event of gas explosion.

- ✓ Integrity and support of the helideck on the living quarters
- ✓ Integrity of the living quarters support to the platform
- ✓ Integrity and support of the deck on the substructure
- ✓ Integrity and foundation of the substructure

Blast loadings due to vapour cloud explosions are obtained from the calculations with the program AutoReaGas™. This program provides pressure history plots in selected gauge points. Triangular pulses for both the positive and the negative part of the overpressure can approximate the pressure pulse.

Since blast loadings are inherently dynamic, dynamic simulation is required except when the pulse duration is greater than say 2 times the lowest natural period of the part of the structure to be analysed. As blast loadings constitute a low probability it is not economically justifiable to design all structures as fully elastic. The function of structural elements must therefore be reviewed. The design approach must allow for plastic deformations and possibly large displacements, in such a way that the overall structural integrity is maintained.

Structural elements, which are required to stay within the elastic limit, shall be designed to either BS5950 or API-RP2A.Those elements, which may develop plastic deformations, shall be designed to remain within the specified deformation limits. In addition, shear stresses shall comply with elastic design requirements as prescribed in the BS5950 or API-RP2A code.

For the design of members subject to tension or bending, strain rate effects may be incorporated to increase the minimum specified yield stress. In the absence of more detailed information an increase of 10% may be taken. The upper limit for the shear stress is 0.5 of the enhanced tensile stress. There is no increase permissible for a member in compression.

Once the pressure profiles are available from AutoReaGas™ structural calculations have to be carried out. Two approaches are available for the design of structural steelwork to resist the blast overpressures. These are set out below. The step-by-step integration of these two methods result in an optimum design for blast impact resistance.

5.2 Simplified Approach

Using a Rayleigh-Ritz method Biggs [4] derived Single Degree of Freedom (SDF) models for beams and plates under dynamic bending loading, for various boundary conditions and elastic or elastic-plastic behaviour. An Excel program has been developed

to carry out the calculations. Figures 17, 18 and 19 depict the dynamic response in displacement as function of time and as spring force as function of displacement.
Figure 19 depicts the reaction forces as function of time.

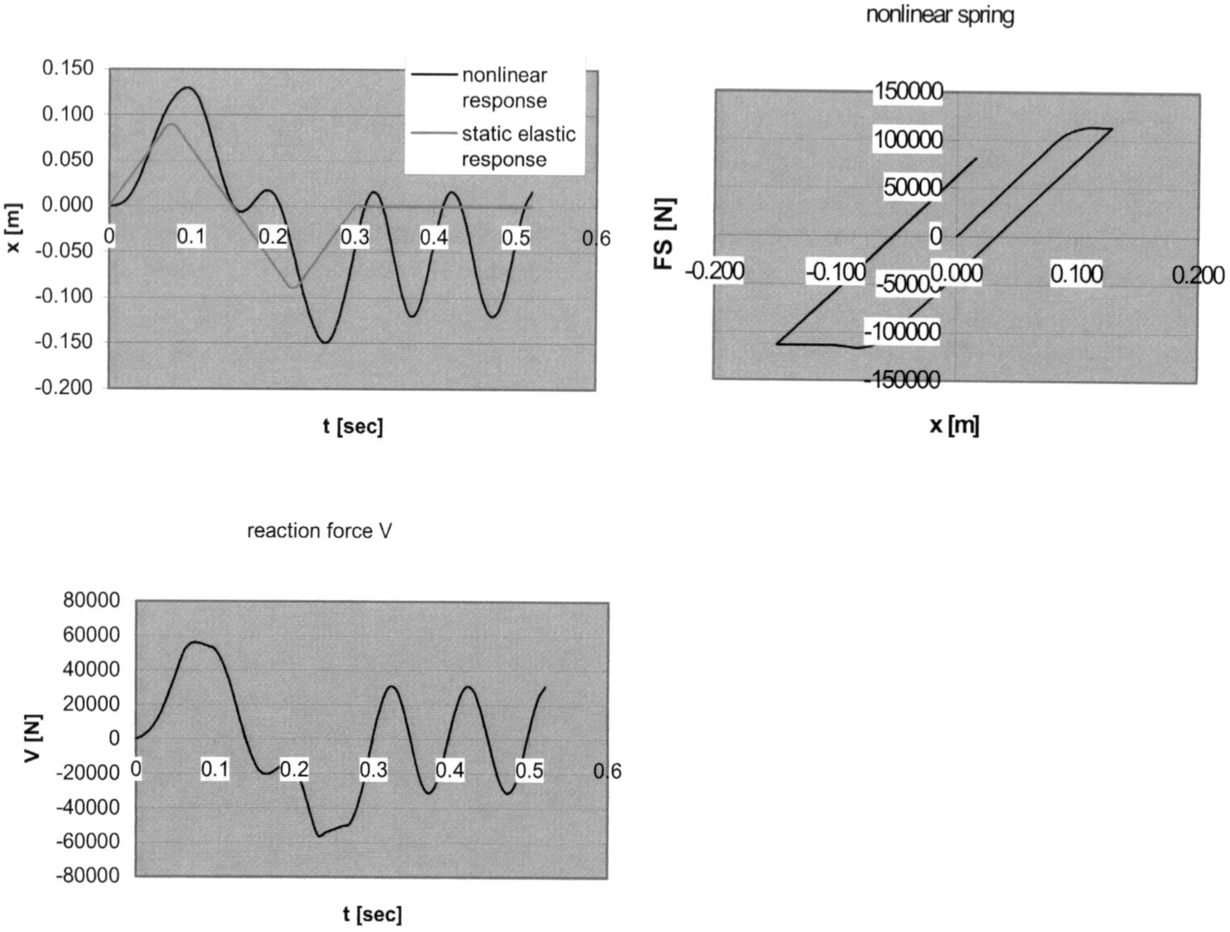

Figures 17, 18 and 19: Dynamic response in displacement

5.3 Non Linear Approach

This approach can in principle be used for all blast calculations, but when applied to the whole or a large part of the structure would lead to unacceptable computer time. The approach is used for parts of the structure that cannot be treated with the simplified approach (e.g. membrane action, interacting vibration modes) or in case of doubt regarding assumptions.

5.4 Global Integrity Analysis

For the global stiffness analysis of the primary structure, the blast loads are applied as static loads transmitted by blast resisting walls and floors in the following way:

- The wall reactions, calculated in the Excel sheets, are applied as point loads at "hard points" of the appropriate deck level in the SACS model.
- The floor loads, calculated in the Excel sheets, are applied as point loads to the truss chords and spine beams in the SACS model.

In addition the effects of drag forces inducing minor axis bending have been checked separately. It shall also be assumed that there is only a single blast event in any compartment at any one time.

5.5 Structural Elements

Decks

All deck beams, excluding truss chords, but including the stringers and deck plates may be considered secondary and are designed to deform plastically, provided that their deformations do not create the risk of serious escalation of hazardous events or affect the stability of primary steel members.

Trusses

All truss members are considered to be of primary importance to the overall structural integrity of topsides facilities. Therefore, trusses shall be designed elastically, using the maximum reactions transferred by deck and wall panels.
Ballistic missile damage to tubular braces can be ignored. The effect of drag forces on truss members is generally insignificant. In case of doubt drag forces may be obtained from the AutoReaGas™ program.

Independent blast walls

These walls may be allowed to deform plastically, provided their deformations do not create the risk of serious escalation of hazardous events. Escalation of events can particularly be generated if improper attention is given to the interfaces between blast walls and the main structure; e.g. under blast loading the blast wall will move horizontally, the lower floor downwards and the upper floor upwards, thus generating a potential problem.

Load carrying blast walls

These walls are designed such that after a blast the platform can still withstand a year sea wave loading.

Enclosures

Enclosures protecting Emergency Shut Down Valves (ESDV's) and other critical valves/equipment/fittings are designed elastically so as to ensure the continued functioning of safety systems in an emergency.

Heavy equipment

The heavy equipment is taken into account as added point masses on plates or beams to count for its effect in the natural frequency.

Equipment and pipe supports

Critical equipment and pipe supports (essential to the safe shutdown and evacuation of the platform) are designed elastically. In addition, unless shielded by primary steel girders, they will experience drag forces due to gas velocities generated by the explosion, which could potentially overturn the equipment concerned, leading to escalation of hazardous events. Therefore, such supports are reinforced to prevent overturning.

Hatches

The hatches are checked with respect to ballistic effects due to blast; in this case special clamps were to be applied to prevent hatches coming out of their support. The vertical upward movement of one of the hatches is depicted in Figure 20.

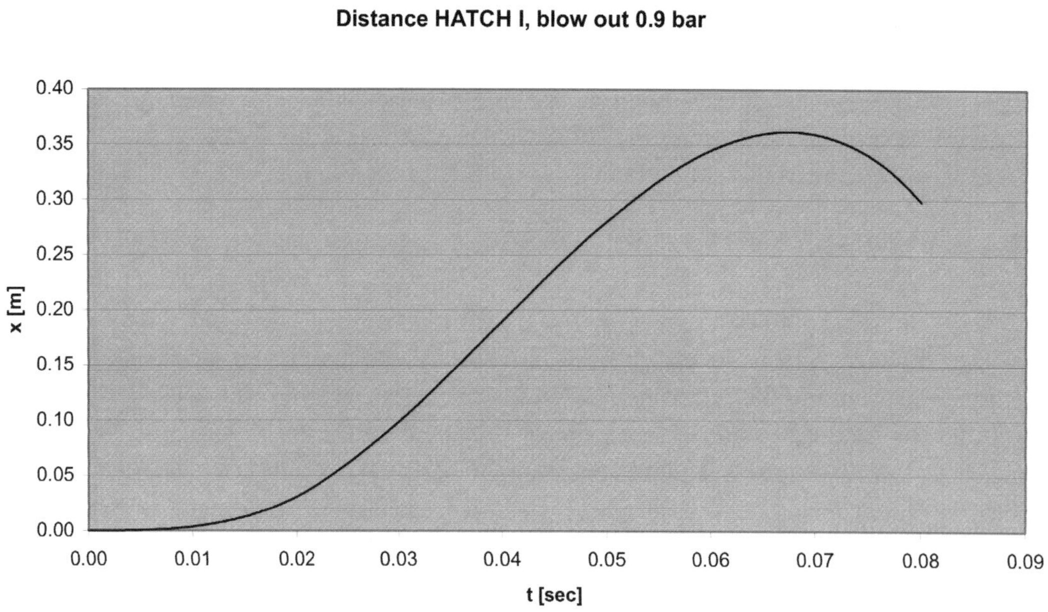

Figure 20: Upward movement of one of the hatches

6. CONCLUSIONS AND RECOMMENDATIONS

In order to get optimal quality control, the gas dispersion, gas explosion and structural impact analyses should be carried out by an integrated team, securing optimal exhange of information between the disciplines involved.

Most likely gas cloud scenarios should be used.

Piping and equipment play an important role in the pressure build up.

The explosions on the production deck produce the largests pressures, also on main and cellar deck.

Underpressures may not be neglected and should be accounted for in the structural analysis.

Typical pressure pulse duration in this case is 100 msec.

REFERENCES

1. Directive 94/9/EC, *On the approximation of the laws of the Member States concerning equipment and protective systems intended for use in potentially explosive atmospheres*, European Parliament and the Council, Brussels, March 23, 1994
2. *Upgraded Guidance For Fire And Explosion Hazards*, UKPOOA/Health and Safety Consortium, London, 2002
3. *Hydrocarbon Leak and Ignition data*, Report No. 11.4/180, E & P Forum, London, 1992
4. Biggs, J.M., *Introduction to Structural Dynamics*, 1964, McGraw-Hill Book Company, New York

THIRD PARTY LIABILITY ESTIMATION

ANALYSIS FOR THE ONSHORE ENERGY INDUSTRY

Alexander John Munnings-Tomes

Marsh Ltd
Tower Place
London, EC3R 5BU
United Kingdom
john.munningstomes@marsh.com

Prepared for Presentation at 38th Annual Loss Prevention Symposium AIChE 2004 Spring Meeting,
New Orleans, April 26-29
Session 4 – Advances in Consequence Modelling I

November 2003

Unpublished

Abstract

For a business to be viable in the long term, companies need to be protected against possible loss scenarios, which can include significant exposure to liability claims following an unfortunate incident such as an explosion, fire, toxic release or environmental damage. This is especially true in societies that have a litigious attitude.

In many cases, potential third party liability exposures can far outweigh other exposures that have been given more attention historically. By having a clear liability exposure analysis, a control and risk financing strategy can be implemented that will ensure that the company assets, image and surrounding community are adequately protected.

A review process has been developed to include an analysis of potential scenarios, which may include one or more of the following: VCE (vapour cloud explosion), BLEVE (boiling liquid expanding vapour explosion),vessel disintegration, flash fires, toxic releases, oil spills and others. Evaluation of these scenarios is based on the use of in-house and commercially available modelling tools, and databases not commonly available in published standards, supported by information available in the public domain. Scenarios are graded in terms of quantified physical damage, impact on third party property, impact on personnel and impact on the environment.

What is Liability and Liability Insurance

"....No.229 If a builder builds a house for a man and do not make its construction firm and the house he has built collapse and cause the death of the owner of the house – that builder will be put to death..." – The Code of Hammurabi, 1700 BC [1].

The Commercial Rationale

The basic principle of liability as established by the common law of England is that a wrong-doer should be responsible for the loss occasioned by his or her wrong-doing. However, it does not follow necessarily that because damage is caused by one person to another, such damage is the act of a wrong-doer. It may have been caused by circumstances which are beyond the control of the person doing the damage, in which case, it cannot be considered to be the act of a wrong-doer. The resulting loss is a misfortune, which must be borne by the person suffering the damage. Equally, that the act causing the damage was unintentional does not free the doer of the act from responsibility automatically. He or she is still liable if it can be shown that the act arose from breach of statute, contract or was due to negligence, on his or her part.

As a wrong-doer will be held responsible for the loss occasioned by his of her wrong-doing and as the wrong may be dealt with by the payment of damages, the discerning person purchases insurance, in order not to be inconvenienced financially, or even bankrupted, by the results of the incident.

The basis for assigning responsibility for such loss is the body of the law, generally common law, which governs the legal rights and duties of all members of society. Recognition and understanding of its obligations allows a commercial operation to manage its exposure to losses.

The laws governing civil wrongs, or torts, whether derived from common law or statutory law, are as numerous and varied as the countries and states in whose jurisdictions they apply, and in a paper such as this, impractical to embrace them all. It is however felt useful to provide some comment to the fundamental principles and practice of liability and liability insurance against a background of the common laws of England and the USA.

Historical Developments

The need for, and development of, liability insurance in England and the USA parallels that development of their respective industries and spread of wealth, as the potential for injury to both employees and public (third parties) alike, grew. Inspite of the principle of a wrong-doer being held responsible for his acts being an ancient one, in early Victorian England it was held against public policy for anyone to purchase insurance to recompense themselves for damages awarded against them as a result of their own negligent acts. The advent of rail travel amongst others is thought to have given rise to accident insurance, from which non-marine liability is thought to have its origins. The first passenger rail accident took place at the opening of the Liverpool to Manchester railway on 15 September 1830, when the Rt. Hon. William Huskisson (Member of Parliament) was rundown by "The Rocket" as he stood on the track. The construction industry recognised the risks involved in its trade, and rudimentary third party liability coverage became available in England in 1880.

The need for liability insurance in the USA took off at a pace upon conclusion of the Civil War (1865), when the industrial power base grew. Employers' liability insurance was introduced in 1886 to meet the demand of employers fearing financial loss because of suits by employees. These were followed by contractors' public liability insurance in 1886, and manufacturers' public liability insurance in 1892. Developments at this time were specific to particular activities, machinery or products. Coverage was provided on a separate policy, excluding all such losses that could be insured by another policy. Excess liability policies became available during the 1920's as claims started to put pressure on the primary policy, typically underwritten by one insurance carrier. These excess policies followed strictly the terms and conditions of the primary policy.

In these early experimental days separate policies were in the interests of insurers and insured alike, although as exposures grew it became increasing complex with the insured dealing with a number of different policies, insurers and brokers. Increasing experience led to a 'scheduled liability approach', which permitted buyers to insure several exposures in one contract, albeit with separate coverage documents for various aspects of a business. The pitfalls of such insurance, with potential for 'gaps' in cover led in the USA in 1941, to the Comprehensive General Liability insurance policy. Such polices were 'accident based' which had been defined by courts of law as an event that was unintended, unpredictable, unforeseen and sudden in nature. In 1966, an 'occurrence' trigger became available, reflecting the reality that a great many liability losses resulted not only from accidents but also a continuous or repeated exposure to the same generally harmful conditions. Further comprehensive changes to this policy took place in 1986.

Industry has many potential areas where injury may result from exposure over a period of time to a hazardous substance such as chemicals, dust (mining) or fibres (asbestos). This in turn has led to both debate and legislation in determining which policy is or isn't responsible for claims made. Pollution and seepage, an area closely linked with the Onshore Energy Industry extended this debate, and with increasing claims on insurers for pollution occurrences with origins many years previously, some 'markets' made the decision to offer liability coverage on a 'claims made' basis only. This covers claims made against the insured during the named policy period only.

There remains today a number of different types of liability policies available such as Directors & Officers (D&O), employment practices, employers' (workers compensation), product, professional indemnity, errors & emissions, medical malpractice, etc. For the main, this paper will continue with a focus on third party exposure associated with operational liabilities.

Basis of Liability

Liability insurance policies afford coverage to an insured for tort (civil wrong) liabilities imposed by common law or assumed by contract. The person (or corporation) committing a tort is called a wrong-doer or a tortfeasor. Liability may arise under the following broad headings:

- Intentional Torts – Including libel, slander, assault & battery, trespass and nuisance, false arrest, wrongful detention, malicious prosecution.

- Negligence – Most prevalent of the unintentional torts, involving the failure of a person (or company) to exercise the degree of care that a reasonably prudent person would have exercised, under similar circumstances, to avoid harming another person.

- Strict Liability – May be imposed, regardless of negligence or intent, as a result of situations which present special danger. This is very much the case for the Onshore Energy Industry and broader processing industries, involving handling of toxic, explosive and flammable materials.

- Vicarious Liability – Under certain circumstances, a person may be held liable for the actions of others, even though there is no negligence or assumption of liability on the part of that person. The clearest example of this is a company's employee whose acts, on behalf of the company, in the course of his employment, give rise to liability.

- Contractual Liability – A tort liability of others transferred or assumed under the terms of a contract or agreement. An example in the energy industry context would be an oil company entering into agreement with a significant number of contractors, who in turn employ sub-contractors. These contracts specify a series of assumptions by one or more parties, whereby the oil company assumes responsibility for any injury or damage caused to a third party even though the assuming party may not have been responsible for the act or the damage. More specifically an example would be a major oil producer taking responsibility for the activities of a small drilling contractor above a specified monetary limit.

Property damage and its potential consequences are relatively easy to project, although liability losses are another matter, as ultimately there is no upper financial limit in a tort lawsuit. Property losses within the context of the energy industry typically involve two parties, the insured (first party) and insurer (second party). The relationship is based on contract, with a claim being referred to as a first-party claim. In contrast, liability claims introduce a third party, a plaintiff not party to the contract (third party). Property damage claims tend to be settled in relatively short periods of time, in contrast to liability losses where delays are common. As already identified a claim may be made a long period after the injury took place or an illness producing condition began to exist. Litigation will invariably delay a claim. Asbestos, once viewed as a 'miracle fibre' is a classic example of the complications associated with liability insurance, both with regard to longevity of claims, and the ever increasing circle of defendants, once limited to direct manufacturers, but now extending to companies using asbestos in its products, installers and retailers.

As a general rule, only legal liability is insurable. It would be impractical and unenforceable to insure against moral liability, i.e. to provide compensation where the insured felt that he or she ought to make a payment to an injured party, although there is no legal obligation to do so.

The Onshore Energy Industry

The Onshore Energy Industry faces many exposures with potential for loss of financial assets due to claims brought by a third party. Strict liability is particularly relevant to the industry (Rylands v. Fletcher, (1868) L.R 3 H.L 330). In this judgement it was held that if someone brings something dangerous onto their land and it escapes, then the party bringing the dangerous object/substance onto the land has a strict liability for injury or damage caused by it. It is noted that such liability is not absolute and dependant upon the defendant having known or being reasonably able to foresee that damage of the type claimed would arise. Clearly the Onshore Energy industry, including in its broadest definition: exploration, gas plants, refineries, petrochemical & downstream chemical facilities, terminals, pipelines and jetties all handle 'dangerous substances'. Whilst we shall review in later sections of this paper in detail some of the technical aspects associated with risks to the industry, we shall for now summarise some of the main concerns that the Onshore Energy Industry presents to the liability underwriter:

- VCE (vapour cloud explosion) – A potentially devastating event, with impact to adjacent third party property, and injuries. In the main, with good cost data the impact of such events in respect of property damage can be reasonably quantified. It is noted that for large manufacturing complexes within a common plot area, it is not uncommon for companies involved to mutually waive their rights of recourse through so-called 'hold harmless' agreements, and rely on their own property damage insurance, even if this damage emanates from outside their own premises.

- Vessel disintegration – for example, often considered in plants involving methanol and ammonia synthesis.

- Extended fires (pool or jet), involving process plant and/or storage areas.

- BLEVE (boiling liquid expanding vapour explosion) and flash fires with potential for injury far beyond site battery limits.

- Missile damage associated with VCE, BLEVE and vessel disintegration.

- Toxic releases, arguably a more prevalent risk in the petrochemical and chemical sector. One need look no further than Bhopal, India, 1984, to see the potential harm from such a release.

- Loss of containment associated with offsite handling and distribution facilities. This includes jetties, road and rail distribution, and pipelines, the routing of which and their proximity to populations and water courses are of particular concern to the insurance industry.

- Land transit risks for road and rail present particular concern, not only for raw material, but also catalysts and consumables such as hydrogen fluoride and in a few remaining geographical areas TEL (tetra-ethyl lead).

- Aircraft re-fuelling has potential for major disaster, where agreements such as the 'Tarbox' agreement provide for liability remaining with the fuel provider, even if a third party is involved with re-fuelling.

- Products – In the oil industry, incidence of claims is relatively low but overall risk is high, particularly in the case of jet-fuel. There have also been a number of recent cases of gasoline contamination of paraffin used in the domestic environment, which has resulted in fatalities and injury to users. In the petrochemical and downstream chemical industry product liability claims are more prevalent, given a more diverse product range, and often more exacting specifications, with widespread consequences in the event of failure. One such example, arising from the 1970's & 80's in the USA involved polybutylene in piping. Damages paid have run in to the billions of dollars.

- Contract Works/Contractors –Often the insured will waive rights of recourse against the contractor and/or sub-contractor for damage arising out of the work. However the policy would deem employees of a contractor/sub-contractor as third parties and therefore covered under a third party policy. Where technical service agreements are entered into between 'operators' and engineering companies it is often common for the engineering company in question to be included as a joint assured under the 'operators' policy, thereby enjoying the same level of protection as the 'operator'.

The Next Step

We have in this opening summary of Liability and Liability Insurance established that it is a more complex 'product' than for example Property Damage, given the introduction of a third party, and potential for litigation. This complexity demands a new approach within the Onshore Energy insurance industry. Before we consider such an approach, let us first provided further emphasis for the need for a defendable methodology by looking at the cost of liability.

The Costs of Liability

Whilst this paper does not intend to provide a detailed analysis of liability insurance pricing drivers and 'market' capacity, issues themselves worthy of a dedicated paper and best left to the broker or underwriter rather than a risk engineer, it is worth considering some general trends.

In recent decades, what has been called by some a litigation explosion has taken place in the USA. Everyone suffering a harm wants to make someone pay, and the obvious truth that some injuries are purely accidental - with no one to blame - has gone out the window.

In recent years, some of this 'compensation culture' has been exported overseas. England now permits lawyers to charge clients only if they win, so it's no-risk litigation for the plaintiff. Cries for tort reform are being heard in yet another English-speaking country, Australia. And, slowly but surely, the problem is creeping up on non-English speaking countries, too.

To protect themselves from being on the wrong end of a claim, the insured will inevitably ask the question "how much is enough?" and in the same breath, "how much can we afford?" gets mixed up with "how much are we willing to afford?"

Marsh publishes an annual Research Report [2], based on a global sample survey, in which insurance buyers are able to compare their limits-buying decisions with those of their peers. Some of the more salient summary observations from this report are below:

- Average cost per US$ 1 million of liability coverage to US firms (all sectors) rose 63.4% in 2003, and 82.2% in Europe.
- Worldwide, all regions are paying more.
- Average limits purchased by US firms (all sectors) declined 9.4% in 2003, with a three year decline of 14.5%. Average limits in Europe fell by 21% in 2003.
- In spite of increasing cost and decline in average limits purchased, many businesses, especially those who have experienced large losses, showed signs that they understand the need to align limits with potential losses. Whilst average decline in purchased limits for US firms that have not suffered a loss of US$ 5 million or greater fell by 21.9% from 2002 to 2003, those that have suffered a loss of US$ 5 million or greater maintained the same level of coverage in spite of the 'hard' market.
- The US industry group purchasing the highest limits on average was Chemicals and Pharmaceuticals, this was the same for Europe. Second highest group was that of Mining and Energy in the US, and Transportation in Europe.
- Excess Liability market capacity fell almost 25% between 2000 and 2002.
- 43.3% of worldwide participants have terrorism exclusions.

A second question that needs to be addressed is that of the cost of incidents to both insured and insurer alike.

To provide some perspective we need to remember that not all accidents/injuries result in an insurance claim, and few of those become lawsuits, with yet fewer still becoming jury verdicts/awards. There is a challenge to the insured to distinguish between the exposures that merit concern and those that do not, particularly given that large losses have a historical lower frequency of occurrence.

In regard to the cost of incidents to the insured and insurer alike it is appropriate that we consider the below table, which considers court awards and out-of-court settlements for 17 countries.

Country	US$ million (includes settlements)
USA	124.0
Switzerland	10.8
Australia	7.3
Germany	7.2
Belgium	5.9
UK	5.8
France	5.5
Canada	5.0
Italy	4.3
Spain	3.4
Hong Kong	3.3
Japan	3.2
Austria	2.0
Sweden	1.9
Norway	1.4
Denmark	0.6
Portugal	0.6

(Source: Swiss Re)

Highest Indemnity Award for a Single Bodily Injury

Whilst an indicator of broad trends, these specific amounts have been generated by widely varying legal systems

The data shown, whilst concluding that geographical location counts, gives little insight into likely range of awards. What is the cost of a life or serious injury then? In short this paper we will not give the answer. In fact the following analysis section only goes as far as to quantify the number of fatalities by 'levels'. It would however be unfair to leave the reader without some further guidance.

In the USA, for a more detailed analysis one can turn to information developed by Jury Verdict Research (JVR). In the United Kingdom, a body by the name of the Judicial Studies Board puts out similar cost data. Though it is authorized by Parliament, its conclusions don't have the force of law.

If we consider death and injury awards, JVR offers some interesting data. Instead of just providing a mean or median verdict, it provides a "verdict probability range," defined as the middle 50% of all verdicts. This leaves 25% of cases settled below this range and 25% above.

Below are the verdict probability ranges for quadriplegia, catastrophic burns, and death. If it seems strange that death should come in third, bear in mind that extended illness or incapacity often generates significant medical expenses.

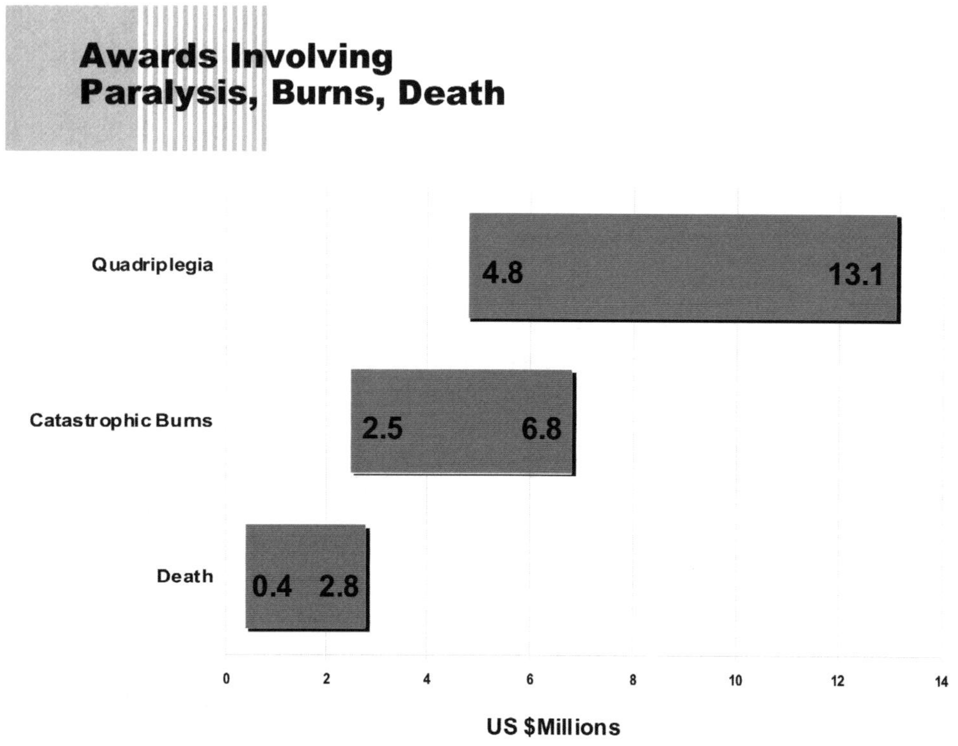

Source JVR 2002

"Verdict probability range," defined as the middle 50% of all verdicts.

It is noted that in the USA, the 'peak probability award' for wrongful death has increased almost threefold in the past ten years, and is now over US$ 4 million per life.

Needless to say the final answer to "how much is enough" is part science and part art. What is not in dispute however is that tort costs are increasing significantly, year after year, in wrongful death and injury cases.

For the Onshore Energy industry, analysis of actual incidents provides us with insight (or rather hindsight) and opportunity for the general industry to learn some lessons from the mistakes of others. From an insurers and insured perspective it also provides an indication of potential financial impact from a loss. The latter in particular, supports the credibility of modelled events, and more pointedly in the case of liability insurance, sets possible precedents.

We shall take a review of some incidents that have occurred in the Onshore Energy industry that are significant from a liability perspective. This review is not intended to be a comprehensive analysis of the incidents, as this in the main has been documented elsewhere and often tabled at Symposiums such as this one. It does however set out to the highlight some of the 'costs' associated with the liability fallout from the incident.

It is emphasised that one cost that is **not typically insurable** and increasingly the larger portion of a liability exposure is that of punitive (or exemplary) damages. These are essentially 'punishments' handed down by the courts, and awarded over and above compensatory damages. There is in effect no ceiling to these damages, and amounts vary significantly based on issues such as number of claims and severity of event for which a defendant has been found liable. Punitive damages alone from the Exxon Valdez (see below) incident amounted to US$ 5 billion at the time of the award. Analysis of awards in the USA, suggests some 4% of cases result in punitive damages being awarded, and whilst this is reasonably stable, there is an increase in severity of award and significant state to state variation.

Incident Review

It is noted that much of the below information has been sourced from the Marsh Loss Database- complied broadly from information within the public domain. Where alternative sources have been used, these are referenced.

Multiple BLEVE – 19 November 1984. This well publicised incident in a suburb of Mexico City serves to remind that it is not always the major sites that pose the major hazard risk from a liability perspective. This incident occurred at an LPG terminal. The root cause has not been passed into the public domain, although what is clear is that an 8" LPG line failed, with catastrophic consequences. Because of the proximity of informal housing, the incident resulted in more than 550 deaths and in excess of 2,000 injuries, with 200 homes destroyed. Actual details of liability payments are not publicly available. Thirteen of the LPG bullets were projected outside of the plant boundary, and many vessels BLEVE'd. LPG droplets were reported as "raining from the sky" which led to numerous fires away from the plant. Above ground firemains damaged in the initial explosion, and relatively low levels of safety system hardware were amongst contributing factors to the seriousness of the incident.

Toxic Gas Cloud Release – 3 December 1984. A release of methyl isocyanate (MIC) from a relief valve on a storage tank at a pesticide plant, Bhopal, India. To this day the actual number of casualties remains estimates only, but majority of sources concur that fatalities are in excess of 3,000 and injuries in excess of 10,000. People continued to die of the effects many years after. A final liability settlement of US$ 470 million was reached. Plant modification without proper analysis of consequences has been attributed as one of main root causes. Claims of sabotage have never been substantiated. The presence of a large informal population adjacent to the plant boundary was a significant contributing cause of the extent of loss. The severity of the incident makes it the worse recorded within the chemical industry.

Refinery Vapour Cloud Explosion – 5 May 1988. A vapour cloud explosion at the refinery in Louisiana. The incident resulted in 7 fatalities, and near to 50 injuries. In 1993 a settlement of US$ 160 million was agreed for over 17,000 remaining third party claims. Settlements and costs in the five years previous are estimated to have exceeded US$ 100 million.

Source: XL Insurance Industry Loss Report.

Platform Loss – 6 July 1988. Whilst not an Onshore loss, this incident is significant for the Energy industry from a liability perspective in many ways. The incident involved explosion, fire and ultimately total loss of a North Sea production platform, with 167 fatalities. The operator settled liability cost of US$ 160 million in money of the day. The incident led to a major industry and government review of operations in the North Sea, with the introduction of a compulsory Safety Case being but one of the legislated requirements resulting from the incident. Later, High Court judgements found the contractors, rather than the operator, were responsible for their own people and equipment.

Oil Spill – 24 March 1989. The nature of the Onshore Energy industry, is that raw materials are transported in bulk by tanker to facilities close to end user. This in turn gives rise to risk of loss of containment of the raw material in close proximity to a shoreline or river course. In this regard the liability claims databases are 'littered' with claims in running into the US$ 00's million. The loss in question, in Prince William Sound, Alaska is significant in the size of punitive damages awarded in addition to other settlements. These punitive damages of US$ 5 billion, reflect the severity of the circumstances of the loss. Cost for clean-up and other third party settlements were significantly in excess of insured cover.

Petrochemical Vapour Cloud Explosion – 23 October 1989. A major vapour cloud explosion at a polymer plant, Pasadena, Texas. The incident resulted in 23 fatalities, and more than 300 injuries. Overall cost was estimated at US$ 1.6 billion in money of the day, of which some US$ 300 million has been estimated as third party liability, although total final settlements are not known. Debris from the explosion was found 6 miles (10km from the plant). Failings in maintenance procedures were amongst the root cause of the incident. The company was fined heavily by OSHA.

Transportation Release – 26 July 1993. This incident involved a release of sulphuric acid from a tank car, allegedly impacting more than 24,000 people in the release area. A total settlement of US$ 180 million was approved by a Supreme Court Judge, with individual payments made under a formula taking into account the severity of each plaintiff's injuries.

Source: XL Insurance Industry Loss Report.

Chemicals Warehouse Fire and River Poisoning – 1 November 1996. A warehouse fire near Basel, Switzerland, containing about 1 300 tonnes of at least 90 different chemicals resulted in widespread damage to the ecosystems in the River Rhine. The majority of these chemicals were destroyed in the fire, but large quantities were introduced into the atmosphere and, into the Rhine River through runoff of fire-fighting water (about 10,000 to 15,000 m^3), and into the soil and groundwater at the site. The exact mass of chemicals entering the Rhine has been estimated at somewhere between 13 and 30 tonnes. Following the accident, the aquatic life in the Rhine was heavily damaged for several hundred kilometres. Several compounds were detected in the sediments of the Rhine after the accident. Third party liability exposure has been estimated at US$ 250 million.

Source: UNEP (except TPL value).

Fatal Tank Explosion – 17 July 2001. This incident involving a single fatality and eight injuries has been recently extensively investigated by the U.S Chemical Safety and Hazardous Investigation Board, and forms the subject of another paper at this very Symposium. The incident is recorded here for three main reasons. First the incident involves an oil refining major, one of the key audiences of this paper. And secondly at the time of writing this report, the settlement reported in September 2003 of US$ 36.4 million to the family of the worker killed, represents to date for 2003 the most significant settlement for a single fatality in the Onshore Energy industry. And in this regard provides a timely reminder as to potential liability exposures. Thirdly given the detailed investigation available, drawing further attention to this can only serve to remind us of the lessons that can be learned from this incident. It is noted that the U.S Chemical Safety and Hazardous Investigation Board investigation found issue with Mechanical Integrity Programme, Engineering Management, Management of Change and Hot Work Systems.

Fertiliser Plant Explosion – 21 September 2001. Less than two weeks after the tragic events at the World Trade Centre, this explosion involving ammonium nitrate at Fertiliser Plant in Toulouse, France received relatively little international media attention. It will however undoubtedly have major ramifications in France and the Europe Union with regard to 'major hazard' legislation within the process industries. The actual cause of the loss remains under investigation, and is subject to difference of opinion between the parties involved. The explosion resulted in 31 fatalities, of which 25 are understood to have been outside of the plant boundaries. Up to 600 injuries have also been reported, with damage to 3,000 homes, of which 500 at the time of the incident were reported as uninhabitable. The site suffered extensive damage, and will not be reopened. Processing of third party claims is ongoing and as such final details are not in the public domain, it is however understood that the claims far exceed level of third party insurance purchased, by as much a six times.

Gas Plant Explosion – 25 September 1998 (deliberately out of date sequence). Reference is made to this incident for as much as the final liability exposure it didn't have, as for the exposure it did. The explosion at a Gas Plant in Victoria, Australia, caused 2 fatalities and injured 8, all on the site. Further, natural gas supplies to the State of Victoria were interrupted for 19 days. The incident prompted a Royal Commission of investigate and significant fines for the 'operator'. It also caused the State of Victoria to fast track the implementation of its Major Hazard Facility Regulations. One recent development in this incident is a ruling this year that the 'operator' was **not liable** for US$ 500 million of economic damages caused as a result of interruption of the natural gas supply, although was found responsible for two minor third party property damage incidents.

Before moving on, and leading on from the Victoria Gas Plant explosion, one theme worth noting from a number of the above is the relationship between large liability exposure incidents and impact on legislation affecting the process industries and their potential third party exposure. In addition to the Victoria Major Hazard Facility Regulations, we have in Europe the Seveso II Directive, named after a 10 July 1976 incident in Seveso, Italy in which dioxin was released following a runaway reaction, causing injury to more than 600 persons. These regulations are applicable to all European Union members, and are implemented via 'competent authorities' in the member states under various guises. In the United Kingdom for example it is the Health and Safety Executive that implement the COMAH (Control of Major Accident Hazard) Regulations. This is a two-tier set of regulations, in which amongst others an assessment of impact (and mitigation/elimination of this impact) to third parties forms a major part. In the USA there is of course the Risk Management Plan 40 CFR 68 Chemical Accident Prevention Provisions, where amongst other things a site is required to a present a worse case release scenario and alternate release scenario analysis. Many other countries are introducing their own bespoke systems aligned with either USA or European models. Parallels can be drawn with such regulatory requirements and the analysis methodology below. Such systems, if proven by audit to be as effective in application as they are comprehensive on paper, also provides a degree of validation to the liability insurer.

Third Party Liability Exposure Analysis

Background

Risk engineers have been employed in the Insurance Industry in significant numbers from the mid-1970's. This, particularly in the London Insurance Market, was triggered by the events in the Flixborough plant on 1 June 1974. Here, a vapour cloud explosion at a caprolactum plant involving cyclohexane, caused 28 fatalities, 36 site injuries, 53 reported offsite injuries, extensive plant damage and third party damage to housing in the neighbouring village. Whilst generally acknowledged as one of the more significant 'wake up' calls to the chemical industry in the United Kingdom, it also alerted insurance underwriters that there was a need for specialists within the insurance industry to understand the detail and risks associated with process plant insurance.

Whilst providing an overall qualitative and sometimes quantitative opinion as the quality of 'risk' associated with a given site, when quantifying actual loss scenarios the insurance risk engineers has focused mainly on property damage and business interruption EMLs (Estimated Maximum Losses). These EMLs are key in the decision making process of an underwriter as to the '% line' they wish to take for a given risk (the main factor here is the size of the EML relative to the underwriting 'capacity' of a particular insurance company). In contrast, liability exposures, historically have been addressed qualitatively only. In many cases, as has been illustrated in the above *Incident Review*, potential third party liability exposures can far outweigh property damage and business interruption exposures that have been given more attention historically.

Within the Marsh Marine and Energy Engineering division we have long recognised the liability potential impact of incidents. We have over the years developed a review process to be included either within the remit of a conventional property damage insurance survey, or by a dedicated liability assessment to analyse the potential liability exposures found with a facilities with the Onshore Energy Industry. Identification has been aided by checklists, developed as part of on internal Engineering Management System.

These exposures may include one or more of the following: vapour cloud explosions, BLEVE, vessel disintegration, flash fires, toxic releases, oil spills and others. More recently, a formal procedure for such analysis has been developed, with recourse to simulation of incidents and quantitative and semi-quantitative calculation of exposure.

The simulation techniques used behind the analysis are in themselves not new. Vapour cloud explosions are modelled on in-house software, SLAM, and other releases on commercially available simulation software. What is new is the interpretation of data and the clarity in which it is presented to the insurance audience. By having a clear liability exposure analysis, a control and risk financing strategy can be implemented that will ensure that the company assets, image and surrounding community are more adequately protected.

The analysis is not intended to be a cumulative assessment of liability exposures as one would find in rigorous quantitative risk assessment. Rather from an insured's and insurer's perspective the method focuses on the potential impact from single catastrophic worse case events. The intention is not to look at quantitative societal or individual exposures, rather to produce a defendable estimate of worse case exposure that can be used for insurance planning purposes. The scenarios developed are not intended to represent or serve as a complete risk assessment. It is however crucial that the site in question has considered the factors leading to the events modelled within an in-house risk assessment, and that a management process has implemented the appropriate corrective actions where 'barriers' have been found to be inadequate. Such a management process should embody as a minimum the following simple principles:

- hazards are recognised and understood
- equipment "fit for purpose"
- systems and procedures are in place to maintain integrity
- competent staff are engaged
- emergency procedures are developed and proven effective
- performance monitoring and continuous improvement is entrenched.

Events modelled to generate the Estimated Maximum Loss (EML) are low probability events, in that the EML is defined as: 'the loss that could be sustained under abnormal conditions with the failure of all protective systems.' No credit is given to the effectiveness of barriers or the impact of emergency response activities. Although the probability is not typically specifically defined, events evaluated must be considered credible and supported by industry precedent.

Third Party Exposure - A Methodology

The calculation of property loss potentials forms a key part of any underwriting submission or client sponsored risk assessment. In the Onshore Energy industry, one of the acknowledged event types that has most potential to cause a loss is the vapour cloud explosion (VCE). Calculation of damage from a VCE demands a methodology, which ensures a degree of consistency across all using it, with the same rules, scientific basis, and engineering principles.. The model (SLAM), used by Marsh was developed inhouse in the mid 90's and is based on the premise that vapour cloud explosions arise through combustion processes which, given the right conditions, generate high burning velocities and hence damaging overpressures. This methodology represented at the time of its launch a major departure from previous insurance industry methods. The SLAM model today is used to produce EMLs from VCE's, aids understanding of exposures during project activity, and forms the basis of the third party exposure from a VCE, both property and human effect.

First and Third Party Property Damage from VCE

The paper does not at this stage intend to review the detailed background of SLAM, this has been done on earlier occasions [3,4]. For the purposes of understanding the following example, the SLAM model is created from a site's plot plan, values are allocated to identified and categorised plant areas from a schedule of insurance and known third party detail. Areas of plant congestion are identified and quantified, and sources of VCE forming material identified together with possible sources of ignition. The model is then run, and allocates damage according to the overleaf table.

Fire & Blast Damage Allocation (%)					
Overpressure Annulus Bar	Process Plant	Heavy Machinery	Buildings Cooling Towers	Tankage	Average Site
> 0.70	100	80	100	100	100
0.70 – 0.35	80	40	100	100	80
0.35 – 0.20	20	0	100	100	20
0.20 – 0.10	5	0	100	50	5
0.10 – 0.05	0	0	50	0	0

SLAM – Allocation of Fire and Blast Damage

Results are represented as per the below.

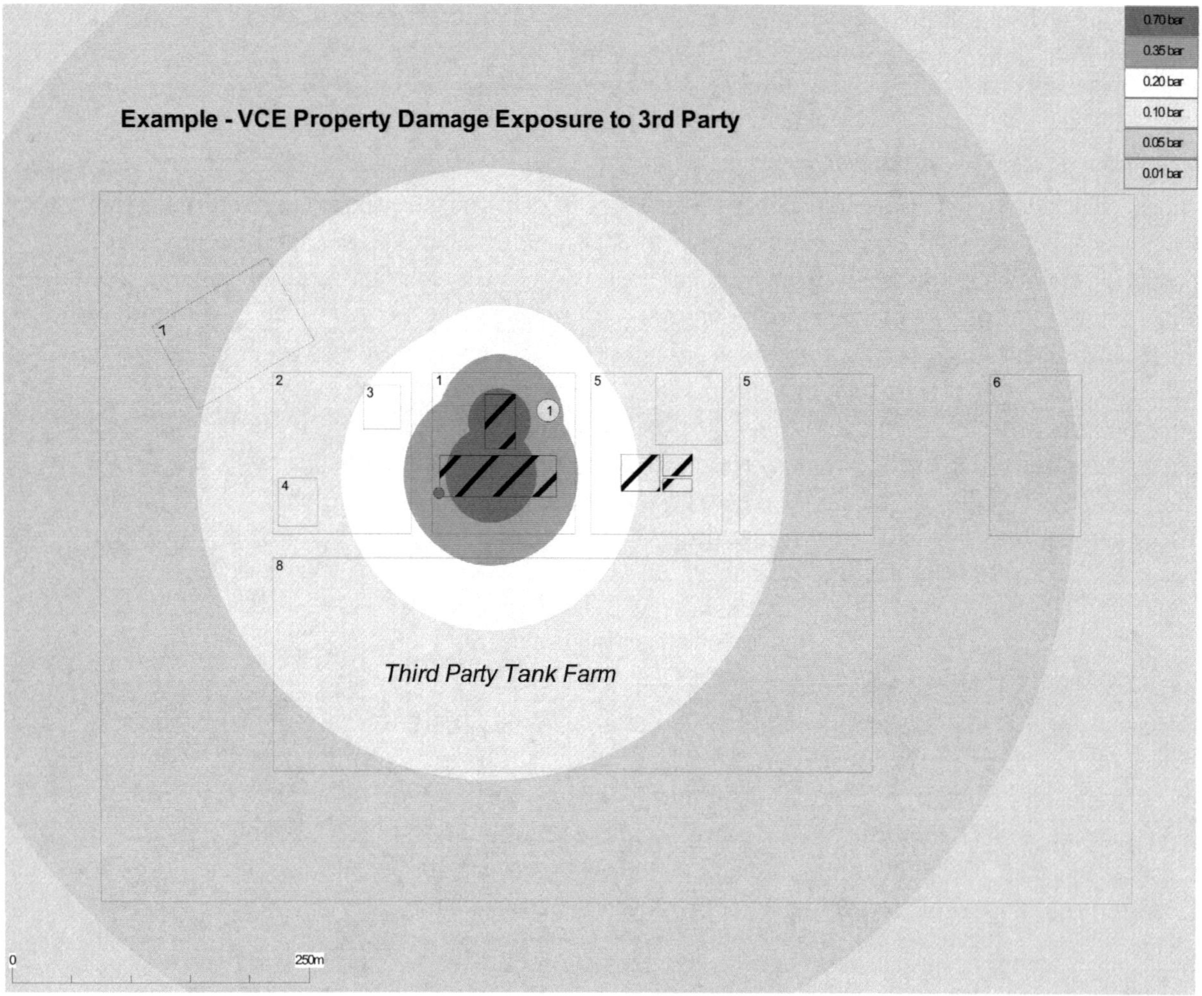

SLAM – Overpressure Diagram

This translates into the following damage estimate for both first and third parties, with the latter in ***italics***:

Unit			Area Exposed (%) Overpressure Zones barg					Total Damage US$ Million
No	Name	Value US$ Million	0.70	0.35	0.20	0.10	0.05	
6	Admin	15.0	-	-	-	-	94.8	7.1
5	Plat	50.0	-	-	18.1	43.7	38.2	2.9
1	CDU	100.0	40.7	52.3	6.9	-	-	84.0
4	HDS CT	5.0	-	-	-	100.0	-	0.3
2	HDS	50.0	-	1.5	41.1	57.1	-	6.2
8	***Tankage***	***80.0***	**-**	***0.4***	***10.1***	***58.6***	***30.9***	***31.8***
Total Value		300.0	40.7	42.8	15.4	26.2	7.1	132.2

Damage Estimates

Data as such in the above table would typically be considered as raw would be subject to additional costs including:

- Inflation arising from variation in values from time of valuation to time of loss
- Inflation arising from rebuild time
- Allowance for temporary facilities, investigation, commissioning and start-up costs and fire fighting expenses.

Other insurable property damage loss scenarios are possible such as jet fires, spill fires, tank fires, and natural hazards, and would be considered as appropriate, although for a typical medium size refinery example, a VCE scenario typically represents the EML event. Care in analysis should be taken to ensure that the worse case third party property damage liability event has been identified, as it is clearly not always concurrent with the worse case first party exposure. For example, if one considers a site with marine facilities, loss of a tanker and crew at a jetty, as a result of the site's negligence, has potential for significant liability exposure.

It is noted that for liability assessment purposes, unless specifically asked otherwise, it is reasonable to assume that no account has been taken of any contractual limitations that may be in force between any third parties on or offsite.

Nor in all cases, is the third party property so easily categorised and valued as in the above example. Under these circumstances, it is necessary to rely on more generic assumptions as detailed in the next section.

Liabilities – Other considerations

In this section we consider the liability implications arising from human and environmental impact. Events typically considered as giving rise to large operational liability exposures for a processing facility, but are not necessarily limited to, are:

- VCE
- BLEVE
- Flash Fires
- Jet Fires
- Toxic Release
- Oil Spills.

Use is made for analysis of these exposures, with exception of VCE, of Det Norske Veritas's software process hazard analysis modelling - PHAST. Meteorological conditions for modelling purposes are taken from site data.

Impact criteria

The demand of the liability exposure modelling is to produce Impact Criteria that are 'fit for purpose'. The model must be easily and consistently applied across of portfolio of risks, with minimal demand for data. In this regard, the below impact/endpoint have been selected.

Event	**Impact/Endpoint**		
	Property	**Human**	**Environmental**
VCE	Use SLAM e.g. 80%+ damage within 0.35 bar overpressure circle. Glass breakage limit at 0.01 bar	0.2 bar overpressure No protection indoors.	Considered minimal
BLEVE	Considered minimal	Thermal dose of 2300 $(W/m^2)^{4/3}s.10^{-4}$ No impact on personnel indoors	Considered minimal
Flash Fires	Considered minimal	Within ½ LFL contour No protection indoors	Considered minimal
Toxic releases	Considered minimal	LC_{50} dose for 10-15 minutes. No protection indoors	Considered minimal
Oil Spills	Considered minimal	Considered minimal	Spill from maximum 30 minute release

Onsite Liability Assessment Impact/Endpoint Criteria

With regard to human impact, points have been selected to approximately represent the 50% fatality point.

Personnel at Risk

Again for the purposes of insurance assessment a number of simplifying assumptions have been made. It is worth re-emphasising that the intent is not to produce quantitative societal or individual exposures, rather to produce a defendable estimate of worse case exposure that can be used for insurance planning purposes. Site populations are generally available with reasonably levels of accuracy. There is no differentiation between own employees and contract staff. Whilst there is a great difference between the on-site population during normal working days and evening and weekends, for an oil refinery, for study purposes, a pessimistic normal working day population is typically chosen. In the absence of breakdown, this is generally estimated as 75% of the total payroll, including contractors. It is noted, that in our experience, for a medium sized oil refinery in the USA/Europe, that on-site personnel outside of the normal working hours typically vary between 5-10% of the total payroll.

It is not typical to consider larger site populations such as present during shutdown activity, as hydrocarbon inventory is more often or not removed at point of peak population. Should however there be ongoing construction activity, for example a major revamp or expansion within the plant boundaries, it would be prudent to consider including numbers from such activity within the totals.

For populations on property adjacent or within the site boundary, headcounts are normally typically available from Safety Case studies. Likewise, such 'off the shelf' assessments will also contain wider population data. In the USA, there are a number of excellent 'web-based' sources of population data, although generally more difficult to find easily within the public domain elsewhere.

Having identified populations at risk, the following broad assumptions are made:

- There is an even distribution of personnel (persons/m^2) through both on-site and off-site areas. Identification of actual plot area for both onsite and offsite areas under consideration will allow simple derivation of this number.
- 85% of persons are indoors at any one time.

Human Exposures

Rather than report actual number of deaths/injuries from what is broad-based assessment tool, we choose to rank potential human exposures based on a banded basis. These are represented below for numbers of people affected for the above given endpoints:

Human Effects (HE)	
Level	**Exposure** 0.2 barg, or Thermal dose of 2300 $(W/m^2)^{4/3}.s.10^{-4}$ or, Within ½ LFL contour, or LC_{50} dose for 10-15 minutes
1	Less than 10 people affected
2	Between 10 and 100 people affected
3	Greater than 100 people affected

Human Exposures Effects

Using the above VCE example, it would be typical in an assessment to consider four overpressure radii as part of the analysis. The first as identified is for human effect purposes at 0.2 barg, with the other three used to assess any additional third party damage considerations that may not have been taken into account in the SLAM model.

The points are:

Overpressure Bar	Radius m	Comments
0.2	196	Human effect end point
0.1	248	Severe housing damage
0.05	492	50% housing damage
0.01	1881	Window breakage limit

SLAM – Overpressure Profile

By example, the area of exposure human exposure resulting from the above equates to approximately 30,000 m^2. This in turn is converted to actual number of persons exposed by multiplication with the identified persons/unit area, and from there a **'Level'** is allocated.

In much the same way, the area of human exposure is calculated for other events and corresponding **'Level'** identified.

It is noted, that based on our experience, a VCE event rarely constitutes the most significant human exposure, assuming a typical medium sized refinery. BLEVE and flash fire both have potential for far reaching exposure beyond plant boundaries on thermal impact alone. Often for a flash fire, the end point needs to be considered carefully given that prediction does take account of the many ignition sources a dispersing cloud may approach.

Environmental Exposures

Likewise for environmental exposures, for which often the main consideration is release of liquid to a shoreline, either on a coastal, lake or river location, we have ranked the potential exposure as a **Category**, rather than specifying actual spill volumes:

Environmental Effects (EE)	
Category	**Exposure**
1	Less than 1,000 barrels spilled
2	Between 1,000 and 1,000,000 barrels spilled
3	Greater than 1,000,000 barrels spilled

Environmental Exposures Effects

At time of submittal of this paper, we were considering re-evaluation of the exposures, on the basis that experience has illustrated Category 2 to be too wide.

It is noted, that the cost for a spill from a tanker in most waters howsoever caused, is in many cases covered by convention ratified by Governments. For example, there is one convention, the Civil Liabilities Convention (CLC) for International Waters, and the Oil Pollution Act (OPA) for US waters. These conventions essentially limit tanker owners liability (except for gross negligence). However, no such limited liability extends to the onshore operator, who if proven negligent could be liable for the full cost of clean up, plus any loss of business, amenity, and punitive damages etc.

One final point in this area, with reference to the earlier *Incident Review* section, it is noted that the 'Basel' incident of 1 November 1996, arose from the environmental impact of the event escalation as a result of the fire-fighting activity.

Loss Estimates Summary Presentation

Below, an example is offered of presentation of end results, based on an assumed coastal medium sized oil refinery, with moderate third party exposure.

First Party Physical Damage EML

Scenario	**US$ million**
VCE from release in CDU Area	250

Worst Case Liability Exposures

	Exposure		
Scenario	**Physical**	**Human**	**Environmental**
Physical Damage – CDU VCE	US$ 50 million	Level 1	minimal
Human Effect - Bleve/Flash Fire	minimal	Level 2	minimal
Environmental - Oil spill on water	minimal	minimal	Category 2

Clearly, for a company with multiple locations of different sites types, site historical development, layout, geographical locations and proximity/function of third parties, one can start to build up an risk register and exposure profile. This in turn becomes the tool by which strategy decisions can be made with a clear understanding of exposures.

By definition, the above profile generated is one of catastrophic exposure, and as such events considered have a low probability of occurrence. There may be some benefit in applying broad probability categories to these events, perhaps using a risk matrix to identify any differentiation. Should lower 'tier' events, i.e. non-catastrophic be considered, probability profiling is definitely encouraged.

Other Exposures – Third Party Business Continuity

One area of analysis not specifically addressed in the above has been that of financial impact with regard to business continuity. We have looked at property damage, human effects and environmental impact but not the financial loss to a third party from a business interruption perspective in the event. Consideration of this would arguably add further value to an analysis. The challenge here is that third parties affected by a modelled event would need to share their own profitability data. This may not always be forthcoming and practical. An alternative could be to consider a level or category approach as has been applied to human effects and environmental impact (as above) based on the type of industry involved and generic profitability data.

Topical Causes

Whilst we have not explored the root cause of a particular scenario, there is one topical cause worthy at this time of individual mention, namely terrorist attack and broader security violations which have potential for wide liability exposure. Security challenges to a facility are dealt with in detail at a dedicated session at this symposium. It is sufficient for this paper, that sites worldwide (and not limited to those where it is compulsory by legislation) should be encouraged to perform a broad Security Vulnerability Analysis (SVA), and then to plan for improvement. Following this analysis, further attempt should be made to integrate security into management systems and then subject security to routine review and audit as with any other system. Security issues should also be addressed in management of change to ensure that it gets due consideration.

As to liability implications arising from a terrorist attack, it is considered imperative as with any risk that has a potential to cause injury/damage to third parties, that the site is able to demonstrate that security exposures have at least been mitigated to as low as reasonably practical (ALARP-principle). Security considerations are of particular interest with regard to ALARP, given that security enhancement works on the basis of displacement, i.e. the potential perpetrator is 'encouraged' to move to the site with lower levels of security in place.

Conclusion

Future & Developing Exposures

What does the future hold for insureds and potential areas of exposure? **Trend Watch** [2] identifies the below as areas in 2002 that provide possible indication as to growing areas for future court awards:

- Human Rights – Of the Top 100 verdicts in 2002 seven came from civil rights cases, in 2001 none made the list. Cases included false imprisonment, race discrimination and defamation charges. Many were brought against individuals, rather than corporations or institutions.

- Intellectual Property – When told in simple terms, juries understand that stealing is stealing, with US$ 641 million awarded in 2002 for patent, copyright and trade secret cases in the Top 100, more than double that of 2001.

- Medical Malpractice – Arguably reaching a crisis point in the USA, with the need for malpractice tort reform pushed to the forefront of domestic-policy. Some argue that the medical-liability system is already broken, few doubt the need for the reform. Thirteen malpractice awards were among the Top 100, with nine of these involving failure in neo-natal care. Caps on damages for pain and suffering are a possible way forward.

- Punitive Damages – If these are a measure of juror anger, then clearly there were a lot of angry jurors in 2002. In comparison of the Top 50 awards for 2001 and 2002, punitive damages were awarded in 22 cases. In 2001, total for the 22 cases was US$ 3.2 billion, in 2002 the figure was US$ 32.6 billion. Clearly the awards may not be just down to anger, but simply that more wrongdoing came to light in 2002, borne out by a 37% increase in compensatory damages in cases where punitive awards were made. Nonetheless the ratio of punitive damages to compensatory damages averaged 29 in 2002, compared to 4 in 2001. Clearly use of average figures can be influenced by extreme cases, as was the case for 2002. However even using the median, an increase in median compensatory award ratio from 2.3 to 4.5 between 2001 and 2002 was seen for punitive awards.

- Railroad Litigation – Awards in 2002 were high, with a number of cases involving the alleged spoilage of evidence by defendants.

Add to this list other emerging/evolving exposures such as 'toxic molds, lead paints, auto liability, electromagnetic fields, SARS, and for the process industries, a 'terrorist threat' and potential liability fall-out, the need for understanding of exposures has arguably never been greater.

Arguably the area above most directly impacting the Onshore Energy industry is that of Punitive Damages, this as previously stated being the one area not normally insurable.

Fundamentals

Ultimately, the best way to protect against potential claims in the first place has to come back to the elementary identification, quantification and control of risks, with a cycle of audit and improvement entrenched. These in turn support regulatory compliance, and a demonstration that a point of ALARP has been reached. From these simple and fundamental 'building blocks' a company is in a far stronger position to start identifying its own Limits of Liability.

A technical, operational risk focussed analysis as illustrated in this paper is but part of the first item in the below general list of what an insured needs to consider when considering liability exposure and its subsequent 'protection':

- **Evaluate and prioritise exposures**
- Consider the impact of exposures particular to the varied geographical locations in which they do business
- Understand exposure to multi-claimant suits
- Consider raising deductibles rather than a reduction of limits
- Consider 'ventilation' of insurance programmes to fund additional catastrophic layers
- Access alternative insurance programme structures and funding mechanisms
- Recognise that occurrence limits are permanent – claims can be made under an occurrence-based policy of long ago
- Choose highly rated carriers with the financial strength to pay future claims
- Devise new or revitalise effective risk control strategies
- Make careful use of benchmarking data.

Insureds should look to secure limits that are reasonably well aligned with experience and informed expectation. Further, it is emphasised that liability insurance limits do not exist in a vacuum, and the ultimate question of "How much is enough?" depends on assessment of every aspect of a business that creates risk and every tool that can be used to reduce risks.

References

1. The Hazards of Life and All That, John Bond, ISBN 0 7503 0360 3.

2. Limits of Liability 2003, Research Report, Marsh.

3. Fuel Gas Explosion Guidelines - practical applications, Barton R.F. - paper presented at I.Chem.E Major Hazards Onshore and Offshore II Conference; Manchester, Major hazards IChem.E Symposium Series No. 139, page 285, 1995.

4. Risk Management: An Improved Approach to Modelling Vapour Cloud Explosions, Barton R.F. – paper presented at 2nd International Conference on Loss Prevention and Safety, The Bahrain Society of Engineers, page 361, 1995.

The Role of ASTM E27 Methods in Hazard Assessment

David Frurip
The Dow Chemical Co.,
Bldg. 1897F
Midland MI 48667
989-636-2446
DJFrurip@dow.com

Larry Britton
Consultant,
69, Percy Road,
Hampton TW12 2JT
England UK
(734) 769-0670
lbritton@neolytica.com

William Fenlon
SHE Technologies
90 Red Hickory Drive
Rochester, New York 14626
(585) 225-6661
WilliamJFenlon@cs.com

John Going
Fike Corp.
704 S. 10th St.
Blue Springs, MO 64013
(816) 229-3405x521
john.going@fike.com

Keith Harrison
Dept. of Chemical Engineering,
University of South Alabama,
Mobile, AL 36688
(251) 460-6160
kharriso@jaguar1.usouthal.edu
sun1.che.usouthal.edu/harris.htm

Jeff Niemeier
Eli Lilly and Co.
Lilly Technology Ctr.
Indianapolis, IN 46285
(317) 276-2066
Niemeier@Lilly.com

Erdem A. Ural*
Loss Prevention Science &
Technologies, Inc.
659 Pearl Street
Stoughton, MA 02072
(781) 344-7656
erdem.ural@lpsti.com
www.lpsti.com

**Author to whom correspondence should be directed.*

Prepared for presentation at the 38th Loss Prevention Symposium
at the 2004 Spring National Meeting of the American Institute of Chemical Engineers.

January 2004

Unpublished

Abstract

Any consequence model is only as good as the selected input data. Accurate reactive chemicals data form the cornerstone of procedures used to assess the hazards associated with commercial chemical production. For over 35 years, the ASTM E27 Committee (Hazard Potential of Chemicals) has issued numerous, widely used consensus standards dealing with diverse testing and predictive procedures used to obtain relevant chemical hazard properties. The decision to issue a standard rests solely with the membership, which consists of representatives from industry, government, consulting firms, and instrument suppliers. Consequently, the procedures are automatically relevant, timely, and widely applicable. The methods developed by E27 and described in this paper cover flammability, ignitability, and reactivity of fuel/air mixtures, thermodynamics, thermal stability, and chemical compatibility. The purpose of this paper is to highlight some of the widely used standards, complemented with hypothetical but relevant examples describing the testing strategy, interpretation, and application of the results. A further goal of this paper is to encourage participation in the standards development process.

I. Introduction

ASTM Committee E27 (Hazard Potential of Chemicals) has been in existence for over 35 years. During that time, its members have developed and approved over 30 consensus standards relating to the fields referred to as "reactive chemicals" and flammability. The Committee was formed because of the need for consistent, scientifically based, accurate, easy to apply, and most of all, *consensus* methodologies for the determination of parameters and properties which allow industry to properly design safe processes. Up until the time of the Committee's formation, many companies had developed "home-grown" test methods which may have served them well but perhaps lacked technical accuracy or general applicability. The synergy of a motivated group of experts, following the well established procedures of a premier consensus standards organization like ASTM (American Society for Testing and Materials) led to the development of a series of useful standards. E27 standards are widely recognized and used in hazard assessment and loss prevention engineering.

The official scope of ASTM Committee E27 is the development and standardization of physical and chemical test methods, nomenclature, and the promotion of knowledge and stimulation of research bearing on the hazard potential of chemicals. The areas of interest to the Committee are the development of test procedures for determining the degree of susceptibility of materials to ignition or release of energy under varying industrial conditions, including flash points, flammability limits, autoignition temperatures, dust explosion limits, and kinetics of decomposition. This includes experimental procedures and estimation methods.

The membership of E27 includes experts from chemical and pharmaceutical companies, academia and process safety testing and consulting firms. These companies and organizations have been willing to share their knowledge of this important area for the synergism which occurs during free and open technical interactions.

Committee E27 Organization
Currently the Committee consists of approximately 80 members and is divided into the following technical sub-subcommittees:

E27.01 Editorial and Nomenclature
E27.02 Thermal Stability and Condensed Phases
E27.04 Flammability and Ignitability of Chemicals
E27.05 Explosibility and Ignitability of Dust Clouds
E27.07 Estimation Methods

Meetings are held twice per year, typically in March and October. Occasionally, meetings are held at the member's sites in order to get a better understanding of how hazard evaluation and the entire Process Safety Management process is implemented by individual companies and organizations. In recent times, these meetings have occurred at Kodak, Fike, CIBA Specialty Chemicals, Dow, Eli Lilly, and the Pittsburgh Research Center (formerly the U.S. Bureau of Mines). More information on upcoming meetings, Committee contacts, and Committee standards can be found under the Technical Committees link at www.astm.org.

Standards Administered by E27

A comprehensive listing of all standards currently administered by E27 is given in Table I. In this listing, we have included some of the citations where the standard may be recommended for use. Also included is a short description of the scope of each standard. The table is organized into three sections representing Flammability and Ignitability Test Methods, Special Test Methods for Dust Clouds, and Thermal Stability and Estimation Methods.

To give the reader a further sense of the general applicability of E27 standards, Figure 1 shows two flowcharts, one for dust explosion protection, and other for thermal stability evaluation. Within each appropriate box relevant ASTM E27 test methods are listed. Clearly, the methods administered by ASTM E27 are well aligned with these and other commonly used hazard/consequence evaluation methodologies.

Survey of ASTM Method Usage

In preparation of this paper, an informal survey was conducted in the fall of 2003, to estimate the frequency of use of ASTM E27 test methods for flammability, ignitability and explosibility. The results were expected to support the general importance of standard methods as well as indicate to the Committee which specific methods were of greatest interest to the user community. A survey form was sent to all of the laboratories in the US known by the Committee to be using these methods. This includes commercial laboratories, government laboratories, chemical manufacturers' laboratories and explosion protection manufacturer's laboratories. Out of a total of 14 information requests, a response was received from 10. The results were tabulated in terms of total number of times a method was used on an annual basis, using the past 3-4 years as a base. The results are listed in Table II by ASTM method number. It appears from Table II that the dust test methods are being used much more frequently than the rest of the rest of the methods. Flash point test method E 502 appears to enjoy frequent use. What seems most remarkable is that the three most frequently used dust cloud test methods E 1226, E 2019 and E 1515, account for 74% of all reported tests.

There are two apparent reasons for this finding. First, the gas and vapor test methods have been around for a long time, and test data have been compiled and published for a large number of pure chemicals (e.g. NFPA 325). Existing mixing rules or correlations (e.g. Kuchta) are considered adequate for most loss prevention applications involving fuel mixtures. On the other hand, a large data base obtained using E 789 for dust clouds is now obsolete, being regenerated using E 1226. The second reason for the large frequency of dust cloud tests is the strong sensitivity of flammability, ignitability, and explosibility values to the particle size distribution of the sample. That is why different powder products made up of the same pure chemical needs to be tested individually.

Although this survey was focused on the areas of flammability and dust explosion hazards, it is expected that a similar survey of the standards in the chemical stability area would show a high frequency of use, as well. As an example, the thermal stability screening test, E 537, is applied several hundred times each year at The Dow Chemical Company alone. Certainly the strong sales and frequent citation of the CHETAH program over the years is a testament to its usefulness.

Paper Organization

This paper seeks to describe a selected number of standards administered by E27. The standards described in this paper are among those methods considered by the authors (all authors are active members of E27) to be generally applicable and widely used in industry. In many cases, a hypothetical example is given to illustrate exactly how the results of the standard might be applied to a real scenario. If any reader is interested in learning additional details about these methods, please read the standard and contact ASTM who can put you in contact with the appropriate E27 subject matter expert.

II. Standards Related To Flammability, Ignitability, And Explosibility Of Fuel/Air Mixtures

These methods are used to assess the gas, vapor and dust explosion hazard properties of materials, to establish safe operating conditions, and to perform consequence analysis for hypothesized accidents. Generally speaking:

- Flammability test methods aim to determine the concentration limits of fuel, air (oxidant), or diluent (suppressant) in the presence of a sufficiently strong ignition source. As will be seen below, this concept can also be extended to temperature or pressure limits of flammability as related to commonly used flash point values.
- Ignitability test methods aim to characterize the minimum "strength" of an ignition source required to ignite the most sensitive fuel/air/diluent mixture.
- Explosibility test methods assume the presence of both the sufficiently strong ignition source, and the worst case mixture, and aim to characterize how fast the combustion reaction can proceed in a particular fuel/air/diluent combination.

II.A. Flammability Test Methods

ASTM E27 Committee maintains and develops separate flammability test standards for dust-air and gas/vapor-air mixtures.

Flammability Limits of Dust Clouds: ASTM E 1515, Minimum Explosible Concentration of Dust Clouds, and ASTM WK1680, Limiting Oxygen Concentration of Dust Clouds.
Two methods (one approved and one in development) are designed to assess flammability limits of dust clouds. The methods address fuel as a limit (E 1515, Minimum Explosible Concentration of Dust Clouds) and oxygen as a limit (WK1680, Limiting Oxygen Concentration of Dust Clouds). The latter method is a working draft document available to sub-committee members. In both cases, when one of the required elements for flame propagation is decreased, a point is reached where flame propagation ceases to occur. Beyond that point, the heat release rate is insufficient to allow self-sustaining flame propagation through the dust cloud. In practice, the minimum explosible concentration (MEC) is determined in the presence of excess oxygen (normally air) while the limiting oxygen concentration (LOC) is determined at the "optimum" dust concentration. It is important to recognize that the optimum dust concentration decreases with decreasing oxygen concentration.

Flammability limits are used to characterize safe operating conditions in industrial processes. The MEC identifies the point at which a dust explosion hazard exists in air. It is matched with the industrial process and what is known and what can be controlled reliably. If the operating conditions and maintenance procedures can be depended upon to never exceed the MEC (or a fraction of the MEC), then the hazard may be considered to be minimal. In the extreme, the test may identify a dust as being inert or non-ignitable at any concentration

Similarly, the LOC method is used to identify safe operating conditions <u>when the oxygen concentration is controlled</u>.

The testing procedure is fundamentally the same as described below for E 1226. In the case of MEC determination, the dust concentration is reduced systematically from an explosible level until no deflagration occurs. The fuel is reduced in 25% steps near the MEC and the pressure is monitored as the indicator of deflagration. When the measured overpressure first drops below 1 bar in duplicate tests, the MEC is determined from the data as the concentration that yields an overpressure of 1 bar or greater **(Cashdollar 1992)**.

When determining the LOC, the oxygen concentration is reduced in 1% steps again until the measured overpressure is less than 1 bar. The dust concentration is varied up and down and the oxygen concentration changed as necessary to find the minimum concentration not resulting in a deflagration. The LOC is then reported as the average of the lowest concentration that deflagrated and the highest concentration that did not **(Going 2000)**.

In both MEC and LOC testing selection of the appropriate igniter energy can be difficult, especially for hard to ignite materials. Importance of differentiating flammability limits from the ignitability limits is discussed in E 1515. Therefore, tests may need to be repeated using different energy igniters. On the other hand, strong ignition sources overdrive a deflagration in a small 20 L chamber, predicting explosibility where the material is not explosible. Such hard to ignite material are tested in larger (1 m^3) chambers with strong ignition source.

As seen in Table II, the MEC method enjoys frequent use. Chapter 6 of NFPA 69 discusses explosion protection by control of dust concentration. Although it does not reference E 1515, this would be the applicable method for a dust cloud. It is identified as the standard method in NFPA 654 and 664 and in FM Data Sheet 7-76. The LOC method has not yet been published and so no direct citations have been made. NFPA 69, Chapter 5, however, addresses deflagration prevention by oxygen control. Safety factors are specified when using LOC data and reference is made to E 2079, Standard Method for Limiting Oxygen Concentration of Gases and Vapors. When approved, the draft LOC method for dusts will be applicable to Chapter 5 of NFPA 69

<u>Example</u>. Prior to installing vents in their dust collector, a company desired to determine if an explosion hazard did exist. An MEC test was conducted on their active ingredient, and a value of 30 g/m^3 was found (see Fig. 2). This is relatively low for a MEC and indicates the high hazard associated with this material. Engineering calculations and material recovery data indicated that the dust concentration in the collector could exceed this level, particularly during pulsing. Consequently, the decision was made to install explosion vents on the dust collector.

Flammability Limits of Vapors and Gases

The **<u>ASTM E 681</u>** test method aims to simulate the behavior of large volumes of test mixture with respect to the lowest and highest fuel concentrations that can propagate a flame through a homogeneous mixture with air (or other oxidant no stronger than air) at atmospheric pressure and given test temperature. For most fuels a 5-L spherical glass test vessel is used. For fuels with large quenching distances that may be difficult to ignite, a special procedure is adopted using a 12-L spherical flask. The gaseous mixtures are subjected to electrical spark ignition source and the absence or the presence of flame propagation is determined visually. Recognizing the subjectivity of the visual flame propagation assessment for near-limit mixtures, some laboratories use video recording systems for subsequent analysis. It is important to note that the energy released by the chemical igniters used in dust cloud flammability testing is almost a thousand times that used in E 681 method. The reason why stronger ignition sources are not used for hard to ignite materials in E 681 is explained as "if too high an energy ignition source is used, all that can be seen is the dissipation of the ignition energy and not the propagation of a flame."

<u>ASTM E 918</u> is designed to determine flammability limits at elevated temperature (up to 200 ^{O}C) and initial pressure of up to 200 psia. The pressure resistant metal test vessel is a vertical cylinder, only 1-L in size, which is much smaller than that used in E 681. However, vessel size effect may not be as important for tests at high initial pressure since the quenching distances decrease with increasing initial pressure. Electrically exploding fuse wire is used as an ignition source. E 918 uses a pressure criterion for flame propagation which can be much less subjective than the visual method. Flame propagation is defined as a combustion reaction that produces at least a 7% rise of the initial absolute pressure.

<u>ASTM E 2079</u> was published recently to test the limiting oxygen (oxidant) concentration of gases and vapors. Being a new method, E 2079 has been designed to address the problems associated with the E 681 and E 918 methods **(Ural 2002)**. E 2079 requires the test vessel to be larger than 4-L, and nearly spherical. Typically a 10 J electrical ignition source, and the 7% pressure rise criterion is used to establish flammability. Where necessary, the use of strong igniters is permitted. Advice on how to modify the pressure criterion for such tests is also provided in the test method. Presently, the sub-committee is developing a new draft standard test method to determine the lower and upper flammability limits, employing the concepts developed in E 2079. A round-robin study of the pressure criteria for flame propagation is also underway.

A review article describing the evolution of flammability test methods was recently published (**Britton 2002a**). It is pointed out that current European test methods using small, vertical test vessels are generating erroneously wide "flammable limits" by needing to adopt an unrealistically stringent ignition criterion. The lower flammable limits (LFLs) of organic fuels in air agree quite well with the predictions of a "heat of oxidation" model **(Britton 2002b)** provided the LFL data are generated in a vessel of adequate size such as required in ASTM E 681. The absolute error caused by "overdriving" the ignition process in small test vessels is far greater at the upper flammable limit (UFL) than at the LFL. This can cause unnecessary expense when the data are applied to large or high-throughput systems operated above the UFL.

Example: Deflagration Prevention by Control of Combustible Concentration

The objective is to operate outside the flammable range (below the LFL or above the UFL) using appropriate safety factors to mitigate errors in either the gas concentration or the flammable limits. In some systems the composition may vary with both position and time while there may be additional instrumentation errors caused by calibration offset and/or time lag. Where such effects make it impractical to operate above the gas mixture UFL, such as in some vent collection headers, it is common practice to add a sufficient concentration of "enrichment gas" to cause the net fuel composition to always be above the UFL, regardless of how the vent stream composition may change. Similarly, some early suppression systems used fuel as the suppressant.

Chapter 6 of NFPA 69 provides that after the appropriate LFL for the combustible components has been determined, addressing all operating conditions, the combustible concentration shall be maintained at or below 25% of the LFL. The exception for gas-phase systems is where automatic instrumentation with safety interlocks is provided, in which case the concentration shall be maintained at or below 60% of the LFL. This method is termed "deflagration prevention by combustible concentration reduction".

An alternative control method described in Chapter 5.7.3 of NFPA 69 permits operation above the UFL. This is a common strategy for atmospheric storage of liquids whose vapor pressures are high enough to maintain vapor concentrations above the UFL under all operating conditions. This method is included under "deflagration prevention by oxidant concentration reduction" in NFPA 69 because arguably, when operating above the UFL the concentration of oxygen available from ambient air is insufficient to produce an ignitable mixture.

Specific guidance is given for near-atmospheric vent collection headers in Chapter 5.7.3.2. Addition of at least 25 vol% of methane or natural gas can be used to render the gas stream nonflammable provided no component of the stream has a UFL greater than that of hydrogen in air (75 vol%) and the vent header does not contain oxygen at a concentration greater than can be derived from ambient air. Marine vent collection headers can also be rendered nonflammable by the addition of enrichment gas but the precise requirements are separately covered in Title 33.CFR. Part 154.

Flash Point Tests

Flash Point is considered a material hazard property, perhaps the most frequently used in fire protection engineering. ASTM defines flash point as "the lowest temperature, corrected to a pressure of 101.3 kPa, at which application of an ignition source causes the vapors of the specimen to ignite under specified conditions of test." As this definition suggests, the flash point values can be significantly influenced by the test apparatus and the test method. Everybody knows that the open cup flash point tests yield generally higher flash point values than the closed cup tests. It is also important to recognize that significant differences may exist even among the values obtained from closed cup test methods. Unfortunately, flash point data reference tables (e.g. NFPA 325) are populated with values of unknown origin, perhaps because of the varying influence of different industries in the consensus standard making processes. For example, NFPA 30 defines flash point as "the minimum temperature of a liquid at which sufficient vapor is given off to form an ignitible mixture with the air, near the surface of the liquid or within the vessel used, as determined by the appropriate test procedure and apparatus specified" in that standard. NFPA 704 on the other hand, does not specify a test method, and defines flash point as "the minimum temperature at which a liquid or a solid emits vapor sufficient to form an ignitable mixture with air near the surface of the liquid or the solid."

ASTM E27 Committee administers two standards solving some of the problems in this arena.

E 502 "Standard Test Method for Selection and Use of ASTM Standards for the Determination of Flash Point of Chemicals by Closed Cup Methods":
Despite being one of the earlier standards of the E27 Committee E 502 is widely used and referenced today (see for example, NFPA 1, NFPA 30, NFPA 35, NFPA 77, NFPA 115, NFPA 385, and NFPA 704). This test method covers the determination of the flash point of liquid and solid chemical compounds flashing from below - 10 to 370°C (16 to 700°F). E 502 employs the procedures and apparatus in ASTM Test Methods D 56, D 93, D 3278, D 3828, and D 3941. It provides additional explanatory notes and procedure modifications not contained in the individual methods. E 502 also permits determination of flash point for solids and highly viscous liquids. For a given fuel viscosity and anticipated flash point range, the user is advised which particular test is suitable.

E 502 offers a valuable discussion providing insight into some of the dangerous or curious results obtained from flash point tests. First, the flash point does not represent the minimum temperature at which a material can evolve flammable vapors. With the exception of certain equilibrium test methods, most flash point tests are run at a finite heating rate, and therefore, vapor concentrations are not representative of equilibrium conditions. Flash point testing employs downward and horizontal propagation of flame. Flame propagation in these directions generally requires slightly higher vapor concentrations than is required for upward flame propagation. The flame is introduced at a finite distance above the liquid surface. Since the vapors are denser than air, the vapor concentration is slightly higher at the liquid surface than at the flame position. There are instances with pure materials where the absence of a flash point does not ensure freedom from flammability. Included in this category are materials that require large quenching diameters (e.g. halogenated hydrocarbons or certain aqueous acetic acid solutions). Such material will not propagate a flame in apparatus the size of a flash-point tester, however, its vapors are flammable and will burn when ignited in a larger process or storage vessel. Some materials having very dense vapors, a narrow range of flammability, or the requirement for being somewhat superheated to burn, will not exhibit a flash point with conventional test methods, but can form flammable vapor - air mixtures if heating and mixing are optimum and the temperatures are raised. In specific instances, contrary to usual behavior, the open-cup flash point may be at a lower temperature than the closed-cup flash point.

Conventional flash point tests can also introduce large errors when the fuel is a mixture or contains flammable impurities. A liquid containing flammable and nonflammable components can evolve flammable vapors under certain conditions while not exhibiting a closed-cup flash point. This phenomenon is observed when a nonflammable component is sufficiently volatile and present in sufficient quantity to inert the vapor space of the closed cup. In many instances, liquids of this type will exhibit an open-cup flash point. In certain cases the material may exhibit no flash point, either open or closed, but when spilled it may become flammable after the nonflammable component evaporated. Therefore, it is important to test samples that have been weathered to an extent comparable to the material in the process stream. It is also important to keep in mind that liquids containing a highly volatile nonflammable impurity, which exhibit no flash point because of the influence of the nonflammable material, may form flammable mixtures if totally flash vaporized in air in the proper proportions. Some mixtures of water and hydrocarbons, or low volatility halogenated hydrocarbons and volatile hydrocarbons, may have low flash points but will not of themselves sustain burning. These materials can present explosion hazards in closed vessels but will not burn as a pool fire out in the open. Therefore for process and storage vessels, E 502 recommends the use of E 1232 which is described below.

E 1232 "Standard Test Method for Temperature Limit of Flammability of Chemicals":
The temperature limit of flammability test measures the minimum temperature at which liquid or solid chemicals evolve sufficient vapors to form a flammable mixture with air under equilibrium conditions. This test is designed to remedy limitations inherent in flash point tests, and yields a result closely approaching the minimum temperature of flammable vapor formation for equilibrium situations in the chemical processing industry such as in process vessels, storage tanks and similar equipment. This test also allows

the use of oxidant/diluent mixtures other than air, which needs to be evaluated in cost-saving partial inerting applications.

The results of this test is deliberately differentiated from the flash point by using the term "lower temperature limit of flammability (LTL)", which is defined as the lowest temperature, corrected to a pressure of 101.3 kPa (760 mm Hg, 1013 mbar), at which application of an ignition source causes a homogeneous mixture of a gaseous oxidizer and vapors in equilibrium with a liquid (or solid) specimen to ignite and propagate a flame away from the ignition source under the specified conditions of test. Unlike flash point, LTL is a true material property in most circumstances. The difference between the flash point and LTL increases as the flash point increases. It is important to recognize that LTL can be tens of degrees Celsius lower than the closed cup flash point.

The test apparatus is similar to that described in E 681. The 5 liter test flask is placed in a heated chamber to ensure a controlled and uniform temperature. A pool of liquid sample is stirred in a closed vessel in an air atmosphere. The vapor-air mixture above this liquid is exposed to an ignition source and the upward and outward propagation of flame away from the ignition source is noted by visual observation. Temperature in the test vessel is varied between trials until the minimum temperature at which flame will propagate away from the ignition source is determined. A striking comparison of this apparatus with the common closed cup flash point test apparatuses can be found in Table III.

Temperature limit of flammability results obtained by E 1232 are consistent with vapor pressure and concentration limit of flammability data. This provides a built in check of results on pure materials that produce vapors obeying the ideal gas law at the test conditions, and leads to a high degree of confidence in the results obtained for mixtures. Figure 3 taken from Bodman 2002 shows the comparison of experimental LTL and UTL values for a number of pure chemicals obtained in accordance with E 1232 with the theoretical values computed from the vapor pressure and flammability limit data.

Since the lower concentration limits of flammability are generally determined at temperatures above the saturation temperature of the vapors (in order to avoid condensation during testing) a slight correction to LFL may be required in the prediction of LTL for pure materials. Certain chemicals (organic acids for example), exhibit a high degree of non-ideal vapor phase behavior, being highly associated in the vapor phase. In order to properly interpret the vapor pressure/volume percent/weight percent/temperature relationships, a knowledge of the degree of non-ideality (association and molecular weight) is necessary. (Concentration limits of flammability for these materials are generally expressed in terms of weight per unit volume.). The facts that E 1232 combines all important aspects and bypasses such difficulties associated with predictions make E 1232 a useful tool for hazard and consequence analysis, particularly for mixtures.

II.B. Ignitability Test Methods

Minimum Ignition Energy (MIE)
ASTM standards E 582 and E 2019 respectively address the lowest energy, stored by a capacitor, that when released as a spark will ignite gas-oxidant and dust cloud-oxidant mixtures. For any given fuel concentration a Minimum Ignition Energy (MIE) can be determined. By testing a range of concentrations, the Lowest Minimum Ignition Energy (MIE) can be determined for the optimum mixture. Observed MIE and MIE values are highly sensitive to the test method, particularly the spark electrode geometry and characteristics of the capacitor discharge circuit. Gas ignition energy standard E 582 uses a simple high voltage spark circuit designed to minimize inductance, stray capacitance and current leakage from the power supply. Dust ignition energy standard E 2019 describes test methods in current use that have been found to yield comparable results; however, unlike E 582 it is a "performance standard" whereby the methodology adopted must produce data within the expected range for a series of reference dusts.

Minimum Ignition Energy of Gas Mixtures (ASTM E 582)
The gas/vapor MIE standard test method ASTM E 582 has been changed little since the 1960s, by which time research conducted by the Bureau of Mines had determined the important features. The test involves production of a spark between electrodes held at an optimum distance within a homogeneous mixture of the

most easily ignitable mixture. The test method also allows determination of the "parallel plate quenching distance", which is directly related to the quenching diameter. Among various provisions of the ASTM E 582 standard method is that the spark circuit should have negligible inductance and low capacitance. The smallest MIE values are typically found using low capacitance circuits with correspondingly low time constants. The MIE is normally assumed equal to the stored energy just sufficient to ignite the optimum mixture at the optimum gap length, although the option exists to report the MIE calculated by integrating the measured voltage and current across the spark gap. In principle this must always yield smaller MIE values owing to energy losses in the circuit. One option not noted in E 582 is the use of pointed electrodes, which can yield somewhat smaller MIE values than the standard flanged electrodes by reducing heat and radical losses from incipient flames. However, deviations from the normal procedure have the disadvantage of generating disparate data sets that do not compare with the set of "standard" values.

For most fuel gases/vapors MIE is below 1 mJ, and therefore ignition control strategies can not be used as the sole means of explosion protection. As an exception to this old and widely accepted rule, commercial airplane fuel tanks are designed so that maximum credible electrical spark energy is one tenth the jet fuel MIE. A paper presented at the last Symposium **(Ural 2003)** discussed how the commercial aviation sector, after a series of deadly fuel tank explosions, is rediscovering this old and widely accepted rule. For low MIE gases and vapors, ignition control is still useful as it reduces the frequency of ignition incidents. The MIE can vary by several orders of magnitude depending on the types and concentrations of fuel and oxidant, plus the temperature and pressure. Further, the operating conditions might be such that the optimum mixture cannot form, in which case some higher MIE value might be taken to represent worst-case conditions for a particular process.

<u>Example: Evaluating the Ignition Hazard of a Gas Mixture</u>
There is no accepted method of calculating the ignition energy of a fuel-air mixture from the values corresponding to the separate components. A possible method for "C+H+O+N" organic fuels has not yet been fully evaluated **(Britton 2002b)** and in many cases, one or more of the fuels may contain other elements such as halogens, or be entirely inorganic. If the mixture contains easily ignited components such as hydrogen, plus hard to ignite components, such as ammonia, ASTM E 582 can be used to determine at what hydrogen concentration the mixture becomes ignitable at an assumed effective energy level. The latter might be taken as 10 mJ to represent corona and brush discharges from plastic surfaces, or 25 mJ to represent sparks from ungrounded personnel **(Britton 1999)**. Alternatively, ASTM E 582 might be used to evaluate the benefits or hazards of changing the oxygen concentration relative to that available from ambient air.

As nitrogen or other inert gas is added to air, its strength as an oxidizer diminishes and the MIE of optimum fuel-oxidant mixtures increases by orders of magnitude. If the maximum effective energy of ignition sources present in a system can be determined, partial inerting might be used to complement existing explosion protection systems. For instance, the frequency of activation of a suppression system might be minimized. Alternatively, the ignition hazard due to an upset in a nitrogen-air mixer serving an inhibited monomer storage tank might be assessed.

As additional oxygen is added to air, for example via decomposition of a peroxide, the MIE of optimum fuel-oxidant mixtures decreases by orders of magnitude. Difficult-to-ignite gases may ignite readily and common solvents may be ignited by corona discharges and other very weak sources. NFPA 53 plus a series of ASTM publications such as G-128 provide detailed information on the hazards of oxygen-enriched systems.

Minimum Ignition Energy of Dust Clouds, ASTM E 2019
ASTM E 2019 describes several variants of apparatus that subject a dust suspension to an electrical spark. Owing in part to the high dependence of dust MIE on physical characteristics such as particle size distribution and (in some cases) moisture content, plus the large ignition energies involved relative to most gases, there has been considerable variability in published "dust ignition energy" data even for common materials such as sulfur. Even lycopodium, a fairly monodispersed pollen dust, has yielded a wide range of values in different apparatus. Apart from physical differences between samples of the "same" dust, the cause of much of the data scatter has been the design of the electrical circuit and the calculation of the

energy released by the spark. It has been found that unlike the case with most gases, the ignitability of dusts is dependent on the spark duration, which is directly affected by the spark generating system. Since there is no "absolute" value for dust MIE it is difficult to assess the accuracy of a particular value except by reference to other values measured for an identical sample. Thus, E 2019 is a performance-based standard in which calibration tests are carried out with at least three different reference dusts.

Four spark generating systems are described. Three of these are high voltage capacitance circuits (10-30 kV) triggered by (a) auxiliary spark in three electrode system (b) moving electrode or (c) slowly increasing the voltage across the gap. The fourth circuit involves energy storage at a lower voltage (< 2.5 kV) and triggering via a low energy, high voltage pulse applied across the gap. This pulse is generated by discharging a small capacitor through a transformer whose secondary is in series with the spark gap.

Since the storage capacitor is normally only partially discharged by the spark, the residual energy is subtracted from the energy initially stored. While some test methods report the MIE, others report the MIE as falling within a series of defined ranges whose ratios increase by a factor of about 3. For example, the MIE of lycopodium might be reported as 20 mJ or 10-30 mJ.

As seen in Figure 4, MIE is strongly affected by the particle size. Aluminum and polyethylene data seen in this figure were taken from Eckhoff 1997, and are reportedly plotted against the mass median particle size of the sample. The zinc stearate data (circles) were generated in an unpublished E27 round robin study among four different labs, all using the slightly modified Eckhoff circuit which is referred in E 2019 as the "Triggering by Auxiliary Spark." In this study, mass mean particle size of each size fraction used in the tests was characterized using a light scattering method. The scatter of the circles in Figure 4 is a good indication of the inter-laboratory reproducibility of the MIE test results. As illustrated by the solid lines in Figure 4, all three data sets of appear to support the theoretical expectation that MIE is proportional to approximately the cube of the particle size.

Examples: Safety of Additives and Special Procedures

A choice of antioxidants for a granular polyolefin production facility may represent different chemical and physical options. One consideration is the MIE of the additive compared with that of the polymer. If the MIE is less than that of the polymer the likelihood of ignition will be increased in accumulation sites such as baghouses. The likelihood of ignition may also increase in bins, especially if the additive MIE is less than 10 mJ. Options include a different additive whose MIE is greater when tested at the same particle size, or to purchase an additive that is coarser. For example, a sub-400 mesh additive whose MIE is less than 10 mJ could be replaced by a coarser additive that contains no material having an MIE less than 20 mJ.

Many companies have adopted policies to restrict handling of dusts at some prescribed MIE and in some cases to inert conveying systems and storage bins. The "action level" adopted varies with operating experience but is typically 10 mJ or less. This is the effective energy normally ascribed to bulking brush discharges that are believed responsible for dust ignition in properly grounded equipment (Britton 1993, Jaeger 1998).

Spontaneous Combustion

A series of ASTM standards address spontaneous combustion of gas mixtures (ASTM E 659), dust clouds (ASTM E 1491) and dust layers (ASTM E 2021). Presently there is no ASTM test addressing spontaneous ignition of bulk solids. The principal application is defining the maximum acceptable surface temperature in a particular area, usually for electrical classification purposes. Other applications include qualitative assessment of hazards inside process equipment rather than external to it (e.g. bucket elevator bearings).

Autoignition of Liquid Chemicals, ASTM E 659

A liquid or adequately volatile solid sample is added to an open-necked glass flask held inside a furnace. The quantity of added sample is systematically varied in order to determine the lowest temperature at which hot-flame ignition occurs (Autoignition Temperature or AIT). Additional test observations are the lowest temperature at which cool flames are observed (Cool Flame Temperature or CFT) and the lowest temperature at which exothermic, non-luminous pre-flame reactions are observed (Reaction Threshold

Temperature or RTT). The AIT is the most widely used of these parameters and is commonly listed in data compilations.

The current test method is ASTM E 659, which uses a spherical 500 ml flask. In 1976 this replaced ASTM D 2155, which used a 200 ml flask. Depending on an excess rate of heat and radical production in the test vessel, AITs differing by 10-20°C are to be expected simply due to flask size differences, with the current method normally yielding lower values. However, another reason for adopting the larger standard (borosilicate) flask was to reduce catalytic wall effects, which could lead to spuriously low AIT values for certain chemicals. Since the compromise of a 500 ml flask size has not been adopted in Europe, two disparate databases continue to be generated.

The following example uses AIT exclusively. The parameter is here synonymous with "ignition temperature" since it is tacitly assumed that hot surfaces in a classified area cannot trap stagnant volumes of gas-air mixture significantly larger than those used in the standard test and that no significant surface catalysis occurs. For chemical process applications, the CFT and RTT parameters may need to be considered. For instance, if a solvent vapor-air mixture is heated in a large atmospheric process vessel, "cool flames" might trigger two-stage ignition at much less than the "standard" AIT. The CFT may be on the order 50-100°C less than the AIT while exothermic pre-flame reactions occur at lower temperatures still. The situation becomes more complicated if pressures other than atmospheric are involved, since the effect of pressure can be highly irregular. In general, expert advice should be taken when assessing such situations.

<u>Example: Applying Equipment "T Code" for Gases and Vapors</u>
Article 500.8 of NFPA 70, also known as the National Electrical Code, provides that "Class I equipment shall not have any exposed surface that operates at a temperature in excess of the ignition temperature of the specific gas or vapor". The ignition temperature is assumed to be equal to the AIT.

AIT data compilations are given in NFPA 325 and NFPA 497. However, these data compilations do not address mixtures and there is no available method for combining individual component data. Also, data compilations often use the lowest published AIT value, which can cause expensive difficulties in cases where a commercial grade chemical (such as hexane) displays a range of AITs depending on the components actually present. Finally, most of the compiled data predate the current ASTM E 659 method and can involve tests conducted in either larger or smaller vessels. Hence it is here assumed that a new AIT determination is conducted using ASTM E 659.

Table 500.8(B) of Article 500.8 of NFPA 70 shows a list of fourteen temperature class (T codes) that are to be marked on nameplates for Class 1 equipment. Each T code in the table has a corresponding "maximum safe operating temperature". For example, code "T1" corresponds to a temperature of 450°C and code T6 to 85°C. Article 500.8 provides that the temperature markings in Table 500.8(B) shall not exceed the ignition temperature of the specific gas or vapor to be encountered.

A particular item of equipment is often available with a range of different T codes depending on the needs of the purchaser. Hence, the AIT of the specific gas, vapor or mixture present in an area can determine which manufacturer or style of equipment is purchased and at what cost.

A similar AIT application is to determine the maximum safe operating temperature of other hot surfaces in a classified area, such as process or utility pipelines. It might be determined that additional insulation is required to reduce the external temperature. Provided large stagnant pockets of gas-air mixture are not trapped by the hot surfaces involved, the AIT is normally a conservative measure of the ignition temperature. Exceptions include certain reactive chemicals whose ignition temperatures are lowered by rusty surfaces.

Dust Cloud Autoignition, ASTM E 1491, and Dust Layer Ignition, ASTM E 2021
These two test methods are entirely different, although in many instances the data are used synonymously for "dust ignition temperature".

The ASTM E 1491 dust cloud AIT test involves blowing dust into a heated furnace set at a predetermined temperature. The dust concentration is systematically varied to find the lowest temperature at which self-ignition occurs at ambient pressure, known as the Minimum Autoignition Temperature (MAIT). A visible flame exiting the furnace provides evidence for ignition. Four different furnaces are described in ASTM E 1491 (0.27-L Godbert-Greenwald Furnace, 0.35-L BAM Oven, 1.2-L Bureau of Mines Furnace and 6.8-L Bureau of Mines Furnace). Each yields somewhat different MAIT data, the largest deviations occurring at the greatest MAIT values. However, the lower MAIT range is of more practical importance and here the agreement is better (for example 265±25°C for sulfur).

The ASTM E 2021 dust layer hot surface ignition temperature test uses a constant temperature hot-plate to heat the dust on one side only (see Figure 5). Routine tests use a 12.7 mm (0.5 inch) thick layer, which might simulate a substantial build-up of dust on the outside of hot equipment. However, since the ignition temperature normally decreases markedly with increased dust layer thickness, the method allows layer thickness to be varied according to the application. This can depend on the user's housekeeping standards.

The dust layer ignition temperature is usually less than the dust cloud MAIT. This is because the duration of the layer test may be hours as opposed to the few seconds that a dust cloud is suspended in a furnace. However, some exceptions occur. One possible reason for the cloud MAIT being less than the layer ignition temperature is if melting occurs. This excludes combustion air from a layer but not from a suspension of dust particles. If the layer ignition temperature is near the softening or melting point, increasing the layer thickness may promote ignition rather than melting.

Owing to the great dependence of dust layer ignition temperature on test conditions, especially layer thickness, data of uncertain origin should not be used for direct comparison purposes. Indeed, many "dust layer ignition temperature" data reported in the literature were determined using a basket held in the so-called "modified Godbert-Greenwald furnace" where a small sample is heated from all sides.

It should be noted that the lowest ignition temperature for a dust corresponds to bulk heating under conditions that cause the dust to be heated from all sides. This is the situation for thick dust layers or accumulations inside heated equipment such as dryers. Neither the dust cloud MAIT nor the dust layer hot surface ignition temperature is appropriate for this scenario. Instead, either scaled isothermal basket tests or adiabatic calorimetry with appropriate mathematical model tying the reaction rate to heat and air flow has been used.

Example: Applying Equipment "T Code" for Dusts

This is similar to the example for gas/vapor T code but with two important differences affecting data application. These involve first, the definition of "ignition temperature", and second, the provision of a maximum allowable surface temperature for dusts that may dehydrate or carbonize.

Article 500.8 of NFPA 70 provides that "The temperature marking specified in 500.8(B) shall be less than the ignition temperature of the specific dust to be encountered. For organic dusts that may dehydrate or carbonize, the temperature marking shall not exceed the lower of either the ignition temperature or 165°C (329°F)."

The user of NFPA 70 is referred to a compilation of "dust ignition temperatures" presented in Table 2-5 of NFPA 499, which states: "The ignition temperature of a dust layer is a function of the type of dust and its physical and chemical properties and is often less than the cloud ignition temperature. The ignition temperature shown in Table 2-5 is the lower of the two". The data compiled in Table 2-5 of NFPA 499 are not derived from consistent test methods and nor were both dust cloud and layer ignition temperatures available in all cases. Hence ideally both ASTM E 1491 and E 2021 should be run to determine the lowest ignition temperature of a representative dust sample, addressing both chemical and physical properties. Assuming that the layer ignition temperature is less than the cloud MAIT, the layer tested should be thick enough to represent "worst credible case" accumulations. There is however no formal guidance with respect to dust layer thickness employed in either NFPA 70 or NFPA 499. This is attributed to the fact that, until recently, the majority of dust layer ignition temperature data were generated using the "modified Godbert-Greenwald furnace," where the layer thickness is not a test parameter.

II.C. Explosibility Test Methods

ASTM E 789 Dust Explosion in a 1.2-Liter Closed Cylindrical Vessel and ASTM E 1226 Pressure and Rate of Pressure Rise for Combustible Dusts.

It is well known that when an atmosphere of a fuel is ignited, a rapid combustion can occur. If this is occurring in a confined space, considerable pressure is experienced from the rapid temperature rise which can be destructive, i.e., an explosion. The requirements for this to occur are well known: fuel, oxidant, ignition, and confinement. The fuel needs to be dispersed and in the case of a solid, must be sufficiently small in size to be dispersed and able to support flame propagation. It is of interest to know how much pressure is developed and how fast that pressure is developed. Figure 6 shows typical large scale test data for the combustion of a dispersed dust in a 10 m^3 vessel. The explosibility parameters P_{max} (maximum pressure) and K_{ST} (maximum rate of pressure rise of the worst case concentration times the cube root of the test volume) are obtained from such measurements.

The determination of a P_{max} and K_{ST} for a material first establishes that it is an explosible dust. In fact, most carbon based organic materials and many elemental metals are explosible if they are of a size that is readily suspended in air. The primary use of the test data P_{max} and K_{ST} is for the design of explosion protection systems: venting, suppression, isolation. Vent designs provide a relief area that will limit damage to the process equipment to an acceptable level. The required vent area is calculated using

equations from NFPA 68 and requires knowledge of the process – volume, temperature, operating pressure, design strength, vent relief pressure- and of the fuel, P_{max} and K_{ST}. The safety of the design is only as good as the explosibility data. Suppression is the active extinguishment of the combustion and again limits the explosion pressure to an acceptable level. Suppression designs require similar process and hazard data in order to determine the hardware requirements such as size, number and location of containers, detection conditions, and the final or reduced explosion pressure. Isolation, the prevention of flame propagation through interconnections, requires the same process and hazard data to determine hardware needs and locations.

The measurement of the explosibility parameters requires the reproducible generation of a near homogeneous dust cloud inside a containment vessel of known volume. This is best accomplished using a spherical test chamber of 1 m^3 or larger. However, a bench scale test method described in E 1226 is also deemed adequate. It is based on a vessel at least 20 liters in volume in which a dust cloud is formed using the discharge of a small cylinder of compressed air. After a prescribed time delay, the highly turbulent dust cloud is ignited using a strong ignition source of known energy. (The two 5 kJ igniters used in these tests by themselves are responsible for 1.6 bar pressure rise in the test chamber.) Pressure is monitored versus time by appropriate transducers and expressed as pressure, P_{ex}, and pressure rate of rise, dP/dt_{ex} (see Figure 6). Dust concentration is varied to determine the maxima of both parameters (see Figure 7). Particle size and moisture are other variables that must be considered. Particle size should be less than 75 μm ensuring a design that is conservative.

Method E 789 uses a smaller test vessel (1.2 liters), a much weaker ignition source and a lower turbulence intensity. These parameters are known to affect the results, therefore the E 789 data should not be used for explosibility assessment and explosion protection designs.

E 1226 is the most frequently used ASTM E27 method. It is cited in numerous NFPA documents (1, 61, 68, 484, 654, 664, 921) and Factory Mutual documents (Data Sheets 7-36, 7-76). The K_{ST} and Pmax data are also used in hazard assessment, forensic analysis, and consequence modeling, to estimate the deflagration burning velocity, flame speeds and induced velocities in dust clouds (e.g. **Ural 1992**, and **Ural 2001**).

Example. A pharmaceutical company is processing an active ingredient through a spray dryer, cyclone and dust collector. The cyclone is used for collection of the bulk of the material and the dust collector retains the fines. It is located outdoors and requires protection by venting. The only material available for testing was collected from the cyclone and gave a P_{max} of 8.1 barg and a K_{ST} of 153 bar.m/s when tested by E

1226. The same material when sieved to sub 200 mesh (< 75μm) and tested gave a P_{max} of 8.3 barg and K_{ST} of 242 bar.m/s. From the dirty volume of the dust collector of 250 ft^3 and a design strength of 3.0 psig, NFPA 68 equation for L/D < 2 was used to calculate the required vent areas using 0.1 bar P_{stat} vents. The required area was 4.5 and 7.2 ft^2 for the coarser and finer material, respectively. Since the finer particles could be expected to be present in the dust collector, the larger vent area was selected for design. One 24"x44" (7.3 ft^2) vent was installed such that the bottom of the filter bags was above the vents.

Explosibility of Gases and Vapors:
Gas explosibility parameter, K_G, is frequently used in deflagration vent or suppression system design. Currently, ASTM does not offer a test method for the determination of this parameter.

III. ASTM E27 Standards Related to Thermal Stability and Estimation Methods

The following section covers some of the ASTM E27 methods used to assess the thermal stability and overall energy content of a material. These properties are determined experimentally via calorimetry and also from theoretical estimations using a purely thermodynamic approach coupled with empirical correlations.

Knowledge of the thermal stability of a material, including mixtures, remains one of the most important types of information in reactive chemicals hazard evaluation. Certainly, the term "thermal stability" encompasses both the kinetic regime (how fast a reaction takes place) and the thermodynamic regime (how much energy is released). Both are necessary to define safe operating limits for a chemical process. An accurate assessment of this property allows one to answer practical questions like:

- **What is a safe storage temperature for this material?**
- **Are two materials safe to mix and store?**
- **If a material or mixture reacted/decomposed, what is the maximum expected temperature?**
- **During a desired chemical process, are there any undesired energetic reactions which may occur at temperature elevated from the normal process temperature?**
- **Might the substance be subject to a strong energy release event (deflagration or detonation)?**

Outlined below are two important ASTM tests useful in assessing the thermal stability properties of chemical substances

Thermal Stability of Chemicals by Differential Scanning Calorimetry, ASTM E 537
Most chemical companies who have active reactive chemicals hazard evaluation laboratories and programs have adopted differential scanning calorimetry (DSC) as their primary screening tool for thermal stability. Certainly DSC is thoroughly discussed and recommended in various widely used monographs in the area of hazard evaluation (for example, see **CCPS 1995; Grewer 1994; Crowl 1990; Yoshida 1987**). In a DSC test, a small sample (typically 0.5-5mg) is encapsulated in an inert sample container with a headspace gas (air or N_2, typically) then heated at a constant temperature ramp rate (typically 5-20 deg. C/min) from ambient conditions to some elevated temperature (typically 300-500 C). The DSC instrument measures the heat flow from a sample (in a differential manner, with respect to an inert reference cell). Of course, the heat flow is only measurable if the heat release rate is above the detection limit of the instrument. Typically, the measured heat flow is plotted as power output as a function of the furnace temperature (see Fig. 1). Events which are accompanied by heat evolution (exotherms) are seen in the output trace as peaks. Endothermic events are observed as deviations in the opposite direction. Integration of these events from a stable, flat baseline allows the quantification of the energy released or absorbed. Implied in this brief explanation is that a "flat" trace indicates that no events are detectable over the temperature range. This is normally a desired result but one has to be quite careful in this case if the material or mixture is to be exposed to elevated temperatures during normal processing (see below) since the DSC becomes less sensitive to some events as the temperature is raised (**Hofelich 2002**).

DSC test results are quite useful for hazard screening, for example:

- The absence of detected activity generally implies a stable material under most conditions. (See **Hofelich 2002** for a detailed discussion of this.)
- The observation of detectable exothermic activity allows one to make decisions about storage and processing or the need for further testing (like ARC or impact sensitivity testing)
- Important parameters, such as Time to Maximum Rate (TMR), may be determined or approximated by straightforward calculations using only the detected onset of the exotherm.
- Heats of chemical reactions may be determined accurately, which allow adiabatic temperature rise estimations due to a runaway reaction.

The advantages of DSC as a screening technique are:

- Small sample size (typically 0.5-5 mg)
- Rapid testing time (typically 1-3 hours)
- Variety of sample encapsulation materials (e.g. metal and glass, which can withstand high pressure)
- Relatively high calorimetric sensitivity (ca. 45 mW/g) and accuracy
- Generally straightforward interpretation of results
- Direct, quantitative kinetics may be obtained from manipulation of the raw data using various data analyses methods

Disadvantages and limitations, include:

- Low activation energy (Ea) events occurring at elevated T may not be detectable by DSC;
- Gas evolving events may not be noted;
- Heterogeneous materials may not be sampled properly due to the small sample size

The ASTM standard E 537 provides practical information about performing a DSC test on a sample. The spirit of the method is to assess the presence or absence of thermal events, the detected onset temperature of these events, and to quantify the heat release from any events. The method describes the minimum required instrumental capabilities for performing the test including furnace specifications, temperature control, etc. (Most commercially available DSC instruments meet the minimum specifications for this test method.) Specific information about instrument calibration is provided along with recommendations for sample size, heating rate, and temperature range. The procedure also specifies what parameters should be reported (detected onset T, extrapolated onset T, peak T, etc.). Finally guidance is given to the contents of the report.

Example: Introduction of a new chemical into a process.

A new chemical, a high molecular weight mononitro substituted organic alcohol (MW=385, MP approx. 100C) is being introduced to a plant at the 55 gal. drum scale. The drum will be stored in a warehouse under ambient conditions not expected to exceed 30 C. This material will be added to a liquid product formulation of polyols at the 1% level as a corrosion inhibitor. As part of the normal Management of Change (MOC) process in the plant, the Reactive Chemicals testing group is consulted about the need for testing. The MSD sheet indicates that the material is stable but, in part, because of the presence of the nitro group, it is decided to screen this material for its thermal stability using DSC. Following the ASTM standard E 537, a small (0.87mg) sample is sealed in a glass capillary under nitrogen and tested to 400 C using a heating rate of 10 deg. C/min. The resulting DSC trace is given in Fig. 8. The endotherm near 100 C is interpreted as the melting point. The exotherm detected near 170 C is not unexpected due to the presence of the nitro group in the molecule.

Do these data convince us that the material is safe to store under the conditions specified? Unfortunately the DSC trace alone, without further kinetic evaluation, is not enough to answer this question with 100% certainty. Certainly the fact that the detected onset T is significantly above the intended storage temperature is working in our favor. Also, the fact that the peak is reasonably narrow indicates a reasonably high activation energy for the decomposition reaction. This also is working in our favor. Using some simple assumptions about the activation energy of the reaction, and knowing the calorimetric sensitivity of the DSC, we can use the detected onset temperature to predict the adiabatic Time to Maximum Rate. This is a very conservative calculation since in drum storage, there is a finite heat loss to the surroundings. In this example, a calculation yields a TMR value on the order of years for this material.

The expected usage rate in on the order of several drums per month. Hence, the decision is made that ambient storage conditions will pose a minimal risk and no further thermal testing is required. (Note that if any autocatalytic behavior of the material is suspected, further isothermal testing may be required.)

Thermal Stability of Materials by Methods of Accelerating Rate Calorimetry (ARC), ASTM E 1981
Similar in intent to E 537 (Differential Scanning Calorimetry), E 1981 also outlines suggested procedures for assessing the thermal stability of chemicals. In contrast to E 537, however, E 1981 is performed under quasi-adiabatic (near equilibrium) conditions, instead of in a scanning mode. A (0.5-10 gram) sample is placed inside a reaction container (typically, a metal sphere), secured inside a calorimeter, heated to a user-selected initial temperature, and allowed to come to equilibrium. A search for evidence of an exothermic reaction is undertaken, indicated by a self-heat rate exceeding a user-selected rate of temperature rise, typically 0.02°C/min.

If no exotherm is detected, the calorimeter increments the temperature of the system 5°C, and the heat-wait-search cycle outlined above is continued. The cycle continues until an exotherm is detected, or the upper temperature limit of the test is reached. If an exotherm is detected, the calorimeter switches to an adiabatic mode, raising the temperature of the surroundings to that of the reaction container, so that no heat is lost. Temperature and pressure data are recorded as a function of time for the duration of the exotherm. These data may then be used to calculate a time to maximum rate, kinetics, heats of reaction, adiabatic temperature rises, and other parameters. The origins of this widely used technique are detailed in the reference (**Townsend 1980**).

A significant difference in the two methods is related to the closed system capability of ARC, which allows any pressure from gaseous decomposition products to be measured. From these measurements, estimates of total gas evolution and rates of pressure rise may be calculated, and applied to process conditions within properly understood restrictions and limitations. Another significant difference lies within the sample size tested: DSC, with its milligram-sized samples, is less likely to obtain a representative sample of a material which may be heterogeneous or contain impurities. ARC, which can employ sample sizes up to 1000 times larger than a DSC, greatly decreasing the probability of obtaining a non-representative sample, is still considered suitable for testing in a laboratory-scale setting. For a more detailed comparison of ARC and other thermal stability test methods, see reference (**Fenlon 1984**).

Other options available to ARC include stirring of the sample within the reaction container, as well as the capability of dosing, shot additions or head space gas sampling in the course of a test. In addition, any desired atmosphere or pressure may be chosen for the reaction container. Although fully automated and computer controlled, one disadvantage of the ARC is the amount of time required for a test: Typically one test per day may be performed, after allowing for cool-down, disassembly and clean-up of the apparatus.

Example: Heated storage scenario for a new material.
In the previous example for the DSC Standard, let's now assume that the quantity of material to be used by the plant has increased from drum quantity to a larger storage tank (1000 kg). Furthermore, the plant is considering heating the material in the storage tank above the melting point in order to make material transfers to the reactor easier via pumping through a heated transfer line. It is decided to run an ARC test on the sample in order to gain a better understanding of the kinetics of the exothermic decomposition reaction observed in the DSC test (Fig. 8). Also, for purposes of relief design, the process owners want to know if there is any gas release associated with the degradation.

Fig. 9 shows the results of an ARC test on a 4g sample, heated from 30 C to 350 C following the guidelines in E 1981. The exothermic reaction of the material is now detected in the more sensitive ARC calorimeter near 150 C accompanied by significant gas generation (pressure data are not shown in Fig. 9). An estimate is made of the adiabatic time to maximum rate from an expected worst case condition (known maximum temperature of the heating medium) of T=140C. This TMR turns out to be only 22 hours and the project team decides there is an unacceptable risk so it is further decided to seek an alternate route for material transfer.

Preparation of a Binary Chemical Compatibility Chart, ASTM E 2012
Managing chemical reactivity hazards is an important requirement of any chemical process safety program. Many serious accidents have occurred because of unintended mixing of materials during storage, processing, and transportation. Often times these accidents are due to a lack of knowledge of the hazard. In other cases, they are due to lack of communication. For example some people within an organization may be well aware of the hazard of mixing certain chemicals, but this information may not be documented in an easy to interpret format. For these reasons a thorough program is necessary for evaluation and communication of chemical reactivity issues. In some cases gathering reactivity information may be required by regulations. The US OSHA Process Safety Management Rule (OSHA 1910.119) requires that reactivity information be gathered on the highly hazardous chemicals listed by the rule. Transportation regulations forbid incompatible materials to be transported together.

There are many possible ways to document reactivity issues, including reports and MSDSs. However, a binary compatibility chart (sometimes called reactivity chart) is ideal for documenting reactivity issues between pairs of chemicals or materials. Such a chart allows concise presentation of reactivity information for all materials used on a process and is useful for operators, process engineers, technical services personnel, and those conducting process hazard analyses and training programs. Another advantage of a compatibility chart is that it forces an evaluation of reactivity for each binary pair.

An example of a binary compatibility chart is shown in Figure 10.

ASTM Committee E27 recently developed E 2012, a Standard Guide for the Preparation of a Binary Chemical Compatibility Chart. This guide addresses a number of issues that are important for the preparation of such charts: accurate assessment of chemical compatibility, the importance of defining the scenario of the mixing event, consideration of reactivity of common cleaning agents and heat transfer fluids, suitable experimental techniques for gathering compatibility information, incorporation of user-friendliness, documentation guidelines, and provision for revisions. This standard was based primarily on a more detailed discussion and description of the issues involved and is recommended to any user of this standard (**Frurip 1997**).

Many sources of reactivity information are available including textbooks, MSDSs, databases, publicly available general compatibility charts, software programs, and knowledgeable personnel within an organization. If information does not exist on a particular binary pair it may be necessary to test. E 2012 provides an overview of information sources, reactivity testing, and data interpretation.

E 2012 recommends that the scenario under consideration be defined in the chart. Examples of variables that should be defined include temperature, confinement, scale, time, atmosphere (e.g., air vs. nitrogen), and heat transfer. These variables have a large influence on whether a mixture poses a hazard or not.

The example chart above uses a simple hazard rating scheme – reactive or not reactive. For some organizations it may be useful to use a more detailed scheme. Also, its recommended that footnotes be used to provide the source of information for each binary mixture. This allows the user to obtain more detail if desired. See E 2012 for more information.

Mixing of more than two materials may be credible and any known hazards of such a combination should be documented. This can be done by listing the third necessary material (often times a catalyst) as a footnote in a binary compatibility chart.

<u>Example: Preparation of a compatibility chart for a process</u>
Due to a recent accident, a company has decided to upgrade its understanding and documentation of reactivity hazards on its process for making its best selling product. A chemist has been assigned to lead this effort. The chemist decides to convert the existing reactivity information into a compatibility chart including the materials used on the process plus water, air, acetone (the cleaning agent), and the heat transfer fluid. She uses ASTM E 2012 to guide development of the chart. She finds that reactivity information is scattered through process development, process hazard analysis, and accident reports. She also finds that no information has been documented for many of the reactivity pairs. She uses Bretherick's

Handbook (**Bretherick 1999**), the NOAA reactivity worksheet (**NOAA 2003**), MSDS's, and performs a literature search to gather existing public information. She complements this with her knowledge of chemistry to fill in most of the cells in the compatibility chart.

She is unsure of the reactivity of certain reaction mixtures within the process, so she has these tested by the company's Process Safety Lab. The testing reveals that a runaway reaction could occur between one of the reagents, bromine, and a reaction mixture. This combination is possible if excess bromine is overcharged in a previous step. Eventually she obtains information on each binary pair.

Equipment and procedural changes are made to prevent the overcharging of bromine. The completed chart is rolled-out at staff and safety meetings. It's used to train new process operators. Subsequently, a change is proposed to dry one of the reagents, benzyl chloride, by passing it over molecular sieves. The change is not made because the chart revealed that this may cause a polymerization reaction.

Due to creation of a compatibility chart the recognition of potential hazards is increased, and the likelihood for injuries, equipment damage, and process shutdown on this important product has been reduced.

CHETAH® a Computer Program for Energy Release Evaluation and Prediction of Chemical Thermodynamic Properties.

Description of the Program: ASTM Committee E27 is responsible for the computer program CHETAH® (ASTM Computer Program for Chemical Thermodynamic and Energy Release Evaluation). It was first introduced in 1974. Since that time the Committee has worked to further develop the technology in the program and the associated databases. Also the Committee keeps CHETAH® up to date with current operating systems and user interfaces. The current version 7.3 was released in 2002 (**CHETAH 2002**).

The CHETAH® program is a unique tool for predicting both thermochemical properties and certain reactive chemicals hazards associated with a pure chemical, a mixture of chemicals, or a chemical reaction. The calculations are made using only information concerning the molecular structure of the components, using the well-accepted Benson's group contribution technique (**Benson 1968**) to predict the important thermodynamic properties. The database of molecular fragments (Benson's groups) used to describe the molecules is believed to be the largest such database in existence, allowing a very large number of possible molecules to be calculated. For convenience the current version of CHETAH® allows the importation of structures drawn using ChemDraw® software. This imported data is then processed without the user needing to choose the molecular fragments that make up the molecule. Also CHETAH® has an extensive database of molecules for which the complete necessary thermochemical data are available from the literature for immediate calculations.

For chemical reactive hazard evaluation, CHETAH® is designed to be a conservative screening tool for use during the early stages of compound synthesis or process development. Energy related hazards associated with handling new chemicals are frequently not known. In addition experimentally determined thermochemical data for new chemicals used for process design and to predict hazards are often not available. For these situations CHETAH® can perform an energy release evaluation indicating the maximum energy of decomposition of a compound or mixture and the likelihood of an energy release hazard. These hazard evaluations are valid for gas, liquid, or solid materials. CHETAH® can also predict certain flammability parameters. However, CHETAH® is not meant to replace the physical testing of materials. CHETAH® should be used to complement experimental results to help identify the need for further testing in the areas of energy release and/or flammability.

CHETAH® can also calculate heats of combustion for pure components and mixtures. Unlike most methods which only apply to compounds composed of six or so elements, the method used in CHETAH® will work with compounds composed of approximately 60 different elements. Combustion products are predicted based on complete combustion.

CHETAH® is also of great utility for predicting ideal gas thermochemical properties of compounds. This would include standard heats of formation, heat capacities, entropies, and free energies. Thermodynamic properties of user specified reactions can also be calculated. Extensive databases are also provided for known thermodynamic properties of gas phase molecules and ionic solids. CHETAH® includes a thermodynamic property prediction method for ionic solids.

In summary CHETAH® can be used to perform the following functions by simple menu selections:

- Classify a material or mixture with respect to its ability to decompose with violence.
- Calculate the enthalpy of combustion for a compound or mixture.
- Calculate thermochemical properties for reactions: ΔCp_{rxn}, ΔH_{rxn}, ΔS_{rxn}, ΔG_{rxn}, logK
- Calculate thermochemical properties for compounds: C_p., S, ΔH_f, ΔG_f, $\log K_f$, free energy function (G-H)/T, H_T-H_{298}
- Estimate lower flammable limits.
- View thermochemical data in CHETAH®'s database.

Example of a CHETAH calculation; an energy release evaluation for 2,4,6-trinitrotoluene (TNT).

Describe the molecule's structure using molecular fragment building blocks called Benson Groups. A form listing the Benson Groups needed is filled out as shown in Figure 1. The groups needed are chosen off a menu listing the groups available.

Alternatively, the molecule may be drawn using the program ChemDraw® and then pasted into CHETAH. If this procedure is followed, CHETAH will automatically identify the Benson groups needed.

Next the user chooses the energy release evaluation icon (the firecracker) above the molecular structure table on Figure 11. The calculation form will appear (as shown in Figure 12). The calculation button is clicked to start the evaluation.

After calculations are completed, an output report will be presented as shown in Figure 13. Note that TNT is found to be a HIGH hazard for energy release using three different criteria: Maximum Heat of Decomposition, the Plosive Density method, and the Overall Energy Release Potential. Other secondary indicators are also calculated and presented. Note that the products from the maximum energy release decomposition are given as well as results from a heat of combustion calculation. A discussion of the various output measures is given in the user manual. The reader may also consult other publications which discuss the use of the program (see References section for a listing of articles related to previous versions of the program). A critical review of the CHETAH hazard criteria may be found in **Shanley 1995**.

IV. Conclusions

ASTM Committee E27 has produced and continues to produce important standard methodologies that are used for hazard evaluation. Dedicated use of these standards offer the following benefits:

- More reproducible data due to well-defined test methods.
- Reduced probability of incorrect data or data interpretation due to warnings on pitfalls that may negatively affect data generation and interpretation.
- The ability to verify adherence to test procedures by comparing test results on standard materials with those published in the Precision and Bias section of the E27 standards.
- Simpler data interpretation and application due to the number of guides (e.g., NFPA) and papers based on data from E27 methods.
- Reduced probability of accidents during testing due to warnings regarding issues that may negatively affect safety.

The E27 standards greatly enhance the ability of an organization to produce accurate, meaningful data. Consequently, those organizations who practice these methods and apply the data to their processes have a competitive edge in that they are able to avoid reactive chemical accidents, fires, and explosions.

New technologies will require modification of existing test methods and development of new ones, and therefore the E27 Committee will continue to be active. Individuals knowledgeable in process safety testing are encouraged to become part of E27. Individuals that participate in E27 are able to influence the modification and development of standards and, perhaps more importantly, are able to establish contacts and interactions with individuals at other organizations. These contacts yield ideas on improvement of process safety testing methods and instrumentation.

As with any volunteer organization, a key to success is to have active, enthusiastic, dedicated, and knowledgeable members participating. The current emphasis within companies to enhance performance in Environmental Health and Safety coupled with the Process Safety Management regulations (OSHA) makes the relevance of ASTM E27 activities more important than ever. Please consider participating in this group by contacting ASTM directly for membership requirements and more information. Certainly, any of the authors of this paper would be happy to talk to you more about the benefits of joining.

V. Acknowledgements

The authors are indebted to ASTM for facilitating and promulgating the E27 standards. The contributions of all ASTM E27 members, past, present, and future are gratefully acknowledged. Mr. Ken Cashdollar of Pittsburgh Research Center kindly collected and tabulated the responses to the ASTM method usage survey and preserved the anonymity of the respondents.

VI. References

ASTM 2003 Vol. 14.02, General Test Methods, (2003); ASTM, West Conshohocken, PA. (2003). Individual standards are also available from http://www.astm.org.

Benson 1968 S.W Benson: "Thermochemical Kinetics", Wiley, New York, Chap. 2. (1968).

Bodman 2002 Bodman, G., Britton, L., Labarge, M., and Ural, E. A.,"A REVIEW OF THE CHARACTERISTIC SCATTER AND BIAS OF ASTM E 1232, Standard Test Method for Temperature Limit of Flammability of Chemicals.," December 6, 2002, available from ASTM as RR: E27-1004..

Bretherick 1999 P. Urben and L. Bretherick, "Bretherick's Handbook of Reactive Chemical Hazards", Butterworth-Heinemann (Elsevier) 1999.

Britton 1993 L. Britton, "Static Hazards Using Flexible Intermediate Bulk Containers for Powder Handling", Process Safety Progress, Vol. 12, No. 4, page 242 (1993).

Britton 1999 L. Britton, "Avoiding Static Ignition Hazards in Chemical Operations", AIChE-CCPS Concept Book (1999).

Britton 2002a L. Britton, "Two Hundred Years of Flammable Limits", Process Safety Progress, Vol. 21, No. 1 (2002).

Britton 2002b L. Britton, "Using Heats of Oxidation to Evaluate Flammability Hazards", Process Safety Progress, Vol. 21, No. 1 (2002).

CCPS 1995 "Guidelines for Chemical Reactivity Evaluation and Application to Process Design", Center for Chemical Process Safety of the American Institute of Chemical Engineers, New York NY, 1995, ISBN 0-8169-0479-0.

Cashdollar 1992 K. Cashdollar and K. Chatrathi, "Minimal explosible dust concentrations measured in 20-L and 1-m3 vessels", Combustion Science and Technology, 87, 157(1992).

CHETAH 2002 CHETAH Version 7.3: The ASTM Computer Program for Chemical Thermodynamic and Energy Release Evaluation, 2002. Copies of the complete program are available from ASTM, 100 Barr Harbor Drive, West Conshohocken, PA 19428-2959. URL http://www.astm.org.

Crowl, 1990 D. A. Crowl and J. F. Louvar, "Chemical Process Safety: Fundamentals with Applications", Prentice Hall (1990).

Eckhoff 1997 R. K. Eckhoff, "Dust Explosions in the Process Industries," Butterworth-Heinemann (1997)

Fenlon 1984 W.J. Fenlon, "A Comparison of ARC and Other Thermal Stability Test Methods", Plant Operations Progress, Vol 3, No. 4, 197-202 (1984).

Frurip 1997 D.J. Frurip, T.C. Hofelich, D.J. Leggett, J.J. Kurland, and J.K. Niemeier, "A Review of Chemical Compatibility Issues," Proceedings of the Annual Loss Prevention Symposium, AIChE, Paper 43c, (1997); see also: T.C. Hofelich, D. J. Frurip, and J. B. Powers, "The Determination of Compatibility via Thermal Analysis and Mathematical Modeling," Process Safety Progress, vol. 13, no. 4, pp. 227-233 (1994).

Going 2000 J. Going, K. Chatrathi and K. Cashdollar, "Flammability limit measurements for dusts in 20-L and 1-m3 vessels", J. Loss Prevention in the Process Industries, 13, 209 (2000)

Grewer 1994 T. Grewer, "Thermal hazards of Chemical Reactions", Elsevier (1994).

Hofelich 2002 Hofelich 2002 T.C. Hofelich and M.S. LaBarge, "On the use and misuse of detected onset temperature of calorimetric experiments for reactive chemicals", Journal of Loss Prevention in the Process Industries, vol. 15, pp 163-168 (2002).

Jaeger 1998 N. Jaeger, T. Hoppe, and J. Terry, International Symposium of Hazards, Prevention and Mitigation of Industrial Explosions, Schaumburg IL (1998).

NFPA XX Various publications and standards from the National Fire Protection Association, Quincy, MA, USA. Also visit www.nfpa.org. Specific NFPA publications and standards mentioned in this paper are: NFPA 53 "Recommended Practice on Materials, Equipment and Systems Used in Oxygen Enriched Atmospheres"
NFPA 69; NFPA 70; NFPA 325; NFPA 497; NFPA 499.

NOAA 2003 Jim Farr's "Chemical Reactivity Worksheet" is a very useful tool for compatibility evaluation and is available for free download from:
http://response.restoration.noaa.gov/chemaids/react.html

Shanley 1995 E. S. Shanley and G. A. Melhem, "A review of ASTM CHETAH 7.0 hazard evaluation criteria", Journal of Loss Prevention in the Process Industries, Vol. 8, No. 5, Pages 261-264 (1995).

Townsend 1980 Townsend, D. I., "Thermal Hazard Evaluation by an Accelerating Rate Calorimeter," Thermochimica Acta 37, 1-30 (1980).

UN, 1999 Recommendations on the Transport of Dangerous Goods, Manual of Tests and Criteria, 3rd Revised Edition, United Nations, 1999, ISBN 92-1-139068-0.

Ural 1992 E.A. Ural, "Dust Entrainability and its Effect on Explosion Propagation in Elongated Structures," Plant/Operations Progress, v. 11, no. 3, pp. 176-181, 1992.

Ural 2001 E.A. Ural, "A Simplified Development Of A Unified Dust Explosion Vent Sizing Formula," Process Safety Progress, v. 20, no.2, pp.136-144, 2001.

Ural 2002 E.A. Ural, "Flammability Potential of Selected Halogenated Fire Suppression Agents and Refrigerants," Process Safety Progress, v. 22, no. 1, pp. 65-73, 2003.

Ural 2003 E.A. Ural, "Airplane Fuel Tank Explosions," presented at the 37th Annual Loss Prevention Symposium, American Institute of Chemical Engineers, New Orleans, LA, March 30 – April 3, 2003.

Yoshida, 1987 T. Yoshida, "Safety of Reactive Systems", Elsevier (1987).

Other Referenced Standards from Part 1 of this Paper:
ASTM D-2155; ASTM E 659; ASTM E 1491; ASTM E 2021; ASTM G-128 "Guide for Control of Hazards and Risks in Oxygen Enriched Systems"
Code of Federal Regulations 33.CFR Part 154.

Reference Material Related to Earlier Versions of CHETAH:

C. A. Davies, D .J. Frurip, E. Freedman, G.R. Hertel, W.H. Seaton, and D. N. Treweek, 'CHETAH 4.4: the ASTM chemical thermodynamic and energy release program,' 2nd edition, ASTM Publication DS 51A (1990).

C. A. Davies, I. M. Kipnis, M. W. Chase, and D. N. Treweek, "The thermochemical and hazard data of chemicals, estimation using the ASTM CHETAH program", ACS Symposium Series No.274, Chemical Process Hazard Review, J. M. Hoffman and D. C. Maser (Eds.), 82-90 (1985).

J. R. Downey, D. J. Frurip, M. S. LaBarge, A. N. Syverud, N. K. Grant, M. D. Marks, B. K. Harrison, W.H. Seaton, D. N. Treweek, and T. B. Selover, "CHETAH 7.0 Users Manuals: the ASTM Computer program for chemical thermodynamic and energy release evaluation", NIST Special Database 16, ASTM, ISBN 0-8031-1800-7, Philadelphia, PA (1994).

D. J. Frurip , E. Freedman and G. R. Hertel, "A new release of the ASTM CHETAH program for hazard evaluation: versions for mainframe and personal computer", Plant/Operations Progress 8, 100-4 (1989).

T. Grewer, D.J. Frurip, and B.K. Harrison, "Prediction of thermal hazards of chemical reactions", Journal of Loss Prevention in the Process Industries 12, 391-398 (1999).

B. K. Harrison, "Computerized screening of chemicals for energy release hazards", Computer Aided Chemical Engineering , (European Symposium on Computer Aided Process Engineering -10, 2000), 799-804, (2000).

Table I. Part A. A list of standards currently administered by ASTM Committee E27. These standards are available from ASTM online (www.ASTM.org) or in print (ASTM 2003).

Table I. Part A Flammability and Ignitability Test Methods		
ASTM Designation	**Title**	**Scope, Citations, Applications, or General Comments**
E 502	Selection and Use of ASTM Standards for the Determination of Flash Point of Chemicals by Closed Cup Methods	Guides in the selection of the appropriate ASTM flash point test method. Repeatability and reproducibility for different test methods. Effect of ambient pressure on flash point. Referenced by NFPA 30 (Classification of Liquids fuels), NFPA 704, NFPA 49, NFPA 325, NFPA 77, and NFPA 497.
E 582	Minimum Ignition Energy and Quenching Distance in Gaseous Mixtures	Tests are performed on specified gas mixtures in a 1 L bomb using capacitance sparks and flanged electrodes. NFPA 77: MIE criteria for various types of static discharge. NFPA 497
E 659	Autoignition Temperature of Liquid Chemicals	Vapor-air mixtures are tested in open-necked 0.5L heated glass flask. NFPA 70 (National Electrical Code): Article 500.8: Specify maximum allowable equipment surface temperature or determine that AIT exceeds equipment "T Code". See also NFPA 497: Table 2-1. NFPA 921: Ch. 14.10.2.11
E 681	Concentration Limits of Flammability of Chemicals	Lower flammable limit (LFL) and upper flammable limit (UFL) in 5L spherical vessel (plus special procedure for gases that are difficult to ignite). NFPA 69: Control of combustible gas concentration by either operating above the UFL (Ch. 5) or below the LFL (Ch. 6). NFPA 921: Ch. 14.10.2.22
E 918	Determining Limits of Flammability of Chemicals at Elevated Temperature and Pressure	This practice covers the determination of the lower and upper concentration limits of flammability of combustible vapor-oxidant mixtures at temperatures up to 200°C and initial pressures up to as much as 1.38 MPa (200 psia).
E 1232	Temperature Limit of Flammability of Chemicals	Minimum and maximum temperatures at which a liquid in equilibrium with its vapor forms a flammable mixture with air at atmospheric pressure. The lower temperature limit of flammability (LTLF) is generally less than the flash point and is more relevant for large process vessels.
E 2021	Hot-Surface Ignition Temperature of Dust Layers	A laboratory procedure to determine the minimum surface temperature capable of igniting a dust layer. NFPA 70 (National Electrical Code): Article 500.8: Specify maximum allowable equipment surface temperature or determine if ignition temperature is greater than equipment "T Code". See also NFPA 499: Table 2-5. NFPA 654: Ch 5-7: Maintain hot surfaces below either 165°C or 80% of dust layer ignition temperature (°C)
E 2079	Limiting Oxygen (Oxidant) Concentration in Gases and Vapors	Limiting oxidant concentration (LOC) in 5L spherical vessel. NFPA 69: Ch. 5: Prevent flame propagation by keeping the oxygen concentration below the LOC.

Table I. Part B. A list of standards currently administered by ASTM Committee E27. These standards are available from ASTM online (www.ASTM.org) or in print (ASTM 2003).

Table I. Part B Special Test Methods for Dust Clouds		
ASTM Designation	**Title**	**Scope, Citations, Applications, or General Comments**
E 789	Dust Explosions in a 1.2-Litre Closed Cylindrical Vessel	Historical value only, **obsolete** for design. Occasionally used as a qualitative screening tool.
E 1226	Pressure and Rate of Pressure Rise for Combustible Dusts	Maximum pressure "P_{max}" and maximum rate of pressure rise (hence K_{St}) in closed vessel of at least 20L volume. NFPA 68: Design of Deflagration Vents. NFPA 69: Design of Containment, Suppression and Isolation Systems. NFPA 484, NFPA 654, NFPA 921: Ch 14.10.2.25
E 1491	Minimum Autoignition Temperature of Dust Clouds	This test method determines the minimum temperature at which a dust cloud will autoignite when exposed to air heated in a furnace at local atmospheric pressure. NFPA 654
E 1515	Minimum Explosible Concentration of Combustible Dusts	Minimum concentration of dust suspended in air (mass per unit volume) that will propagate a flame. NFPA 69: Ch 6: Control of Combustible Dust Concentration. NFPA 77, NFPA 484, NFPA 654, NFPA 664.
E 2019	Minimum Ignition Energy of a Dust Cloud in Air	Methods for minimum ignition energy (MIE) of dust clouds in air. NFPA 61, NFPA 77: MIE criteria for various types of static discharge plus recommendations for different dust handling situations, such as FIBC selection. NFPA 654, NFPA 921
WK1680	Limiting Oxygen (Oxidant) Concentration of Combustible Dust Clouds	In preparation (method will be similar to E 1515). Will apply to NFPA 69, NFPA 484, NFPA 654

Note: WK1680 refers to the ASTM draft standard currently being developed by the E27.05 sub-Committee.

Table I. Part C. A list of standards currently administered by ASTM Committee E27. These standards are available from ASTM online (www.ASTM.org) or in print (ASTM 2003).

Table I. Part C Thermal Stability and Estimation Methods		
ASTM Designation	**Title**	**Scope, Citations, Applications, or General Comments**
E 476	Thermal Instability of Confined Condensed Phase Systems (Confinement Test)	Used to determine the temperature at which a chemical or mixture of chemicals, confined initially as a solid or liquid will "start" a reaction, generating appreciable heat when subjected to a programmed temperature increase.
E 487	Constant-Temperature Stability Of Chemical Materials	Used to assess the constant-temperature stability of chemical materials that undergo exothermic reactions for solids, liquids, or slurries of chemical substances.
E 537	Thermal Stability Of Chemicals By Differential Scanning Calorimetry	Widely used in industry to screen for thermal stability of a material. Used in NFPA 704: Determination of instability rating via instantaneous power density
E 680	Drop Weight Impact Sensitivity Of Solid-Phase Hazardous Materials	Comparison of impact sensitivity of sample against "standard" materials such as picric acid.
E 698	Arrhenius Kinetic Constants for Thermally Unstable Materials	This method covers the determination of the overall kinetic parameters for exothermic reactions.
E 1231	Calculation of Hazard Potential Figures-of-Merit for Thermally Unstable Materials	This practice covers the calculation of hazard potential figures-of-merit for exothermic reactions, including: Time-to-thermal-runaway, Critical half thickness, Critical temperature, Adiabatic decomposition temperature rise, Explosion potential, Shock sensitivity, Instantaneous power density, and NFPA instability rating.
E 1981	Assessing the Thermal Stability of Materials by Methods of Accelerating Rate Calorimetry	Another widely used method in industry for assessing the thermal stability of a material. May be used in NFPA 704: Determination of instability rating via instantaneous power density
E 2012	Preparation of Binary Chemical Compatibility Chart	This guide provides an aid for the preparation of Compatibility charts. It reviews a number of issues that are critical in the preparation of such charts: accurate assessment of chemical compatibility, suitable experimental techniques for gathering compatibility information, incorporation of user-friendliness, and provision for revisions. See separate section in this paper.
E 2046	Reaction Induction Time by Thermal Analysis	This test method describes the measurement of Reaction Induction Time (RIT) of chemical materials that undergo exothermic reactions with an induction period.
	CHETAH	The Chemical Thermodynamic and Energy Release Program is not an official E27 Standard but it's development has been coordinated by E27 for many years. This program is used widely to estimate reaction heats and to provide an indication of the tendency of a material to hazardous energy release. See separate section of this paper.

Table II. Survey Results of ASTM E27 Method Usage.
ASTM Method WK1680 is not a currently published standard, but a working draft.

ASTM Method	Title	Average Annual Use among the 10 responding US labs
Subcommittee E27.04 Flammability and Ignitability		
E 502	Selection and Use of ASTM Standards for the Determination of Flash Point of Chemicals by Closed Cup Methods	105
E 582	Minimum Ignition Energy and Quenching Distance in Gaseous Mixtures	1
E 659	Autoignition Temperature of Liquid Chemicals	13
E 681	Concentration Limits of Flammability of Chemicals	16
E 918	Determining Limits of Flammability of Chemicals at Elevated Temperature and Pressure	6
E 1232	Temperature Limit of Flammability of Chemicals	2
E 2021	Hot-Surface Ignition Temperature of Dust Layers	29
E 2079	Limiting Oxygen (Oxidant) Concentration in Gases and Vapors	7
Subcommittee E27.05 Dust Cloud Ignitability & Explosibility		
E 789	Dust Explosions in a 1.2-Litre Closed Cylindrical Vessel	28
E 1226	Pressure and Rate of Pressure Rise for Combustible Dusts	469
E 1491	Minimum Autoignition Temperature of Dust Clouds	48
E 1515	Minimum Explosible Concentration of Combustible Dusts	123
E 2019	Minimum Ignition Energy of a Dust Cloud in Air	284
E 2021	Hot-Surface Ignition Temperature of Dust Layers	29
WK1680	Limiting Oxygen (Oxidant) Concentration in Dust Clouds	19

Table III. Comparison of E 1232 test apparatus
with commonly used closed cup flash point test apparatuses.

Method/Apparatus	D 56 Tag Closed	D 93 Pensky-Martens	D 3828 Setaflash	E 1232 LTL
Test Volume shape	Cylinder	Cylinder	Cylinder	Sphere
Nominal Vapor space volume (ml)	67	46	19	5000
Sample volume (ml)		72	2 or 4	50 or more
Stirrer	No	Yes	No	Yes
Heating Rate	finite	finite	equilibrium	equilibrium
Ignition source	flame	flame	flame	spark
Ignition location	top	top	top	below center

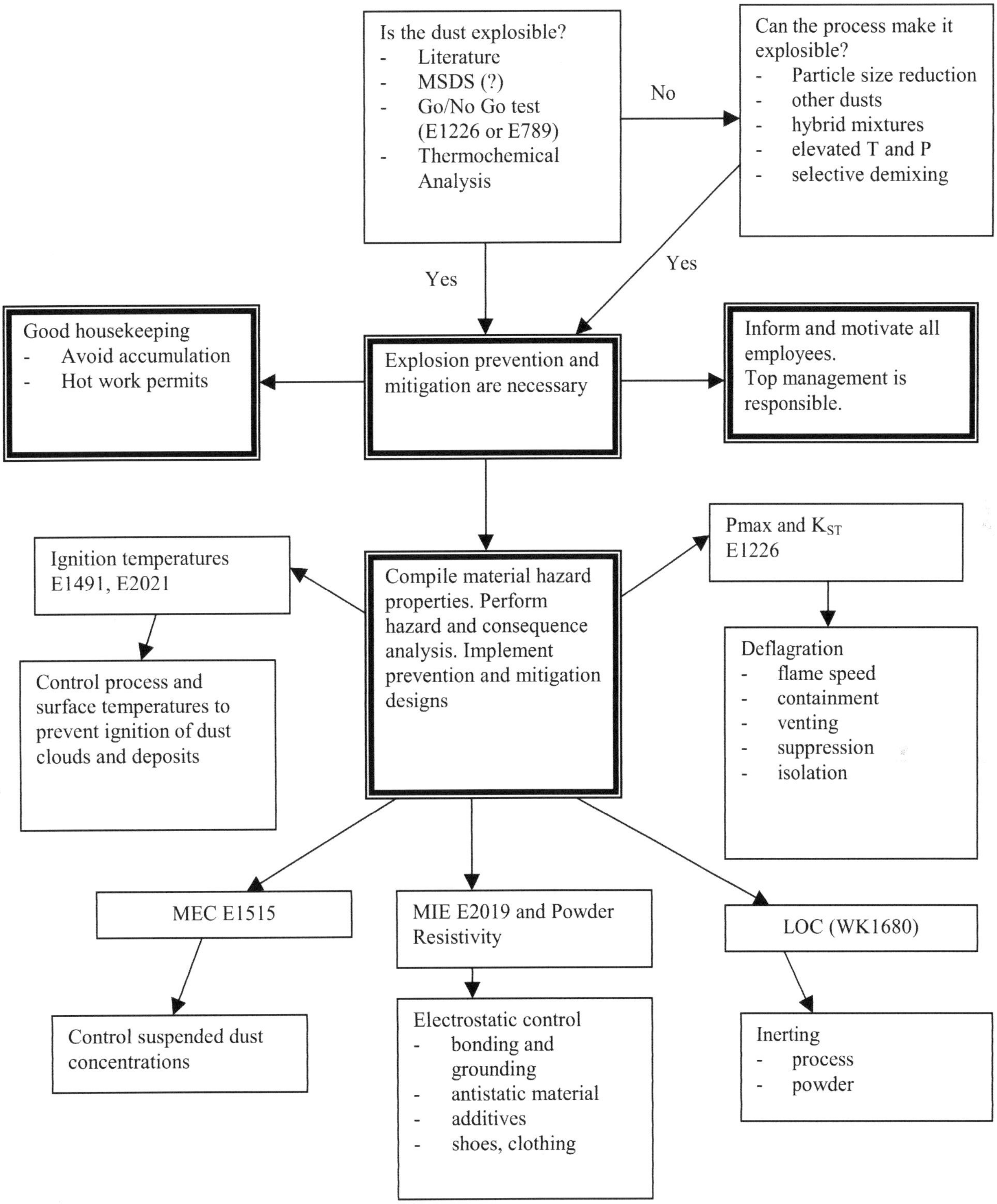

Figure 1a. Dust Explosion Protection Philosophy (Adapted from Eckhoff)

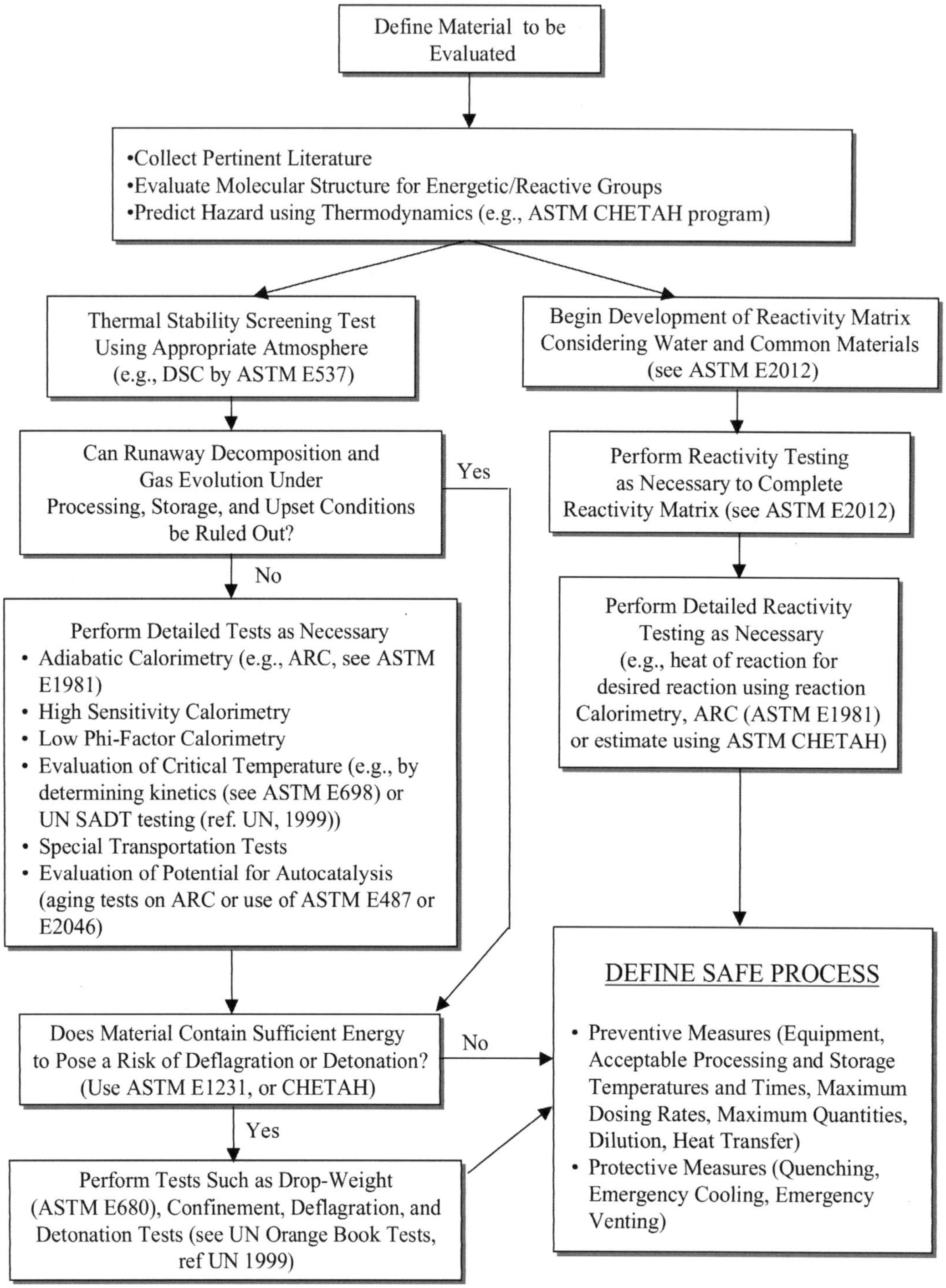

Figure 1b. General Scheme for Evaluation of Thermal Stability and Reactivity Hazards. Adapted from "Guidelines for Chemical Reactivity Evaluation and Application to Process Design" (CCPS 1995).

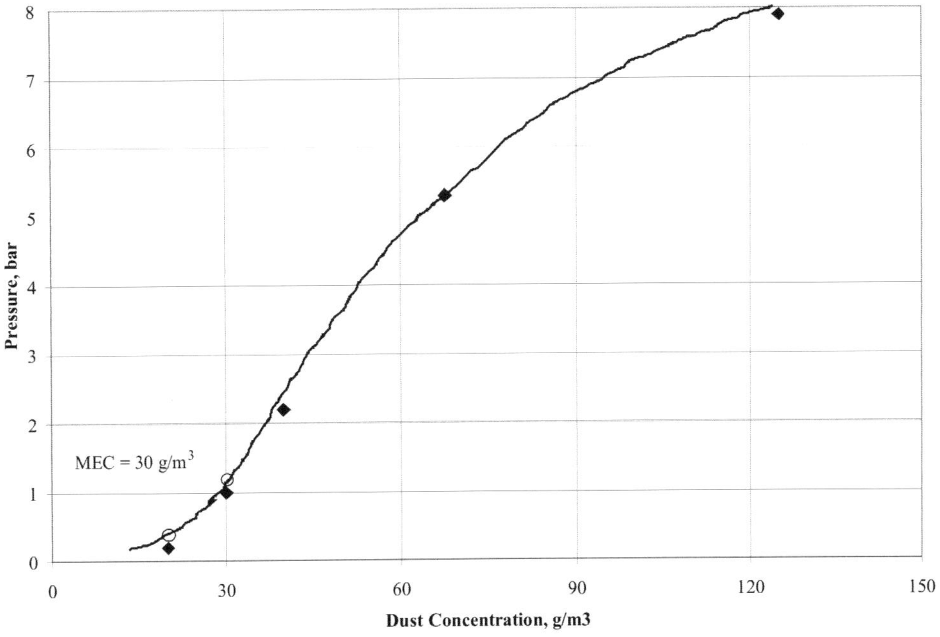

Figure 2. Typical test results for the determination of the MEC for a dust. In this hypothetical case, the MEC is determined to be 30 g/m3.

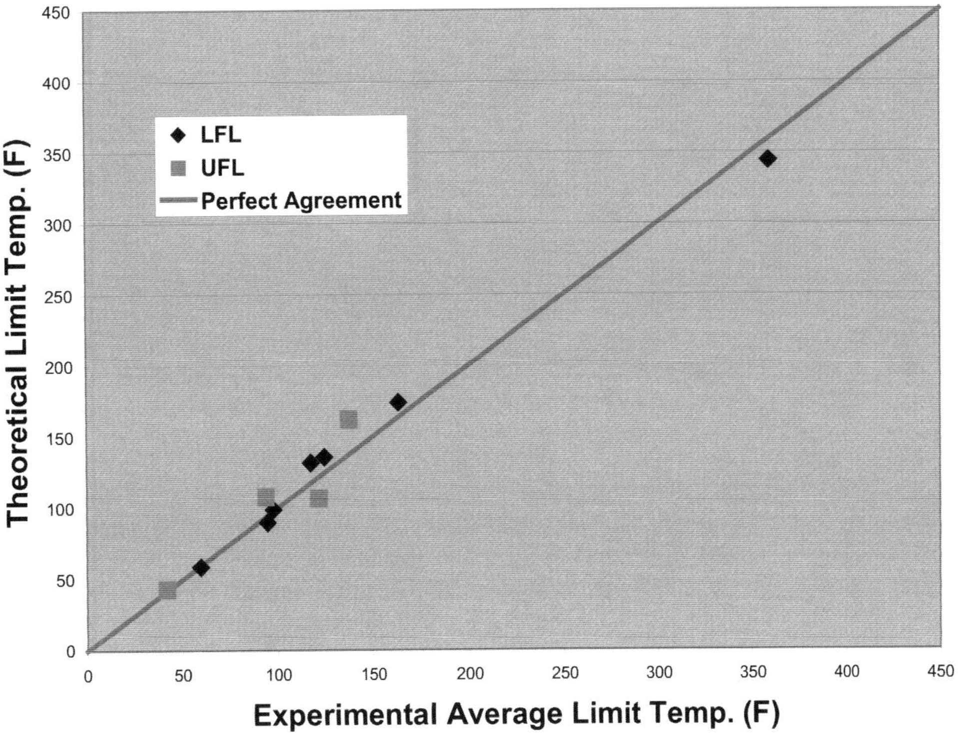

Figure 3. Comparison of experimental LTL and UTL values with theoretical values

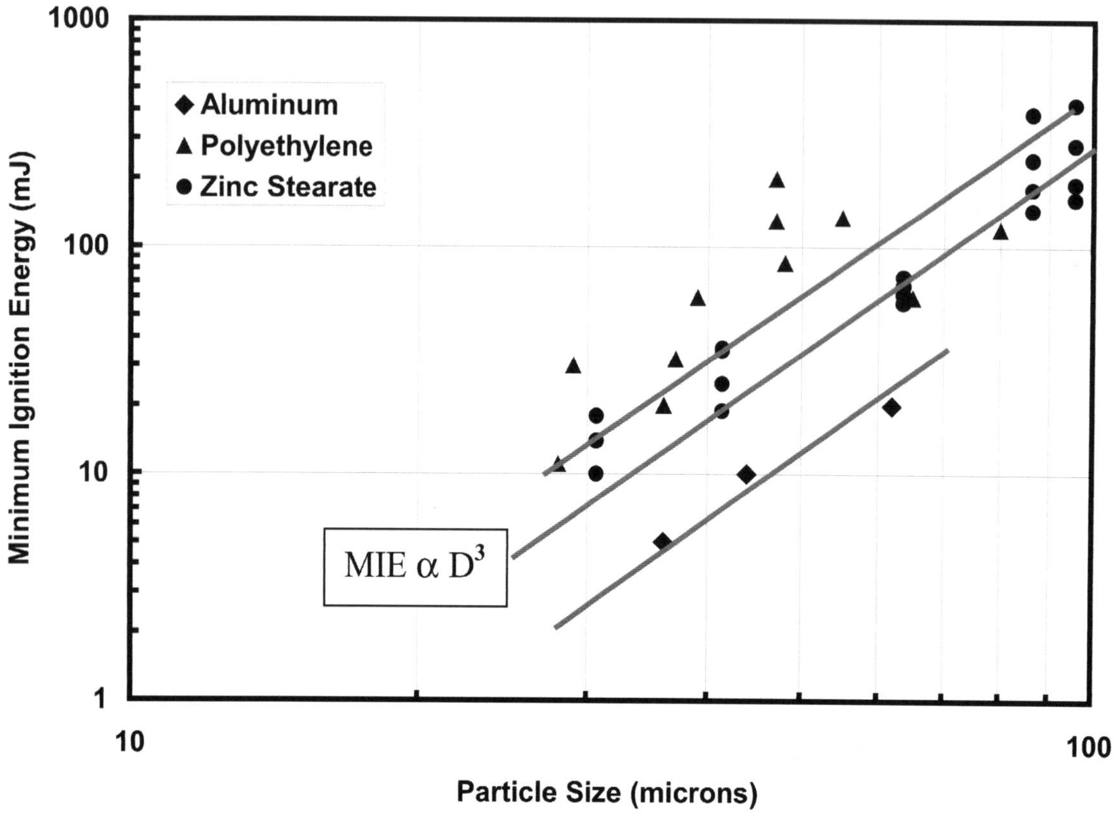

Figure 4 Round Robin MIE data showing the effect of particle size.

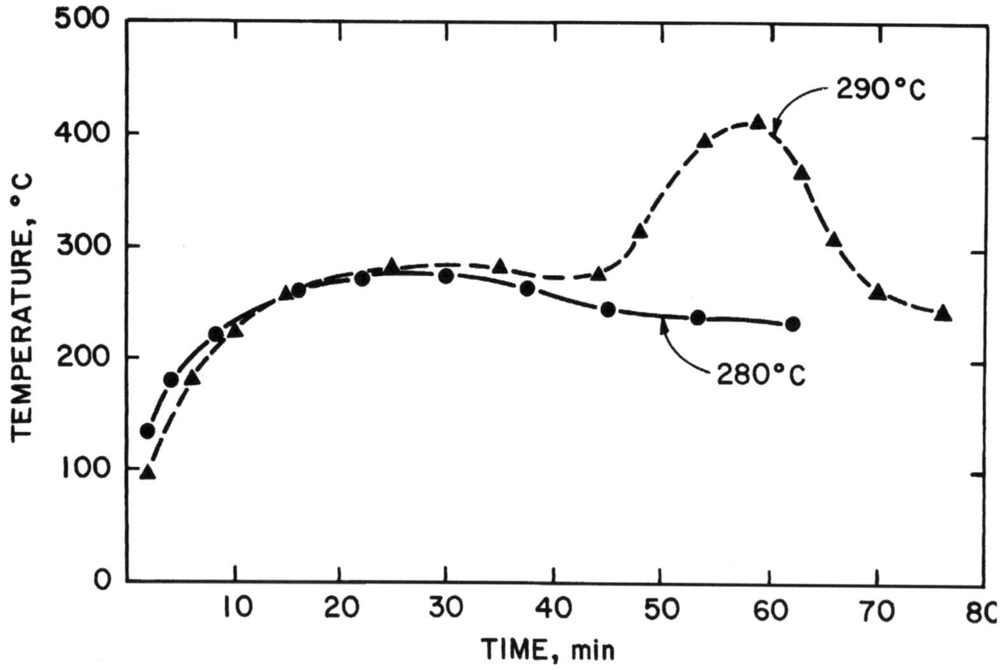

Figure 5. Typical results of the dust layer hot surface ignition temperature test described in ASTM E 2021.

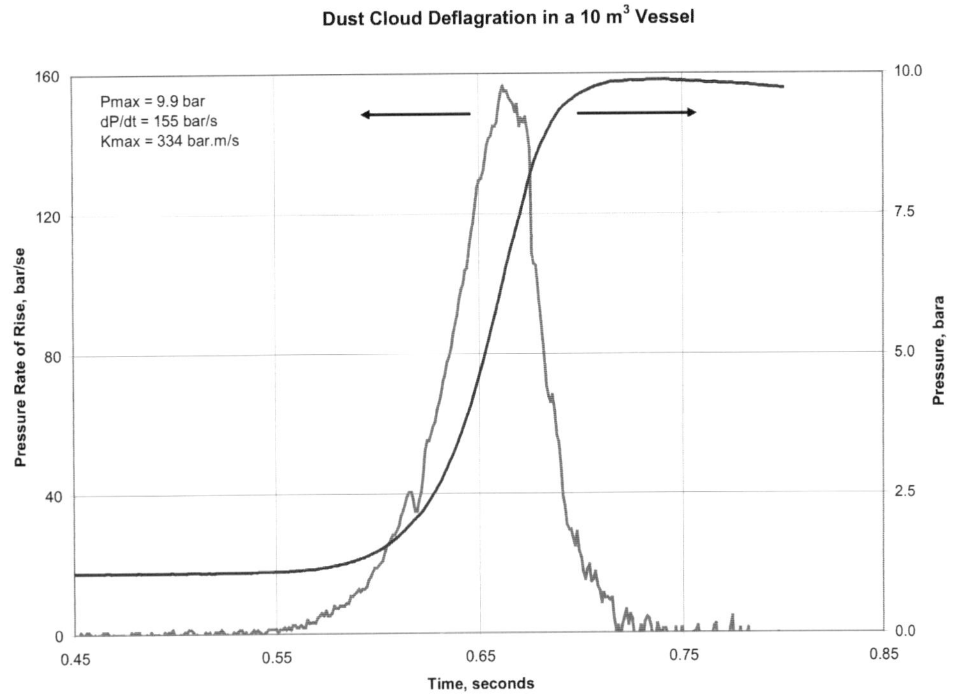

Figure 6. Typical pressure and pressure rate versus time behavior for the combustion of a dispersed dust in a 10 m^3 vessel.

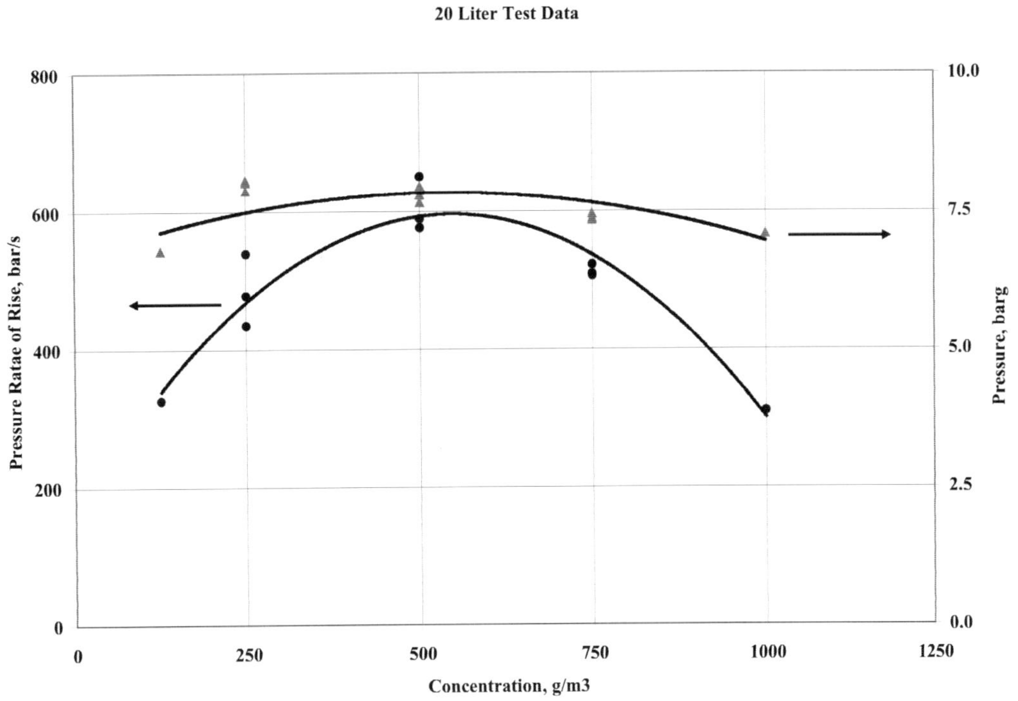

Figure 7. Typical pressure rate and pressure versus dust cloud concentration in a 20 L vessel.

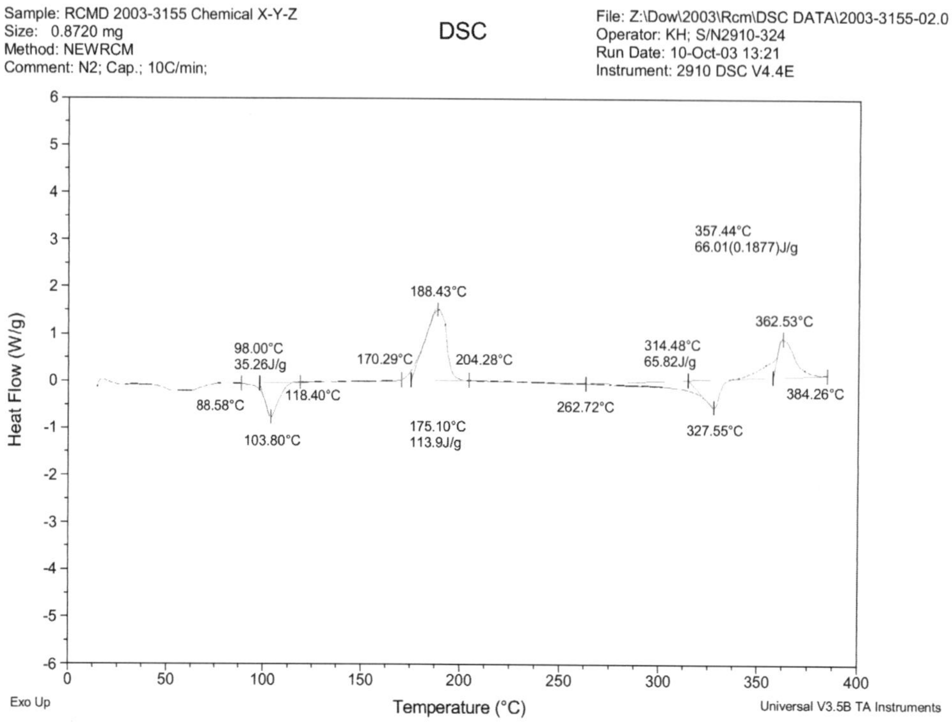

Figure 8: DSC trace of a new additive material. The endotherm near 100 C is interpreted as the melting point. The exotherm detected near 170 C is not unexpected due to the presence of the nitro group in the molecule.

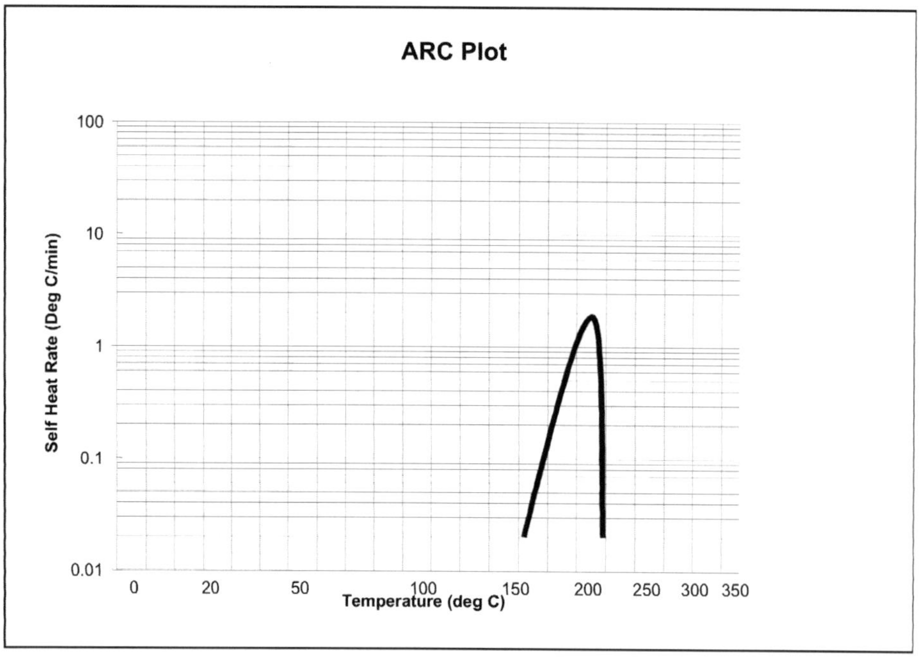

Figure 9. A hypothetical heat rate versus temperature plot for an ARC run on the material specified in the example. The ARC heat rate plot is typically plotted as a log of the temperature rate versus an inverse 1/T (T in kelvins) scale. (Note: the hypothetical data shown here have already been corrected for the so-called phi factor: see E 1981 for a description.)

1	Hydrochloric Acid	1					
2	Sulfuric Acid	R	2				
3	Acetic Acid	?	R	3			
4	Ethanol	NR	R	NR	4		
5	Ethylenediamine	R	R	R	NR	5	
6	Water	R	R	NR	NR	R	6

Legend:

R	Reactive under the stated scenario - incompatible
NR	Non-Reactive under the stated scenario - compatible
?	Unknown – assume incompatible until further information is obtained

Figure 10. Hypothetical Compatibility Chart for Process Y at Site X (*example of possible format - not for use*), developed by John Doe, last update 12/99. Scenario: Ambient temperature, adiabatic, non-vented conditions. Maximum contact time: 4 hours. Definition of incompatibility: Adiabatic temperature rise greater than 25°C, or a gassy reaction.

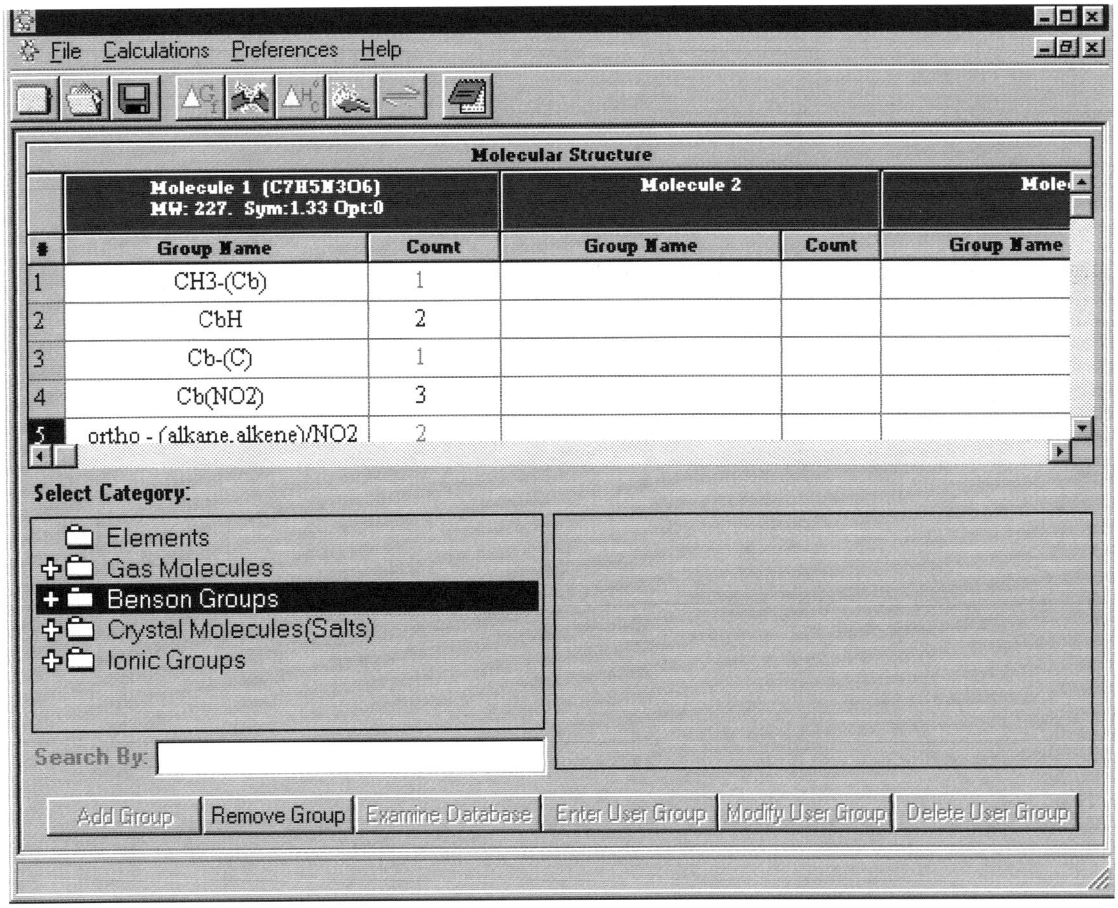

Figure 11. TNT (2,4,6-Trinitrotoluene) molecule built with Benson groups

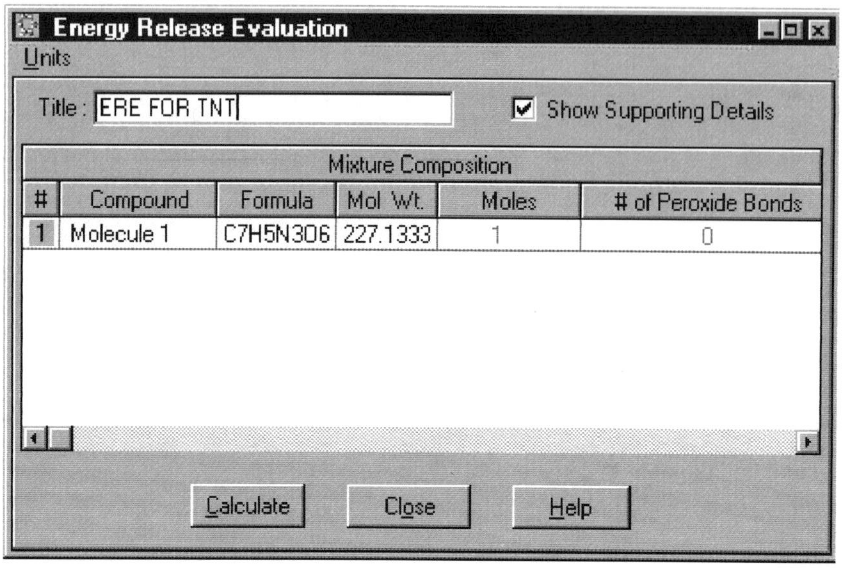

Figure 12. ERE calculation form.

ASTM PROGRAM CHETAH 7.3
COMMITTEE E27 CHEMICAL THERMODYNAMIC AND ENERGY APPRAISAL PROGRAM

ENERGY RELEASE EVALUATION

Compound Name : Molecule 1
Formula : C7H5N3O6
Molecular Weight : 227.133
Amount : 1 Mole(s)
Heat of Formation at 25 C : 9.830 kcal/mol

Hazard Classification Section:

Primary Results:

Criterion	Value	Units	Hazard Classification(#1)
Maximum heat of decomposition (#2)	-1.404	kcal/g	HIGH
Over-all Energy Release Potential	-1.221		HIGH
Net Plosive Density(#4)	0.547		PLOSIVE

Secondary Results:
The following criteria are parameters of the over-all ERE calculation and serve to enhance the fit for particular classes of compounds, but are not generally useful for hazard analysis out of context:

Criterion	Value	Units	Hazard Classification(#1)
Fuel Value - Heat of Decomposition	-2.174	kcal/g	HIGH
Oxygen Balance(#3)	-73.963	g O2/100g	HIGH
CHETAH ERE Criterion 4	213.234	kcal^2/gmole.gm	HIGH
Total Number of Peroxide Bonds	0.000		

Warning: These ratings only apply to hazards associated with strong shock. This does not imply the absence of other hazards.

Notes:
#1 This evaluation was developed to classify a composition as able or not able to decompose with violence, if subjected to the proper conditions. Information on the interpretation of the hazard classification criteria used by CHETAH may be found in the CHETAH documentation (ASTM publication DS-51A) and in the User's Manual.
#2 For decomposition products shown.
#3 Experience has shown that the oxygen balance criterion is useful only for compounds composed of the elements C, H, N, and O. It was derived for explosives, especially nitrates [W.C.Lothrop and G.R Handrick, Chem.Rev.1949,44,419-45], for example it is not valid for water. Units are grams additional O2 per 100g reactants needed for complete combustion.
#4 Sum of auxoplosive and plosophoric weights per gram of mixture.(J. Chem. Ed. v66, A137 (1989)

Decomposition Products (chosen to maximize heat of decomposition):

Moles	State	Species	
5.250	ref-cr	C	Carbon
1.500	ref-gas	N2	Nitrogen
1.750	gas	CO2	Carbondioxide
2.500	gas	H2O	Water

Heat of Combustion Section

Fuel Value (Net Heat of Combustion):

Mass Basis		Mole Basis	
-3.578	kcal/g	-812.680	kcal/mol
-14.970	kJ/g	-3400.253	kJ/mol
-6440.377	Btu/lb		

Combustion Products (Chosen for Fuel Value and Net Heat of Combustion):

Moles	State	Species	
1.500	ref-gas	N2	Nitrogen
7.000	gas	CO2	Carbondioxide
2.500	gas	H2O	Water

Groups Present in this Compound:-

Description	Count
CbH	2
Cb-(C)	1
Cb(NO2)	3
CH3-(Cb)	1
Ortho - (alkane,alkene)/NO2	2

Figure 13. Output for Energy Release Evaluation (ERE) of TNT.

An Update to the Baker-Strehlow-Tang Vapor Cloud Explosion Prediction Methodology Flame Speed Table

Adrian J. Pierorazio, J. Kelly Thomas, Quentin A. Baker, and Donald E. Ketchum

Baker Engineering and Risk Consultants, Inc.
San Antonio, Texas
AdrianP@BakerRisk.com
Prepared for presentation at the 38th Loss Prevention Symposium
at the 2004 Spring National Meeting of the American Institute of Chemical Engineers.

January 2004

Unpublished

Abstract
The Baker-Strehlow-Tang vapor cloud explosion (VCE) blast load prediction methodology utilizes flame speed as a measure of explosion severity. In previous publications, guidance has been presented for selecting flame speeds as a function of congestion, confinement, and fuel reactivity. These recommended values were based on empirical data available from the literature. Over the last five years, a series of medium-scale VCE tests have been conducted through a joint industry program to better understand vapor cloud explosions and to allow a more accurate definition of the flame speed applicable to a given combination of congestion, confinement, and fuel reactivity. These tests have demonstrated that the previously published flame speeds are not conservative for all configurations for the case of no confinement (3-D flame expansion). This paper provides an overview of the tests along with an update to the flame speed table where the previously published guidance was not conservative.

Introduction
Vapor cloud explosion (VCE) prediction methodologies can be organized into three broad categories: simplified (point source), phenomenological, and numerical. Because each method has its own set of advantages and disadvantages, they are commonly used to address significantly different types of problems. Simplified models are used for many on-shore plant analyses. These models are not the best choice for blast load prediction within or very close to the explosion source because they do not account for the fine details of equipment layout. However, the areas of interest for most on-shore facilities (e.g., occupied buildings) are usually at a significant enough distance from the explosion sources that this shortcoming is not an issue. The simplified models permit analysts to perform assessments more quickly than numerical or phenomenological models, while still providing reasonably accurate results at the areas of interest, consequently offering a significant cost advantage over the more time-consuming approaches.

Background
The three most widely used simplified VCE blast load prediction models are the TNT equivalent method, the TNO multi-energy method[1], and the Baker-Strehlow-Tang (BST) method[2, 3, 4]. All three methods use non-dimensionalized blast curves to predict the blast load for a given source energy and standoff distance. The methodologies differ only in the number and type of curves used. The TNT equivalent model has one pressure and one impulse curve and inherently assumes that all VCEs are detonations that behave like a condensed-phase high explosive. This assumption represents a gross simplification, and this method is no longer widely used. The TNO multi-energy method provides ten numerically derived curves for both pressure and duration. These curves span a range of severities from mild deflagrations to detonations, with the curves evenly spaced based on their maximum pressures. The applied impulse can be estimated from the pressure and duration data provided by the curves. The Baker-Strehlow-Tang method uses a continuum of numerically determined pressure and impulse curves that are based on the Mach number of the VCE flame front relative to a stationary point of reference. Duration can be calculated from pressure and impulse.

Since all three of these simplified methods are based on looking up values from numerically derived non-dimensional blast curves that provide pressure, impulse, and duration, the main difference between the methods is the means of selecting which curve to use. In the TNT methodology, there is only one curve and, therefore, only one option. The TNO methodology can be applied by using severity number 7 unless there is a good reason to use a different value[5], or by using one of several methodologies (GAME, company internal methodologies, etc.) that assign severity based on unit size (number of floors) or congestion level. The BST method provides guidance on selecting a flame speed based on broad categories of congestion (obstacle density), confinement (degrees of freedom of expansion), and fuel reactivity (based on laminar burning velocity).

The accuracy of any of these methods is limited by the ability to select an appropriate curve for a particular plant geometry. The original BST flame speed table was produced based on published experimental data. At the time of publication, much of the published experimental data was determined using small-scale apparatus. It was recognized that some of these small-scale tests might not be ideal for application to a full-scale plant; however, the data were used because no alternative was available. The Explosion Research Cooperative (a joint industry program) initiated an extended series of experiments to refine the functional relationship between flame speed and the degree of congestion and confinement along with the flammable gas mixture reactivity. These tests have been performed over a wide range of congestion levels (low, medium and high) and degrees of confinement (three-dimensional flame expansion, two-dimensional flame expansion with varying aspect ratios, and mixed two- and three-dimensional expansion). Tests have been conducted with near-stoichiometric methane-air, propane-air, and ethylene-air mixtures, which represent low, medium, and high reactivity mixtures, respectively. A limited number of tests have also been performed with other fuels as well as with lean or rich mixtures. The participating companies of the Explosion Research Cooperative agreed to release this update to the flame speed table in order to ensure that the data available in the published literature is conservative.

The Baker-Strehlow-Tang Methodology

The complete Baker-Strehlow-Tang method was first published at the 28th Loss Prevention Symposium in 1994[2] shortly after the development of a correlation for determining maximum flame speed in a VCE. Since that time, the Baker-Strehlow method has been used extensively in VCE hazard assessments in refineries and chemical plants. The goal of the original study in which the methodology was developed was to achieve an objective methodology to provide consistent prediction of VCE blast effects.

The VCE blast curves developed by Strehlow were chosen for the original Baker-Strehlow methodology because blast curves are selected based on flame speed, which affords the opportunity to use empirical data in the selection. Tang and Baker subsequently developed a new set of VCE blast curves, which were adopted in 1999[4] and the methodology was renamed Baker-Strehlow-Tang. The Baker-Strehlow-Tang blast curves are presented in Appendix A.

Determination of the energy term is based on the size of the flammable cloud within confined and congested portions of a plant. Multiple blast sources can emanate from a single release. Fuel reactivity, confinement and obstacle density influence the reaction rate as mentioned above.

Test Description

The congested region for these tests was constructed of modular cubic sections. The length, width, and height of each cube are 6 feet (1.8 meters). The congestion is provided by a regular array of vertical circular tubes (2" schedule 40 pipe). A 4x4 array of tubes per cube was used for low congestion, a 7x7 array represented medium congestion, and 11 rows of alternating 4 and 7 tubes were used for high congestion (see Figure 1). The corresponding pitch-to-diameter, area blockage, and volume blockage ratios are provided in Table 1. The corner tubes of the congestion cubicle were used as substitutes for four of the tubes in each congestion pattern.

Table 1: Congestion Levels Utilized in Tests

Congestion Level	Pitch to Diameter Ratio	Area Blockage Ratio (%)	Volume Blockage Ratio (%)
Low	7.6	13	1.5
Medium	4.3	23	4.3
High	3.1	23	5.7

Sixteen such cubes were arranged in a 2x8 pattern for these tests to provide an elongated length-to-width aspect ratio that is representative of many on-shore facilities. An illustration and photograph of the test rig in this configuration are shown in Figure 2 and Figure 3, respectively. The test rig was configured without any confinement (i.e., no wall or roof sections) for all of the tests reported in this paper.

A near-stoichiometric fuel-air mixture was employed in these tests. Methane was used as the low reactivity fuel, propane as the medium reactivity fuel, and ethylene as the high reactivity fuel. The fuel gas was dispersed through the test rig using a distributed set of venturi mixing devices and its concentration was monitored using an online gas analyzer. A thin (0.001 inch) plastic tent was placed around the rig to facilitate development of the flammable gas mixture. The mixture completely filled the congested region, but did not extend beyond it. Weights holding down the bottom of the plastic tent were removed just before ignition to minimize the impact of the tent on flame propagation. The flammable gas mixture was ignited using a low-energy source in the center of the rig at ground level.

An array of pressure transducers was placed inside and outside the rig at distances of up to 300 feet along both the long and short axis centerlines and diagonally from one corner. Figure 4 shows a typical pressure transducer distribution. High-speed video cameras were positioned outside the rig to provide flame front position recordings. An array of ionization probes was placed along the long axis centerline to track the position of the flame front for selected tests.

Test Results

The pressure-time histories for each pressure transducer were analyzed to determine the peak side-on pressure and impulse at each location. The results were plotted against predicted pressure and impulse versus distance for a variety of flame speeds. The flame speed was iterated until the best match with the data was obtained (see Figure 5 for an example of such a fit). Since pressure and impulse predictions outside congestion are the primary objective of the Baker-Strehlow-Tang methodology, greater emphasis was placed on the best match of prediction to test data outside congestion.

A secondary check of the flame speed was performed next. The flame speed yielding the best fit was compared to the flame front location over time as measured by the ionization probes and/or high-speed video. The best fit for subsonic flame speeds (deflagrations) essentially represented the average flame front speed in the rig (see Figure 6). This result shows that the flame speed that is the best fit for prediction of pressure and impulse is also a good fit to the measured flame speeds.

The most likely reason that these flame speeds are higher than the ones originally used for the flame speed table is due to the scale of the experiments. The original experiments were less than 6 feet (1.8 meters) in their longest dimension for many cases and thus, did not have sufficient length for flame acceleration. The current set of experiments was conducted with a rig size approaching that of actual process equipment, so that these results are more applicable to typical industrial plants. Furthermore, the flame speed data given in the following section have been scaled up to account for the maximum size of a typical industrial plant in order to provide additional margin.

During the course of the testing, a detonation-to-deflagration transition (DDT) was observed during a medium congestion, unconfined ethylene test. The DDT caused extensive damage to the test rig and associated systems. After repairs were made, the test was repeated to confirm the result. The details of this portion of the testing and the justification for concluding that the cloud underwent a DDT are discussed in Thomas et al (2003)[6].

Test Results

A new flame speed table (see Table 2) was produced based on the test program discussed in the preceding sections. It is recommended that this flame speed table be used for all BST blast load predictions since it corresponds to a scale representative of typical chemical processing plants.

Table 2: Revised Baker-Strehlow-Tang Methodology Flame Speed Table

Confinement	Reactivity	Congestion		
		Low	Medium	High
2-D	High	0.59	DDT	DDT
	Medium	**0.47**	0.66	**1.6**
	Low	0.079	0.47	0.66
2.5-D	High	**0.47**	**DDT**	**DDT**
	Medium	**0.29**	**0.55**	**1.0**
	Low	0.053	**0.35**	**0.50**
3-D	High	**0.36**	**DDT**	**DDT**
	Medium	**0.11**	**0.44**	**0.50**
	Low	0.026	**0.23**	**0.34**

Notes: (1) Bold values have been updated based on the current set of experiments.
(2) 2.5-D values are the simple average between 2-D and 3-D values.

It is important to note that this new flame speed table includes the 2.5-D category that was put forward by Baker et al (1997) to be used in cases where the confinement is made of either a frangible panel or by a nearly solid confining plane (e.g., pipe rack where the pipes are almost touching). As described, the 2.5-D values are obtained by taking a simple average between the 2-D and 3-D confinement values for the same congestion and fuel reactivity. The 1-D entries have been deleted since the maximum flame speed achieved in true one-dimensional expansion conditions (i.e., a pipe) is a function of the length-to-diameter ratio of the pipe in addition to pipe geometry (elbows, tees, etc.), fuel reactivity and congestion level. Many fuels are able to undergo a DDT in a 1-D geometry if the combination of length-to-diameter ratio and obstacle density are sufficiently high. As a result, the use of a single number to represent all length-to-diameter ratios is overly simplified and a more detailed analysis is recommended for all such cases.

Acknowledgements

The tests described in this paper were performed under the sponsorship of the Explosion Research Cooperative, an ongoing joint industry research program organized by Baker Engineering and Risk Consultants, Inc. (BakerRisk). The Explosion Research Cooperative is comprised of companies in the petrochemical and chemical industries with a strong commitment to process safety. The Cooperative has supported VCE testing and model development by Baker Engineering and Risk Consultants, Inc. over the past several years, and this support is gratefully acknowledged. The VCE experiments were carried out with the help of Roland Ramirez, Greg Burton, Jeremy McElroy, Kenny Martin, Martin Goodrich, and Massimiliano Kolbe. The authors also acknowledge the contributions of Ming Jun Tang to the VCE test program.

Recommendations for Future Work

While the tests discussed in this paper have increased the understanding of VCE phenomena and contributed greatly to enhancing the BST blast load predictive methodology, it is recognized that there are many relevant questions these tests do not address. The Explosion Research Cooperative continues to support VCE testing research. The Cooperative invites companies with an interest in VCE blast loads to join with them and participate in these efforts.

Figures

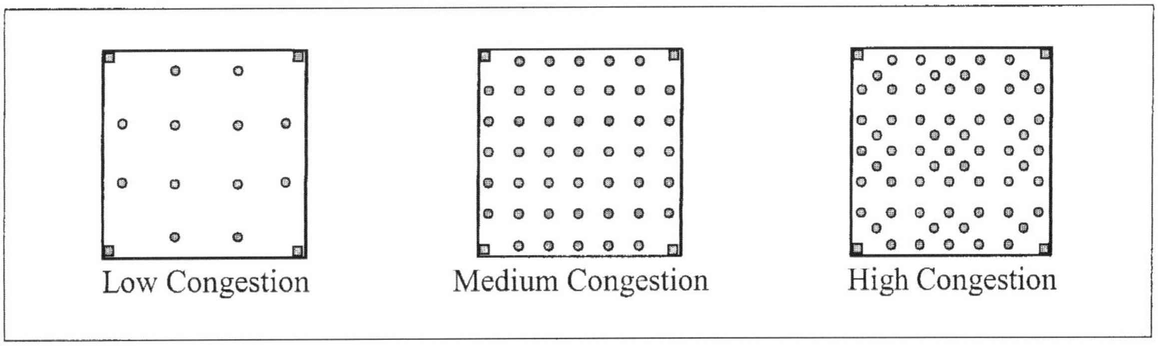

Figure 1: Congestion Patterns Used

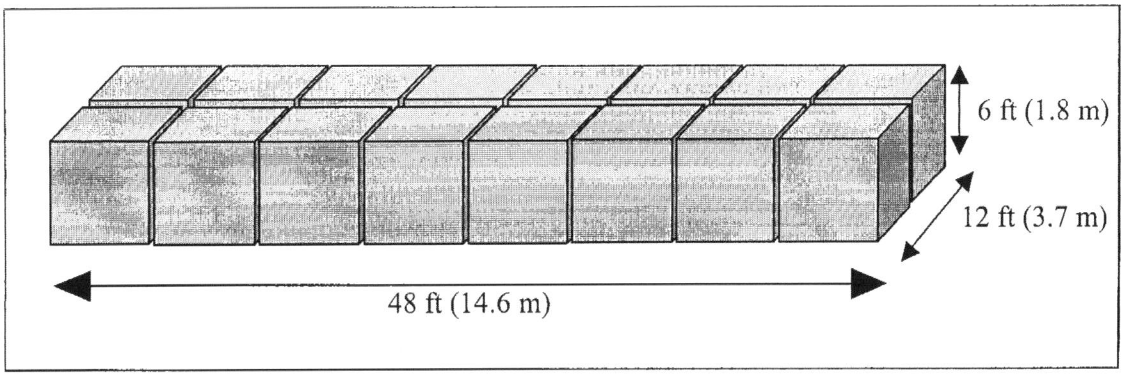

Figure 2: Illustration of test cube arrangement

Figure 3: Photograph of test rig

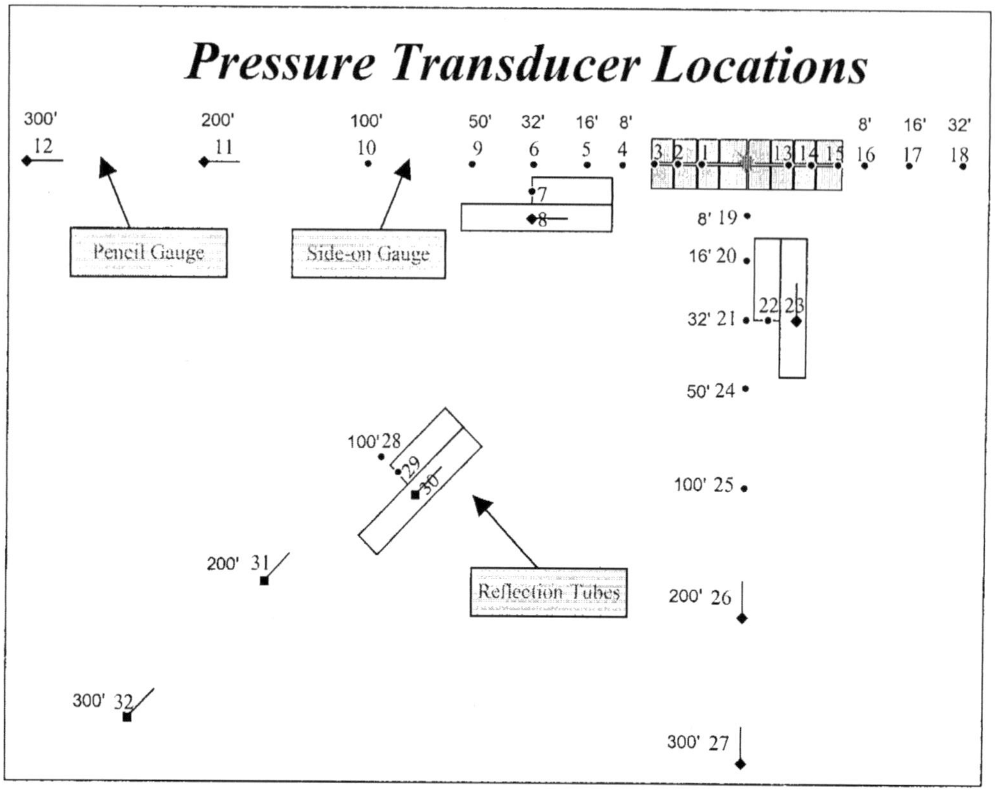

Figure 4: Pressure Transducer Locations

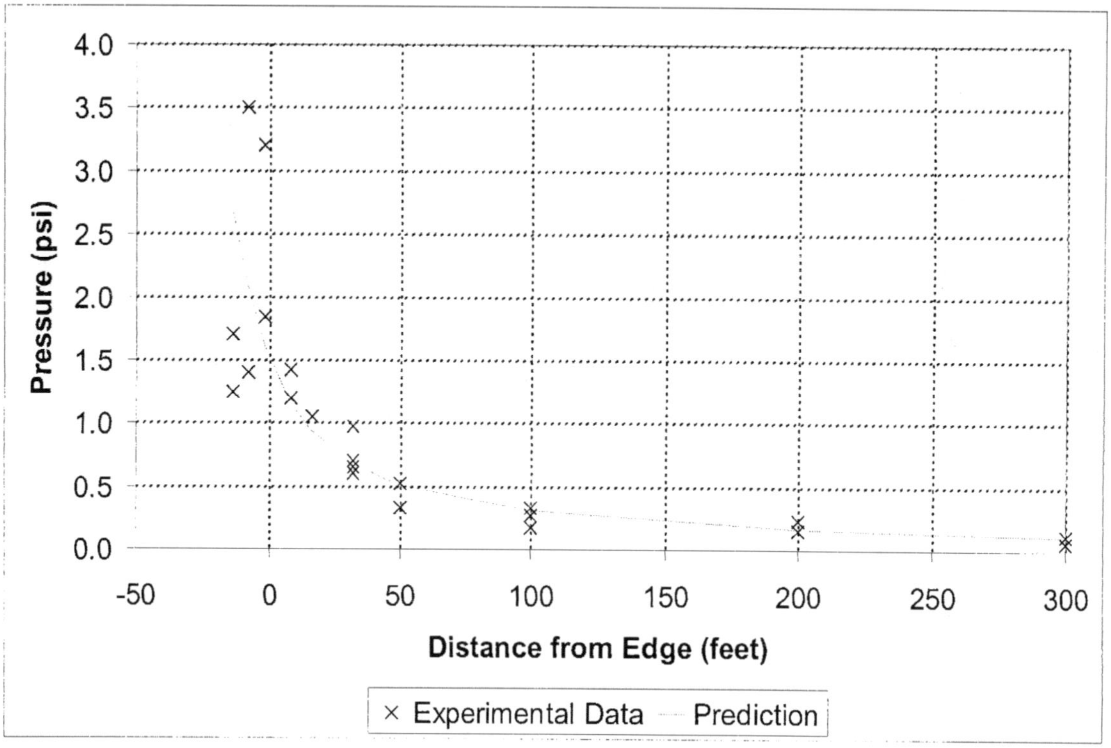

Figure 5: Sample fit of Flame Speed to 3-D High Congestion Propane Pressure Data

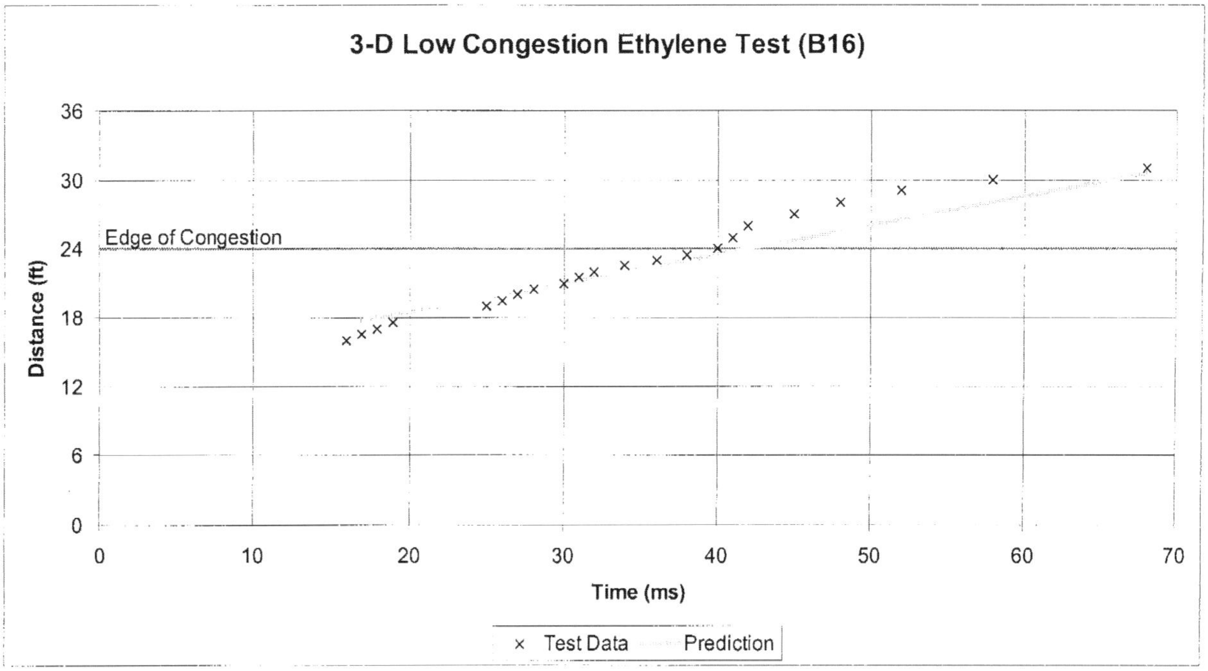

Figure 6: Comparison between predicted and actual flame front location for a deflagration

References

1) TNO Yellow Book, 3rd ed., Apeldoorn, TNO, The Netherlands, 1997.

2) Baker, Q. A., M.J. Tang, E.A. Scheier and G.J. Silva (1994) "Vapor Cloud Explosion Analysis," 28th Annual Loss Prevention Symposium.

3) Baker, Q. A., C.M. Doolittle, G.A. Fitzgerald and M.J. Tang (1997) "Recent Developments in the Baker-Strehlow VCE Analysis Methodology," 31st Annual Loss Prevention Symposium.

4) Tang, M.J., Baker, Q.A., "A New Set of Blast Curves from Vapor Cloud Explosions," 33rd Loss Prevention Symposium, American Institute of Chemical Engineers, Paper 29e, March 14-18, 1999.

5) Crowl, D. A., Understanding Explosions, Center for Process Safety of the American Institute of Chemical Engineers, New York, 2003

6) R.A. Strehlow, R.T. Luckritz, A.A. Adamczyk and S.A. Shimp, "The Blast Wave Generated By Spherical Flames", *COMBUSTION AND FLAME*, 35: 297-310, 1979

7) Thomas, J.K., A.J. Pierorazio, M. Goodrich, M. Kolbe, Q.A. Baker and D.E. Ketchum (2003), "Deflagration to Detonation Transition in Unconfined Vapor Cloud Explosions," Center for Chemical Process Safety (CCPS) 18th Annual International Conference & Workshop, Scottsdale, AZ, 23-25 September 2003.

8) Tang, M.J and Baker, Q.A., "Predicting Blast Effects From Fast Flames", 32nd Loss Prevention Symposium, AIChE March 1998

9) Tang, M.J., Cao, C.Y., and Baker, Q.A., "Blast Effects From Vapor Cloud Explosions", International Loss Prevention Symposium, Bergen, Norway, June 1996

Appendix A. Baker-Strehlow-Tang Blast Curves

The Baker-Strehlow-Tang blast curves are constructed as scaled blast wave properties versus scaled distance and are presented as families of curves with the flame Mach number as the parameter. The flame Mach number is the apparent flame speed divided by the ambient sound velocity. The blast properties and the distance are in non-dimensional coordinates, as shown in figures on the following pages.

According to Sach's scaling, following non-dimensional parameters are used in Baker-Strehlow-Tang blast curves.

$$\overline{P^+} = \frac{p_{max}^{+} - p_0}{p_0} \qquad \overline{P^-} = \frac{\left|p_{max}^{-} - p_0\right|}{p_0}$$

$$\overline{I^+} = \frac{i^+ a_0}{E_t^{1/3} p_0^{2/3}} \qquad \overline{I^-} = \frac{i^- a_0}{E_t^{1/3} p_0^{2/3}}$$

$$\overline{t^+} = \frac{t^+ a_0}{\left(E_t / p_0\right)^{1/3}} \qquad \overline{t^-} = \frac{t^- a_0}{\left(E_t / p_0\right)^{1/3}}$$

$$\overline{t_a} = \frac{t_a a_0}{\left(E_t / p_0\right)^{1/3}} \qquad \overline{U} = \frac{u_{max}}{a_0}$$

$$\overline{R} = \frac{R}{\left(E_t / p_0\right)^{1/3}}$$

where p_0 is atmospheric pressure;
a_0 is the sound velocity at ambient conditions;
p_{max}^{+} is the maximum of positive peak absolute pressure;
p_{max}^{-} is the maximum of negative peak absolute pressure;
R is stand-off distance;
E_t is the total energy release from the explosion source;
i^+ is positive specific impulse;
i^- is negative specific impulse;
t^+ is time duration of positive phase;
t^- is time duration of negative phase;
t_a is the arrival time of wave front;
u_{max} is the maximum of flow velocity.

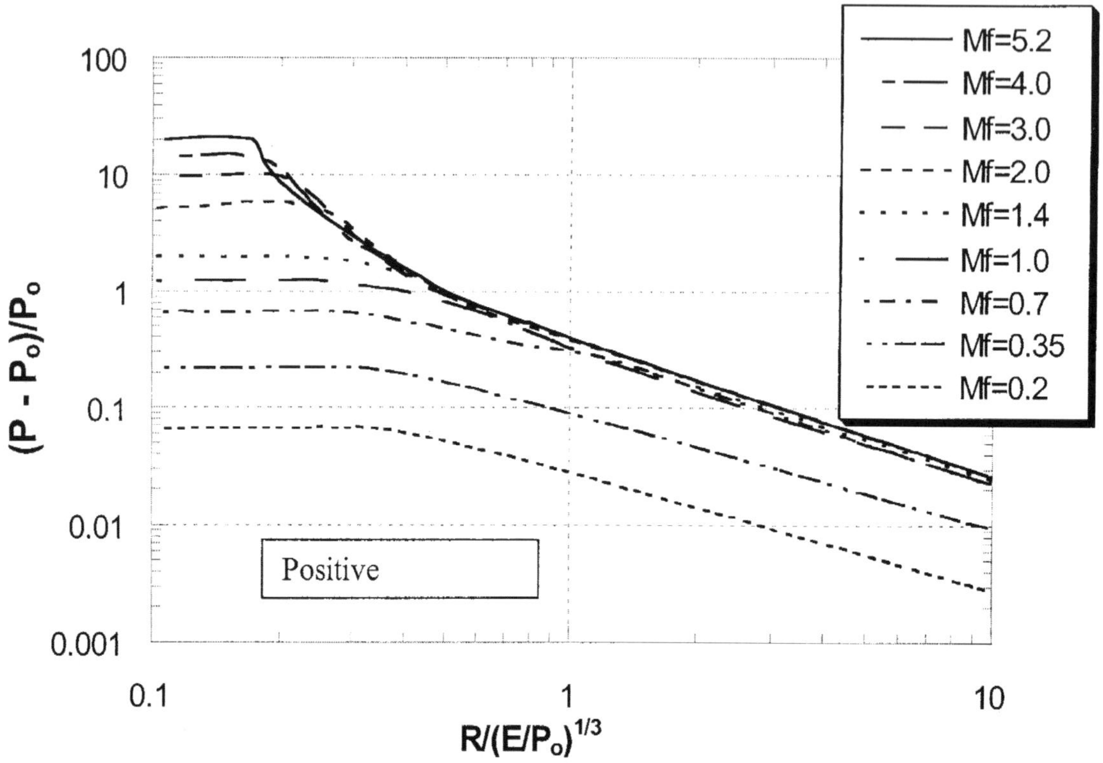

Figure A1 Positive Overpressure vs Distance for Various Flame Speeds

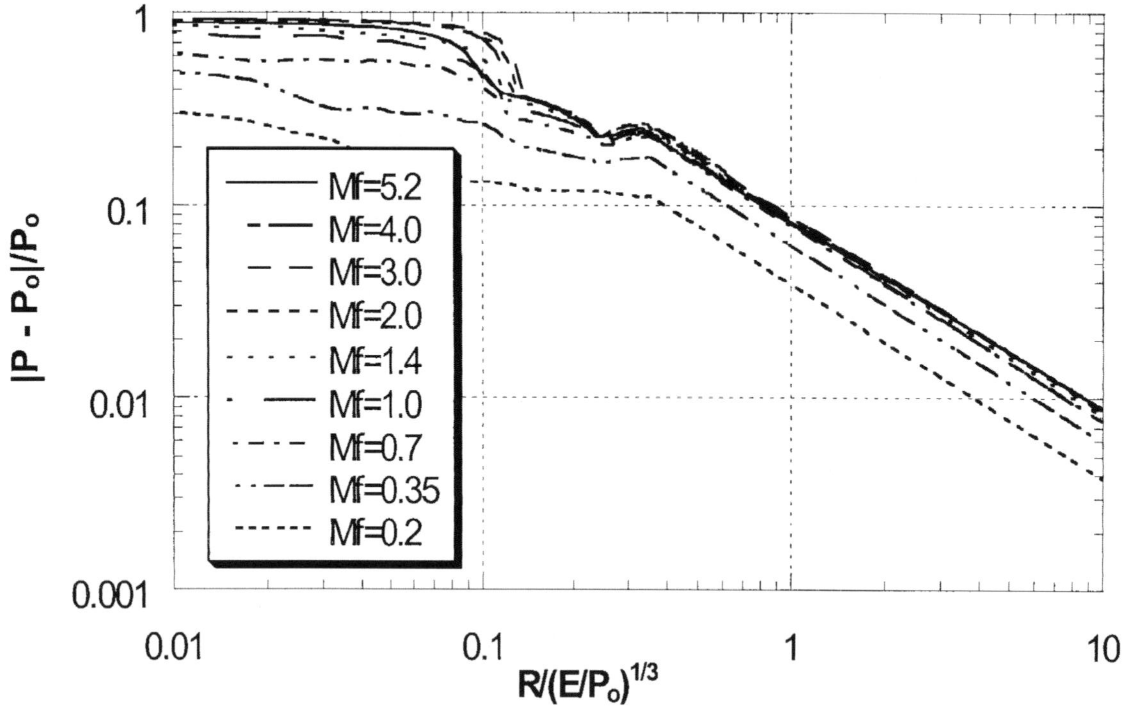

Figure A2 Negative Overpressure vs Distance for Various Flame Speeds

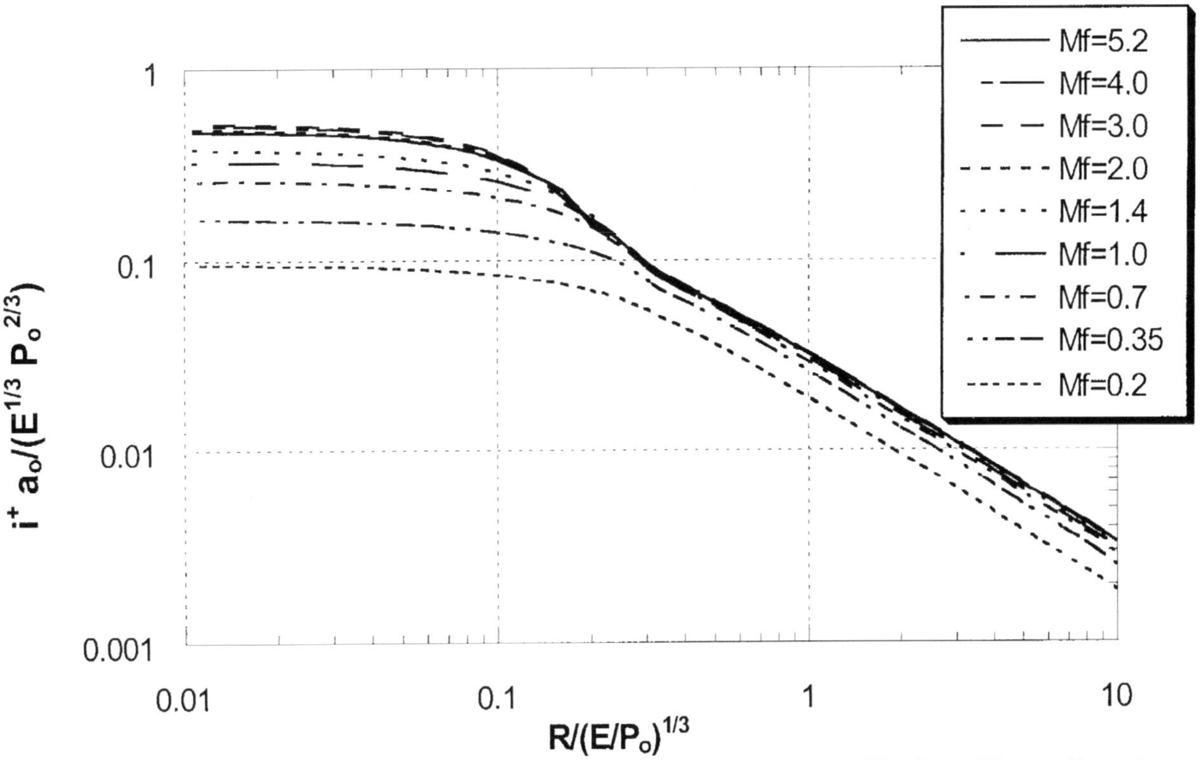

Figure A3 Positive Impulse vs Distance for Various Flame Speeds

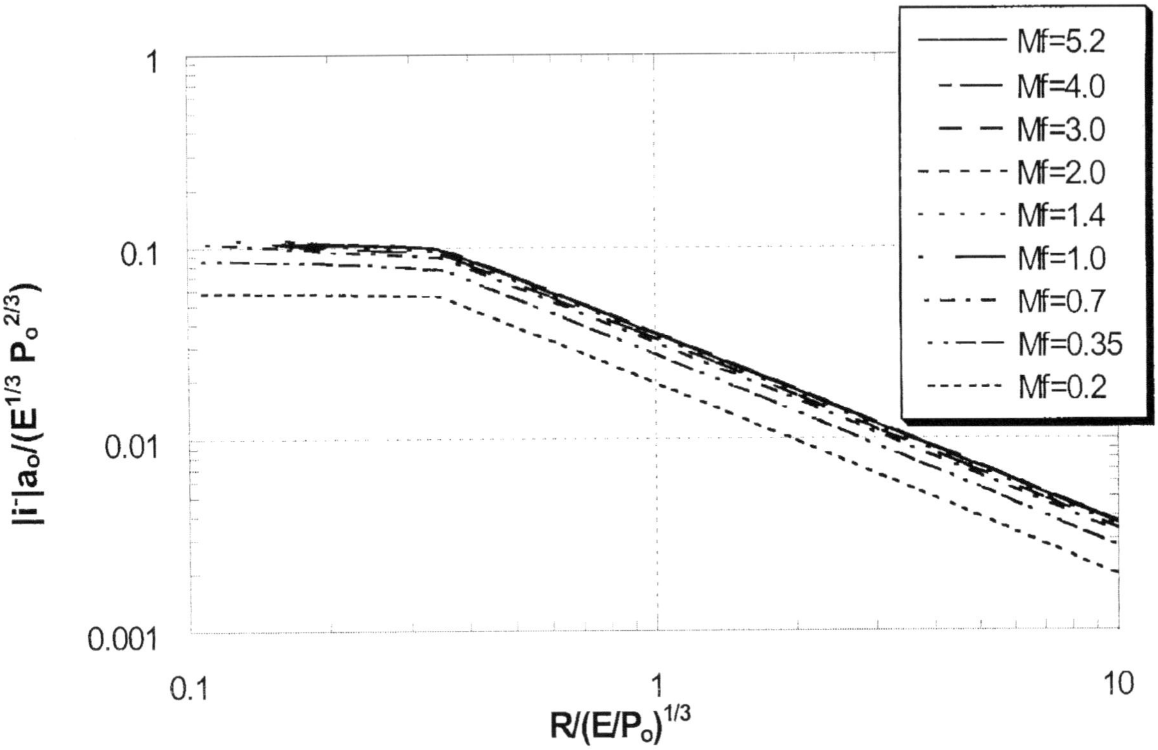

Figure A4 Negative Impulse vs Distance for Various Flame Speeds

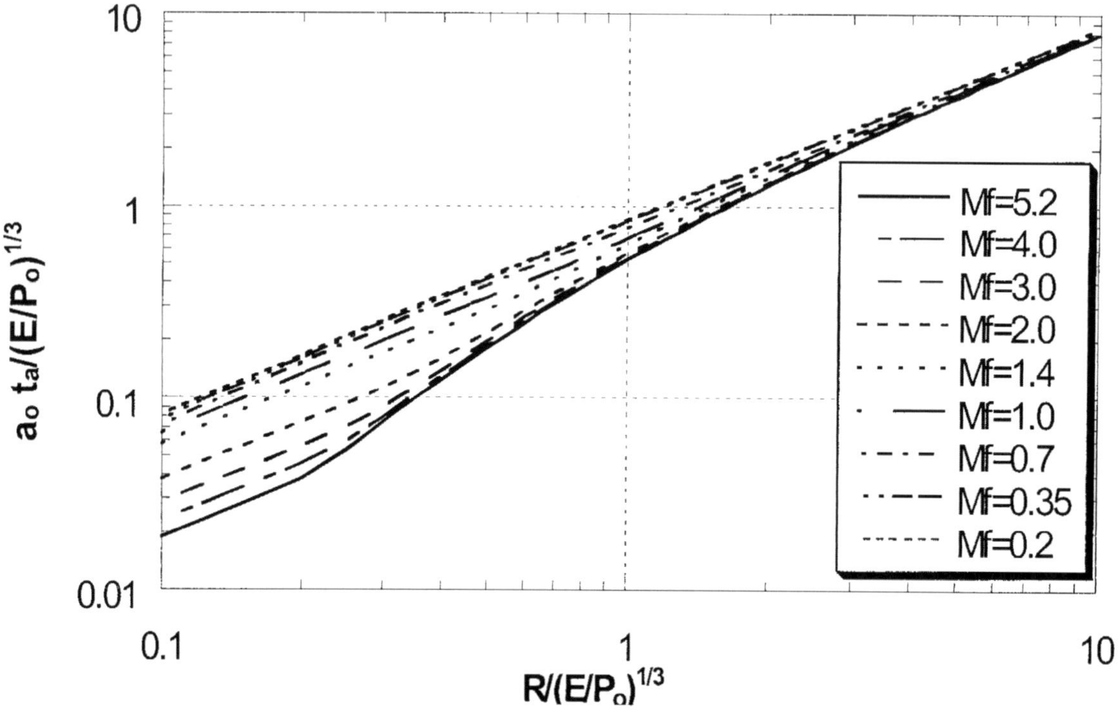

Figure A5 Arrival Time vs Distance for Various Flame Speeds

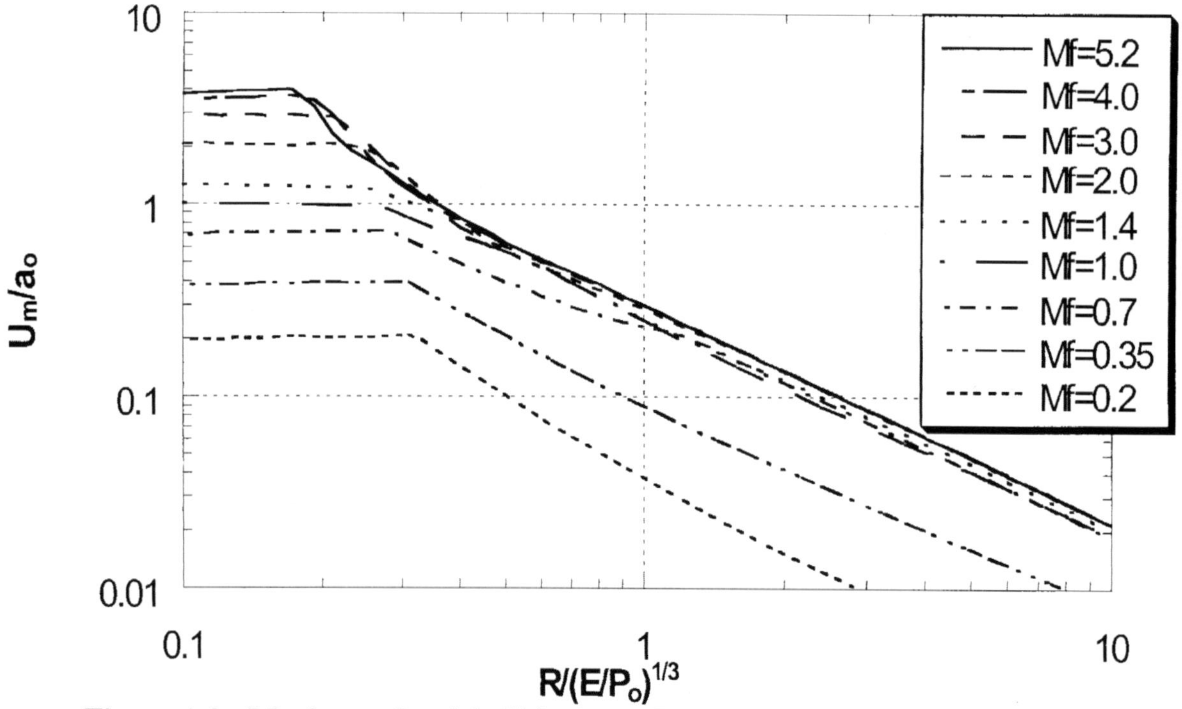

Figure A6 Maximum Particle Velocity vs Distance for Various Flame Speeds

Reference:

Tang, M.J., Baker, Q.A., "A New Set of Blast Curves from Vapor Cloud Explosions," 33rd Loss Prevention Symposium, American Institute of Chemical Engineers, Paper 29e, March 14-18, 1999.

New Methods for Estimating Sonic Gas Flow Rates in Pipelines

Jason M. Keith and Daniel A. Crowl
Department of Chemical Engineering
Michigan Technological University
Houghton, MI 49931

Abstract

This paper presents a detailed review and analysis of sonic gas flow in pipelines, adding considerably more detail to the analysis, particularly for long pipelines. Our results show that 1) the mass flow rate is asymptotic as the velocity head pipe loss increases 2) the asymptotic value is identical for both adiabatic and isothermal conditions and 3) a maximum is found in the gas flow, although this maximum is near the asymptotic value. A graphical method for isothermal flows and a simple, shortcut formula is presented using asymptotic analysis which accurately estimates gas flow rates in long pipelines under both adiabatic and isothermal conditions. Process safety applications are provided.

Compressible Flow Theory

There have been many studies of the flow of compressible gases in pipelines in textbooks (Churchill, 1980; Levenspiel, 1984; Crowl and Louvar, 2002), in articles (Levenspiel, 1977; Cochran, 1996; Farina, 1997; Kumar, 2002), and technical documents (Crane, 1986). Two limiting cases, adiabatic and isothermal, are often considered. Adiabatic flow conditions assume flow through an insulated pipe. These conditions are usually valid for short pipelines since there is little heat transfer to or from the gas. Isothermal flow conditions assume flow through a pipe held at a uniform temperature; these conditions are commonly assumed when studying the flow of a gas in an uninsulated pipeline. Most natural gas pipelines are considered isothermal.

The easier case of isothermal flow is presented first. The literature (Dodge and Thompson, 1937; Crane, 1986) states that flow under isothermal conditions is applicable to most industrial conditions, an example being a lengthy pipeline submerged under water. (Note: This is almost contradictory since most industrial pipelines are not submerged in water.) The analysis begins with a mechanical energy balance about a fluid element of length dx within a horizontal pipe of diameter D. The shear stress at the exterior surface of the fluid element is balanced by the sum of the pressure head and kinetic energy of the flowing gas, both acting over the cross-section of the fluid element:

$$\tau \pi D dx = -\frac{\pi D^2}{4} dP - u\rho \frac{\pi D^2}{4} du \tag{1}$$

In industrial practice, long piping networks may not be horizontal. However, the contribution of gravitational potential to the mechanical energy balance is generally negligible. Equation (1) can be rewritten in terms of the mass velocity $G = u\rho$ (which is also called the mass flux) and the Fanning friction factor f from the relationship $\tau = f\rho u^2/2 = fG^2/2\rho$, yielding:

$$\frac{f}{2}dx = -\frac{\rho D}{4G^2}dP + \frac{D}{4}\frac{d\rho}{\rho} \tag{2}$$

This can be integrated over the length of the pipe from $x = 0$ to L, where the inlet pressure and density at $x = 0$ is $P = P_1$ and $\rho = \rho_1$ and the exit pressure and density at $x = L$ are given by $P = P_2$ and $\rho = \rho_2$. For an ideal gas ($\rho = PM/RT$), the result is:

$$\frac{4fL}{D} = -\frac{M}{G^2 RT}\left(P_2^2 - P_1^2\right) + 2\ln\left(\frac{\rho_2}{\rho_1}\right) \tag{3}$$

Suppose there is a gas at temperature T entering the pipe at a velocity u_1 and pressure P_1. The velocity can be recast in dimensionless form by dividing by the speed of sound, a, which is described by the thermodynamic derivative:

$$a^2 = \left(\frac{\partial P}{\partial \rho}\right)_s = \frac{\gamma RT}{M} = \frac{\gamma P}{\rho} \tag{4}$$

The entering Mach number, Ma_1, is given by u_1/a. The inlet density is calculated from the ideal gas law as $\rho_1 = P_1 M/RT$.

At the exit of the pipe, the velocity is u_2, and the corresponding Mach number is $Ma_2 = u_2/a$. Mass conservation states that the mass velocity, G, denoted by the product of the density and velocity, ρu, must be a constant at steady-state. Thus, $\rho_1 u_1 = \rho_2 u_2$. The exiting density can now be calculated as $\rho_2 = \rho_1(u_1/u_2)$ which can be written as $\rho_2 = \rho_1(Ma_1/Ma_2)$ in terms of the entering and exiting Mach numbers. Because the pressure is proportional to the density, the exiting pressure P_2 can also be determined as $P_2 = P_1 (Ma_1/Ma_2)$. The mass velocity can also be written in terms of the Mach number at the inlet and exit conditions:

$$G = \rho_1 u_1 = \rho_1 Ma_1 a = Ma_1 P_1\sqrt{\frac{\gamma M}{RT}} = \rho_2 u_2 = \rho_2 Ma_2 a = Ma_2 P_2\sqrt{\frac{\gamma M}{RT}} \tag{5}$$

Using these relationships, the integrated mechanical energy equation given by Equation 3 can be rewritten in terms of the Mach number at inlet and outlet conditions:

$$\frac{4fL}{D} = \frac{1}{\gamma}\left(\frac{1}{Ma_1^2} - \frac{1}{Ma_2^2}\right) + 2\ln\left(\frac{Ma_1}{Ma_2}\right) \tag{6}$$

Under choked flow conditions, the speed of the exiting gas reaches a maximum (as shown by Churchill (1980)), such that the exiting Mach number $Ma^2{}_2 = 1/\gamma$. This gives:

$$\frac{4fL}{D} + \ln\left(\frac{1}{\gamma Ma_1^2}\right) + \left(1 - \frac{1}{\gamma Ma_1^2}\right) = 0 \tag{7}$$

such that the exit velocity is $u_2 = a/\gamma^{1/2}$, the exit pressure is $P_2 = P_1 Ma_1 \gamma^{1/2}$ and the exit density is $\rho_2 = \rho_1 Ma_1 \gamma^{1/2}$.

A more complicated situation exists under adiabatic flow conditions because the temperature changes as the gas flows towards the tube exit. Thus, the gas velocities at the inlet and exit become:

$$u_1 = Ma_1 a = Ma_1 \sqrt{\frac{RT_1\gamma}{M}} \qquad \text{and } u_2 = Ma_2 a = Ma_2 \sqrt{\frac{RT_2\gamma}{M}} \tag{8}$$

Since $G = \rho_1 u_1 = \rho_2 u_2$, a relationship between the inlet and outlet Mach number, temperature, and pressure follows directly from Equation 8:

$$Ma_1^2 P_1^2 T_2 = Ma_2^2 P_2^2 T_1 \tag{9}$$

Since the gas temperature is changing, the conservation of thermal energy also needs to be solved:

$$udu = -dh = -C_p dT \tag{10}$$

which states that there is a balance between kinetic energy and enthalpy, where C_p is the heat capacity per unit mass. Integrating from the inlet (where $u = u_1$ and $T = T_1$) to any point in the pipe (where u and T are arbitrary), one obtains:

$$\frac{u^2}{2} - \frac{u_1^2}{2} = -C_p (T - T_1) \tag{11}$$

Substituting $G = u\rho$ and the ideal gas law for the temperature, $T = PM/\rho R$:

$$\frac{G^2}{2}\left(\frac{1}{\rho^2} - \frac{1}{\rho_1^2}\right) = \frac{\gamma}{\gamma - 1}\left(\frac{P_1}{\rho_1} - \frac{P}{\rho}\right) \tag{12}$$

Substituting the velocities from Equation 8 into Equation 11 with u and T being evaluated at the exit (such that $u = u_2$ and $T = T_2$) and after some manipulation yields:

$$\frac{T_2}{T_1} = \frac{Y_1}{Y_2} \qquad \text{where } Y_i = 1 + \frac{\gamma - 1}{2} Ma_i^2 \tag{13}$$

This relationship, when substituted into Equation 9, yields:

$$\frac{P_2}{P_1} = \frac{Ma_1}{Ma_2}\sqrt{\frac{Y_1}{Y_2}} \tag{14}$$

The ideal gas law can then be used to determine the density ratio:

$$\frac{\rho_2}{\rho_1} = \frac{P_2 T_1}{P_1 T_2} = \frac{Ma_1}{Ma_2}\sqrt{\frac{Y_2}{Y_1}} \tag{15}$$

Solving Equation 12 for P and differentiating with respect to ρ gives:

$$dP = \left[\frac{G^2}{2}\frac{\gamma-1}{\gamma}\left(\frac{1}{\rho_1^2}+\frac{1}{\rho^2}\right)+\frac{P_1}{\rho_1}\right]d\rho \tag{16}$$

which can be substituted in for dP in Equation 2 and integrated from the inlet to the outlet to yield:

$$\frac{4fL}{D} = \frac{\gamma+1}{\gamma}\ln\left(\frac{\rho_2}{\rho_1}\right)+\left[1-\left(\frac{\rho_2}{\rho_1}\right)^2\right]\left(\frac{\gamma-1}{2\gamma}+\frac{P_1\rho_1}{G^2}\right) \tag{17}$$

Equation 12, using exit conditions (where $\rho = \rho_2$ and $P = P_2$), can be rewritten in terms in temperature and pressure to give an expression for the mass velocity:

$$G^2 = \frac{2\gamma M}{(\gamma-1)R}\frac{T_2 - T_1}{\left(\frac{T_1}{P_1}\right)^2 - \left(\frac{T_2}{P_2}\right)^2} \tag{18}$$

Equation 18 can be inserted into Equation 17 to give the following expression:

$$\frac{\gamma+1}{\gamma}\ln\left(\frac{P_1T_2}{P_2T_1}\right)-\frac{\gamma-1}{2\gamma}\left(\frac{P_1^2T_2^2-P_2^2T_1^2}{T_2-T_1}\right)\left(\frac{1}{P_1^2T_2}-\frac{1}{P_2^2T_1}\right)+\frac{4fL}{D}=0 \tag{19}$$

Equation 19 can also be written in terms of the inlet and outlet Mach numbers:

$$\frac{\gamma+1}{2}\ln\left(\frac{Ma_2^2Y_1}{Ma_1^2Y_2}\right)-\left(\frac{1}{Ma_1^2}-\frac{1}{Ma_2^2}\right)+\gamma\frac{4fL}{D}=0 \tag{20}$$

Under choked flow conditions, the maximum velocity of the exit gas is at the speed of sound (as shown by Churchill (1980)), such that the Mach number is unity. This simplifies Equation 20 to the form:

$$\frac{\gamma+1}{2}\ln\left[\frac{2Y_1}{(\gamma+1)Ma_1^2}\right]-\left(\frac{1}{Ma_1^2}-1\right)+\gamma\frac{4fL}{D}=0 \tag{21}$$

such that the temperature ratio $T_2/T_1 = 2Y_1/(\gamma+1)$, the pressure ratio $P_2/P_1 = Ma_1(2Y_1/(\gamma+1))^{1/2}$, and the density ratio $\rho_2/\rho_1 = Ma_1((\gamma+1)/2Y_1)^{1/2}$.

The assumptions that are inherent in the above analysis are:

1. Either isothermal or adiabatic conditions along the entire pipe length.
2. The fluid is an ideal gas
3. Potential energy contributions, Δz, are negligible compared to kinetic energy and pressure head / compressibility
4. Constant compressibility ratio, γ, along the entire pipe length
5. Constant friction factor, f, along the entire pipe length.

The validity of each of these assumptions will be discussed in detail later.

Solution Techniques for Choked Flow Conditions

In this section, solution procedures for estimating mass flow rates under choked conditions will be outlined. These include trial-and-error methods, graphical methods, and a new shortcut method formulated using asymptotic theory.

Trial-and-Error Methods

For problems involving the estimation of choked isothermal flow rate conditions, the feed temperature T, feed pressure P_1, pipe length L, and inside diameter D are known. In order to computer the mass velocity, one needs to first determine the value of the Fanning friction factor f. Generally, it is safe to assume fully developed turbulent flow at high Reynolds numbers when considering choked flow. Then, the friction factor is given by:

$$\frac{1}{\sqrt{f}}=4\log\left(3.7\frac{D}{\varepsilon}\right) \tag{22}$$

where ε is the pipe roughness. Equation 7 is solved first using a trial and error procedure to determine the Mach number at the inlet:

$$\frac{4fL}{D}+\ln\left(\frac{1}{\gamma Ma_1^2}\right)+\left(1-\frac{1}{\gamma Ma_1^2}\right)=0 \tag{23}$$

The mass velocity is then determined from Equation 5:

$$G = \frac{\dot{m}}{A} = Ma_1 P_1 \sqrt{\frac{\gamma M}{RT}} \tag{24}$$

Under adiabatic conditions, Equation 21 must be solved iteratively:

$$\frac{\gamma+1}{2}\ln\left[\frac{2Y_1}{(\gamma+1)Ma_1^2}\right]-\left(\frac{1}{Ma_1^2}-1\right)+\gamma\frac{4fL}{D}=0 \tag{25}$$

The mass velocity is then determined from Equation 24 with $T = T_1$:

$$G = \frac{\dot{m}}{A} = Ma_1 P_1 \sqrt{\frac{\gamma M}{RT_1}} \tag{26}$$

Graphical Methods - Lupo Correlation (Crowl, 1999)

Crowl (1999) suggests using the Darcy formula to solve for the mass velocity under adiabatic flow conditions:

$$G = \frac{\dot{m}}{A} = Y\sqrt{\frac{2\rho_1(P_1-P_2)}{\sum K}} \tag{27}$$

where Y is a unitless gas expansion factor and ΣK is the sum of frictional losses, including pipe friction, entrance and exit losses, and fittings, which are written in terms of an equivalent friction factor given by $\Sigma K = 4fL/D$.

For adiabatic flows, the gas expansion factor Y is determined by equating Equations 26 and 27 to yield:

$$Y = Ma_1\sqrt{\frac{\gamma\sum K}{2}\frac{P_1}{(P_1-P_2)}} \tag{28}$$

The graphs from Figures 1 and 2 are determined by repeated trial-and-error or direct solutions of Equation 25, given a value for ΣK and for γ. Then the exit pressure is determined from $P_2 = P_1 Ma_1 (2Y_1/(\gamma+1))^{1/2}$ and the gas expansion factor, Y, is determined from Equation 28.

A similar graphical procedure, previously unpublished, can be developed for isothermal conditions. The only differences to the procedure is that Equation 23 is used to determine Ma_1, and now $P_2 = P_1 Ma_1 \gamma^{1/2}$. Figure 3 shows the resulting pressure ratio and Figure 4 shows the gas expansion factor. As expected, the pressure ratio is independent of the heat capacity ratio γ, which affects the temperature of the gas during adiabatic conditions.

The procedure for determining the mass velocity, given the type and diameter of the pipe, the pipe length, the type of gas, and the upstream and downstream pressures, is as follows:

1. Given the inlet conditions, determine the sum of friction terms ΣK
2. Determine the pressure ratio $(P_1 - P_2)/P_1$ as a function of ΣK from Figure 1 or 3
3. Determine the gas expansion factor Y as a function of ΣK from Figure 2 or 4
4. Calculate the mass velocity from Equation 27

An analytical prediction of the gas expansion factor and sonic pressure drop shown in Figures 1 through 4 can be made by fitting the numerical data to a curve of the following form: $\ln Y = A (\ln K)^3 + B (\ln K)^2 + C (\ln K) + D$. This is done using the "Normal Equations" (Leighton, 1997) by letting $z = \ln Y$ and $t = \ln K$. Thus the system we are trying to solve, in matrix form, can be written as:

$$\begin{pmatrix} z_1 \\ z_2 \\ \cdot \\ \cdot \\ z_N \end{pmatrix} = \begin{pmatrix} 1 & t_1 & t_1^2 & t_1^3 \\ 1 & t_2 & t_2^2 & t_2^3 \\ \cdot & \cdot & \cdot & \cdot \\ \cdot & \cdot & \cdot & \cdot \\ 1 & t_N & t_N^2 & t_N^3 \end{pmatrix} \begin{pmatrix} D \\ C \\ B \\ A \end{pmatrix}$$

In matrix notation, this can be written as $\underline{z} = \underline{\underline{A}}\underline{x}$ where $\underline{x}$ is the column vector of unknown constants.

For any vector $\underline{x}$, the error or residual between the actual data $\underline{z}$ and the computed data is given by: $\underline{r} = \underline{z} - \underline{\underline{A}}\underline{x}$. The goal is to find the unknown fitting parameters (the vector $\underline{x}$) to minimize the square of the residual, given by:

$$\underline{r}^T \underline{r} = (\underline{z} - \underline{\underline{A}}\underline{x})^T (\underline{z} - \underline{\underline{A}}\underline{x}) = \underline{z}^T \underline{z} - \underline{z}^T \underline{\underline{A}}\underline{x} - \underline{x}^T \underline{\underline{A}}^T \underline{z} + \underline{x}^T \underline{\underline{A}}^T \underline{\underline{A}}\underline{x}$$

This function is minimized by taking its gradient with respect to $\underline{x}$ and equating it to zero, to yield:

$$\underline{\underline{A}}^T \underline{\underline{A}}\underline{x} = \underline{\underline{A}}^T \underline{z}$$

which can be solved for $\underline{x}$ by simple matrix inversion techniques. Applying this to the data in Figures 1-4 yields the Lupo correlations, which are summarized in Table 1 for adiabatic flow and in Table 2 for isothermal flow.

Inspection of Figures 1 and 3 show that as the loss term ΣK increases, the pressure drop ratio reaches an asymptotic value of $(P_2-P_1)/P_1 = 1$. Also, Figures 2 and 4 show that the gas expansion factor Y reaches an asymptotic value of about 0.7 for both the adiabatic and isothermal cases. The asymptotic value is also evident from the graphs of Crowl and Louvar (2002). It is also interesting to note that under all conditions, our analysis identified a maximum value for Y as a function of ΣK. Table 3 shows the location of this maximum value. For adiabatic flow, the maximum is a function of the heat capacity ratio, as shown in Figure 5. For isothermal flow, the maximum is independent of the heat capacity ratio, but interestingly, it is the same as for the $\gamma = 1.0$ case in adiabatic flow. In all cases, this maximum is within 3% of the asymptotic value. We will now analytically obtain these asymptotic limits and use them to develop a simple shortcut formula for predicting choked gas flows.

Shortcut Method

The goal of the shortcut method is to develop a simple formula to predict the maximum flowrates through very long pipelines. Under isothermal conditions the inlet Mach number Ma_1 is determined from Equation 7, where the frictional pipe losses ΣK are balanced by kinetic energy, $[\ln (1/\gamma Ma_1^2)]$, and compressibility $[1-1/\gamma Ma_1^2]$:

$$\sum K + \ln\left(\frac{1}{\gamma Ma_1^2}\right) + \left(1 - \frac{1}{\gamma Ma_1^2}\right) = 0 \tag{29}$$

For very small inlet gas velocities (and thus small Ma_1), the gas compressibility term grows more rapidly (scaling as Ma_1^{-2}) than the kinetic energy (scaling as $-2 \ln Ma_1$). From a physical standpoint, the gas expands as the pressure changes down the pipeline while the kinetic energy is relatively unimportant. In fact, as the frictional pipe losses approach infinity, the inlet Mach number is expected to vary as:

$$Ma_1 = \frac{1}{\sqrt{\gamma \sum K}} \tag{30}$$

A similar argument can be made during adiabatic conditions, where the inlet Mach number can be computed from Equation 21

$$\gamma \sum K + \frac{\gamma+1}{2} \ln\left[\frac{2Y_1}{(\gamma+1) Ma_1^2}\right] + \left(1 - \frac{1}{Ma_1^2}\right) = 0 \tag{31}$$

Comparing Equation 29 with Equation 31 shows that the only difference between the isothermal and adiabatic model are the kinetic energy terms. This is because the change in temperature within the pipe affects the fluid density (and hence kinetic energy). Nevertheless, as the frictional pipe losses grow, Equation 30 can be used to sufficiently describe its relationship with the inlet Mach number Ma_1. Thus, for large pipelines this theory predicts that there is no fundamental difference between adiabatic and isothermal flows! The literature claims that "real gas flow" lies in between those predicted by the isothermal and adiabatic models (Churchill, 1980; Levenspiel, 1984; Crowl and Louvar, 2002).

Nevertheless, one can insert Equation 30 into the relationship between inlet and outlet pressure: $P_2 = P_1 Ma_1 \gamma^{1/2}$ for isothermal conditions or $P_2 = P_1 Ma_1 (2Y_1/(\gamma+1))^{1/2}$ for adiabatic conditions to show, for large pipelines when $\Sigma K \to \infty$, $P_2 = 0$, such that the exiting gas has no gauge pressure (and thus cannot speed up by expansion). Inserting this relationship, along with Equation 30, into Equation 28 gives a simplified formula for the asymptotic value for the gas expansion:

$$Y = \frac{1}{\sqrt{2}} = 0.7071 \tag{32}$$

which can subsequently be inserted into Equation 27 to give a simple, analytical estimate for the gas flowrate under sonic flow conditions, which does not require using cumbersome figures or trial-and-error solutions:

$$\dot{m} = A\sqrt{\frac{\rho_1 P_1}{\sum K}} \tag{33}$$

The asymptotic value of Y and the pressure drop ratio are evident in Figures 1-4.

The validity of the five assumptions following Equation 21 will now be discussed in detail. The first concerns itself with having adiabatic or isothermal conditions along the entire pipe length. However, the literature has suggested that "real" pipe flow lies somewhere in between the adiabatic and isothermal limits (Churchill, 1980; Levenspiel, 1984; Cochran, 1996; Crowl and Louvar 2002). The results presented here, in particular the derivation of Equation 33, shows that this is quantitatively true, especially for long pipelines. Therefore, maintaining an entire pipe under isothermal conditions may not be necessary for the theory to hold.

The second assumption addresses whether or not the pressure-volume relationship of the expanding gas can be described by the ideal gas equation of state, and depends mainly on the upstream gauge pressure P_1. In many industrial situations, the upstream pressure is dictated by a storage tank pressure or reactor operating pressure. On the other hand, the downstream gauge pressure is equal to P_1 minus the pressure drop ΔP with the pressurized discharge entering the atmosphere. However, during choked flow conditions the pressure drop in the pipeline ΔP approaches P_1 for very long pipelines (where $\Sigma K >> 10$, as seen from Figures 1 and 3). Thus, the exit gauge pressure approaches zero. Under these circumstances, the upstream pressures can be high enough to suggest the possibility of non-ideal behavior. In fact, for very long pipelines simulation of Equation 2 suggests there is a small pressure drop over most of the pipe. Near the pipe exit, the pressure drop is much higher as the gas is now expanding at a rapid rate. Therefore this is the most restrictive assumption of the theory. Nevertheless, ideal gas solutions are usually used as a back-of-the-envelope tool for predicting the pressure-volume relationship. The ideal gas law thus can be used here to provide a similar tool for predicting choked flow.

The assumption of constant friction factor will now be discussed. Simulation of Equation 2 shows that the local gas Mach number (and hence the velocity) at any position x slowly increases until near the pipe exit, where it quickly increases to a value of unity. Although the velocity is changing along the pipe length, its rate of change is small and a constant friction factor is probably a good assumption.

The remaining two assumptions concerning horizontal pipes and constant gas compressibility ratio will now be discussed. In many industrial situations, pipelines are not horizontal; however, even for cases with significant potential energy changes these losses can be lumped into an effective ΣK value. The heat capacity ratio is a material property and is relatively independent of pressure. Therefore, the assumption of constant heat capacity ratio should not affect the solution.

Special Cases

Two special cases of choked flow occur in practice which can use the asymptotic solution of Equation 33. In the first case, the pipe diameter and length are known, and the mass flow is required. This is equivalent to the rupture of a pipeline at a specified location. In the second case, the mass flow is specified over a fixed distance and the pipeline diameter required to transport the gas is required. In both cases it is assumed that the frictional loss is due entirely to the pipe friction due to the pipe length.

For the first case, for a circular pipe $A = \pi D^2 / 4$ and assuming that the frictional losses are due entirely to the pipes length results in $\sum K = 4fL / D$. Substituting into Equation 33 results in

$$\dot{m} = \frac{\pi}{8}\sqrt{\frac{\rho_1 P_1 D^5}{fL}} \tag{34}$$

Furthermore, by the ideal gas law ($\rho_1 = P_1 M / RT_1$) it follows that,

$$\dot{m} = \frac{\pi}{8}\sqrt{\frac{P_1^2 M D^5}{RT_1 fL}} \tag{35}$$

Equation 35 is used to determine the mass flow for the first case.

For the second case, the solution can be found directly from Equation 34. Solving for D,

$$D = \sqrt[5]{\frac{64\dot{m}^2 fL}{\pi^2 \rho_1 P_1}} \tag{36}$$

If we further substitute the ideal gas law for the upstream density, then,

$$D = \sqrt[5]{\frac{64\dot{m}^2 fLRT_1}{\pi^2 MP_1^2}} \tag{37}$$

The solution to Equation 37 is by trial and error since the friction factor, *f*, depends on the pipe diameter.

Some Examples

We now use these results to estimate mass flow rates under sonic conditions for two examples. The first problem is number 4-14 from the second edition of Crowl and Louvar's (2002) textbook on chemical process safety.

Example 1 A laboratory apparatus uses nitrogen at 250 psig. The nitrogen is supplied from a cylinder, through a regulator, and to the apparatus via 15 feet of ¼-inch ID drawn copper tubing. If the tubing separates from the apparatus, estimate the flow of nitrogen from the tubing. The nitrogen in the tank is at 75 ^{o}F. Assume that γ= 1.4 and a friction factor of f = 0.00355, which is estimated for fully developed turbulent flow. The gas density at the supply pressure is determined from the ideal gas law and is 1.2846 lb_m/ft^3.

Solution: Adiabatic flow is assumed. The ¼-in ID is equivalent to 2.08 x 10^{-2} ft, and the cross-sectional area of the tube can be calculated as 3.41 x 10^{-4} ft^2. Thus the friction losses can be calculated as $\Sigma K = 4fL/D$ = (4 x 0.00355 x 15 ft) /(2.08 x 10^{-2} ft) = 10.24. A (lengthy!) trial and error computation using Equation 25 yields Ma_1 = 0.231.

Equation 26 can be used to compute the mass flowrate:

$$\dot{m} = AMa_1P_1\sqrt{\frac{\gamma M}{RT_1}} = (3.41\times10^{-4}\,\text{ft}^2)(0.231)\times(250+14.7)\left(\frac{\text{lb}_\text{f}}{\text{in}^2}\right)\left(\frac{144\ \text{in}^2}{\text{ft}^2}\right)$$

$$\times\sqrt{\frac{(1.4)(32.17\ \text{ft-lb}_\text{m}/\text{lb}_\text{f}\text{-s}^2)(28\ \text{lb}_\text{m}/\text{lbmol})}{(1545\ \text{ft-lb}_\text{f}/\text{lbmol-}^\circ\text{R})(75+460)(^\circ\text{R})}} = 0.118\ \text{lb}_\text{m}/\text{s}$$

An almost identical answer can be obtained using Equation 27 and the charts in Figures 1 and 2 or the Lupo correlation in Table 1. For this case the pressure drop ratio $(P_1\text{-}P_2)/P_1$ = 0.794 and Y = 0.70.

Alternatively, substituting into Equation 33 yields the asymptotic solution:

$$\dot{m} = A\sqrt{\frac{\rho_1 P_1}{\sum K}} = (3.41\times10^{-4}\ \text{ft}^2)$$

$$\times\sqrt{\frac{(1.2846\ \text{lb}_\text{m}/\text{ft}^3)(32.17\ \text{ft-lb}_\text{m}/\text{lb}_\text{f}\text{-s}^2)(250+14.7)(\text{lb}_\text{f}/\text{in}^2)}{10.24}} = 0.134\ \text{lb}_\text{m}/\text{s}$$

which has a difference of 13% from the rigorous solution. Clearly, the velocity head pipe loss of 10.24 is too small for the asymptotic solution to apply. Nevertheless, it provides a good estimate of the mass flow, which is within reasonable accuracy for process safety applications.

It is worth mentioning that if $\Sigma K = 50$, the full solution of Equation 26 for adiabatic flow gives a mass flow rate of 0.058 lb_m/s while the asymptotic theory of Equation 33 yields a mass flow rate of 0.061 lb_m/s, an error of 3.9%. Furthermore, increasing ΣK to 100 and 500 give an error of 2.2% and 0.2%, respectively. Thus, the asymptotic theory becomes more accurate with increasing the pipe length.

Example 2_You wish to construct a pipeline to deliver 1 kg/s of nitrogen to a location 10 km away. The pipeline is constructed from commercial steel pipe. If the upstream pressure of the nitrogen is 50 barg, what is the required pipeline diameter? Assume a temperature of 25°C, that nitrogen is an ideal gas, and the gas exits at choked conditions.

Solution: The asymptotic solution, provided in Equation 37, applies for this very long pipeline. The procedure is as follows:

1. Guess a pipe diameter, *D*.
2. Use the Equation 22 for fully developed turbulent flow to determine *f*.
3. Compute a new diameter using Equation 37.
4. Continue steps 1 through 3 until guessed *D* equals computed *D*.

The procedure converges rapidly if the computed *D* is used as the next guess.

For this case, P_1 = 50 barg = 51 bara = 5.1×10^6 kg/m s.

Guess *D* = 10 cm = 0.1 m. For commercial steel pipe, ε =0.046 mm. From Equation 22, *f* = 0.004098. Substituting into Equation 37

$$D = \sqrt[5]{\frac{64\dot{m}^2 fLRT_1}{\pi^2 MP_1^2}}$$

$$D = \sqrt[5]{\frac{(64)(1\ \text{kg/s})^2(0.004098)(10{,}000\ \text{m})(8.3143\ \text{kPa}\cdot\text{m}^3/\text{kg-mole}\cdot\text{K})[(1000\ \text{kg/m}\cdot\text{s})/(1\ \text{kPa})](298\ \text{K})}{(3.14)^2(28\ \text{kg/kg-mole})(5.1\times10^6\ \text{kg/m}\cdot\text{s})^2}}$$

D = 0.0618 m = 6.18 cm

Now guess *D* = 6.18 cm. Then *f* = 0.00457, and the computed *D* = 6.32 cm
Now guess *D* = 6.32 cm. Then *f* = 0.00455 and the computed *D* = 6.32 cm.

The answer is $D = 6.32$ cm. For this problem $4fL/D = 2880$ so the asymptotic solution does apply.

Discussion and Conclusions

This paper has presented significantly more detail on sonic gas flow than has been presented elsewhere, particularly for the case of a long pipeline. In this case, both the adiabatic and isothermal equations show maximum and asymptotic values in the gas expansion parameter. The asymptotic values for the adiabatic and isothermal cases are the same. A shortcut formula is derived from the asymptotic values, enabling easy calculation of the flow.

The key to the success of the shortcut formula lies mainly in the Y parameter, and somewhat in the pressure ratio. Figures 1-4 show that for pipelines with $\Sigma K = 5$ the gas expansion parameter, Y, is within 5% or less of the asymptotic value of 0.707. In addition, the pressure ratio is within 30% or less of the asymptotic value of 1.0. For $\Sigma K =$ 10 the error in the gas expansion factor and pressure ratio are 2% and 20%, respectively. As most industrial pipelines have large ΣK values due to its length, the asymptotic equation is very good at predicting the maximum gas flow rate. The argument of isothermal vs. adiabatic flow is also unimportant.

The equations presented here are useful for consequence modeling of releases of gases. A simple, direct solution is possible rather than a complicated trial and error procedure.

References

Churchill, S., *Inertial Flows*, Etaner Press, Thornton, PA, pp. 53-58, 1980.

Cochran, T. W., "Calculate the Pipeline Flow of Compressible Fluids," *Chem. Eng.*, Vol. 103, No 2, 115, 1996.

Crane Co, *Flow of Fluids Through Valves, Fittings, and Pipe,* Technical Paper No. 410, New York, NY, 1986.

Crowl, D. A. and Louvar, J. F., *Chemical Process Safety: Fundamentals with Applications*, Prentice Hall PTR, Upper Saddle River, NJ, pp. 136-151, 2002,.

Dodge, R. A. and Thompson, M. J., *Fluid Mechanics*, McGraw-Hill Book Company, New York, NY, pp. 193, 288, and 407, 1937,.

Farina, I. H., "Critical Length Helps Calculate Compressible Flow," *Chem. Eng.*, 104n2, 88 , 1997.

Kumar, S., "Spreadsheet Calculates Critical Flow," *Chem. Eng.*, Vol. 109, No. 10, 62, 2002.

Leighton, D. T., "Lecture Notes on Computer Methods for Chemical Engineers," University of Notre Dame, 1997.

Levenspeil, O., "The Discharge of Gases from a Reservoir through a Pipe," *AIChE J.*, 23, 402, 1977.

Levenspeil, O., *Engineering Flow and Heat Exchange*, Plenum Press, New York, NY, pp.41-47, 1984.

Nomenclature

a	Speed of sound
A	Area
C_P	Heat capacity
D	Diameter
F	Fanning friction factor
G	Mass flux
h	Enthalpy
K	Velocity head losses
L	Length
$\dot{m}$	Mass flow
M	Molecular weight
Ma	Mach number
P	Pressure
R	Ideal gas constant
T	Temperature
u	Velocity
x	Distance
Y	Expansion factor defined by Equation 28

Symbols

ε	Pipe roughness
γ	Heat capacity ratio
ρ	Density
τ	shear stress

Table 1. Correlations for the expansion factor Y and the sonic pressure drop ratio $(P_1 - P_2)/P_1$ as a function of the pipe loss ΣK for adiabatic flow conditions. The equation used to fit the functions is of the form $\ln Y = A (\ln K)^3 + B (\ln K)^2 + C (\ln K) + D$.

Function value	***A***	***B***	***C***	***D***
Expansion factor Y	0.0009	-0.0189	0.1125	-0.5295
Sonic pressure drop ratio $\gamma = 1.2$	0.0009	-0.0308	0.2610	-0.7248
Sonic pressure drop ratio $\gamma = 1.4$	0.0011	-0.0302	0.2384	-0.6455
Sonic pressure drop ratio $\gamma = 1.67$	0.0013	-0.0287	0.2310	-0.5633

Table 2. Correlations for the expansion factor Y and the sonic pressure drop ratio $(P_1 - P_2)/P_1$ as a function of the pipe loss ΣK for isothermal flow conditions. The equation used to fit the functions is of the form $\ln Y = A (\ln K)^3 + B (\ln K)^2 + C (\ln K) + D$.

Function value	***A***	***B***	***C***	***D***
Expansion factor Y	0.0003	-0.0080	0.0611	-0.4558
Sonic pressure drop ratio $\gamma = 1.2$	0.0007	-0.0237	0.2409	-0.7678
Sonic pressure drop ratio $\gamma = 1.4$	0.0007	-0.0237	0.2408	-0.7677
Sonic pressure drop ratio $\gamma = 1.67$	0.0007	-0.0237	0.2407	-0.7677

Table 3. Maximum values in the expansion factors for adiabatic and isothermal sonic flows.

Adiabatic Flow:		
Heat Capacity Ratio γ	Velocity Head Loss ΣK	Expansion Factor Y
1.0	56.3	0.7248
1.1	64.0	0.7277
1.2	72.5	0.7210
1.3	81.3	0.7195
1.4	90.0	0.7182
1.5	101.1	0.7172
1.6	112.0	0.7162
1.7	123.5	0.7154
1.8	135.8	0.7147
1.9	148.7	0.7141
2.0	162.4	0.7135
Isothermal Flow (all values of γ):	56.3	0.7248

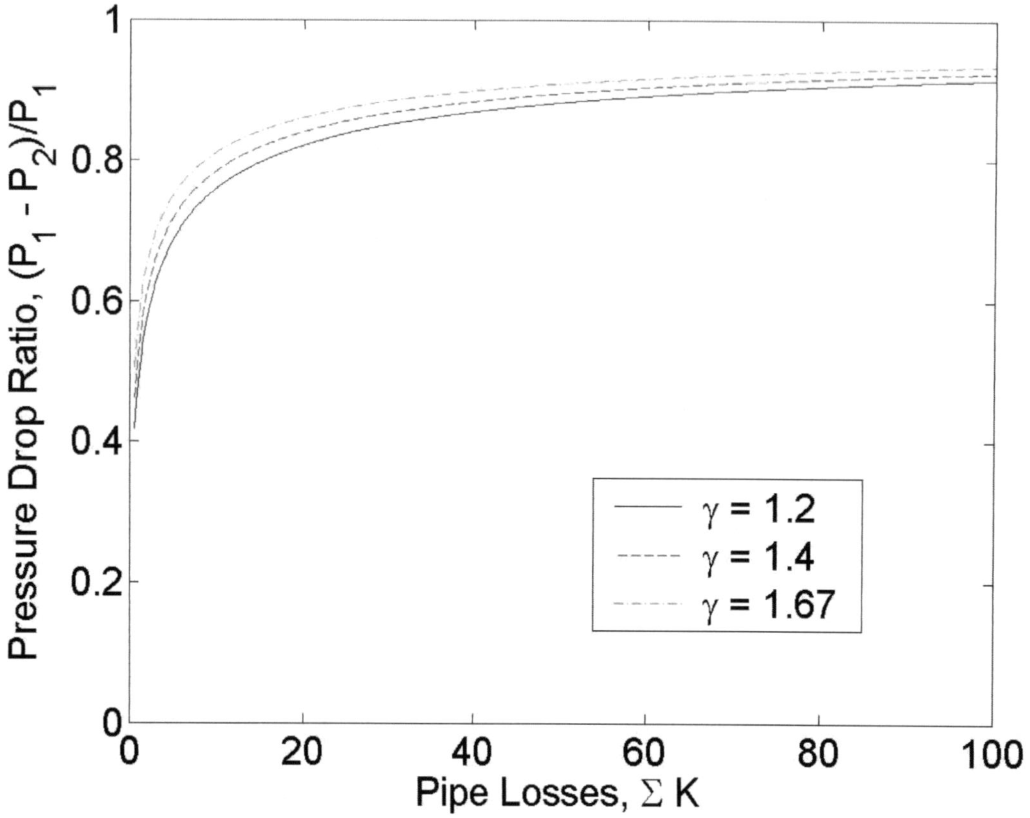

Figure 1. Pressure drop ratio as a function of the excess velocity head pipe losses for adiabatic flow.

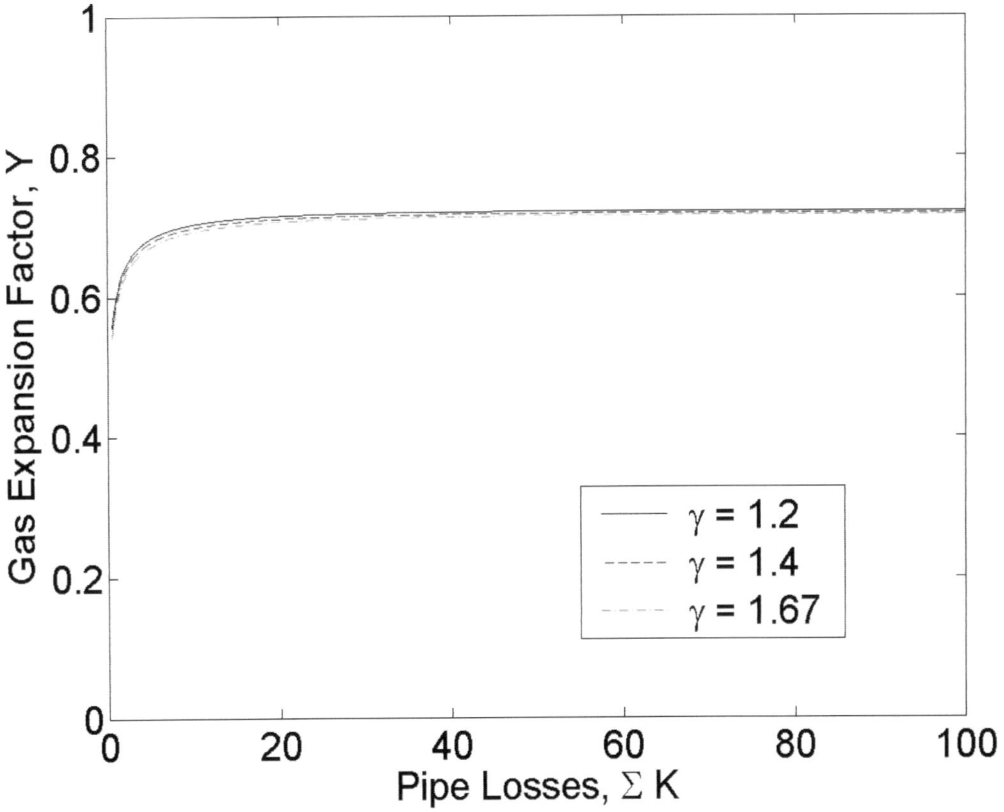

Figure 2. Gas expansion factor as a function of the excess velocity head pipe losses for adiabatic flow.

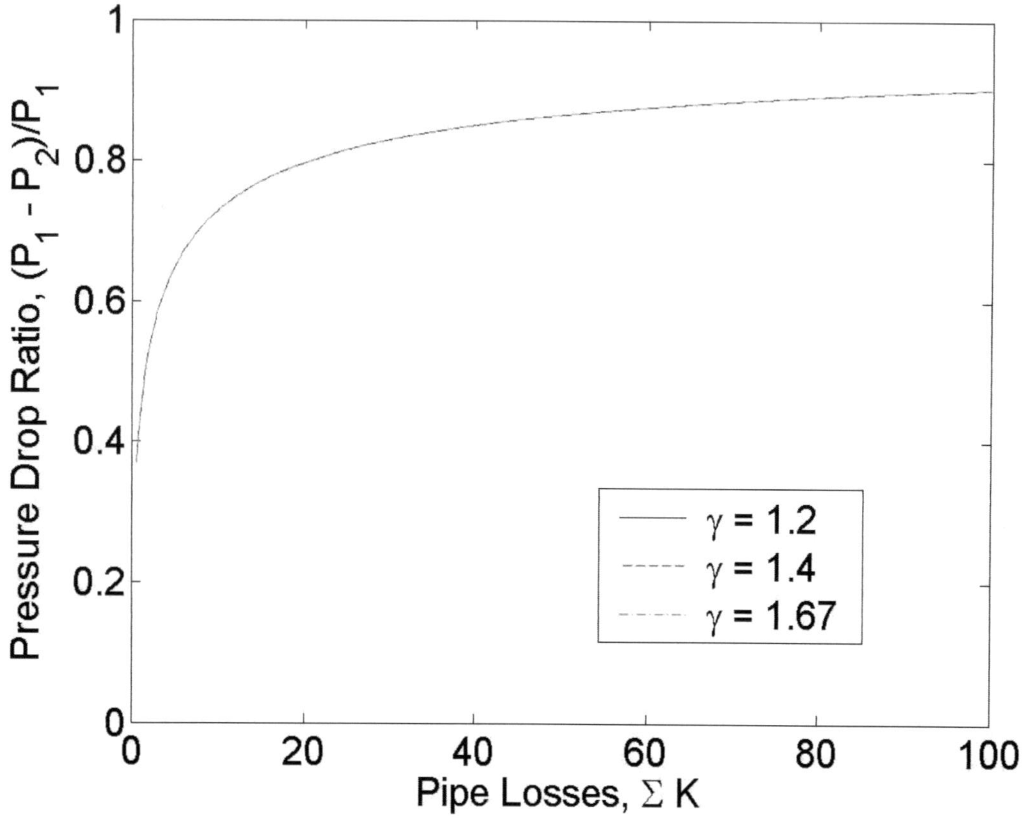

Figure 3. Pressure drop ratio as a function of the excess velocity head pipe losses for isothermal flow. The results are independent of the heat capacity ratio.

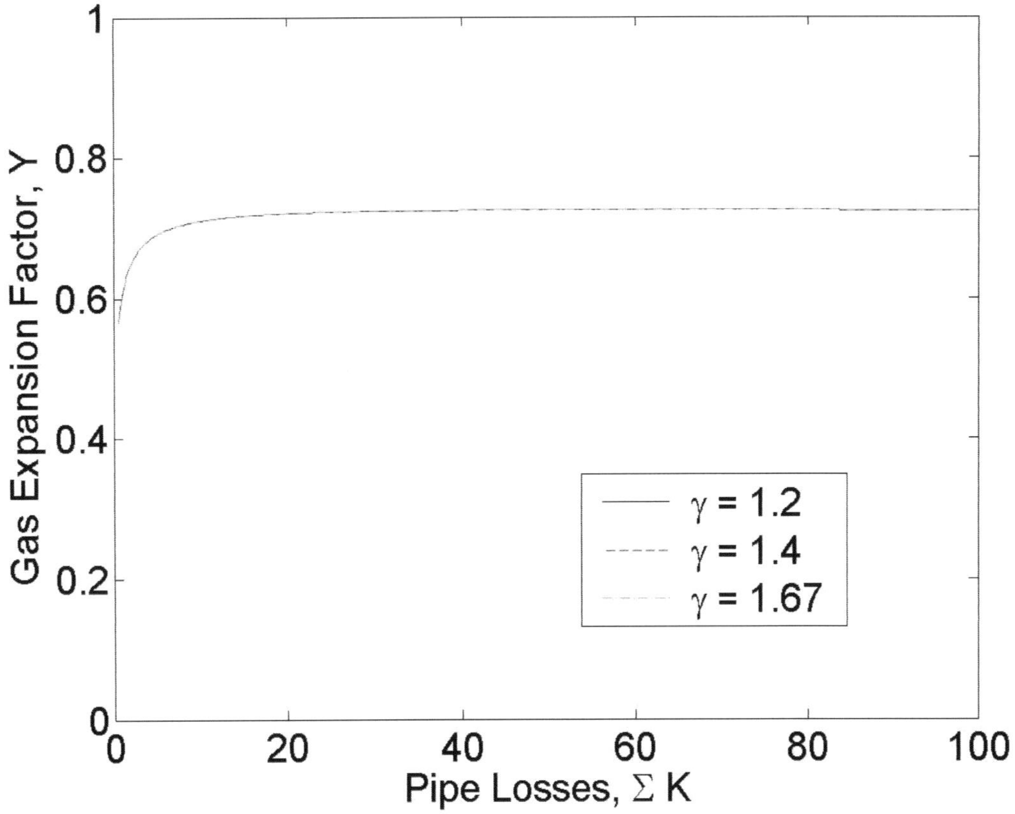

Figure 4. Gas expansion factor as a function of the excess velocity head pipe losses for isothermal flow. The results are independent of the heat capacity ratio.

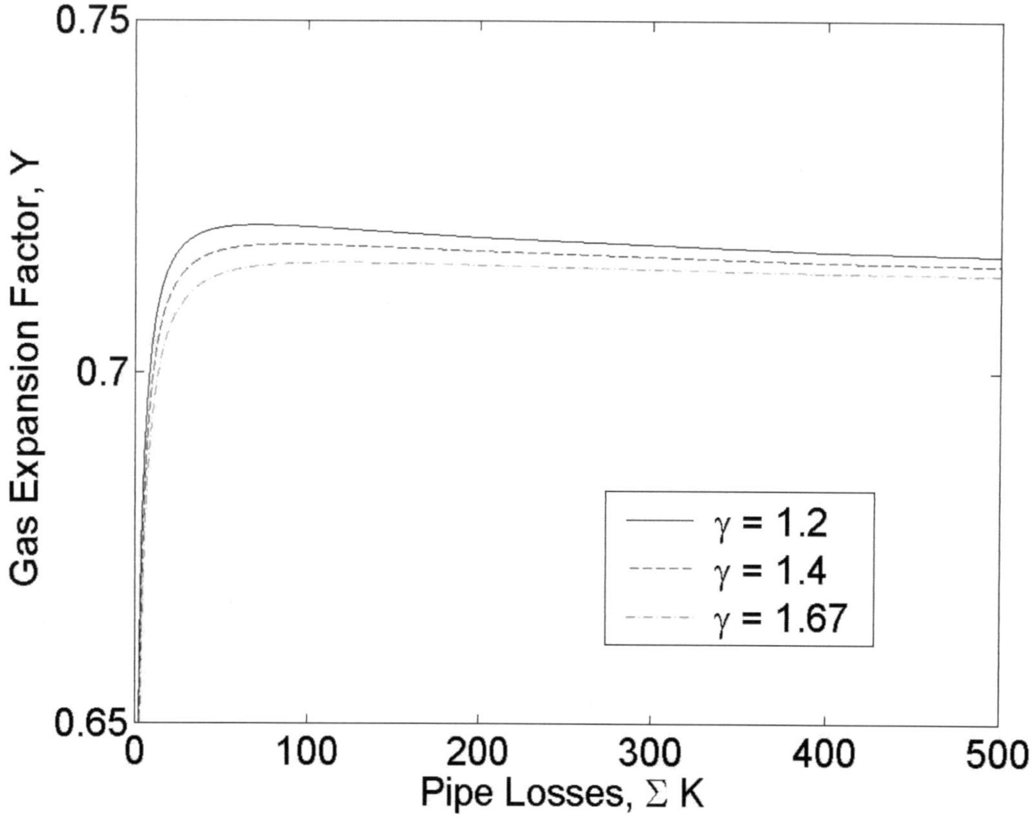

Figure 5. Maximum and asymptotic gas expansion factors for adiabatic flow.

MODELING OF HAZARDOUS RELEASES FROM OIL & GAS WELL BLOWOUTS, AND PIPELINE LEAKS

by: Mohammed Ayyoubi

Paper LPS-2b

AUTHOR INFORMATION

Mohammed J. Ayyoubi is a Loss Prevention Engineer with Saudi Aramco (in Dhahran, Saudi Arabia) and has over 14-years of experience in conducting process safety, consequence, and risk analysis studies. Mohammed received his Industrial and Safety Engineering degrees from Texas A&M University. Before joining Saudi Aramco, Mohammed worked for Bechtel Corporation's Petroleum and Chemical Headquarters in Houston, Texas. At Saudi Aramco's Technical Services Unit, Mohammed is responsible for conducting and reviewing a variety of safety, consequence, and risk analysis studies for Aramco projects around the world. Also, as a "National Fire Protection Association" (NFPA) board *Certified Fire Protection Specialist,* Mohammed current responsibilities extend to cover a variety of fire prevention and protection projects. Mohammed is part of a Saudi Aramco team that combines theoretical modeling knowledge with practical plant experience.

1. ABSTRACT

The world's population continues to increase and expand beyond its original boundaries. As a result, emerging populations have encroached on existing oil and gas wellhead areas, and on pipeline right-of-ways. In addition, new wells are being drilled and new pipelines constructed near existing populations. As a result, the potential for hazardous impacts to nearby populations from wellhead and pipeline uncontrolled releases has increased.

Consequence modeling of wellhead blowouts and pipeline leaks is a complex task, and requires the consideration of several important factors. Simplistic consequence modeling tools and substandard approximation methodologies are not fit-for-purpose to properly model this complex phenomenon, and may result in underestimation of the potential hazards associated with wellhead and pipeline uncontrolled releases. This may result in unaccounted exposures (to personnel, assets, environment, and to the surrounding public). Calculations will then indicate risk values that are lower than actual risk levels. The opposite: consequence modeling overestimation of hazards and their associated risks may result in unnecessarily spending of funds in an attempt to prevent or mitigate (in reality) low risk events. Thus, the need for more accurate consequence modeling tools and proper modeling methodologies has become more important than ever before.

The physical and chemical characteristics of wellhead and pipeline fluids, and the physical configurations of their containment systems are critical factors to the accurate dispersion modeling of hazardous release events from such systems. This paper discusses some of these primary factors and illustrates why they are critical as well as some of the how to (and how <u>not</u> to) model such events. Case studies and examples illustrate some of the factors and modeling methodologies.

The factors and consequence modeling analysis methodologies discussed by this paper are applicable to other type fluids, and other type releases (for example, pressurized liquid releases, superheated liquids, and two-phase releases from pressure vessels/ equipment, plant piping, and relief valves).

2. INTRODUCTION

Underestimation of hazards from wellhead and pipeline uncontrolled releases may result in un-accounted for exposures, and actual risk levels that are higher than the modeled (calculated) risks. This may cause a "false" sense of security and a dangerous perception that the associated risks with wellhead and pipeline hazardous release events are lower than actual. This "false" sense of security may be shattered when a major incident with large catastrophic consequences occurs. The opposite: overestimation of the same hazards may result in unnecessarily spending of funds in an attempt to prevent or mitigate (in reality) low risk events.

Selected factors influencing the dispersion modeling of toxic and flammable hazards from wellhead and pipeline leaks are discussed and their importance illustrated. Some of these factors include: 1) the different types of released fluids (i.e., phases) from such leaks, 2) the mixture-composition of each of the phases, 3) the mass release rates of the different phases, 4) their time-transient nature, 5) the total mass released following a leak, 5) and the effects from wellhead and pipeline systems' physical configurations and features (for example, length and diameter of the pipeline and the presence of isolation valves). Examples of other factors influencing the dispersion modeling of toxic and flammable hazards that are not discussed by this paper include atmospheric conditions (e.g., wind speed, stability, relative humidity), and obstacles along the dispersion path. Selected case studies illustrate the potential significance of some of these factors and how they can significantly affect dispersion modeling results.

This paper excludes the *likelihood* aspect of wellhead and pipeline uncontrolled releases and primarily addresses the *"consequence"* portion associated with the dispersion of toxic and flammable fluids as a result. This paper does not address the potential explosion hazards associated with these release events.

2.1. Primary Exposures from Wellhead and Pipelines Hazardous Releases

Exposures to toxic gas hazard and flammable hazards are the two main concerns associated with uncontrolled hydrocarbon releases from oil and gas well blowouts and from pipeline leaks. Toxic gas hazards are often caused by the presence of toxic material(s) (such as Hydrogen Sulfide) within the dispersed wellhead or pipeline leaked-fluids. Flammable hazards are caused by the ignition of the leaked fluids.

Dispersing hydrocarbon fluids may result in personnel exposures to different degrees of flammable/ toxic hazards. The degree and downwind extent of the flammable or toxic hazards depends on internal factors of the release such as the wellhead or pipeline fluid-mixture composition, initial fluid temperature and pressure, normal flow rate, duration of normal flow, duration of the release, and available inventories. External factors such as the release hole-size, atmospheric release conditions, and the physical-nature of the surroundings (for example the roughness of the terrain) play an important role in determining the degree (extent) of the hazard as well.

Flammable-hazards to personnel and/or to structures consist of higher than acceptable heat-radiation-level exposures from spreading liquid-pool fires, flash-fires, or jet fires; and exposure to overpressures from vapor-cloud-explosions.

In general, exposure to toxic hazards present a more complex problem to model, as the impact needs to be assessed by considering *both* the calculated magnitude and the duration of toxic exposures, often referred to as "exposure-dose".

Following a major wellhead or pipeline uncontrolled release, neighboring public communities may be exposed to a number of different hazards. Uncontrolled releases from wellhead or pipeline systems containing fluid streams with toxic substances (components) typically present the most severe consequences and potential risk to the public. The reason is that toxic hazards tend to reach further distances from the release point than flammable hazards, and may adversely impact members of the general public far beyond a wellhead location or a pipeline corridor. This is primarily due to the low allowable toxic exposure levels of certain hydrocarbons. For example, Hydrogen Sulfide has an Emergency Response Planning Guideline ERPG-2 concentration value of only 30 parts-per-million. ERPG-2 is defined as the maximum airborne concentration below which it is believed that nearly all individuals could be exposed for up to one hour without experiencing or developing life-threatening health effects. Hydrogen Sulfide's low toxic concentration value of 30 ppm is apparent when compared to its Lower Flammable Limit (LFL) concentration set at approximately 4.0% percent or 40,000 ppm. This means that following a release; a Hydrogen-Sulfide LFL-cloud could be diluted approximately ten fold and still be over the toxic ERPG-2 concentration of 30 ppm. Depending on atmospheric conditions, durations, and physical characteristics of hazardous releases; the toxic concentrations of concern (such as ERPG-2) can travel for very long downwind distances (sometimes up to several kilometers). Thus, toxic hazards may impact public targets that are far from the hazard's point of origin.

2.2. Basics of Dispersion Modeling

In order to accurately assess the magnitude of either toxic or flammable hazards, two important steps are required. First, determine the *source-term*, which are mainly the *form (phases and compositions of the released materials)*, the *total amount* released, and the *rate of release* of the leaked hydrocarbon fluids. Second, the source-term information is input into *dispersion analysis* algorithms in order to predict the downwind extent of possible toxic and flammable hazards. Industry uses the term "consequence modeling analysis of toxic/ flammable hazards" to be the sum of both above steps.

A modeling end-point is a term used to define the lowest concentration of interest for a component or a mixture where the dispersion modeling algorithms will terminate the calculation. For example, a modeler may select the LFL as an end-point to model a flammable type hazard. To model a toxic hazard, a modeler may select an end-point equal to a component's ERPG-2 value.

Flammable hazards extents are often modeled to LFL concentrations or fractions thereto, depicting the flammable zones of such releases. It is common to model flammable hazards to one-half the lower flammable limit (1/2 LFL) in order to account (and provide a safety cushion) for variability in modeling accuracies and changes in atmospheric dispersion conditions.

In order to determine a range of physiological responses (e.g., nuisance/irritant, recoverable injuries, unrecoverable injuries, and death), toxic hazards extents are typically modeled over a wide range of concentration end-points. For each physiological response a corresponding toxic-component concentration end-point is modeled. Commonly, the modeled end-point will depend on the expected exposure time from the hazardous release event.

Since the first "*source-term determination*" step directly feeds into the second "*dispersion analysis*" step, (i.e., step one's output is the second step's input), the accurate determination of the first step is critical to the accuracy of the second, and therefore to the overall consequence modeling results. The common saying amongst computer-professionals *"garbage-in equals garbage-out"* summarizes the point above.

Some consequence modeling packages integrally perform the above two steps (back-to-back). Others perform only the dispersion portion and require the source term data as input into their dispersion routines.

3. THE TYPES AND IMPORTANCE OF SOURCE TERM MODELING

3.1. Source Term Types and Dispersion Behavior

Leaks from gas wells or from gas pipelines will take only a single vapor-phase form.

However, leaks from liquid or two-phase (both liquids and vapor) systems (including wells and pipelines) may take one or several of the following four distinctly different forms (phases): i) vapor, ii) aerosolized small liquid droplets, iii) large aerosol droplets, or iv) heavy liquid. A description of each of the four phases is detailed below.

i) *Vapor Phase:* Dispersion of vapor-phase materials normally is one of two types. First is momentum jet type dispersion where the velocity of the released vapor (primarily due to process conditions and exit hole diameter) is higher than the surrounding atmospheric wind speed. Second, is a passive type (sometimes referred to as a Gaussian) dispersion where the released vapors velocity is equal to or lower than the velocity of the surrounding wind. High-pressure releases typically start as momentum-jet and gradually slow down to Gaussian type dispersions some distance away from the release point.

   Flashed-vapors (sometimes referred to by petroleum engineers as associated gas) are typical from leaks of hydrocarbon liquid mixtures containing light-end components, particularly under high temperature and pressure conditions. Upon a hazardous release, no significant differences exist between the physical dispersion behavior of vapor phase materials from pure gas systems and of flashed-vapor phase materials. Both the pure-vapor phase and the flashed-vapor phase will tend to initially disperse as momentum jet cloud.

ii) *Aerosolized small liquid droplets (mist):* These are the small droplets of liquid that remain airborne for sometime following a release. They are too small to settle (rainout) onto the substrate (e.g., the ground). During high-pressure releases, this phase can travel long distances while gradually evaporating into the air carrying with it toxic and flammable components before the mist-like droplets are completely vaporized and depleted.

   The primary dispersion mechanism for the vapors generated by this phase is similar to the vapor phases discussed in (i) above.

iii) *Large aerosol droplets:* These are the larger droplets that are not small enough to remain airborne through out the entire dispersion and their total mass will not be completely depleted by evaporation. Only portions of these droplets will flash-evaporate in the atmosphere along their "fall-path" trajectory to the substrate. The remaining droplets' mass will fall onto the substrate forming a liquid pool, which may further evaporate and produce a different kind of vapor cloud (for example, a heavy-type gas cloud). The flash-evaporated portion of the larger-aerosol-particles primary disperse as part of the overall momentum jet vapor cloud, which already contains the initial flashed vapors, and the mist vapors in (i) and (ii) above. The amount of flash evaporation from large aerosols depends on internal factors (for example, the aerosol mixture composition and starting temperature). External factors that have an effect include: the release elevation and orientation, and how long the large droplets have to travel before settling on the substrate, and on certain atmospheric conditions such as atmospheric temperature and humidity. Eventually, the remaining portions of the large aerosol droplets (mainly containing heavier-end components) fall to the substrate where they will primarily form a puddle, and potentially disperse in a manner similar to the heavy liquid phase mechanism discussed in (iv) below.

The significance of phases (ii) and (iii) above is that the aerosol particles, both small and large, increase the density of the overall momentum jet cloud, and take longer to mix with the surrounding air to dilute below concentrations of concern (i.e., end-points). Typically, this will result in a heavier than air cloud that will disperse further downwind and in a downwardly manner resulting in grade-level impact. For example, a lighter than air momentum jet cloud from a natural gas release will normally lift-off close to the source, while a momentum jet cloud from a pressurized NGL release will typically have a longer downwind reach, a larger footprint, and will disperse near grade level.

iv) *Heavy Liquid Phase:* Higher molecular weight components that do not initially flash or flash-evaporate typically end up forming a liquid pool on the substrate surface. Vaporization from the liquid pool normally results in heavy gas cloud dispersion behavior. These heavy clouds are different from momentum-jet releases in that they typically stay close to grade and are passively dispersed by the wind since they have no driving force of their own. However, heavy gas clouds tend to be different than Gaussian type clouds in that they are denser, thicker, slower in motion, and harder to dilute.

Heavy-gases resulting from evaporation from a liquid-pool surface follow different dispersion mechanisms than either a Momentum Jet/ or Gaussian type clouds.

The sequence of events described by fire fighters and well-blowout control individuals from *Boots and Coots International Well Control, Inc.,* based on their actual experience in capping and extinguishing wellhead blowouts and fires, is as follows:

1) *"Hydrocarbons are released near wellhead."*
2) *"Some liquids flash evaporate, forming an aerosol of liquid droplets and vapor."*
3) *"Heavier hydrocarbon liquids that do not flash evaporate fall and pool around the well and release vapors."*

3.1.1. Example to illustrate the Significance of Source Term Determination

When performing uncontrolled release modeling of wellhead and pipeline hazardous releases, it is important to properly determine the types, compositions, release rates, and amounts of the hydrocarbon mixture break-ups explained above. There are a number of reasons that make this initial task critical.

First, when a hydrocarbon mixture is released, and if conditions warrant the breaking-up of the mixture into more than a single phase, the original mixture's composition-profile is not likely to resemble the composition profiles of any of the resulting phases. Thus, each of the four potentially released phases discussed earlier are likely to have different compositions from one another. This means different molecular weights, densities, and physical and chemical properties among the different phases. As a result, upon a mixture's release into the atmosphere, some of these phases will behave differently from one another and from the original mixture.

The following example illustrates this phenomenon: a sample hydrocarbon wellhead fluid mixture of ten components at 205 °F degrees-Fahrenheit and 1,500 pounds-per-square-inch (absolute) pressure is released into the atmosphere from a 6-inch diameter wellhead production tubing. The tubing is 4,550 feet long (distance between surface release point and the reservoir).

Table 3-1 illustrates the calculated composition profiles and release rates of the different resulting phases, and illustrates how the composition of each phase varies from the original mixture and from the other phases.

TABLE 3-1

Example: Calculated* composition profile break-ups and flow rates for a hazardous oil wellhead blowout

Well-head Fluid Components	**Pre-Release Reservoir Fluid Composition in Mole-Percent (%)**	Composition of Post-Release Phases in Mole-Percent (%)			
		Flashed vapor phase (%)	***Evaporated vapor phase (%)***	***Aerosol liquid phase (%)***	***Liquid pool phase (%)***
C1(Methane)	*1*	*16*	*1*	*0*	*0*
C3 (Propane)	*2*	*17*	*6*	*1*	*0*
C4 (n-Butane)	*1*	*7*	*6*	*1*	*0*
C5 (n-Pentane)	*2*	*9*	*13*	*2*	*1*
C6 (n-Hexane)	*3*	*7*	*15*	*3*	*2*
C7 (n-Heptane)	*12*	*12*	*33*	*12*	*11*
C10 (n-Decane)	*60*	*6*	*19*	*63*	*67*
C_{15} (Heavy Hydrocarbon)	*11*	*0*	*1*	*11*	*13*
C_{22} (Heavy Hydrocarbon)	*6*	*0*	*0*	*6*	*6*
H2S (Hydrogen Sulfide)	*2*	*26*	*6*	*1*	*0*
	100%	***100%***	***100%***	***100%***	***100%***
Calculated Release Rate	***150 lb/sec***	***17 lb/sec. (11%)***		***87 lb/sec. (58%)***	***46 lb/sec. (31%)***

** Tabulated results are courtesy model output from " CANARY® by Quest" Consequence Modeling Package.*

Examining the results of the model-run summarized in Table 3-1, we can see that due to the high release temperature and pressure the lighter-end (low molecular weight) component hydrocarbons (for example, methane and propane) are mostly released in the flashed-vapor form. Heavier hydrocarbons, such as Butane *(C4)* through Pentane *(C7),* are mainly released as evaporated-vapors from small liquid aerosol particles (initially in the form of a mist). Decane (C10) and heavier hydrocarbons are mostly released as either large aerosol particles (which eventually settle on the substrate), or as heavier spilled liquids onto the substrate

So, why is this multi-phase, multi-component, multi release-rate behavior determination significant? The answer is relatively simple. The different released phases are likely to act and behave somewhat independently from one another. Phases behaving similarly will act uniformly (forming one plume) with specific set of potential hazards that are <u>different</u> from the hazards formed by other plumes. So, which of the plumes will result in the worst downwind hazard? For example, is it the momentum jet plume formed by the flashed and the evaporated vapor phases, or is it the heavy gas plume formed by the slow evaporation of the liquid pool phase? The only way to answer this question is by knowing what are all the phases formed, what is the composition and release rate of each phase, and which phases will act uniformly to form a single plume.

Typically, flashed vapors and small-aerosolized liquid droplets (mist), along with evaporated-vapors from the small and large aerosol droplets will disperse as one momentum jet plume. The heavier liquid aerosols will settle on the substrate forming a liquid pool (following a travel trajectory from the release point to the substrate surface). Components remaining in the liquid phase following a release will form liquid pools (puddles). The substrate surface will then have both phases (the large aerosols that settled on surface and the heavier liquid phases). This liquid puddle may slowly "cook" and "weather off" generating a heavier (i.e., denser) type cloud (due to heat exchange with the substrate and surroundings). This heavier gas cloud is likely to have completely different dispersing characteristics, concentration profiles, and will develop and reach its maximum extent <u>over a different time span</u> than the momentum jet cloud. Therefore, it is very important to know the components of the multi-phases, the total amount released from each phase, and the release rate of each phase across the leak duration.

In summary, either the momentum jet cloud or the heavy gas cloud will produce a dominant toxic or flammable hazard (i.e., one with longer downwind distances, or larger hazardous foot-prints to flammable or toxic modeling concentration end-points of concern). Thus, without proper source term determination prior to subsequent data input into the dispersion algorithms, one should have little confidence that the final consequence modeling results will closely resemble reality. If they occasionally do, then we have probably got the correct answer for all the wrong reasons. Unfortunately, this is not likely to happen all the time, nor do we know when it will happen and when it will not.

4. MAIN FACTORS INFLUENCING THE FORMATION OF AEROSOLS AND THEIR IMPORTANCE

The formation of aerosols from *leaked fluids* is significant under high-pressure release conditions, and in releases where there is some degree of superheat [i.e., when a component(s) within a mixture, or the entire mixture is at temperature(s) above its atmospheric boiling point]. Superheated conditions are typical of wellhead fluids and some pipeline releases.

In his handbook *Loss Prevention in the Process Industries (Volume I, 1991),* Frank P. Lees states, "When a superheated liquid is released from a containment, a proportion flashes off…" "In practice it frequently happens that there is in addition a significant amount of spray formation caused by the sudden release of pressure and the violent boiling of the liquid. This spray vaporizes rapidly in the air, from which it receives its latent heat of vaporization. The proportion of the liquid vaporized is thus considerably increased."

Some consequence modeling methodologies depend on simplistic process thermodynamic flash calculations and separate the released oil wellhead or pipeline fluids into only two phases: a flashed vapor phase and a liquid phase (without proper consideration of aerosols, or with rough empirical "fudge" factors to account for aerosol formation, evaporation, and settlement). These methodologies may then use the vapor phase as the source term input into a momentum jet type dispersion algorithm for calculation of downwind distances to hazardous concentration end-points. The liquid phase portion is often ignored by these methodologies. Another problem is: other phases (including evaporated vapors from aerosol particles, the small aerosolized liquid droplets that are too small to settle and will travel as mist with the momentum jet cloud, and the larger aerosol particles that will eventually settle on the substrate and may later evaporate) are not considered by these methodologies. Typically, these methodologies come into play due to limitations of the consequence modeling tools themselves. The final result is an *under* or *over*estimation (potentially by several orders of magnitude) of the flammable and toxic dispersion hazards.

The prediction of aerosol formation as a result of liquid releases is of particular importance for a number of reasons. Some are listed below:

- Aerosols liquid particles are generally heavier than vapors and when released along with the vapors, the resulting plumes will become denser and heavier (due to the presence of liquid) than when compared to only vapor-phase clouds. The result is a plume that is dispersing in the downward direction with slower air entrainment from the surroundings. This translates to longer dispersion distances to flammable or toxic modeling end-points. Also, plumes that are heavier than air will travel in a negatively buoyant manner (i.e., close to ground if the release takes place near ground, or slump to ground if the release takes place at an elevation above ground). These heavier clouds will therefore tend to have more effect on people since they will travel and disperse at (or towards) grade level, which is where people are normally present. Lighter than air (positively buoyant) vapor clouds tend to lift-off quickly following a release and have less potential effects on people in the far-field (i.e., at longer distances from the leak source).

- Second, flammable and/or toxic components of interest (for example, toxic H2S) may be fully or partially dissolved in the released mixture's aerosol particles, and may gradually evaporate along the dispersion path of a momentum jet release. This may yield significantly different dispersion results compared to the immediate sudden/ instantaneous flashing (or lack there of) of these components (as assumed by the simple flash calculation methodologies). Thus, the simplistic-modeling methodologies may result in calculations showing quick dilution and shorter calculated distances to components of interest end-point concentrations. The simplistic methodologies and the tools they utilize would then have underestimated the actual hazards, and the resulting risks.

5. SUMMARY

When a liquid or a two-phase mixture is released from a pipeline or wellhead, the original mixture is broken into several phases. Determining what the resulting phases are, the composition of the phases, and their release rates and total amounts is referred to as source-term-determination. Output from the source term determination process feeds into the dispersion algorithms to calculate downwind distances to flammable and toxic end-points of concern.

Proper source term determination is very critical to accurate dispersion analysis and overall consequence modeling results. When modeling wellhead and pressurized pipeline releases, consequence modeling tools and methodologies should be able to calculate and consider the resulting multi-phases, including the various aerosol phases and their compositions and respective release rates.

Dispersion analysis should take into account aerosol formation and aerosol dispersion since the heavier aerosol plumes tend to travel further distances and at grade level where people are normally present. Aerosols may both provide a physical carrying mechanism of flammable and toxic components (as part of a heavier aerosol cloud) and a mechanism for flash evaporation of toxic and flammable substances that are dissolved as part of the aerosol particles. When compared to pure vapor or flashed-vapor releases, aerosol releases tend to be characterized by longer distances to dilution to concentration end-points. Thus, aerosol clouds tend to yield longer distances to the same modeling end-point when compared to dispersing vapor clouds.

Thus, proper source term determination of aerosol phases resulting from high pressure and/or superheated releases (common to liquid well blowouts and high pressure pipeline releases), followed by the utilization of this information in aerosol dispersion analysis will better predict potential hazards. It will reduce the possibility of dispersion analysis underestimation of hazards, and the under-prediction of risks.

Simplistic consequence modeling tools and substandard approximation methodologies are not fit-for-purpose to properly model the complex phenomenon of wellhead and pipeline releases.

5.1. Case Study

A simple two-component mixture contains 50 mole percent methane and 50 mole percent pentane. The computed mixture phase-envelop is illustrated in figure 4-1.

FIGURE 4-1

Computed Methane-Pentane Mixture Phase Diagram

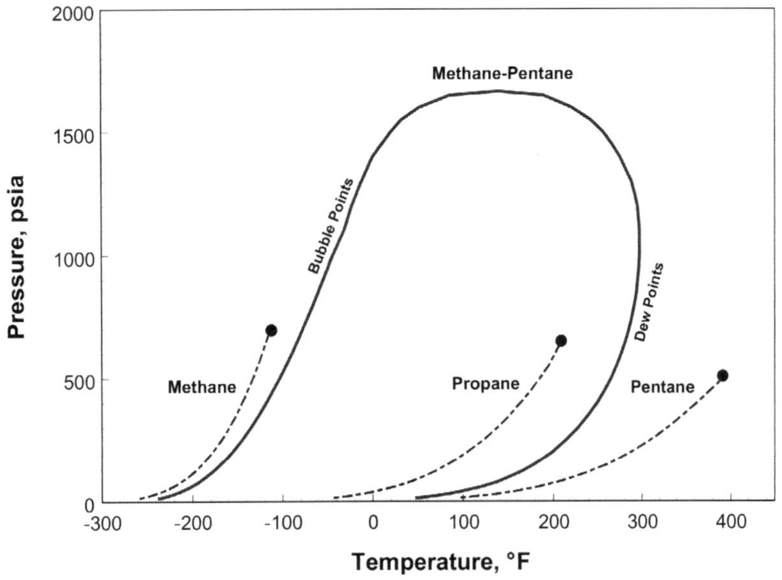

The area to the right of the phase-envelop reflects where the mixture is completely vapor phase. The area to the left of the phase-envelop reflects where the mixture is completely liquid phase. Within the phase diagram the mixture exist as two phase (i.e., the change from vapor to liquid or vise-versa does not happen at one temperature as with pure components (e.g., methane, propane, and pentane, as illustrated in Figure 4-1). For a mixture at a constant (below critical) pressure, there is a temperature range that must be traversed to complete the phase change from all-liquid to all-vapor or vise-versa.

A 6-inch pipeline contains the 50-50% *(C1-C5)* mole fraction mixture at various temperatures, pressures, and flow conditions. Upon a wellhead blowout or a pipeline full bore-rupture, the three-dimensional histogram in Figure 4-2 illustrates the original mixtures' break-up phases as a function of starting temperature and pressure conditions. Calculations indicate that the original mixture's phase fraction break-ups are maintained by changing the normal flow rate, and that they seem to be primarily a function of changes in the mixture's initial temperature and pressure conditions. For wellhead blowouts and pipeline releases, the normal flow rate is the rate dictated by a wellhead Absolute-Open-Flow (AOF) conditions, or a pump's performance curve, respectively. At

1,000-psia pressure, the mixture's normal flow rate is assumed at (150)-lb/sec. Ninety-three (93)-lb/sec normal flow rate is assumed at 100-psia pressure, and a (13)-lb/sec normal flow rate is assumed at (25)-psia.

FIGURE 4-2

Example of *C1-C5* Mixture Break-up Phases as a Function of Different Initial Temperature & Pressure Conditions

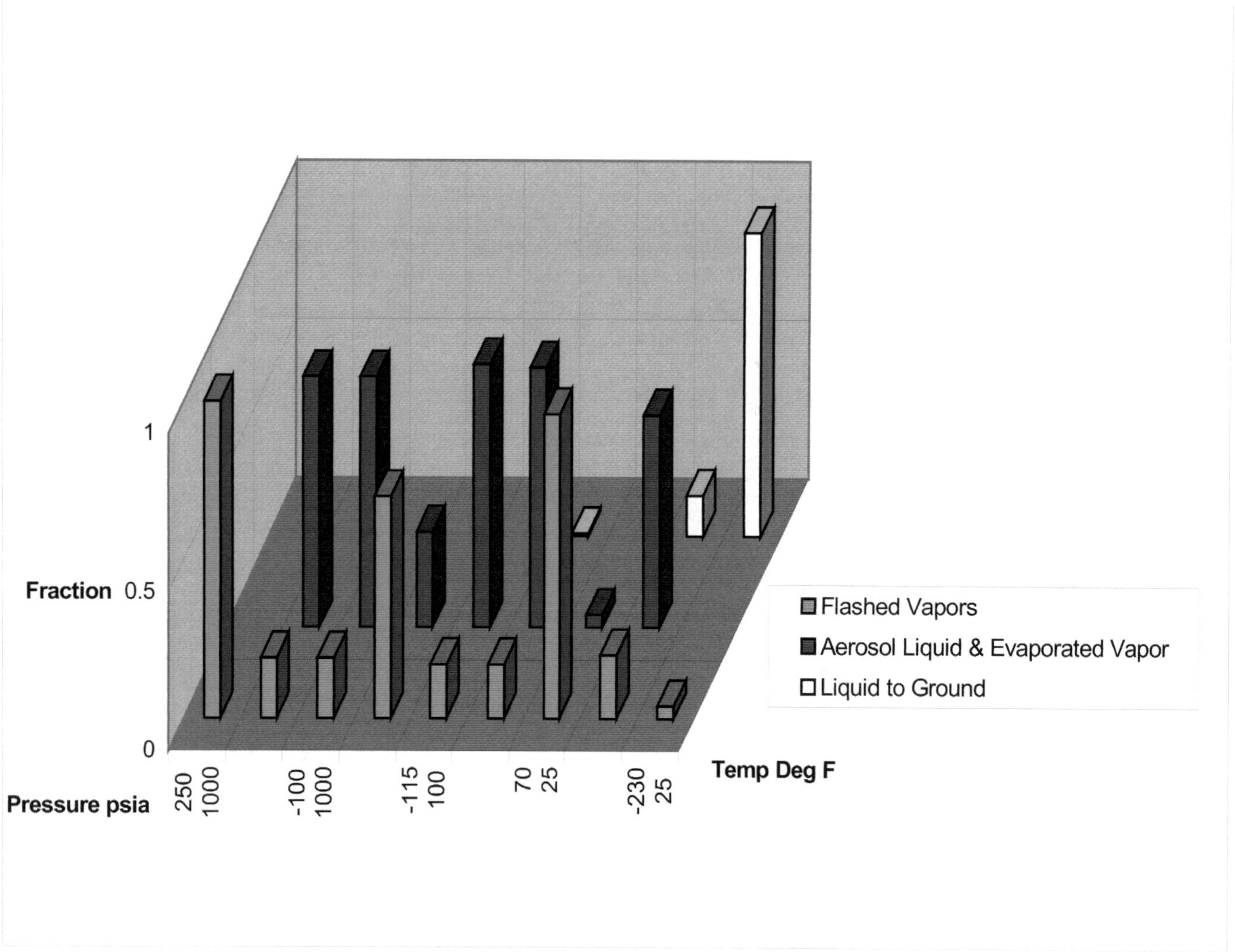

Comparing the low temperature and pressure release conditions (e.g., 25 psia and -230 °F), calculations indicate that high vapor and aerosol generation rates are noticed at relatively medium-to-high pressure ranges (100 psia and above), and higher temperatures (superheat or at temperatures near or above the mixture's bubble point).

For the C1-C5 modeled <u>two-component</u> mixture, small variations in either temperatures or pressures can have significant differences in the resulting source-term release or phases. This situation can get considerably more complex as the number of components in the released mixture increase (e.g., a typical oil wellhead fluid may easily have ten-plus individual components). Also, pipelines commonly transport multi-component hydrocarbon mixtures rather than pure single components.

6. THE IMPORTANCE OF DETERMINING SOURCE-TERM RELEASE RATE VARIATIONS AS A FUNCTION OF TIME (I.E., TRANSIENT NATURE OF PIPELINE RELEASES)

Long flow/ trunk tubing strings of oil and gas wellheads (for example between a wellhead choke and an inlet separator), and large-bore diameter pipelines can hold large hydrocarbon inventories. This is particularly true at higher pressures.

Following a large release from a transport pipeline system, the fluids packed at high pressures in these piping systems will initially blow-down at high rates over a period of time (sometimes lasting for tens of seconds). This is particularly true in the case of high-pressure gas transport pipelines. Then, the release rate will reach a "steady-state" flow condition. Examples of steady state conditions include flows from pumps and compressors in liquid and gas pipelines, respectively.

Depending on factors such as the length and diameter of a piping system, initial systems' pressure, and the type and phase (vapor, liquid, or two phase) of the released materials; the amount of time it may take a pipeline system to blow-down its initial inventory to reach steady state flow conditions may last for a considerable duration.

For large size hazardous releases from large diameter pipelines, the initial blow-down periods are typically characterized by very high flow rates. In some instances, the high blow-down rates during the initial periods of a release may produce potential hazardous vapor clouds with larger footprints than the ones produced later under steady-state flow conditions. Often, only one of the two flow regimes will typically produce a controlling worst-case toxic/ flammable release scenario. Thus, it is very important *not* to ignore the variable blow-down flow rates during the initial periods of a release. The case study below illustrates by example.

6.1. Toxic Hazards

The issue of source-term release rate variation as a function of time is of particular importance when assessing toxic hazardous effects since component toxicity levels are measured in terms of dose (i.e., the concentration to which individuals are exposed to a toxic substance, and the time of that exposure). To illustrate, the following is a description of the events that will take place following a toxic hazardous release from a long, high pressure gas transport pipeline: "several minutes of very high pipeline blow-down flow rates prior to reaching steady-state flow rate condition will typically result in a toxic hazardous plume that resembles the shape of a tadpole (i.e., fat and wide head at the front-end leading the cloud downwind, and a thinner longer tail section lagging behind). The front, wide-head portion typically resembles the hazardous plume that is generated by the initial very high release rates during the initial system blow-down stages (similar to a "smoke -puff" from a cigar). The lagging, thinner/

narrower portion of the cloud resembles the hazardous part formed as a result of steady state release rate conditions."

Which one of these two plumes will then yield the worst toxic hazardous exposure scenario? The answer depends on several factors, including: the type and temperature/ pressure conditions of the released materials; the duration of the release; the piping/ tubing systems' physical characteristics and configurations (e.g., length, diameter, isolation means); and on thc population density and distribution around the hazardous release point. The following case study illustrates some aspects of this phenomenon.

6.1.1. Case Study

A population center is located approximately 750-meters away from a 45 kilometer long, 32" (inch) gas pipeline. The pipeline is operated at approximately 365-(psia) pressure and 110-°F. The gas contains toxic H_2S (at 12% mole-percent concentration). The pipeline is assumed to have automatic isolation valves that are evenly spaced every seven kilometers. Upon a large leak (e.g., full-bore rupture) the calculated high initial blow-down mass flow rates, and the lower long-term steady-state release rate are plotted as a function of time in Figure 5-1.

FIGURE 5-1

Case Study (Toxic Release): Hazardous Release Mass Flow Rate as A Function of Time

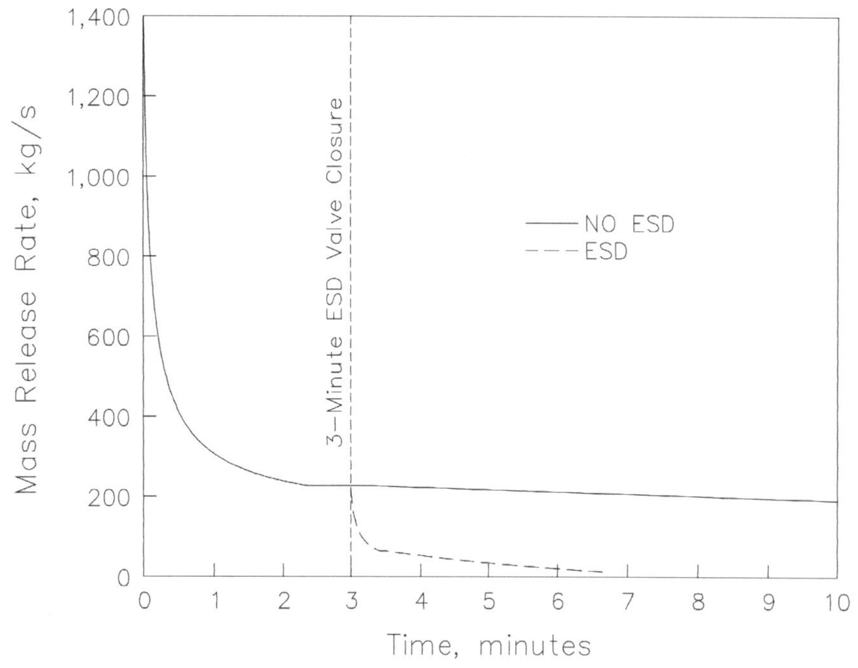

As stated previously, an important parameter in the definition of the hazard zones associated with a release of gas from the pipeline is the variable escape rate of gas. The larger releases will quickly depressurize the line and, as the line pressure drops, the mass flow out of the opening will be reduced.

The initial high flow rate quickly drops off as the pressure inside the pipeline decays. After a period of time, approximately one minute, the flow out of the pipeline would drop off to the normal line flow rate (if this rate could be maintained). In this case, the pipeline normal flow rate can be maintained provided mechanical drivers (e.g., compressors) continue to operate following the rupture.

The effect of having operating ESD valves in place can be seen in Figure 5-1. If it is assumed the pipeline rupture is detected within seconds of occurrence, the valves along the pipeline could be actuated within seconds, or several minutes. Also, time needs to be allowed for the 32" isolation valves to close from the time they are actuated. An estimated full three minutes are assumed to elapse between when the rupture occurs and when the valves are completely closed.

Once the valves are closed, the distance between the closed ESD valves determines the mass flow rate out of the rupture. The closer the valves, the less mass there is for release; thus, the pressure in the line decreases faster. Figure 5-1 shows the effect of having ESD valves. With the assumption that the valves are closed three minutes after the rupture occurs, the flow starts to decay at time equal three minutes. From Figure 5-1, it is clear that the 7-km spacing of ESD valves results in a rapid decay of the gas flow once the valves close.

The important point to be made with Figure 5-1 is that the maximum rates at which the gas can escape the rupture occurs during the first minute or so after the release. The flow rate drops rapidly as the line is depressured. From Figure 5-1, it can be seen that the ESD valves have nothing to do with the initial flow rate. For this case, ESD valves have no effect in limiting the initial large release rates during the initial periods. The value of the ESD valves is to limit the time over which the lower steady-state normal flow rate occurs.

Consequence modeling calculations indicate that the short-lived, long, tadpole-head shape toxic plume(s) generated by the initial, high blow-down flow rates may lethally impact the population center located 750-meters away from the pipeline. However, the longer duration, shorter, and thinner shape toxic plume generated by the lower steady-state flow rate does not pose a lethal threat to the population center. This case study presents a classical example where ESD valves have little to no value in reducing toxic effects on targets of interest following a large gas release, and where the long-term steady-state flow rate does not result in the worst toxic release scenario. In this example, it is rather the initial stages of the hazardous release (and its transient stages) that are more critical and yield the worst toxic hazard to the population.

6.2. **Flammable Hazards**

With regards to the effects of source-term release rate variations on flammable leaks, it is common to use the higher initial release rates to model flammable hazards (ultimately resulting in longer flammable vapor clouds). Typically, as flow rates increase, the footprint and/or downwind extents of the resulting flammable vapor clouds increase. Larger footprint flammable clouds have a higher likelihood of finding an ignition source. A general presumption is that the probability of finding an ignition source increases in an exponential fashion as the release rate increases. For flammable-type releases, hazardous outcomes have less dependency on time (compared to toxic releases). It is a common and a conservative approach to use the high initial blow-down release rates for modeling flammable hazards, since the higher rates will typically yield further downwind (larger size) flammable plumes.

7. CONCLUSIONS

The physical and chemical characteristics of wellhead and pipeline released fluids, and the physical configurations of their containment systems are critical to the accurate dispersion modeling of hazardous release events from such systems.

Some of the major factors that play an important role in assessing the flammable and toxic hazards associated with wellhead and pipeline uncontrolled releases include: 1) properly determining the different phases resulting from the original mixture break-up, the composition profiles and thermodynamics of each of the released phases (including vaporization from, and settlement of liquid aerosols), 2) determining the mass release rate for each of the phases, 3) their time-transient nature, and 4) taking into account the mechanical features and physical configurations of the wellhead and pipeline systems.

For large size hazardous releases (particularly from large diameter gas pipelines) the initial blow-down periods are typically characterized by very high flow rates lasting for several tens of seconds. In some instances, the very large blow-down rates during the initial periods of a release may produce hazardous vapor clouds with larger footprints than the ones produced later under steady-state flow conditions. Often, only one of the two flow regimes will typically produce a controlling worst-case toxic/ flammable release scenario. Thus, it is very important not to ignore the *variable* blow-down flow rates (i.e., transient behavior) during the initial period of a hazardous release. This issue is of particular importance when assessing toxic effects since component toxicity levels are measured in terms of time of exposure and concentration (i.e., dose).

Consequence modeling of wellhead blowouts and pipeline leaks is a complex task, and requires the consideration of several important factors. Simplistic consequence modeling tools and substandard approximation methodologies are not fit-for-purpose to properly model this complex phenomenon, and may result in underestimation of the potential hazards associated with wellhead and pipeline uncontrolled releases. This may result in unaccounted exposures (to personnel, assets, environment, and to the surrounding public). Calculations will then indicate risk values that are lower than actual risk levels. The opposite: consequence modeling overestimation of hazards and their associated risks may result in unnecessarily spending of funds in an attempt to prevent or mitigate (in reality) low risk events. Thus, the need for more accurate consequence modeling tools and proper modeling methodologies has become more important than ever before.

The factors and consequence modeling analysis methodologies discussed by this paper are applicable to other type fluids, and other releases (for example, pressurized liquid releases, superheated liquids, and two-phase releases from pressure vessels/ equipment, plant piping, and relief valves).

REFERENCES

American Industrial Hygiene Association, *"Emergency Response Planning Guidelines criteria,"* 2003.

David W. Johnson, and John L. Woodward, *"RELEASE, A Model with Data to Predict Aerosol Rainout in Accidental Releases,"* Center for Chemical Process Safety of the American Institute of Chemical Engineers, New York 1999.

Cornwell, John B., *"The Importance of Multiphase and Multicomponent Modeling in Consequence and Risk Analysis,"* ASSE-MEC, 6th Professional Development Conference and Exhibition, 2003.

Coots, Mathews (Boots and Coots Consultants), and Flak, L (former John Wright Co. employee), *"Part 9-Firefighting. The Mechanics of Oil/Gas Fires, Meltdown and Secondary Damage, Water/Chemical/Explosive Extinguishing Methods and Considerations for Voluntary Ignition."* Internet Article.

Lees, Frank P., *"Loss Prevention in the Process Industries"* Butterworth-Heinemann Ltd, 1991.

Center for Chemical Process Safety of the American Institute of Chemical Engineers, *"Chemical Process Quantitative Risk Analysis,"* New York, 1989.

Determining Fractionator Pressure Relief Loads Using Dynamic Simulation

Freeman Self, Bechtel and Glenn Dissinger, AspenTech

Prepared for presentation at the
AIChE 2004 Spring National Meeting - 38th Loss Prevention Symposium
Session T007 "Advances in Consequence Modeling II

April 25-29, 2004

Unpublished

Abstract

Dynamic simulation has developed into one of the most powerful tools for determining relief loads. Nowhere is this more evident than in establishing the relief flow from fractionating towers, one of the most thorny of all relief calculations.

Fractionators generally have large relief loads and significant flows to a disposal system and it is beneficial to calculate the loads accurately. Although dynamic simulation has been used successfully for years, it gained prominence in the late 1990s with the maturity of high-level programs only incrementally more complicated than steady-state simulation. One of the most important benefits is the reduction in contingency usually assigned to the relief load. Additionally, the impacts of control systems, drivers, heat exchange, and the changes in relief and inventory over time are logically displayed. With the advantages of dynamic simulation, it is no longer necessary to rely on "hand" or "estimation" methods that approximate properties and fluid traffic in the column. Even though some estimation methods are theoretically sound, they are inherently inaccurate due to the treatment of inventory and static assumptions for phase equilibrium, heat transfer, pressure drop and hydraulics.

The presentation will culminate with a HYSYS demonstration of a fractionator, which illustrates the changes in inventory and pressure leading to the relief scenario.

What is Dynamic Simulation?

Dynamic simulation enables one to gain a detailed understanding of the transient behavior of a new or existing process and its control system. Process equipment are rigorously modeled using differential and algebraic equations resulting from mass and energy balances, phase equilibria, heat transfer, and hydraulics. These equations are integrated over time to predict how the process and its equipment will respond to various disturbances. Applications include:

- design and troubleshooting of control systems
- improving the flexibility and robustness of processes to handle a variety of disturbances
- safety studies including pressure relief and reactor runaway
- design and optimization of batch processes
- simulation and design of transient operations as load following and startup / shutdown sequences

Why Dynamic Simulation?

There are several major reasons to use dynamic simulation rather than steady-state simulation for determining pressure relief loads. A look at these will illustrate the engineering and financial attractions.

- Reduces margin of error. Since estimation methods, by definition, utilize assumptions that add more variability to the calculation, they inherently have a greater margin of error. In actual practice, this margin of error results in an engineering contingency thus adding costs to the design. Dynamic simulation is able to remove this contingency and thus lower engineering costs.

• <u>May lower absolute relief loads</u>. A smaller relief load will result in considerable economic benefit, both for the fractionator relief valve and header system. Based on a limited number of services, a survey by AspenTech found that on the average the relief load as calculated by dynamic simulation, as compared to estimation methods, was lower 90% of the time and the average reduction was approximately 50%. This result is similar to other individual studies. A strong caveat is that no effort was made to rationalize the estimation methods and part of this reduction could possibly be due to a difference in basis.

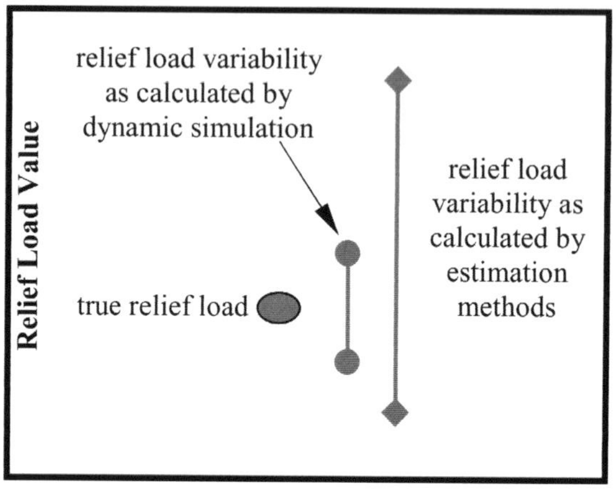

On the other hand, it is important to recognize that there is no guarantee that this result will be true for all comparisons. But due to the inherent variability of estimation methods and resulting conservatism, it is reasonable to expect that dynamic simulation may produce lower relief loads.

• <u>Shows relief changes over time</u>. It is this feature that particularly enhances understanding of the relief event. At any point in time, the user may change the process dynamics and witness the transient response. The relationship of parameters to changes in equipment and instrumentation are easily identified. For example, typical parameters tracked are pressure, overhead and reboiler temperatures, column holdup accumulation, and flows especially streams that are pumped or compressed. These parameters may be plotted for easy visualization, with multiple plots on the screen. Dynamics are also important to determine the timing of reliefs from different fractionators under global scenarios such as power failure.

• <u>Analyzes impacts of control systems, drivers, heat exchange, inventory, pump curves</u>. The peripheral equipment of fractionators will often influence the relief load. Dynamic simulation allows the designer to quickly understand the impacts of the equipment operation, which is especially valuable at the design constraints. The response of the control system to a change in the process or control is provided. As the unit pressures, the changes in feed & pump-around flows are calculated. Heat exchangers are rerated as the process changes. And the impacts of multiple unit interactions (e.g., heat integration, relief rate suppression) may be analyzed.

• <u>Resolves estimation methods</u>. Presently there are several methods for calculating fractionator relief loads, but due to their different defining assumptions all may give widely varying answers. However, dynamic simulation offers excellent accuracy and results in reasonable completion time.

Comparison of Execution

Dynamic simulation compares very favorably with estimation methods. The results are subject to reduced variability since the simulation is more rigorously calculated, and there are fewer assumptions that may promote wide ranges of results. This translates to a more predicable time estimate. Additionally, the time required for assessment and verification of assumptions and results is reduced.

Dynamic Simulation	Typical Estimation Method
Simulation completion requires approximately 30 - 50 hours per fractionator. Schedule is sufficiently predicable.	Calculation completion requires approximately 20 - 70 hours per fractionator. Schedule is subject to more unpredictability.
Variability in relief flow result is low, due to rigor of calculation.	Variability in relief flow is high, and results may have differences over 50% depending on even simple assumptions.
Complex hydraulics are difficult to handle unless programmed into simulation.	Problems that are difficult to handle include the amount of fluid accumulation (holdup), steam stripping, and the selection of the relief molecular weight.

Are Results More Rigorous?

Whenever dynamic simulation is discussed, the subject of rigor is invariably raised. Many citations in the literature demonstrate the success of dynamic simulation as checked against field data. The majority of tests are done for small process perturbations on training models. On the other hand, data for large perturbations during actual field relief upsets are infrequent and often qualitative. There are four points to address.

- Flashes are rigorous using exact vapor-liquid equilibrium. This is a standard feature of process simulators but an area that must be approximated using estimation methods. For example, steam strippers or fractionators with streams of widely varying enthalpies may pose complications in properly mixing the stream enthalpies if rigorous flashes are not performed.

- Temperature profile is accurate. Accurate heat and mass transfer is required for correct phase equilibrium and is furnished by dynamic simulation. The temperature profile is also important in understanding the progress of the relief.

- Flows and Inventories are modeled. The proper description of inventory is one important parameter in obtaining proper fractionator relief loads. For example, relief determinations due to loss of feed are very difficult by estimation methods since the material balances and inventories can only be roughly established for most situations. In general, estimating the traffic at the bottom of the fractionator is a critical consideration but a source of large variability for estimating methods, although it easily handled by dynamic simulation.

- Pressure drops / hydraulics are calculated. Pressure drops and hydraulics are required for correct flows and inventories. Dynamic simulation uses standard correlations for pressure drops and hydraulics. As will be shown by the example, the effects of vapor traffic influence liquid traffic down the column and liquid inventory in the fractionator bottoms. With insufficient bottoms holdup, the relief event subsides.

Requirements for Dynamic Simulation?

The precision of dynamic simulation imposes several extra requirements over steady-state methods. Equipment parameters are typically available from the process design or manufacturers drawings. Additional process modeling skills and attention are also required.

- Requires more equipment information. Equipment information is needed, especially volumes, distributor details, and liquid levels on trays. If such information is not available, HYSYS Dynamics has sizing algorithms for many equipment, including drums, fractionator diameters, and tray sizing. Heat exchanger duty may be set using various methods, including a set duty, a simple rate equation, or rigorous heat exchanger rating. Stages are best modeled using actual stages with tray efficiency. On the other hand, certain parameters need not be accurately specified. For instance, detail control valve information is not required if it is assumed that the control valve is operated in the manual mode and the valve position does not change.

- Requires HYSYS Dynamics and fractionation expertise. Fractionation experience is certainly helpful in planning the simulation and evaluating the results. But dynamic simulation programs are becoming increasingly user-friendly and conversion from a steady-state simulation to dynamics is generally easily performed.

- Must consider pressure / flow requirements. Unlike steady-state simulation, the user must think in terms of pressures and flows. For example, stream pressures need not balance in a steady-state simulation for convergence. However, all flows and pressures must match in a dynamic simulation.

History of Dynamic Simulation

Dynamic simulation of fractionators was introduced in the 1940s using analogue techniques and later large digital computers. In the 1980's, early work using dynamic simulation was performed to model reactor runaway scenarios. These often used programs where one had to meticulously build up the partial differential equations that described the system and specialists skilled in programming and process engineering were required. The 1990s saw a rapid development in advanced simulators and the growing use of dynamic simulation for pressure relief studies. Current simulators are completely different programs than their predecessors and involve an order of magnitude less time to execute a model. Today model libraries in dynamic simulation software enhance the ease of pressure relief studies.

Documentation

Documentation is an essential issue for relief studies. Surprisingly, documentation of a dynamic simulation is understandable, concise, and complete. The structure of the simulation is part of the input options. Annotations may be added. Graphs of any of the several thousand parameters are possible, although only a few important parameters are shown in the following example. The typical documentation may include these elements.

• Reference to the steady-state simulation establishes the initial basis.
• Equipment and instrumentation details may be taken directly from the simulation.
• The Case Summary may be formatted using various techniques, depending on the application. For example, a schematic with initiating and subsequent events is a convenient method. Alternatively, a list from the program inputs may suffice.
• The Case Outputs are effectively presented as graphs of the parameters over time. These are provided by the program.

Reflux Failure Case Study

A case study will illustrate the strength of dynamic simulation. The study was done using HYSYS Dynamics, one of two dynamic simulation programs available from AspenTech. In addition to HYSYS Dynamics, AspenTech also produces the program Aspen Dynamics.

Figure 1
Fractionator Schematic and Dynamic Response

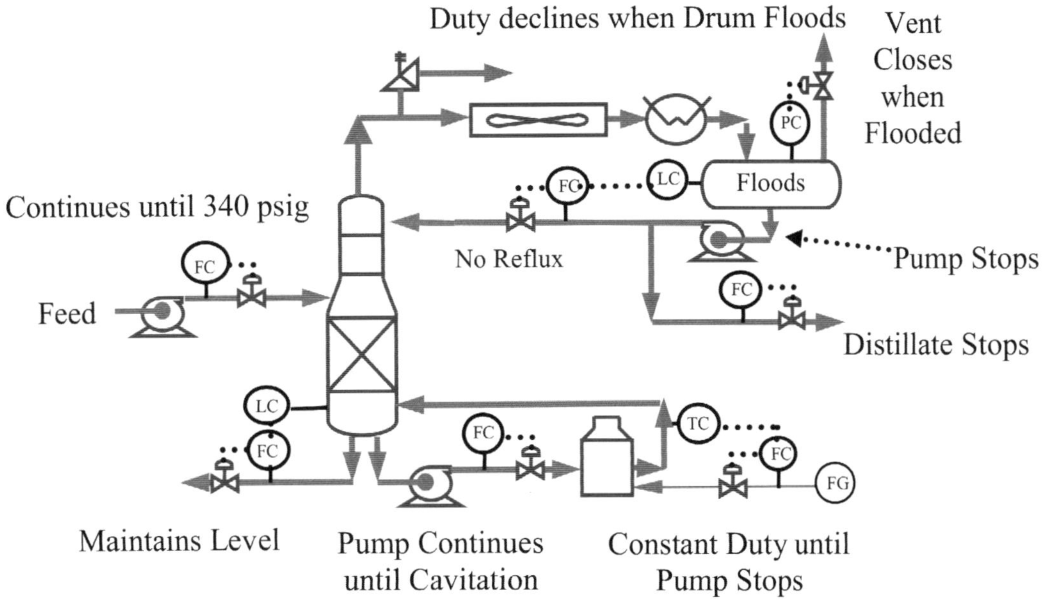

The evaluation of the relief event due to failure of reflux will illustrate several aspects of the relief load. The fractionator is a reboiled via pump-around through a fired heater. The condensers include both a water cooler and air cooler. The feed is pumped. The initiating event for the case is the stoppage of the reflux pump, resulting in no reflux or distillate product. This in turn will ultimately result in loss of cooling and blockage of the overhead vapor line, resulting in a relief.

After the reflux flow stops, the vapor from the fractionator top tray will still continue to flow overhead and condense, causing the reflux drum level to slowly increase. As shown in Figure 2, the assumption of a manual position for the overhead vent valve results in a slight column depressuring. The loose control action of the bottoms product level control valve also is illustrated.

When the reflux drum floods, the overhead vent is blocked. With no outlet for the vapor and heat still being introduced by the reboiler, the column pressure now starts to increase. The level in the bottom of the column is maintained at the normal liquid level by the liquid product control valve.

As the pressure rises to 340 psig (just below the relief pressure), the feed pump ceases to have sufficient pressure to enter the column. With the feed no longer replenishing the liquid in the bottom of the tower, the liquid product valve goes closed while the liquid holdup inventory starts to rapidly diminish as the tower pressure increases. Since the feed had a "cooling" effect on the liquid inventory, loss of feed causes the pressure to increase more rapidly.

Figure 2
Pressure and Holdup Over Total Time Period

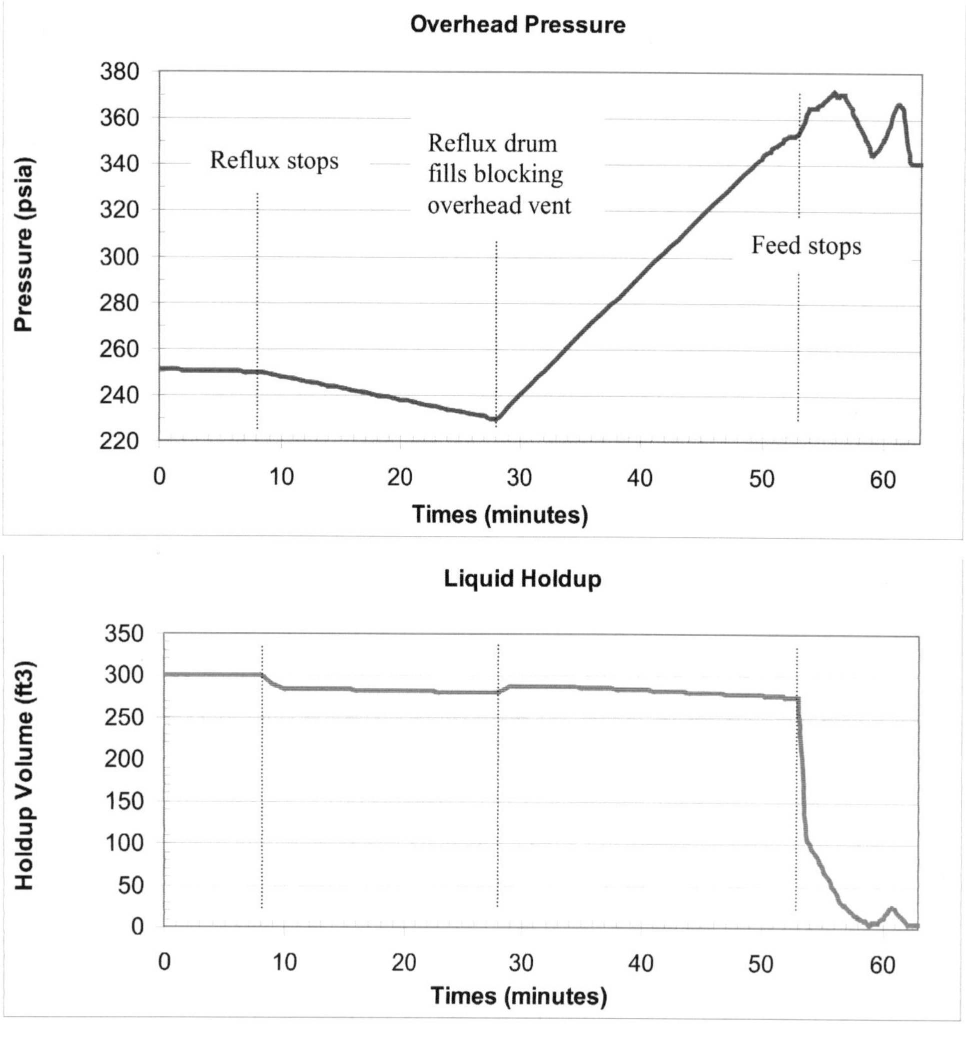

When the tower reaches the relief valve set pressure of 350 psig (364.7 psia), the relief valves open. Refer to Figures 2 and 3. The pressure in the column, as well as the relief rate, continue to increase as the liquid holdup is depleted.

The liquid holdup at the bottom of the tower is finally consumed, and the reboiler pump and reboiler duty stops, triggering a rapid fall in relief load and pressure.

But without traffic flow up the column, liquid is allowed to flow from the top trays to the bottom of the column to partially refill the bottom liquid inventory.

Figure 3
Relief Flow, Pressure, Holdup

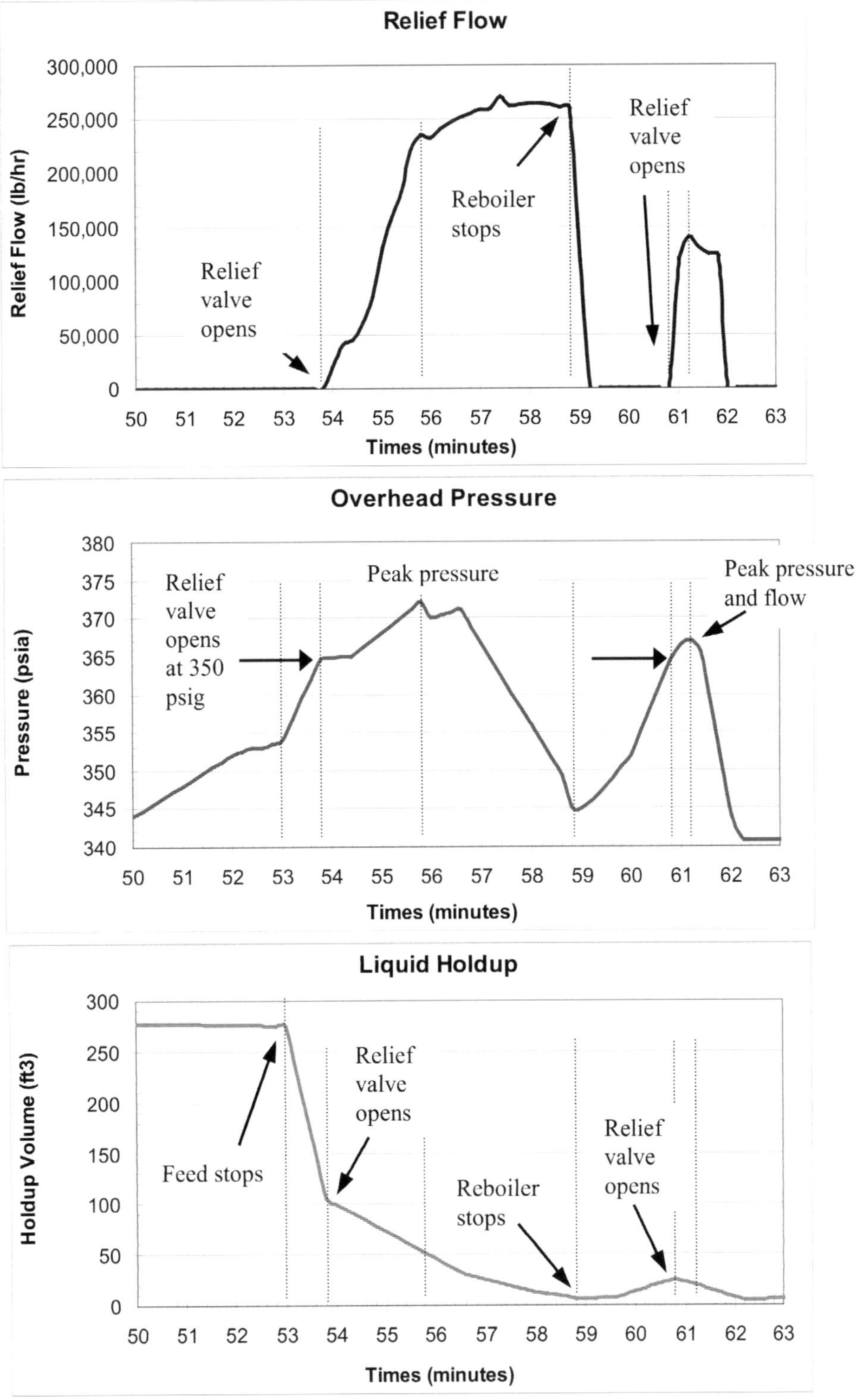

This in turn assumes that the reboiler pump and reboiler start to re-introduce heat and the tower pressure again increases to 340 psig, when the relief valve reopens. This point corresponds to the temporary maximum in liquid holdup.

Subsequently, the holdup is depleted and the relief valve closes.

The complete scenario unfolded over a period of less than one hour from stopping of the reflux pump to the final relief.

However, the time from the reflux drum flooding to the opening of the relief valve was less than 30 minutes. It is to be noted that these relief times are indicative only of this particular fractionator and will be different depending on the individual circumstances, including initial liquid inventory and action of the feed stream.

USES OF FIRE DYNAMICS SIMULATOR V4 FOR LARGE SCALE/INDUSTRIAL INCIDENTS

Jason E. Floyd
Hughes Associates, Inc.
3610 Commerce Dr., Suite 817
Baltimore, MD 21227-1652

Prepared for
38th Annual Loss Prevention Symposium
AIChE 2004 Spring National Meeting
April 26-29, 2004
Advances in Consequence Modeling

Unpublished

ABSTRACT

Consequences of large scale/industrial incidents can be costly both in terms of external impacts to the surrounding communities and on site losses, which include personnel casualties, damage to facilities, and facility downtime during recovery. The use of computational fluid dynamics (CFD), such as Fire Dynamics Simulator (FDS) v4, a large eddy simulation CFD program being developed at the National Institute of Standards and Technology (NIST), can aid in reducing these costs in both the facility design process and in the post-accident recovery process. Unique aspects that may not be adequately addressed by existing fire codes may hamper design of fire protection systems and concepts for some industrial facilities. The use of CFD can aid in avoiding under design of active and passive protection systems to ensure adequate protection. Environmental recovery from an incident will be done with some resource limitations. CFD can aid the resource allocation by predicting the path and fallout of incident related emissions. FDS v4 will be used to show examples of CFD predictions in both design and consequence analysis.

FIRE DYNAMICS SIMULATOR

FDS is computation fluid dynamics tool under development by the NIST Building and Fire Research Laboratory (BFRL) for the purpose of predicting the heat and mass transfer resulting from combustion (McGrattan, et al., 2003a and 2003b). The current release of FDS is version 4. FDS solves a low-Mach number form of the Navier-Stokes equations on a multi-block, Cartesian grid with either a large eddy simulation (LES) or direct numerical solution (DNS) approach for turbulence. Combustion is computed using either a mixture fraction or a one-step, finite-rate reaction method. Radiation heat transfer is computed using a finite volume method. FDS also contains submodels for predicting droplet dynamics including transport and evaporation that can be used for either suppression systems or fuel sources. The following subsections will briefly discuss the each of aforementioned code capabilities.

Hydrodynamics

FDS solves a low-Mach number form of the Navier-Stokes equations (Rehm and Baum, 1978). For most types of fire hazards the primary driving force is fire-generated buoyancy, which results in low speed flows. By eliminating pressure waves, a single pass, Fast Fourier Transform (FFT) solver can be used for the pressure equation which allows for fast execution times as compared to CFD codes that use an iterative or matrix solution pressure solver. The FDS equations for conservation of mass, momentum and energy are given below:

Mass

$$\frac{\partial \rho Y_i}{\partial t} + \boldsymbol{u} \cdot \nabla \rho Y_i = -\rho Y_i \nabla \cdot \boldsymbol{u} + \rho D \nabla Y_i + \dot{m}_i''' \quad (1)$$

Momentum

$$\frac{\partial \boldsymbol{u}}{\partial t} + \boldsymbol{u} \times \omega + \nabla \mathrm{H} = \frac{1}{\rho}\left((\rho - \rho_\infty) \boldsymbol{g} + \boldsymbol{f} + \nabla \cdot \tau \right) \quad (2)$$

Energy

$$\nabla \cdot \boldsymbol{u} = \frac{1}{\rho c_p T}\left(\nabla \cdot k \nabla T + \nabla \cdot \sum_i \int c_{p,i} dT \rho D_i \nabla Y_i - \nabla \cdot \boldsymbol{q}_r + \dot{q}''' \right) + \left(\frac{1}{\rho c_p T} - \frac{1}{p_0} \right) \frac{dp_0}{dt} \quad (3)$$

In the above equations $\boldsymbol{g}$ is the acceleration due to gravity, $\boldsymbol{f}$ is a force term resulting from droplet drag or other external forces, ρ is density, Y is a species mass fraction, i is a species, $\boldsymbol{u}$ is velocity, T is temperature, p_0 is the background pressure, $\boldsymbol{q}_r$ is the radiative flux vector, k is conductivity, τ is the viscous stress tensor, c_p is the specific heat, H is a pressure term, t is time, ω is vorticity, D is diffusivity, $\dot{m}'''$ is a mass source term, and $\dot{q}'''$ is an energy source term which includes the convective heat release rate and energy gain or loss due to droplet interactions with the gas.

FDS uses one of two turbulence models, DNS and LES. DNS requires a highly resolved grid, on the order of 1 mm, and it is thus, not practical for use in most large scale/industrial scenarios where length scales are tens of meters to kilometers. As such it will not be discussed further. The large eddy simulation model uses the Smagorinski model (Smagorinski, 1963). This model assumes that turbulence below the grid resolution can be modeled by an effective viscosity whose magnitude is related to the local deformation, related to the Dissipation Function. When doing an LES computation the viscosity used in generating τ in Equation 2 is replaced by:

$$\mu_{LES} = \rho (C_s \Delta)^2 \left(2 (def\ \boldsymbol{u}) \cdot (def\ \boldsymbol{u}) - \frac{2}{3} (\nabla \cdot \boldsymbol{u})^2 \right), \quad (4)$$

where C_s is a constant, Δ is a length scale related to the grid size, and $def\ \boldsymbol{u}$ is the deformation tensor.

The computational grid used by FDS is a multi-block, Cartesian grid. The multi-block method divides the computational grid into one or more regions, or blocks, each with its own grid (Prassad, Patnaik, and Kailasanath, 2000). Information is exchanged between grids at coincident boundaries. This method allows for three potential paths for speed improvement. First, in geometries such as an L-shaped compartment, a large number of grid cells can exist in a region where there is no fluid flow. Computational time and memory is wasted in tracking these regions. Multi-block allows one to eliminate such dead zones. Second, the accuracy of the LES method is dependent upon resolving the dominant eddies in the flow field. In the case of fire driven flows the dominant eddy size within the combusting fire plume is generally smaller than the dominant eddy size in regions where only smoke movement is occurring. Multi-block could be used to define a region of higher resolution around the fire while reducing the resolution in more remote sections of the domain. Third, since information is only exchanged at grid boundaries, each grid is in effect an entirely separate simulation. This leads to an easy method of parallelization by assigning each grid to a separate processor.

Combustion

FDS contains two combustion models. The first is a single-step, finite-rate model. This model is only appropriate for use when doing a DNS calculation, thus it will not be discussed further. The second model is a mixture fraction combustion model (Kuo, 1986). The mixture fraction, denoted as Z, is a dimensionless quantity that represents the fraction of mass that was originally fuel. Thus, a cell containing ambient air will have Z=0, and a cell containing pure fuel will have Z=1. A special quantity of mixture fraction is defined for a mixture representing stoichiometric conditions called Z_F. Combustion in the FDS mixture fraction model is assumed to be infinitely fast with a single equilibrium state, and it occurs at the surface where $Z=Z_F$. Z can also be defined by the following equation (Mahalingam, et al., 1999) where: *F* is fuel and O_2 is oxygen, *I* indicates the incoming fuel stream and *8* indicates ambient:

$$Z = \frac{sY_F - \left(Y_{O_2} - Y_{O_2}^{\infty}\right)}{sY_F^I + Y_{O_2}^{\infty}}; s = \frac{\nu_{O_2} w_{O_2}}{\nu_F w_F} \tag{5}$$

If we assume a generic hydrocarbon fuel that produces a fixed level of CO and soot per unit mass of fuel consumed we can write the following equation where *?*, which can vary from 0 to 8, indicates the degree of excess oxygen:

$$\begin{aligned} C_xH_y + \xi\left(\frac{x\left(1+\frac{1}{2}\chi_{CO}\right)}{1+\chi_S+\chi_{CO}} + \frac{y}{4}\right)(O_2 + 3.76N_2) \rightarrow \mathrm{Max}[0, 1-\xi]C_xH_y \\ + \mathrm{Min}[1,\xi]\left(\frac{x}{1+\chi_S+\chi_{CO}}\right)CO_2 + \mathrm{Min}[1,\xi]\frac{y}{2}H_2O + \mathrm{Min}[1,\xi]\left(\frac{x\chi_{CO}}{1+\chi_S+\chi_{CO}}\right)CO \\ + \mathrm{Min}[1,\xi]\left(\frac{x\chi_S}{1+\chi_S+\chi_{CO}}\right)C + \mathrm{Max}[0,\xi-1]\left(x+\frac{y}{4}\right)O_2 + \xi\left(x+\frac{y}{4}\right)3.76N_2 \end{aligned} \tag{6}$$

Note that as ? varies from 0 to 8 the mixture fraction will vary from 1 to 0 (e.g. ? =0 indicates pure fuel or Z=1). By mapping ? onto Z a series of state relationships for each of the combustion products can be created as shown in Figure 1 below for propane with assumed CO and soot yields of 0.1 and 0.15 respectively. The fact that Z, a single scalar quantity, can be used as a surrogate for all the combustion product species means that only one Equation 1 is required rather than a separate Equation 1 for each combustion product species.

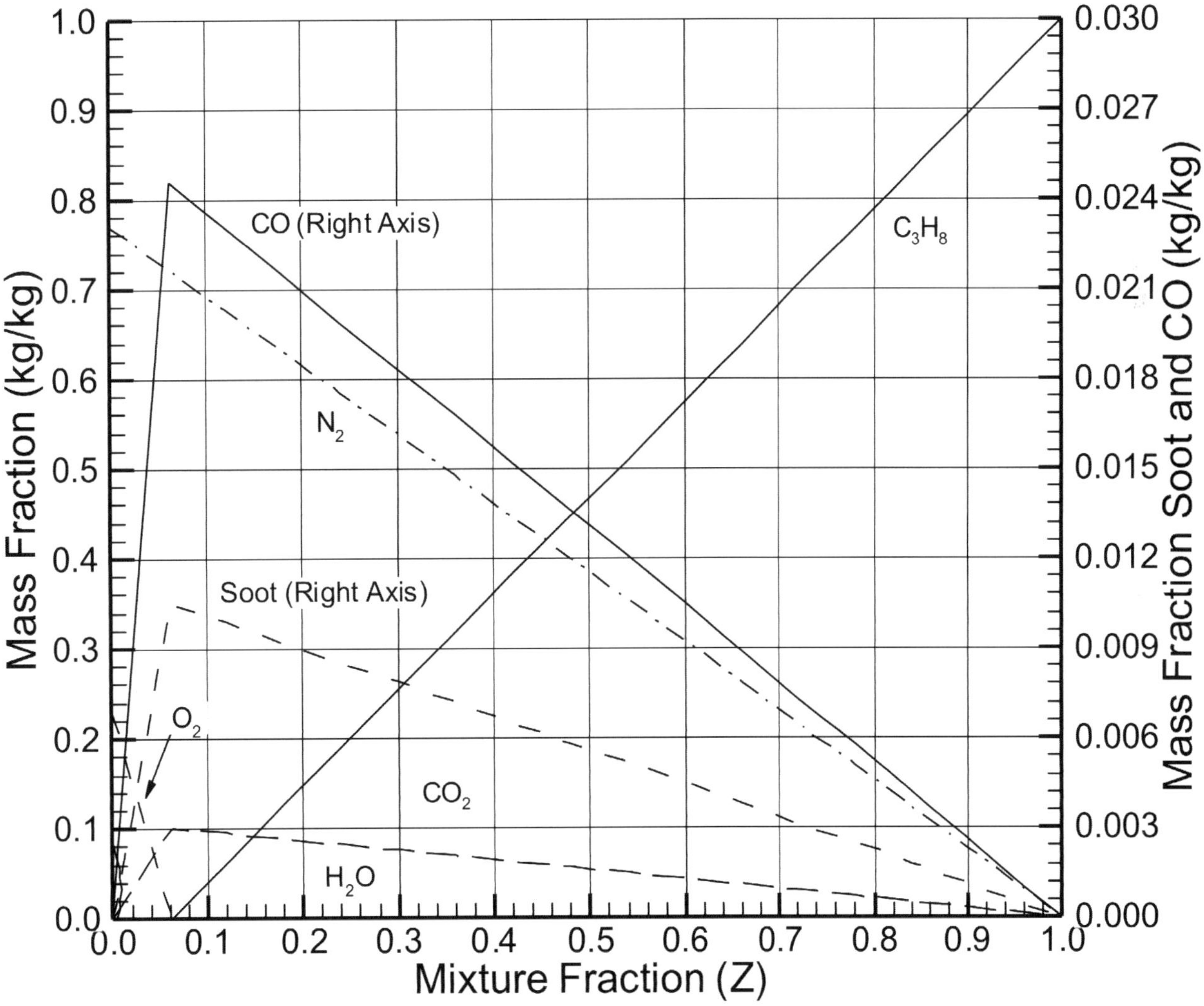

Figure 1: Mixture Fraction State Relationships

With the assumption of an infinitely fast reaction with a single equilibrium state, the local rate of combustion is mere a function of how quickly fuel and oxygen are transported across the stoichiometric surface, $Z=Z_F$ (Floyd, et al., 2003). By applying the state relationships determined by Equation 6, an expression for the oxygen consumption can be derived which when multiplied by a heat of combustion based on oxygen consumption results in the heat release rate. This equation is shown below where $\vec{n}$ is the normal vector to the stoichiometric surface.

$$\dot{m}''_{O_2} = -\frac{dY_{O_2}}{dZ}\rho D \nabla Z \cdot \vec{n}\Big|_{Z=Z_F} \qquad (7)$$

Radiation

FDS uses a gray-gas, finite-volume method for computing radiative heat transfer (Raithby and Chui, 1990). This method is obtained by taking the gray-gas, radiation transport equation (RTE) (Siegel and Howell, 1992) and dividing into a series of equations each representing a differential angular direction. Typically about 100 angles are chosen. The RTE and the derived finite volume equation are shown below.

$$\hat{s} \cdot \nabla I(\vec{x}, \hat{s}) = \kappa(\vec{x})\left[I_b(\vec{x}, \hat{s}) - I(\vec{x}, \hat{s})\right] \qquad (8)$$

$$\sum_{m=1}^{6} A_m I_m^L \int_{\Omega^L} \hat{s} \cdot \hat{n}_m d\Omega = \kappa_{ijk}\left(I_{b,ijk} - I_{ijk}^L\right)V_{ijk} d\Omega^L \qquad (9)$$

Where $\hat{s}$ is a path vector, $\hat{n}$ is the normal vector to a grid cell face, I is the radiant intensity, I_b is the radiation source term, κ is the absorption coefficient, $\vec{x}$ is a position vector, A_m is the cell area of face m, m is one of the six faces of a grid cell, ijk is a grid cell, V_{ijk} is the volume of grid cell ijk, $d\Omega$ is a differential angle, and L indicates an angular direction. The absorption coefficient, κ, for a grid cell is a function of the grid cells temperature and gas composition. FDS uses RADCAL (Grosshandler, W, 1993) to compute the absorption coefficient for each grid cell. Since the possible species compositions are limited to the mixture fraction state relationships, a table of κ as a function of temperature and mixture fraction can be pre-computed.

Special treatment is required for the source term, I_b in a typical LES calculation. The radiation emitted by a hot gas can be given by the Stefan-Boltzman law, which relates the energy emission to the fourth power of the gas temperature. In an LES calculation the gas temperature of a grid cell may not properly represent the true radiation emission of the cell. For a grid cell without combustion the Stefan-Boltzman law can be used with reasonable confidence. In a grid cell with combustion it is likely that the fraction of the grid cell that represents the true region of active combustion is only a small portion of the entire grid cell volume. The temperature computed for the grid cell is the result of the high temperature region where fuel is combusting averaged with the possibly much lower temperature of the surrounding gas. In other words in an LES calculation with large grid cells, it is not likely that the cells with combustion occurring will have combustions temperatures. Using the Stefan-Boltzman law would significantly under predict the radiant source term. Instead in a grid cell with combustion FDS will compute the radiant source term using the Stefan-Boltzman law, but will use the greater of the computed source term or a user specified radiative fraction, which defaults to 35 % (Karlsson and Quintiere, 2000).

Equation 9 is solved by determining the upwind direction for the current angle and propagating the incoming boundary condition (Hostikka, McGrattan, and Hamins, 2002) from the upwind direction through the computational domain. No costly matrix inversion is required. Equation 9 must be solved for each grid cell for each direction. To do this for all angles at every time step would consume significant computational resources. However, in general the hydrodynamic time step is going to be much shorter than the speed at which bulk temperatures and the size of the combustion region will be changing, i.e. the source term does not vary greatly between time steps. Thus, to reduce computational time FDS updates only a subset of the angles at each time step. Furthermore, it will also skip time steps. The default is to update one fifth of the angles every three time-steps.

Droplets

FDS contains a series of submodels for determining the interaction between liquid droplets and the gas phase. These were originally intended for use in predicting the effects of water based suppression systems (e.g. sprinkler and mist systems), but have been extended to include combustible droplets (e.g. fuel sprays). The droplet submodels include a routine for inserting droplets, a routine for predicting the velocity and position of droplets along with momentum transfer to the gas phase, and a routine for determining droplet evaporation, which includes absorption due to radiation.

FDS uses a superdrop, Lagrangian particle approach for computing effects to a dispersed droplet phase. In this approach a modest number of drops are tracked by FDS with each drop representing a much larger number of drops. Drops are inserted by first converting a user specified angular distribution of droplet velocity and droplet massflux into a probability distribution of angular location and velocity. An angular location and its corresponding velocity are randomly chosen along with a droplet size which is determined by a user specified Rosin-Rammler distribution (Chan, 1994). The drop is then weighted to obtain the correct net massflux from the injection nozzle. Droplet position is then tracked by updating a droplet momentum and position equation for each droplet at each time step. A drag coefficient force is computed for each drop based on its relative velocity to the gas and combined with a gravitational force. This is then used to compute the droplet's acceleration which is used to update its position. Newton's third law is used to transfer the droplet's force term to the gas through the ***f*** term in Equation 2.

Mass and energy transfer from droplets to the gas phase is computed semi-empirically using a quasi-equilibirum method. First a maximum evaporation rate is determined from a correlation that is a function of the local vapor mass fraction and the droplet equilibrium mass fraction, which is determined using the Clausius-Clapeyron equation. The correlation is given below where m_d is the droplet mass, r_d is the droplet radius, Y_d is the equilibrium vapor mass fraction, Y_g is the current vapor mass fraction in the cell containing the drop, r is the gas density, D is the gaseous diffusion for water vapor into air, and *Sh* is the droplet Sherwood number (Cheremisinoff, 1986).

$$\frac{dm_d}{dt} = -2\pi r_d \, Sh \, \rho D \ln\left(\frac{Y_g - 1}{Y_d - 1}\right) \tag{10}$$

This maximum rate is limited by the total energy available for evaporation given by the droplet enthalpy and any convective and radiative heat transfer between the droplet and its surroundings. Convective heat transfer is computed using a correlation for heat transfer to a drop. Radiative heat transfer is computed by modifying Equation 9 to account for the droplet absorption and scattering. This modified equation is shown below:

$$\hat{s} \cdot \nabla I(\vec{x},\hat{s}) = -[\kappa_d(\vec{x}) + \overline{\sigma}_d(\vec{x})] I(\vec{x},\hat{s}) + \kappa_d(\vec{x}) I_{b,d}(\vec{x},\hat{s}) + \frac{\overline{\sigma}_d(\vec{x})}{4\pi} U(\vec{x}), \quad (11)$$

where $\overline{\sigma}_d$ is a scattering term, U is the total integrated radiant intensity, and the source term and absorption coefficients in Equation 9 now include droplet components which are based on the mean droplet diameter and temperature within a grid cell. As with RADCAL computed absorption coefficient, the droplet scattering and absorption parameters are tabulated in advance as functions of the droplet mean diameter using the MieV code (Wiscombe, 1979).

EXAMPLES OF USE

Three examples of using FDS for industrial applications are given in the following three subsections. The examples are an evaluation of the effectiveness of a sprinkler system, a fire in a flammable liquid storage tank farm, and a flammable liquid spray fire.

Evaluation of Sprinkler System Effectiveness

One potential application for FDS and its droplet model is to reduce the need for large scale testing of sprinkler based fire protection systems (McGrattan, Hamins, and Forney, 2000). A typical industrial application these systems is protection of rack storage in warehouses. Sprinkler nozzles are tested in large-scale facilities in tests mimicking as-installed configurations and anticipated fire threats. If an end user of a sprinkler system needs to install the system in a facility or with a fuel load that is significantly different than tested configurations, additional testing may be required to ensure confidence in the system's performance.

As an example a large retail chain is planning to construct a series of warehouse style stores (such as a Home Depot ® or Bed Bath and Beyond ®). Full scale testing has been performed of various arrangements of commodity with an overhead sprinkler system to demonstrate that the planned system is capable of meeting its life safety and asset protection goals. After testing is completed the retail chain changes the store design and increases the height of the ceiling. It is obvious that an increase in ceiling height will result in an increased time to sprinkler activation and a larger fire size. To repeat the large scale testing to account for the new overhead height would be a significant expense in both time and money. Instead one could use FDS to predict fire conditions at the time of sprinkler activation and from that determine if the change in facility design will prevent the system from performing.

Such an activity is a three-part analysis. The first part of the analysis is to create a model of the original tests and demonstrate that FDS can reproduce the experimental results. The second part of the analysis is to modify the model for the new boundary conditions. Provided that changes to the model do not introduce significant new phenomena or flow changes, one can be confident in the accuracy of the predictions for the new prototype. The final part of the analysis is to compare the model results and determine if the new prototype still meets the desired performance characteristics within the anticipated model error. There will be predictive errors associated with any type of numerical simulation, and the first part of the analysis should provide an indication of the magnitude of the potential error. If the end result is success or failure of the design within the margin of error, then further large scale testing is not required.

As an example of such a calculation a mockup of a rack storage layout is shown in Figure 2 where group A plastic commodity is stored in a mix of rack storage and floor pallets. The overhead is 6.7 m and it contains an arrangement of sprinkler nozzles with 4 m spacing. A fire is postulated to start near the floor at the intersection of the rack storage. An FDS computation of this layout predicts a sprinkler activation time of 6.6 minutes with a peak fire size of 1 MW, shown in Figure 3. If the overhead height is increased to 8 m, FDS predicts an activation time of 9.2 minutes with a peak fire size of 1.5 MW. A comparison of the two scenarios is shown in Figure 4. Depending on the required success criteria, the suitability of the current sprinkler layout for the new store design could be determined without the need for further testing.

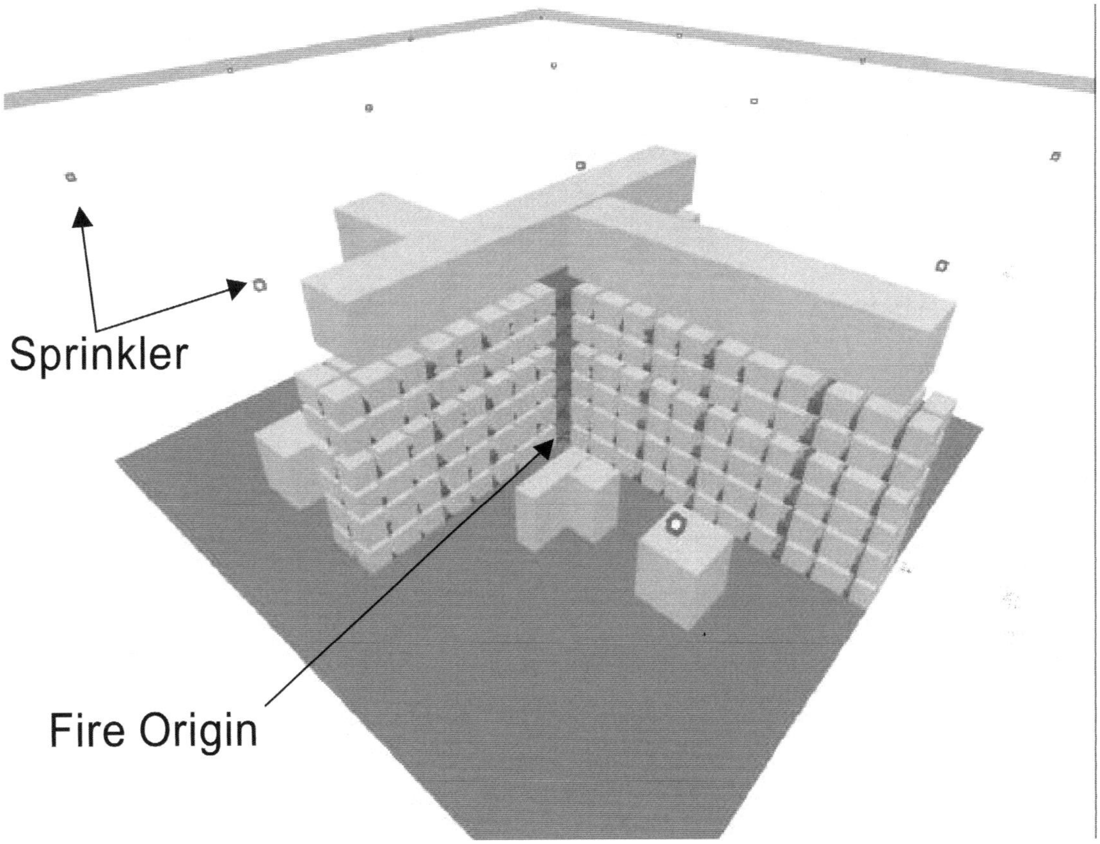

Figure 2: Example 1, Sprinkler System Effectiveness Layout

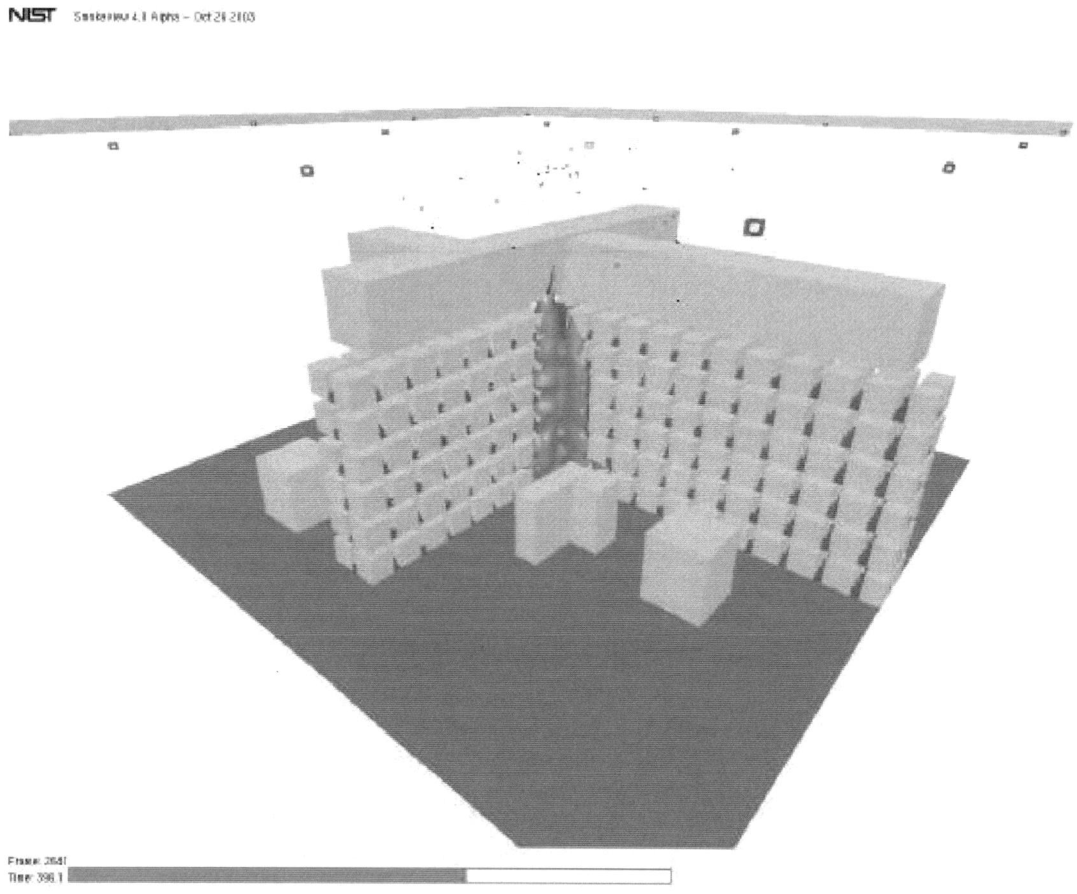

Figure 3: Example 1, 1st Sprinkler Activation for 6.7 m Overhead

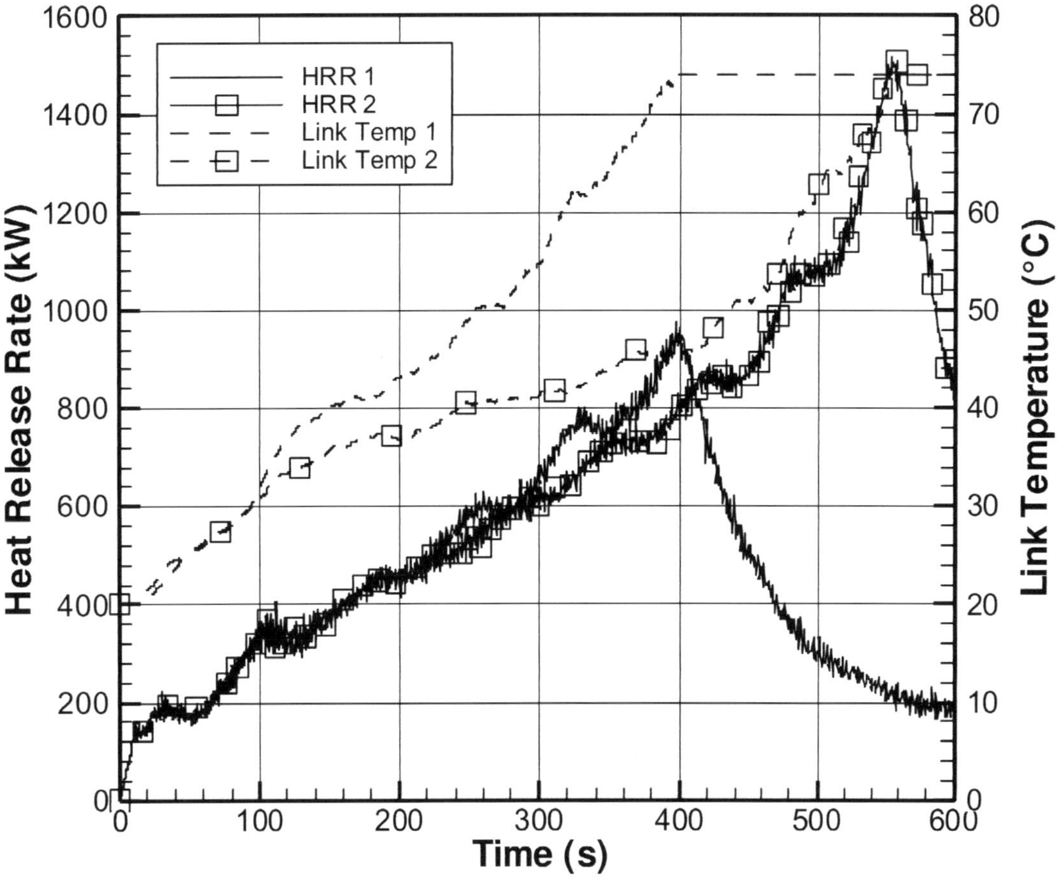

Figure 4: Example 1, 1st Sprinkler Activation for 6.7 m Overhead

Flammable Liquid Storage Tank Farm

This example predicts the impact of a fire in a flammable liquid storage tank upon other tanks located in a tank farm. FDS allows the user to specify wind profiles typical of large scale outdoor environments and also allows the specification of a an atmospheric lapse rate, accounts for atmospheric stratification. Combined these give the ability to predict the smoke plume dynamics from large outdoor fires (Baum and McGrattan, 1999, and Walton, et al., 1993). In Figure 5 below is a tank farm composed of 9 storage tanks with each tank 84 m in diameter and 27 m tall. Each tank is set within a containment trench that is 9 m deep. Two scenarios are run, both postulating the center tank of an edge is on fire. The first scenario is a quiescent atmosphere, and the second scenario imposes a wind profile of 5 m/s at the top of the burning tank. A series of calculations such as this, especially if the input files accounted for the local topography and wind conditions, could be used to optimize the tank farm layout to reduce the risk of secondary ignitions (by precluding the possibility of critical heat fluxes being reach) and could determine the downwind dispersion of combustion products (McGrattan, et al., 1994).

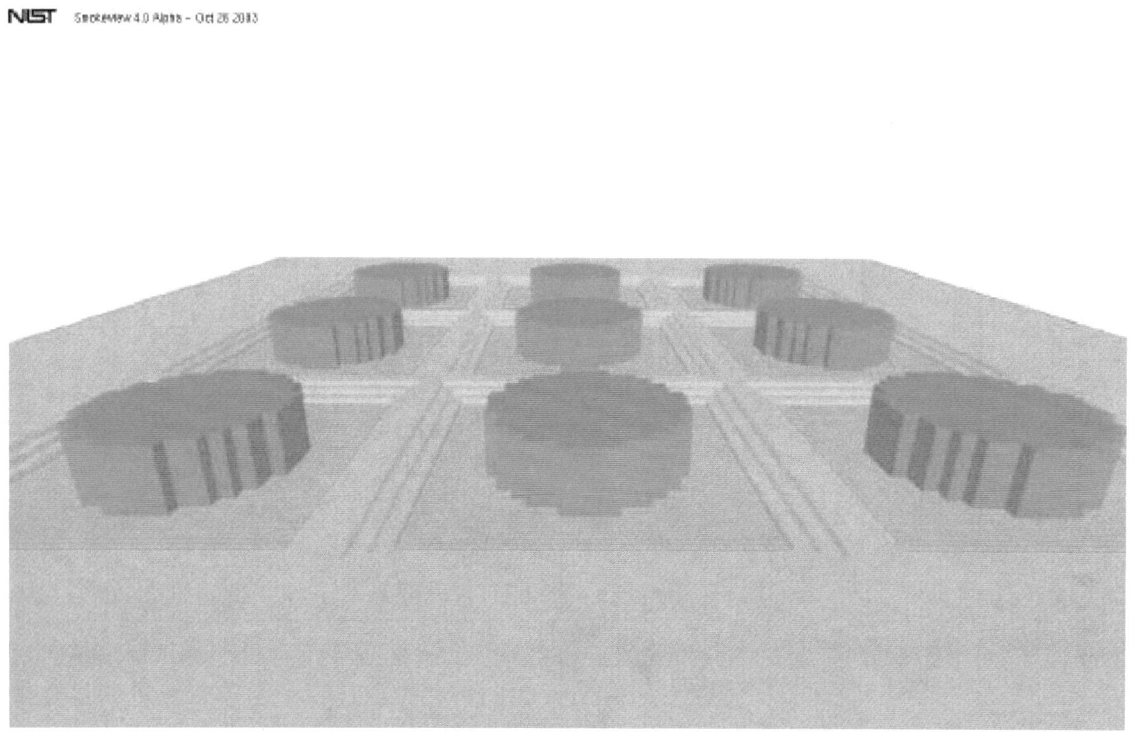

Figure 5: Example 2, Tank Farm Layout

Computational results for the first scenario are shown in Figure 6. Displayed are the fire and smoke along with the heat to the tanks. For this scenario we can see that the primary threat to the adjacent tank is along its side and the edge of the top nearest the fire. This is not unexpected as these are the regions with the largest view factors to the fire plume.

Figure 6: Example 2, Tank Farm Fire with no Wind

In the second scenario the wind blown fire plume is pushed towards the top of the neighboring tank. This results in increased view factors to the plume, which acts to increase the radiative heat flux by over 50 % to the neighboring tank. If the computational domain were to be extended, through the use of mutiblock grids, this calculation would also indicate the downwind extent and concentrations of the fire plume. With a multiblock approach the fire plume could be tracked over many kilometers with a minimal consumption of computational resources. Since the topography can be specified in the same manner as any other flow obstacle, the effects of terrain can be accounted for in a manner not possible with traditional windblown, Gaussian plume models (McGrattan, Baum, and Rehm, 1997). This information combined with toxicity information on the tank contents and their combustion products could aid in the creation of emergency plans and evacuation areas.

Figure 7: Example 2, Tank Farm Fire with 5 m/s Wind

Spray Fire

A typical fire hazard present in industrial facilities is pressurized flammable liquids. These can take the form of hydraulic fluid, solvents, or lubricating oils. A pipe leak can result in a combustible fuel spray, which can contain the concurrent hazards of both the fire and airborne droplets of a potentially toxic substance. The third scenario demonstrates the combination of the mixture fraction combustion model with the droplet model. For this scenario a single nozzle used for a heptane spray burner was attached to a stand and pointed in the horizontal direction. The nozzle was fed with 3.5 bar of heptane and the spray ignited. Figure 8 shows a still from a video of the test (McGrattan, 2003). The measured nozzle characteristics (droplet size, distribution, and geometry) where used to create a sprinkler description file for FDS, and the test was simulated.

Figure 8: Example 3, Spray Fire Experiment

Figure 9 shows the FDS computed fire and droplet distribution where the extents of the computational domain are approximate equal to the area of the floor covering seen in Figure 8. In this simulation the as the droplets evaporate they create fuel which burns according to the mixture fraction model. The radiative and convective feedback between the burning gas and the droplets determines the size of the fire and the amount of drops that fallout from the fire without burning (a phenomena which was observed, but not measured). This figure indicates that FDS is accurately capturing the dynamics of this spray fire as both the extent and shape of the flame seen in the experiment is reproduced.

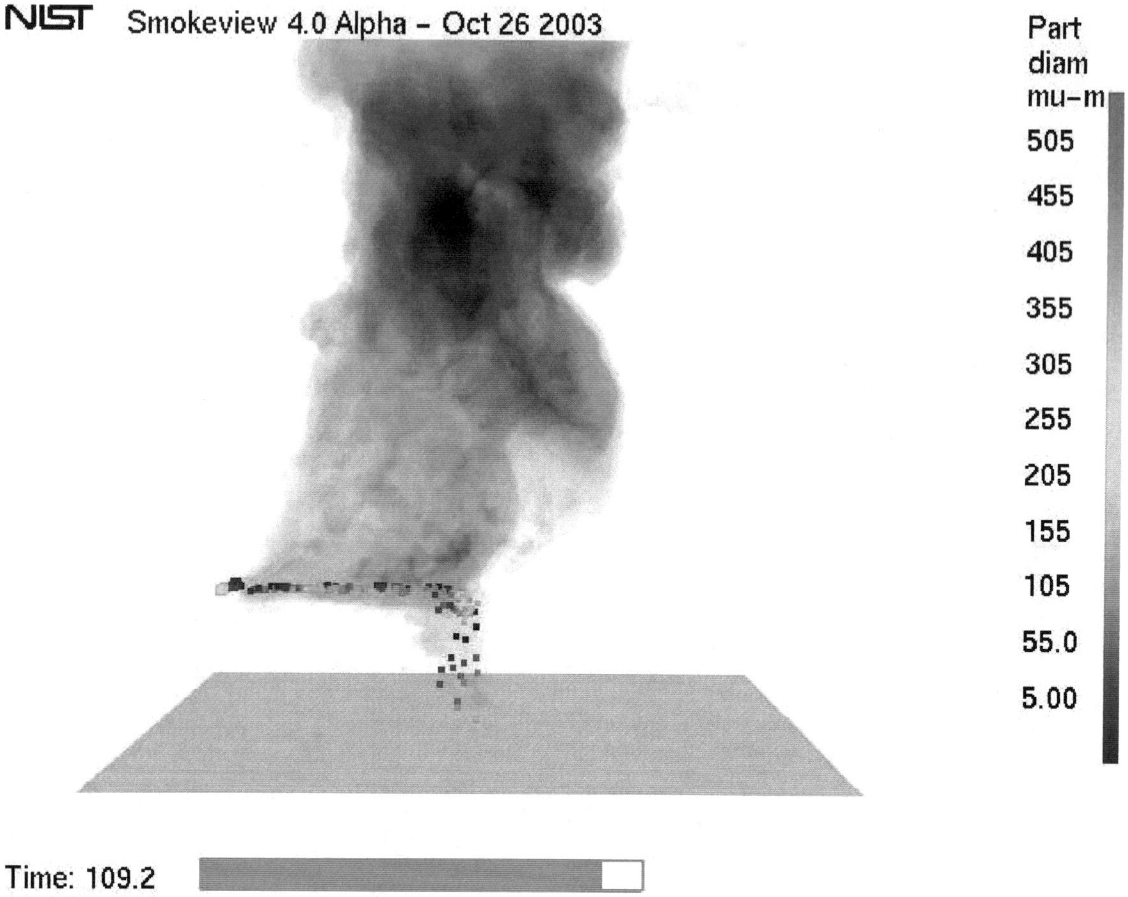

Figure 9: Example 3, Spray Fire Simulation

With this benchmark established, a second case was run where a wall was placed in that path of the fuel spray. In Figure 10, the results of this computation are shown including the predicted heat flux to the wall. Were this wall to represent a flammable object or a heat sensitive electric or mechanical device, the predicted heat flux could be used to evaluate the risk to that object from the spray fire. If the wall represented a shield for other equipment the flux on the far side of the wall could be used to determine the effectiveness of the wall as a passive safety feature.

Figure 10: Example 3, Spray Fire Against a Wall Simulation

CONCLUSIONS

There are many large scale/industrial incident scenarios that involve a fire or the release of chemical compounds into the atmosphere. Provided the fire consequences are not dependent upon pressure (e.g. explosions) or the chemical release can be posed as an ideal gas or droplet stream, FDS can be used to predict the consequences of the incident including thermal hazards and the time dependent, downwind spread of contaminants.

Each of these computations required under a man-day of time for setup, which includes creation of the input geometry and specification of any sprinkler specific inputs. The longest calculation, the retail storage and sprinkler scenario with over 1 million nodes, took approximately two days to execute on a 2 GHz machine. Contrast this with the man-months of time required to design, plan, construct, and execute a series of large-scale experiments, and the potential of CFD as a tool in design and consequence analysis can easily be seen.

As with any analysis tool the use of a CFD tool such as FDS for large-scale incidents has its limitations. For example accurate prediction of flame spread and accurate prediction of fire suppression effects (not activation due to smoke/temperature sensors) on a fire are difficult with very large grids as these phenomena a governed by processes at very small length and timescales. However, with an appropriate understanding of the limitations of the tool, there are many problems that FDS can aid in solving and understanding.

REFERENCES

Baum, H. and McGrattan, K., "Simulation of Oil Tank Fires," *Proceedings of Interflam '99*, Interscience Communications Ltd., Edinburgh, Scotland, 1999.

Chan, T., "Measurements of Water Density and Droplet Size Distribution of Selected ESFR Sprinklers," *Journal of Fire Protection Engineering*, Vol. 6(2), 1994, pp. 79 – 87.

Cherisinoff, N., *Encyclopedia of Fluid Mechanics, Volume 3: Gas-Liquid Flows.* Gulf Publishing Co., Houston, TX, 1986.

Floyd, J., McGrattan, K., Hostikka, S., and Baum, H., "CFD Fire Simulation Using Mixture Fraction Combustion and Finite Volume Radiative Heat Transfer," *Journal of Fire Protection Engineering*, Vol. 13 (1), 2003, pp. 11-36.

Grosshandler, W., "RADCAL: A Narrow-Band Model for Radiation Calculations in a Combustion Environment," National Institute of Standards and Technology, NIST Technical Note 1402, 1993.

Hostikka, S., McGrattan, K., Hamins, A., "Numerical Modeling of Pool Fires Using LES and Finite Volume Method for Radiation," *Proceedings of the 7th International Symposium*, Evans, D. editor, International Association of Fire Safety Science, Worcester, MA 2002.

Karlsson, B. and Quintiere, J., *Enclosure Fire Dynamics*, CRC Press, 2000.

Kuo, K, *Principles of Combustion*, New York, NY, John Wiley and Sons, 1986.

Mahalingam, S., et al., "Analysis and Numerical Simulation of a Nonpremixed Flame in a Corner," *Combustion and Flame*, Vol. 118, 1999, pp. 221-232.

McGrattan, K., et al., "Fire Dynamics Simulator (Version 4)- Technical Reference Guide," National Institute of Standards and Technology, NISTIR 6783, 2003a.

McGrattan, K., Forney, G., et al., "Fire Dynamics Simulator (Version 4) - User's Manual," National Institute of Standards and Technology, NISTIR 6784, 2003b.

McGrattan, K., Private communication, 2003.

McGrattan, K., Forney G., and Hamins, A., "Modeling of Sprinkler, Vent and Draft Curtain Interaction," *Proceedings of the 6th International Symposium*, Curtat, M. editor, International Association of Fire Safety Science, Poitiers, France, 2000.

McGrattan, K., Baum, H., and Rehm, R., "Smoke Plumes from Large Fires," *Proceedings of the UJNR 13th Joint Panel Meeting, Vol. 2*, NISTIR 6030, US/Japan Government Cooperative Programs on Natural Resources, 1997.

McGrattan, L., Baum, H., Walton, W., and Trelles, J., "Smoke Plume Trajectory From in Situ Burning of Crude Oil in Alaska: Field Experiments and Modeling of Complex Terrain," National Institute of Standards and Technology, NISTIR 5958, 1994.

Prassad, K., Patnaik, G., and Kailasanath, K., "Advanced Simulation Tool for Improved Damage Assessment 1) A Multiblock Technique for Simulating Fire and Smoke Transport in Large Complex Enclosures," US Naval Research Laboratory, NRL/MR/6140—00—8428, 2000.

Raithby, G. and Chui, E. "A Finite-Volume Method for Predicting Radiant Heat Transfer in Enclosures with Participating Media," Journal of Heat Transfer, Vol. 112, 1990, pp. 415-423.

Rehm, R. And Baum, H., "The Equations of Motion for Thermally Driven, Buoyant Flows" *Journal of Research of the NBS*, Vol. 83, 1978, pp. 297-308.

Siegel, R. and Howell, J., *Thermal Radiation Heat Transfer 3rd Edition,* Hemisphere Publishing Corp., Philadelphia, PA, 1992.

Walton, W., Evans, D. McGrattan, K., et al., "In Situ Burning of Oil Spills: Mesoscale Experiments and Analysis," *16th Artic and Marine Oil Spill Program Technical Seminar, Vol. 2*, Environment Canada, Edmonton, Alberta, Canada, 1993.

Wiscombe, W., "Advances in Technique and Fast, Vector-Speed Computer Codes," National Center for Atmospheric Research (NCAR), 1979.

A complete description of an advanced evaporation model for accidentally released hazardous liquids from land.
Part IA – Theoretical background.

Dr. F. Kootstra, Ing. B.F.P. van het Veld and Ir. G. Opschoor

Netherlands Organisation for Applied Scientific Research,
TNO Environment, Energy and Proces Innovation,
Department of Industrial Safety,
Laan van Westenenk 501,
P.O. Box 342,
7300 AH Apeldoorn,
The Netherlands,
Phone: +31 55 5493208,
Fax: +31 55 5493390,
Web: http://www.tno.nl,
E-mail: F.Kootstra@mep.tno.nl

Prepared for presentation at the 38th Annual Loss Prevention Symposium, AIChE 2004 Spring National Meeting, New Orleans, Louisiana, April 26 – 29, 2004.

ABSTRACT

In two papers the theoretical background of a completely integrated evaporation model will be described for all kinds of possible pools. In the first paper the evaporation of pools from land (Part 1A) and in the second paper the evaporation of non-soluble, floating pools from water (Part 1B) will be described for as well boiling as non-boiling liquids. The outflow of the evaporating liquid on land or water can be continuous as well as instantaneous. The spreading of the liquid is considered until a minimum layer thickness has been reached, or until the borders of the bund have been reached on land, or the confinement of the water surface (rivers). The mass- and heat balance need to be simultaneously solved for the evaluation of the temperature and height of the pool in time after (or during) the release. For the evaporation from water the formation of ice is considered in case of a confined water surface (rivers), and the different boiling regimes in case of an open water surface (sea). Within the evaporation model the following environmental- and weather conditions are taken into account: several subsoils for evaporation from land, humidity, wind speed and cloud cover in the air, the influence of position in the world and date (day:month:year).

INTRODUCTION

All over the world large amounts of chemicals are being transported, produced and stored for use in e.g. chemical plants. Sometimes during the transportation over land or water, or during the storage of these chemicals at the site, accidents happen. Consequently the chemical is being released under the local storage conditions or the transportation conditions. Such an accidental release can causes risk to the people and the environment depending upon e.g. the toxicity or the flammability of the chemical. Due to legislation the manufacturers and transporters of these chemicals are obliged to certain criteria to minimise the risk associated with the production or transportation of these chemicals. For being able to calculate the risk of such an accidental release of the transported or stored chemical, one needs to determine the physical effect of the accidental release of such a chemical, cause the risk associated with a certain event is defined as the (physical) effect of the event times the frequency of the occurance of the event. In case the chemical is being released under such conditions that it is in it's liquid form, for such cases the following two papers give a complete description of the theoretical background to determine the physical effects associated with the accidental release of the consequent evaporation process of the time-dependent evaporating, spreading liquid pool. As can be seen from the flow chart in Figure 1, a seperation should be made between the physical description of the evaporation process of liquids from land and water.

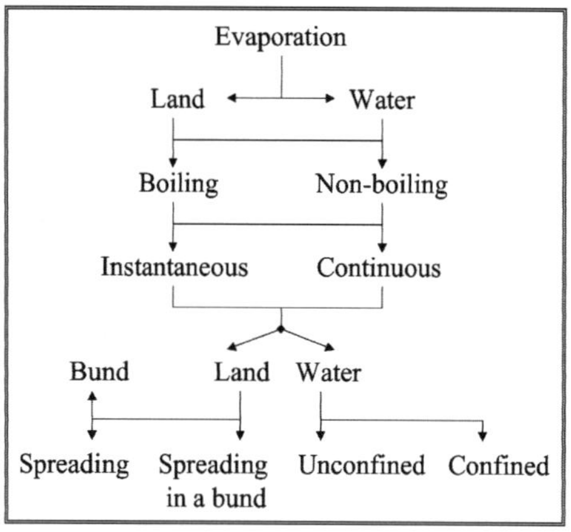

Figure 1: Flow chart for the evaporation process

Therefore, the first paper describes the evaporation process for pools from land (Part 1A) and the second paper describes the evaporation process for non-soluble, floating pools from water (Part 1B). Next to the seperation of the evaporation process from land or water, the following distinctions should also be made (as can be seen from the flow chart in Figure 1), between a boiling or a non-boiling pool, between an instantaneous or continuous release of the liquid and, for evaporatoin from land, between a spreading pool or a spreading pool until a certain bund and, for evaporation from water, between a confined or unconfined water surface. Working out of all the above distinctions results in the following eighteen (to be considered) cases in the evaporation process, as shown in Table 1.

	Non-boiling liquid		Boiling liquid	
Outflow upon land	*Instantaneous*	*Continuous*	*Instantaneous*	*Continuous*
Bunds	(1)	(*)	(6)	(**)
Spreading	(2)	(4)	(7)	(9)
Spreading in bunds	(3)	(5)	(8)	(10)
Outflow upon water	*Instantaneous*	*Continuous*	*Instantaneous*	*Continuous*
Unconfined (open water)	(15)	(17)	(11)	(13)
Confined (rivers/canals)	(16)	(18)	(12)	(14)

(*) Same as case (5)

(**) Same as case (10)

Table 1: The eighteen separate cases for the evaporation process from land and water.

The organisation and lay-out for the description of the theoretical background of a completely integrated evaporation model in the two papers is the following: in the first paper the background is given for the evaporation from land (Part IA), and in the second paper from water (Part IB). Thus, as can be seen from Table 1, the first ten cases (1-10) are treated in the first paper, and the last eight cases (11-18) are dealt with in the second paper. The practical implications of the first two papers will be described in another third paper [Kootstra: 2004], in which the use and application of the advanced evaporation models will be shown and demonstrated in several practical example calculations.

A list of the symbols can be found at the end of the paper in which all symbols throughout the equations in the paper are given together with a short description, the unit and the numbers of the equations in which the symbol is used.

EVAPORATION FROM LAND FOR NON-BOILING LIQUIDS

Dangerous chemicals/liquids which are accidentally released on land are considered to be non-boiling when they have a normal boiling point (T_b) (under ambient conditions) that is higher than the temperature of the subsoil (T_s) (on which the chemical is spilled) and/or the ambient temperature (T_a).

The evaporation of a non-boiling liquid pool on land depends mainly on the rate at which the vapour above the pool is removed by the wind. Therefore, in general for non-boiling liquid pools, the mass transfer by diffusion will be the limiting factor in the evaporation proces for a non-boiling liquid pool on land.

Case 1 : Instantaneous release in a bund

Evaporation mass flow rate

According to [Kawamura: 1987] the vaporisation mass flow rate $q_v^{'}(t)$ for a non-boiling pool on land can be estimated by the following equation:

$$q_v^{'}(t) = q_v^{''}(t) \cdot A_p \tag{1}$$

Note: A_p is not time-dependent because of an instantaneous release in a bund

In equation (1), $q_v^{''}(t)$ is the vaporisation mass flux which can be calcultated according to:

$$q_v^{''}(t) = \frac{k_m \cdot P_v\left(T_p(t)\right) \cdot \mu_i}{R \cdot T_p(t)} \tag{2}$$

In equation (2), k_m is the mass transfer coefficient

$$k_m = 0.004786 \cdot \left(u_{w,10}\right)^{0.78} \cdot d_p^{-0.11} \cdot Sc^{-0.67} \tag{3}$$

and Sc is the Schmidt number

$$Sc = \upsilon_a / D_a \approx 0.8 \tag{4}$$

Heat balance

In general the heat-balance for an evaporating pool reads [Yellow Book: 1997]:

$$\frac{d\left(h_p(t)\cdot \rho_l \cdot C_{p,l} \cdot T_p(t)\right)}{dt} = \Sigma H(t) - q_v''(t)\cdot L_v\left(T_p(t)\right) \tag{5}$$

If we assume change in the specific heat, the liquid density and the height of the pool to be small compared to the change in the pool temperature in time, the heat-balance results in:

$$h_p(t)\cdot \rho_l \cdot C_{p,l} \cdot \frac{dT_p(t)}{dt} = \Sigma H(t) - q_v''(t)\cdot L_v\left(T_p(t)\right) \tag{6}$$

And thus the change in the pool temperature can be evaluated with:

$$dT_p(t) = \frac{\Sigma H(t) - q_v''(t)\cdot L_v\left(T_p(t)\right)}{h_p(t)\cdot \rho_l \cdot C_{p,l}} \cdot dt \tag{7}$$

Therefore the pool temperature changes in a time-step dt according to:

$$T_p(t+dt) = T_p(t) + dT_p(t) \tag{8}$$

In equation (5), the sum of the heat sources $\Sigma H(t)$ comprises:

$$\Sigma H(t) = H_{subsoil}(t) + H_{air}(t) + H_{solar-radiation} + H_{(long-wave)-radiation} \tag{9}$$

in which:

1. $H_{subsoil}(t)$: the heat flux as conducted from the subsoil, which is given by:

$$H_{subsoil}(t) = \frac{\lambda_s \cdot (T_{subsoil} - T_p(t))}{\sqrt{a_s \cdot \pi \cdot t}} \tag{10}$$

The numerical values for the thermal conductivity (λ_s) and thermal diffusivity (a_s) from the subsoil is given in Table 1 for various subsoils.

Subsoil	λ_s [J/msK]	$a_s \cdot 10^{-7}$ [m^2/s]
Isolation concrete	0.207	2.5
Light concrete	0.418	2.5
Heavy concrete	1.3	5.9
Clinkers	0.7	4.2
Average subsoil 8% moist	0.9	4.3
Dry sandy subsoil	0.3	2.0
Wet sand 8% moist/clay	0.6	3.3
Wood	0.2	1.6
Gravel	2.5	11
Carbon steel	46	128

Table 1: Heat conduction properties for various subsoils.

2. $H_{air}(t)$: the heat flux as conducted by air, which is given by:

$$H_{air}(t) = k_{h,a} \cdot (T_{air} - T_p(t)) \tag{11}$$

The heat transfer coefficient $k_{h,a}$ follows from:

$$k_{h,a} = \frac{Nu \cdot \lambda_a}{d_p} = \frac{(0.037 \cdot \mathrm{Pr}_a^{1/3} \cdot \mathrm{Re}^{4/5}) \cdot \lambda_a}{d_p} \tag{12}$$

with:

$$\mathrm{Pr}_a = \upsilon_a / D_a \approx 0.786 \tag{13}$$

and:

$$\mathrm{Re} = \frac{\rho_a \cdot u_{w,10} \cdot d}{\eta_a} \tag{14}$$

3. $H_{solar-radiation}$: the solar radiation, which is given by:

$$H_{solar-radiation} = (c_1 \cdot \sin\chi - c_2) \cdot (1 - 0.75 \cdot N^{3.4}) \tag{15}$$

The solar elevation χ (in rad.) is calculated by:

$$\chi = \arcsin\left(\sin(SD) \cdot \sin\left(\frac{\pi}{180}\varphi\right) + \cos(SD) \cdot \cos\left(\frac{\pi}{180}\varphi\right) \cdot \cos\alpha \right) \tag{16}$$

in which SD (in rad.) is the solar declination according to:

$$SD = \arcsin\left(0.398 \cdot \sin(SL)\right) \tag{17}$$

and α is the smallest angle through which the earth must turn to bring the meridian of the given location directly under the sun, it is calculated according to:

$$\alpha = \psi_w \cdot \frac{\pi}{180} + 0.043\sin(2 \cdot SL) - 0.033\sin\left(\frac{\pi}{180} \cdot J_n\right) + \pi \cdot \frac{t_u}{(12-1)} \tag{18}$$

In the equation (17) and (18), SL (in rad.) is solar longitude according to:

$$SL = 4.871 + \frac{\pi}{180} \cdot J_n + 0.033\sin\left(\frac{\pi}{180} \cdot J_n\right) \tag{19}$$

and J_n the Julian day number, which is defined as the number of days elapsed since the first of January. It is calculated as follow: from a date in the form *DD-MM-YYYY*, it is first examined if *YYYY* is a leap year. And if it is a non-leap year, and up to Feb. in all years, J_n is found by:

$$J_n = DD + X_{MM-1} \tag{20}$$

and in leap years after Feb., J_n is found by:

$$J_n = DD + X_{MM-1} + 1 \tag{21}$$

Thereby using Table 2.

MM	1	2	3	4	5	6	7	8	9	10	11	12
	Jan	Feb	Mar	Apr	May	Jun	Jul	Aug	Sep	Oct	Nov	Dec
X	031	059	090	120	151	181	212	243	273	304	334	365

Table 2:

4. $H_{(long-wave)-radiation}$: the heat flux as

$$H_{(long-wave)-radiation} = \frac{(1-e)\cdot H_{solar-radiation} + c_3 \cdot T_{air}^{\ 6} - \sigma \cdot T_{air}^{\ 4} + c_4 \cdot N}{1.12} \tag{22}$$

Values for Albedo e for various subsoils can be found in Table 3.

Subsoil	Albedo e
Soils (Dark, wet – Light, dry)	0.05 – 0.40
Desert	0.20 – 0.45
Grass (Long [1.0 m. – Short [0.02 m.]])	0.16 – 0.26
Agricultural crops	0.18 – 0.25
Orchards	0.15 – 0.20
Forests	
Decious (bare-leaved)	0.15 – 0.20
Coniferous	0.05 – 0.15
Water	0.03 – 1.00
Snow (Old – Fresh)	0.40 – 0.95
Ice	
Sea	0.30 – 0.45
Glacier	0.20 – 0.40
Tarmac, Asphalt	0.05
Buildings	0.20 – 0.45

Table 3: Albedo for various subsoils from [Byrne: 1992]

Mass balance

In general the mass-balance for an evaporating pool reads:

$$\frac{dM_p(t)}{dt} = -q_v^{'}(t) = -q_v^{''}(t)\cdot A_p \tag{23}$$

The surface area of the pool is constant due to an instantaneous release in a bund, which gives us for the mass-balance ($M_p(t) = \rho_l \cdot V_p(t)$):

$$\rho_l \frac{dV_p(t)}{dt} = -q_v^{"}(t)\cdot A_p \rightarrow$$
$$\rho_l \cdot A_p \frac{dh_p(t)}{dt} = -q_v^{"}(t)\cdot A_p \rightarrow \frac{dh_p(t)}{dt} = -q_v^{"}(t)/\rho_l \tag{24}$$

Therefore the height of the pool changes in a time-step dt according to:

$$h_p(t+dt) = h_p(t) + dh_p(t) \tag{25}$$

Evaporation process in time

Starting from an initial pool at it's initial temperature $T_p(t_0)$, and a pool surface area (A_p) equal to the bunded area which contains an amount of liquid equal to the amount which was instantaneous released, $M_p(t_0)$. The evaporation of the pool is followed in time by updating (evaluating) after specific time steps dt, the temperature, $T_p(t)$, and height, $h_p(t)$, of the pool. The evaporation process of the non-boiling liquid, which was instantaneously released in a bunded area, is followed until the pool has totally evaporated, or before (at a specific time, t_{study}) if requested by the user.

Case 2 : Instantaneous release, spreading

Surface area of the pool

Compared to the situation in case 1: Instantaneous release in a bund, the difference in this case comes from the fact that the surface area A_p of the pool is not a constant value, as determined by the surface area of the bund, anymore.

In the case of a spreading pool, the surface area of the pool becomes time-dependent, and is denoted as $A_p(t)$:

$$A_p(t) = \pi \cdot r_p^2(t) \tag{26}$$

and the volume of the pool $V_p(t)$ is calculated by:

$$V_p(t) = A_p(t)\cdot h_p(t) = \pi \cdot r_p^2(t)\cdot h_p(t) \tag{27}$$

The volume of the pool at the time of instantaneous release, $t=0$, can be evaluated according to:

$$V_p(0) = V_{p,0} = M_{p,0} / \rho_l \tag{28}$$

Spreading of the pool

The spreading of the pool for an instantaneous released liquid on flat, non-permeable subsoil can be evaluated by means of the following equation:

$$\frac{dr_p(t)}{dt} = \sqrt{C'' \cdot g \cdot \left(h_{p,0} - h_p(t)\right)} \tag{29}$$

The volume of the spreading pool at a specific time t can be approximated by equation (27) (hereby neglecting the evaporation of the pool for the time interval $t \in [0,t]$). By substituting $h_p(t)$ from equation (27) in (29), and after some simple algebra we find for the radius of the spreading pool at a specific time t:

$$r_p(t) = r_{p,0} + \sqrt{C'' \cdot g \cdot h_{p,0}} \cdot t \tag{30}$$

The linear dependence of the pool spreading with respect to time is confirmed by experimental results in reference [Jacobs: 1985] and theoretically derived in reference [Verhagen: 1985] from the Daburst principle. The spreading of pool for times $t < t_e$ is evaluated by equation (30). The time t_e is the spreading time of the pool until the minimum layer thickness $h_{p,\min}$ has been reached.

Derivation of the spreading time t_e (minimum layer thickness)

The evaporated quantity $q_v(t)$ of the pool at a specific time t can be evaluated by the following relation:

$$q_v(t) = \int_0^t q_v''(t) \cdot A_p(t)\,dt = \int_0^t q_v''(t) \cdot \pi \cdot r_p^2(t)\,dt \tag{31}$$

If we insert for the evaporation flux $q_v''(t)$ the following (approximate) equation

$$q_v''(t) = \frac{\Sigma H(t)}{L_v(T_p(t))} \approx \frac{H_{subsoil}(t)}{L_v(T_p(t))} = \frac{\lambda_s \cdot \left(T_{subsoil} - T_p(t)\right)}{L_v(T_p(t)) \cdot \sqrt{a_s \cdot \pi \cdot t}} = \frac{e}{\sqrt{t}} \tag{32}$$

with:

$$e = \frac{\lambda_s \cdot \left(T_{subsoil} - T_p\left(t\right)\right)}{L_v\left(T_p\left(t\right)\right) \cdot \sqrt{a_s \cdot \pi}} \tag{33}$$

In equation (32), it is assumed that the evaporation rate of the pool is in the initial stages of the evaporation process mainly controlled by the rate at which heat can get into the pool from the subsoil, which is only valid in case of a 'boiling' evaporating pool. In principle the equations (32) and (33) can be extended for the proper use in case of a non-boiling liquid by not making the assumption $\Sigma H\left(t\right) \square H_{subsoil}\left(t\right)$. Equation (33) reads in this case:

$$e = \frac{\lambda_s \cdot \left(T_{subsoil} - T_p\left(t\right)\right)}{L_v\left(T_p\left(t\right)\right) \cdot \sqrt{a_s \cdot \pi}} + \frac{\left(H_{air} + H_{solar-radiation} + H_{(long-wave)-radiation}\right)}{L_v\left(T_p\left(t\right)\right)} \tag{34}$$

Consequently we find for equation (31):

$$q_v\left(t\right) = e \cdot \pi \cdot \int_0^t \frac{r_p^2\left(t\right)}{\sqrt{t}} dt = \frac{2}{5} \cdot e \cdot \pi \cdot C'' \cdot g \cdot h_{p,0} \cdot t^{5/2} + 2 \cdot e \cdot \pi \cdot r_{p,0}^2 \cdot \sqrt{t} \tag{35}$$

The evaporation rate $q_v^{'}\left(t\right)$ of the pool becomes equal to:

$$q_v^{'}\left(t\right) = e \cdot \pi \cdot C'' \cdot g \cdot h_{p,0} \cdot t^{3/2} + \frac{e \cdot \pi \cdot r_{p,0}^2}{\sqrt{t}} \tag{36}$$

If we now let the pool spread until a minimum thickness layer $h_{p,\min} = h_p\left(t_e\right)$ is reached, the time t_e therefore required can be calculated from the mass-balance:

evaporated quantity at t_e + mass in the pool at t_e = mass of the instantaneous released liquid at t_0 (37)

which gives:

$$q_v\left(t_e\right) + \rho_l \cdot \pi \cdot r_p^2\left(t_e\right) \cdot h_{p,\min} = \rho_l \cdot V_{p,0} \tag{38}$$

equation (38) should be solved by iteration:

$$\frac{2C''e\pi g h_{p,0}}{5\rho_l} t_e^{5/2} + \frac{2e\pi r_{p,0}^2}{\rho_l} \sqrt{t_e} + C''\pi g h_{p,0} h_{p,\min} t_e^2 = V_{p,0} - \pi r_{p,0}^2 h_{p,\min} \tag{39}$$

Assuming that, after reaching the minimum layer thickness $h_{p,\min}$ at time t_e, the layer thickness to be constant means that after time t_e the pool will shrink.

The derivation of the radius of the shrinking pool for times $t > t_e$ can be evaluated by the following relation based on the mass-balance:

$$\rho_l \cdot \frac{d}{dt}\left(V_p(t)\right) = \rho_l \cdot \frac{d}{dt}\left(A_p(t) \cdot h_{p,\min}\right) = \rho_l \cdot \frac{d}{dt}\left(\pi \cdot r_p^2(t) \cdot h_{p,\min}\right) = -q_v^{'}(t) \qquad (40)$$

Assuming the following relation for the evaporation rate $q_v^{'}(t)$:

$$q_v^{'}(t) = \frac{e^* \cdot \pi \cdot r_p^2(t)}{\sqrt{t}} \qquad (41)$$

in which e^* is chosen in such a way that the function $q_v^{'}(t)$ is continuous at time t_e, by:

$$e^* = \frac{q_v^{'}(t_e) \cdot \sqrt{t}}{\pi \cdot r_p^2(t_e)} \qquad (42)$$

We find by substitution of equation (41) in (40):

$$\frac{d}{dt}\left(\pi \cdot r_p^2(t) \cdot h_{p,\min}\right) = -\frac{e^* \cdot \pi \cdot r_p^2(t)}{\rho_l \cdot \sqrt{t}} \qquad (43)$$

with the initial condition that at time t_e the radius of the pool is $r_{p,e} = r_p(t_e)$.

The solution of equation (43) gives:

$$r_p^2(t) = r_{p,e}^2 \cdot e^{\left[-\frac{2 \cdot e^*}{\rho_l \cdot h_{p,\min}}\left(\sqrt{t} - \sqrt{t_e}\right)\right]} \qquad (44)$$

By combining equation (41) and (44) we find for the evaporation rate $q_v^{'}(t)$:

$$q_v^{'}(t) = \frac{e^* \cdot \pi \cdot r_{p,e}^2 \cdot e^{\left[-\frac{2 \cdot e^*}{\rho_l \cdot h_{p,\min}}\left(\sqrt{t} - \sqrt{t_e}\right)\right]}}{\sqrt{t}} \qquad (45)$$

And for the evaporated quantity $q_v(t)$ after the time t_e the following applies:

$$q_v(t) = \int_{t_e}^{t} q_v^{'}(t) \cdot dt = \int_{t_e}^{t} \frac{e^* \cdot \pi \cdot r_{p,e}^2 \cdot e^{\left[-\frac{2 \cdot e^*}{\rho_l \cdot h_{p,\min}}\left(\sqrt{t} - \sqrt{t_e}\right)\right]}}{\sqrt{t}} \cdot dt \qquad (46)$$

which gives after integration:

$$q_v(t) = \pi \cdot r_{p,e}^2 \cdot \rho_l \cdot h_{p,\min} \cdot \left(1 - e^{\left[-\frac{2 \cdot e^*}{\rho_l \cdot h_{p,\min}}\left(\sqrt{t} - \sqrt{t_e}\right)\right]}\right) \qquad (47)$$

Mass balance

The mass-balance for an evaporating pool reads in case of a spreading pool:

$$\frac{dM_p(t)}{dt} = -q_v^{'}(t) = -q_v^{''}(t)\cdot A_p(t) \tag{48}$$

which gives for the change in the volume of the spreading pool, $dV_p(t)$:

$$\frac{dV_p(t)}{dt} = \frac{-q_v^{''}(t)\cdot \pi \cdot r_p^2(t)}{\rho_l} \tag{49}$$

Therefore the volume of the pool changes in a time-step dt according to:

$$V_p(t+dt) = V_p(t) + dV_p(t) \tag{50}$$

Compared to case 1: Non-boiling, instantaneous released liquid in a bund, the difference in equation (49) is found in the evaluation for the vaporisation mass flux a specific time *t*. The vaporisation mass flux $q_v^{''}(t)$ follows from:

$$q_v^{''}(t) = \frac{k_m(t)\cdot P_v(T_p(t))\cdot \mu_i}{R\cdot T_p(t)} \tag{51}$$

In which the mass transfer coefficient $k_m(t)$ is now time-dependent, because the radius of the pool is time-dependent in case of a spreading pool, $r_p(t)$. The mass transfer coefficient is now found according to:

$$k_m(t) = 0.004786\cdot(u_{w,10})^{0.78}\cdot(2\cdot r_p(t))^{-0.11}\cdot Sc^{-0.67} \tag{52}$$

Heat balance

The heat-balance for an evaporating pool reads in case of a spreading pool:

$$\frac{d\left(h_p(t)\cdot \rho_l \cdot C_{p,l}\cdot T_p(t)\right)}{dt} = \Sigma H(t) - q_v^{''}(t)\cdot L_v(T_p(t)) \tag{53}$$

Assuming that the change in the specific heat, the liquid density and the height of the (spreading) pool to be small compared to the change in the pool temperature in time, the change in the pool temperature is given by:

$$dT_p(t) = \frac{\Sigma H(t) - q_v^{''}(t)\cdot L_v(T_p(t))}{h_p(t)\cdot \rho_l \cdot C_{p,l}}\cdot dt \tag{54}$$

And the pool temperature changes in a time-step δt according to:

$$T_p(t+dt) = T_p(t) + dT_p(t) \tag{55}$$

Compared to the case of a non-boiling liquid within a bund, which is instantaneously released, the differences in equation (54) are found in:

(a) the sum of the heat sources,

(b) the vaporisation mass flux, and

(c) the height of the pool a specific time t.

(a) The sum of the heat sources $\Sigma H(t)$ comprises:

$$\Sigma H(t) = H_{subsoil}(t) + H_{air}(t) + H_{solar-radiation} + H_{(long-wave)-radiation} \tag{56}$$

In which the heat flux as conducted by air, $H_{air}(t)$, is given by:

$$H_{air}(t) = k_{h,a}(t) \cdot (T_{air} - T_p(t)) \tag{57}$$

The heat transfer coefficient $k_{h,a}(t)$ is now also time-dependent and follows from:

$$k_{h,a}(t) = \frac{Nu \cdot \lambda_a}{2 \cdot r_p(t)} = \frac{\left[0.037 \cdot \mathrm{Pr}_a^{1/3} \cdot (\mathrm{Re}(t))^{4/5}\right] \cdot \lambda_a}{2 \cdot r_p(t)} \tag{58}$$

with:

$$\mathrm{Re}(t) = \frac{\rho_a \cdot u_{w,10} \cdot 2 \cdot r_p(t)}{\eta_a} \tag{59}$$

(b) The vaporisation mass flux $q_v''(t)$ changes as shown above, according to the equations (51) and (52).

(c) The height of the pool $h_p(t)$ follows from:

$$h_p(t) = \frac{V_p(t)}{A_p(t)} = \frac{V_p(t)}{\pi \cdot r_p^2(t)} \tag{60}$$

In which the volume $V_p(t)$ and radius $r_p(t)$ of the pool at a specific time t, is evaluated to the equations (50) and (30)/(44).

Evaporation process in time

Starting from an initial pool at it's initial temperature $T_p(t_0)$ which contains an amount of liquid equal to the amount which was instantaneous released, $M_p(t_0) = M_{p,0}$, the dimensions of the pool like: surface area; $A_p(t)$ and radius; $r_p(t)$ can be calculated after making an assumtion concerning the dimensions of the pool at the time of instantaneous release, $t=0$. For the volume of the pool at $t=0$, we get: $V_{p,0} = A_{p,0} \cdot h_{p,0} = \pi \cdot r_{p,0}^2 \cdot h_{p,0}$.

E.g. it is possible to make one of the following choices:

- $A_{p,0}$ and $h_{p,0}$ dependent upon the dimensions of the vessel.
- choose $h_{p,0}$: e.g. $h_{p,0} = 10 \cdot h_{p,\min} = 0.05$ m.

Example:

Variable	**Value**
Chemical name:	Acrylonitrile
$M_{p,0}$ (mass released at $t = 0$):	50.000 kg
T_b (boiling point chemical):	350.5 K
ρ_l (liquid density chemical) at 293 K:	807 kg/m^3
$h_{p,0}$ (height pool at $t = 0$):	0.05 m [fixed]
$V_{p,0}$ (volume pool at $t = 0$):	62 m^3
$A_{p,0}$ (surface area pool at $t = 0$):	1239 m^2
$r_{p,0}$ (radius pool at $t = 0$):	19.86 m

The evaporation of the pool is followed in time by updating (evaluating) after specific time steps dt, the temperature, $T_p(t)$, and volume, $V_p(t)$, of the pool. The evaporation process of the non-boiling, spreading liquid, which was instantaneously released, is followed until the pool has totally evaporated, or before (at a specific time) if requested by the user.

Case 3 : Instantaneous release, spreading in bunds

Spreading of the pool

Compared to the situation in case 2: Instantaneous release, spreading, the difference in this case comes from the fact that the radius of the pool is limited to a maximum value, $r_{p,\max}$ as determined by the surface of the bunded area, A_{bund} by:

$$r_{p,\max} = \sqrt{\frac{A_{bund}}{\pi}} \tag{61}$$

Therefore during the time of spreading of the pool from $t = 0$ till $t = t_e$, and as long as the pool has not totally evaporated yet. I.e. for $t \in [0, t_e]$ and $V_p(t) > 0$, the radius of the pool is calcalated as:

$$r_p(t) = \text{Min}\left[(r_{p,0} + \sqrt{C^{"} \cdot g \cdot h_{p,0}} \cdot t), (r_{p,\max}) \right] \tag{62}$$

For times after the spreading time of the pool $t > t_e$, but still within the time of study $t < t_{study}$, and as long as the liquid in the pool has not totally evaporated yet. I.e. for $t > t_e$ and $t < t_{study}$ and $V_p(t) > 0$, the radius of the shrinking pool is calculated according to:

$$r_p(t) = \text{Min}\left[\left(\sqrt{\frac{V_p(t)}{\pi \cdot h_p(t)}} \right), (r_{p,0} + \sqrt{C^{"} \cdot g \cdot h_{p,0}} \cdot t), (r_{p,\max}) \right] \tag{63}$$

In equation (63), the height of the pool for times $t > t_e$ is calculated according to:

$$h_p(t) = \frac{V_p(t)}{A_{bund}} \tag{64}$$

In case $h_p(t)$ becomes smaller than $h_{p,\min}$, the height of the pool $h_p(t)$ is set to $h_{p,\min}$. Implying that from that moment on the evaporating pool will be shrinking, $A_p(t) < A_{bund}$.

Case 4 : Continuous release, spreading

Release rate

Compared to the situation in case 2: Instantaneous release, spreading, the difference in the case for a continuous release comes from the fact that the release of the liquid is now over a certain period of time. The release rate q'_{out}, is assumed to be constant over the complete outflow time t_s.

Spreading of the pool

The spreading of a pool for an continuous released liquid on flat, non-permeable subsoil can be evaluated by means of the following differential equation:

$$r_p^2(t) \cdot \frac{dr_p^2(t)}{dt} = \frac{2}{\pi} \cdot g \cdot C^{"} \cdot q'_{out} \cdot t^2 \tag{65}$$

Equation (65) is derived from the law of the 'conservation of impulse' for the spreading pool based upon the Damburst principle and simplified up to the two most dominating terms:

- the change of liquid mass in the pool, and
- the hydrostatic pressure between the liquid pool and the surrounding area.

Equation (65) is valid for times smaller than the outflow time $t < t_s$, and it is assumed that the continuous outflow rate q'_{out} is constant in the interval $t \in [0, t_s]$.

Solution of equation (65) gives for the radius of the spreading pool $r_p(t)$ at a time $t \le t_s$:

$$r_p(t) = \left[\frac{4 \cdot C'' \cdot g \cdot q'_{out}}{3 \cdot \pi} \right]^{1/4} \cdot t^{3/4} \tag{66}$$

For times $t > t_s$ the radius of the spreading pool $r_p(t)$ is found by:

$$r_p^2(t) = \frac{1}{16} \cdot r_{p,s}^2 \cdot \left(21 \cdot \left(\frac{t}{t_s} \right)^2 - 18 \cdot \frac{t}{t_s} + 13 \right) \tag{67}$$

in which continuity in $r_p(t)$ is assured by finding $r_{p,s}$ in equation (66) when substituting $t = t_s$.

Mass balance

The mass-balance for the evaporating pool reads in this case for times $t \le t_s$:

$$\frac{dM_p(t)}{dt} = -q'_v(t) + q'_{out} = -\left(q''_v(t) \cdot A_p(t) \right) + q'_{out} \tag{68}$$

which gives for the change in the volume of the spreading pool, $\delta V_p(t)$:

$$\frac{dV_p(t)}{dt} = \frac{-\left(q''_v(t) \cdot \pi \cdot r_p^2(t) \right) + q'_{out}}{\rho_l} \tag{69}$$

For times $t > t_s$, the release rate q'_{out} becomes zero and should be left out of the equations (68) and (69). Consequently the volume of the pool changes in a time-step dt according to:

$$V_p(t+dt) = V_p(t) + dV_p(t) \qquad (70)$$

In case of a continuous release, the volume of the pool at the time, $t = 0$, is:

$V_p(0) = V_{p,0} = 0$.

Heat balance

The evaluation of the heat-balance in case of a continuous released spreading liquid is exactly the same as described in case 2: instantaneous release, spreading.

Evaporation process in time

The evaluation of the radius, the height, the volume, the temperature, the evaporation flux, the evaporation mass flow rate, and the evaporated mass from the pool is evaluated until:

- the total mass which has been released into the pool during the outflow time t_s has totally evaporated, or
- a specific time, t_{study} as requested by the user.

When the minimum layer thickness, $h_{p,\min}$ is reached before the pool has totally evaporated the evaluation of the pool evaporation is continued with at a constant value for the height of the pool $h_p(t)$ at it's minimum layer thickness:

$$h_p(t) = h_{p,\min} = 0.005 \text{ m} \qquad (71)$$

And the radius of the pool is evaluated according to:

$$r_p(t) = \sqrt{(V_p(t))/(\pi \cdot h_{p,\min})} \qquad (72)$$

Case 5 : Continuous release, spreading in bunds

Spreading of the pool

Compared to the situation in case 4: Continuous release, spreading, the difference in this case comes from the fact that the radius of the pool is limited (by the surface of the bunded area, A_{bund}) to a maximum value, $r_{p,\max}$. Therefore for times smaller than the outflow time, $t \in [0, t_s]$ the radius of the evaporating pool is calculated according to:

$$r_p(t) = \mathrm{Min}\left[\left(\sqrt{\left[\frac{4 \cdot C'' \cdot g \cdot q'_{out}}{3 \cdot \pi}\right]^{1/4} \cdot t^{3/4}} \right), \left(r_{p,\max} \right) \right] \tag{73}$$

For times after the outflow time of the pool $t > t_s$, but still within the time of study $t < t_{study}$, and as long as the height of the pool is higher than the minimum layer thickness. I.e. for $t < t_{study}$ and $h_p(t) \geq h_{p,\min}$, the radius of the pool is calculated according to:

$$r_p(t) = \mathrm{Min}\left[\left(\sqrt{\frac{1}{4} \cdot r_{p,s}^2 \cdot \sqrt{\left(21 \cdot \left(\frac{t}{t_s}\right)^2 - 18 \cdot \frac{t}{t_s} + 13 \right)}} \right), \left(r_{p,\max} \right) \right] \tag{74}$$

After the minimum layer thickness has been reached but still within the time of study and a certain amount of liquid in the pool, i.e. for $t < t_{study}$ and $V_p > 0$, the radius of the shrinking pool is evaluated according to:

$$r_p(t) = \mathrm{Min}\left[\left(\sqrt{\frac{V_p(t)}{\pi \cdot h_{p,\min}}} \right), \left(\sqrt{\frac{1}{4} \cdot r_{p,s}^2 \cdot \sqrt{\left(21 \cdot \left(\frac{t}{t_s}\right)^2 - 18 \cdot \frac{t}{t_s} + 13 \right)}} \right), \left(r_{p,\max} \right) \right] \tag{75}$$

EVAPORATION FROM LAND FOR BOILING LIQUIDS

Dangerous chemicals/liquids which are accidentally released on land are considered to be boiling when they have a normal boiling point (T_b) (under ambient conditions) that is higher than the temperature of the subsoil (T_s) (on which the chemical is spilled) and/or the ambient temperature (T_a).

At the first moments after the accidental release, the boiling rate of the pool is mainly controlled by the rate rate at which heat from the surroudings can be transferred into the pool from the subsoil, thereby supplying the latent heat of vaporisation for the spilled (boiling) chemical.

At later stages, the other heat sources, like the solar heat flux and the heat flux as conducted by air, become equally important to the heat flux from the subsoil and control the boiling rate of the pool.

Case 6 : Instantaneous release in a bund

Heat balance

In a simple approach for a boiling liquid, the temperature of the pool is at a constant value ($dT_p(t)/dt = 0$), the boiling temperature T_b, and therefore the heat-balance from equation (5) reduces to the following equation:

$$0 = \Sigma H(t) - q_v''(t) \cdot L_v(T_b) \qquad (76)$$

From equation (76), the vaporisation mass flow rate $q_v'(t)$ can now easily be found:

$$q_v'(t) = q_v''(t) \cdot A_p = \frac{\Sigma H(t)}{L_v(T_b)} \cdot A_p \qquad (77)$$

In equation (77), the surface area of the pool is determined by the surface of the bunded area $A_p = A_{bund}$, and the sum of the heat sources: $\Sigma H(t)$ from equation (9), reads in case of a boiling pool:

$$\Sigma H(t) = \frac{\lambda_s \cdot (T_{subsoil} - T_b)}{\sqrt{a_s \cdot \pi \cdot t}} + k_{h,a} \cdot (T_{air} - T_b) + H_{solar-radiation} + H_{(long-wave)-radiation} \qquad (78)$$

Mass balance

As derived before in case 1 for a non-boiling, instantaneous released liquid in a bund, the change in the height of the pool in time, $dh_p(t)$ follows from the mass balance as given in equation (24). And the height of the pool in time can be followed during the evaporation process by equation (25).

Evaporation process in time

Starting from a boiling pool at boiling temperature T_b, and a pool surface area (A_p) equal to the bunded area (A_{bund}), which contains an amount of liquid equal to the instantaneously released amount at t_0; $M_p(t_0)$. The evaporation of the pool is followed in time by updating (evaluating) after specific time steps dt, the height, $h_p(t)$ of the pool. Thereby assuming that pool 'remains' boiling, at it's boiling temperature T_b. The evaporation process of the constantly boiling liquid, which was instantaneously released in a bunded area, is followed until the pool has totally evaporated, or before (at a specific time, t_{study}) if requested by the user.

Pool temperature in time

The assumption that the pool temperature is (and remains) at a constant value, the boiling temperature (T_b) of the liquid, is only valid in case the temperature difference (ΔT) between the evaporating pool and the subsoil (T_s) and/or ambient temperature (T_a) is high enough (and remains high enough) to supply the pool with the heat (ΣH) necessary for the evaporation process ($q_v''(t) \cdot L_v(T_b)$). In case this is not the fact and the heat supply to the pool (through the heat as supplied by the subsoil ($H_{subsoil}$) and/or air (H_{air})) becomes smaller than the heat necessary for the evaporation process, $\Sigma H < q_v''(t) \cdot L_v(T_b)$, the temparture of the pool will drop below it's boiling point and the evaporating pool will subtract the heat necessary for the evaportion process from itself, and consequently thereby lowering it's temperature.

Case 7 : Instantaneous release, spreading

Surface area of the pool

As described previously in case 2 for a non-boiling, instantaneously released spreading liquid , the surface area of the evaporating pool is in case of a spreading pool, time dependent and denoted as: $A_p(t)$. The equations (26), (27) and (28), as derived before also apply in this case.

Spreading of the pool

As described previously in case 2 for a non-boiling, instantaneously released spreading liquid, the spreading of the boiling evaporating pool can also be evaluated by the equations (29) and (30) as derived previously.

Spreading time t_e (minimum layer thickness)

As described previously in case 2 for a non-boiling, instantaneously released spreading liquid, the solution of the spreading time t_e (until a minimum layer thickness $h_{p,\min}$ has been reached) can be found by the iterative solution of equation (39), in which in the case of a boiling liquid the approximation is valid, that in the initial stages of the evaporation process the rate at which heat can get into the pool is mainly controlled by the heat input from the subsoil. Therefore equation (33) can be used for the value of e.

Mass & heat balance

Assuming that the heat supply to the evaporating (initially) boiling pool is high enough during the evaporation process, the pool temperature can be considered constant ($dT_p(t)/dt = 0$), at the value of it's boiling temperature T_b. Consequently from the heat balance in equation (5), the evaporation flux $q_v^{"}(t)$ can be calculated according to the following equation:

$$q_v''(t) = \frac{\Sigma H(t)}{L_v(T_b)} \tag{79}$$

in which the sum of the heat sources: $\Sigma H(t)$, is calculated according to equation (78), with the only differences that the heat transfer coefficient $k_{h,a}(t)$ becomes time-dependent in case of a spreading evaporating pool. From equation (79),the evaporation flux $q_v'(t)$ can be found according to:

$$q_v'(t) = q_v''(t) \cdot A_p(t) = q_v''(t) \cdot \pi \cdot r_p^2(t) \tag{80}$$

And the evaporated quantity $q_v(t)$ from the pool at a time $t + dt$ can be approximated by:

$$q_v(t + dt) = q_v(t) + q_v'(t) \cdot dt \tag{81}$$

Evaporation process in time

Starting from an initial pool at it's boiling temperature T_b which contains an amount of liquid equal to the amount which was instantaneous released, $M_p(t_0) = M_{p,0}$, the dimensions of the pool like: surface area; $A_p(t)$ and radius; $r_p(t)$ can be calculated in time after making an assumption concerning the dimensions of the pool at the time of instantaneous release, $t = 0$. If we make the following assumptions for the initial height of the pool, $h_{p,0}$ and the initial volume of the pool, $V_{p,0}$:

- $h_{p,0} = 10 \cdot h_{p,\min} = 0.05$ m, and
- $V_{p,0} = M_{p,0} / \rho_l$,

we find the radius, $r_{p,0}$ and area $A_{p,0}$ of the intial pool by:

$$r_{p,0} = \sqrt{\frac{V_{p,0}}{\pi \cdot h_{p,0}}} \qquad A_{p,0} = \pi \cdot r_{p,0}^2 \tag{82}$$

The evaporation of the pool is followed in time by evaluating after specific time steps dt, the height, $h_p(t)$, of the pool through the update of the mass balance, because the height of the pool at a time *t* follows directly from the mass balance:

$$h_p(t) = \frac{M_{p,0} - q_v(t)}{\pi \cdot r_p^2(t) \cdot \rho_l} \tag{83}$$

In advance of the evaporation calculation, the spreading time t_e, for the pool to spread until a minimum thickness layer $h_{p,\min} = h_p(t_e)$ is reached, can be estimated. It can be estimated from the mass-balance as given in the equation (37)-(39) in which for the value of e the approximated value for a boiling liquid can be substituted as given in equation (33). In the evaluation of the evaporation process, after reaching the spreading time t_e, the evaporating pool will be shrinking (equation (44)). The evaporation process of the boiling, spreading liquid, which was instantaneously released, is followed until the pool has totally evaporated, or before (at a specific time) if requested by the user.

Case 8 : Instantaneous release, spreading in bunds

Spreading of the pool

Compared to the situation in case 7: Instantaneous release, spreading, the difference in this case comes from the fact that the radius of the pool is limited to a maximum value, $r_{p,\max}$ as determined by the surface of the bunded area, A_{bund} as given in equation (61). Therefore during the time of spreading of the pool from time $t = 0$ until $t = t_e$, and as long as the evaporated amount of liquid from the pool is less than the instantaneous released amount of liquid into the pool at time $t = 0$. I.e. for $t \in [0, t_e]$ and $q_v(t) < M_{p,0}$, the radius of the pool is calculated accoring to:

$$r_p(t) = \text{Min}\left[(r_{p,0} + \sqrt{C'' \cdot g \cdot h_{p,0}} \cdot t), (r_{p,\max}) \right] \tag{84}$$

The pool will spread until a minimum layer thickness $h_{p,\min} = h_p(t_e)$ has been reached at time t_e, the spreading time of the pool. The calculation of the spreading time t_e is done by the iterative solution of equation (39). Assuming that, after reaching $h_{p,\min}$ at time t_e, the layer thickness to be constant implies that the pool will shrink for times $t > t_e$. In case the spreading pool reaches the bund before $t = t_e$, i.e. before reaching the minimum layer thickness $h_{p,\min}$, the spreading immediately stops and t_e is in this case determined by the bunded area. For times after the spreading time of the pool $t > t_e$, but still within the time of study $t < t_{study}$, and as long as the evaporated amount of liquid from the pool is less than the instantaneous released amount of liquid into the pool at time $t = 0$. I.e. for $t > t_e$ and $t < t_{study}$ and $q_v(t) < M_{p,0}$, the radius of the shrinking pool is calculated according to:

$$r_p(t) = \mathrm{Min}\left[(\sqrt{r_{p,e}^2 \cdot e^{\left[-\frac{2 \cdot e^*}{\rho_l \cdot h_{p,\min}} \left(\sqrt{t} - \sqrt{t_e} \right) \right]}}), (r_{p,\max}) \right] \tag{85}$$

Case 9 : Continuous release, spreading

Release rate

Compared to the situation in case 7: Instantaneous release, spreading, the difference in the case for a continuous release comes from the fact that the release of the liquid is now over a certain period of time. The release rate $q_{out}^{'}$, is assumed to be constant over the complete outflow time t_s.

Spreading of the pool

As described previously in case 4 for a non-boiling, continuous released spreading liquid, the spreading of the boiling evaporating pool can also be evaluated by the equations (66) and (67) as derived previously.

Heat balance

From the heat balance in equation (76) for a boiling pool, it's possible to derive the equations for the evaporation rate and quantity for a continuous release by inserting the spreading in equations (66) and (67). The evaporated quantity $q_v(t)$ at a specific time t can be calculated according to:

$$q_v(t) = \int_0^t q_v^{''}(t) \cdot \pi \cdot r_p^2(t)\, dt \;\square\; e \cdot \pi \cdot \int_0^t \frac{r_p^2(t)}{\sqrt{t}} dt \tag{86}$$

when assuming that the boiling rate in the initial stage of the evaporation process is mainly controlled by the rate at which heat can get into the liquid pool from the subsoil. This means that the following approximation can be made for the evaporation flux $q_v^{''}(t)$ in equation (86) in case of a boiling pool:

$$q_v^{''}(t)=\frac{\Sigma H(t)}{L_v(T_b)} \square \frac{H_{subsoil}(t)}{L_v(T_b)}=\frac{\lambda_s\cdot(T_{subsoil}-T_b)}{L_v(T_b)\sqrt{a_s\cdot\pi\cdot t}}=\frac{e}{\sqrt{t}} \tag{87}$$

with e as given in equation (33), in which $T_p(t)$ is replaces by the boiling temperature T_b of the pool. By inserting equation (66) in equation (86) gives for times $t\le t_s$:

$$q_v(t)=e\cdot\left[\frac{3\cdot\pi^3\cdot C^{''}\cdot g\cdot q_{out}^{'}}{16}\right]^{1/2}\cdot t^2 \tag{88}$$

and for times $t>t_s$ is:

$$q_v(t)=e\cdot\pi\cdot r_{p,s}^2\cdot\sqrt{t_s}\cdot\left\{\left(\frac{t}{t_s}\right)^2\cdot B_{\frac{t_s}{t}}\left(\frac{5}{2},\frac{1}{2}\right)+\sqrt{\left(\frac{t}{t_s}-1\right)}\left(2\cdot\frac{t}{t_s}+\frac{7}{5}\cdot\left(\frac{t}{t_s}-1\right)^2\right)\right\} \tag{89}$$

in which $B_{\frac{t_s}{t}}\left(\frac{5}{2},\frac{1}{2}\right)$ is the incomplete Beta function, which in general is of the form:

$$B_x(a,b)=\int_0^x \tau^{a-1}\cdot(1-\tau)^{b-1}\cdot d\tau \tag{90}$$

In this specific case it is:

$$B_{\frac{t_s}{t}}\left(\frac{5}{2},\frac{1}{2}\right)=\int_0^{\frac{t_s}{t}} \tau^{3/2}\cdot(1-\tau)^{-1/2}\cdot d\tau \tag{91}$$

Working out equation (91) gives the following result:

$$B_{\frac{t_s}{t}}\left(\frac{5}{2},\frac{1}{2}\right)=\frac{1}{8}\cdot\frac{2\cdot\left(\frac{t_s}{t}\right)^2+4\cdot\left(\frac{t_s}{t}\right)^3-6\cdot\frac{t_s}{t}+3\cdot\sqrt{-\left(\frac{t_s}{t}\right)^2+\frac{t_s}{t}}\cdot\arcsin\left(2\cdot\frac{t_s}{t}-1\right)}{\sqrt{\frac{t_s}{t}}\cdot\sqrt{1-\frac{t_s}{t}}}+\frac{3}{16}\cdot\pi \tag{92}$$

From the eqauations (88) and (89), we find for the evaporation rate $q_v^{'}(t)$ for times $t\le t_s$:

$$q_v^{'}(t) = e \cdot \left[\frac{3 \cdot \pi^3 \cdot C^{"} \cdot g \cdot q_{out}^{'}}{4} \right]^{1/2} \cdot t \tag{93}$$

and for times $t > t_s$:

$$q_v^{'}(t) = \frac{2 \cdot e \cdot \pi \cdot r_{p,s}^2}{\sqrt{t_s}} \cdot \left(\frac{t}{t_s}\right) \cdot B_{\frac{t_s}{t}}\left(\frac{5}{2}, \frac{1}{2}\right) + \frac{1}{2} \cdot e \cdot \pi \cdot r_{p,s}^2 \cdot \frac{\sqrt{t_s}}{t} \cdot \sqrt{\left(\frac{t}{t_s} - 1\right)} \left(7 \cdot \left(\frac{t}{t_s}\right)^2 - \frac{t}{t_s} + 2\right) \tag{94}$$

Mass balance

From the mass balance the heigt of the pool can be found and is calculated according to:

$$h_p(t) = \frac{V_p(t)}{A_p(t)} = \frac{M_p(t)}{\pi \cdot \rho_l \cdot r_p^2(t)} = \frac{\left(\rho_l \cdot q_{out}^{'} \cdot t\right) - q_v(t)}{\pi \cdot \rho_l \cdot r_p^2(t)} \tag{95}$$

in which for times $t > t_s$ in equation (95) the term $\left(\rho_l \cdot q_{out}^{'} \cdot t\right)$ should be replaced by the total mass released into the pool during the outflow time t_s.

Evaporation process in time

The evaluation of the radius $r_p(t)$, the height $h_p(t)$, the evaporation flux $q_v^{"}(t)$, the evaporation mass flow rate $q_v^{'}(t)$, and the evaporated mass $q_v(t)$ from the pool is evaluated according to the above equation until:

- the total mass which has been released into the pool during the outflow time t_s has totally evaporated, or
- a specific time, t_{study} as requested by the user.

In case the minimum layer thickness $h_{p,\min}$ is reached before the pool has totally evaporated, the evaluation of the pool evaporation process for times greater than t_e is according to the equations as derived before in case 7 for a boiling spreading liquid which was instantaneous released. For times $t > t_e$, the height of the pool $h_p(t)$, is kept at the constant value of the minimum layer thickness: $h_p(t) = h_{p,\min} = 0.005 \text{ m}$, and as derived before in case 2, the radius of the pool is evaluated according to equation (44). And the evaporation rate $q_v^{'}(t)$ follows from equation (45). The evaporated quantity

$q_v(t)$ is calculated according to equation (47). From the equations (44) and (45), the evaporation flux $q_v^{''}(t)$ can consequently be found as:

$$q_v^{''}(t) = \frac{q_v^{'}(t)}{A_p(t)} = \frac{q_v^{'}(t)}{\pi \cdot r_p^2(t)} \tag{96}$$

Case 10 : Continuous release, spreading in bunds

Spreading of the pool

Compared to the situation in case 9: Continuous release, spreading, the difference in this case comes from the fact that the radius of the pool is limited (by the surface of the bunded area, A_{bund}) to a maximum value, $r_{p,\max}$. Therefore during the outflow time of the liquid into the pool in case of a continuous release from time $t = 0$ until $t = t_s$, and as long as the thickness of the pool is higher than the minimal layer thickness of 5 mm. I.e. for $t \in [0, t_s]$ and $h_p(t) \geq h_{p,\min}$, the radius of the pool is calculated according to:

$$r_p(t) = \text{Min}\left[\left(\sqrt{\left[\frac{4 \cdot C^{''} \cdot g \cdot q_{out}^{'}}{3 \cdot \pi} \right]^{1/4}} \cdot t^{3/4} \right), \left(r_{p,\max} \right) \right] \tag{97}$$

For times after the outflow time of the liquid into the pool $t > t_s$, but still within the time of study $t < t_{study}$, and still as long as the thickness of the pool is higher than the minimal layer thickness $h_{p,\min}$, and the evaporated liquid from the pool is less than the amount of released liquid into the pool during the outflow time t_s. I.e. for $t > t_s$ and $t < t_{study}$ and $h_p(t) \geq h_{p,\min}$ and $q_v(t) < M_p$, the radius of the pool is calculated according to:

$$r_p(t) = \text{Min}\left[\left(\sqrt{\frac{1}{4} \cdot r_{p,s}^2 \cdot \sqrt{\left(21 \cdot \left(\frac{t}{t_s} \right)^2 - 18 \cdot \frac{t}{t_s} + 13 \right)}} \right), \left(r_{p,\max} \right) \right] \tag{98}$$

At a certain moment t, the thickness of the pool becomes lower than the minimum layer thickness of 5 mm. I.e. for $t > t_s$ and $t < t_{study}$ and $h_p(t) \in \left[0, h_{p,\min} \right]$ and $q_v(t) < M_p$, than the pool thickness is kept at the constant value $h_p = h_{p,\min}$, and the radius of the pool is found from:

$$r_p(t) = \mathrm{Min}\left[\left(\sqrt{\frac{M_p - q_v(t)}{\pi \cdot h_{p,\min} \cdot \rho_l}}\right), \left(\sqrt{\frac{1}{4} \cdot r_{p,s}^2 \cdot \sqrt{\left(21 \cdot \left(\frac{t}{t_s}\right)^2 - 18 \cdot \frac{t}{t_s} + 13\right)}}\right), \left(r_{p,\max}\right) \right] \quad (99)$$

CONCLUSIONS

In this first paper, in a series of two, the theoretical background of a completely integrated evaporation model has been described for all kinds of possible pools evaporating from land (Part 1A). In the subsequent paper in the series, the subject will be on evaporating pools from water for non-soluble, floating liquids on water (Part 1B). For the situation on land, the evaporation process has been worked out and described for as well boiling as non-boiling liquids. The (accidental) outflow of the evaporating liquid is described for as well continuous as instantaneous releases. The spreading equations for the liquid on land are considered until a minimum layer thickness has been reached, or until the borders of the bund are reached. The mass- and heat balance are simultaneously solved during the outflow, spreading and evaporation process for the evaluation of the temperature and the height of the pool over time during (or after) the release.

The evaporation model takes the following environmental- and weather conditions into account: several subsoils, humidity, wind speed, cloud cover, the influence of position in the world and date (day:month:year).

LIST OF SYMBOLS

Symbol	Description	Unit	Equation
$q_v^{'}(t)$	vaporisation mass flow rate at specific time *t*	kg/s	1,23,36,40,41,44,46,48,68,76,80,81,93,96
$q_v^{''}(t)$	vaporisation mass flux at specific time *t*	kg/m^2s	1,2,5,6,7,23,24,31,32,48,49,51,53,54,68,69, 76,77,79,80,86,87,96
A_p	surface area of the pool (instantaneous release in a bund)	m^2	1,23,24,77
k_m	mass transfer coefficient (time-independent)	m/s	2, 3
$P_v(T_p(t))$	vapour pressure at pool temperature at specific time *t*	N/m^2	2,51
μ_i	molecular weight of substance *i*	kg/mol	2,51
R	universal gas constant	J/molK	2,51
$T_p(t)$	temperature pool at specific time *t*	K	2,5,6,7,8,32,33,34,51,53,54,55,57
$u_{w,10}$	wind speed at 10 m. height	m/s	3,14,52,59
d_p	diameter of the pool (instantaneous release in a bund)	m	3,12,14
Sc	Schmidt number	-	3,4,52
υ_a	kinematic viscosity of air	m^2/s	4,13
D_a	diffusion coefficient of air	m^2/s	4,13
$h_p(t)$	height of the pool at specific time *t*	m	5,6,7,24,25,27,29,53,54,60,63,64,71,83,95
ρ_l	density of the liquid pool	kg/m^3	5,6,7,24,28,38,39,40,43,44,45,46,47,49,53, 54,69,83,85,95,99
$C_{p,l}$	specific heat of the liquid	J/kgK	5,6,7,53,54
$\Sigma H(t)$	sum of the heat sources at specific time *t*	J/m^2s	5,6,7,32,53,54,56,76,77,78,79,87
$L_v(T_p(t))$	latent heat of vaporisation at pool temperature	J/kg	5,6,7,32,33,34,53,54
$H_{subsoil}(t)$	heat as conducted from the subsoil	J/m^2s	9,10,32,33,34,56,87
λ_s	thermal conductivity of the subsoil	J/msK	10,32,33,34,78,87
$T_{subsoil}$	temperature of the subsoil	K	10,32,78,87
a_s	thermal diffusivity of the subsoil	m^2/s	10,32,33,34,78,87
$H_{air}(t)$	heat as conducted by air	J/m^2s	9,11,34,56,57
$k_{h,a}$	heat transfer coefficient to the atmosphere (time-independent)	J/m^2sK	11,78
T_{air}	temperature of the air (ambient temperature)	K	11,22,57,78
Nu	Nusselt's number	-	12,58
λ_a	thermal conductivity of air	J/msK	12,58
$\Pr_a$	Prandtl's number for air	-	12,13,58
Re	Reynold's number (time-independent)	-	12,14
ρ_a	density of air	kg/m^3	14,59
η_a	dynamic viscosity of air	Ns/m^2	14,59
$H_{solar-radiation}$	heat from solar radiation	J/m^2s	9,15,34,56,78

c_1	constant = 990	J/m²s	15
χ	solar elevation	rad	15,16
c_2	constant = 30	J/m²s	15
N	cloud cover (between 0 and 1)	-	15,22
SD	solar declination	rad.	16
α	angle (see text)	°	16,18
φ	latitude of the location	°N	16
SL	solar longitude	rad.	17,18,19
ψ_w	western longitude	°W	18
J_n	Julian day number	-	18,19,20,21
t_u	universal time	hours	18
$H_{(long-wave)-radiation}$	heat by long-wave radiation	J/m²s	9,22,34,56,78
e	Albedo number	-	22
c_3	constant = 5.31.10^{-13}	J/m²sK^6	22
σ	Stefan-Boltzmann constant = 5.67.10^{-8}	W/m²K^4	22
c_4	constant = 60	J/m²s	22
$M_p(t)$	liquid mass in the pool at specific time *t*	kg	23,48,68,95
$V_p(t)$	volume of the pool at at specific time *t*	m³	24,27,40,49,50,60,63,64,69,70,72,75,95
$A_p(t)$	surface area of the pool at specific time *t*	m²	26,27,31,40,48,60,68,80,95,96
$r_p(t)$	radius of the pool at specific time *t*	m	26,27,29,30,31,35,40,41,43,44,49,52,58, 59,60,62,63,65,66,67,69,72,73,74,75,80,83, 84,85,86,95,96,97,98,99
$V_{p,0} = V_p(t_0)$	volume of the liquid pool at time *t*=0	m³	28,38,39,82
$M_{p,0} = M_p(t_0)$	liquid mass in the pool at time *t*=0	kg	28,83
C''	emperical constant = 1.08 (land)	-	29,30,35,36,39,62,63,65,66,73,84,88,93,97
g	acceleratioin of gravity	m/s	29,30,35,36,39,62,63,65,66,73,84,88,93,97
$h_{p,0}$	height of the pool at time *t*=0	m	29,30,35,36,39,62,63,82,84
$r_{p,0}$	radius of the pool at time *t*=0	m	30,35,36,39,62,63,82,83,84
$q_v(t)$	evaporated quantity of the liquid pool at a specific time *t*	kg	31,35,46,47,81,83,86,88,89,95,99
$q_v(t_e)$	evaporated quantity of the liquid pool at time t_e	kg	38
$r_p(t_e) = r_{p,e}$	radius of the pool at spreading time t_e	m	38,42,44,45,46,47,85
$h_{p,\min}$	minimum height of the pool at time $t_{e)}$ (minimum layer thickness)	m	38,39,40,43,44,45,46,47,71,72,75,85,99
t_e	spreading time to reach minimum layer thickness	s	37,39,44,45,46,47,85
$q_v'(t_e)$	vaporisation mass flow rate at time t_e	kg/s	42
$k_m(t)$	time-dependent mass transfer coefficient	m/s	51,52
$k_{h,a}(t)$	time-dependent heat transfer coefficient to the atmosphere	J/m²sK	57,58
$\mathrm{Re}(t)$	time-dependent Reynold's number	-	58,59
$r_{p,\max}$	maximum radius of the pool as determined by the bund	m	61,62,63,73,74,75,84,85,97,98,99

A_{bund}	surface area of the bund	m^2	61,64
q'_{out}	continuous outflow rate	kg/s or m^3/s	65,66,68,69,73,88,93,95,97
t_s	outflow time of a continuous release	s	67,74,75,89,92,94,98,99
$r_{p,s}$	radius of the pool at at ouflow time $t=t_s$	m	67,74,75,89,94,98,99
$L_v(T_b)$	latent heat of vaporisation at boiling temperature T_b	J/kg	76,77,79,87
$A_{p,0}$	surface area of the pool at time $t=0$	m^2	82
$B_x(a,b)$	Incomplete Beta function	-	89,90,91,92,94

REFERENCES

[Kawamura: 1987] P.I. Kawamura and D. Mackay, "*The evaporation of volatile liquids*" Journal of hazardous materials, 15 (1987) 343-364.

[Yellow Book: 1997] Yellow Book, third edition 1997, *Methods for the calculation of physical effects, resulting from releases of hazardous materials (liquids and gases).* Directorate-General of Social Affaires and Employment, CPR 14E third edition, The Hague 1997.

[Byrne: 1992] A.J. Byrne, S.J. Jones, S.C. Rutherford, G.A. Tickle and D.M. Webber, "*Description of ambient atmospheric conditions for the computer code DRIFT*" SRD/HSE R553, SRD, Culcheth, United Kingdom (1992).

[Jacobs: 1985] F.A.M.H. Jacobs, "*Experimenteel onderzoek naar de spreiding van instantaan vrijkomende vloeistoffen op land*" TNO-PML Report 1985.

[Verhagen: 1985] Th.L.A. Verhagen, "*Spreiding van instantaan of continu vrijkomende, tot vloeistof gekoelde gassen op land en op water*" TNO-PML Report 1985.

[Kootstra: 2004] F. Kootstra, B.F.P. van het Veld and G. Opschoor, "*Use and application of the advanced evaporation model for liquids from land and water (non-soluble, floating)*" to be submitted.

A complete description of an advanced evaporation model for non-soluble, floating liquids from water. Part IB – Theoretical background.

Dr. F. Kootstra, Ing. B.F.P. van het Veld and Ir. G. Opschoor

Netherlands Organisation for Applied Scientific Research,
TNO Environment, Energy and Proces Innovation,
Department of Industrial Safety,
Laan van Westenenk 501,
P.O. Box 342,
7300 AH Apeldoorn,
The Netherlands,
Phone: +31 55 5493208,
Fax: +31 55 5493390,
Web: http://www.tno.nl,
E-mail: F.Kootstra@mep.tno.nl

Prepared for presentation at the 38th Annual Loss Prevention Symposium, AIChE 2004 Spring National Meeting, New Orleans, Louisiana, April 26 – 29, 2004.

ABSTRACT

In two papers the theoretical background of a completely integrated evaporation model will be described for all kinds of possible pools. In the first paper the evaporation of pools from land (Part 1A) and in the second paper the evaporation of non-soluble, floating pools from water (Part 1B) will be described for as well boiling as non-boiling liquids. The outflow of the evaporating liquid on land or water can be continuous as well as instantaneous. The spreading of the liquid is considered until a minimum layer thickness has been reached, or until the borders of the bund have been reached on land, or the confinement of the water surface (rivers). The mass- and heat balance need to be simultaneously solved for the evaluation of the temperature and height of the pool in time after (or during) the release. For the evaporation from water the formation of ice is considered in case of a confined water surface (rivers), and the different boiling regimes in case of an open water surface (sea). Within the evaporation model the following environmental- and weather conditions are taken into account: several subsoils for evaporation from land, humidity, wind speed and cloud cover in the air, the influence of position in the world and date (day:month:year).

INTRODUCTION

All over the world large amounts of chemicals are being transported, produced and stored for use in e.g. chemical plants. Sometimes during the transportation over land or water, or during the storage of these chemicals at the site, accidents happen. Consequently the chemical is being released under the local storage conditions or the transportation conditions. Such an accidental release can causes risk to the people and the environment depending upon e.g. the toxicity or the flammability of the chemical. Due to legislation the manufacturers and transporters of these chemicals are obliged to certain criteria to minimise the risk associated with the production or transportation of these chemicals. For being able to calculate the risk of such an accidental release of the transported or stored chemical, one needs to determine the physical effect of the accidental release of such a chemical, cause the risk associated with a certain event is defined as the (physical) effect of the event times the frequency of the occurance of the event. In case the chemical is being released under such conditions that it is in it's liquid form, for such cases the following two papers give a complete description of the theoretical background to determine the physical effects associated with the accidental release of the consequent evaporation process of the time-dependent evaporating, spreading liquid pool. As can be seen from the flow chart in Figure 1, a seperation should be made between the physical description of the evaporation process of liquids from land and water.

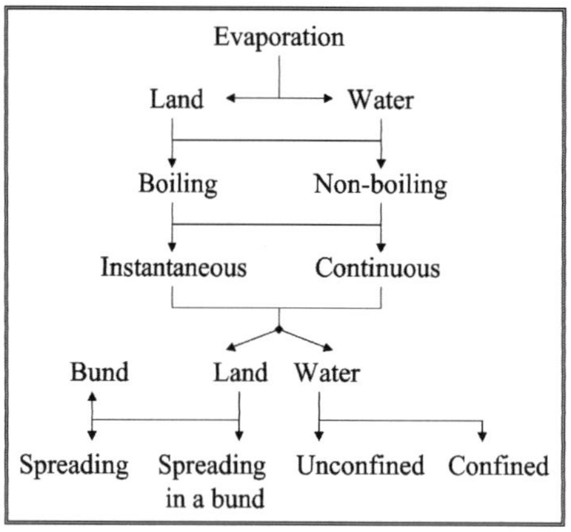

Figure 1: Flow chart for the evaporation process

Therefore, the first paper describes the evaporation process for pools from land (Part 1A) and the second paper describes the evaporation process for non-soluble, floating pools from water (Part 1B). Next to the seperation of the evaporation process from land or water, the following distinctions should also be made (as can be seen from the flow chart in Figure 1), between a boiling or a non-boiling pool, between an instantaneous or continuous release of the liquid and, for evaporatoin from land, between a spreading pool or a spreading pool until a certain bund and, for evaporation from water, between a confined or unconfined water surface. Working out of all the above distinctions results in the following eighteen (to be considered) cases in the evaporation process, as shown in Table 1.

	Non-boiling liquid		Boiling liquid	
Outflow upon land	*Instantaneous*	*Continuous*	*Instantaneous*	*Continuous*
Bunds	(1)	(*)	(6)	(**)
Spreading	(2)	(4)	(7)	(9)
Spreading in bunds	(3)	(5)	(8)	(10)
Outflow upon water	*Instantaneous*	*Continuous*	*Instantaneous*	*Continuous*
Unconfined (open water)	(15)	(17)	(11)	(13)
Confined (rivers/canals)	(16)	(18)	(12)	(14)

(*) Same as case (5)

(**) Same as case (10)

Table 1: The eighteen separate cases for the evaporation process from land and water.

The organisation and lay-out for the description of the theoretical background of a completely integrated evaporation model in the two papers is the following: in the first paper the background is given for the evaporation from land (Part IA), and in the second paper from water (Part IB). Thus, as can be seen from Table 1, the first ten cases (1-10) are treated in the first paper, and the last eight cases (11-18) are dealt with in the second paper. The practical implications of the first two papers will be described in another third paper [Kootstra: 2004], in which the use and application of the advanced evaporation models will be shown and demonstrated in several practical example calculations.

A list of the symbols can be found at the end of the paper in which all symbols throughout the equations in the paper are given together with a short description, the unit and the numbers of the equations in which the symbol is used.

EVAPORATION FROM WATER FOR BOILING LIQUIDS

Evaporation process for a boiling liquid from an unconfined water surface

Within the assumption of no-ice formation it's assumed that the heat supply within the two-dimensional infinite water surface to the evaporating pool is fast enough, not to cause the freezing of the water below the evaporating pool. Consequently, after the outflow of a liquefied gas on water, a boiling process begins, due to the temperature difference between the water and the released liquid. The boiling process can be separated in three boiling regimes as depicted in Figure 1:

- nucleate boiling,
- transition boiling and
- film boiling.

The trajectory 0-1, in Figure 1, is characterized by nucleate boiling. The increase of the temperature difference between the liquid and the water generates the appearance of vapor bubbles around the nuclei, on the contact area between liquid and water. These bubbles become bigger in size due to the heating and finally come loose. This bubble formation increases along with the increase of the temperature difference between water and liquid. The trajectory of nucleate boiling is limited, in its upper part, by the temperature difference ΔT_1 between liquid and water, whereby the heat-flow is than maximal. Beyond the temperature difference ΔT_1 (between liquid and water), the number of bubbles which form increases in such a way that it exerts a disturbing influence on the heat transfer from water to the liquid. The bubbles, then, form a steadily improving insulation layer between liquid and water. The resulting decrease of heat-transfer between water and liquid produces a decrease in evaporation rate of the liquid. This period, represented in Figure 1 by the trajectory 1-2, is called meta-stable (transition) boiling. At a temperature difference between liquid and water higher than a given value (in Figure 1: ΔT_2, the so called Leidenfrost-point) the liquid becomes completely separated from the water by a vapor film. The trajectory after this point is called film boiling. In literature a lot of attention is given to nucleate boiling and film boiling. The meta-stable boiling regime is often not taken into consideration due to its complexity. During meta-stable boiling, a certain portion of the water surface is wetted by the liquid during a short time (nucleate boiling), while thereafter a vapor layer will be formed between water and the boiling liquid. In order to calculate the heat-flow during meta-stable boiling, the relations for nucleate boiling and film-boiling should first be known.

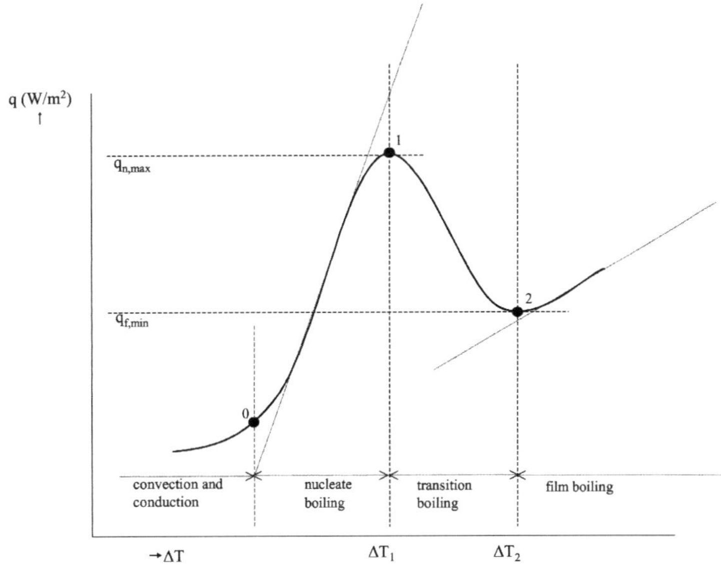

Figure 1: The boiling regimes in case of a boiling liquid.

The heat-flux during the nucleate boiling regime is given by the empirical formula of Grigoryev *et al* [Grigoryev: 1973]:

$$q_n = \frac{4.1 \cdot \left\{ 1 + 10 \left[\frac{\rho_{v,s}}{\rho_{vl} - \rho_{v,s}} \right]^{\frac{2}{3}} \right\}^3 \cdot \Delta T^3}{\gamma \cdot T_b \cdot \left\{ \left(\frac{\sqrt{\upsilon_{vl}}}{\lambda_{vl}} \right) + \left[\frac{10}{\sqrt{\rho_w \cdot c_w \cdot \lambda_w}} \right] \right\}^2 \cdot \left\{ 1 + 10 \cdot \sqrt{\frac{\rho_{vl} \cdot c_{vl} \cdot \lambda_{vl}}{\rho_w \cdot c_w \cdot \lambda_w}} \right\}} \tag{1}$$

The maximum heat-flux during nucleate boiling is according to Kutateladze [Kutateladze: 1952]:

$$q_{n,\max} = 0.16 \cdot h_v \cdot \sqrt{\rho_{v,s}} \cdot \left[\gamma \cdot g \cdot \left(\rho_{vl} - \rho_{v,s} \right) \right]^{\frac{1}{4}} \tag{2}$$

And the heat-flux during film boiling is given by [Kalinin: nyp]:

$$q_f = 0.18 \cdot \lambda_v \cdot \Delta T \cdot \left[\frac{g}{\upsilon_v \cdot a_v} \cdot \left(\frac{\rho_{vl}}{\rho_{v,f}} - 1 \right) \right]^{\frac{1}{3}} \tag{3}$$

To find the temperature difference at which the heat flow is maximal, ΔT_1 can be calculated by setting equation (1) equal to (2), which gives:

$$\Delta T_1 = \sqrt[3]{\frac{q_{n,\max} \cdot \left(\gamma \cdot T_b \cdot \left\{ \left(\frac{\sqrt{\upsilon_{vl}}}{\lambda_{vl}} \right) + \left[\frac{10}{\sqrt{\rho_w \cdot c_w \cdot \lambda_w}} \right] \right\}^2 \cdot \left\{ 1 + 10 \cdot \sqrt{\frac{\rho_{vl} \cdot c_{vl} \cdot \lambda_{vl}}{\rho_w \cdot c_w \cdot \lambda_w}} \right\} \right)}{4.1 \cdot \left\{ 1 + 10 \left[\frac{\rho_{v,s}}{\rho_{vl} - \rho_{v,s}} \right]^{\frac{2}{3}} \right\}^3}} \tag{4}$$

For ΔT_2 the relation as given by Kalinin in [Kalinin: 1973] is used:

$$\Delta T_2 = \left(T_c - T_{vl} \right) \cdot \left[0.16 + 2.4 \cdot \left(\frac{\rho_{vl} \cdot c_{vl} \cdot \lambda_{vl}}{\rho_w \cdot c_w \cdot \lambda_w} \right) \right]^{\frac{1}{4}} \tag{5}$$

The heat flux in the transition boiling regime is now calculated by the following interpolation [Kalinin: 1975]:

$$q_t = q_n \cdot f + q_f \cdot (1 - f) \tag{6}$$

in which:

$$f = \left(1 - \frac{\Delta T - \Delta T_1}{\Delta T_2 - \Delta T_1} \right)^7 \tag{7}$$

Using the equation (1) through (7) for several liquefied gasses spreading on water gives the results as shown in Table 1. The evaporation fluxes m'' are calculated according to:

$$m'' = \frac{q_t}{h_v} \tag{8}$$

Liquid	Formula	T_b [K]	ΔT_1	ΔT_2	$q_{n,\max}$ [kW/m^2]	$q_{f,\min}$ [kW/m^2]	m'' *10^{-3} [kg/m^{2}s]
Propane	C_3H_8	231.1	18.75	105	287	20	33
Butane	C_4H_{10}	272.6	20.35	114	301	23	(780) 18
1,3-Butadiene	$CH_2CHCHCH_2$	268.7	21.29	118	326	27	(630) 21
1-Butene	$CH_2CHCH_2CH_3$	266.9	20.39	114	311	23	(510) 24
Propene	C_3H_6	225.5	19.25	108	298	23	36
Isobutene	$CH_2C(CH_3)_2$	266.2	21.85	114	310	20	(530) 25
Ethane	C_2H_6	184.6	14.89	94	269	18	43

Ethene	C_2H_4	169.4	13.85	90	265	16	46
Ethylene oxide	CH_2OCH_2	283.6	31.72	149	502	31	22
Ammonia	NH_3	239.8	22.17	195	727	91	148

Table 1: Calculated evaporation fluxes during transition boiling of several boiling liquids spreading on an unconfined water surface. (water temperature: T_w=293 K)

It appears in Table 1 that the calculated evaporation rates for liquids with a boiling temperature near the water temperature is extremely high, and the heat flux necessary for such high evaporation rates cannot be delivered by the water. These calculated evaporation rates are given in brackets in Table 1. The heat to these liquids is assumed to be provided by the convective heat flow in water, as given by [Opschoor: 1977, Opschoor: 1980]:

$$q_w = 0.85 \cdot \lambda_w \cdot \left[\frac{g \cdot \alpha \cdot \Delta T^4}{a_w \cdot \upsilon_w} \right]^{\frac{1}{3}} \tag{9}$$

and the evaporation fluxes m'' are calculated according to:

$$m'' = \frac{q_w}{h_v} \tag{10}$$

Evaporation process for a boiling liquid from a confined water surface

Strong ice-formation is a consequence of a spillage of a liquefied gas on a confined water surface. The formation of an ice layer will strongly influence the evaporation of the released liquid. After the spillage on a confined water surface, the liquid will spread until the complete water surface area has been covered by the released liquid. The evaporation for this 'spreading' period of time can be calculated as formulated before in case of an unconfined water surface. At later stages, it will be assumed that water boundary is of circular shape, and that the outflow takes place in the center. After the coverage of the complete water surface by the released liquid, first film boiling or meta-stable boiling will take place, and a solid ice layer will be formed. During the formation of the ice layer the temperature difference between the released liquid and the ice/water will decrease, and after some time nucleate boiling will take place. The heat-flow from the increasing ice layer to the spreading pool during nucleate boiling is modelled (first equality), and approximated (second equality) by the following relation:

$$q_i(t) = \frac{\lambda_{ice} \cdot \Delta T^*}{\delta_{ice}(t)} = \left(\rho_w \cdot h_w + \frac{\rho_{vl} \cdot c_{ice} \cdot \Delta T^*}{2} \right) \cdot \frac{d\delta_{ice}(t)}{dt} \tag{11}$$

with the initial condition that at $t = 0 = t_0$: $\delta_{ice}(t_0) = 0$.

The solution of equation (11) gives for the layer thickness of the ice:

$$\delta_{ice}(t) = \sqrt{\frac{4 \cdot \lambda_{ice} \cdot \Delta T^* \cdot t}{(2 \cdot \rho_w \cdot h_w) + (\rho_{vl} \cdot c_{ice} \cdot \Delta T^*)}} \tag{12}$$

In the derivation of equation (12) ΔT^* is assumed constant, which is formally only valid in case $T_{liquid} = T_b$ and remains at the boiling temperature.

If the heat-flow due to the freezing of the water and the cooling down of the ice can be assumed equal to the evaporation flux $m''(t)$ of the liquid which is present on top of the ice layer, than by combining equation (11) with this assumption gives:

$$q_i(t) = m''(t) \cdot h_v = \frac{\lambda_{ice} \cdot \Delta T^*}{\delta_{ice}(t)} \tag{13}$$

and for the evaporation flux $m''(t)$ of the liquid on the ice layer we find:

$$m''(t) = \sqrt{\frac{(2 \cdot \rho_w \cdot h_w + \rho_{vl} \cdot c_{ice} \cdot \Delta T^*) \cdot \lambda_{ice} \cdot \Delta T^*}{4 \cdot h_v^2 \cdot t}} \tag{14}$$

Hereby it is assumed that a layer of ice is already present, in reality a layer of ice will be formed after a certain amount of time, if the water surface is sufficiently calm.

From experiments with LPG [Reid: 1978], ethane [Drake: 1975], and LNG [Opschoor: 1977], times for the formation of a solid layer of ice (by spreading of a boiling liquid over a confined water surface) τ_e are known: 2/3, 10, 20 for respectively LPG, ethane, and LNG. Looking at these results there appears to be a linear relation between the time required for the formation of a solid layer of ice τ_e [s] for a liquid spreading on a confined water surface and the boiling temperature T_b [K] of that liquid. This linear relation is given by:

$$\tau_e = -0.14 \cdot T_b + 35.58 \tag{15}$$

Equation (15) will give a negative value for τ_e for chemicals with a boiling point higher than 254.14 K. For these chemical the value of τ_e should be set equal to zero. When considering the evaporation of a boiling liquid on a confined water surface, a distinction is made in times before, and after, the formation of a solid layer of ice.
From experimental results [Opschoor: 1977] it is found that for the first period of time $0 \le t \le \tau_e + 1$, the evaporation flux can be found by the following linear relation:

$$m''(t) = M \cdot t \tag{16}$$

and for the second period of time (nucleate boiling on an increasing solid ice layer), the evaporation flux is calculated by:

$$m''(t) = \frac{M \cdot (\tau_e + 1)}{\sqrt{t - \tau_e}} \tag{17}$$

in which:

$$M = \sqrt{\frac{\left(2 \cdot \rho_w \cdot h_w + \rho_{vl} \cdot c_{ice} \cdot \Delta T^*\right) \cdot \lambda_{ice} \cdot \Delta T^*}{4 \cdot h_v^2 \cdot (\tau_e + 1)^2}} \tag{18}$$

The calculated evaporation fluxes with the equations (16) and (17) are in good agreement with the experimental findings in [Reid: 1978] and [Drake: 1975], as was shown graphically in [Opschoor: 1980].

Now with help of the calculated values for M and τ_e, it is possible to calculated the evaporation flux $m''(t)$ for the liquid considered. Consequently it is possible to calculate the evaporation of this boiling liquid for as well an instantaneous as continuous release of such a liquid on a confined water surface with the equation given below.

Case 11: Instantaneous release on an unconfined water surface

In the evaporation model which describes simultaneously the spreading and evaporation of boiling liquids on an open water surface for an instantaneous spill, it has been assumed that the heat of vaporisation comes primarily from the water below the spilled boiling liquid, and, during the spreading of the boiling liquid on the open water surface, no solid ice layer is being formed between the liquid and the water.

Spreading of the pool

The equations governing the spreading of a fluid make use of the concept of the equilibrium of the spreading and resisting forces and the continuity equation. For the spreading holds (gravitational force):

$$F_g = \frac{\pi \cdot \rho_{vl} \cdot g \cdot h_p^2(t) \cdot r_p(t) \cdot (\rho_w - \rho_l)}{\rho_w} \tag{19}$$

And for the inertia force holds:

$$F_t = -\frac{\pi \cdot \rho_{vl} \cdot h_p(t) \cdot r_p^2(t)}{C^{"}} \cdot \frac{d^2 r_p(t)}{dt^2} \tag{20}$$

Setting equation (19) equal to (20) gives the following differential equation:

$$r_p(t) \cdot \frac{d^2 r_p(t)}{dt^2} = -\frac{C^{"} \cdot h_p(t) \cdot g \cdot (\rho_w - \rho_{vl})}{\rho_w} \tag{21}$$

In general the volume of the spreading liquid can be written, when not taking the evaporation into account, as:

$$V_p(t)\left[= V_{p,0}\right] = \pi \cdot r_p^2(t) \cdot h_p(t) \tag{22}$$

Substitution of the height of the pool, $h_p(t)$, in equation (21) and solving the differential equation gives for the radius of the pool:

$$r_p^2(t) = \sqrt{\frac{4 \cdot C^{"} \cdot g \cdot (\rho_w - \rho_{vl}) \cdot V_{p,0}}{\pi \cdot \rho_w}} \cdot t + r_{p,0}^2 \tag{23}$$

The evaporation process

The evaporated quantity of the liquid at a time t can be calculated according to:

$$m(t) = \int_0^t m^{"} \cdot \pi \cdot r_p^2(\tau) d\tau \tag{24}$$

The values for the evaporation flux m'' should be calculated according to equation (8) or (10). Substitution of equation (23) in (24) and integration gives for the evaporated quantity at a specific time t:

$$m(t) = \left[m'' \cdot \sqrt{\frac{\pi \cdot C'' \cdot g \cdot (\rho_w - \rho_{vl}) \cdot V_{p,0}}{\rho_w}} \right] \cdot t^2 + \left[m'' \cdot \pi \cdot r_{p,0}^2 \right] \cdot t \tag{25}$$

and for the evaporation rate m' at a specific time t:

$$m'(t) = \left[2 \cdot m'' \cdot \sqrt{\frac{\pi \cdot C'' \cdot g \cdot (\rho_w - \rho_{vl}) \cdot V_{p,0}}{\rho_w}} \right] \cdot t + \left[m'' \cdot \pi \cdot r_{p,0}^2 \right] \tag{26}$$

To calculate the time t_e at which the total pool has evaporated (neglecting the formation of a minimum layer thickness), we insert the total initial mass $\left(= \rho_{vl} \cdot V_{p,0}\right)$ in the quadratic equation (25), the time t_e can than be approximated by (only the quadratic term in t is taken into account):

$$t_e = \left[\frac{\rho_{vl}^2 \cdot \rho_w \cdot V_{p,0}}{\pi \cdot C'' \cdot g \cdot (\rho_w - \rho_{vl}) \cdot m''^2} \right]^{\frac{1}{4}} \tag{27}$$

The corresponding maximum radius of the pool $r_{p,\max}$ is found by substitution of equation (27) in equation (23), and we find (only the linear term of t is taken into account):

$$r_{p,\max} = \left[\frac{16 \cdot C'' \cdot g \cdot (\rho_w - \rho_{vl}) \cdot \rho_{vl}^2 \cdot V_{p,0}^3}{\pi^3 \cdot \rho_w \cdot m''^2} \right]^{\frac{1}{8}} \tag{28}$$

Untill the time t_e has been reached the height of the pool $h_p(t)$ can be calculated according to:

$$h_p(t) = \frac{V_p(t)}{A_p(t)} = \frac{M_p(t)}{\pi \cdot \rho_{vl} \cdot r_p^2(t)} = \frac{(\rho_{vl} \cdot V_{p,0}) - m(t)}{\pi \cdot \rho_{vl} \cdot r_p^2(t)} \tag{29}$$

in which $\rho_l \cdot V_{p,0}$ is the amount of liquid which has been instantaneously released at $t = 0$.

Case 12: Instantaneous release on a confined water surface

Spreading of the pool

Immediately after the start of the outflow of the boiling liquid over the confined water surface, the liquid will start spreading until the entire water surface is completely covered by the liquid. This spreading time, t_0, can be determined from equation (23) and is given by:

$$t_0 = \frac{\left(A_b - \pi \cdot r_{p,0}^2\right)\sqrt{\rho_w}}{\sqrt{4 \cdot \pi \cdot C'' \cdot g \cdot (\rho_w - \rho_{vl}) \cdot V_{p,0}}} \tag{30}$$

in which A_b is the surface area of the confined water surface. The radius of the pool from $t = 0$ until $t = t_0$ is calculated with equation (23).

The evaporation process

The amount of liquid which has evaporated from $t = 0$ until $t = t_0$ can be calculated from equation (25) in which the equations (16) and (17) should be substituted for $m''(t)$.

The amount of liquid which has evaporated when the complete water surface is covered follows from equation (25), in which the time t_0, as deterimed in equation (30), is substituted:

$$m_0 = m(t_0) = \left[m''(t_0) \cdot \sqrt{\frac{\pi \cdot C'' \cdot g \cdot (\rho_w - \rho_{vl}) \cdot V_{p,0}}{\rho_w}}\right] \cdot t_0^2 + \left[m''(t_0) \cdot \pi \cdot r_{p,0}^2\right] \cdot t_0 \tag{31}$$

For the time interval $t \in [0, t_0]$ the height of the pool follows from:

$$h_p(t) = \frac{W_0 - m(t)}{\pi \cdot \rho_{vl} \cdot r_p^2(t)} \tag{32}$$

and the evaporation rate from:

$$m'(t) = A_b \cdot m''(t) \tag{33}$$

After the time t_0, the non-spreading period begins. For the time interval $t \in [t_0, t_e]$ the mass-balance looks like:

$$W_0 - m_0 = \int_{t_0}^{t_e} A_b \cdot m''(t) dt \tag{34}$$

The mass balance from equation (34) determines the time t_e at which all liquid has evaporated. Substitution of the equations (16) and (17) for the evaporation fluxes $m''(t)$, into the mass balance, we find for times $t_0 \le t_e < \tau_e + 1$:

$$W_0 - m_0 = \int_{t_0}^{t_e} A_b \cdot M \cdot t dt \tag{35}$$

and for times $t_e \ge \tau_e + 1$:

$$W_0 - m_0 = \int_{t_0}^{\tau_e+1} A_b \cdot M \cdot t dt + \int_{\tau_e+1}^{t_e} \frac{A_b \cdot M \cdot (\tau_e + 1)}{\sqrt{t - \tau_e}} dt \tag{36}$$

The solution of the integral in equation (35) gives for $t_0 \le t_e < \tau_e + 1$:

$$\begin{gathered} W_0 - m_0 = \int_{t_0}^{t_e} A_b \cdot M \cdot t dt = \left[\frac{A_b \cdot M \cdot t^2}{2} \right]_{t_0}^{t_e} = \frac{A_b \cdot M \cdot t_e^2}{2} - \frac{A_b \cdot M \cdot t_0^2}{2} \Rightarrow \\ 2(W_0 - m_0) = A_b \cdot M \cdot t_e^2 - A_b \cdot M \cdot t_0^2 = A_b \cdot M \left(t_e^2 - t_0^2 \right) \Rightarrow t_e = \sqrt{\frac{2 \cdot (W_0 - m_0)}{A_b \cdot M} + t_0} \end{gathered} \tag{37}$$

and the solution of the integrals in equation (36) gives for $t_e \ge \tau_e + 1$:

$$W_0 - m_0 = \int_{t_0}^{\tau_e+1} A_b \cdot M \cdot t dt + \int_{\tau_e+1}^{t_e} \frac{A_b \cdot M \cdot (\tau_e + 1)}{\sqrt{t - \tau_e}} dt =$$

$$W_0 - m_0 = \left[\frac{A_b \cdot M \cdot t^2}{2} \right]_{t_0}^{\tau_e+1} + \left[2 \cdot A_b \cdot M \cdot (\tau_e + 1) \cdot (t - \tau_e)^{1/2} \right]_{\tau_e+1}^{t_e} =$$

$$W_0 - m_0 = \left[\frac{A_b \cdot M \cdot (\tau_e + 1)^2}{2} - \frac{A_b \cdot M \cdot t_0^2}{2} \right] + \left[2 \cdot A_b \cdot M \cdot (\tau_e + 1) \cdot (t_e - \tau_e)^{1/2} - 2 \cdot A_b \cdot M \cdot (\tau_e + 1) \right] = \quad (38)$$

...

$$t_e = \left[\frac{2 \cdot (W_0 - m_0) - A_b \cdot M \cdot \left((\tau_e + 1)^2 - t_0^2 \right)}{4 \cdot A_b \cdot M \cdot (\tau_e + 1)} \right]^2 + \tau_e + 1$$

For time $t \in [t_0, t_e]$ the radius of the pool has a constant value which is determined by the confined water surface A_b and equals:

$$r_p(t) = r_p = \sqrt{\frac{A_b}{\pi}} \quad (39)$$

The evaporation rate, $m'(t)$, and the amount of liquid which has evaporated from the pool during this time period, $m(t)$, follow from the evaporation flux $m''(t)$ which is calculated according to the equations (16) or (17). The evaporation rate follows from:

$$m'(t) = A_b \cdot m''(t) \quad (40)$$

and the evaporated amount of liquid from the pool at a time $t + \Delta t$ can be approximated by:

$$m(t + \Delta t) = m(t) + m'(t) \cdot \Delta t \quad (41)$$

Case 13: Continuous release on an unconfined water surface

Spreading of the pool

In case of a continuous outflow, the following differential equation is found:

$$-\frac{\pi \cdot \rho_w}{C'' \cdot g \cdot (\rho_w - \rho_{vl})} \cdot r_p^3(t) \cdot \frac{d^2 r_p(t)}{dt^2} = V' \cdot t \quad (42)$$

in which $V^{'}$ is the outflow rate. In equation (42), the duration of the outflow time, t_s, should be substituted on the right hand site for t, in case $t > t_s$.

Solving equation (42), gives for the radius of the pool for times $t \le t_s$:

$$r_p(t) = \left[\frac{16 \cdot C^{''} \cdot g \cdot (\rho_w - \rho_{vl}) \cdot V^{'}}{3 \cdot \pi \cdot \rho_w}\right]^{\frac{1}{4}} \cdot t^{\frac{3}{4}} \tag{43}$$

and for times $t > t_s$:

$$r_p(t) = r_{p,s} \cdot \left[\frac{3 \cdot t}{2 \cdot t_s} - \frac{1}{2}\right]^{\frac{1}{2}} \tag{44}$$

in which $r_{p,s}$ is $r_p(t_s)$ as found by inserting t_s in equation (43).

The evaporation process

The total evaporated quantity of the liquid at a time t can be calculated according to:

$$m(t) = \int_0^t m^{''} \cdot \pi \cdot r_p^2(\tau) d\tau \tag{45}$$

in which the correct values for the evaporation flux $m^{''}$ should be substituted. In the case of a continuous release on open water, we find for times $t \le t_s$:

$$m(t) = \frac{2 \cdot \pi \cdot m^{''}}{5} \cdot \left[\frac{16 \cdot C^{''} \cdot g \cdot (\rho_w - \rho_{vl}) \cdot V^{'}}{3 \cdot \pi \cdot \rho_w}\right]^{\frac{1}{2}} \cdot t^{\frac{5}{2}} \tag{46}$$

and for times $t > t_s$:

$$m(t) = \frac{m^{''} \cdot \pi \cdot r_{p,s}^2 \cdot t_s}{20} \cdot \left\{15 \cdot \left[\frac{t}{t_s}\right]^2 - 10 \cdot \frac{t}{t_s} + 3\right\} \tag{47}$$

Consequently the evaporation rates $m^{'}(t)$ can be found by, for times $t \le t_s$:

$$m^{'}(t) = \pi \cdot m^{''} \cdot \left[\frac{16 \cdot C^{''} \cdot g \cdot (\rho_w - \rho_{vl}) \cdot V^{'}}{3 \cdot \pi \cdot \rho_w}\right]^{\frac{1}{2}} \cdot t^{\frac{3}{2}} \tag{48}$$

and times $t > t_s$:

$$m'(t) = \frac{m'' \cdot \pi \cdot r_{p,s}^2}{2} \cdot \left\{ 3 \cdot \frac{t}{t_s} - 1 \right\} \tag{49}$$

To calculate the time t_e at which the total pool has evaporated, we insert the total mass released at $t = t_s \left(= \rho_l \cdot V' \cdot t_s \right)$ in equation (47), the time t_e can than be approximated by (only the quadratic term in t is taken into account):

$$t_e = \left[\frac{20 \cdot V' \cdot t_s^2 \cdot \rho_{vl}}{15 \cdot m'' \cdot \pi \cdot r_{p,s}^2} \right]^{\frac{1}{2}} \tag{50}$$

The corresponding maximum radius of the pool $r_{p,\max}$ can be found by substitution of equation (27) in equation (44), and we find (when only taken the linear term of t into account):

$$r_{p,\max} = \left[\frac{21 \cdot r_{p,s}^6 \cdot V' \cdot \rho_{vl}}{10 \cdot \pi \cdot m''} \right]^{\frac{1}{8}} \tag{51}$$

For times $t \le t_s$, the height of the pool $h_p(t)$ should be calculated according to:

$$h_p(t) = \frac{V_p(t)}{A_p(t)} = \frac{M_p(t)}{\pi \cdot \rho_{vl} \cdot r_p^2(t)} = \frac{\left(\rho_{vl} \cdot V' \cdot t \right) - m(t)}{\pi \cdot \rho_{vl} \cdot r_p^2(t)} \tag{52}$$

in which $\rho_l \cdot V' \cdot t$ is the amount of liquid which has been released until the time t. For times $t > t_s$, the height of the pool follows from:

$$h_p(t) = \frac{V_p(t)}{A_p(t)} = \frac{M_p(t)}{\pi \cdot \rho_{vl} \cdot r_p^2(t)} = \frac{\left(\rho_{vl} \cdot V' \cdot t_s \right) - m(t)}{\pi \cdot \rho_{vl} \cdot r_p^2(t)} \tag{53}$$

in which $\rho_l \cdot V' \cdot t_s$ is the total amount of liquid which has been released at the time t_s.

Case 14: Continuous release on a confined water surface

Spreading of the pool

After the start of a continuous outflow of the boiling liquid on the confined water surface, the liquid will start spreading. The moment t_0 at which the water surface A_b is completely covered by the liquid can be determined for $0 \le t_0 \le t_s$ from equation (43), and for times $t_0 > t_s$ from equation (44). This gives for times $0 \le t_0 \le t_s$:

$$t_0 = \left[\frac{16 \cdot \pi^3 \cdot C'' \cdot g \cdot (\rho_w - \rho_{vl}) \cdot V'}{3 \cdot \rho_w} \right]^{-\frac{1}{3}} \cdot A_b^{\frac{4}{3}} \tag{54}$$

and for times $t_0 > t_s$:

$$t_0 = \frac{2 \cdot A_b^2 \cdot t_s + A_{p,s}^2 \cdot t_s}{3 \cdot A_{p,s}^2} \tag{55}$$

in equation (54), A_b is the surface area of the confined water surface, and in equation (55), $A_{p,s} = \pi \cdot r_{p,s}^2 = \pi \cdot r_p^2(t_s)$, t_s is the duration of the outflow, and $r_{p,s}$ is the radius of the pool at the moment the continuous outflow stops as calculated with equation (43).

Evaporation process

The amount of liquid which has evaporated before the complete water surface has been covered can be calculated with equation (46) for times $t \le t_s$, and with equation (47) for times $t > t_s$ when t_0 is substituted.

The determination of the time t_e at which the total amount of liquid has evaporated is similar to the way it was calculated in case 16: instantaneous release on a confined water surface: and results now for times $t_0 \le t_e < \tau_e + 1$ in:

$$t_e = \sqrt{\frac{2 \cdot (W_0 - m_0)}{A_b \cdot M}} + t_0 \tag{56}$$

and for times $t_e \geq \tau_e + 1$ in

$$t_e = \left[\frac{2 \cdot (W_0 - m_0) - A_b \cdot M \cdot \left((\tau_e + 1)^2 - t_0^2 \right)}{4 \cdot A_b \cdot M \cdot (\tau_e + 1)} \right]^2 + \tau_e + 1 \qquad (57)$$

in which W_0 is given by:

$$W_0 = \rho_{vl} \cdot V' \cdot t_s \qquad (58)$$

For the period in which the complete water surface is covered by the liquid $t > t_0$, the evaporation rate $m'(t)$ can be found by equation (40), and for times $t < t_0$ the evaporation rate follows from:

$$m'(t) = \pi \cdot r_p^2(t) \cdot m''(t) \qquad (59)$$

The evaporated amount of liquid from the pool at a time $t + \Delta t$ can consequently be approximated by equation (41).

Case 15: Instantaneous release on an unconfined water surface

In case of an accidental release of a liquid on an unconfined water surface, the liquid is considered non-boiling when the normal boiling point (T_b) (under ambient conditions) is higher than the water temperature (T_w) (on which the chemical is spilled) and/or the ambient temperature (T_a). The evaporation mass flow rate from a non-boiling liquid on an unconfined water surface will mainly depend upon:

1. the rate at which the vapour above the pool is removed by the wind. (mass transfer coeffient).
2. the sum of the heat fluxes which is supplied to the evaporating pool. (water, solar, air).

Evaporation mass flow rate

The evaporation mass flow rate $q_v^{'}(t)$ is generally given by:

$$q_v^{'}(t) = q_v^{''}(t) \cdot A_p(t) = q_v^{''}(t) \cdot \pi \cdot r_p^2(t) \tag{60}$$

in which $q_v^{''}(t)$ is the evaporation flux, and $r_p(t)$ the radius of the unconfined spreading pool on the water surface. For the evaporation flux the relation by [Kawamura: 1987] is applied which reads:

$$q_v^{''}(t) = \frac{k_m(t) \cdot P_v(T_p(t)) \cdot \mu_i}{R \cdot T_p(t)} \tag{61}$$

with the time-dependent mass transfer coefficient:

$$k_m(t) = 0.004786 \cdot (u_{w,10})^{0.78} \cdot (2 \cdot r_p(t))^{-0.11} \cdot Sc^{-0.67} \tag{62}$$

Spreading of the pool

The radius of the spreading pool is evaluated according to:

$$r_p^2(t) = \sqrt{\frac{4 \cdot C^{''} \cdot g \cdot (\rho_w - \rho_{vl}) \cdot V_{p,0}}{\pi \cdot \rho_w}} \cdot t + r_{p,0}^2 \tag{63}$$

as found before in case of a boiling liquid which is the solution of the differential equation as given in equation (21). In equation (63) the density of the liquid ρ_{vl} is evaluated at the actual pool temperature $T_p(t)$, which was, in case of a boiling pool, evaluated at the boiling temperature T_b of the liquid. The radius of the pool at $t = 0$, $r_{p,0}$ in equation (63) is found according to:

$$r_{p,0} = \sqrt{\frac{V_{p,0}}{\pi \cdot h_{p,0}}} \tag{64}$$

in which the height of the pool at $t = 0$ is assumed the have a value of e.g.:

$$h_{p,0} = 10 \cdot h_{p,\min} = 0.05 \text{ m}. \tag{65}$$

Heat balance

The temperature of the spreading pool $T_p(t)$ in equation (61) is evaluated according to:

$$T_p(t+dt) = T_p(t) + dT_p(t) \tag{66}$$

in which the temparature change $dT_p(t)$ is found from the heat balance as showed before in [Kootstra2: 2004], it reads:

$$dT_p(t) = \frac{\Sigma H(t) - q_v''(t) \cdot L_v(T_p(t))}{h_p(t) \cdot \rho_{vl} \cdot C_{p,l}} \cdot dt \tag{67}$$

The sum of the heat sources $\Sigma H(t)$ is in case of an evaporation from a water surface found by:

$$\Sigma H(t) = H_{water}(t) + H_{air}(t) + H_{solar-radiation} + H_{(long-wave)-radiation} \tag{68}$$

in which the heat flux as supplied by the water surface $H_{water}(t)$ can be found by:

$$H_{water}(t) = \frac{\lambda_w \cdot (T_{water} - T_p(t))}{\sqrt{a_w \cdot \pi \cdot t}} \tag{69}$$

From [PTZ: 49 Ed.] the value for the thermal conductivity of (sea-)water (3% salt at T_w=293 K) has been taken: $\lambda_w = 0.6$ W/mK [=J/msK]. The value for the thermal diffusivity of (sea-)water is now calculated according to:

$$a_w = \frac{\lambda_w}{\rho_w \cdot C_{p,w}} = \frac{0.6 \text{ [J/msK]}}{1.024 \cdot 10^3 \text{ [kg/m}^3\text{]} \cdot 3.93 \cdot 10^3 \text{ [J/kgK]}} = 1.49 \cdot 10^{-7} \text{ [m}^2\text{/s]} \tag{70}$$

in which the values for the density of water ρ_w, and the specific heat of water $C_{p,w}$, are also taken from [PTZ: 49 Ed.]. The equations for the heat fluxes as supplied by air, solar and long-wave radiation can be found in [Kootstra2: 2004].

Mass balance

The volume of the spreading pool $V_p(t)$ is evaluated according to:

$$V_p(t+dt) = V_p(t) + dV_p(t) \tag{71}$$

in which the volume change $dV_p(t)$ is found from the mass balance as showed before in [Kootstra2: 2004], it reads:

$$\frac{dM_p(t)}{dt} = -q_v^{"}(t) \cdot A_p(t) \Rightarrow dV_p(t) = \frac{-q_v^{"}(t) \cdot \pi \cdot r_p^2(t)}{\rho_{vl}} \cdot dt \tag{72}$$

The spreading of the pool on water is evaluated with equation (63) until the height of the pool becomes equal to the minimum pool thickness: $h_{p,\min}$. After reaching this value, the radius of the pool is calculated according to:

$$r_p(t) = \sqrt{\frac{V_p(t)}{\pi \cdot h_{p,\min}}} \tag{73}$$

Case 16: Instantaneous release on a confined water surface

Pool dimensions

For an instantaneous liquid released on a confined water surface it is assumed that the confined water surface, A_p, at $t = 0$ is completely filled with the total released amount of liquid at $t = 0$, $M_{p,0}$. The radius of the evaporating pool is assumed circular and therefore simply given by:

$$r_p = r_{p,0} = r_p(t) = \sqrt{\frac{A_p}{\pi}} \tag{74}$$

The height of the pool at $t = 0$ is consequently found by:

$$h_{p,0} = \frac{\left(\frac{M_{p,0}}{\rho_{vl}}\right)}{A_p} \tag{75}$$

Mass & heat balance

The evaluation in time of the height $h_p(t)$ and temperature $T_p(t)$ of the non-boiling, evaporating, confined pool, which has been released instantaneously, are (in general) according to:

$$X_p(t+dt) = X_p(t) + dX_p(t) \tag{76}$$

- The change in the height of the pool $dh_p(t)$ is found from the mass balance as showed before in [Kootstra2: 2004], it reads:

$$\begin{aligned} &\frac{dM_p(t)}{dt} = \left(-q_v^{'}(t) = \right) - q_v^{"}(t) \cdot A_p \rightarrow \rho_{vl} \frac{dV_p(t)}{dt} = -q_v^{"}(t) \cdot A_p \rightarrow \\ &\rho_{vl} \cdot A_p \frac{dh_p(t)}{dt} = -q_v^{"}(t) \cdot A_p \rightarrow dh_p(t) = -\left(\frac{q_v^{"}(t)}{\rho_{vl}}\right) \cdot dt \end{aligned} \tag{77}$$

  in which the evaporation flux $q_v^{"}(t)$ is the one by [Kawamura: 1987] as described in the previous case (case 15).

- The temparature change $dT_p(t)$ is found from the heat balance as showed before in [Kootstra2: 2004], and given in equation (67), in which the sum of the heat sources $\Sigma H(t)$ is the same as described in the previous case (case 15).

Case 17: Continuous release on an unconfined water surface

Pool dimensions

In case the liquid is released continuously over a certain period of time, the release rate $q_{out}^{'}$, during the outflow time t_s, is assumed constant. For the spreading of the pool on an unconfined water surface, the equations as derived before in case 13, can be used. The radius of the pool for times $t \le t_s$ can be found by equation (43), and for times $t > t_s$ by equation (44), in which $r_{p,s}$ is $r_p(t_s)$ as found by inserting t_s in equation (43). The spreading of the pool on water is evaluated with the equations (43) and (44) until the height of the pool reaches a minimum pool thickness: $h_{p,\min}$. After reaching this value, the radius of the pool is calculated according to equation (73).

Mass & heat balance

The evaluation in time of the volume $V_p(t)$ and temperature $T_p(t)$ of the evaporating, spreading pool are according to the general equation as given before in equation (76).

- The volume change $dV_p(t)$ is found from the mass balance as showed before in [Kootstra2: 2004], and is given by:

$$dV_p(t) = \frac{-\left(q_v^{"}(t)\cdot\pi\cdot r_p^2(t)\right)+V'}{\rho_{vl}}\cdot dt \tag{78}$$

  in which the evaporation flux $q_v^{"}(t)$ is the one by [Kawamura: 1987] as described in case 15: instantaneous release on an unconfined water surface. For times $t > t_s$, the release rate V' becomes zero and should be left out of equation (78).

- The temparature change $dT_p(t)$ is found from the heat balance as showed before in [Kootstra2: 2004], and given in equation (67).

Case 18: Continuous release on a confined water surface

Pool dimensions

For the (initial) spreading of the pool on a confined water surface, the equations as derived before in case 13, can be used. The radius of the pool for times $t \leq t_s$ can be found by equation (43), and for times $t > t_s$ by equation (44), in which $r_{p,s}$ is $r_p(t_s)$ as found by inserting t_s in equation (43). The spreading of the pool on water is evaluated with the equations (43) and (44) until the confined water surface is completely covered by the pool or the height of the pool reaches a minimum pool thickness: $h_{p,\min}$.

In case the borders of the confined water surface are reached by the spreading pool before layer thickness of the pool becomes equal to the minimum pool thickness, from that moment the radius of the pool has the constant value as deterimed by the confined water surface and can be evaluated by equation (74)

In case the layer thickness of the spreading pool becomes equal to the minimum pool thickness before the spreading pool reaches the borders of the confined water surface, from that moment, the radius of the pool is calculated according to equation (73).

Mass & heat balance

The evaluation in time of the volume $V_p(t)$ and temperature $T_p(t)$ of the evaporating pool is the same as described in the previous case (case17: continuous release on a unconfined water surface), with the exception that in case of a confined water surface after a certain moment the maximum radius of the pool will be reached, $r_{p,\max}$, as determined by the confinement of the water surface.

CONCLUSIONS

In this second paper, in a series of two, the theoretical background of a completely integrated evaporation model has been described for all kinds of possible pools evaporating from water in case of non-soluble, floating liquids on water (Part 1B). In the first paper in the series, the subject was on evaporating pools from land (Part 1A). For the situation on water, the evaporation process has been worked out and described for as well boiling as non-boiling liquids. The (accidental) outflow of the evaporating liquid is described for as well continuous as instantaneous releases. The spreading equations for the liquid on water are considered until a minimum layer thickness has been reached, or until the confinement of the water surface (rivers) is reached. The formation of ice is considered in case of a confined water surface (rivers), and the different boiling regimes in case of an open water surface (sea). The mass- and heat balance are simultaneously solved during the outflow, spreading and evaporation process for the evaluation of the temperature and the height of the pool over time during (or after) the release.

The evaporation model takes the following environmental- and weather conditions into account: several subsoils, humidity, wind speed, cloud cover, the influence of position in the world and date (day:month:year).

LIST OF SYMBOLS

Symbol	Description	Unit	Equation
q_n	heat flux in the nucleate boiling regime	W/m^2	1,6
q_f	heat flux in the film boiling regime	W/m^2	3,6
q_t	heat flux in the transition boiling regime	W/m^2	6,8
q_w	convective heat flux from water to the liquid	W/m^2	9,10
$q_{n,\max}$	maximum heat flux during nucleate boiling	W/m^2	2,4
$\rho_{v,s}$	density of the evaporated liquid at saturation temperature	kg/m^3	1,2,4
$\rho_{v,f}$	density of the evaporated liquid by an average temperature of liquid and water	kg/m^3	3
ΔT	temperature difference between liquid and water	K	1,3,7
γ	surface tension	N/m	1,2,4
T_b	boiling point of the liquid	K	1,4,15
T_c	critical temperature of the liquid	K	5
T_{vl}	temperature of the liquid	K	5
h_v	evaporation heat of the evaporated liquid	J/kg	2,8,11,13,14,18
g	acceleration of gravity	m/s^2	2,3,9,19,21,23,25,26,27,28,30,31,42,43,46,48, 54,63
ρ_w	density of water	kg/m^3	1,4,5,11,12,14,18,19,21,23,25,26,27,28,30,31, 42,43,46,,48,54,63,70
c_w	specific heat of water	J/kgK	1,4,5
λ_w	heat conduction coefficient of water	W/mK	1,4,5,9,69,70
ρ_{vl}	density of the liquid	kg/m^3	1,2,3,4,5,11,12,14,18,19,20,21,23,25,26,27,28 29,30,31,32,42,43,46,48,50,51,52,53,54,58,63 67,72,75,77,78
c_{vl}	specific heat of the liquid	J/kgK	1,4,5
υ_{vl}	kinematic viscosity of the liquid	m^2/s	1,4
λ_{vl}	heat conduction coefficient of the liquid	W/mK	1,4,5
λ_v	heat conduction coefficient of the evaporated liquid	W/mK	3
υ_v	kinematic viscosity of the evaporated liquid	m^2/s	3
a_v	thermal diffusivity of evaporated liquid	m^2/s	3
m''	evaporation rate	kg/m^2s	8,10,24,25,26,27,28,45,46,47,48,49,50,51
a_w	thermal diffusivity of water	m^2/s	9
υ_w	kinematic viscosity of water	m^2/s	9
α'	expansion coefficient of water	1/K	9
$q_i(t)$	heat flux at a specific time *t*	W/m^2	11,12
λ_{ice}	heat conduction coefficient of ice	W/mK	11,12,13,14,18
$\delta_{ice}(t)$	layer thickness of ice	m	11,12,13

ΔT^*	temperature difference between liquid and ice	K	11,12,13,14,18
h_w	sublimation heat of water	J/kg	11,12,14,18
c_{ice}	specific heat of ice	J/kgK	11,12,14,18
W_0	initial quantity of the instantaneously released liquid at t=0	kg	32,34,35,36,37,38,56,57,58
m_0	amount of liquid which has evaporated at $t=t_0$	kg	31,34,35,36,37,38,56,57
t_0	time until the complete water surface is covered with liquid	s	30,31,37,38,54,55,56,57
t_e	time until the total amount of liquid has evaporated	s	27,37,38,50,56,57
A_b	surface area of the confined water surface	m^2	30,33,34,35,36,37,38,39,40,54,55,56,57
$m''(t)$	evaporation flux of the liquid at time t	kg/m^2s	13,14,16,17,33,34,40,59
$m'(t)$	evaporation rate of the liquid at time t	kg/s	26,33,40,48,49,59
$m(t)$	evaporated mass from the liquid pool at time t	kg	24,25,29,32,41,45,46,47,52,53
$h_p(t)$	height of the pool at time t	m	19,20,21,22,29,32,52,53,67,77
τ_e	time required for the formation of a solid ice layer	s	15,17,18,36,38
F_g	gravitational force	N	19
F_t	inertia force	N	20
$r_p(t)$	radius of the liquid pool at time t	m	19,20,21,22,23,24,29,32,39,42,43,44,45,52, 53,59,60,62,63,72,73,74,78
C''	emperical constant = 2 (water)	-	20,21,23,25,26,27,28,30,31,42,43,46,48,54,63
$V_p(t)$	volume of the liquid pool at time t	m^3	22,29,52,53,71,72,73,77,78
$V_{p,0}$	volume of the liquid pool at t=0	m^3	23,25,26,27,28,29,30,31,63,64
$r_{p,0}$	radius of the pool at t=0	m	23,25,26,30,31,63,64,74
$h_{p,0}$	height of the pool at t=0	m	64,65,75
$M_p(t)$	mass in the pool at time t	kg	29,52,53,72,77
$A_p(t)$	surface area of the pool at time t	m^2	29,52,53,60,72
$r_{p,\max}$	maximum radius of the pool	m	28,51
V'	continuous outflow rate	kg/s of m3/s	42,43,46,48,50,51,52,53,54,58,78
t_s	outflow time of a continuous release	s	44,47,49,50,53,55
$q_v'(t)$	vaporisation mass flow rate at specific time t	kg/s	60,77
$q_v''(t)$	vaporisation mass flux at specific time t	kg/m^2s	60,61,67,72,78
$k_m(t)$	time-dependent mass transfer coefficient	m/s	61,62
$P_v(T_p(t))$	vapour pressure at pool temperature at specific time t	N/m^2	61
$T_p(t)$	temperature pool at specific time t	K	61,66,67,69
μ_i	molecular weight of substance i	kg/mol	61
R	universal gas constant	J/molK	61
$u_{w,10}$	wind speed at 10 m. height	m/s	62

Sc	Schmidt number	-	62
$\Sigma H(t)$	sum of the heat sources at specific time t	J/m^2s	67
$L_v(T_p(t))$	latent heat of vaporisation at pool temperature	J/kg	67
$C_{p,l}$	specific heat of the liquid	J/kgK	67
$H_{water}(t)$	heat as conducted from the water surface	J/m^2s	68,69
$H_{air}(t)$	heat as conducted by air	J/m^2s	68
$H_{solar-radiation}$	heat from solar radiation	J/m^2s	68
$H_{(long-wave)-radiation}$	heat by long-wave radiation	J/m^2s	68
T_{water}	temperature of the water surface	K	69
a_w	thermal diffusivity of the water surface	m^2/s	69
$C_{p,w}$	specific heat of the water	J/kgK	70

REFERENCES

[Kootstra: 2004] F. Kootstra, B.F.P. van het Veld and G. Opschoor, "*Use and application of the advanced evaporation model for liquids from land and water (non-soluble, floating)*" to be submitted.

[Kawamura: 1987] P.I. Kawamura and D. Mackay, "*The evaporation of volatile liquids*" Journal of hazardous materials, 15 (1987) 343-364.

[Kootstra2: 2004] F. Kootstra, B.F.P. van het Veld and G. Opschoor, "*A complete description of an advanced evaporation model for accidentally released hazardous liquids from land. Part IA – Theoretical background.*" to be submitted to AIChE 2004.

[Grigoryev: 1973] V.A. Grigoryev, Yu. M. Pavlov, and E.V. Ametistov, *Teploenergetika* **9**:57 (1973).

[PTZ: 49 Ed.] Poly Technisch Zakboek, 49e druk, ISBN: 90-6228-358-6.

[Kutateladze: 1952] S.S. Kutateladze, "Heat transfer in condensation and boiling", *Energia*, Moscow (1952).

[Kalinin: nyp] E.K. Kalinin, I.I. Berlin, V.G. Karavayev, V.V. Kostyuk, and E.M. Nosova, *Inzhenerno-fizicheskii Zh.*, to be published.

[Kalinin: 1973] E.K. Kalinin, I.I. Berlin, *Inzhenerno-fizicheskii Zh.* **25**(2):205 (1973).

[Kalinin: 1975] E.K. Kalinin, I.I. Berlin, V.V. Kostyuk, and E.M. Nosova, *Advances in Cryogenic Engineering* **21:**273 (1975).

[Opschoor: 1977] G. Opschoor, "Investigation into the spreading and evaporation of LNG spilled on water", *Cryogenics*, Vol. **17**, No. 11, pp. 629-633 (1977).

[Opschoor: 1980] G. Opschoor, "Investigation into the evaporation of liquefied gasses spilled on water", Bureau Industriële Veiligheid TNO, Rijswijk, Ref. nr. 80-05624 (April 1980).

[Reid: 1978] R.C. Reid, and K.A. Smith, "Behavior of LPG on water", *Hydrocarbon Processing*, Vol. **58**, No. 4 (1978).

[Drake: 1975] E.M. Drake, A.A. Jeje, and R.C. Reid, "Transient boiling of liquid cryogens on a water surface", *Int. J. Heat Mass Transfer*, Vol. **18**, pp. 1361-1368 (1975).

U.S. CHEMICAL SAFETY AND HAZARD INVESTIGATION BOARD

INVESTIGATION REPORT

HYDROGEN SULFIDE POISONING

(2 Dead, 8 Injured)

GEORGIA-PACIFIC
Naheola Mill
PENNINGTON, ALABAMA
JANUARY 16, 2002

KEY ISSUES
REACTIVE HAZARD IDENTIFICATION
HYDROGEN SULFIDE SAFETY
EMERGENCY RESPONSE

REPORT NO. 2002-01-I-AL
ISSUE DATE: JANUARY 2003

Abstract

This investigation report examines a hydrogen sulfide poisoning incident that occurred on January 16, 2002, at the Georgia-Pacific Naheola mill in Pennington, Alabama. Two contractors were killed, and eight were injured. County paramedics reported symptoms of hydrogen sulfide exposure. This report identifies the root and contributing causes of the incident and makes recommendations on reactive hazard identification, hydrogen sulfide safety, and emergency response.

The U.S. Chemical Safety and Hazard Investigation Board (CSB) is an independent Federal agency whose mission is to ensure the safety of workers, the public, and the environment by investigating and preventing chemical incidents. CSB is a scientific investigative organization; it is not an enforcement or regulatory body. Established by the Clean Air Act Amendments of 1990, CSB is responsible for determining the root and comtributing causes of accidents, issuing safety recommendations, studying chemical safety issues, and evaluating the effectiveness of other government agencies involved in chemical safety. No part of the conclusions, findings, or recommendations of CSB relating to any chemical incident may be admitted as evidence or used in any action or suit for damages arising out of any matter mentioned in an investigation report (see 42 U.S.C. § 412[r][6][G]). CSB makes public its actions and decisions through investigation reports, summary reports, safety bulletins, safety recommendations, special technical publications, and statistical reviews. More information about CSB may be found at www.chemsafety.gov.

Salus Populi Est Lex Suprema
People's Safety is the Highest Law

Information about available publications may be obtained by contacting:
U.S. Chemical Safety and Hazard Investigation Board
Office of Prevention, Outreach, and Policy
2175 K Street NW, Suite 400
Washington, DC 20037
(202) 261-7600

CSB publications may be purchased from:
National Technical Information Service
5285 Port Royal Road
Springfield, VA 22161
(800) 553-NTIS or
(703) 487-4600
Email: info@ntis.fedworld.gov
For international orders, see: www.ntis.gov/support/cooperat.htm.

For this report, refer to NTIS number PB2003-101293

Acronyms and Abbreviations

AFPA	American Forest and Paper Association
AIChE	American Institute of Chemical Engineers
ANSI	American National Standards Institute
ASSE	American Society of Safety Engineers
ATSDR	Agency for Toxic Substances and Disease Registry
CCPS	Center for Chemical Process Safety
CFR	Code of Federal Regulations
CPR	Cardiopulmonary resuscitation
CSB	U.S. Chemical Safety and Hazard Investigation Board
DHHS	U.S. Department of Health and Human Services
DOT	U.S. Department of Transportation
EMS	Emergency Medical Services
ERT	Emergency response team
H_2S	Hydrogen sulfide
HAZCOM	Hazard Communication Standard (OSHA)
Hazwoper	Hazardous Waste Operations and Emergency Response Standard (OSHA)
IBEW	International Brotherhood of Electrical Workers
IChemE	Institution of Chemical Engineers
IDLH	Immediately dangerous to life or health
MMG	Medical Management Guidelines
MOC	Management of change
MSDS	Material safety data sheet
NaSH	Sodium hydrosulfide
NIOSH	National Institute of Occupational Safety and Health
OSHA	Occupational Safety and Health Administration
PACE	Paper, Allied-Industrial, Chemical & Energy Workers International Union

Acronyms and Abbreviations (cont'd)

ppb	Parts per billion
ppm	Parts per million
PPSA	Pulp and Paper Safety Association
PSM	Process Safety Management (OSHA)
SCBA	Self-contained breathing apparatus
SCT	Secretariat of Transport and Communications (Canada)
TC	Transport Canada
U.S.C.	United States Code
WTTP	Wastewater treatment plant

Executive Summary

ES.1 Introduction

On January 16, 2002, highly toxic hydrogen sulfide (H_2S) gas leaked from a sewer manway at the Georgia-Pacific Naheola mill in Pennington, Alabama.[1] Several people working near the manway were exposed to the gas. Two contractors from Burkes Construction, Inc., were killed. Eight people were injured–seven employees of Burkes Construction and one employee of Davison Transport, Inc. Choctaw County paramedics who transported the victims to hospitals reported symptoms of H_2S exposure.

Because of the serious nature of this incident, the U.S. Chemical Safety and Hazard Investigation Board (CSB) initiated an investigation to determine the root and contributing causes and to issue recommendations to help prevent similar occurrences. CSB defines a reactive incident as:

> A sudden event involving an uncontrolled chemical reaction–with significant increases in temperature, pressure, or gas evolution–that has caused, or has the potential to cause, serious harm to people, property, or the environment.

Based on this definition, the incident that occurred at the Georgia-Pacific Naheola mill is a reactive chemical incident.

. . . the incident that occurred at the Georgia-Pacific Naheola mill is a reactive chemical incident.

CSB was assisted in this investigation by the Agency for Toxic Substances and Disease Registry (ATSDR), an agency of the U.S. Department of Health and Human Services (DHHS). One of the key functions of ATSDR is to respond to emergency releases of hazardous substances that affect public health.

[1] The mill began operation in 1958 as the Marathon Southern Division of the American Can Company. It was acquired in 1982 by James River Corporation. In 1997, James River merged with Fort Howard Corporation to form Fort James Corporation. Georgia-Pacific acquired Fort James Corporation in November 2000. The mill now operates as Fort James Operating Company, a fully owned subsidiary of Georgia-Pacific Corporation. This report describes events and history at the Naheola mill and does not necessarily differentiate among owners.

ES.2 Incident

Burkes Construction employees were working on a construction project at the Naheola mill in the vicinity of the tank truck unloading station, where various chemicals could be unloaded. Sodium hydrosulfide (NaSH) was being unloaded on January 15–16.

The unloading station consists of a large concrete pad sloped to a collection drain. A shallow concrete pit containing unloading pumps and associated process piping is located directly next to the pad and collection drain. This pit–commonly referred to as the oil pit–collects rainwater, condensate, and occasionally spilled chemicals from the unloading station. Due to environmental concerns about oil from the fuel oil pumps getting into the mill effluent, the drain valve from the oil pit to the acid sewer was locked closed.

The job required Burkes employees to work in or near the oil pit, which–at the time of the incident on January 16–contained liquid. Those interviewed estimated that it was typical for approximately 5 gallons of NaSH to collect in the oil pit from various sources (pump leaking, flushing unloading lines, etc.) during each offloading of a tank truck.

Fifteen tank trucks of NaSH had unloaded in the 24 hours prior to the incident.

Fifteen tank trucks of NaSH had unloaded in the 24 hours prior to the incident. Consequently–though the material in the pit was mainly water–it also contained NaSH from the unloading of the 15 trucks. To avoid having the construction crew stand in the fluid-filled pit, an operator opened a valve to drain the oil pit;[2] after 5 minutes, the valve was closed and relocked.

In the same area, three Davison Transport tank trucks arrived carrying NaSH. With the assistance of two Georgia-Pacific operators, one of the truck drivers connected his vehicle to the unloading hose. Witnesses estimated that when the connection was made, up to 5 gallons of NaSH spilled to the collection drain. (The tank truck, however, was not actually unloaded.)

[2] There was a procedure for periodically checking the oil pit and draining it if necessary. Although the oil pit was not full, the operator followed procedure when draining the pit.

On the day of the incident, sulfuric acid was being added to the acid sewer to control pH downstream in the effluent area. NaSH from the oil pit and the collection drain drained to the sewer and reacted with the sulfuric acid to form H_2S. Within 5 minutes, an invisible cloud of H_2S gas leaked through a gap in the seal of a manway in the area of the Burkes Construction workers. Two contractors near the manway were killed by H_2S poisoning; seven other Burkes employees and one Davison Transport driver were injured due to H_2S exposure.

NaSH from the oil pit and the collection drain drained to the sewer and reacted with the sulfuric acid to form H_2S.

Seven of the injured contractors were driven in private vehicles to Thomasville Infirmary in Thomasville, Alabama. Choctaw County Emergency Medical Services (EMS) transported three other victims (including the two fatally injured) to hospitals in Meridian, Mississippi. The clothing of one victim was completely removed and placed in a bag; the clothing of the other two victims was not removed.[3]

Two contractors near the manway were killed by H_2S poisoning; seven other Burkes employees and one Davison Transport driver were injured due to H_2S exposure.

The six Choctaw County paramedics who transported the victims reported symptoms of H_2S exposure; however, the two paramedics who removed the clothes of their patient reported milder symptoms. All of the County paramedics were medically evaluated and then released.

ES.3 Key Findings

1. H_2S was not identified as a hazard in the immediate area of the mill where the incident occurred. For this reason, there were no monitors, alarms, or warning signs in the area.

2. Modifications to the acid sewer over a period of several years included connections to the chlorine dioxide sewer, to the sewer from the truck unloading area, and to the containment area known as the oil pit. When these changes were made, the chemicals that could be added to the sewer and their interactions with other chemicals were not identified, nor were formal hazard evaluations or management of change (MOC) analyses conducted.

[3]The shirt of one victim was removed to facilitate medical procedures.

Georgia-Pacific did not require detailed H_2S safety training for those working in this area of the mill.

Beginning on the morning of January 16, sulfuric acid–used to control the pH of the mill wastewater–was continuously added to the acid sewer. . . . the volume of liquid in the acid sewer was lower than normal; and, consequently, the concentration of acid was high.

No hazard review or MOC analysis was performed when the oil pit and the collection drain from the truck unloading area were connected to the acid sewer.

3. Georgia-Pacific did not require detailed H_2S safety training for those working in this area of the mill. The contractors working on the day of the incident had only a basic awareness of H_2S and its hazards.

4. Beginning on the morning of January 16, sulfuric acid–used to control the pH of the mill wastewater–was continuously added to the acid sewer. Because the chlorine dioxide unit that emptied into the sewer line was not running at the time of the incident, the volume of liquid in the acid sewer was lower than normal; and, consequently, the concentration of acid was high.

5. During the truck unloading process, several potential sources of NaSH could leak and drain through the oil pit or collection drain to the acid sewer–such as material released from connecting and disconnecting the trucks, from flushing the loading line, or from a leaking pump seal or packing. In the 24 hours prior to the incident, 15 tank trucks of NaSH were unloaded, and one was connected to begin unloading. This activity resulted in NaSH collecting in the oil pit and draining to the sewer.

6. The manufacturer's material safety data sheet (MSDS) for NaSH states that this substance will generate H_2S gas if it contacts acid. For large spills, it recommends that runoff be prevented from entering sewers or drains.

7. The Naheola mill did not apply the principles of process safety management to truck unloading or to the acid sewer.[4]

    - Hazard information about NaSH, available on the MSDS, was not incorporated into mill procedures or training.
    - No hazard review or MOC analysis was performed when the oil pit and the collection drain from the truck unloading area were connected to the acid sewer.

[4] The principles of process safety management are incorporated into the Occupational Safety and Health Administration (OSHA) Process Safety Management (PSM) Standard (29 CFR 1910.119); however, even a facility that is not subject to the OSHA regulation should apply the good practices of process safety. The NaSH truck unloading area at the Naheola mill was not part of a process covered by the PSM Standard.

8. The manway involved in the incident was originally an unsealed open grate. The sewer was modified to convert it to a closed sewer, and the manway was eventually outfitted with a fiberglass cover and sealed.

The manway involved in the incident was originally an unsealed open grate.

9. Interviewed employees had observed leaking chlorine dioxide from the fiberglass manway on previous occasions and recalled repairs that were sometimes documented by work orders. These events were not reported as near-miss incidents, nor were the causes of the leaks formally investigated. The Georgia-Pacific corporate investigation policy–which requires investigating near-miss incidents, including releases of hazardous materials–was not yet in place at the Naheola mill.

The victims . . . were not decontaminated at the scene.

10. The victims were removed from the incident scene and taken to the mill first-aid station prior to setup of the incident command system. They were not decontaminated at the scene. Mill guidelines did not provide for decontamination at the first-aid station.

11. The clothing of one of the three victims transported to hospitals was removed, which was not the case with the other two men. The six Choctaw County paramedics who evacuated the men from the mill first-aid station reported symptoms consistent with H_2S exposure.

The ATSDR Medical Management Guidelines do not recognize that victims . . . [could] pose a medical risk to responders.

12. The ATSDR Medical Management Guidelines (MMG) do not recognize that victims of H_2S gas exposure may release H_2S, which can pose a medical risk to responders (ATSDR, 2001).

ES.4 Root Causes

1. **Good engineering and process safety practices were not followed when joining the drain from the truck unloading station and the oil pit to the acid sewer.**
   - Neither the chemicals that could be introduced into the sewer nor the hazards of their interactions were identified.

No formal hazard review or MOC analysis was conducted when connecting sewer lines from the tank truck unloading and chlorine dioxide areas to the acid sewer.

- No formal hazard review or MOC analysis was conducted when connecting sewer lines from the tank truck unloading and chlorine dioxide areas to the acid sewer. Consequently, no scenarios leading to the possible release of H_2S were identified, nor were warning devices placed in the area.

2. **There was no management system to incorporate hazard warnings about mixing sodium hydrosulfide (NaSH) with acid into process safety information.**

Operating procedures . . . did not warn of the hazard of mixing NaSH with acids or the hazard of allowing NaSH to enter sewers.

- Design information for projects involving NaSH did not specify the hazard of mixing NaSH with acid.
- Operating procedures for NaSH tank truck unloading and oil pit operations did not warn of the hazard of mixing NaSH with acids or the hazard of allowing NaSH to enter sewers.
- Mill personnel were not trained on the specific hazards of NaSH, such as handling spilled material or keeping it separate from acid.

ES.5 Contributing Causes

1. **The fiberglass manway was not adequately designed or sealed to ensure that the sewer remained closed.**

The manway was originally an open grate. It was modified by adding a fiberglass cover and sealed. Those interviewed recalled prior occasions when chlorine dioxide escaped from the fiberglass manway; they also recalled repairs that were sometimes documented in work orders, but these events were not reported or investigated as incidents. Thorough incident investigation would have provided an opportunity to recognize that the fiberglass manway was not appropriately designed to provide and maintain an airtight seal on the sewer.

. . . the fiberglass manway was not appropriately designed to provide and maintain an airtight seal on the sewer.

2. **The injured contractors did not have adequate training to understand the hazards of hydrogen sulfide (H_2S).**

H_2S training should include specific instruction on the importance of wearing proper protective equipment prior to attempting rescue.

ES.6 Recommendations

Georgia-Pacific Corporation

1. Conduct periodic safety audits of Georgia-Pacific pulp and paper mills in light of the findings of this report. At a minimum, ensure that management systems are in place at the mills to:

   - Evaluate process sewers where chemicals may collect and interact, and identify potential hazardous reaction scenarios to determine if safeguards are in place to decrease the likelihood or consequences of such interactions. Take into account sewer system connections and the ability to prevent inadvertent mixing of materials that could react to create a hazardous condition.
   - Identify areas of the mill where hydrogen sulfide (H_2S) could be present or generated, and institute safeguards (including warning devices) to limit employee exposure. Require that personnel working in the area are trained to recognize the presence of H_2S and respond appropriately. Update emergency response plans for such areas, to include procedures for decontaminating personnel exposed to toxic gas.
   - Apply good engineering and process safety principles to process sewer systems. For instance, ensure that hazard reviews and management of change (MOC) analyses are completed when additions or changes are made where chemicals could collect and react in process sewers. (Such principles may be found in publications from the Center for Chemical Process Safety [CCPS].)

2. Communicate the findings and recommendations of this report to the workforce and contractors at all Georgia-Pacific pulp and paper mills.

Georgia-Pacific Naheola Mill

1. Evaluate mill process sewer systems where chemicals may collect and react to identify potential hazardous reaction scenarios to determine if safeguards are in place to decrease the likelihood or consequences of such interactions. Evaluate sewer connections and ensure that materials that could react to create a hazardous condition are not inadvertently mixed, and that adequate mitigation measures are in place if such mixing does occur.

2. Establish programs to comply with recommendations from manufacturers of sodium hydrosulfide (NaSH) regarding its handling, such as preventing it from entering sewers because of the potential for acidic conditions.

3. Establish programs to require the proper design and maintenance of manway seals on closed sewers where hazardous materials are present.

4. Identify areas of the plant where hydrogen sulfide (H_2S) could be present or generated, and institute safeguards (including warning devices) to limit personnel exposure. Institute a plan and procedures for dealing with potential H_2S releases in these areas, and require that anyone who may be present is adequately trained on appropriate emergency response practices, including attempting rescue. Require contractors working in these areas to train their employees on the specific hazards of H_2S, including appropriate emergency response practices.

5. Update the Naheola mill emergency response plan to include procedures for decontaminating personnel who are brought to the first-aid station. Include specific instructions for decontaminating personnel exposed to H_2S so that they do not pose a secondary threat to medical personnel.

Agency for Toxic Substances and Disease Registry (ATSDR)

Evaluate and amend as necessary the ATSDR Medical Management Guidelines to consider the risk to responders posed by the exposure to victims of high levels of hydrogen sulfide (H_2S) gas. Specify procedures for adequate decontamination. Communicate the results of this activity to relevant organizations, such as the American Association of Occupational Health Nurses.

Burkes Construction, Inc.

Train your employees on the specific hazards of hydrogen sulfide (H_2S), including appropriate emergency response practices, in areas where Georgia-Pacific has identified this material as a hazard.

Davison Transport, Inc.

Communicate the findings and recommendations of this report to those employees who haul or handle sodium hydrosulfide (NaSH).

American Forest and Paper Association (AFPA)

International Brotherhood of Electrical Workers (IBEW)

Paper, Allied-Industrial, Chemical & Energy Workers International Union (PACE)

Pulp and Paper Safety Association (PPSA)

Communicate the findings and recommendations of this report to your membership.

1.0 Introduction

1.1 Background

On January 16, 2002, hydrogen sulfide (H_2S) gas generated in a sewer leaked from a gap in the seal of a manway at the Georgia-Pacific Naheola mill in Pennington, Alabama. Several people working near the manway were exposed to the gas. Two contractors from Burkes Construction, Inc., were killed. Seven employees of Burkes Construction and one employee of Davison Transport, Inc., were injured. The six Choctaw County paramedics who transported the victims to hospitals reported symptoms consistent with H_2S exposure.

1.2 Investigative Process

The U.S. Chemical Safety and Hazard Investigation Board (CSB) examined physical evidence at the site, conducted interviews, and reviewed relevant documents. CSB exercised a Memorandum of Understanding with the Agency for Toxic Substances and Disease Registry (ATSDR) to assist in investigating emergency and medical response to the incident.

1.3 Georgia-Pacific Naheola Mill

The Georgia-Pacific Naheola mill is located in Pennington, Alabama, approximately 125 miles north of Mobile and 150 miles southwest of Birmingham. Pennington is also located 50 miles east of Meridian, Mississippi.

The mill began operation in 1958 as the Marathon Southern Division of the American Can Company. It was acquired in 1982 by James River Corporation. In 1997, James River merged with Fort Howard Corporation to form Fort James Corporation. Georgia-Pacific acquired Fort James Corporation in November 2000. The mill now operates as Fort James Operating Company, a fully owned subsidiary of Georgia-Pacific Corporation.

The Naheola mill produces over 650,000 tons of paper, paperboard, and pulp annually. Approximately 1,475 employees work at the mill. The Paper, Allied-Industrial, Chemical & Energy Workers International Union (PACE) and the International Brotherhood of Electrical Workers (IBEW) represent some of the mill employees.

1.4 Overview of Naheola Mill

The pulping liquor is recycled, and fresh chemicals are occasionally added to maintain the proper liquor chemistry. Sodium hydrosulfide is one of these makeup chemicals.

1.4.1 Kraft Process

The Naheola mill uses the kraft process to produce pulp.[1] In this process, wood chips are treated with a liquor that chemically breaks them down into pulp. Kraft pulping liquor is made up of sodium hydroxide and sodium sulfide. The pulping liquor is recycled, and fresh chemicals are occasionally added to maintain the proper liquor chemistry. Sodium hydrosulfide (NaSH) is one of these makeup chemicals.

The pulp is bleached using chlorine dioxide, caustic, and oxygen. The chlorine dioxide is generated onsite. Some of the pulp is dried and sold as a final product. The remaining pulp is processed primarily into tissue, towels, and paperboard.

The overall pH of the effluent is affected by various mill operations. If the pH is low, caustic is added in the wastewater treatment area; if the pH is high, sulfuric acid is added.

1.4.2 Waste Treatment Process

The Naheola mill contains a process sewer network that collects waste; the process sewers join at a mixing basin in the effluent treatment area.

The pH[2] of the untreated mill effluent is maintained between 7 and 9 to ensure that the waste is efficiently treated. The overall pH of the effluent is affected by various mill operations. If the pH is low, caustic is added in the wastewater treatment area; if the pH is high, sulfuric acid is added.

[1] Pulp is the fibrous material derived from wood. It is the main raw material in making paper.

[2] pH is a measure of the acidity or alkalinity of a solution. It is numerically equal to 7 for neutral solutions; it increases with decreasing alkalinity and decreases with increasing acidity.

The acid is manually added in a sewer line, commonly referred to as the acid sewer, which originates in the chemical area where the acid tank is located (Figure 1). Drainage from the truck unloading station flows to the acid sewer and mixes with other mill waste streams in the mixing basin. Wastewater flows from the mixing basin to one of two clarifiers. The streams from the clarifiers are combined at a lift station prior to additional treatment and discharge to the Tombigbee River.

Drainage from the truck unloading station flows to the acid sewer and mixes with other mill waste streams in the mixing basin.

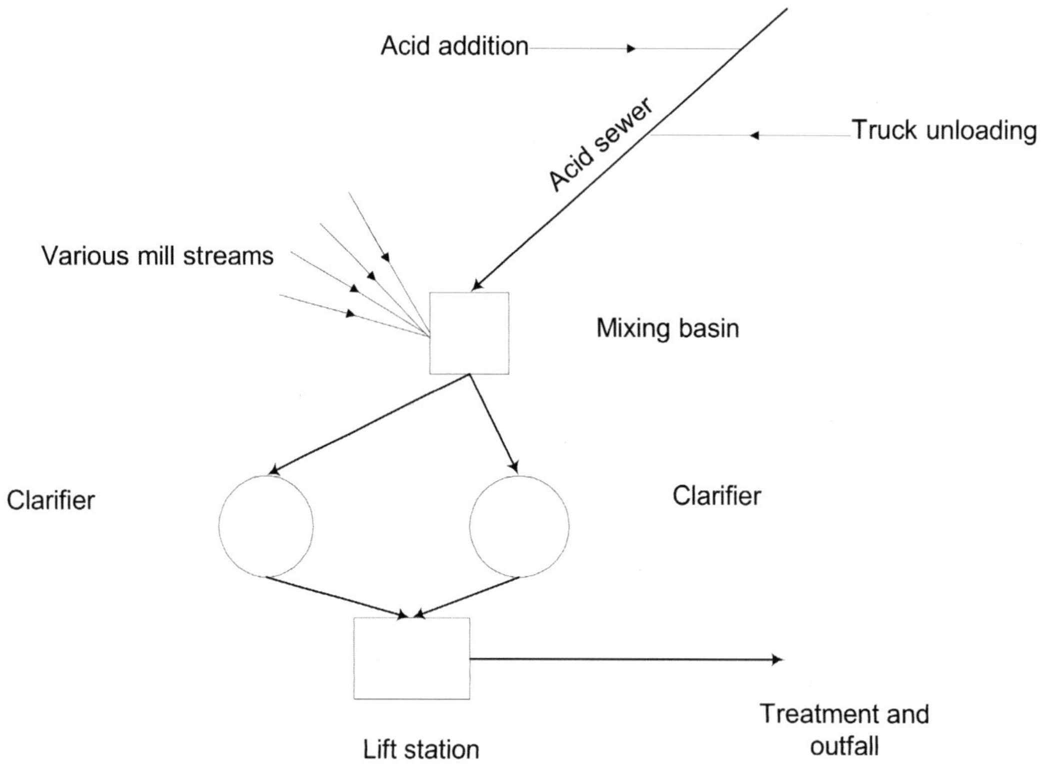

Figure 1. Naheola mill waste treatment process.

2.0 Description of Incident

2.1 Background

2.1.1 Sewer Operation

On January 16, the pH of the wastewater at the mill effluent treatment plant was high. To ensure efficient wastewater treatment and to comply with environmental permits, sulfuric acid was added to the acid sewer per normal procedure, as described in Section 1.4.2. Acid was continuously added to the sewer on the day of the incident; at the same time, the chlorine dioxide generator was not running, which resulted in a lower than normal flow of water through the sewer. This combination of acid volume and low water flow accounted for the high concentration of acidic water in the sewer.

To ensure efficient wastewater treatment and to comply with environmental permits, sulfuric acid was added to the acid sewer per normal procedure . . .

[The] combination of acid volume and low water flow accounted for the high concentration of acidic water in the sewer.

2.1.2 Tank Truck Unloading

NaSH[3] is used in the kraft process to make up pulping liquors. It is delivered by tank truck and stored onsite. The Naheola mill may go several months without a delivery and then bring in several tank trucks in a short span of time to replenish the supply. NaSH is delivered to an unloading station located in a typically unoccupied area near the maintenance shops, between the chemical area and the wastewater treatment area. Fuel oil and caustic are unloaded in the same area.

NaSH is delivered to an unloading station located in a typically unoccupied area near the maintenance shops . . .

The tank truck unloading station (Figure 2) is located on a large concrete pad sloped to a collection drain. The drain is routed through two normally open valves (valves 2 and 3) to the acid sewer. The process piping and various unloading pumps are located on a shallow curbed concrete containment area, directly next to the pad and collection drain (Figure 3). This area is commonly referred to as the oil pit (Figure 4).

[3] The Naheola mill used a solution of 30 percent NaSH in water.

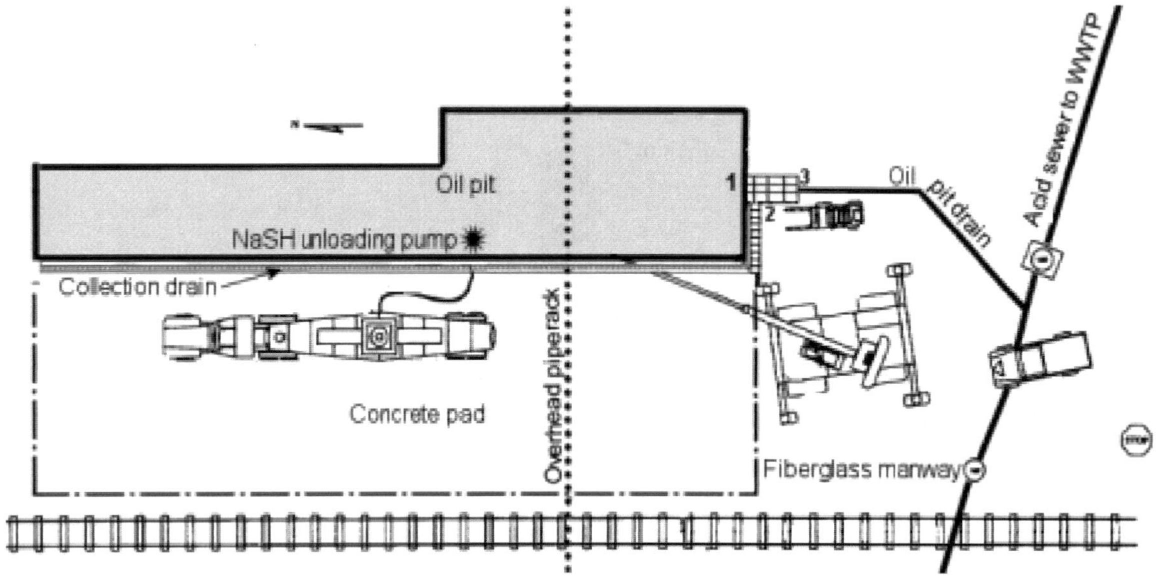

Figure 2. Tank truck unloading station.

Figure 3. Collection drain next to tank truck unloading station.

Figure 4. Oil pit and adjacent tank truck unloading station.

The oil pit (with a maximum depth of 20 inches) collects rainwater, condensate, and incidentally spilled chemicals from the tank truck unloading station. It is emptied through two valves (see also Figure 5). The drain valve (valve 1) on the pit is kept closed and locked due to environmental concerns about oil getting into the mill effluent; valve 3 is kept open. Per procedure, operators periodically inspect the oil pit; if no oil is present, they unlock valve 1 and drain the pit to the acid sewer.

The oil pit . . . collects rainwater, condensate, and incidentally spilled chemicals from the tank truck unloading station.

Figure 5. Oil pit control valves.

Personnel interviewed estimate that it is typical for approximately 5 gallons of NaSH to collect in the oil pit from various sources (pump leaking, flushing unloading lines, etc.) during each offloading of a tank truck. NaSH spilled during hose connections to the truck flows through the collection drain to the acid sewer. Material lost from the pump or piping during the unloading process accumulates in the oil pit until drained.

At the time of the January 16 incident, the mill was replenishing its NaSH inventory. In the 24 hours preceding the release of H_2S gas, 15 truckloads of NaSH were unloaded and a sixteenth was being prepared for unloading.

In the 24 hours preceding the release of H_2S gas, 15 truckloads of NaSH were unloaded and a sixteenth was being prepared for unloading.

2.1.3 Construction Work

Burkes Construction was contracted to replace supports on an overhead piperack located near the fiberglass manway . . .

Burkes Construction was contracted to replace supports on an overhead piperack located near the fiberglass manway; the piperack crossed over the tank truck unloading station and the oil pit, as shown in Figure 2. As the supports were removed, they were placed on the side of the road, directly in front of the manway.[4] On the day of the incident, the contractors were working in the area of the tank truck unloading station.

2.2 Incident Description

At approximately 3:15 pm on January 16, the Burkes Construction employees finished their afternoon break and prepared to return to work (see the timeline in Appendix A). They were standing on level ground in an open area adjacent to the tank truck unloading station, near the fiberglass manway.

. . . the Burkes Construction employees . . . were standing on level ground in an open area adjacent to the tank truck unloading station, near the fiberglass manway.

The piperack work required some Burkes employees to be in or near the oil pit, which contained liquid at the time; the material was primarily water and NaSH that had collected during unloading of the 15 tank trucks.[5]

The Burkes contractors asked a Georgia-Pacific operator to drain the oil pit. After inspecting for oil, the operator opened valve 1; the pit drained for approximately 5 minutes, and then the valve was closed.

Meanwhile, with the assistance of two Georgia-Pacific operators, the Davison truck driver connected his vehicle to the unloading hose. Witnesses estimated that when the connection was made, up to 5 gallons of NaSH spilled to the collection drain. Unloading had not actually begun.

[4] There is a possibility that one of the supports contacted the manway and compromised its seal.

[5] Those interviewed estimated that–during the unloading process–it is typical for up to 5 gallons of NaSH to collect in the oil pit from various sources (i.e., pump leaking, flushing unloading lines, etc.).

Also on the same day, sulfuric acid was being added to the acid sewer to control pH downstream in the effluent area. The NaSH draining to the acid sewer from the oil pit and the collection drain around the truck unloading station reacted with sulfuric acid (present in the sewer) to form H_2S.

The NaSH draining to the acid sewer from the oil pit and the collection drain . . . reacted with sulfuric acid (present in the sewer) to form H_2S.

Within 5 minutes of the material draining from the oil pit and collection drain to the acid sewer, an invisible cloud of highly toxic H_2S gas leaked through a gap in the seal of the manway near the Burkes Construction workers. Three contractors (A, B, and C in Figure 6) were immediately overcome by H_2S and fell down. Another contractor (D) passed out as he rushed to aid a coworker (A). Simultaneously, one contractor (I) left the area, and four others (E, F, G, and H) walked to fresh air. One of these men (F) passed out momentarily before he and another contractor (E) realized that their coworkers had fallen; the two returned to the area to assist in removing the others to fresh air. Contractor (F) passed out a second time while attempting to assist his coworkers. Georgia-Pacific operators working in the vicinity of the NaSH tank truck saw the Burkes employees falling and went to the control room to report the incident.

Within 5 minutes of the material draining . . . to the acid sewer, an invisible cloud of highly toxic H_2S gas leaked through a gap in the seal of the manway . . .

Two Burkes employees (B and C) in the immediate area of the manway were killed. Seven other Burkes employees and one Davison truck driver (J) were injured.

Two Burkes employees in the immediate area of the manway were killed.

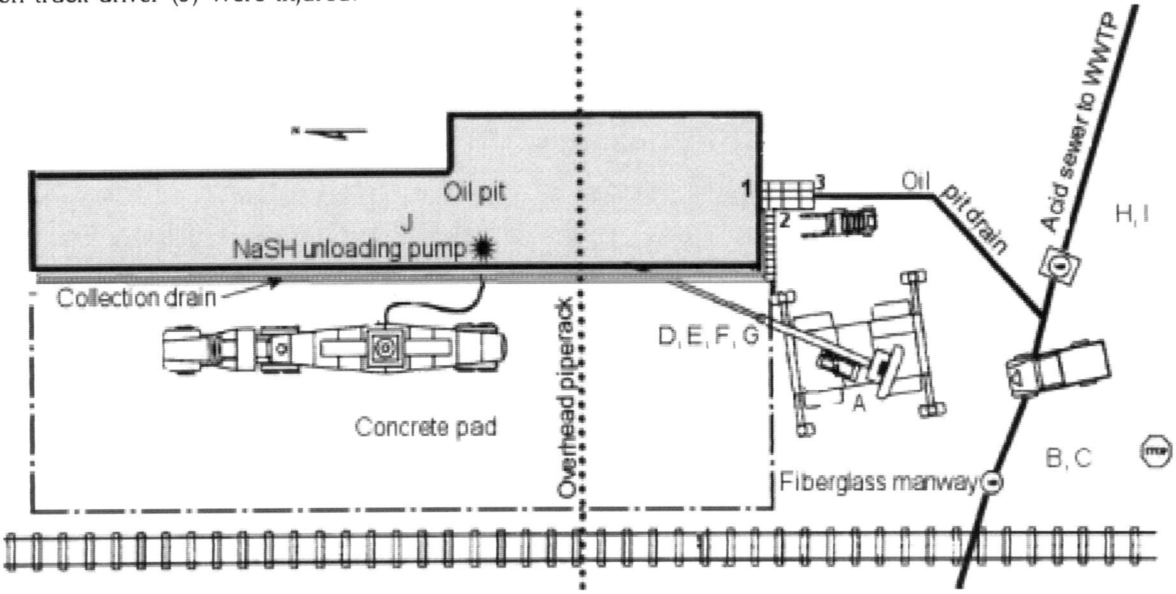

Figure 6. Approximate locations of the 10 victims ("A" through "J").

2.3 Incident Response

The injured were not decontaminated at the scene or at the first-aid station.

The six paramedics who transported the victims to hospitals . . . all described a strong odor in the ambulance bays.

Two of the paramedics reported less severe exposure symptoms–probably because their patient was exposed to a lower level of H_2S and his clothing had been removed.

The injured Burkes employees assisted each other and carried their fatally injured coworkers from the area closest to the release, as described in Section 2.2. The mill emergency response team (ERT) was notified. Others who came to the area performed cardiopulmonary resuscitation (CPR) on the two fatally injured men, who were transferred by mill ambulance to the first-aid station.[6]

The other injured personnel were transported by pickup trucks and motorized carts. The injured were not decontaminated at the scene or at the first-aid station. The incident commander arrived shortly after the victims had left the scene. He and the ERT evacuated the area, began search/rescue, set up zones, and began air monitoring.

Four Choctaw County ambulances–staffed with two paramedics each–arrived at the mill first-aid station. Two ambulances transported the fatally injured men, and a third ambulance transported the injured truck driver.[7] Two victims were taken to Rush Hospital and one to Riley Hospital, both located in Meridian, Mississippi, a 45-minute drive. Burkes and Georgia-Pacific employees transported the other injured Burkes employees to nearby Thomasville Infirmary, where six were admitted and one was sent to a hospital in Mobile because of a previous heart condition.

The six paramedics who transported the victims to hospitals in Meridian all described a strong odor in the ambulance bays. They opened windows and turned fans on to reduce the odor. After delivering the victims to the hospitals, the paramedics reported symptoms consistent with H_2S exposure. They were medically evaluated and released.

Two of the paramedics reported less severe exposure symptoms–probably because their patient was exposed to a lower level of H_2S and his clothing had been removed.

[6] The mill first-aid station is located near the front gate of the mill, a considerable distance from the incident scene.

[7] The fourth ambulance remained at the scene on standby and did not transport patients.

3.0 Analysis of Incident

3.1 Good Engineering Practices

The sewer line from the oil pit to the acid sewer was installed as part of a project to direct water from various storage pits to a process sewer. The work was completed in 1995 when the mill was owned by James River Corporation.

CSB did not find any procedures that described the engineering process. However, the James River capital appropriations request contained a Project Impact Statement checklist. The checklist asked general safety and environmental questions, focusing mainly on regulatory compliance. There was no detailed information on chemicals that could be present in the oil pit and drain to the acid sewer–nor was there a management of change (MOC) analysis or formal hazard review.[8]

There was no detailed information on chemicals that could be present in the oil pit and drain to the acid sewer–nor was there a management of change analysis or formal hazard review.

The NaSH material safety data sheet (MSDS) and manufacturer's information explain the hazards of NaSH and state that its interaction with acid will produce H_2S. For large spills, it recommends that runoff be prevented from entering sewers or drains. Another NaSH manufacturer's product literature cautions against allowing NaSH to drain to sewers because their contents are often acidic. If James River had identified the chemicals that could potentially drain to the acid sewer and completed a proper hazard review, it would likely have realized the risk of allowing NaSH to enter the sewer.

The NaSH material safety data sheet and manufacturer's information explain the hazards of NaSH and state that its interaction with acid will produce H_2S.

Drawings of the new sewer connections were made for construction. However, it was not James River's practice to update overall mill sewer drawings to reflect modifications. Therefore, at the time of the incident, no drawing of the mill sewer system accurately portrayed its layout.

If James River [Corporation] had identified the chemicals that could potentially drain to the acid sewer and completed a proper hazard review, it would likely have realized the risk of allowing NaSH to enter the sewer.

Good engineering practices during the conceptual design phase of a capital project such as this include identifying all chemicals that could be directed to the sewer, their hazards, and the hazards of their interactions.[9] The design information found for this project did not list NaSH or other substances that collected in the oil pit and could

[8] A hazard review is a systematic process to investigate hazards, assess potential consequences, and establish a design and operating basis for safety.

[9] The American Institute of Chemical Engineers (AIChE) Center for Chemical Process Safety (CCPS) describes good engineering practices for various phases of a capital project (CCPS, 1989).

The design information found for this project did not list NaSH or other substances that collected in the oil pit and could enter the acid sewer.

enter the acid sewer. Identification of the chemicals and their hazards, and a proper hazard review, would have identified the likely consequences and also the modifications required to eliminate or control the hazard. The Naheola mill did not follow these practices.

Process sewers are an integral part of a waste collection, treatment, and disposal process that presents unique hazards. Additionally, the various effluents collected present their own hazards (e.g., mixing and cross-reactions). There is considerable documentation of incidents that have occurred in sewers and drains. Appropriate techniques used for other chemical processes should also be applied to the design and hazard evaluation of process sewers (Lees, 1996; CCPS, 1995; Kletz, 1999).

3.2 Principles of Process Safety Management

Process safety management is the application of management systems to control hazards and to ensure the safety of a process. The use of such systems is considered good practice for handling chemical processes.[10]

The inadvertent mixing of incompatible chemicals is not uncommon. However, the Naheola mill had no formal reactive hazard management system to identify and control reactive hazards.

3.2.1 Reactive Hazard Management

NaSH from the tank truck unloading station was released to the acid sewer, where it inadvertently reacted with sulfuric acid to produce H_2S. The inadvertent mixing of incompatible chemicals is not uncommon. However, the Naheola mill had no formal reactive hazard management system to identify and control reactive hazards.

[10] The principles of process safety management are incorporated into the Occupational Safety and Health Administration (OSHA) Process Safety Management (PSM) Standard (29 CFR 1910.119); however, even a facility that is not subject to the OSHA regulation should apply the good practices of process safety management.

The first step in a reactive hazard management program is to determine whether a material has reactive hazards (CCPS, 2001). Reactive hazard information is found in literature and on some MSDSs. The MSDS collected by the mill from its NaSH supplier indicated that this substance readily reacts with acid to form H_2S–which clearly indicates a potential reactive hazard.

After the identification of a reactive hazard, precautions should be taken to mitigate the hazard. If such precautions were in place at the Naheola mill, the oil pit containing NaSH would likely not have been tied into the acid sewer.

The MSDS collected by the mill from its NaSH supplier indicated that this substance readily reacts with acid to form H_2S–which clearly indicates a potential reactive hazard.

If [mitigative measures] were in place at the Naheola mill, the oil pit containing NaSH would likely not have been tied into the acid sewer.

3.2.2 Process Safety Information

The NaSH MSDS contains warnings about the hazards of combining NaSH solution with acid. It notes that the reaction will produce toxic H_2S. The MSDS also specifies that large spills should be cleaned up using absorbent material, and that the substance should be prevented from entering sewers or natural waterways. Additionally, it notes that the headspace of a tank truck could contain H_2S vapors that evolve from the solution. The NaSH MSDS was available in the control room and at the tank truck unloading station.

The mill used a chemical approval form to collect information on all chemicals brought onsite. Although the form for NaSH contained information about its hazards, none of this information was incorporated into operating procedures or training. Process safety information should be incorporated into policies, procedures, and programs. (CCPS, 1995b) Had this been done, employees would likely have understood the hazards of NaSH, including the manufacturer's recommendation that spills be cleaned up with absorbent material instead of being washed into a sewer.

Although the [chemical approval] form for NaSH contained information about its hazards, none of this information was incorporated into operating procedures or training.

3.2.3 Incident Investigation

Interviewed employees reported prior instances of chlorine dioxide leaking from the manway involved in the January 16 incident. . . . however, these events were not reported as incidents or investigated.

Incident investigation is an important element of a safety management system. It allows companies to assess and correct underlying causes of near misses and accidents (CCPS, 1989). Georgia-Pacific has a corporate policy for investigating incidents.

Interviewed employees reported prior instances of chlorine dioxide[11] leaking from the manway involved in the January 16 incident. Some employees recalled repairs, and at least some of these were documented by work orders to reseal the manway; however, these events were not reported as incidents or investigated. The Georgia-Pacific corporate incident investigation policy–which requires investigation of near-miss incidents, including releases of hazardous materials–had not yet been put in place at the Naheola mill.

Because of the design of the sewer, maintenance personnel were unable to ensure that the manway was adequately sealed even after repairing it.

Because of the design of the sewer, maintenance personnel were unable to ensure that the manway was adequately sealed even after repairing it. If earlier incidents had been investigated, it is likely that management would have determined that the manway required design modification.

3.3 Hydrogen Sulfide Safety

H_2S is a colorless, extremely flammable toxic gas. Although it has a characteristic "rotten egg" odor detectable at concentrations as low as 0.5 part per billion (ppb), olfactory fatigue[12] occurs rapidly and at low concentrations. Exposure to concentrations of 500 parts per million (ppm) causes loss of consciousness, and exposure to concentrations as high as 700 ppm can cause immediate death.

The majority of fatalities associated with H_2S exposure occur in confined spaces . . . In this case, however, the H_2S poisoning occurred in a flat unconfined area.

The majority of fatalities associated with H_2S exposure occur in confined spaces, such as sewers, waste dumps, sludge plants, tanks, and cesspools (ATSDR, 1999). In this case, however, the H_2S

[11]Employees identified chlorine dioxide by its color; it is a greenish-yellow manufactured gas. Breathing air containing chlorine dioxide gas may cause nose, throat, and lung irritation.

[12]Olfactory fatigue refers to cessation of the sense of smell.

poisoning occurred in a flat unconfined area. When identifying areas with potential H_2S hazards, both confined and unconfined spaces should be considered.

According to ATSDR, the men who died were likely exposed to H_2S concentrations greater than 700 ppm.

According to ATSDR, the men who died were likely exposed to H_2S concentrations greater than 700 ppm. The injured victims were exposed to lower, but still dangerous, concentrations. Some of the injured reported smelling an odor for an instant, but then their sense of smell was deadened.

H_2S was not identified as a hazard in the typically unoccupied area where the incident occurred. . . . Personnel had only their sense of smell to indicate [its] possible presence . . . however, smell is not a reliable indicator because the gas causes olfactory fatigue.

H_2S was not identified as a hazard in the typically unoccupied area where the incident occurred. There were no monitors or alarms to warn of a release. Personnel had only their sense of smell to indicate the possible presence of H_2S; however, smell is not a reliable indicator because the gas causes olfactory fatigue.

The Burkes employees had a general awareness of H_2S. More detailed safety training would have covered how to identify H_2S and how to respond in an emergency. Emergency training should instruct individuals to leave the area and not to attempt rescue of others unless they themselves are trained and wearing the proper protective equipment.

Emergency training should instruct individuals to leave the area and not to attempt rescue of others unless they themselves are trained and wearing the proper protective equipment.

Per Georgia-Pacific policy, Burkes construction workers wore the Scott Speed Evac disposable mouth bit respirator (model 90 AG) on their belts and were trained in its use. These respirators are used for escape in situations where unexpected releases result in atmospheres containing low levels of chlorine, chlorine dioxide, sulfur dioxide, H_2S, or hydrogen chloride. They are not intended for use in immediately dangerous to life or health (IDLH) atmospheres.

Several Burkes employees tried to use the respirators, but H_2S levels were too high for the equipment to be effective. Only a self-contained breathing apparatus (SCBA) would have provided adequate protection at the exposure levels present on January 16.

The Naheola mill did not adequately identify areas that contain–or have the potential to contain–dangerous levels of H_2S gas. All areas of H_2S hazard should have been identified; and personnel, including contractors, should have been trained on how to respond in the event of an H_2S release.

3.4 Emergency Response

The mill ambulance arrived at the scene and removed the injured prior to arrival of the incident commander. Mill personnel did not remove the victims' clothing at the first-aid station because there was no visible contamination.

Paramedics were initially told that the victims had been exposed to either chlorine dioxide or H_2S. . . . [They] had only the [DOT] Emergency Response Guidebook to refer to for treatment of these chemical exposures.

Injured personnel were not decontaminated–neither at the scene of the incident nor at the mill first-aid station. The mill's emergency response plan contains procedures for decontamination, but it is a function of the incident command system. The mill ambulance arrived at the scene and removed the injured prior to arrival of the incident commander. Mill personnel did not remove the victims' clothing at the first-aid station because there was no visible contamination.

At the first-aid station, mill personnel continued to aid the victims until the Choctaw County Emergency Medical Services (EMS) team arrived. The first-aid station was staffed by five registered nurses[13] who provided 24-hour coverage; two nurses were on duty at the time of the incident. When the County paramedic units arrived at the mill, they took over treatment of the most critical victims.

The first Choctaw County ambulance arrived 3 to 5 minutes after being notified (and 19 minutes after the incident began). Paramedics were initially told that the victims had been exposed to either chlorine dioxide or H_2S. The victims were treated for their symptoms.

The responding paramedics had only the Emergency Response Guidebook (DOT/TC-SCT, 2000) to refer to for treatment of these chemical exposures. This guidebook is intended to be used by first responders to hazardous material transportation incidents. It contains only cursory information on first aid and decontamination, and advises that contaminated clothing and shoes be removed. The two paramedic teams that did not remove the victims' clothing stated that they saw no obvious signs of contamination.

The six paramedics stated that during the 40- to 50-minute trips to hospitals, they noticed a strong odor and opened the vehicle windows and turned on the exhaust fans. The paramedic team transporting the truck driver initiated decontamination of the victim by removing his clothing.

[13] The plant nurses are members of the American Association of Occupational Health Nurses.

Each of the paramedics was evaluated for symptoms of H_2S exposure, including burning throat and eyes and nausea. The paramedic team that removed the truck driver's clothing had less severe symptoms than the other two crews. Collectively, the paramedics may have experienced fewer symptoms if the clothes of all the victims had been removed at the scene or at the first-aid station.

The paramedics did not have hazardous material training, nor were they trained on use of the Emergency Response Guidebook (DOT/TC-SCT, 2000)–though they did refer to it. The paramedics should have received more detailed information on H_2S at the scene.

The ATSDR Medical Management Guidelines (MMG) were designed for emergency medical personnel (ATSDR, 2001); however, the responders were not aware of this guidance at the time of the incident. The MMG on H_2S exposure states that responders are not at risk when they assist a victim of H_2S gas exposure.

However, H_2S is released from the clothing and skin, and via gas that expires from the lungs. The threat to responders may be greater when the situation occurs in a confined space, such as an ambulance. Through its participation in this investigation, ATSDR has concluded that persons who are exposed to elevated levels of H_2S (i.e., greater than 500 ppm) may pose a threat to responders who come into contact with them.

Each of the paramedics was evaluated for symptoms of H_2S exposure, including burning throat and eyes and nausea. The paramedic team that removed the truck driver's clothing had less severe symptoms than the other two crews.

The ATSDR Medical Management Guidelines . . . [state] that responders are not at risk when they assist a victim of H_2S gas exposure.

. . . H_2S is released from the clothing and skin, and via gas that expires from the lungs.

ATSDR has concluded that persons who are exposed to elevated levels of H_2S (i.e., greater than 500 ppm) may pose a threat to responders . . .

4.0 Regulatory Analysis

4.1 OSHA Process Safety Management

The OSHA PSM Standard (29 CFR 1910.119) is intended to prevent or minimize the consequences of a catastrophic release of toxic, reactive, flammable, or explosive chemicals. Paper mills are not exempt from the standard, which applies to processes[14] containing more than a threshold quantity of any one of 137 OSHA-listed highly hazardous chemicals or substances classified by OSHA as flammable. Substances are listed based on their toxic or reactive properties. OSHA lists H_2S as a highly hazardous chemical; NaSH is not listed in the PSM Standard.

OSHA lists H_2S as a highly hazardous chemical; NaSH is not listed in the PSM Standard.

The PSM Standard covers certain processes at the Naheola mill; however, the area where the incident occurred was not part of a PSM-covered process. Although OSHA lists H_2S as a highly hazardous chemical, it is not present in sufficient quantities[15] or–in this case–is not part of an OSHA-defined covered process. There were no other OSHA PSM highly hazardous chemicals in the area of the release.

. . . the area where the incident occurred was not part of a PSM-covered process.

The Naheola mill applied the site process safety management plan only to areas of the facility that were covered by the PSM Standard. Section 3.2 discusses the elements of a process safety management program that are causally related to this incident.

4.2 Hazard Communication

The OSHA Hazard Communication (HAZCOM) Standard (29 CFR 1910.1200) requires companies to provide their employees with information on the hazardous chemicals to which they may be exposed.

If the good engineering and process safety management practices outlined in Section 3.0 had been followed, mill personnel would likely have determined that H_2S could possibly be present at the location of the incident. Good practice would have required the

[14] OSHA defines a process as any activity involving a highly hazardous chemical, including any use, storage, manufacturing, handling, or onsite movement of such chemicals, or any combination of these activities.

[15] The threshold quantity for H_2S is 1,500 pounds.

communication of H_2S hazards to Naheola mill employees; similarly, Burkes Construction would have to communicate the hazards of H_2S to its employees. The HAZCOM Standard requires that this be done by a formal communication program, to include labels, MSDSs, and training.

4.3 Hazardous Waste Operations and Emergency Response

The OSHA Hazardous Waste Operations and Emergency Response (Hazwoper) Standard (29 CFR 1910.120) details requirements for emergency response activities associated with the release of hazardous substances. The Naheola mill is subject to Hazwoper requirements; among other things, the standard mandates specific emergency response training.

The Hazwoper Standard allows for an exemption if employees will be evacuated in the event of a release, which applies to Burkes Construction. Accordingly, it was not bound to meet the Hazwoper requirements.

5.0 Root and Contributing Causes

5.1 Root Causes

1. **Good engineering and process safety practices were not followed when joining the drain from the truck unloading station and the oil pit to the acid sewer.**

   - Neither the chemicals that could be introduced into the sewer nor the hazards of their interactions were identified.
   - No formal hazard review or MOC analysis was conducted when connecting sewer lines from the tank truck unloading and chlorine dioxide areas to the acid sewer. Consequently, no scenarios leading to the possible release of H_2S were identified, nor were warning devices placed in the area.

Neither the chemicals that could be introduced into the sewer nor the hazards of their interactions were identified.

2. **There was no management system to incorporate hazard warnings about mixing sodium hydrosulfide (NaSH) with acid into process safety information.**

   - Design information for projects involving NaSH did not specify the hazard of mixing NaSH with acid.
   - Operating procedures for NaSH tank truck unloading and oil pit operations did not warn of the hazards of mixing NaSH with acids or the hazards of allowing NaSH to enter sewers.
   - Personnel were not trained on the specific hazards of NaSH, such as handling spilled material or keeping it separate from acid.

Personnel were not trained on the specific hazards of NaSH, such as handling spilled material or keeping it separate from acid.

5.2 Contributing Causes

1. **The fiberglass manway was not adequately designed or sealed to ensure that the sewer remained closed.**

   The manway was originally an open grate. It was modified by adding a fiberglass cover and sealed. Those interviewed recalled prior occasions when chlorine dioxide escaped from the fiberglass manway; they also recalled repairs that were sometimes documented in work orders, but these events were not recorded or investigated as incidents. Thorough incident investigation would have provided an opportunity to recognize that the

Those interviewed recalled prior occasions when chlorine dioxide escaped from the fiberglass manway . . . but these events were not recorded or investigated as incidents.

H_2S training should include specific instruction on the importance of wearing proper protective equipment prior to attempting rescue.

fiberglass manway was not appropriately designed to provide and maintain an airtight seal on the sewer.

2. **The contractors injured during the incident did not have adequate training to understand the hazards of hydrogen sulfide (H_2S).**

H_2S training should include specific instruction on the importance of wearing proper protective equipment prior to attempting rescue.

6.0 Recommendations

Georgia-Pacific Corporation

1. Conduct periodic safety audits of Georgia-Pacific pulp and paper mills in light of the findings of this report. At a minimum, ensure that management systems are in place at the mills to:

   - Evaluate process sewers where chemicals may collect and interact, and identify potential hazardous reaction scenarios to determine if safeguards are in place to decrease the likelihood or consequences of such interactions. Take into account sewer system connections and the ability to prevent inadvertent mixing of materials that could react to create a hazardous condition. (2002-01-I-AL-R1)
   - Identify areas of the mill where hydrogen sulfide (H_2S) could be present or generated, and institute safeguards (including warning devices) to limit personnel exposure. Require that personnel working in the area are trained to recognize the presence of H_2S and respond appropriately. Update emergency response plans for such areas to include procedures for decontaminating personnel exposed to toxic gas. (2002-01-I-AL-R2)
   - Apply good engineering and process safety principles to process sewer systems. For instance, ensure that hazard reviews and management of change (MOC) analyses are completed when additions or changes are made where chemicals could collect and react in process sewers. (Such principles may be found in publications from the Center for Chemical Process Safety [CCPS].) (2002-01-I-AL-R3)

2. Communicate the findings and recommendations of this report to the workforce and contractors at all Georgia-Pacific pulp and paper mills. (2002-01-I-AL-R4)

Georgia-Pacific Naheola Mill

1. Evaluate mill process sewer systems where chemicals may collect and react to identify potential hazardous reaction scenarios to determine if safeguards are in place to decrease the likelihood or consequences of such interactions. Evaluate sewer connections

and ensure that materials that could react to create a hazardous condition are not inadvertently mixed, and that adequate mitigation measures are in place if such mixing does occur. (2002-01-I-AL-R5)

2. Establish programs to comply with recommendations from manufacturers of sodium hydrosulfide (NaSH) regarding its handling, such as preventing it from entering sewers because of the potential for acidic conditions. (2002-01-I-AL-R6)

3. Establish programs to require the proper design and maintenance of manway seals on closed sewers where hazardous materials are present. (2002-01-I-AL-R7)

4. Identify areas of the plant where hydrogen sulfide (H_2S) could be present or generated, and institute safeguards (including warning devices) to limit personnel exposure. Institute a plan and procedures for dealing with potential H_2S releases in these areas, and require that anyone who may be present is adequately trained on appropriate emergency response practices, including attempting rescue. Require contractors working in these areas to train their employees on the specific hazards of H_2S, including appropriate emergency response practices. (2002-01-I-AL-R8)

5. Update the Naheola mill emergency response plan to include procedures for decontaminating personnel who are brought to the first-aid station. Include specific instructions for decontaminating personnel exposed to H_2S so that they do not pose a secondary exposure threat to medical personnel. (2002-01-I-AL-R9)

Agency for Toxic Substances and Disease Registry (ATSDR)

Evaluate and amend as necessary the ATSDR Medical Management Guidelines to consider the risk to responders posed by exposure to victims of high levels of hydrogen sulfide (H_2S) gas. Specify procedures for adequate decontamination. Communicate the results of this activity to relevant organizations, such as the American Association of Occupational Health Nurses. (2002-01-I-AL-R10)

Burkes Construction, Inc.

Train your employees on the specific hazards of hydrogen sulfide (H_2S), including appropriate emergency response practices, in areas where Georgia-Pacific has identified this material as a hazard. (2002-01-I-AL-R11)

Davison Transport, Inc.

Communicate the findings and recommendations of this report to those employees who haul or handle sodium hydrosulfide (NaSH). (2002-01-I-AL-R12)

American Forest and Paper Association (AFPA)

Communicate the findings and recommendations of this report to your membership. (2002-01-I-AL-R13)

International Brotherhood of Electrical Workers (IBEW)

Communicate the findings and recommendations of this report to your membership. (2002-01-I-AL-R14)

Paper, Allied-Industrial, Chemical & Energy Workers International Union (PACE)

Communicate the findings and recommendations of this report to your membership. (2002-01-I-AL-R15)

Pulp and Paper Safety Association (PPSA)

Communicate the findings and recommendations of this report to your membership. (2002-01-I-AL-R16)

By the

U.S. Chemical Safety and Hazard Investigation Board

Carolyn W. Merritt
Chair

John S. Bresland
Member

Gerald V. Poje, Ph.D.
Member

Isadore Rosenthal, Ph.D.
Member

Andrea Kidd Taylor, Dr. P.H.
Member

November 20, 2002

7.0 References

Agency for Toxic Substances and Disease Registry (ATSDR), 2001. *Managing Hazardous Material Incidents (MHMI)*, Volume III, U.S. Department of Health and Human Services (DHHS), Public Health Service.

ATSDR, 1999. *Toxicological Profile for Hydrogen Sulfide*, DHHS, Public Health Service.

American National Standards Institute, Inc. (ANSI), 2001. *American National Standard Accepted Practices for Hydrogen Sulfide (H_2S) Safety Training Programs*, ANSI/ASSE Z390.1-1995 (R2001), American Society of Safety Engineers, May 8, 2001.

Center for Chemical Process Safety (CCPS), 1995a. *Guidelines for Safe Storage and Handling of Reactive Materials*, American Institute of Chemical Engineers (AIChE).

CCPS, 1995b. *Process Safety Documentation*, AIChE.

CCPS, 1989. *Guidelines for Technical Management of Chemical Process Safety*, AIChE.

Kletz, Trevor, 1999. *Hazop and Hazan*, Fourth Edition, Institution of Chemical Engineers (IChemE).

Lees, Frank, 1996. *Loss Prevention in the Process Industries*, Reed Educational and Professional Publishing Ltd.

National Institute for Occupational Safety and Health (NIOSH), 1977. *Criteria for a Recommended Standard, Occupational Exposure to Hydrogen Sulfide*, U.S. Department of Health, Education, and Welfare, Public Health Service, May 1977.

U.S. Department of Transportation (DOT)/Transport Canada (TC)-Secretariat of Transport and Communications (SCT, 2000). *2000 Emergency Response Guidebook.*

APPENDIX A: Incident Timeline

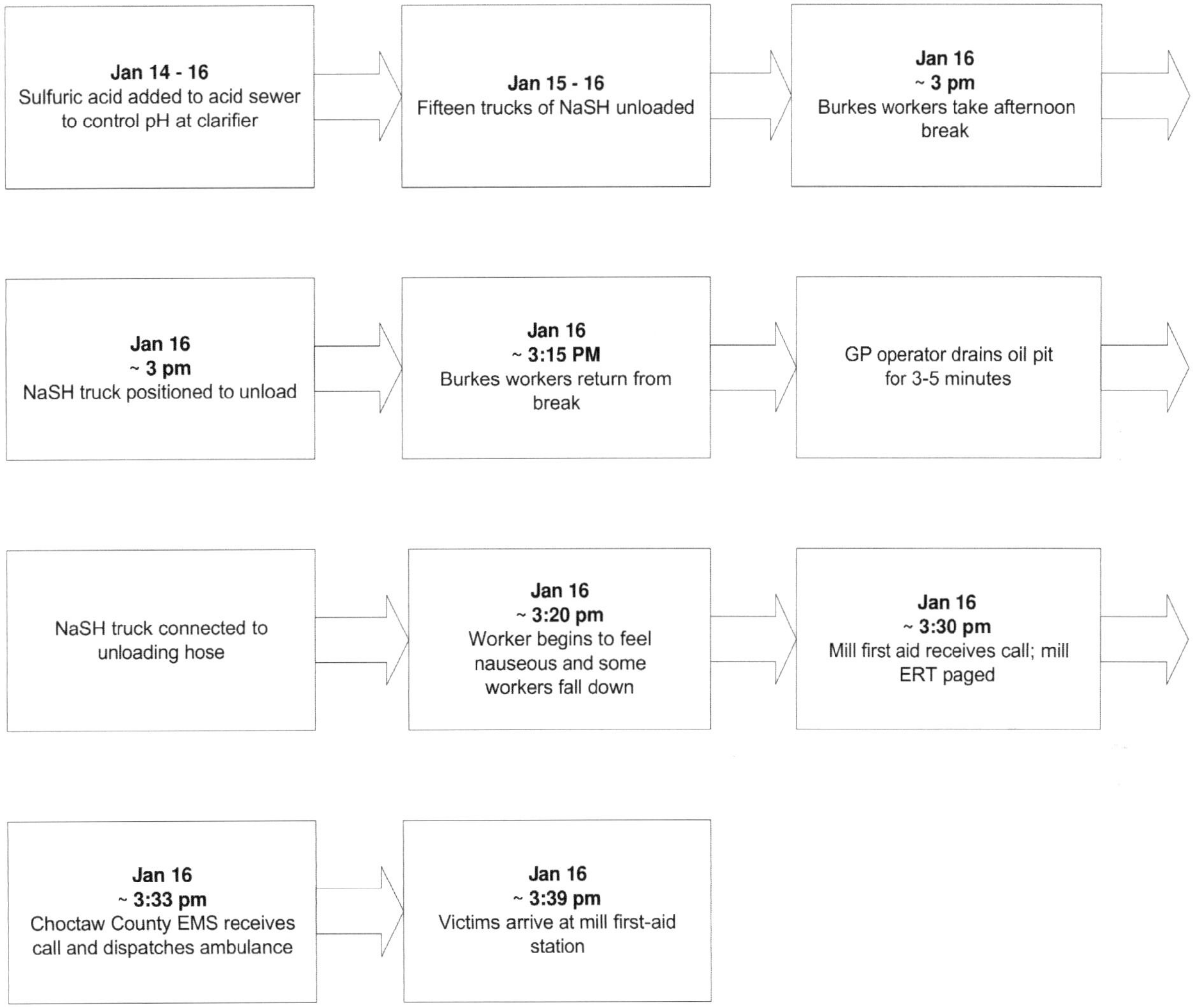

APPENDIX B: Logic Diagram

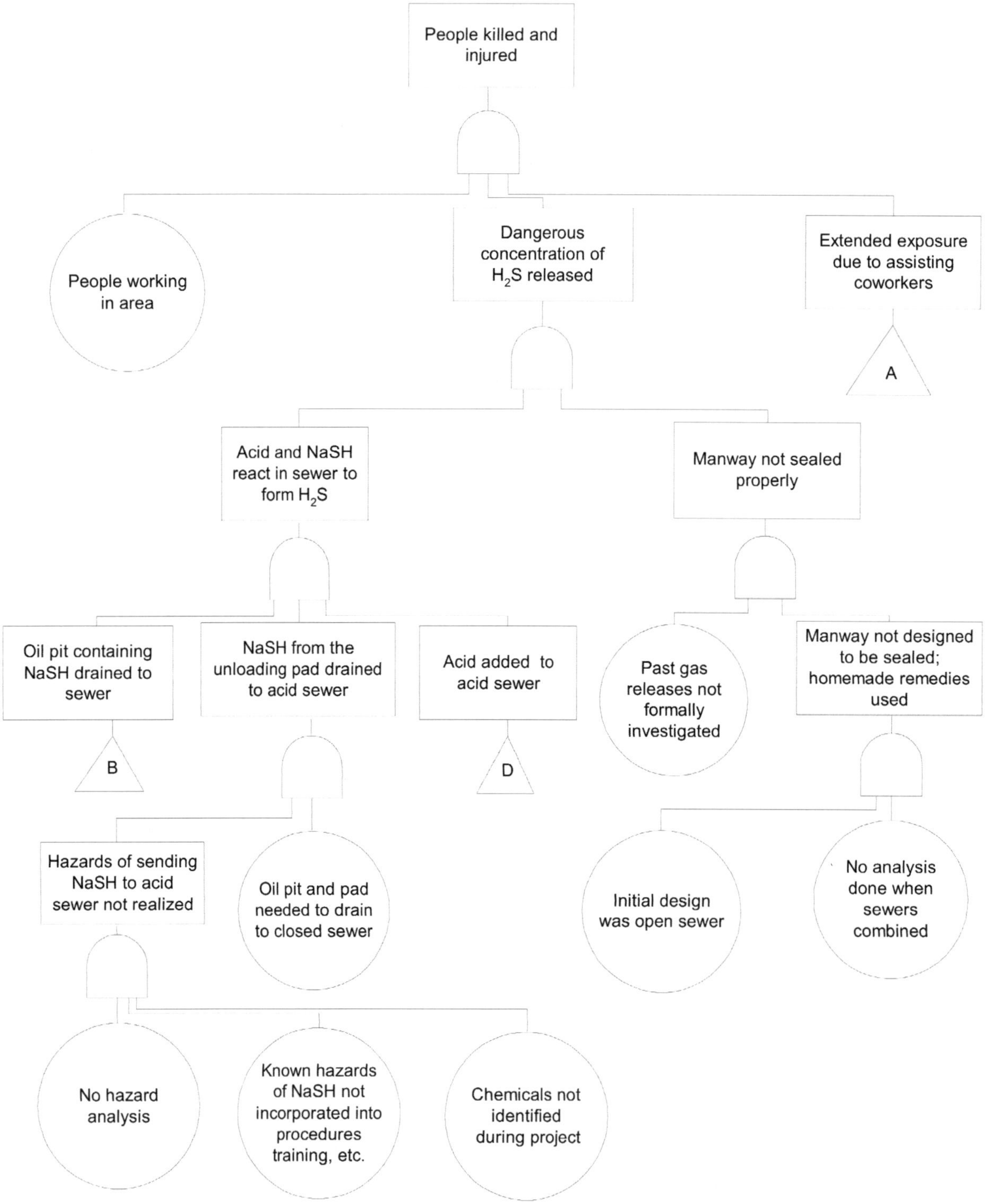

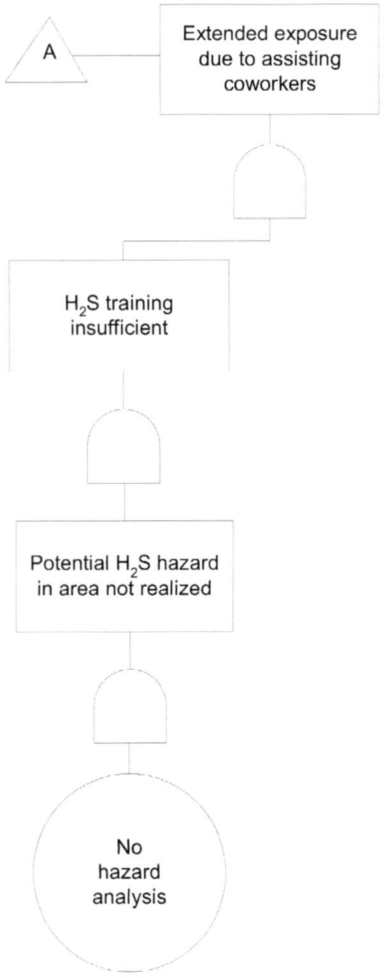
A
Extended exposure due to assisting coworkers
H2S training insufficient
Potential H2S hazard in area not realized
No hazard analysis

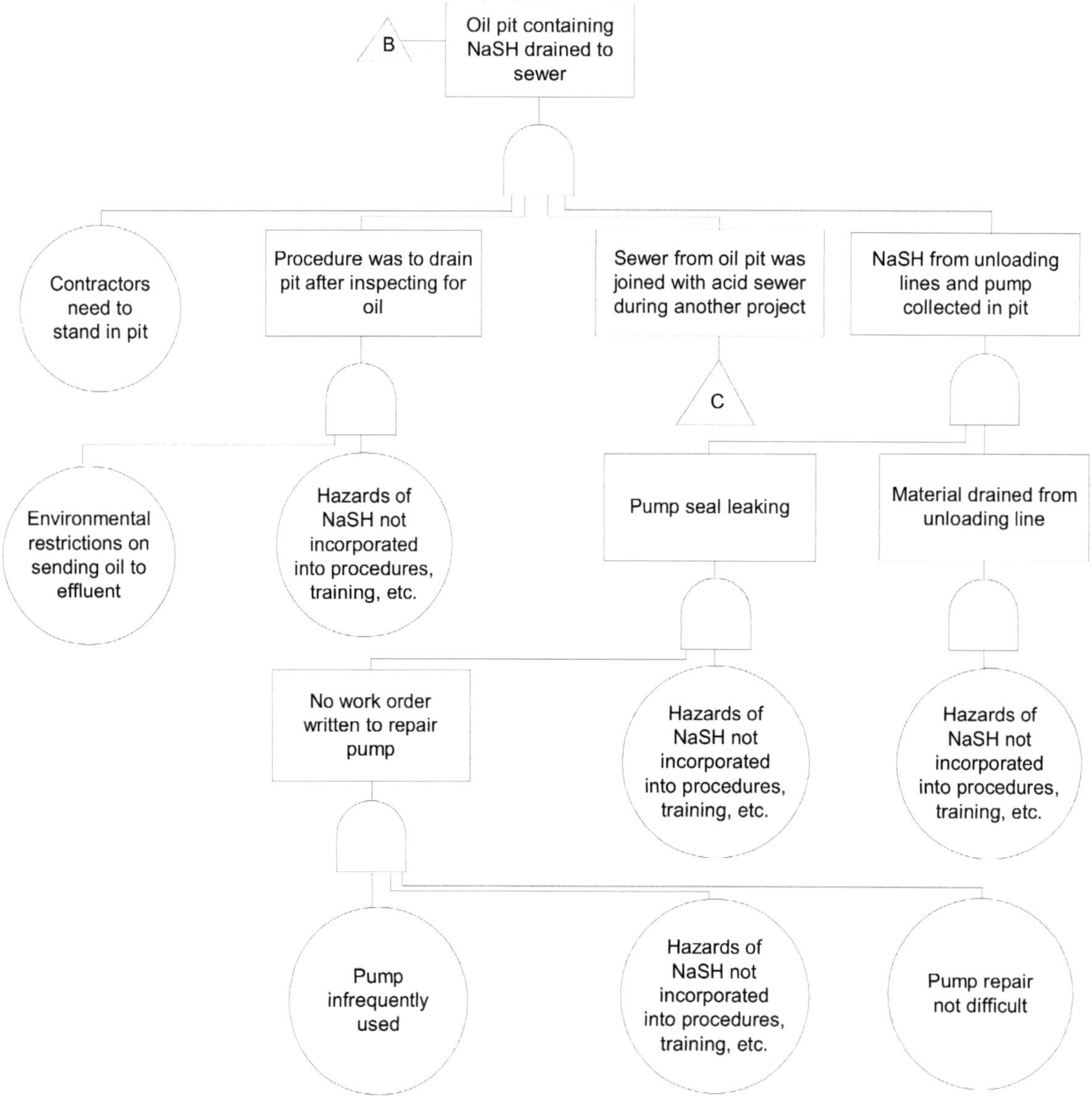
B
Oil pit containing NaSH drained to sewer
Contractors need to stand in pit
Procedure was to drain pit after inspecting for oil
Sewer from oil pit was joined with acid sewer during another project
NaSH from unloading lines and pump collected in pit
C
Environmental restrictions on sending oil to effluent
Hazards of NaSH not incorporated into procedures, training, etc.
Pump seal leaking
Material drained from unloading line
No work order written to repair pump
Hazards of NaSH not incorporated into procedures, training, etc.
Hazards of NaSH not incorporated into procedures, training, etc.
Pump infrequently used
Hazards of NaSH not incorporated into procedures, training, etc.
Pump repair not difficult

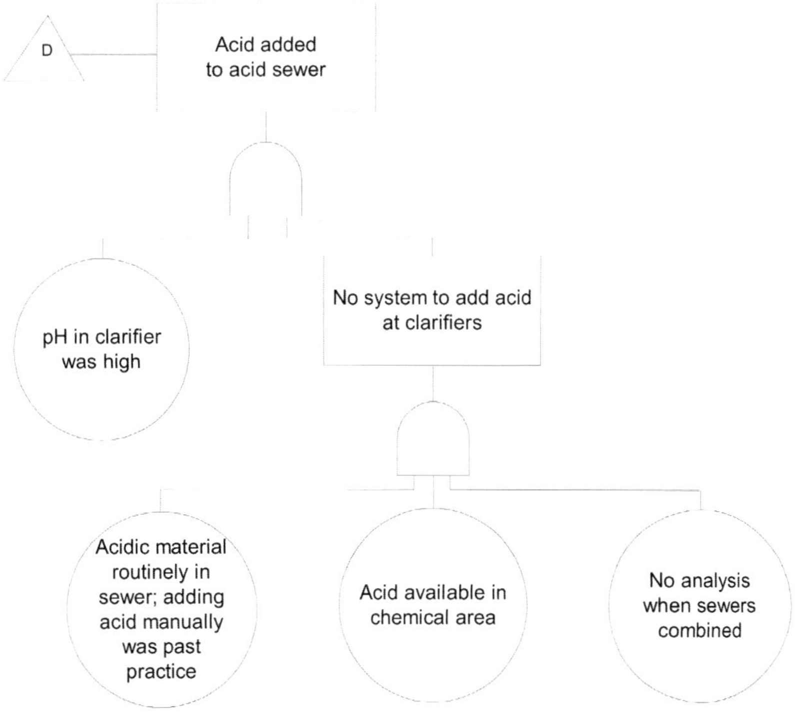
D
Acid added to acid sewer
pH in clarifier was high
No system to add acid at clarifiers
Acidic material routinely in sewer; adding acid manually was past practice
Acid available in chemical area
No analysis when sewers combined

Chlorine Dioxide: Green or Mean? An Oxidation Under Biphasic Conditions

B. O'Reilly, F. Okuniewicz, R. Schild, S. Y. Wang, L. Rossano, R. Shapiro, D. Chen, S. Kiang

Bristol-Myers Squibb Pharmaceutical Research Institute
Process Research & Development
One Squibb Drive
New Brunswick, New Jersey 08903-0191

Abstract

A procedure using sodium chlorite / chlorine dioxide as an oxidant was developed for a multi-kilo campaign to prepare a pharmaceutical intermediate. During process optimization and process hazards evaluation, a laboratory fire highlighted the need for an in-depth hazards analysis on the generation and use of chlorine dioxide. A procedure using a dual-cell OPTEK spectrophotometer to continuously monitor chlorine dioxide concentration in the reactor head-space and reaction mixture allowed for the evaluation and control of potential process hazards. A review of the safety and engineering issues associated with chlorine dioxide oxidations in organic synthesis is presented.

Introduction

A primary goal of chemical reaction engineering is the ability to reliably *predict and control* what will occur in a given chemical process. In the pharmaceutical industry, as project planning schedules become more aggressive, the associated timelines have become shorter and less flexible. Hence, the need to evaluate complex reactions with potential safety issues in a short timeframe has become one of the most challenging aspects of process development. To meet these challenges, rapid project development must incorporate an understanding of the factors that may lead to potential hazards and evaluate whether a safe operating window exists.

This paper describes the engineering and safety issues encountered during the development of a process to prepare a novel pharmaceutical intermediate on a multi-kilo scale. In this process, sodium chlorite was used in the presence of bleach to generate chlorine dioxide (ClO_2) to selectively oxidize a furan ring to the corresponding carboxylic acid:

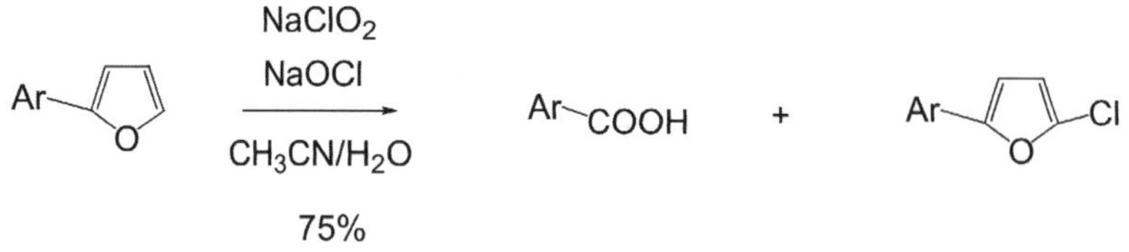

Compound A   Product P   Undesired chlorofuran

ClO_2 was chosen as the oxidant to avoid the waste generation and cleaning issues associated with a more typical permanganate oxidation. The reaction with ClO_2 goes to completion under milder conditions than with permanganate and the process is volume efficient. The reagent is also compatible with the aromatic substituents.

ClO_2 is a toxic and an inherently unstable oxidant[1,2]. In all industrial uses of ClO_2, the instability of gaseous chlorine dioxide has been a major concern for safe operation on scale[2, 3]. The peculiar nature of ClO_2 stems from the potential for free radical initiated chain reaction mechanism of ClO_2 auto-decomposition which can occur with an induction period[4,5]. The incompatible nature of this reagent with reducing agents, common solvents, and many metals (or materials of construction) could easily initiate a powerful exothermic event or even a thermite reaction[2, 4, 5]. When ClO_2 ignites, the extremely high deflagration rate often leads to fire or explosion without warning.

While data is available for safe execution of ClO_2 oxidations in aqueous systems, very little literature data exists for execution in non-aqueous systems. In order to make the process safe for execution on scale, a thorough evaluation of the thermochemistry of the reacting system is needed, and a method for real-time monitoring of the ClO_2 concentrations in the headspace and liquid is needed to insure safe operating conditions. Thus, two primary goals of this study were:

- To evaluate the safe operation of this oxidation process.
- To establish the use of visible spectrometry as a method for monitoring the headspace and solution concentrations of ClO_2 in this process.

Experimental

Thermochemistry

The thermochemistry of the oxidation was investigated using a RC-1 calorimeter. The RC-1 measured heat of the reaction was –2915 kJ/mole of Compound A with a calculated adiabatic temperature rise of 99°C. In the event of a runaway reaction, the system would reach reflux conditions.

Initial Reaction Procedure: Batch Charges, HPLC Analysis

Compound A and acetic acid (3 mL/g) in acetonitrile is treated with aqueous sodium hypochlorite maintaining the temperature at 2-5 °C. Aqueous sodium chlorite (14 eqs) is then added gradually maintaining batch temperature at 2-5 °C. The headspace gas is monitored by HPLC for ClO_2 concentration. When the reaction is complete, as determined by HPLC analysis of the solution, residual chlorine and ClO_2 are removed by subsurface sparging to a metabisulfite scrubber. After quenching the reaction to neutralize any remaining oxidants and testing with starch iodide for residual ClO_2, the work-up proceeds in a normal fashion.

Modified Reaction Procedure: Inverse Addition with Sawtooth Charges, Spectroscopic Analysis

Laboratory development work demonstrated advantages in selectivity and yield when using a modified charge mode. Two streams are prepared: Stream A, containing Compound A and acetic acid in acetonitrile, and Stream B, containing aqueous sodium chlorite solution. After charging 20% of stream A to the reactor, cooling and initiating the reaction with sodium hypochlorite, Stream B is added while maintaining temperature below a predetermined level. Incremental charges of 20% each of the two streams were made in a 'sawtooth' charge pattern while maintaining temperature at 2-5 °C. A continuous purge of nitrogen was added to the headspace to maintain the ClO_2 level below 3% (3 fold below the lower explosion limit of 10% in air).

An Optek-Danulat, Inc. C4000 Spectrometer with a 10 mm optical path was used to determine the overhead concentrations of ClO_2 in the headspace off-gas (430 nm) and the liquid reaction mixture (470 nm). The spectrometer was chosen as it was able to distinguish between chlorine dioxide and chlorine gas and provide real-time concentration data. The gas sensing cell was placed between the reactor and the dry trap (Figure 1). Liquid samples were analyzed off-line, after separation of the two phases, or in real-time as a biphasic mixture in a circulating loop.

Results and Discussion

Batch Reactions, HPLC Analysis

The first two experiments (Run 1 and 2) compared HPLC results to those obtained by spectroscopic analysis. The HPLC results were found to be very sensitive to sampling technique and the duration between sampling and analysis. HPLC results were not reproducible and varied widely. In all cases, overhead gas concentrations of ClO_2 (Figure 2) determined by spectroscopic analysis were greater than those determined by HPLC analysis.

Spectroscopic analysis of liquid concentrations was performed after phase separation of the sample aliquots. Batch maximum ClO_2 values are reported for the acetonitrile phase in Table 1. In the first run, the ClO_2 concentration in the acetonitrile phase exceeded the maximum measurable concentration (~ 45 g/L) twenty minutes after the completion of the chlorite charge. The aqueous phase was concentration was 12 g/L.

In the second run, the ClO_2 concentration in the acetonitrile phase at the end of the sodium chlorite solution addition (18.1 g/L) was less than half that of a sample taken 35 minutes later (40 g/L). By comparison, the aqueous liquid concentration was slightly higher at the end of the chlorite addition (8.9 g/L) than in the sample taken 35 minutes later (7.4 g/L).

This is evidence of significant variation of ClO_2 concentration with rate of addition. As with the off-gas determinations, we found the liquid concentrations determined by HPLC analysis to be lower and less reliable than the concentrations of ClO_2 determined by spectroscopic analysis.

In these first two runs, the coolant flow (~ 1.0 LPM) was insufficient to maintain the batch temperature below 5°C and the off-gas concentration maximums were 3.66% and 3.72 % respectively. These values were greater than the target limit of 3%. The elevated batch temperatures and ClO_2 concentrations (Table 1) were cause for concern.

The second two experiments (Runs 3 and 4) compared the overhead gas ClO_2 concentrations produced by the sawtooth charging mode to the first two experiments. In these runs, coolant flow was increased by an order of magnitude and a circulating loop was installed to measure the biphasic liquid ClO_2 concentration over time. As a result, the observed batch temperatures in Run 3 was held to less than 2°C and the peak ClO_2 gas concentration did not exceed 1.7 % (Figure 3). The reaction selectivity was improved in Run 3 (ratio of product to chlorofuran of 28).

For the sawtooth charging in Run 4, the chlorite charge (Stream B) was made first and Stream A was added to it. Here, gas concentrations were slightly lower than for Run 3. The reaction selectivity decreased significantly in Run 4 (ratio of product to chlorofuran of 0.2) indicating the importance of addition order.

As expected, changing the process to a semi-batch process with headspace purging gave a more controllable process and enhanced reaction selectivity.

Table 1. Four oxidation experiments in which spectroscopic analysis was utilized to determine chlorine dioxide liquid (g/L) and off-gas concentrations (%).

Run No.	T_{jacket} °C	Glycol LPM	$T_{bat,max}$ °C	$[ClO_2]_g$ %	$[ClO_2]_l$ g/L	Charge profile and duration
1	0	1.0	14.0	3.66	> 44	Batch, Chlorite added over 3.0h
2	0	1.0	14.8	3.72	40.0	Batch, Chlorite added over 3.0h
3	-10	10	1.8	1.69	9.46	Sawtooth, Stream B to A over 3.5 h
4	-5	10	4.8	1.47	4.68	Sawtooth, Stream A to B over 6.5 h

Notes:
T_{jacket} refers to the setpoint of the chiller circulating the glycol for cooling.
$T_{bat,max}$ refers to the maximum temperature observed during the course of the reaction.
$[ClO_2]_g$ refers to the maximum concentration of the overhead gas measured at 430 nm.
$[ClO_2]_l$ refers to the maximum concentration of the liquid measured at 470 nm:
During Runs 1 and 2, the liquid analyzed off-line was the acetonitrile phase.
During Runs 3 and 4, the liquid analyzed *in situ* was the biphasic mixture.
Stream A is the solution of Compound A (20 g) and acetic acid (30 mL) in acetonitrile (170 mL).
Stream B is a solution of sodium chlorite (100 g of 80% $NaClO_2$; 850 mmol, 14 equivalents) in water (190 mL).
Spectroscopic measurements were made using an Optek-Danulat C4000 spectrometer (10 mm path).

Fire during Post-Experiment Cleanup

A trace of sodium chlorite solution evaporated on the side of a laboratory reactor and during cleaning, a tiny solid flake landed on a paper towel. The towel instantly burst into flame and was reduced to ash in a few seconds.

The handling of sodium chlorite solutions, which can become a percussive solid when dried on oxidizable materials, must also be addressed.

Equipment Issues and Hazards

This chemistry was intended for a general purpose pharmaceutical pilot plant reactor, glass lined with an overhead condenser at +5° or -30°C, mild agitation (retreating blade) and nitrogen sweep. Because ClO_2 (b.p. = 11 °C) could condense, there could be areas where liquid ClO_2 might accumulate and dead spots where vapor concentrations might exceed safe limits. Liquid ClO_2 is extremely unstable and potentially explosive. Mixing of the two phase reaction was also a concern as this will impact the distribution of ClO_2 in each phase and heat transfer. In a power loss situation with no agitation or coolant circulation, the biphasic mixture could warm and release the ClO_2 resulting in a large exotherm. Because of the high energy of ClO_2, relief venting requirements, relief direction, monitoring instruments, and interlocks must be carefully addressed.

The safe operating window for ClO_2 generation as a function of time [duration], pressure and temperature in aqueous systems is available[5], but similar information for the biphasic system under study was not found. However, applying these criteria shows that reaction temperature and ClO_2 concentration must be tightly controlled. In a regime with temperatures greater than 40°C and or vapor concentrations of ClO_2 >30-120 mm Hg partial pressure, spontaneous combustion of ClO_2 can occur. A thorough understanding of the thermochemistry of the system, reaction kinetics, the heat removal capabilities of the reactor, feed control, mixing, and relief venting is needed to safely carry out this process.

Conclusions

The proposed oxidative process using ClO_2 entailed both a high probably and high consequences in the catastrophic event. The following categories are within our management strategy for reducing risk, whether directly reducing frequency or reducing the consequence of potential accidents:

1. Inherent: Considering the design of inherently safer operations. In this case, replacing ClO_2 with a less hazardous substance and synthesis seems inevitable.
2. Passive: There is no process or equipment design that would guarantee eliminating the hazards involved.

3. Active: Engineering controls such as safety interlocks, purging, relief devices, and emergency shutdowns would not be effective for protecting the people and plant while operating this oxidation process.
4. Procedural: Assuring safety is more important then other priorities such as quality, schedule, and cost. The development team came to the conclusion that to pursue alternative synthetic routes was the correct option. Therefore administrative control was employed to discontinue the use of this oxidation chemistry.

Acknowledgements

The authors would like to acknowledge the contributions from C. Ray, R Wethman and M. Oberholzer.

References

[1] Chlorine Dioxide: Chemistry and Environmental Impact of Oxychlorine Compounds , Masschelein, W.J., Ann Arbor Science Publishers, Inc. 1979; Chem. Eng. News, 1950, 28, 611; National Institute of Standards & Technology Webbook (http://webbook.nist.gov/chemistry/).

[2] Occupational Health Guideline for Chlorine Dioxide, U.S. Department of Health and Human Services, NIOSH, September 1978.

[3] Bretherick's Handbook of Reactive Chemical Hazards, 6th Edition, Butterworth-Heinmann Ltd., 1999.

[4] Simoyi, R.H. et. al., Chem. Eng. News, 1993, 71 (12), 4; Zeitschrift fur Physikalische Chemie, 214, 4, 533-535 (2000).

[5] Cowley, G., Loss Prevention Bulletin, 1993, (113), 1.

Figure 1. Apparatus for the oxidation of Compound A.

The spectrometer gas cell is inserted between the reactor and the dry trap. The liquid cell is placed in a circulation loop fitted with a three-way Teflon valve for sampling.

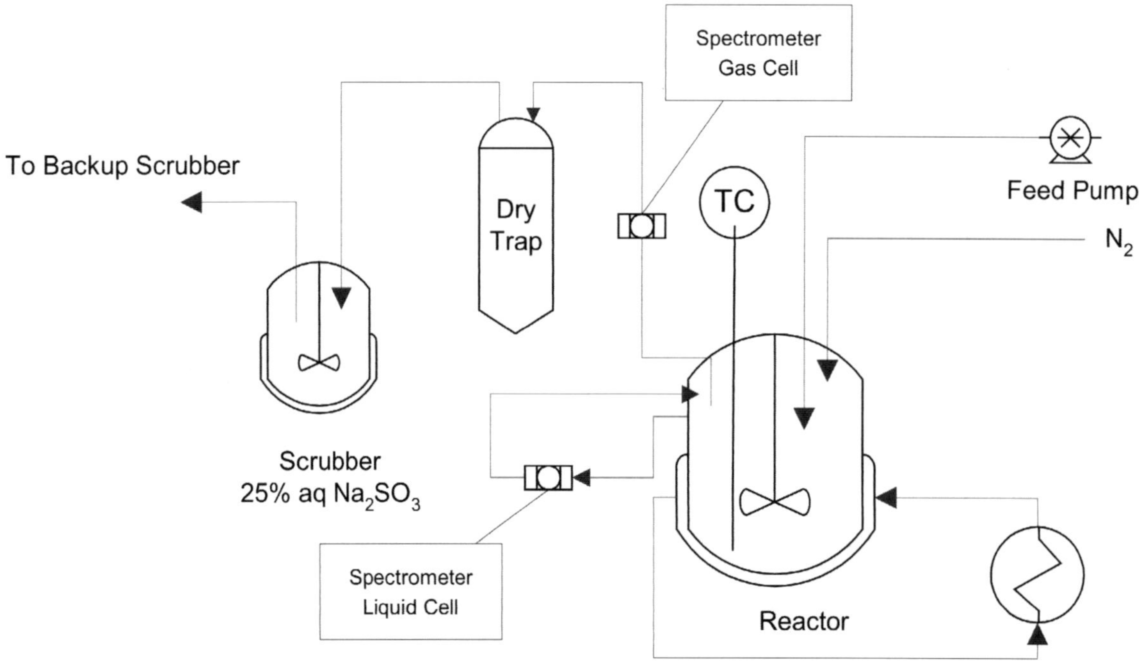

Figure 2. Concentrations of ClO2 (v/v %) in the overhead gas versus reaction time for Runs 1-4.

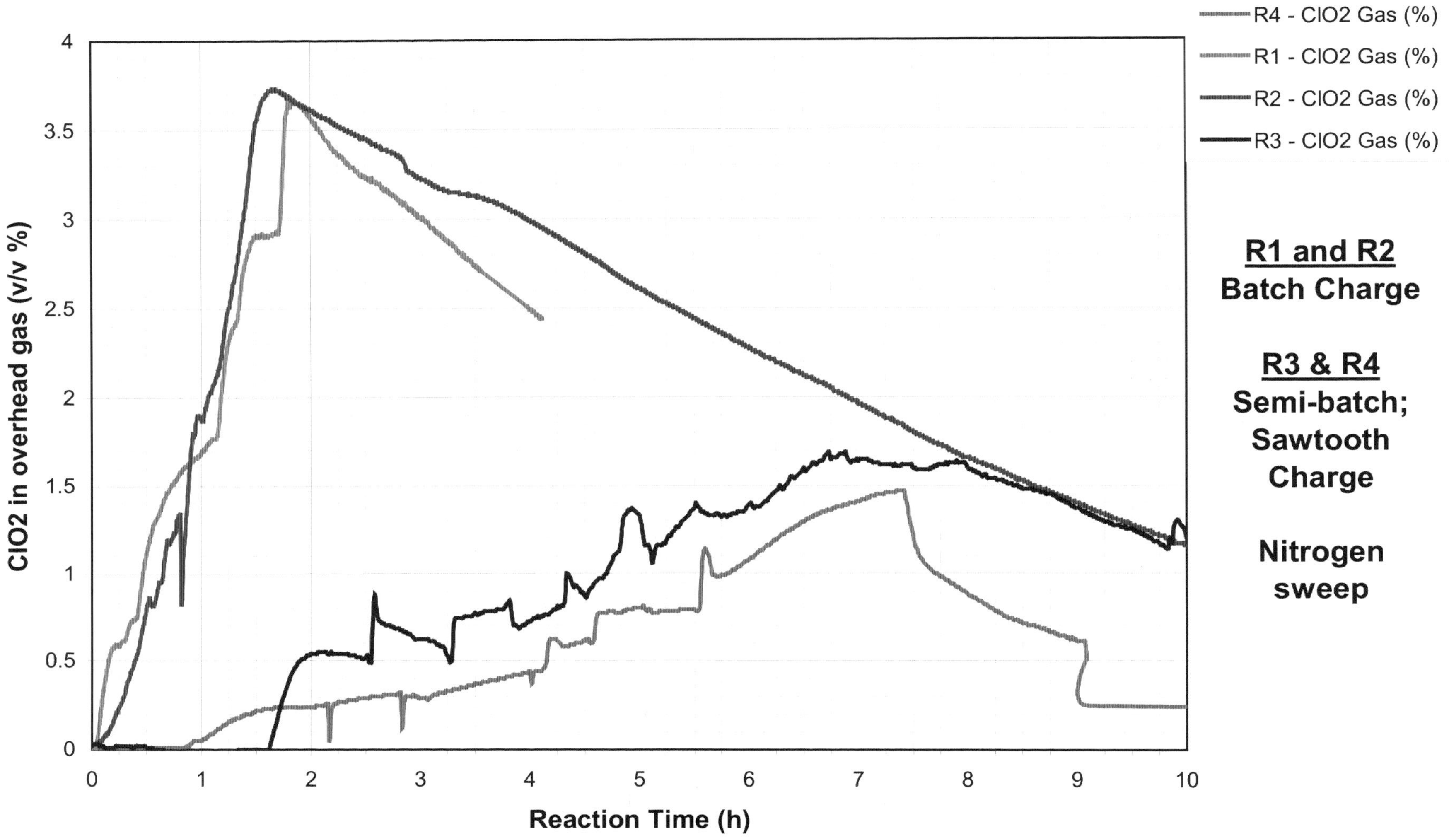

<u>Figure 3</u>. Run 3 concentrations of ClO2 (v/v) in the overhead gas and combined liquid phase versus batch temperature, jacket temperature and reaction time.

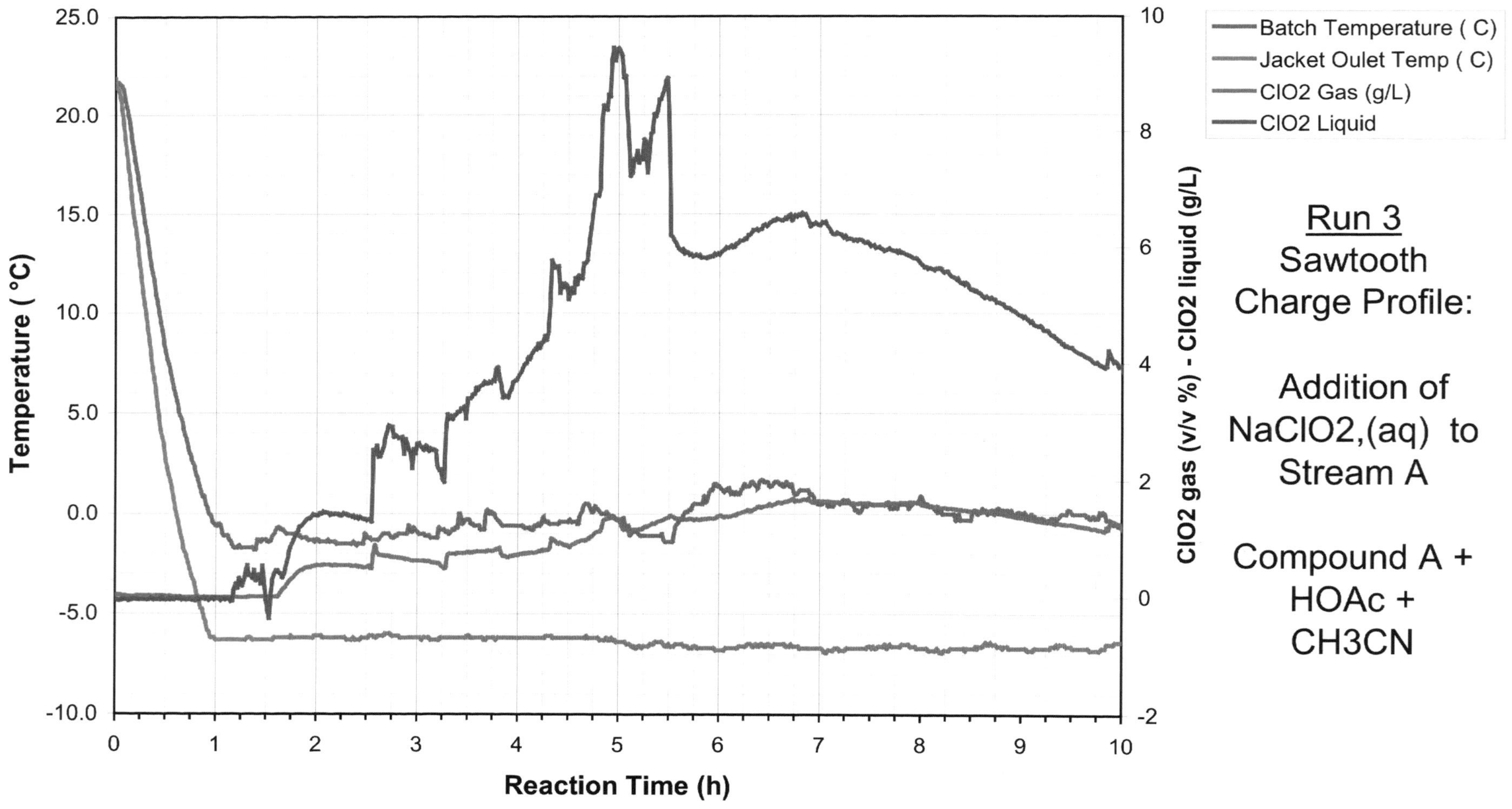

PROCESS SAFETY/LOSS PREVENTION IN PROCESSING OF TANTALUM POWDER

Louis S. Henderson, III*
Manager, Process Engineering / Capital Projects

Melissa A. Perry
Process Engineer, Powders Manufacturing
Cabot Supermetals
County Line Road
Boyertown, PA 19512

Tomoo Izumi
Managing Director, Technology & Production
Cabot Supermetals
Koyo Building
Tokyo 105-0011, Japan

*** Author to whom all correspondence should be addressed.**

Prepared for Presentation at
The 38th Annual Loss Prevention Symposium
Session 1: Fire, Explosion, and Reactive Hazards
April 26 – 29, 2004

ABSTRACT

Tantalum powder has been utilized extensively in the electronics industry for production of capacitors since the 1960's. As consumer product requirements have driven towards smaller capacitors, Ta powder has been pushed towards higher surface areas and smaller particle sizes. Tantalum metal powders can oxidize rapidly and very exothermically in air. This presents unique challenges in material handling in production and use of fine Ta powders. There have been incidents in the past of fires in dust collectors and in milling processes that create newly exposed Ta metal surfaces. However, with the behavior of these powders understood, systems can be designed that mitigate the potential for fire incidents. This paper will review fundamental information on ignition energy and limiting oxygen concentration (LOC), and well as design issues for safe processing.

INTRODUCTION

Tantalum is a transition metal with an Atomic Number of 73. The widest use of tantalum today is in capacitors for the electronics industry. Approximately 51% of purified tantalum produced is used in capacitors (Michaluk, et al., 2000). A capacitor contains tantalum powder which has been compressed, sintered, and treated electrochemically to produce a desired thickness oxide layer on the surface. Tantalum metal surfaces exposed to air quickly oxidize to form Ta_2O_5, a reaction that releases a significant amount of heat. When the tantalum is in the form of a fine powder with a high specific surface area, the oxidation reaction can lead to a rapid temperature rise of the particles.

Tantalum as it exists in nature is present as an oxide, the more common mineral forms being tantalite [$(Fe^{+2},Mn^{+2})Ta_2O_6$] and wodginite [$Mn^{+2}(Sn^{+4},Ta)Ta_2O_8$)]. Companies that mine tantalum ore therefore do not have to deal with tantalum-associated fire hazards. Companies that convert ore to a purified tantalum powder encounter fire hazards, as do companies that form the powders into capacitors. These powders have particles in the size range of 0.1 to 100 microns depending on the step in processing, and the product grade/capacitance target being produced. The Ta particles will have a complex particle morphology to achieve high surface areas. An SEM photomicrograph of an example tantalum particle is shown in Figure 1.

About 16% of tantalum production is purified via melting in electron beam furnaces and formed into ingots that weigh as much as 3,600 pounds. The remainder of tantalum is consumed as tantalum oxide, alloy additives, and in carbides. Ta ingots are subsequently used to produce mill stock products, such as bar, rod, wire (for capacitors), sheet for process equipment, and more recently for sputtering targets in the production of integrated circuits. These large finished ingots and other mill stock products have little surface area for reaction (~ $2x10^{-6}$ m^2/gm) compared to tantalum powder used in capacitors which have surface areas that can exceed 1.2 m^2/gm. Therefore, mill products do not present the same safety issues. However, there have been incidents in the ingot forming stages, where a "green" ingot of compressed, sintered powder has created a fire.

CHEMISTRY OF TANTALUM REACTIONS WITH AIR AND WATER

Oxidation of tantalum metal proceeds rapidly, even at low temperatures, according to Equation (1) below:

$$2Ta\ (s) + 5/2\ O_2\ (g) \Rightarrow Ta_2O_5(s) \qquad \Delta H_{298K} = -489\ kcal/gm\text{-}mole \qquad (1)$$

The heats of reaction for tantalum have been calculated using FactSage software, and are plotted as a function of temperature in Figure 2. Tantalum metal powder is black, while the pure oxide is white, making it easy to determine if any material has oxidized. Inerting the processes in which new tantalum surfaces are created, such as milling, has prevented fires from occurring since it removes oxygen "leg" from the fire triangle. But Reaction (1) must be allowed to proceed under controlled conditions before exposing the material to air. The controlled process to oxidize the surface is commonly referred to as passivation.

At elevated temperatures, tantalum also reacts with nitrogen, where two forms of tantalum nitride can be generated according Equations (2) and (3):

$$Ta\ (s) + 1/2\ N_2\ (g) \Rightarrow TaN(s) \qquad \Delta H_{573K} = -\ 47.5\ kcal/gm\text{-}mole \qquad (2)$$

$$2\ Ta\ (s) + 1/2\ N_2\ (g) \Rightarrow Ta_2N(s) \qquad \Delta H_{573K} = -\ 64.7\ kcal/gm\text{-}mole \qquad (3)$$

Since this reaction is exothermic, and nitriding will affect product quality, nitrogen gas is generally not used for inerting tantalum metal, but rather argon or helium gas. In some low temperature operations, nitrogen can be used to minimize oxygen exposure.

Tantalum has such a strong affinity for oxygen that it will remove it from water, giving off heat and liberating a large volume of hydrogen according to Equation (4) below:

$$2\ Ta(s) + 5\ H_2O\ (g) \Rightarrow Ta_2O_5\ (s) + 5\ H_2\ (g) \qquad \Delta H_{398K} = -\ 198.4\ kcal/gm\text{-}mole \qquad (4)$$

The hydrogen then becomes a fire/explosion hazard and will certainly be ignited in the presence of hot tantalum particles and air according to Equation (5):

$$1/2\ O_2\ (g) + H_2\ (g) \Rightarrow H_2O\ (g) \qquad \Delta H_{398K} = -\ 58\ kcal/gm\text{-}mole \qquad (5)$$

Water is therefore not used as a fire extinguishing media for tantalum, since it will generate additional fuel and spread the fire.

Adiabatic and Actual Temperature Rise

Hertzberg(1992) and Cashdollar(1994) studied the combustion of a variety of metal dusts, including tantalum. They computed the stoichiometric concentration of Ta in air to be 1268 g/m^3 (7 moles of Ta and 8.75 moles of oxygen per cubic meter), and at this concentration the maximum adiabatic temperature rise would be 3290°K. The adiabatic temperature rise is calculated by:

$$\Delta T_{ad} = \Delta H_{oxid} / C_p \qquad (6)$$

Where ΔT_{ad} is the adiabatic temperature rise, ΔH_{oxid} is the heat of oxidation in Reaction (1), and C_p is the heat capacity of the tantalum dust and air mixture. This calculation assumes 100% conversion of the tantalum to the oxide. Also, this does not account for the additional reactions of tantalum with nitrogen (Reaction (2) and (3)), which would generate an even higher temperature, but also leads to lower pressure rise potential. Clearly, tantalum fires burn extremely hot, but actual measurements show values well below ΔT_{ad}. Measurements of ΔT_{ad} reported by Izumi(1999) are on the order of 1800 - 2000°C as shown in Figure 3. A tantalum fire is an extremely hot smoldering fire that is self-fueling and propagates at about 7 mm per second. Fires are sometimes detected by observing metal processing equipment that has a cherry red exterior.

Temperature for Spontaneous Ignition

We have observed that high surface area tantalum particles without an oxide layer on the surface will spontaneously ignite, even at less than 50°C. For tantalum powders that do have a layer of Ta_2O_5 on the surface, higher temperatures are required for ignition. The measurements in Figure 4 show that as capacitance is increased (which is associated with increasing specific surface area and decreasing particle size), the ignition temperature drops from 540°C to 450°C.

Limiting Oxygen Concentration

The limiting oxygen concentration (LOC) for combustion of tantalum powder has been reported to be approximately 6% for finished powders (Izumi), and this value was shown to be relatively insensitive to particle capacitance target (for 20,000 to 80,000 CV powders). NFPA 484 (2002) reports values of 2% for primary particles, and measurements by Chilworth indicate it may be less than 2%, the low limit capability of the test. Inerting systems with continuous monitoring and control of oxygen levels should be maintained at 60% of the LOC to comply with NFPA 69 standards for Explosion Prevention Systems. In some cases, providing some low level of oxygen in the surrounding gas can provide a route for passivation.

Minimum Ignition Energy and Electrostatic Properties

The minimum energy for ignition (MIE) decreases to very low levels as particle size is decreased, as shown in Figure 5. Coarser particles have MIE's on the order of 2.0 mJ, while the fine screen cut is 0.5 mJ. The minimum ignition energy for tantalum powder has been reported as low as 0.2 mJ (Matsuda, 2000) for very fine tantalum particles taken from a dust collector system.

Tantalum is a valve metal, and as such has the ability to store and discharge energy – exactly its function in a capacitor. When tantalum powder passes over conductive surfaces, like stainless steel, no measurable surface potential is generated. However, when non-conductive materials, such as glass or plastic are involved, surface potentials of 0.4 to 0.7 kV are generated, depending on ambient humidity (NFPA 484). Tantalum has moderate chargeability. Proper grounding and bonding of equipment as well as continuity checks are critical to prevent static charge build-up.

Pressure Rise and Maximum Pressure

There are several studies on the rate of pressure rise and ultimate pressure in tantalum dust explosions, and tantalum falls into the category of a weak explosion hazard. Relative to most other combustible metals like aluminum or magnesium, or polyethlyene, the pressure rise with tantalum is relatively slow. In the work of Cashdollar, the maximum pressure was 3 bar. However, with higher surface area powders, pressures of 6 bar have been measured (NFPA 484). Tantalum can consume both oxygen and nitrogen at high particle temperatures, which will clearly reduce the pressure rise rate and ultimate pressure. No studies looked at these competing reactions in a fire scenario. But the significant reaction rate of Ta with nitrogen above 600°C is well-documented (Clancy (2000), Miller (1959)).

SOME CASES OF TANTALUM FIRES

This section describes some of the more common modes by which fires have occurred in the past in handling/processing of tantalum powder.

- Dust Collectors – Dust collection systems are used to remove fine tantalum particles that are generated by various processing operations in either tantalum powder production or in capacitor manufacturing. These systems meet the requirements of industrial hygiene, minimize dust accumulations in the area (which is a fire hazard), and enable recovery the high value of tantalum, which has a value in excess of $8/ounce. There have been several instances of fires occurring in dry baghouse operations, and in one case there was a fatality (Matsuda). In the event Matsuda describes, one worker was killed and another severely injured as they were emptying Ta dust from the dust collector into a polyethylene bag. It was concluded that the fine dust had very low ignition energy of <0.2 mJ, and that electrostatic discharge caused the ignition. Note that particles typically drawn into these systems are much finer than the product, and these particles tend to hang up in ductwork and can be easily charged. If the concentration of dust becomes reaches the explosive limit, it takes very little energy to ignite the tantalum particles.

Interestingly, in all dust collector fire situations reported, the fabric bags were completely burned, but there was no deflagration in the baghouse. These baghouses were of the sheetmetal/box construction design which have very minimal pressure rating, yet there was no rupture of the dust collector walls. Generally, wet dust collectors are preferred for tantalum service. An example of a typical wet dust collector is shown in Figure 6. It is critical to have adequate purging of the vapor space and frequent removal of the collected sludge to prevent hydrogen accumulation per Reaction (4). It uses copious amounts of water to saturate the tantalum, and prevents fires from occurring. These have proven to be very safe, and also very reliable. If a dry dust collector must be used, there are some critical guidelines to follow described in NFPA 484, mainly to insure no accumulation of dust in the collector chamber, dust is to be collected in grounded metal containers, and the design must prevent condensation.

- Milling Systems – Systems that mill tantalum continuously generate fresh tantalum surface area that is very reactive with air. It is critical that these systems provide for controlled passivation of these fresh surfaces. The typical practice is to blanket the milling device in argon or helium, monitoring to insure that oxygen levels are controlled. If high oxygen levels are detected, milling equipment should be shut down automatically. The milling equipment itself is an ignition source due to the potential for metal-to-metal contact and sparking. The milled material is allowed to passivate downstream under more controlled conditions.

- Mixing where Water is Present – There have been situations where wetted tantalum powder was being processed in mixing equipment without an inert gas in the vapor space. Hydrogen was generated, which subsequently ignited and caused a deflagration in the mixing device. In one case, an individual was severely burned. When such operations are conducted, it is imperative that the vapor space is blanketed with argon, and that hydrogen evolution is monitored, controlled, and safely purged if in enclosed equipment. This is particularly critical if new metal surfaces are being generated due to the mixing process.

- Ta Powder Drying Systems - There are operations where wet tantalum powder needs to be dried. In some cases, the drying process takes place under air, and so surface area can be passivated as water is removed. Nonetheless, NFPA guidelines recommend that the drying temperature not exceed 80°C. On the other hand, when vacuum dryers are used, higher drying temperatures are acceptable, but air can not be introduced until the tantalum powder temperature has dropped to less than 80°C. Systems are also installed to inject argon into the dryer when high temperature conditions are detected.

- Dust Accumulations – Situations have been observed where a fire is initiated in a piece of equipment, and the hot burning tantalum particles dispersed through the area, igniting any accumulated layers of tantalum dust. Dusts that are allowed to accumulate in a work area are a fire hazard and also dictate that more stringent electrical classification requirements must be met (Article 502, Class II Locations of NFPA 70, National Electric Code). This is why spills and accumulations must be avoided, and cleaned up promptly. A typical industrial vacuum cleaner that meets the requirements for flammable dusts when properly grounded is shown in Figure 7. This vacuum cleaner is explosion proof, and has very tight dust control from the blower. It is also recommended to empty the chamber on a routine basis and not allow a large inventory of tantalum dust to accumulate and remain in the vacuum.

PREVENTING AND EXTINGUISHING TANTALUM FIRES

NFPA introduced in 2002 a Standard 484 for Combustible Metals, Metal Powders, and Metal Dusts. This provides recommended practices for tantalum in Chapter 7. These standards should be followed to reduce the potential of tantalum fires.

Safeguarding against tantalum fires involves many levels of protection. Inerting is one form of safeguarding against a fire since, as discussed, it removes oxygen from the fire triangle. To inert a system, it must first be flooded with an inert gas and measured for oxygen content. If the oxygen content is too high, the equipment cannot start until a low oxygen concentration is achieved. Once achieved, continuous monitoring should occur. In the case of argon as an inerting agent, the monitoring point should be at the highest elevation in the process since argon is heavier than air. If the monitoring shows a concentration above the LOC limit, the equipment should automatically shut down to prevent any fresh tantalum surfaces from being created. The system should flood with argon and retest. Preventive maintenance should be done on all parts of the system from instrumentation and calibration checks to clearing blockages and eliminating build-up of material in equipment and processing areas.

Emergency stops have been a useful tool in a fire case. An emergency stop is an easily accessible button that stops the equipment and places it in a safe state, which could mean inert flooding, and stops any associated scrubber or dust collector. Another safeguard is to have thermocouples in dust collector ductwork that detects abnormal heat. If a high temperature were observed, it would act the same as an emergency stop.

Proper personal protection equipment should also be worn in case a fire does occur. Fire retardant suits, static dissipative shoes, hard hat, and safety glasses should be required as a minimum for processing tantalum powder. Safety awareness and training are key elements to tantalum safety and loss prevention. The operators who run the equipment every day are the fist link to upset conditions and should also know what to do in case of a fire. A burn area or shed removed from processing and personnel should be created as a safe place where burning tantalum can be taken if necessary.

The best extinguishing agents employed in tantalum processing are either noble gases, like argon or helium gas, or Class D dry agents, like MET-L-X. Use of an automated inert gas blanketing systems with proper controls in a piece of process equipment when a high temperature is detected is very effective. In situations where a fire occurs in a room, MET-L-X powder is applied to extinguish the fire. The material is then removed to a safe area. MET-L-X is basically sodium chloride salt with some additives which melts and smothers the fire (Ansul Incorporated).

Administrative controls should also be employed to avoid fires. The minimization of ignition sources is also key to preventing fires, so use of hot work permit systems, control of heat sources, managing and controlling change, electrical classification, and good training and work instructions are used. Proper maintenance and housekeeping are essential good practices for safety in tantalum manufacturing.

SUMMARY

Over the past eight years, the understanding of requirements for safe handling of tantalum powder has improved dramatically. Much of the learning is reflected in the new NFPA 484 Standard. Concurrently, plant processing systems have been upgraded to mitigate the fire hazards. Table I summarizes the key physical properties and safety parameters of tantalum powders. Some key points to remember in design of systems processing tantalum powder:

1. It is reactive with air, and is a fuel in the fire triangle.
2. Tantalum is an oxygen "getter" and will remove oxygen from water and generate hydrogen. Therefore, water is not used as extinguishing media, nor should it be used in processing at high temperatures without proper inerting systems.
3. The finer the particle size, the lower the ignition energy and the lower the ignition temperature. Particle size distribution impacts these parameters more than the capacitance of the powder.
4. Tantalum is reactive with nitrogen, therefore true inerting is done with noble gases like argon or helium.

5. Milling equipment generates fresh tantalum metal surfaces that are reactive with air, and oxygen levels should be monitored and controlled.
6. Since very fine particles are accumulated in dust collection systems, and these have very low ignition energies, wet dust collectors are preferred to dry collection systems.

Fine tantalum powders are processed very safely with proper attention to system design that reflects the issues discussed in this report and the standards of NFPA 484.

ACKNOWLEDGEMENTS

We would like to thank Cabot Supermetals for their permission to publish this work, and their commitment for continuous improvement in workplace safety.

LITERATURE CITED

Ansul Incorporated**, Fighting Fires with Dry Powder Agents**, White Paper 1013 (1998).

Cashdollar, K.L., **Flammability of Metals and Other Elemental Dust Clouds**, Process Safety Progress, Vol. 13, No. 3, p. 140 - 145 (1994).

Clancy, Donald J., **Statistical Modeling of the Effect of Nitrogenation on the Electrical and Physical Properties of Tantalum,** Tantalum-Niobium International Study Center Annual Conference, San Francisco, October 24 (2000).

Hertzberg,M, I.A. Zlochower, K.L. Cashdollar, **Metal Dust Combustion: Explosion Limits, Pressures, and Temperatures,** 24th International Symposium on Combustion, The Combustion Institute, Pittsburgh, p. 1827 (1992).

Izumi, Tomoo, **Safety Data of Tantalum Powders,** Denkai Chikudenki Hyoron [Electrolytic Condenser Investigation Society], 50(1), 73-81 (1999).

Factsage Software, Center for Research in Computational Thermochemistry, Ecole' Polytechnique (Universite' de Montreal), Quebec, Canada.

Matsuda, Toei and Mizuki, Yamaguma, **Tantalum Dust Deflagration in a Bag Filter Dust-Collecting Device,** Journal of Hazardous Materials, A77, 33-42 (2000).

Michaluk, C., R.O. Burt, and D.P. Lewis, **Tantalum 101: The Economics and Technology of Tantalum Sputtering Targets**, Semiconductor International, Reed Elsevier, July (2000).

Miller, G., **Tantalum and Niobium,** Butterworths Scientific Publications, London (1959).

NFPA 484**, Standard for Combustible Metals, Metal Powders, and Metal Dusts,** Chapter 7, Tantalum (2002).

NFPA 69, **Standard on Explosion Prevention Systems**, Chapter 5(2002).

Table I
Physical Properties and Safety Parameters of Tantalum Powders

Atomic Symbol	Ta	Reference
Molecular Weight	180. 95	
Atomic Number	73	
Specific Gravity	16.65 gms/cc	
Bulk Density – High Cv Powder	2 – 5 gms/cc	
Melting Point	2,996°C	
Boiling Point	5,425 °C	
Ignition Temperature in Air	440 - 540 °C	Izumi
LOC*	2 – 6% Oxygen	Chilworth, NFPA
MIE*	0.2 (Ultra Fine Powder) >500 (mJ)	Matsuda, Izumi, NFPA 484
Maximum Allowable Drying Temperature in Air	80°C	NFPA 484
Stoichometric Dust Concentration	1268 gms/m^3	Hertzberg
Lower Dust Explosivity Limit*	160 – 165 gms/ m^3	NFPA 484
Kst*	149 bar-m/sec	NFPA 484
Minimum Conveying Velocity in Dust Collection Ductwork	1365 meters/min	NFPA 484
Recommended Fire Extinguishing Media	Class D for Metal Fires MET-L-X	NFPA 484 Ansul
Recommended Inerting Agents	Argon, Helium	NFPA 484

* These parameters are sensitive to the particle size, size distribution, and surface area (BET). Have tried to represent the worst case scenario.

Figure 1
SEM Photomicrograph of Tantalum Powder Particles

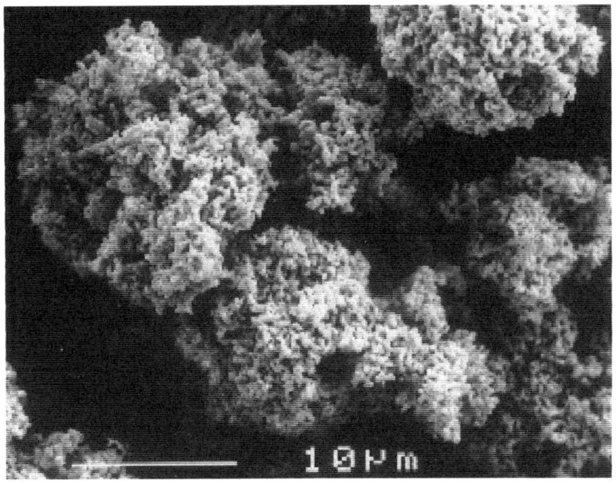

Figure 2
Heats of Tantalum Reactions

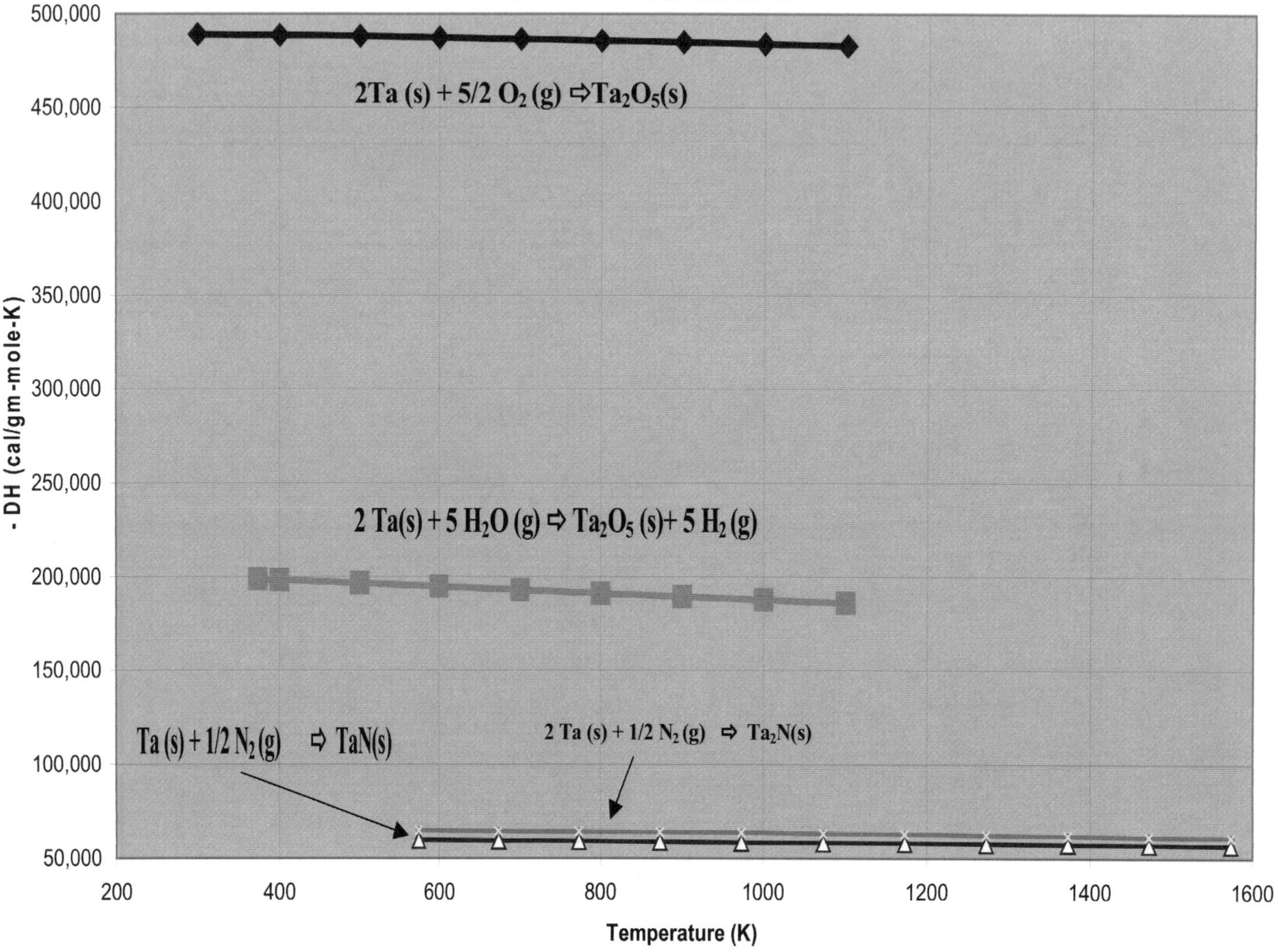

Figure 3
Burning Speed and Maximum Temperature of Ta Combustion
(Izumi, 1999)

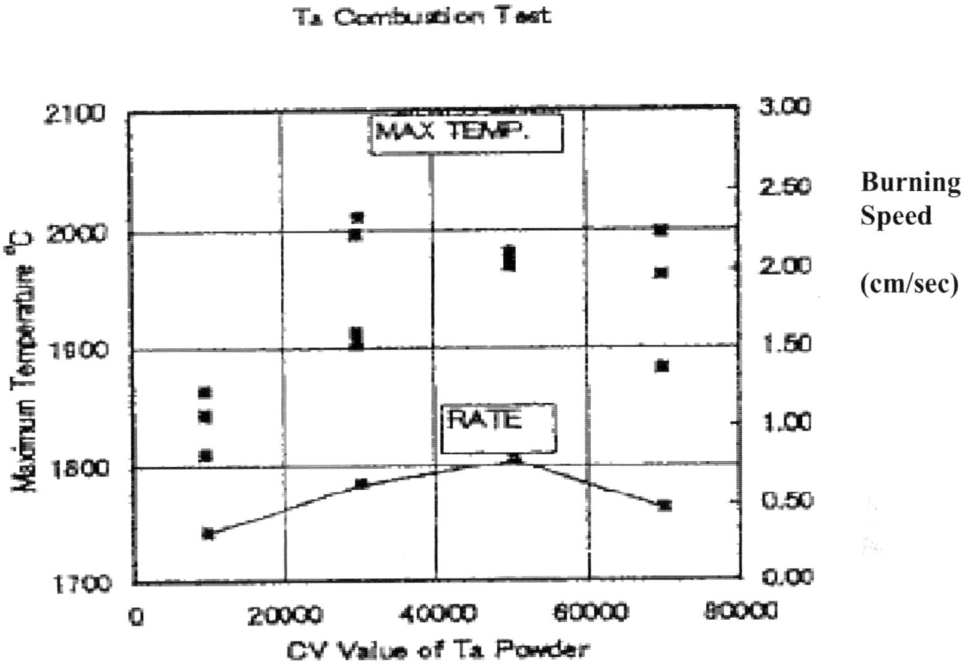

Capacitance of Powder Evaluated (μFV/g)

Figure 4
Relation Between Ignition Temperature and CV Levels of Ta Powder
(Izumi, 1999)

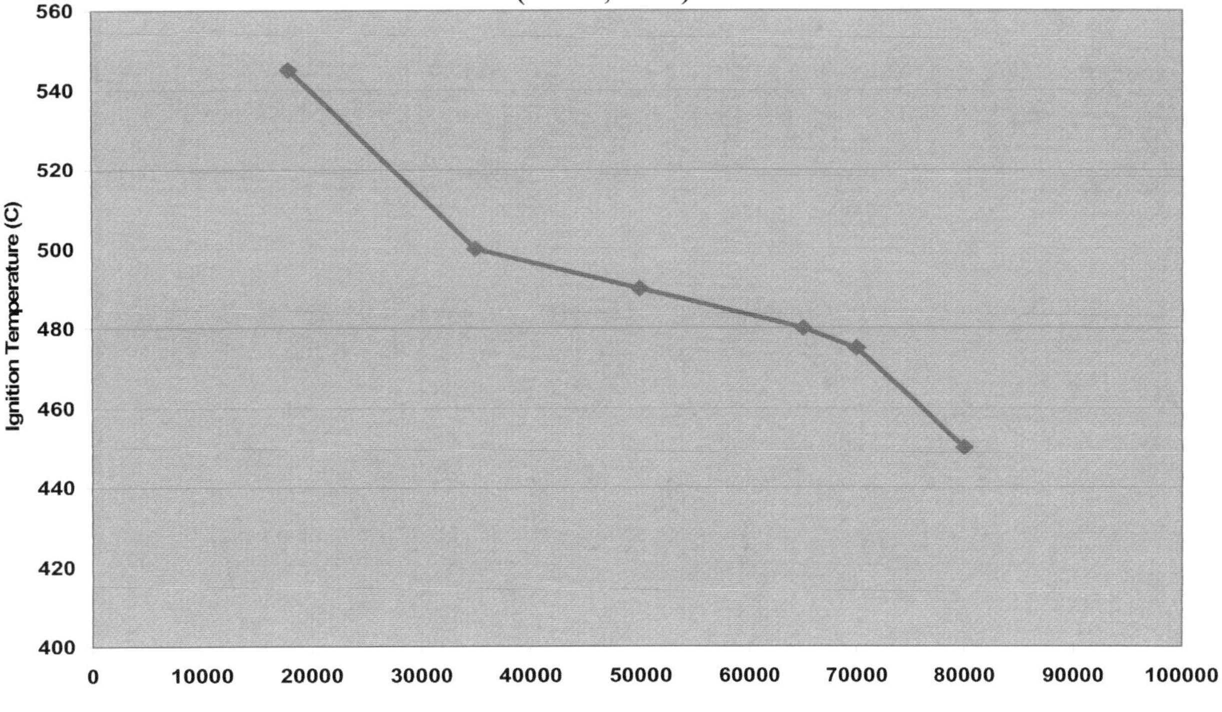

Powder Capacitance (μFV/gm)

Figure 5
The Effect of Tantalum Particle Size Range on MIE
(Izumi, 1999)

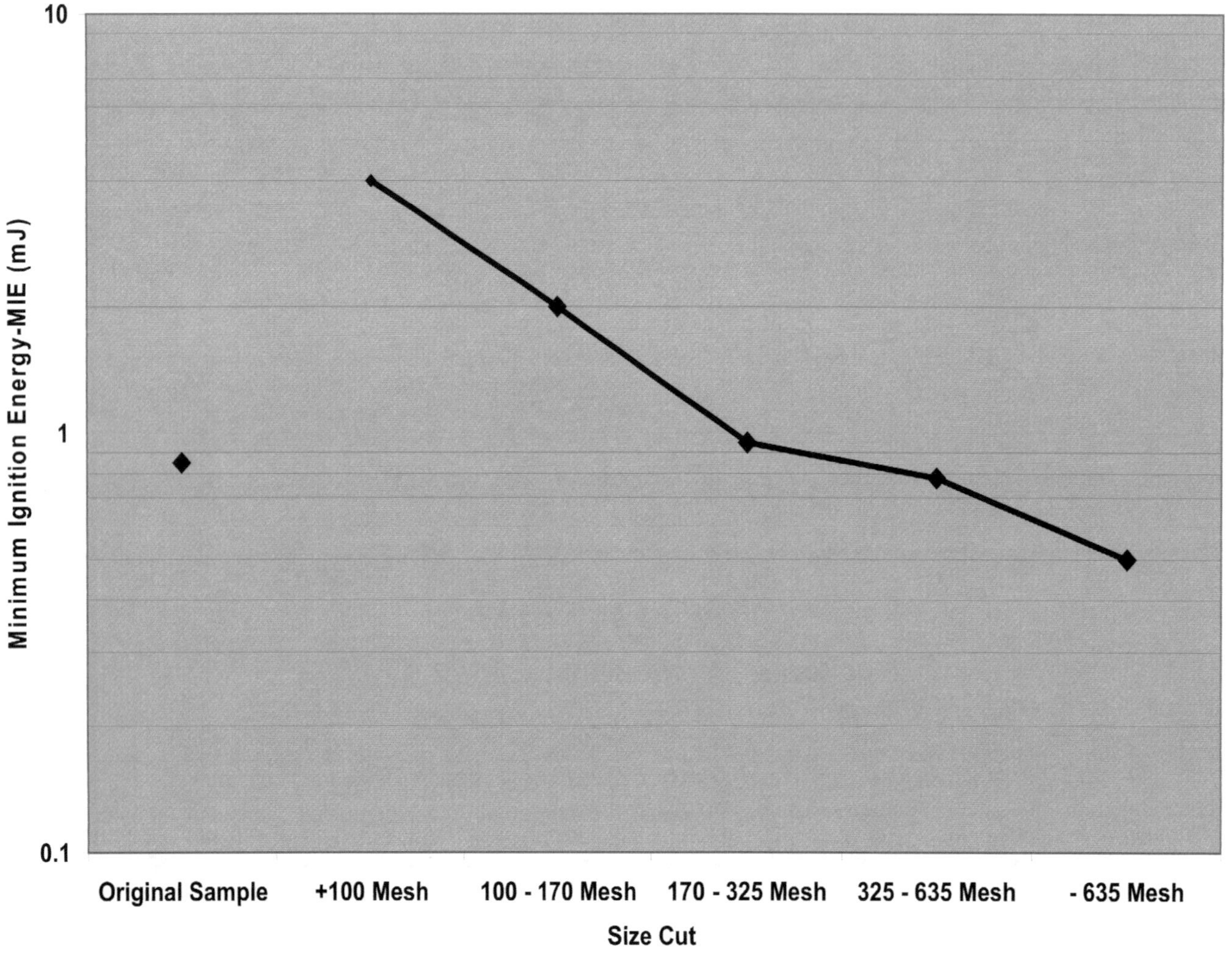

Figure 6
Wet Dust Collector Used in Ta Dust Service - Type N RotoClone
(Photo Courtesy of AAF International – www.aafintl.com)

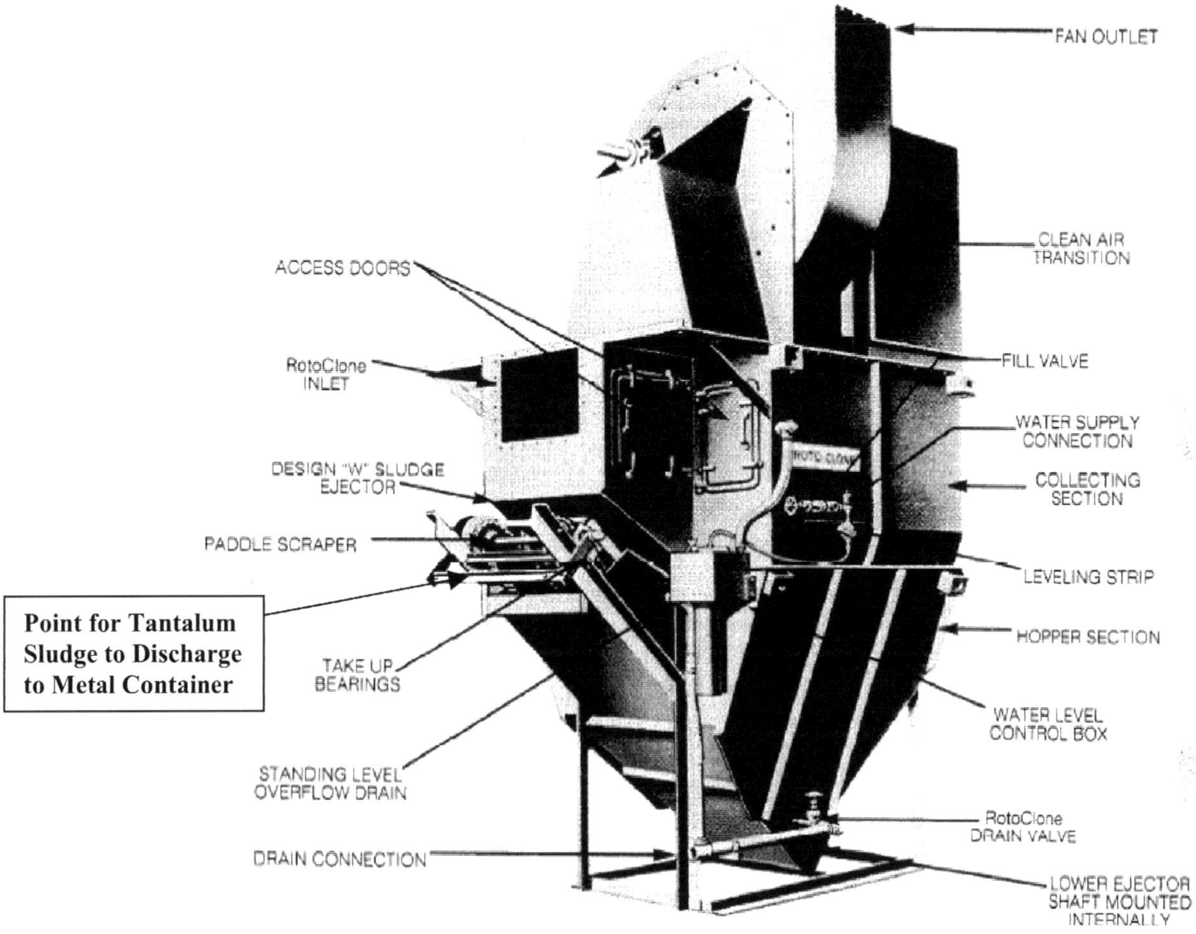

Figure 7
Industrial Vacuum Cleaner – Explosion Proof/Dust Ignition Proof
(Photo Courtesy of Tiger-Vac Inc. / www.tiger-vac.com)

Modeling of Dust Explosions in Industrial Processes:
Use of Explosion Simulations (FLACS, DESC) For Explosion Isolation Design

by

Jef Snoeys, Fike Europe BV, John Going, Fike Corporation,
Olav Roald Hansen and Brian Wilkins, GexCon AS

Prepared for Presentation at the 38th Annual Loss Prevention Symposium,
April 25 – 29, 2004
New Orleans, LA
Fire, Explosion and Reactive Hazards Session

December 2003

Unpublished

Modeling of Dust Explosions in Industrial Processes:
Use of Explosion Simulations (FLACS, DESC) For Explosion Isolation Design

by

Jef Snoeys, Fike Europe, John Going, Fike Corporation,
Olav Roald Hansen and Brian Wilkins, GexCon

Fike Europe BV: Herentaals, Belgium
Fike Corporation:Blue Springs Missouri, USA
GexCon AS: Postboks 6015, N-5892 Bergen, Norway

Abstract

A gas or dust explosion that travels from one vessel through a pipe into a second vessel may lead to serious damage due to pressure piling or jet ignition effects. One must therefore use explosion isolation to prevent the flames from propagating from one vessel to another. Examples of active explosion isolation methods are mechanical isolation and injection of chemical barriers. Passive methods also exist including flame diverters and quenching devices installed in the flowline. The active barriers will need to be installed at a certain distance from the exit of the first vessel (X_{min}) to manage to activate and close before the flames arrive. It is also important that the distance is not too large, as this may lead to damaging pressures or premature activation (chemical barrier). Maximum distances, X_{max}, are therefore also often specified.

It is a challenge to predict the minimum and maximum isolation distances for a general industrial process. The vessel in which the explosion starts may be equipped with vents opening at a pressure similar to activation pressures for the isolation system. The ignition may take place anywhere in the first vessel. Proper installation distance may depend on vent area and opening pressure, ignition location, detection criteria, pipe diameter and dust reactivity. In this study, the CFD-tool FLACS has been applied to generate guidance on installation distances for explosion isolation systems. Multiple scenarios are generated and compared to experimental test data. The good correlation that was found provides encouragement for the future use of explosion simulators in product development, product application and process safety assessment.

1. INTRODUCTION

Primary explosion protection measures are used to safeguard process equipment against the risk of an explosion by containment, venting and/or explosion suppression.
Explosions will also propagate from the vessel to other plant areas through fuel-laden pipes. A propagating explosion will cause the dust or gas to pre-compress; it will also generate increased turbulence resulting in flame jet ignitions. These will increase the explosion characteristics and therefore seriously affect other interconnected plant explosion protection systems. Therefore all explosion protection measures should be combined with explosion isolation methods.
Explosion isolation is an active technique, which prevents flame and/or an explosion pressure wave from propagating via connecting pipes or ducts into other parts of apparatus or plant areas.

In order for a fire or explosion to occur in a pipe, a fuel and oxidizer mixture within the flammable limits must be exposed to an ignition source of sufficient strength. Once ignited, a combustible mixture is capable of supporting deflagrations and possibly detonations as it propagates through the pipe. These three types of combustion propagation phenomena in pipes – fires, deflagrations and detonations – must be clearly identified and understood in order to conduct adequate process safety design.

The principle factors influencing these phenomena in a connected system are:

- Fuel parameters: fuel-oxidizer composition
- Initial pressure, temperature and turbulence level (flow field)
- Pipe diameter and length
- Piping system components (elbows, etc.) and internal geometry in vessels
- Connected systems (vessel-pipe-vessel.)

These factors and phenomena were extensively discussed in papers 'Design for Isolation of Flame Propagation in an Oxygen Atmosphere' (1), presented at the American Institute of Chemical Engineers, 37th National Meeting, Process Plant Safety Symposium New Orleans, LA March 31-April 2, 2003 and "Flame Propagation in Industrial Scale Piping" (2).

Industry standards such as NFPA 69, VDI 2263 lists several explosion isolation techniques used to prevent the spreading of explosions throughout a process. Isolation can be achieved in different ways – chemically (by introducing extinguishing powder in the pipe, quenching the flame) or mechanically (by means of a mechanical block such as a slide valve).

In order to establish any type of barrier, passive or responsive, mechanical or chemical, the flame and pressure in space and time need to be defined and the process of propagation clearly understood. The active barriers will need to be installed at a certain distance from the exit of the first vessel (X_{min}) in order to activate and close before the flame arrives. It is also important that the distance is not too large, as this may lead to damaging pressures or premature activation (chemical barrier). Maximum distances, X_{max}, are therefore also often specified.

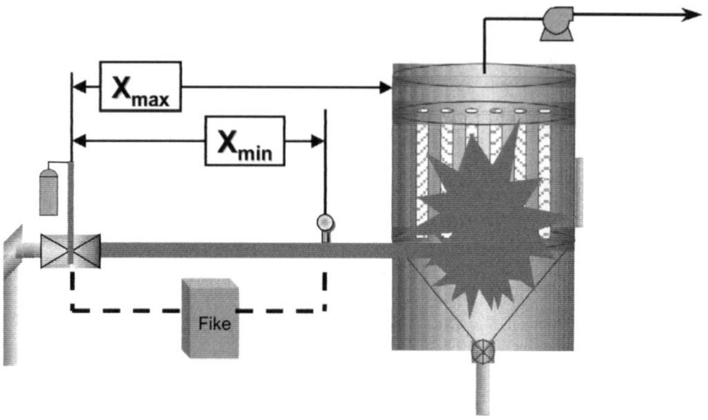

Figure 1 Minimum (X_{min}) and maximum (X_{max}) distances to establish an explosion isolation barrier

Although considerable data is available about gas- and dust explosion propagation, the application of explosion isolation remains complicated. The currently industry guides do not provide methods with which one is able to predict the necessary measures for safe isolation design.

Furthermore, it is a challenge to predict the minimum and maximum isolation distances for any given industrial application. The vessel in which the explosion originates may be equipped with explosion relief vents opening at pressures close to, equal to or even below the required pressure to activate the active explosion barrier system. The ignition may take place anywhere in the first vessel. Correct installation distances depend on vent area and opening pressure, ignition location, detection criteria, pipe diameter and fuel reactivity. The presence of elbows other components will further complicate the situation because the flame will accelerate The availability of a reliable prediction method for flame acceleration in pipes would help dimensioning a safe isolation system for such situations.

2. FLACS EXPLOSION SIMULATOR AND DESC PROJECT

2.1 Gas Explosion Simulation – FLACS.

The FLACS gas explosion and dispersion simulator has been developed at GexCon (CMI/CMR) with support mainly from the offshore industry and to some extent European Commission supported projects. The development started in 1980, and through the years mature models that are well validated against experiments at various scales have developed. Validated CFD-explosion tools like FLACS are used extensively for risk assessments and hazard evaluation offshore. For onshore plants and processes FLACS is also used, and can provide valuable advice on how to optimize design and safety solutions to minimize explosion hazards and risk or maintain the safety level at a lower design cost. Compared to less accurate simpler screening tools widely used by land based industry, FLACS can give much more accurate predictions on potential size and reactivity of gas cloud from a release, and also assess in greater detail explosion pressures (both distribution in time and space for a range of input parameters). One major difference compared to the simplified tools, is that with FLACS actual geometry layout is taken into account when predicting the dispersion and explosion. In Figure 2 some examples of output from simulations are shown both for explosions offshore, gas dispersion, far-field blast propagation, explosion in closed connected vessels and dispersion from a LNG-spill.

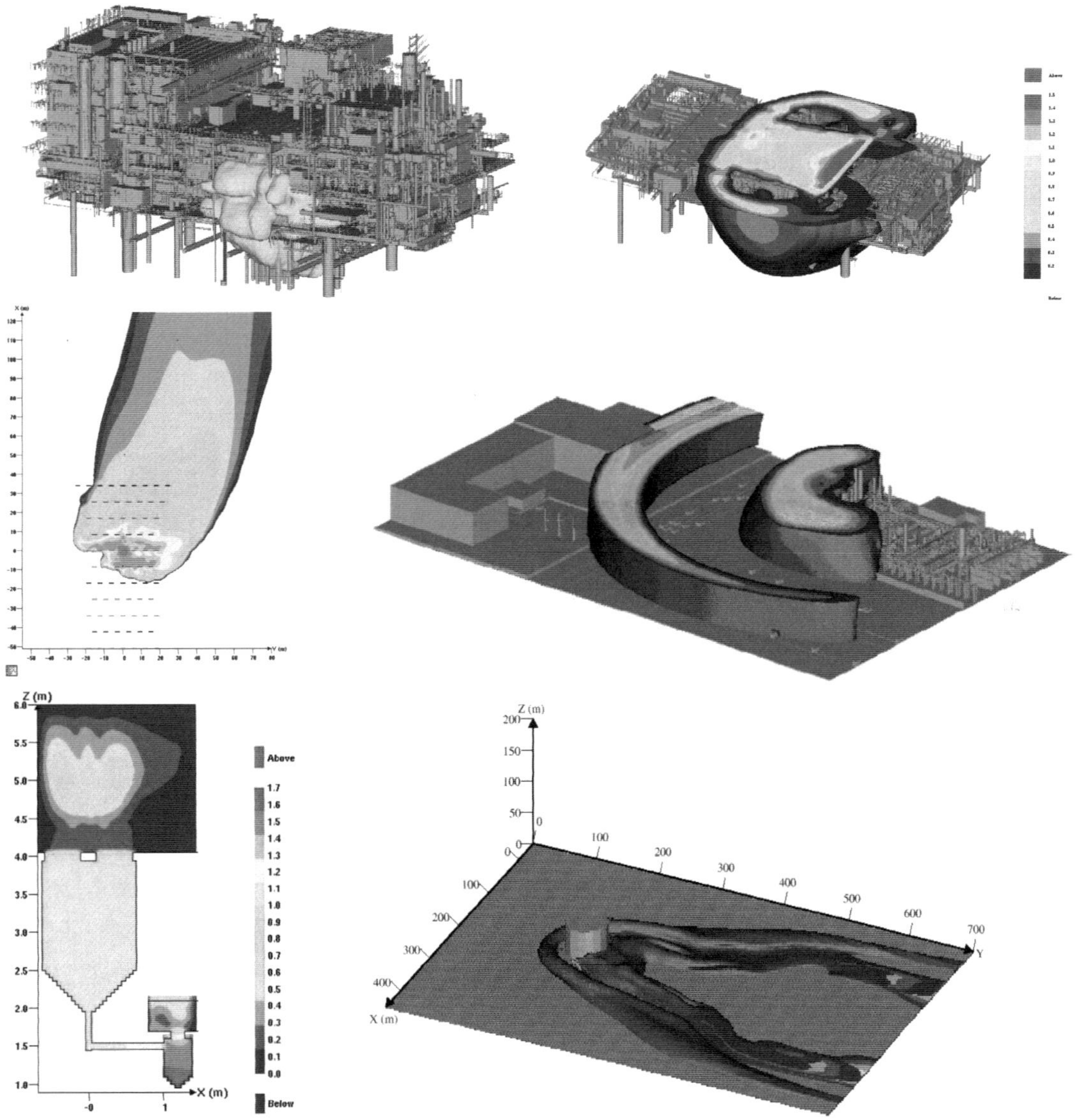

Figure 2 The CFD-tool FLACS was developed for offshore explosion risk assessments (top), but the physical models can also simulate a range of other situations like atmospheric gas dispersion and blast propagation (middle), vented explosions inside process equipment and LNG-releases (lower).

Over the last 10 years FLACS has gradually been used more frequently to estimate explosions in vessels and pipes for guidance on vent sizes and locations. Compared to venting guidelines, FLACS can take into account effects like actual shape of equipment and internal geometry details, actual flow and turbulence patterns, overpressures and elevated temperatures, actual concentration and reactivity of flammable material, exact vent location(s) and shape, opening pressure and inertia, as well as multi-vessel situations.

2.2 Dust Explosion Simulation – DESC.

Early in 2002, a European Commission supported project DESC (Dust Explosion Simulation Code) was started with 11 participants. HSL is coordinating the project; further participants are TNO, Inburex, FSA, Fraunhofer-ICT, Øresund Safety Advisors, Technical Universities of Delft and Warsaw, possible end users Hahn&Co and Lyckeby, in addition to GexCon. Through its cooperation with GexCon and TU Delft, Fike Corporation is also participating actively in the project. The aim of the project was to produce a CFD-tool that could predict industrial dust explosions. The CFD-package FLACS was to be used as basis for the development.

Once this CFD-tool for dust explosions is available, the potential benefits from it are many. Within days, experienced users will be able to predict the outcome of as many as a hundred explosion scenarios. Within minutes, one will be able to evaluate the effect of changing important parameters such as size (area) or location of explosion vent(s) as well as evaluate secondary hazards such as safety zones for personnel (exposure to flame/pressure) and surrounding equipment (blast effects).

In the absence of a simulation code, the options to industry have been;

- **Simplified tools or venting guidelines**
  Such guidelines are easy and quick to use but will very often have an inappropriate physical description of the phenomena, and will usually ignore effects such as process flow, turbulence, geometrical shape of vessels and pipes, internal geometry and vent arrangement (e.g. duct), etc. As a result of these limitations, the solution will have a tendency to become over-dimensioned, an example being unrealistically large vent areas if the effect of vent duct length and shape are not adequately handled. The 'quick' solution therefore becomes expensive and non-optimal in the end.
- **Downsized idealized experiments**
  When experiments are performed, for instance to prove conformity to standards, these are often performed at a smaller scale rather than full scale, with idealized shape of vessel and a limited number of ignition locations only. In many cases, these simplifications are questionable, and effects as seen in the experiments may not be applicable to real processes studied. Such experiments usually will require planning long time in advance, running costs are high and the flexibility is limited during execution.

- **Full-scale experiments**
  Full-scale experiments involving real process equipment strong enough to resist numerous explosion tests will be in most cases very expensive and time consuming.

2.3 Validation of Simulation by Experiment.

Experiments however are still required to validate the performance of simulation models. A very powerful approach is to perform numerous simulations before doing experiments. For several years now, GexCon has been able to benefit from having access to FLACS CFD-tool prior to performing experiments. In this way, a large number of potential experiments can be simulated, both to identify worst-case ignition location, optimal vent configurations, hazardous tests that would damage the test site, and more. Furthermore, the simulations may be used to identify the best locations for mounting pressure transducers and flame detectors. Once experiments are performed, only a limited number of pressure transducers and flame observation locations may be available and the measured data may not be sufficient to describe the observed physics adequately. A comparison with simulations can therefore increase the understanding of the experiments in the post-processing phase. Finally, comparison to simulations is also a good way to check the quality of experiments. For gas explosion experiments FLACS simulations have frequently identified weaknesses in the instrumentation that required improvement.

2.4 Application of Simulators – DESC.

With all the above elements in mind, it is obvious that simulations with a reliable prediction tool such as DESC in combination with experiments will give a significantly added value compared to doing experiments alone. Using the DESC tool directly towards real process situations, for evaluation of explosion hazard and optimization of protection systems, is equally important. Examples of such applications include:

- **Certification processes**

Two newly introduced European directives on explosive atmospheres and equipment used herein, the ATEX directives, require companies to perform risk assessments on explosive hazardous areas (1999/92/EC)(3). It also requires equipment manufacturers to (re-) assess (existing) equipment which will be used in these areas (94/9/EC)(4).
The European (EN) standard on isolation systems (CEN/TC 305/WG 3/SG 6) (5) requires explosion isolation barrier validation tests to be carried out, and assess the method of system design. The current status of the standard however does not provide a procedure to conduct the assessment. In the absence of harmonized standards, the manufacturer should seek advise from a Notified Body in order to validate the product/system under certification.

The validation process will typically consist of:

1. Validating the ability of a given barrier type to stop explosion propagation and determine the limits of application with respect to duct size, fuel type, and pipe/vessel geometry.
2. Defining application rules to make it possible for an engineer to design an isolation barrier system for a specific industrial application. This process requires a method to

calculate the number of isolators, select the type of agent, select an appropriate detection system and finally calculate the detector and barrier placement required to guarantee timely detection and barrier activation.

With regard to the first ATEX directive (3), a simulation tool will be of significant value when estimating the consequences and risk of explosions at different places in a facility. One can avoid unnecessary spending of resources, and also detect situations where simplified guidelines would underestimate the risk. In combination with assumed probabilities and costs of damage, one will be able to define where resources for risk reduction can be spent most effectively. The second directive has lead to a significant amount of explosion (re)testing of equipment. Much of this could be avoided and accelerated if a CFD-simulation tool was to be used in combination with experiments as described above.

- **Vent design in complicated situations**

Venting guidelines such as NFPA-68 and VDI3673 are not always effective and appropriate to be used when designing explosion venting systems. Due to the simplicity of the formula, the validity becomes limited if there is internal geometry in the vessel, if jet ignition from a pipe is possible, if the aspect ratio exceeds 5, if vent ducts are involved etc. . In many cases where the physics cannot be described properly by the venting guidelines, very conservative design will result. A reliable simulation tool will be able to provide a better approximation.

- **Change of process conditions**

In the process industry, and other places handling dust, process conditions may change during the lifetime of the installation. One may for instance change the particle sizes, operating concentrations, flow speed, physical process details, temperature or oxygen content in the air etc. Once a simulation model for the system is established, an evaluation of the possibly increased explosion consequences can be done quickly. After that, further risk-reduction measures can be evaluated and decided which are the most appropriate to be executed.

➢ **Multi-vessel configurations and isolation systems**

Very often process flows consider dust cloud being transported from one vessel through a pipe into another vessel. An explosion isolation system will prevent damage to the connected vessels; the CFD-tool will help to determine the most appropriate installation positions, required detection logics, etc. This will be described further in the following section.

3. SIMULATION OF ISOLATION SYSTEMS

Fike, as a manufacturer of explosion isolation devices will often be in a situation where guidance is needed on how to design an isolation system between protected (vented) vessels. As described in the introduction, several parameters will influence this process, and it is not likely that a simplified guidance will be able to pick up the necessary physics properly. For this reason it was decided to evaluate the use of the FLACS simulation tool for the design of isolation systems. Three different vessel volumes were simulated, $2m^3$, $10m^3$ and $250m^3$. For each of the volumes, explosions with different vent areas, pipe diameters, dust reactivities and ignition locations were simulated.

Since FLACS is a gas explosion tool, a modification had to be done to represent combustion of dusts with different reactivities. Calibration tests in a $2m^3$ sphere were carried out before proper dust reactivity in the simulations was established. A total of 90 scenarios were then simulated for the $2m^3$ situation. The parameter variations were as follows below. Similar variations were used for the $10m^3$ and $250m^3$ volumes.

- Dust reactivity: Kst 100, 200 and 500 bar.m/s
- Pipe diameter: 4" and 10"
- Vent diameter: 200, 500 and 700mm
- Ignition location: Center (1), Far (2), Pipe (3), Worst (4) and Best (5)

The ignition location 3 corresponds to the entry of the pipe. In Figure 2 the simulated geometry is drawn illustrating the ignition locations and measurement points for flame detection and pressure.

The vent will open at a pressure difference of 125 mbarg, whereas the isolation system will be activated at 100 mbarg pressure in the vessel. As can be seen in Figure 3, there is a flame detector for each meter through the pipe. The flame speed is assumed to be effected by the pressure development in the vessel, and it is believed that a grid resolution of 13cm is acceptable. If flame acceleration in the pipe without pressure in the vessel pushing the flame was dominant, a somewhat better grid resolution would have been chosen.

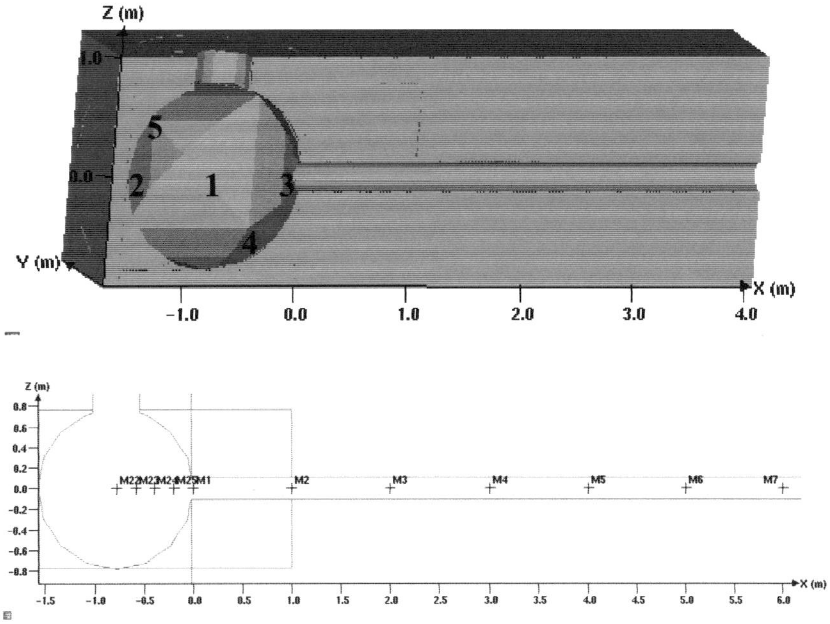

Figure 3 Upper picture shows illustration of the geometry with a 500mm vent upwards and a 10" pipe to the right (pipe length is 20m). Ignition locations are illustrated. In the lower plot the flame detection positions are indicated.

Based on the 90 simulations, tables were generated for flame arrival times relative to the time of detection (rather than time of ignition). These tables would depend on whether the detection criterion is based on pressure or flame detection at pipe entry.

In Figure 4, different simulations are compared for the situation with a very reactive dust Kst = 500 bar.m/s, 500mm diameter vent and 10" pipe. We observe that ignition at the entry of the pipe is the most challenging situation initially. Twenty ms after pressure detection of the explosion, the flame is already 2m into the pipe. If activation and closing times are above 40 ms, the ignition position called "Worst-case ignition" becomes more challenging. Sixty ms after detection, the flames are already 10m into the pipe and the flame speed is of the order 600 m/s. Changing to flame detection at pipe entry will not improve the situation much.

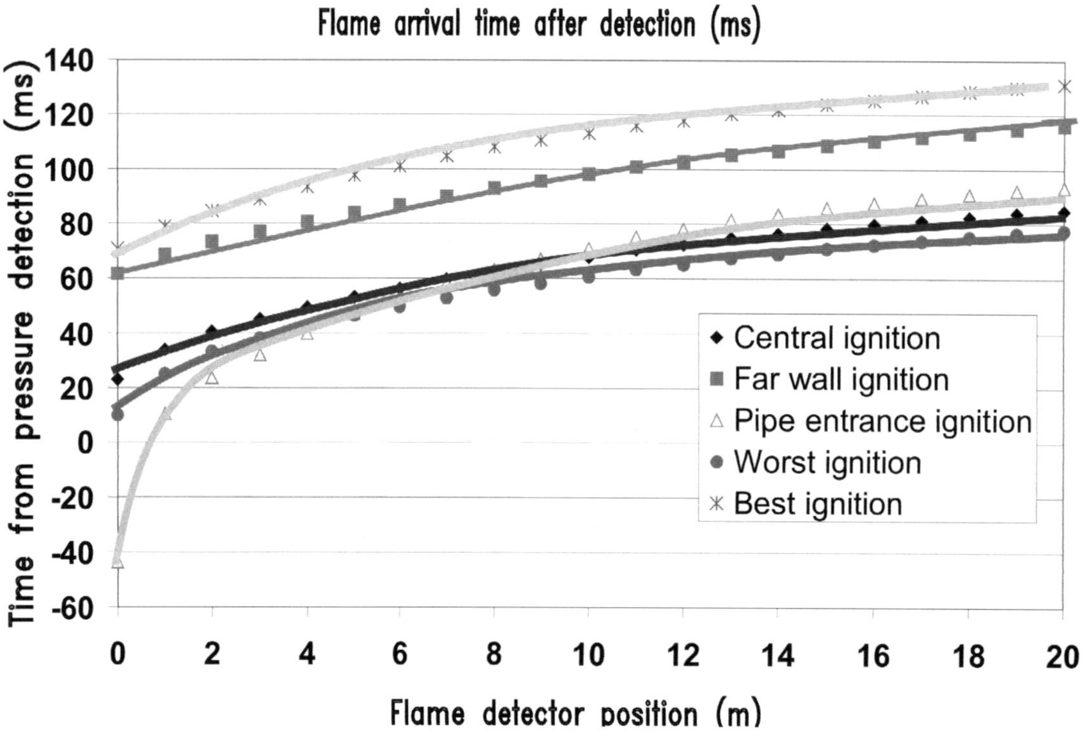

Figure 4 Example of flame propagation predictions for Kst=500 bar m/s, vent diameter of 500mm and 10" pipe.

In Figure 5 another situation is shown, in which a Kst =100 bar.m/s dust burns into a 10" pipe and the vent opening upwards has a diameter of 200mm. This time the flame propagation into the pipe is slower. For ignition at the pipe entrance, the flame is 2m into the pipe before the isolation activation criteria is reached. Within 50ms from detection the flame is 3 meter into the pipe. With a smaller pipe (2") the flame will reach 3m within 30ms. From Figure 5 another effect can be clearly seen. If the initial flame propagation is too slow, it may take some time before the flames reach the pipe. With the "Far Wall" ignition location, it takes more than 250ms from detection by pressure until the flame enters the pipe: with a smaller pipe (2") this took almost 500ms. Due to this variation, it

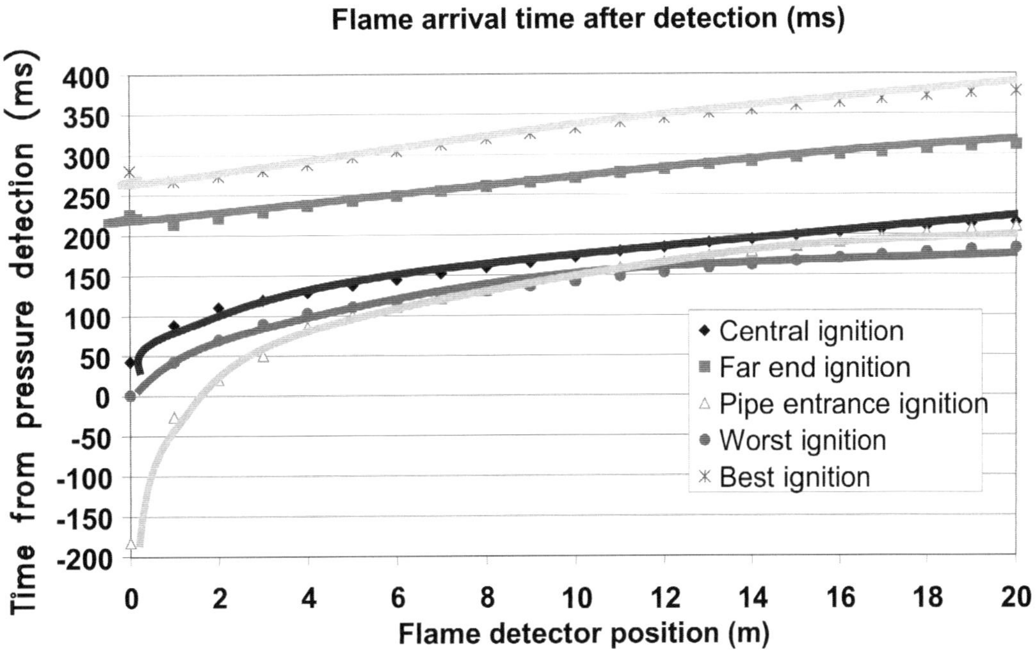

Figure 5 Example of flame propagation predictions for Kst=100 bar m/s, vent diameter of 200mm and 10" pipe.

is important to make sure the agent concentration is sustained over this time period. Maximum flame speed for this latter situation is about 125 m/s.

In Figure 6, the flame propagation in one particular simulation is illustrated. It is seen that the flame accelerates through the pipe in this situation, with increasing pressures along the pipe.

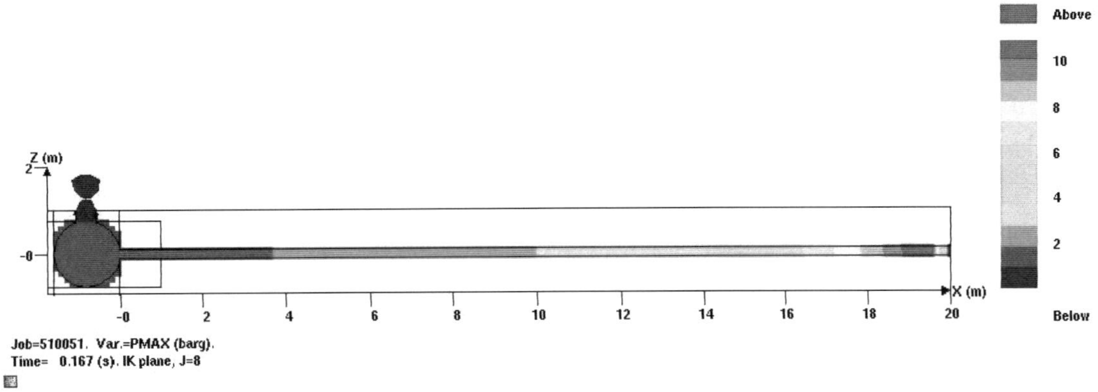

Figure 6 Predicted pressure development through the pipe for the Kst=500 bar m/s with 500mm vent and 10" pipe. Pressure is given in barg (1 barg = 14.5 psi)

The simulator allows the building of multiple scenarios at various conditions, e.g. ignition location. Figure 7 shows the effect on flame propagation at three different ignition points.

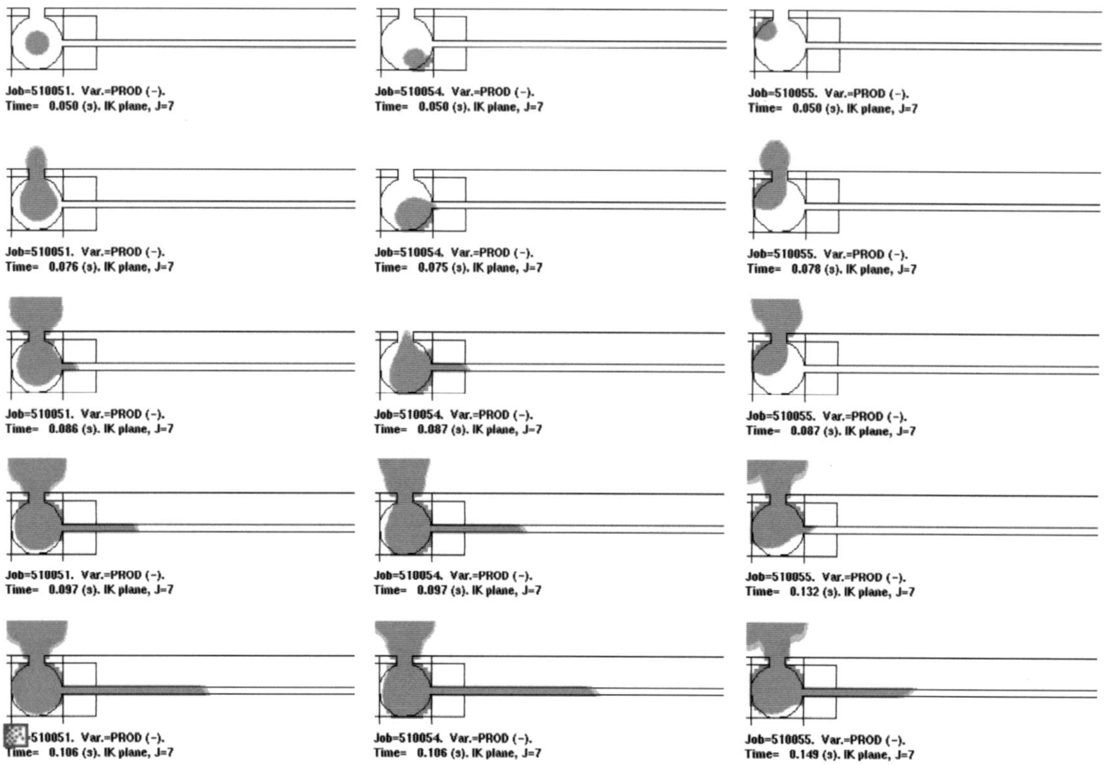

Figure 7 Predicted flame development for three different ignition locations, central ignition (left column), worst ignition (middle column) and best ignition (right column). All cases are for the Kst =500 bar.m/s with 500mm vent and 10" pipe. Effect of strong ignition by chemical igniter has not been modeled; a faster start-up of flame may therefore be expected in the experiments.

4. EXPERIMENTAL

Fike Corporation performed numerous experiments prior to the simulation activities. The test setup can be seen in Figure 8. A 2m^3 sphere is vented upwards with a 500mm vent, and is also connected to a 10" pipe. In the experiments a hybrid mixture of 4.00% propane and 100g/m^3 corn starch was applied to obtain Kst of about 500 bar m/s. Three different ignition locations, Centre, Far wall and Pipe interface, were applied.

Instrumentation consisted of the following;

- **Data Acquisition.** The data acquisition system consisted of a PC for test initiation, a Nicolet Odyssey for multichannel data collection, and a Horiba infrared detector for gas analysis. Test control was accomplished using Lab-Tech data acquisition software and a personal computer. This part of the system was responsible for initiating the test including activating the dust injectors and firing the chemical igniters. Real time data acquisition was accomplished using a Nicolet Odyssey. It is a high-speed data acquisition system used for sampling up to16 channels simultaneously at a rate of 1,000 hertz.

- **Flame Detectors**. Texas Instrument TSL 250 photodiodes were used to detect the arrival of the flame front. The TSL 250 is a light to voltage optical sensor with a built in amplifier and has a spectral response between 300 nm and 1100 nm with a response time of 360 microseconds. The peak response is at approximately 800 nm.
- **Pressure Monitor**. The pressure transducers used were Validyne DP-215 variable reluctance pressure transducers with a pressure range of 0-200 psia.
- **Pressure Detection**. EPACO is the Fike system that consists of a ceramic membrane pressure detector (CEREx) and a control panel (EPC). The detector is calibrated to function between –4.35 and +4.35 psi. It generates a 4-20 ma signal with a response time of <1ms. The panel converts the detector signal into pressure and interprets the results in terms of threshold pressure and pressure rate of rise.
- **Optical Detector**. A commercial optical detector was placed on the pipeline 0.7 feet from the interconnection. The unit, Spectrex 20/20F, uses ultraviolet (0.185-0.260 micron) and infrared (2.5-3 micron) detection and has a reported response time of 5 ms

Figure 8 Illustration of Fike test vessel/pipeline

Experimental figures similar to Figure 4 were generated which allowed the comparison of test data to simulation scenarios. From the flame arrival data, velocity figures were also generated. Figure 9 shows a schematic representation of the test setup and Figures 10 and 11 present the actual flame arrival times and velocities for the vented hybrid deflagration.

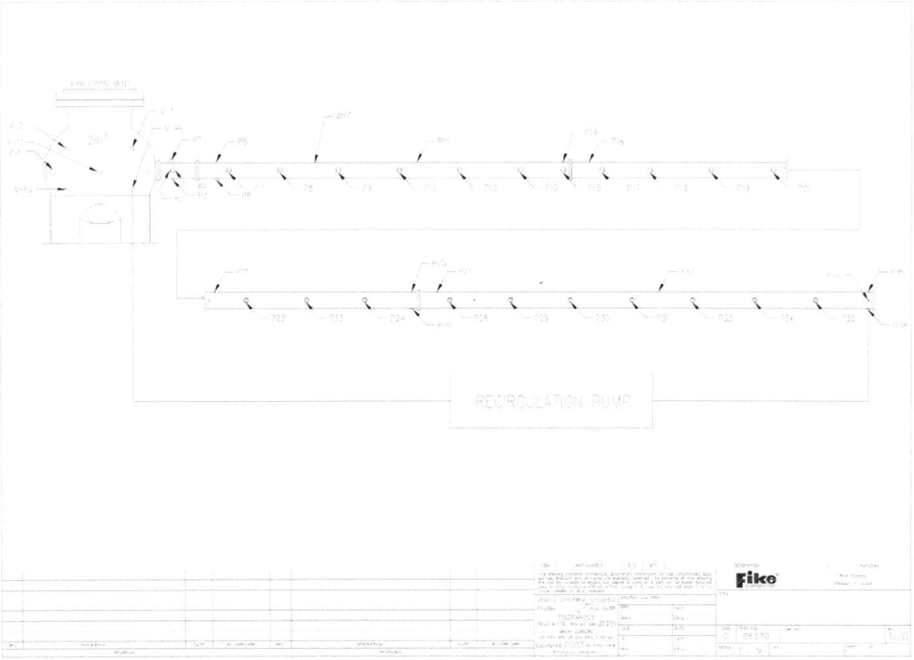

Figure 9. Schematic of Test Layout

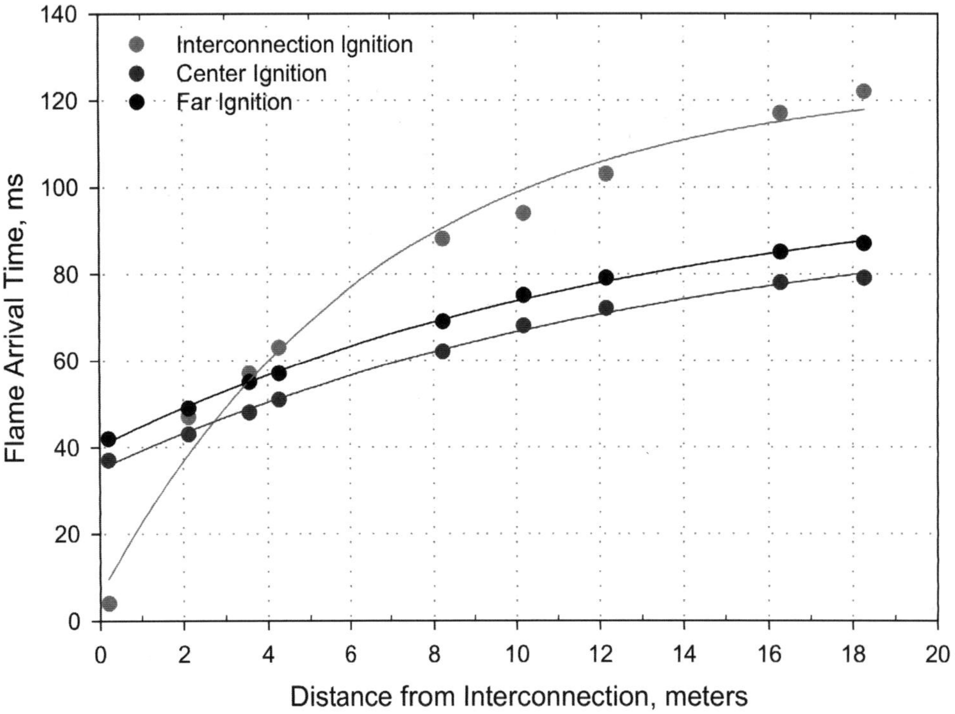

Figure 10. Flame Arrival Times for Hybrid Deflagration; Kst 500 bar.m/s, 10" pie and vent opening of 500 mm

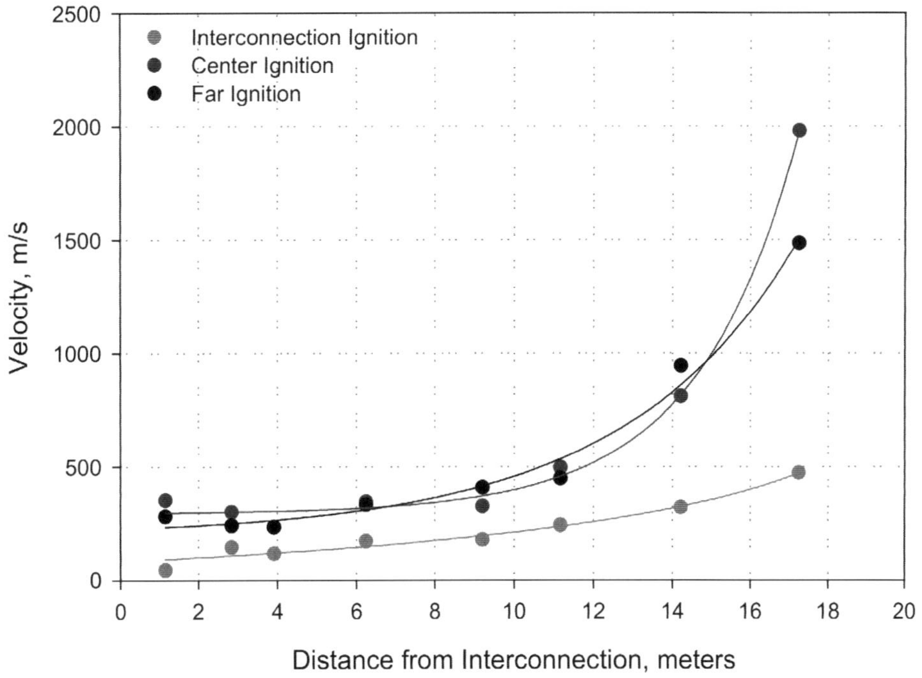

Figure 11 Velocities for Hybrid Deflagration; Kst 500 bar.m/s, 10" pipe and vent opening of 500 mm

5. COMPARISONS OF SIMULATIONS AND EXPERIMENTS

As mentioned before, the objective of the simulations was to perform a technical validation of the tests performed by FIKE, and to identify a number of scenarios which have not been tested experimentally, due to practical limitations. The study of other scenarios is necessary in order for an adequate assessment to be made without imposing impractical restrictions or limitations on the use of the explosion isolation systems. These scenarios were simulated with the explosion simulator FLACS and DESC module.

The simulation results, in terms of flame arrival times, explosion pressures and estimated detection and firing times, are compared to results from chemical isolation function tests that have been performed and then assessed to evaluate the likely functionality of the chemical isolators for a range of application beyond that actually tested.

In Figure 12 a comparison between experiments and simulations is shown for time versus flame position in pipe.

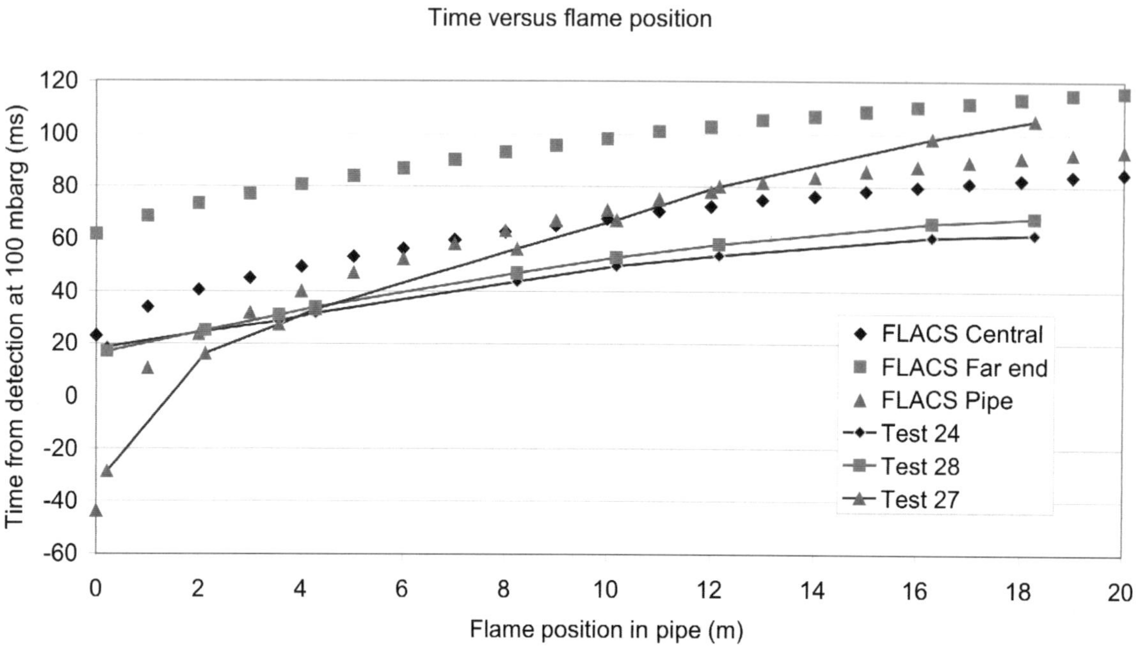

Figure 12 Comparison between Fike Corporation experiments and explosion calculations with FLACS for the case with a high reactivity fuel Ks t=500 bar m/s, 10" pipe and vent opening of 500mm.

The outcome of the comparison is as follows (only 3 ignition locations applied):

Central ignition: The simulated flame is somewhat slower to enter the pipe and burns through the initial parts of the pipe slightly faster than predicted. The maximum flame speeds in the two cases are similar. Deviation could be due to fragmenting of the chemical spark (which will give the flame a shorter path to the pipe) or non-ideal vent that will release at a higher pressure than expected.

Far end ignition: Except for a significant difference in arrival time for the flame, the two curves for flame arrival look very similar. Fragmenting of the chemical spark at ignition may throw an ignition source from the ignition region to a location nearer the pipe entry. The comment about non-ideal vent opening may also apply for this case.

Pipe entrance ignition: The curves from predictions and experiments look very similar.

In Figure 13 and Figure 14 comparisons are shown for the pressure curves in the center of the vessel as well as some locations into the pipe. Comparing the pressure curves one can see that the rate of pressure increase initially is somewhat lower in the simulation than in the experiment. It is possible that this has to do with the use of the chemical igniter in the experiments, which has not been represented properly in the simulations. For the central ignition test, the Pmax in the vessel is 10-15% lower in the simulation than in the experiment, while for ignition in the pipe entrance, the Pmax values are quite similar. Looking at the pressure in the pipe, the curves are similar in some aspects such as the three peaks seen for central ignition, however the timing and magnitude may differs some. Whether a better representation of the chemical igniter in the simulations (or better the use of spark ignition in the experiments) will make the curves more similar remains to be investigated. The correlation between the curves is considered satisfactory for the intended use.

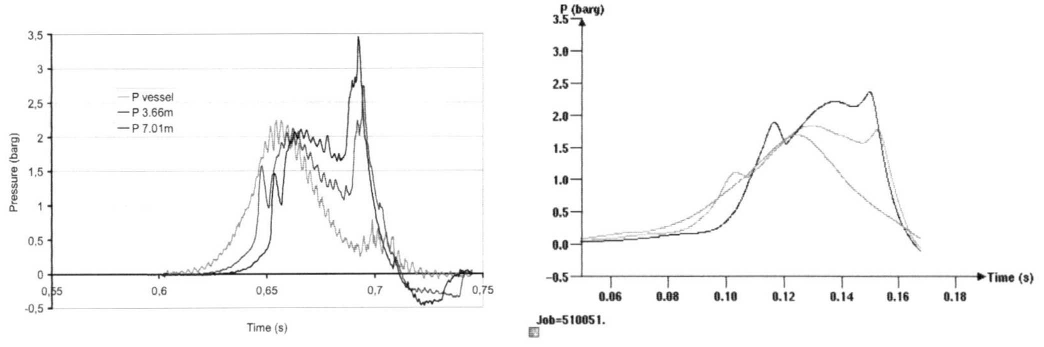

Figure 13 Comparison between explosion pressures in experiment (left) and predicted explosions (right) for the Kst = 500 bar m/s explosion with central ignition.

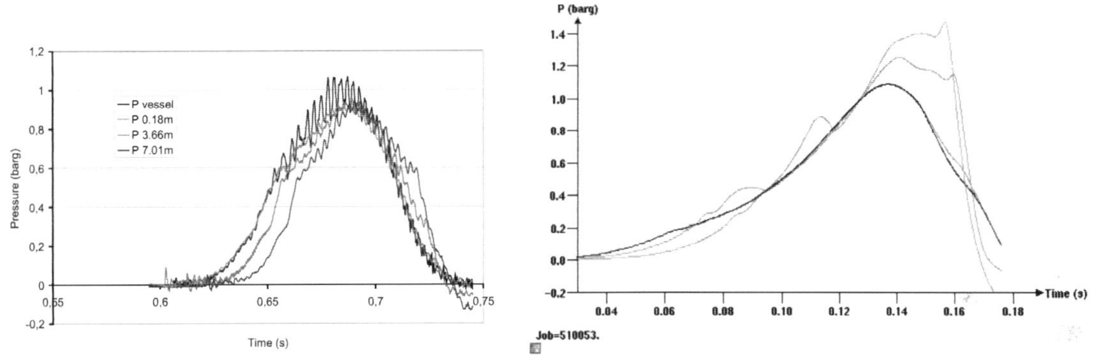

Figure 14 Comparison between explosion pressures in experiment (left) and predicted explosions (right) for the Kst = 500 bar.m/s explosion with ignition at pipe entrance.

For all scenarios, as much as information from experiments was used to check the accuracy of the simulations. For all vessel/fuel combinations, experimental data was available to perform sensitivity checks on the modeling of the chemical reactions. This was especially important because FLACS was used to model dust explosions. The simulations proved to be sufficiently accurate to predict the phenomena under study, allowing the prediction of flame propagation from vented vessels into and through pipes. During the simulation process, small deviations were identified between experiments and simulations, resulting from the difference between 'idealized process simulation' and the practical limitations that limit the 'reproducibility' of experiments. These deviations are discussed and commented in the next chapter.

6. DISCUSSION

The intended use of the simulations is as general guidance on location of a system for chemical isolation between vessels. For a given situation, with a vessel that is not too different from the idealized test setup this library of simulations could be used to identify optimal and safe location for the installation of a chemical barrier. If the system to be protected deviates too much from the idealized tests, for instance due to geometry shape, internal geometry, expected turbulence level or higher opening pressure of the rupture disk, it will be recommended to model the exact system before deciding upon the design of the chemical barrier.

The simulations were performed with the FLACS gas explosion simulator modified slightly to reflect dust explosion properties. The dust explosion simulator DESC, currently being developed, is expected to be a better tool for this type of simulations, as dust properties will be modeled more in detail.

The validity of the predictions will, as already indicated, depend on a number of parameters. These are for instance:

Vessel shape. If the vessel shape deviates strongly from the simplified shape of the sphere the predictions may be not valid. One may for instance have a situation where the initial flame is accelerated into the pipe before generating detection pressure levels in the vessel, because the volume of the vessel is large compared to the pipe area. This mechanism could be achieved with ignition at pipe entry and an early bend in the pipe that will accelerate the flame. One may also have such a situation if the flame is entering the vessel from one pipeline as a jet ignition, and then rapidly enter the pipe to be protected.

Type of ignition. As already discussed the type of ignition is also important for the validity of the simulation results. As an extreme, a very strong ignition source could initiate a detonation spontaneously, which it will be difficult to be protected against. If the possibility for a strong jet ignition into the vessel from another pipe is to be considered, this is more challenging than the situation that has been simulated, and would require dedicated calculations. Even the chemical igniter used in this study proved to give somewhat stronger initial flame acceleration, and probably a fragmentation of ignition source leading to multiple ignition locations, compared to the assumptions in the simulations. A model for chemical igniters of different strength is being developed as a part of the DESC program, if this effect should be taken into account it could be done at a later stage.

Vent opening pressure. In the experiments reported in this study a Teflon-sheet was used as the vent. These sheets are possibly somewhat less reliable with regard to precise opening pressure, relative to commercial vents. Interpreting pressure curves from the test simulations, it seems that the opening pressure of the panels may have varied from about 100 mbar to about 250 mbar in different tests. An increased yielding pressure of the vent can make the flame acceleration into the pipe more severe. To be able to use the simulation results for design of chemical barriers, it is important that the opening pressure of the vents is the same as used in the simulations (125 mbar).

Kst not fundamental parameter. The reactivity of the dust is categorized in terms of Kst value. This may be the best measure of reactivity we have available, but is far from a fundamental variable. Using Kst value, one assumes that the turbulent burning velocity of dust is a scale independent parameter. A serious problems with this is obtaining exactly the same turbulence field ahead of the flame in a large-scale situation that is obtained in a small scale situation (1 m^3 or in particular the 20 l vessel). A phenomenon like radiation is also expected to depend on scale and thus curvature of the flames. To be able to use the simulation results from a study like this, one must make sure that the pre-ignition turbulence in the real situation can not be stronger, i.e. leading to a higher turbulent burning velocity, than is assumed as initial condition in this study. If so, one should repeat the calculations with the higher level of turbulence.

Dust properties. Dust alone may behave differently than the flammable mixture used in this study (hybrid propane/corn starch). In a real situation one should assume that some of the dust settles or agglomerates in the pipe or vessel. It is likely that the concentration of dust will vary, and that the reactivity will be lower than assumed worst-case in some locations.

In the previous section, FLACS simulations have been compared to experiments. In general there is good correlation and similarities between the simulations and experiments, but some aspects with the simulation deviate. Examples are:

- Stronger initial flame acceleration and initial pressure increase are seen in the tests
- Flame accelerates more rapidly into pipe in some tests compared to simulations
- Some deviation in timing and magnitude of pressure peaks

No systematic study has yet been performed to find the reasons for the deviations. The most obvious candidate is however the chemical igniter used as ignition source. With an easily ignitable hybrid mixture with a relatively low dust concentration (100 g/m^3) it is expected that fragmentation of ignition source may lead to multiple ignition locations and rapid initial flame acceleration. It is also assumed that potential variation in the opening pressure of the Teflon sheets used as for venting may have influenced the experiments to some extent. A higher opening pressure will increase the flame acceleration into the pipe.

With the dust explosion tool DESC being developed, an even more suitable tool for this kind of calculations is expected to appear. It will be possible to model the activation of physical isolation devices and probably also inert powders, water mist or gas.

7. CONCLUSIONS

A simulation study has been performed with the CFD-simulator FLACS to produce guidance on installation distances for chemical barriers to be installed as explosion isolation between vessels. A range of different situations has been simulated in order to generate a library that can be used for reference when designing a real situation. The most critical goal is to avoid that the flame can reach beyond the location of the chemical barrier prior to a successful activation. For situations deviating significantly from the idealized situation simulated, it will be recommended to perform a separate study to identify optimal installation distance for the chemical barrier.

The simulations are benchmarked against hybrid propane/corn-starch experiments from the Fike Corporation test site in a $2m^3$ spherical vessel with a 10" 18.9m pipe and a 500mm diameter vent opening. A reasonably good correlation is seen, the observed deviations can probably mainly be explained by the use of a chemical igniter (possibly fragmenting and leading to multiple ignition) in the tests.

The good correlation that was found provides encouragement for the future use of explosion simulators in product development, product application and process safety assessment.

ACKNOWLEDGEMENTS

The authors want to thank the European Commission for the support of the DESC project for development of a dust explosion simulator, as well as the various companies and universities contributing in the DESC consortium.

REFERENCES

1. Going, J. and Snoeys, J., "Design for Isolation of Flame Propagation in an Oxygen Atmosphere", 37th National Meeting, Process Plant Safety Symposium New Orleans, LA March 31-April 2, 2003.
2. Chatrathi, K., Going, J., Grandestaff, B., "Flame Propagation in Industrial Scale Piping", Process Safety Progress, 20, 286, 2001
3. Directive 1999/92/EC of the European Parliament and of the Council of 16 December 1999: on minimum requirements for improving the safety and health protection of workers potentially at risk from explosive atmospheres.
4. Directive 94/9/EC of the European Parliament and Council of 23 March 1994: On the approximation of the laws of the Member States concerning equipment and protective systems intended for use in potentially explosive atmospheres
5. Working document CEN/TC 305/WG 3/SG 6 N 11, Explosion Isolation Systems, June 2003.

Experimental Study on Flammable Gas Explosions induced by Semi-Spherical Obstructions

BI Mingshu, LI Shengjuan
(1 School of Chemical Engineering, Dalian University of Technology, Dalian 116012, China)
(2 School of Power Engineering, University of Shanghai for Science and Technology, Shanghai 200093, China)

Abstract Forecasting for the probability of an explosion is essential in the design of plants. The mechanisms which enhance explosion overpressures need to be established with some certainty in order to ensure that all aspects of the safe design of structures and processes and the safe protection of personnel are taken into account. In present paper, a series of tests were carried out to study the explosion pressure fields of flammable gas clouds built-in semi-spherical obstructions with circular space. The pressure-time histories of explosion process were recorded by a data-acquisition system composed of pressure transducers, an A/D transformer, a data-acquisition card and a computer, with dynamic responding time less than 0.001s. The clouds consisted of acetylene and air with an acetylene volume ratio of 13.3%. The diameters of gas cloud ranged from 50mm to 250mm, the diameters of the obstructions was ranging between 20mm and 60mm, the width of the circular grating was ranging between 15mm and 60mm, and the interspaced ratio was from 20% to 75%. Based on the regression of experimental data and deviation analysis, a fitting equation is given among the cloud deflagration pressure, the cloud radius, the distance from ignition point and the obstruction characteristics. According to the idea of the multi-energy method, a general fitting equation of explosion overpressure is given by combining parameters of boundary conditions, the mixture reactivity and the scale. This fitting relation can be applied to predict other gas cloud explosion field confined by other obstructions.

Key words: flammable gas cloud; deflagration; explosion; obstructions; Overpressure

During the manufacture, storage, transportation of flammable gases or low boil point liquids, flammable gas cloud explosions[1] have caused enormous economic loss and heavy personnel casualty. Therefore, it is very necessary to investigate the mechanism, which enhances explosion overpressure, in order to ensure that all aspects of safe design and the safe protection of personnel are taken into account.

The experimental research on flammable gas explosions was initiated from the unconfined gas explosions, whose overpressure and degree of destroy are little. However, the overpressure will increase greatly when there are obstructions in the gas cloud and sometimes detonation can occur [2,3]. Many researches are focused on the effects of obstructions on explosion. Harrison[4], MERGE[5] and DISCO[6] carried out experiments with multi-layer fences, frame array and pipes of many rows respectively. These experiment results reveal the overpressure mechanism of gas cloud explosions by nature. The role of turbulence is well-established as a mechanism for increasing burning velocity by fragmenting the fame front and increasing the fame surface area and burning rate. It is clear that the propagating flame front is likely to interact directly with the obstructions. However, The quantity description of such interaction is poorly understood.

This study aims to analyze the effects of built-in obstructions on the flame propagation and the variation of the explosion characteristics. In order to describe the built-in obstructions, the experiment study focused on the axis symmetrical semi-sphere obstructions, which are put in the hemispheric gas cloud. Then a quantity equation will be given by combining these test data and the ones from references, which can be applied to predict other gas cloud explosion field confined by other obstructions.

1. Experimental Method

The experiments were carried out in an outdoor base of an area of 4000m^2. A semi-spherical indoor experimental field was built in order to decrease the interference of exterior environment. A confined hemisphere membrane of 0.2mm thickness polythene plastic charged the premixed acetylene-air with an acetylene volume ratio of 13.3% which is the most dangerous concentration confirmed by experiments. The diameters of gas cloud ranged from 250mm to 1250mm, the diameters of the obstructions was between 20mm and 60mm, the width of the circular grating was between 15mm and 60mm□and the interspaced ratio was from 20% to 75%. The schematic diagram of experimental system is shown in Fig.1. The pressure-time histories of explosion process were recorded by a data-acquisition system composed of pressure transducers, an A/D transformer, a data-acquisition card and a computer, with dynamic responding time less than 0.001s. The measurement range and precision of pressure transducers are –5~30kPa and 0.25% respectively. The data-acquisition card has 8 channels with 200 kHz per channel whose precision was 0.02%. An ignition electrode that met ISO6184 and NFPA68 was fixed at the center of gas cloud. Its ignition energy is about 100mJ that was more than that of general industrial ignition source.

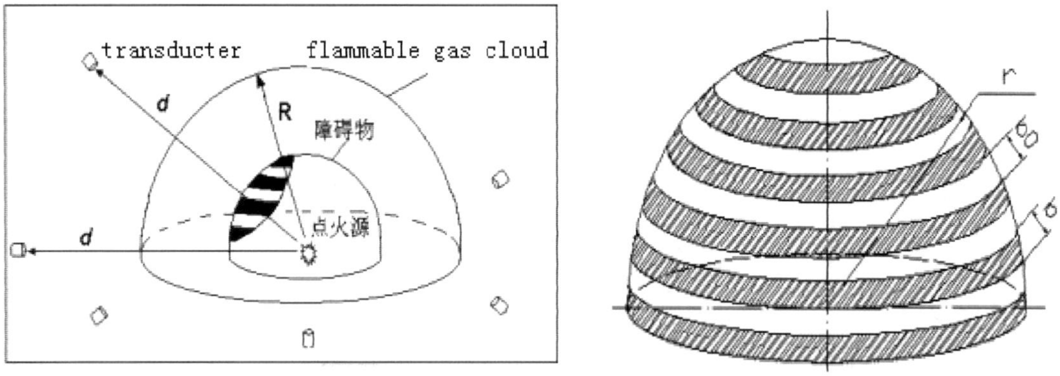

□a□Schematic diagram of experimental system (b) Obstruction schematic diagram

Fig.1 Experimental system schematic diagram

2 Results and Discussion

2.1 The parameter combination and correlation

More than 300 times tests were carried out. The influence factor and level are listed in table 1. Fig 2 shows the experimental results. Based on the regression of experimental data and deviation analysis, the fitting equation is obtained between the explosion overpressure and its influential factors

$$\Delta p = \alpha \cdot \frac{r^2}{b\phi^2}\frac{R}{d} \tag{1}$$

Table 1 The Experiment Factor and Level

	R/m	r/m	bo/mm	b/mm	d/R
1	0.25	0.10	15	15	1.4
2	0.50	0.20	30	30	1.8
3	0.75	0.25	45	45	2.2
4	1.00	0.30	60	60	3
5	1.25				5.4
6					7

Equation (1) can be transformed into dimensionless form

$$\Delta\overline{p} = \frac{\eta}{\overline{H}} \tag{2}$$

Where the dimensionless explosion overpressure $\Delta\overline{p} = \Delta p / p_0$; the dimensionless obstruction parameter $\overline{H} = \left(\frac{E}{p_0}\right)^{1/3} \cdot \frac{b\phi^2 d}{r^2 R}$; η is a constant depending on the property of flammable gas, the concentration and the characteristics of the obstruction. For present experimental conditions, η=0.046.

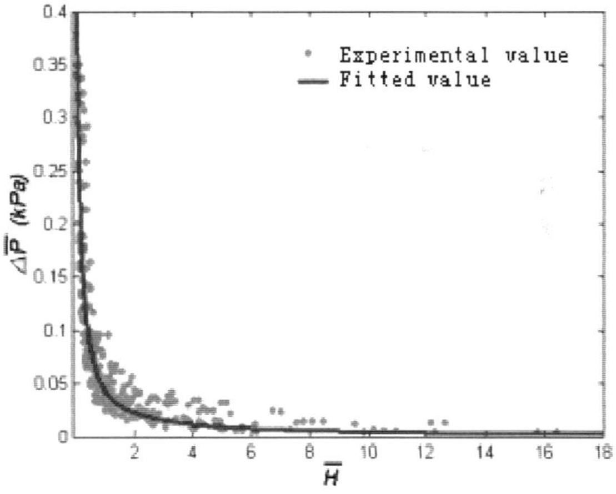

Fig 2 Relativity of fitting equation to experimental data

2.2 Application of multi-energy method

According as equation (1), it is obvious that the factors including obstruction dimension and interspaced ratio have remarkable influence on explosion overpressure. Because of the block of semi-sphere obstructions, regular two-dimensional symmetrical obstructions, the explosion overpressure is growing at exponential law during its development. With reference to the set of confined conditions in Multi-energy Model, in our experiments the overpressure and involved confined factors can be described as followings.

$$\Delta p \propto \left[\frac{(r\cdot\theta)^2 (R/d)}{b}\right]^m \tag{3}$$

where: $(r\cdot\theta)$ is the portion of volume occupied by obstructions, m;

The flammable gas explosions can cause turbulent combustion under the influence of boundary conditions. If on the basis of acoustics, a quadratic relation between overpressure and flame speed is assumed, an approximate relation between the maximum explosion overpressure, laminar burning speed, and scale of the following form can be postulated

$$\Delta p \propto S_l^{2.7} \cdot S_C^{0.7} \tag{4}$$

Based on the regression of experimental data, the fitting equation is obtained between the explosion overpressure and its influential factors

$$\Delta p = a \cdot \left[\frac{(r\cdot\theta)^2 (R/d)}{b}\right]^m \cdot S_l^{2.7} \cdot b^{0.7} \tag{5}$$

In present experimental condition, $a = 119.4\mathrm{kg}\cdot\mathrm{s}^{0.7}/\mathrm{m}^{5.4}, m = 1$. The postulated combination of parameters on confined conditions is

$$L = \frac{(r\cdot\theta)^2 (R/d)}{b} \cdot S_l^{2.7} \cdot b^{0.7} \tag{6}$$

The combination of parameters was evaluated for all experiments in the present experimental data and correlated with the experimentally observed explosion overpressures. The correlation is graphically represented in Fig3. It shows that the correlation is better. To examine the effectiveness of fitting equation, the square deviation analysis result is listed in table 3.

Table 3 Equation fitting and deviation analysis

Fitting equation	Regressing square deviation sum SS_f	Surplus square deviation sum SS_e	$F = \frac{SS_f/1}{SS_e/(n-2)}$	$F_{0.01}(1, n-2)$
$\Delta p = 119.4 \cdot L$	22543.6	3501.09	1693.46	6.72

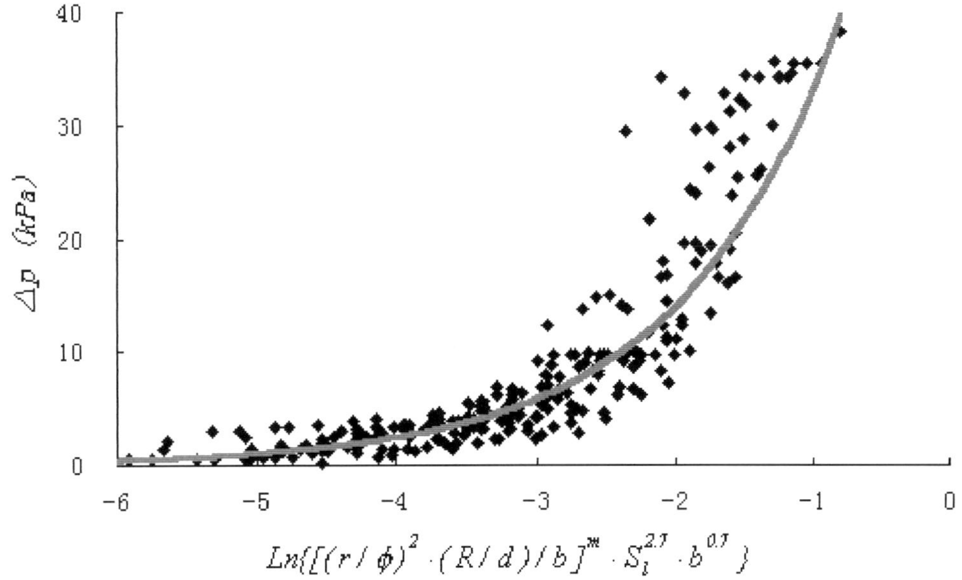

Fig 3 Relativity of fitting equation to experimental data

Other test data have been reported such as Harrison and Eyre[4,7], Hjertager[8] and Mercx[5]. Fig.4 shows their sketch of obstructions. Combining all these experimental data, the results are shown in Fig.5. It shows a good agreement among the four sets of experiment dada on this confined parameter combination. The graphs may well serve as a basis for making a choice of charge strengths in the application of the Multi-Energy method to realistic problems.

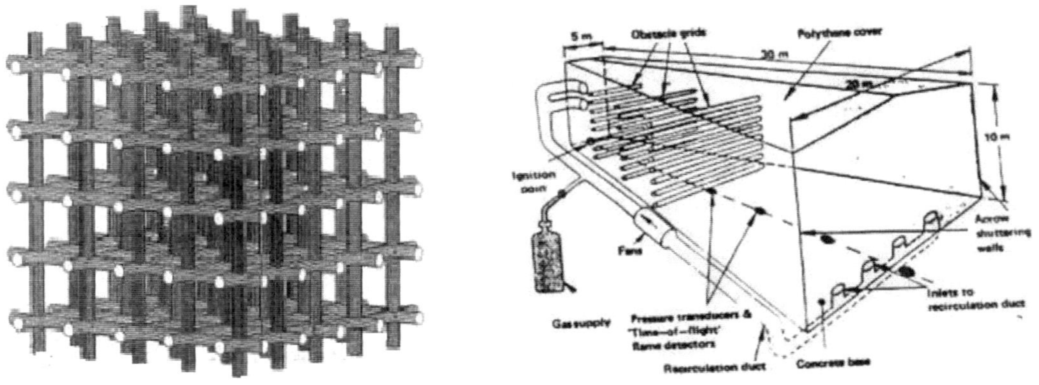

(a)MERGE test□Mercx□ (b) SHELL tests□Harrison,Eyre□

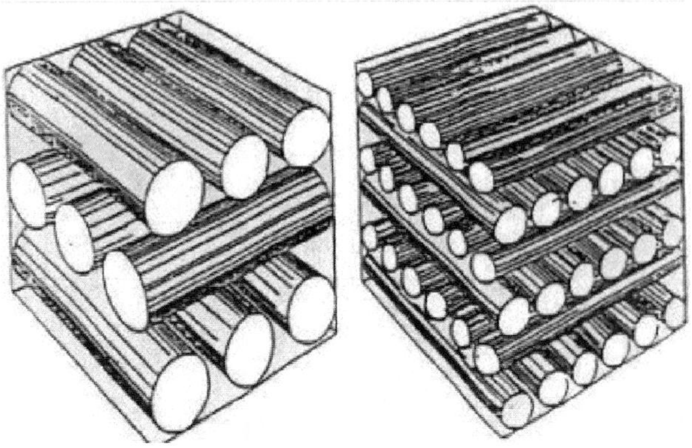

(c)CMR tests（Hfertager）

Fig. 4 Obstruction sketch reported in reference

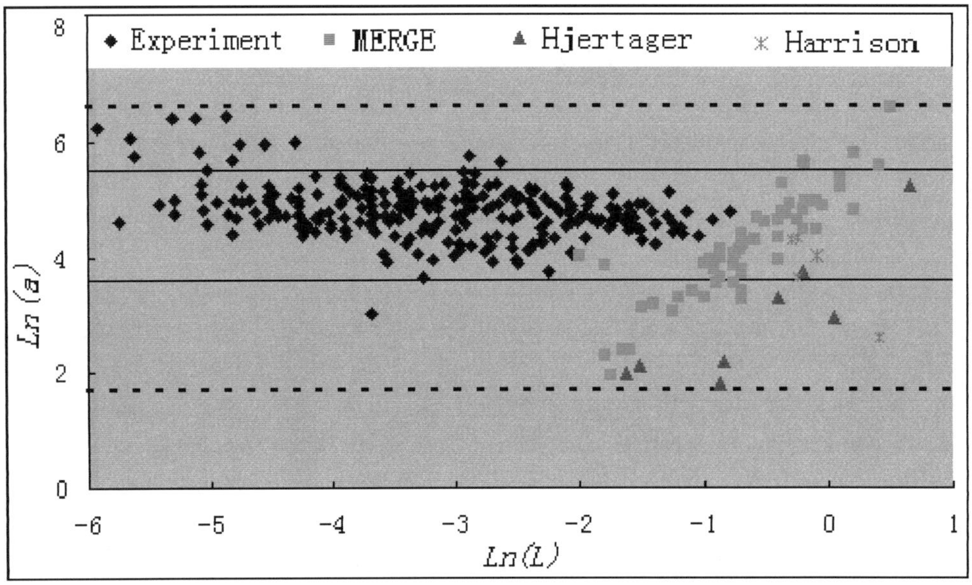

Fig. 5 Observed explosion overpressure of four kinds of obstruction conditions

4 Conclusions

(1) A series of tests were carried out to study the explosion pressure fields of flammable gas clouds built-in semi-spherical obstructions with circular space. It is obvious that the factors including barrier dimension and interspaced ratio have remarkable influence on explosion overpressure. The explosion overpressure increases with the rising of barrier radius and decreases with the increasing of interspaced ratio of barrier.

(2) Based on the regression of experimental data and deviation analysis, the fitting equation is obtained between the dimensionless explosive overpressure and its influential factors:

$$\Delta\overline{p} = \frac{\eta}{\overline{H}}$$

Where η is a constant depending on the property of flammable gas, the concentration and the characteristics of the obstruction. For present experimental conditions, η=0.046.

(3) According to the idea of the multi-energy method, a general fitting equation of explosion overpressure is given by combining parameters of boundary conditions, the mixture reactivity and the scale

$$\Delta p = a \cdot \left[\frac{(r/\phi)^2 (R/d)}{b} \right]^m \cdot S_l^{2.7} \cdot b^{0.7}$$

This fitting relation can be applied to predict other gas cloud explosion field confined by other obstructions.

NOMENCLATURE

a	a fitting constant□kg·s$^{0.7}$/m$^{5.4}$
b	the width of single grating of obstructions□mm
C	the mixture concentration□%
d	the distance of measuring point from the ignition point□m
$\overline{H}$	the dimensionless confined parameter
L	the confined parameter combination
m	a fitting constant
p_0	the atmospheric pressure□Pa
Δp	the explosion overpressure□kPa
$\Delta\overline{p}$	the dimensionless explosion overpressure
r	the radius of obstructions□m
R	the radius of gas cloud□m
S_C	the scale, m
S_l	the laminar burning velocity of the fuel–air mixture□m/s
α	the coefficient
ϕ	void percentage of obstructions□%
θ	block percentage of obstructions□ $\theta = 1 - \phi$
η	the coefficient

REFERENCES

1. Dag Bjerketvedt, Jan Roar Bakke, Kees van Wingerden. Gas Explosion Handbook. Journal of Hazardous Materials, 1997,52□1-150
2. Phylaktou H, Andrews G E.The Acceleration of Flame Propagation in a Tube by an Obstruction. Combustion and Flame, 1991, 85□363-379
3. Masri A R, Ibrahim S S, Nehzat N, Green A R. Experimental study of premixed flame propagation over various solid obstructions. Experimental thermal and fluid science, 2000, 21: 109-116
4. Harrison A J and Eyer J A□The effect of obstruction arrays on the combustion of large premixed gas/air clouds□Combust□and Tech□1987, 52□121-137
5. Mercx W P M, van den Ber A C, Hayhurst C J□Robertson N J, Moran K C. Developments in vapour cloud explosion blast modeling. Journal of Hazardous Materials, 2000,71:301–319
6. Berg A C Van den, Eggen J B M M. Guidance for the Application of the Multi-Energy Method. The Second International Specialist Meeting on Fuel-Air Explosions. Norway: Berger, 1996
7. Harrison A J, Eyre J A. Vapour cloud explosions-the effects of obstructions and jet ignition on the combustion of gas clouds. Proc. 5th Int. Symp. On Loss Prevention and Safety Promotion in the Process Industries, Cannes, France, 1986:1-38
8. Hjertager B H. Explosion in obstruction vessels. Course on Explosion Prediction and Mitigation. University of Leeds, UK, 1993,7: 28-30

Inerting for Explosion Prevention

Walt Frank, P.E.

ABSG Consulting Inc.
5301 Limestone Road, Suite 210
Wilmington, DE 19898
Phone: (302) 239-0496 e-mail: wfrank@absconsulting.com

Prepared for Presentation at

American Institute of Chemical Engineers
38th Annual Loss Prevention Symposium
Fire, Explosion and Reactive Hazards

New Orleans, LA

April 26 – 29, 2004

Abstract

Oxidant concentration control, or inerting, is a commonly used technique for preventing fires and explosions in the process industries. While simple in concept, the details of implementing an inerting system are not always straightforward. This paper is intended to relate a broad range of guidance relative to the design and implementation of inerting systems and to provide a number of caveats addressing some of the more common stumbling points. Special emphasis is placed on the safety hazards associated with the uses of inert gases. A perspective on the challenge of inerting vessels with large height-to-diameter ratios is provided.

1. Introduction

Fires and explosions are among the more frequent and most damaging events experienced in the process industries. This paper addresses inerting for the prevention of such events. It is not meant to imply that inerting is some universally applicable panacea to solve all of our combustion woes. It is suggested, however, that greater understanding and broader use of inerting could lessen the human, capital, and business costs of these events about which we read almost daily in the newspapers, in the trade press, and on the Internet.[1]

Figure 1 illustrates the three components that must be present for a fire or explosion to occur. These are: (1) a fuel at a concentration above its lower flammable limit (LFL) and below its upper flammable limit (UFL), (2) an oxidant at a suitable concentration, and (3) an ignition source of sufficient energy to initiate the combustion of the fuel/oxidant mixture.

Figure 1: Fire Triangle

Removal or effective control of any of the three sides of this fire triangle would prevent the fire or explosion from occurring, and each side should be considered in strategizing an approach to safe system operation. However, practical limitations on the degree of control attainable are often encountered.[2]

For example, in a process handling a volatile, flammable liquid, it is not possible to prevent the formation of a flammable vapor concentration in the ullage above the liquid if the liquid must be processed or stored at a temperature at or moderately above the liquid's flash point. Similarly, the formation of combustible dust/air mixtures often

---

[1] Readers seeking sources of recent incident data should refer to Industrial Fire World (IFW, 2004), the U.S. Chemical Safety and Hazard Investigation Board (CSB, 2004), and SAFTENG (2004).

[2] This paper addresses system safety from the standpoint of preventing the fire or explosion. Other protection options exist for mitigating the effects of such an event should it occur (e.g., designing process equipment to withstand the maximum explosion pressure or relieving the explosion pressure to a safe location). NFPA 68 (2002) and NFPA 69 (2002) provide detailed guidance on the gamut of available options.

cannot be avoided in processes handling combustible particulate solids (e.g., in the head space of a silo during receipt of pneumatically transferred solids).

Our ability to control the ignition source side of the fire triangle is also limited, typically due to the ubiquitous nature of ignition sources in the process environment. A number of National Fire Protection Association (NFPA) publications provide guidance on ignition source controls. A partial listing includes the following:

- NFPA 77 (2000) provides guidance on controlling static electricity
- NFPA 70 (2000), NFPA 497 (1997), and NFPA 499 (1997) address electrical equipment as potential ignition sources
- NFPA 654 (2000) provides guidance on eliminating ignition sources in the handling of combustible particulate solids, such as (but not limited to) frictional heating and mechanical impact sparks

While NFPA requires efforts to eliminate ignition sources, reliance upon ignition source controls alone is not permitted as a primary means of explosion prevention, in recognition of the difficulty of ensuring that all potential ignition sources have been identified and removed from the process environment.

This paper deals with the third side of the fire triangle, through control of the oxidant concentration. As will be discussed in detail, there is a limiting oxidant concentration (LOC) below which combustion cannot occur in a specified fuel/oxidant mixture. The oxidant concentration can be reduced below the LOC through the addition of an inert gas to the system. While this process of inerting is commonly used to prevent fires and explosions, practical limitations also exist for this approach to safety. Some of the more significant limitations will be addressed in this paper.

This paper focuses primarily on inerting for combustion prevention within the process industries. Specific applications include, but are not limited to:

- Displacing flammable atmospheres in preparation for maintenance activities
- Establishing an inert environment before flammables are added to the process
- Maintaining a noncombustible atmosphere inside the process on a continuing basis

There are many other reasons for which an inert gas might be introduced into equipment. A partial list includes:

- Excluding moisture or air to:
  - prevent corrosion or other deterioration of out-of-service equipment
  - protect the quality (i.e., prevent degradation) of sensitive products
  - prevent exothermic reactions with air- or water-reactive materials
- Using inert gases as firefighting agents
- Excluding air during certain types of welding
- Cooling or preventing contaminant entry into bearings and mechanical seals (where compressed air cannot be tolerated for any of the reasons cited above)

- Using low temperature inert gas (e.g., from a cryogenic source) to cool thermally sensitive materials during mechanical processing such as grinding or mixing (may also be intended to provide inerting for fire and explosion protection of such operations)

The design of inert supply systems and controls to address the first item in the above list (i.e., excluding moisture or air to prevent corrosion, etc.) may, in many situations, be very similar to the design of systems for combustion prevention, and much of the guidance in this paper would be applicable. The other uses of inerting listed above will not be explicitly addressed in this paper.[3]

Regardless of the intent, all applications involving the introduction of inert gas to the processing environment share a common concern: the potential for personnel injury, perhaps death, due to exposure to oxygen-deficient environments either inside of equipment or outside of equipment due to leakage of the inerts. The hazards of handling inerts will be addressed in detail in this paper.

2. Definition of Terms

There are many common sources of guidance for the design and operation of inerting systems and, as might be expected, some variety in the use of terminology. For the purposes of this paper, the following terms will have these meanings:

Inerting will be used generically to address any situation "by which a combustible mixture is rendered non-ignitable by addition of an inert gas..." (NFPA 69, 2002). The full NFPA definition includes the addition of noncombustible dust to combustible particulate solids to reduce the explosion hazards of the system. This form of inerting is beyond the scope of this paper.

Inert gas means a "nonflammable, nonreactive gas that renders the combustible material in a system incapable of supporting combustion" (NFPA 69, 2002). As will be seen, the most commonly used "inerts" are not totally "nonreactive," and appropriate consideration must be given to chemical incompatibilities when selecting inerts.

Combustible means "capable of undergoing combustion" (NFPA 69, 2002). This generic usage of the term should not be confused with the more quantitative usage in NFPA 30 (2003), defining the difference between flammable and combustible liquids. Our focus will be on "airborne" mixes of oxidants and combustible fuels, where the fuels could be present as either (1) flammable gases, (2) vapors produced from flammable or combustible liquids heated above their flash points, (3) combustible dusts, or (4) liquid droplets (mists) produced by mechanical dispersion of liquids or the condensation of vapors.

Purging refers to the short-term introduction of an inert gas to displace or dilute the contents of a closed system such as a pipe or vessel. The contents to be displaced or diluted could either be a fuel (e.g., removing flammable vapors from a vessel prior to maintenance), an oxidant (e.g., removing air from a vessel prior to introduction of combustible materials during startup), or a

[3] See NFPA 12 (2000) for guidance on the use of carbon dioxide for fire suppression purposes.

mixture of fuel and oxidant. A key consideration in purging is that the inert gas flow may only be maintained until the desired change in concentration has been affected in the closed system.

Blanketing (or padding) refers to a "technique of maintaining an atmosphere that is either inert or fuel-enriched" in a closed system (NFPA 69, 2002). Blanketing differs from purging in two respects. First, the desired composition change is intended to be maintained in the closed system for an indefinite period of time. Second, the NFPA definition includes the intentional addition of fuel to prevent combustion by keeping the system fuel-rich. NFPA, in effect, presents this as a special case of "inerting."

Displacement means a "form of purging in which replacement of one substance by another is accomplished without appreciable mixing" inside the closed system and is discussed in detail by the Compressed Gas Association (CGA, 2001a).

Dilution means a "form of purging in which replacement of one substance by another is accomplished with appreciable mixing" inside the closed system (CGA, 2001a).

3. Why Inerting Works

To better understand why inerting works to prevent combustion (and, perhaps more importantly, to better understand those situations where inerting will not work), it may be helpful to look at the physical principles underlying inerting.

Combustion involves a propagating reaction. More explicitly, in a flame the combustion occurs in a reaction zone (flame front) that is moving forward through the unburned fuel/oxidant mix ahead of it.[4] The combustion reaction releases heat, much of which goes to increasing the temperature of the combustion products (the adiabatic flame temperatures for stoichiometric mixes of hydrocarbons in air, initially at ambient temperature, are on the order of 2300 °K). Additional energy is lost to the surroundings as radiant heat.

For the flame to propagate, enough heat must be transferred to the unburned fuel/oxidant mixture just ahead of the flame front to raise the temperature of the mixture to the initiation temperature for the combustion reaction. Figure 2 illustrates the temperature profile just

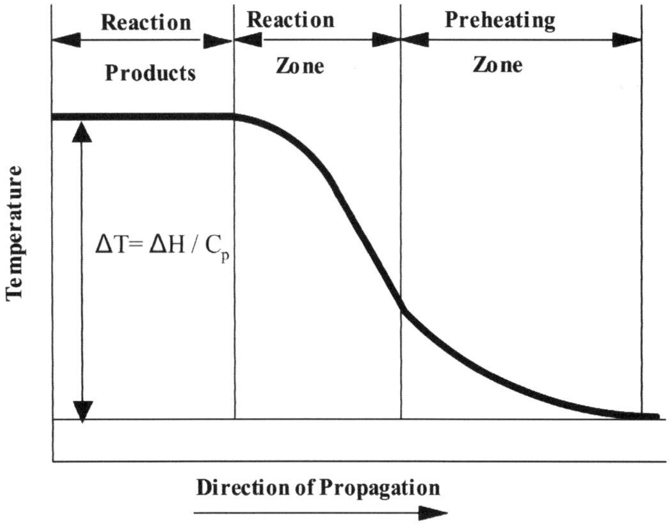

Figure 2: Flame Front Temperature Profile

[4] The phrase "moving forward" must be taken in a relative sense. For example, in a Bunsen burner flame, the flame is essentially stationary, with the fuel/air mixture moving toward the flame.

ahead of and behind a propagating flame front.[5]

Consider the combustion of propane in a stoichiometric mix with air:[6]

$$C_3H_8 + 5\ O_2 + 19\ N_2 \rightarrow 3\ CO_2 + 4\ H_2O + 19\ N_2$$

The 19 moles of nitrogen (N_2) that do not participate in the combustion reaction increase the amount of heat that must be added to the reactants to bring them up to the initiation temperature. This nitrogen, which becomes part of the combustion products, continues to act as a heat sink, lowering the flame temperature and decreasing the rate at which heat can be transferred forward into the unburned gases ahead of the flame front. Doubling the amount of nitrogen in the unburned gas mix to approximately 38 moles of N_2 per mole of propane will provide a heat sink great enough to prevent the propagation of the flame. Studies have demonstrated that, for commonly used thermally stable inert gases such as N_2 or carbon dioxide (CO_2), the heat sink effect is adequate to explain the flame extinction mechanism (Ewing et al., 1984). With gases, the heat sink is due to the sensible heat capacity of the gas molecules. If water vapor were being used as the inert, the heat required to vaporize any water droplets present would further add to the heat sink.[7]

In the above example, additional N_2 added to a stoichiometrically balanced mixture of fuel and oxidant provided an increased heat sink. An additional effect can come into play in other systems. Consider an enclosed tank containing a flammable liquid, with an air atmosphere above the liquid. The concentration of vapor in the ullage of the tank will be determined by the vapor pressure of the liquid at the temperature in the tank. As nitrogen is added to the tank, air (and the 21 vol % oxygen it contains) will be displaced, and the oxygen (O_2) concentration in the ullage will decrease. Since the liquid will continue to assert its vapor pressure, the ratio of fuel to O_2 will increase. For many flammables, this will reduce the reactivity of the system because excess fuel not having O_2 for combustion will provide the same heat sink effect as an inert. There are few flammables, however, whose saturation vapor concentration in air occurs on the lean side of the stoichiometric concentration. Such flammables (e.g., ethanol) will pass though the stoichiometric concentration, producing a more combustible gas mix before the "richening" effect described above occurs. This matter will be discussed in further detail in Section 9.

For inerting to be successful, conditions must be such that the inert can effectively draw heat away from the propagating reaction front, quenching the reaction. The example of a premixed gaseous fuel/oxidant flame described above is clearly such a case. Conversely, inerting the ullage in a vessel provides no protection against the decomposition of thermally unstable liquids

---

[5] Figure 2 illustrates the temperature profile for a flame propagating as a deflagration that, by definition of the term, is moving toward the unburned gases at a speed less than the speed of sound in the unburned gases. In a deflagration, the unburned gases ahead of the flame are warmed by convective and radiative heat transfer. In a detonation, the flame front is moving at a speed greater than the speed of sound in the unburned gases. There is an abrupt, rather than gradual, temperature increase immediately ahead of the flame front as the gases are compressed by the rapidly moving flame front.

[6] The 19 moles of nitrogen, relative to the 5 moles of oxygen required for combustion, reflect the nominal 79 N_2 to 21 O_2 mole ratio for air, rounded to the nearest integer value.

[7] Other agents used for flame suppression rather than inerting are often thermally unstable. For such agents, there can be a significant heat sink associated with their endothermic decomposition.

or solid particulates. While this "experiment" has undoubtedly been run in industry, the inert cannot function as intended since it is not in contact with the bulk of the reacting species, and heat transfer to the inert cannot occur at a sufficient rate to quench the decomposition reaction in the bulk liquid or bulk solid.

While the examples used in this section have focused on the prevention of combustion reactions, it is not meant to imply that combustion is the only chemistry that inerts inhibit. For example, nitrogen is typically added to ethylene oxide (EO) storage tanks to prevent the vapor phase decomposition of EO. Vessels are maintained in the nondecomposable region by controlling the nitrogen pressure in the EO vapor. An industry publication provides guidance for determining the required total pressure, EO plus N_2, as a function of the liquid temperature (Dow Chemical et al., 1999).

4. Sources of Inerts

The most commonly used inert gases are N_2 and CO_2, but other gases and vapors such as argon (Ar), helium (He), steam, and exhaust gases from combustion devices are also used. The choice of the most appropriate inert for a given application must be based upon such factors as (Cunliff, 2001; Grossel and Zalosh, 2004):

- Cost
- Availability
- Reliability of supply
- Effectiveness in reducing the fire and explosion hazards (see Section 8)
- Potential for contamination of the product by the inert gas
- Other chemical compatibility concerns with either the product or materials of construction (see Section 5)

Traditionally, inerts have been obtained from sources such as high pressure gas cylinders or tube trailers or through evaporation of cryogenic liquids from bulk tanks. More specifically:[8]

- Liquid N_2 can be obtained in quantities ranging from a few liters to thousands of gallons, in containers such as dewers, cryogenic liquid cylinders, and bulk cryogenic storage tanks resupplied from transport trucks or railcars (Air Products, 1998a)
- Gaseous N_2 is shipped in high-pressure cylinders, tubes, and tube trailers at pressures as high as 7,500 psig (Air Products, 1998b)
- CO_2 is typically shipped and stored as a liquefied compressed gas. Bulk CO_2 tanks are refrigerated to maintain the liquid at a pressure of less than 305 psig and can be refilled from transport trucks or railcars. CO_2 is also shipped in high-pressure cylinders (816 psig at 20 ^{o}C) and in smaller cryogenic cylinders at a reduced temperature and a pressure of approximately 300 psig (Air Products, 1998c)

[8] The data presented here represent the practices of one particular supplier, which was selected because of the volume of information presented on the supplier's Web site. The data are meant to be representative and not comprehensive of general industry practices. Practices of other suppliers may differ. It is not the intent to endorse any particular supplier.

Gases obtained from these sources tend to be of very high purity, often greater than 99.99% pure, especially for gases obtained from cryogenic sources. As discussed in Section 8, such high purities are typically not required for inerting provided for combustion avoidance.

Other sources of inerts include (NFPA 69, 2002; FM Global, 2000):

- Onsite cryogenic air separation plants
- Pipeline supply from remote sources
- Gas generators burning or catalytically oxidizing a hydrocarbon to produce an oxygen-deficient product gas
- Combustion product gases from process furnaces, boilers, or internal combustion engines
- Steam
- Nitrogen produced by the air oxidation of ammonia
- Nitrogen produced by removal of oxygen from air using pressure swing adsorption (PSA) or membrane separation units
- Fuel gases such as methane or natural gas

The number and variety of commercially available systems for onsite generation of inert gases border on staggering. An Internet search for "nitrogen generation system" turned more than 100 relevant hits. Table 1 provides a small sampling of the types, capacities, and performance features of the systems encountered, with cryogenic separation, PSA, and membrane systems being the most prevalent. Descriptions and comments on some of the technologies follow.

Table 1: Sampling of Commercially Available Nitrogen Generation Systems

Method	Capacity (scfh)	Purity (vol %)	Pressure (psig)
PSA	12,000 to 3,600,000	up to 99.99	
PSA	200 to 1,200	99.5	
PSA	1,200 to 4,000	99.5	
Membrane	12,000	99.90	
Membrane	300 to 9,000	99.50	100 to 5000
PSA	300 to 9,000	99.99	100 to 5000
PSA	575,000	95 to 99.99	100
Membrane	11,000	95	100
Membrane	5,500	98	100
Membrane	39,000	95	5000
PSA	5,000 to 100,000	99.9	
Membrane	5 to 500,000	99.5	
Cryogenic air separation	2,000 to 500,000	99.999999	
Membrane	300 to 120,000	95 to 99.999	

<u>N_2 Produced by Pressure Swing Adsorption (Michael, 1997)</u>. PSA uses a carbon molecular sieve (CMS) that will adsorb both O_2 and N_2, but adsorbs O_2 at a faster rate. Air, at moderate pressures (e.g., 100 to 150 psig), is passed through a vessel containing the CMS. The CMS preferentially adsorbs the O_2, and the air exiting the CMS bed is consequently enriched in N_2 content. The purity of the N_2 can be controlled by controlling the residence time in the bed which, correspondingly, determines the production rate for the unit. At least two adsorber beds are required. While one is online producing N_2, the other is taken offline for regeneration. During regeneration, the vessel is depressurized to ambient pressure, which releases the adsorbed O_2 from the CMS bed.

Purities of 99.9% or higher are attainable. If even lower O_2 concentrations are required, catalytic deoxygenation equipment can be added, in which residual O_2 is reacted with hydrogen to produce water.

Variations of this technology include vacuum pressure swing absorption (VPSA) and vacuum swing absorption (VSA). Both use a vacuum pump to regenerate the offline bed. VPSA and VSA differ in that the feed pressure for the online VSA bed is only slightly greater than ambient.

N_2 Produced by Membrane Separation (Michael, 1997). Membrane separation units take advantage of the fact that certain polymeric materials will selectively pass O_2 and water vapor molecules at a higher rate than N_2 molecules. Polymeric membranes are formed into hollow fibers through which air is passed at pressures typically greater than 150 psig. O_2 and water vapor molecules pass through (permeate) the membrane, leaving the gas inside the fiber enriched in N_2 content. As with PSA, slowing down the flow rate through the membrane (increasing the residence time) increases the purity of the N_2 product. Purities of up to 99.5% are commonly attainable. Unit capacity can be increased by providing more fibers in parallel.

N_2 Produced by Cryogenic Air Separation (Michael, 1997). These onsite units are scaled-down versions of the cryogenic distillation plants used by bulk producers of liquid N_2 and liquid O_2. The distillation process can produce N_2 at purities much higher than can be attained with membrane or PSA units.

Each of the three separation technologies discussed above produce a by-product stream that is enriched in O_2. O_2 and O_2-enriched air are powerful oxidizers and must be handled carefully to prevent the creation of fire and explosion hazards. "Waste" O_2 must be discharged to a safe location.[9] CGA (2003c) provides guidance on the safe installation and operation of PSA and membrane units.

Cryogenic plants have the inherent hazard of handling liquid O_2 in the bottoms of the distillation columns. While this technology has an enviable record of safe operation, several damaging explosions have been reported for cryogenic air separation units (e.g., De Graaf et al., 2000). Particular care must be exercised to prevent the introduction of organics, or other oxidizable substances, in the feed to the cryogenic plant.

Inert Gas Generators Involving Combustion. There is a wide variety of systems in which hydrocarbon fuels are burned in air to consume all or most of the O_2, producing a gas stream that, after further treatment, is suitable for use in inerting. FM Global (2000) describes inert gas generators burning fuel oil or natural gas and producing a stream containing 9 to 12 vol % CO_2, about 85 vol % N_2, and some carbon monoxide (CO). Residual O_2 concentrations are reported to be on the order of 8 vol % for systems burning fuel oil and less than 1 vol % with natural gas.

Another reference describes the combustion of natural gas or propane under fuel-rich conditions to ensure a nearly O_2-free gas stream, albeit with the production of significant amounts of CO (Gullickson, 1996). NFPA 69 mentions systems using the catalytic oxidation of fuels to produce inert gas streams, but the author was unable to locate a source for such systems suitable for

---

[9] Information on the hazards of oxygen-enriched atmospheres is provided in NFPA 53 (1999) and Air Products (2003a).

industrial-scale applications. Finally, combustion exhaust gases from process furnaces, boilers, or internal combustion engines can be used as a source of reduced-O_2 gases for inerting purposes.

The use of combustion-based inert gas generators appears to be most common in the marine industry, where the shipboard use of pure inerts such as N_2 or CO_2 is not practical due to their cost, the cost of transporting them, and their unavailability for resupply at some ports. Combustion gases, either flue gas from the ship's propulsion system or gases from a dedicated inert gas generator, are used for inerting of compartments handling flammable or combustible cargo. U.S. Coast Guard (USCG) regulations (USCG, 1998) and international maritime safety rules (SOLAS, 1974) require or establish inerting requirements based upon the nature and quantity of cargo aboard ship. USCG guidance requires that the O_2 content in cargo tanks be no more than 5 vol % for crude oil carriers, but notes that a lower O_2 concentration may be required for "certain chemicals in product carriers" (USCG, 2000).

Whether shipboard or land-based, combustion systems will produce a hot, dirty gas stream that must be treated before it can be used for inerting purposes. Soot may be present if the combustion was fuel-rich. Sulfur compounds, likely to be present in the off-gas, could react with moisture, forming acids that could corrode equipment. To address these concerns, the USCG requires that the gas stream be washed in a water scrubber. A demister on the scrubber outlet is required to remove entrained water droplets (USCG, 2000). While not required by the USCG, additional treatment to reduce the moisture content of the inert gases may be required for some applications.

The O_2 content of the inert gas stream can be anticipated to be variable and should be monitored, alarmed, and interlocked as required by the application. The USCG requires an alarm if the O_2 concentration reaches 8 vol %.

The American Bureau of Shipping (ABS) provides additional guidance on maritime inert gas systems, which may have relevance to the design and operation of industrial systems (ABS, 2004).

Steam. With proper care, steam can be used as an inert. However, issues related to the use of steam include the following:

- Equipment must be maintained at an elevated temperature to limit the rate at which the steam condenses. FM Global (2000) recommends a minimum temperature of 160 ^oF. The Compressed Gas Association (CGA, 2000) cautions against the use of steam in: (1) systems where brittle materials (such as cast iron) may be stressed by thermal expansion, (2) systems with close clearances where high temperatures may cause permanent warping or maladjustment, and (3) systems where pipe coatings or plastic materials may be damaged by high temperatures. Protections for personnel to prevent thermal burns from equipment may be required
- Condensation of steam inside the equipment leaves in question the concentration of steam present (CGA, 2000). NFPA recommends a steam supply rate of at least 2.5 lb/min/100 ft^3 (NFPA 69, 2002)
- Some equipment or equipment supports may not have the strength to support a significant load of condensate, and provisions must be made for removal of condensate from the

inerted equipment. Steam should not be used in systems where residual water cannot be tolerated when the process is started back up (CGA, 2000)

- Controls must be implemented to prevent pulling a vacuum on the equipment if the steam supply is interrupted and the steam condenses as the equipment cools. This could either pull air into the equipment, defeating the inerting, or collapse equipment not having an adequate vacuum rating (CGA, 2000; NFPA 69, 2002)

Fuel Gases. While it may seem counterintuitive that adding fuel to a system could inert the system and make it safer, the goal here is to displace O_2 and make the system fuel-rich. As mentioned previously, fuel not associated with sufficient O_2 for combustion acts as a heat-absorbing inert. Both NFPA 69 (2002) and USCG (1999) recognize the use of methane or natural gas for this purpose. See Section 8.

Supply Option Selection. Articles periodically published in technical and trade journals provide guidance on determining the supply option most appropriate to particular situations (e.g., Michael, 1997; McMullen and Hojsak, 2003; Ward, 1994; and Hardenburger, 1992). Such articles usually address N_2 supply options and contrast bulk cryogenic liquid tanks, onsite cryogenic separation units, PSA units, and membrane separation units. Factors commonly cited as important in this decision include:

- Required N_2 purity
- Required delivery pressure
- Peak demand flow rate
- Variations in usage pattern (demand)

Generally, extremes with respect to purity, pressure, and peak flow rate, or wide variations in demand rate, mitigate for bulk liquid systems. However, alternative technologies continue to evolve, and the various options should be reviewed with suppliers to determine the most cost-effective supply. Remember that purity costs, and 99.9999% N_2 may not be required for most inerting purposes. See Section 8.

Other considerations (such as chemical incompatibilities) that may dictate the choice of which inert gas to use are discussed in the next section.

5. Chemical Hazards Associated with Inerts (Compatibility Issues)

While N_2 and CO_2 may act as inerts with respect to combustion reactions, they are far from being chemically inert. If they were chemically inert, chemistry textbooks would be much thinner – if there were anyone around to write or read them! Only the noble gases (e.g., Ar and He) can, for all practical purposes, be regarded as true inerts.[10]

Table 2 lists chemical incompatibilities for N_2 and CO_2, as reported in Bretherick (1999) and Pohanish and Greene (2003). Data for CO are also included because appreciable concentrations of CO can be present in gas streams from combustion-based inert gas generators. Incompatibility

[10] Chemists have demonstrated the formation of compounds between the noble gases and certain metals, such as uranium.

information for steam (water) is not presented because such information is so universally available elsewhere.

Table 2: Chemical Incompatibilities for CO_2, CO, and N_2		
Carbon Dioxide	**Carbon Monoxide**	**Nitrogen**
Strong bases, amines, anhydrous ammonia	Oxidizers: oxygen, chlorine dioxide, halogens, dinitrogen oxide, peroxydisulfuryl chloride	Lithium (at ambient temperature)
Lithium, potassium, sodium, Na-K alloy	Potassium, sodium	
Magnesium, beryllium	Nitrogen trifluoride	Titanium will burn in N_2
Aluminum, chromium, manganese, titanium, and zirconium [a]	Cesium monoxide, copper perchlorate, potassium, oxygen difluoride, sodium [c]	Ozone [d]
Uranium, cerium	Aluminum with aluminum halides	
Acrolein	Iron oxide, silver oxide	
Aziridine	Bromine pentafluoride, bromine trifluoride, chlorine trifluoride	
Metal acetylides		
Aluminum halides		
Certain metal hydrides [b]		
Cesium monoxide		
Sodium peroxide, barium peroxide		
Sodium carbide		

[a] These metals are ignitable and explosive when suspended and heated in CO_2.
[b] React explosively with CO_2 when burning.
[c] Form explosive compounds that are heat-, spark-, impact-, or water-sensitive.
[d] Explosive oxidation.

Additional chemical incompatibility issues to consider include the following:

- CO_2 forms carbonic acid in water and can pose a corrosion concern for some materials of construction
- The presence of acidic sulfur compounds in the off-gas from combustion-based inert gas generators was discussed in the previous section
- CO is a strong reducing agent and a flammable gas in its own right

An indirect compatibility issue can be associated with the use of cryogenic nitrogen as a source of N_2 gas. Because N_2 boils at a lower temperature than O_2 (-196 °C versus –183 °C), air can condense on the outside of liquid N_2-bearing piping. The liquid that drops off of the piping will be enriched in O_2 and can pose an enhanced fire or explosion risk in the vicinity of the equipment.

Clearly, chemical incompatibility concerns should be considered in selecting the most appropriate inert for a particular application.

6. Physical Hazards Associated with Inerts

A variety of physical hazards presented by the various inert gases are discussed below.

High Temperature. The high temperature concerns associated with steam and untreated off-gas from combustion-based inert gas generators were discussed in Section 5.

Low Temperature. The atmospheric boiling points for N_2, CO_2, He, and Ar are -196, -79, -269, and -186 °C, respectively. While storage temperatures inside pressurized tankage would be somewhat higher than these values, releases from containment would flash evaporate, producing a mix of liquid and vapor (or, in the case of CO_2, solid and vapor) at the atmospheric boiling point.[11] The potential for cryogenic burns must be considered in specifying operating and maintenance procedures and personal protective equipment requirements.

Cryogenic temperatures can cause embrittlement of many materials of construction (e.g., carbon steel) and must be considered in the design of inert gas delivery systems. Suitable engineered controls should be provided to ensure that operational upsets do not allow the migration of cryogenic liquids to piping or equipment not designed to withstand such low temperatures.

The hazards associated with the condensation and fractional distillation of air on the outside of equipment containing liquid N_2 and other cryogenic liquids with boiling points less than that of O_2 were discussed in the previous section.

High Pressure. Cryogenic liquids produce a very large volume of gas when they evaporate (e.g., 1 volume of liquid N_2 produces 694 volumes of gas at 20 °C) (Air Products, 1999). Containers such as transport and storage vessels (typically supplier-owned and maintained) are provided with appropriate overpressure relief to address this hazard. Similar consideration, however, must be given to piping systems (which may be designed, installed, and maintained by the operating facility). An additional concern is the hydrostatic pressure that can be produced if cryogenic liquids are trapped in a liquid-full system. Absent a vapor space to allow liquid expansion, extremely high pressures can be produced. Pressure relief devices must be installed in sections of equipment where cryogenic liquids might become trapped between closed valves. Air Products (2003b) provides precautionary guidance with respect to manifolding of liquid-containing cylinders. A temperature difference of only a few degrees between cylinders can cause gas from the warmer cylinder to migrate through the manifold to the cooler cylinder where it could condense and, given sufficient time, fill the cylinder. Cylinder rupture could result if the full cylinder was valved closed.

Considering the large liquid-to-gas expansion ratio, consideration should be given to applying controls to the quantity of cryogenic liquid stored inside tight enclosures or buildings. The asphyxiation hazard associated with inert gases is addressed in the next section.

As noted previously, portable containers of high pressure inert gases can operate at pressures of thousands of pounds per square inch. Suitable precautions are required to protect containers and

[11] Temperatures below the atmospheric boiling point could be experienced if there was appreciable evaporative cooling due to the evaporation of liquid droplets entrained in the cloud.

associated regulators and piping from damage. Refer to CGA (2000 and 2003a) and Air Products (2000) for guidance.

Static Electricity. The use of high pressure CO_2 for inerting poses a concern for potential static electricity hazards. CO_2 converts directly to a solid if the liquid is depressurized below 61 psig (Air Products, 1998c). Consequently, high discharge rates can produce CO_2 "snow" which can, because it is moving at a high velocity, generate static electric charge. Incendive sparks (5 to 15 mJ at 10 to 20 kV) have been reported (Bretherick, 1999).

Air Products (1999) provides a good summary of the hazards associated with the handling of cryogenic liquids.

7. Health Hazards Associated with Inerts

The inert gases that we use because they will not support combustion are double-edged swords because they also will not support life. Injuries and fatalities to personnel who either enter inerted equipment, or fall victim to releases of inerts from equipment, are all too frequent events.

The Medical Details. Many people, whose only frame of reference may be falsely based upon a knowledge of how long they can hold their breath, do not understand the physiological processes that occur when someone breathes an O_2-free (or nearly so) atmosphere. Air Products (1998d) provides a very concise description of the physiological processes leading to death from hypoxia (i.e., insufficient supply of oxygen to the body tissues). An abridged version of this description, written from a chemical engineering perspective, follows.

During normal respiration, blood exchanges CO_2 for O_2 in the lungs. A few breaths of an O_2-free atmosphere essentially empties the lungs of O_2, reversing the O_2 concentration gradient between the blood and the alveoli (gas-containing sacs) in the lung. As a result, oxygen is actually stripped from the blood. O_2-deficient blood then returns to the brain, which is extremely sensitive to O_2 deficiency, and swiftly causes unconsciousness and coma. The victim may not recognize any symptoms warning of the predicament, or may not be able to reason or take protective action before unconsciousness occurs. Loss of consciousness occurs within about 10 seconds of the first breath, followed by death within 2 to 4 minutes.

BP (2003) provides a similar estimate of the rapidity with which coma and death would be expected to occur.

The above explanation puts into perspective the rule of thumb that only one or two breaths of O_2-deficient atmosphere are required to cause death from hypoxia. Just removing the victim from O_2-free atmosphere is not sufficient. Resuscitation through the administration of O_2 is required to prevent death (CSB, 2003).

Less severe, but still potentially significant effects occur from breathing atmospheres that have reduced, but nonzero, O_2 concentrations. Table 3 relates O_2 concentrations to physiological effects (BP, 2003). The table reinforces the Occupational Safety and Health Administration (OSHA) requirement (OSHA, 2000) that the O_2 concentration in a confined space must be at least 19.5 vol %, unless air-supplied respiratory equipment is used, and cannot be greater than 23 vol %.[12, 13]

Table 3: Physiological Effects of Reduced O_2 Atmospheres

O_2 (vol %)	Effects
23.5	Maximum "Safe Level" (23 vol % is often the High level alarm of most O_2 detectors)
21	Typical O_2 concentration in air
19.5	Minimum "Safe Level" (19 vol % is often the Low level alarm of most O_2 detectors)
15-19	First sign of hypoxia. Decreased ability to work strenuously. May induce early symptoms in persons with coronary, pulmonary, or circulatory problems
12-14	Respiration increases with exertion, pulse up, impaired muscular coordination, perception, and judgment
10-12	Respiration further increases in rate and depth, poor judgment, lips blue
8-10	Mental failure, fainting, unconsciousness, ashen face, blueness of lips, nausea, vomiting, inability to move freely
6-8	6 minutes, 50% probability of death 8 minutes, 100% probability of death
4-6	Coma in 40 seconds, convulsions, respiration ceases, death

Physical exertion increases oxygen demand and may result in oxygen deficiency symptoms at higher oxygen concentrations (CGA, 2001b).

In addition, commonly available data for the physiological effects of reduced O_2 atmospheres assume a sea-level location. As anyone who has spent time in the mountains will affirm, breathing is much harder at high altitudes. Even though the concentration of O_2 does not decrease (air still has its normal 21 vol % O_2), the amount of O_2 available per volume (per lungful) decreases as the atmospheric pressure and partial pressure of O_2 decrease with increasing altitude (Curtis, 1999). While the guidance in Table 3 should be applied cautiously for facilities at significant altitudes, it should be noted that OSHA, in its Respiratory Protection Standard, accepts 19.5 vol % as a safe O_2 concentration up to an altitude of 8,000 ft (OSHA, 1998).

An asphyxiant is a chemical (either a gas or vapor) that can cause death or unconsciousness by suffocation (BP, 2003). A simple asphyxiant is a chemical, such as N_2, whose effects are caused by the displacement of O_2 in air. These are the effects illustrated in Table 3. A chemical asphyxiant works by interfering with the body's ability to absorb or transport O_2 to the tissues. A relevant example of a chemical asphyxiant is CO which, as noted previously, can be present in inert gas streams produced by combustion.

[12] This paper does not address in detail procedures for safe confined space entry. Readers should consult OSHA (2000), the American National Standards Institute (ANSI, 2003), Hodson (2001), and BP (2003).

[13] An upper limit on O_2 concentration is imposed to protect against the enhanced flammability hazards associated with O_2-enriched atmospheres. Refer to Air Products (2003a), NFPA 53 (1999), and CGA (2003a) for additional information on the hazards of such atmospheres.

Table 4: Effects of Exposure to CO

Concentration, (ppm)	Effect
50	Threshold limit value for no adverse effects even with 6-8 hours exposure
200	Possible mild headache after 2-3 hours
400	Headache and nausea after 1-2 hours
800	Headache, nausea, and dizziness after 45 minutes; collapse and possible unconsciousness after 2 hours
1,000	Loss of consciousness after 1 hour
1,600	Headache, nausea, and dizziness after 20 minutes
3,200	Headache and dizziness after 5-10 minutes; unconsciousness after 30 minutes
6,400	Headache and dizziness after 1-2 minutes; unconsciousness and danger of death after 10-15 minutes
12,800	Immediate physiological effects; unconsciousness and danger of death after 1-3 minutes

CO is an acutely toxic gas that interferes with the ability of the hemoglobin in red blood cells to carry O_2 throughout the body. It is odorless and colorless, and thus provides no immediate warning properties to those exposed. Table 4 summarizes the effects of exposure to CO, showing that significant health effects can result from concentrations measured in parts per million (ppm) (Meidl, 1970). Clearly, the toxic effects of CO would have to be considered in evaluating potential health effects from exposure to inert gas streams produced by combustion.

Table 3 showed the health effects caused by simple asphyxiants which, as defined above, only act by displacing O_2 from air. Examples would include N_2, He, and Ar. It should be noted that CO_2 acts neither as a simple asphyxiant (like N_2) nor a chemical asphyxiant (like CO).

Evolution has programmed the body to respond to high levels of CO_2 because removal of CO_2 is part of the metabolic process. Consequently, the body's response to CO_2 is markedly different than that for simple asphyxiants. The normal concentration of CO_2 in air is on the order of 300 ppm (0.03 vol %). At CO_2 concentrations greater than about 1 vol % the body, not being able to distinguish between internally generated CO_2 and environmental CO_2, begins to breathe more rapidly. Respiration rate increases as the CO_2 concentration increases, producing what some authors have referred to as "asphyxia panic." Physiological effects at higher concentrations are shown in Table 5.

Table 5: Effects of Exposure to CO_2

CO_2 (vol %)	Effect
1	Slight increase in breathing rate
2	Breathing rate increases to 50% above normal. Prolonged exposure can cause headache and tiredness
3	Breathing increases to twice the normal rate and becomes labored. Weak narcotic effect. Impaired hearing, headache, increase in blood pressure and pulse rate
4-5	Breathing increases to approximately four times the normal rate, symptoms of intoxication become evident and slight choking may be felt
5-10	Characteristic sharp odor noticeable. Very labored breathing, headache, visual impairment, and ringing in the ears. Judgment may be impaired, followed within minutes by loss of consciousness
50-100	Unconsciousness occurs more rapidly above 10 vol % level. Prolonged exposure to high concentrations may eventually result in death from asphyxiation

Readers should not attempt to draw any quantitative correlation between Tables 3 and 5. While CO_2 is an asphyxiant at high concentrations, it has other pronounced effects at lower concentrations, as shown in Table 5. The effects of exposure to 5 vol %

CO_2 are not the same as the effects of exposure to air whose O_2 concentration has been reduced by 5 vol %. In fact, tests described by the National Institute for Occupational Safety and Health (NIOSH, 1976) have shown that breathing a mix of 30 vol % CO_2 and 70 vol % O_2 will rapidly cause unconsciousness; therefore, the effect is not due to an insufficiency of O_2.

To summarize the potential health effects associated with commonly available inerts:

- N_2, Ar, and He act as simple asphyxiants, displacing O_2 from air. Their effects at increasing concentrations can be determined based upon the information presented in Table 3
- CO_2, at low concentrations, has a pronounced effect on the body that cannot be attributed to simple asphyxiation alone. While high concentrations of CO_2 will displace enough O_2 from the air to produce O_2 deficiency, the health effects from the CO_2 exposure, as shown on Table 5, may be predominant
- In estimating the health effect from exposure to combustion-generated inert gases, reference should first be made to Table 4 to determine any health effects due to CO exposure, followed by reference to Table 5 for CO_2 exposure effects, followed by reference to Table 3 to determine any simple asphyxiation effects (presuming that neither of the other tables shows a predominant health effect)

Finally, with respect to steam used as an inert, repeated tests have shown that exposure to water at a concentration of 100% is likely to be fatal in as little as a few minutes (Anon, 2002).

As noted in the Introduction, the process and related industries have experienced, and continue to experience, significant numbers of injuries and fatalities due to personnel exposure to inerts. All of the commonly used inerts are colorless and odorless and offer no immediate warning signs to those potentially exposed. Exposures can occur inside of intentionally inerted equipment (e.g., due to failures of isolation procedures or confined space access controls). Exposures can occur outside of equipment due to leaks or intentional releases of inert gases. Examples exist where personnel outside of equipment have been rendered unconscious by such releases, only to fall into the inerted equipment through the opening from which the inert issued.

This paper will not relate any specific case histories. Both BP (2003) and the U.S. Chemical Safety and Hazard Investigation Board (CSB, 1998 and 2003) have published compilations of case histories and investigation reports that should be required reading for anyone having responsibility for designing, operating, or maintaining an inerting system. Understandably, due to the predominant use of N_2 as an inert, the BP and CSB publications focus on N_2. The learnings presented, however, are readily extensible to other inerts.

In its safety bulletin on the hazards of nitrogen asphyxiation, the CSB identified 80 nitrogen asphyxiation deaths and 50 injuries occurring in 85 incidents between 1992 and 2002. Many were "caused by inadequate knowledge of the hazard, or inadvertent use of nitrogen rather than breathing-air delivery systems" (CSB, 2003). It is very likely that these figures are underreported and the incidence of asphyxiation fatalities is greater than yet measured.

A particularly tragic aspect of this issue is the number of would-be rescuers who rush into an inerted atmosphere in an attempt to save a colleague, only to fall victim themselves to the hazard.

The CSB estimates that 10% of the fatalities result from such heroic, but ill-considered, actions. Air Products estimates that over 50% of the workers who die in confined spaces are attempting to rescue others (Air Products, 1998a).

Readers should consult OSHA (2000), ANSI (2003), Hodson (2001), and BP (2003) for guidance on addressing confined space hazards.

8. How Inert Is Inert Enough?

If we focus on the goal of fire and explosion prevention, the obvious question for someone designing an inerting system is "How low do I need to reduce the O_2 concentration?"

The primary guidance for use in inerting to prevent fires and explosions is contained in NFPA 69 (2002). In this document, NFPA defines the LOC as the "concentration of oxidant below which a deflagration cannot occur in a specified mixture."[14]

Determining the LOC. The LOC is specific to a particular fuel and varies depending upon the inert used. To gain a full appreciation of the LOC, it is necessary to have an understanding of the use of flammability diagrams to depict combustible gas compositions.

Figure 3 shows the grid for a triangular flammability diagram. Every point on the diagram represents a triplet of concentrations as indicated by the legends on the axes: fuel, air, and additional nitrogen added as an inert. The point represented by the large dot in the center of the diagram indicates a mixture of fuel, air, and N_2, with concentrations of 40, 30, and 30 vol %, respectively.
There is a handy feature to be aware of with respect to navigating around a flammability diagram. Note the three straight lines drawn from the dot to each of the vertices of the figure. Each vertex represents a point where the composition is 100 vol % of one of the three components. If we follow the line up to the 100 vol % fuel vertex, the compositions along that line have a constant ratio of nitrogen to air. As the fuel concentration increases, the air and nitrogen concentrations decrease, but at a constant ratio. Analogous relationships exist for the other two straight lines going to the 100 vol % air and 100 vol % nitrogen vertices.

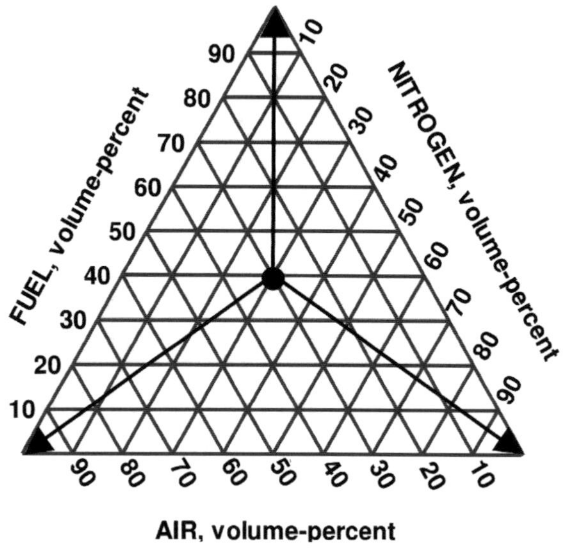

Figure 3: Triangular Flammability Diagram

[14] Other terms that appear in the literature include minimum oxygen (or oxidant) concentration (meaning – to permit combustion) and maximum oxygen (or oxidant) concentration (meaning – to prevent combustion), both abbreviated MOC.

Figure 4 shows a flammability diagram for anhydrous methanol.[15] The shaded area depicts all of the triplet combinations of methanol, air, and N_2 that are flammable at 25 °C and normal atmospheric pressure. This region is bordered on the bottom by a nearly horizontal line representing the LFL, and on the top by a sloping line, representing the UFL. The LFL in air is 7.3 vol %, and the UFL in air is 36 vol % (as read off the left-hand axis of the figure). Traversing the flammability region is a straight line representing all of the mixture compositions for which there is a stoichiometrically balanced ratio of fuel to air. Since the stoichiometric line (C_{st}) passes through the 100 vol % N_2 vertex on the diagram, the ratio of fuel to air is constant.

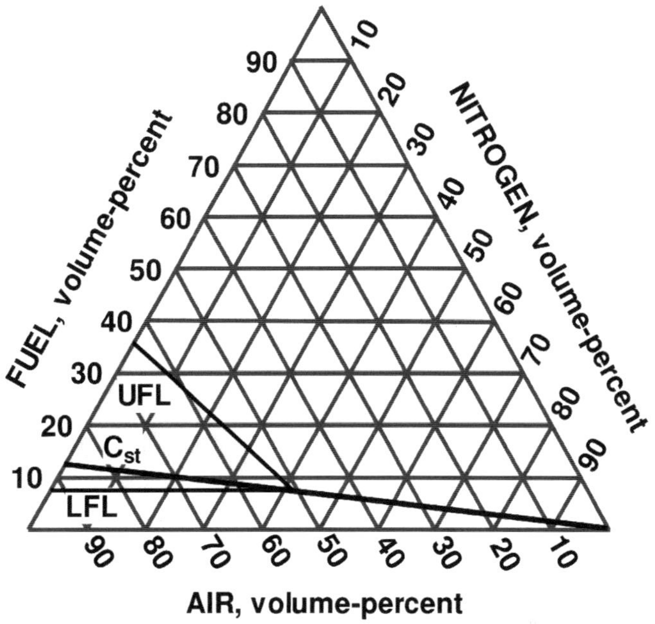

Figure 4: Flammability Diagram for Methanol

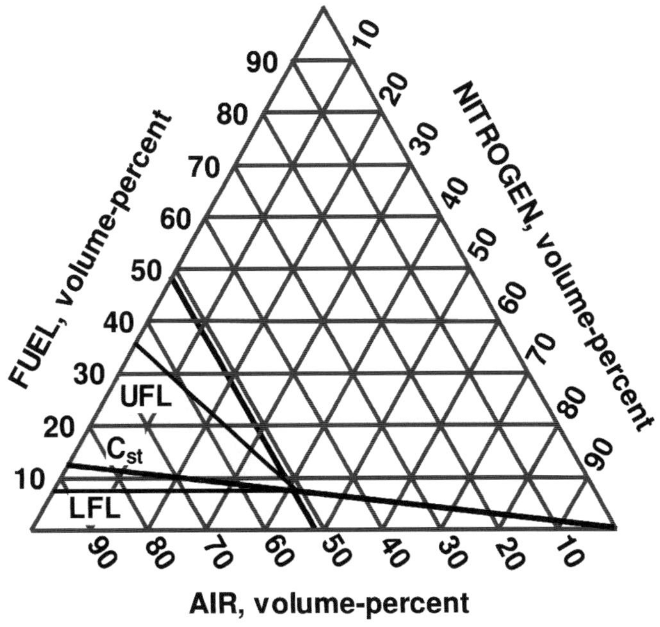

Figure 5: Flammability Diagram with LOC Line

Figure 5 shows the same diagram, with the addition of a line drawn at a constant air concentration, passing near to or through the intersection of the LFL and C_{st} lines. This line describes all points on the diagram having an air concentration of about 51 vol %, as read off the bottom axis.

Close inspection reveals that there are no composition points to the left of and above this line that intersect the flammability region of the diagram. Or, expressed another way, all of the compositions to the left of and above this line are noncombustible. Expressing this yet another way, no mixture composition having less than 51 vol % air is combustible.

To label this the LOC line is a bit of a

[15] The flammability diagram is only intended to be illustrative and should not be used for the purposes of establishing safe operating limits.

misnomer since the line identifies an air concentration. However, recalling that air is 21 vol % O_2, the LOC for methanol as read off this diagram is 10.71 vol %.[16] As we will see shortly, the experimental value for methanol is about 10 vol %.

The above approach illustrates how the LOC can be graphically determined if a flammability diagram is available. Crowl (2003) describes approaches for approximating the flammability diagram, given the UFL and the LFL for the fuel, either in air or in O_2. In point of fact, this approach was used to construct the flammability diagrams above. A less time-consuming approach, also described by Crowl, involves finding the LOC algebraically.

The stoichiometric equation for the combustion of methanol in air is:

$$\mathbf{CH_4O + 1.5\ O_2 + 5.6\ N_2 \rightarrow CO_2 + 2\ H_2O + 5.6\ N_2}$$

What we wish to solve algebraically is what we solved graphically before – the intersection of the stoichiometric line and the LFL line. From the above chemical equation we know that the slope of the stoichiometric line is 1.5 moles of O_2 per mole of methanol. Next, we assume that the LFL line is horizontal (i.e., that the LFL is constant, a relatively safe assumption). In this case the LFL is 7.3 vol %, which is equivalent to 7.3 moles methanol/100 moles of mixture.

The LOC can then be calculated as:

$$\mathbf{LOC = \left(\frac{7.3\ moles\ methanol}{100\ moles\ mixture}\right)\left(\frac{1.5\ moles\ O_2}{mole\ methanol}\right) = \frac{10.95\ moles\ O_2}{100\ moles\ mixture} = 10.95\ vol\%}$$

The algebraic approach yields an LOC of 10.95 vol %. Unfortunately, both the graphical and algebraic approaches yielded answers that were on the nonconservative (high) side of the experimental value of 10 vol %. Experience has shown that LOCs estimated in this manner are generally within about 1 vol % of the experimentally determined LOCs, but are about equally likely to overestimate or underestimate the LOC.

It is not often that we have the luxury of dealing with a simple system involving a single fuel. The LOC for a mixture of fuels can be estimated if it is assumed that the relative proportions of the various fuels are fixed. The approach is based upon the application of Le Châtelier's Rule, which has traditionally been used to estimate the LFL or UFL of mixtures of fuels (Crowl, 2003).

---

[16] Flammability diagrams can also be drawn with O_2 concentration on the bottom axis. Each point on the diagram would reflect a triplet of fuel, O_2, and N_2 concentrations. On such a diagram, the N_2 would be the sum of the N_2 present in the air, plus any additional N_2 added for inerting. A little bit of arithmetic would allow the flammability envelope to be redrawn to accommodate the new scales. A line of constant O_2 concentration drawn through the nose of the flammability region would indicate the LOC in vol % O_2 on the bottom axis.

If the concentrations and individual LFLs are known for each fuel in the mix, the LFL for the mix can be estimated from the following equation:

$$LFL_{mix} = \frac{1}{\sum_{i=1}^{n} \frac{y_i}{LFL_i}}$$

In the equation, y_i is the mole fraction of the ith fuel and LFL_i is the LFL for that fuel. The fuel mole fractions are calculated on a fuel-only basis (i.e., ignoring any air or inert).

Consider the following system:

$$2C_3H_8 + 3C_4H_{10} + 29.5\ O_2 + 111\ N_2 \rightarrow \text{products}$$

Given that the LFL for propane and butane are 2.1 and 1.8 vol %, respectively, we can estimate the LFL for the mix as:

$$LFL_{mix} = \frac{1}{\frac{0.4}{2.1} + \frac{0.6}{1.8}} = 1.91 \text{ v/o}$$

From the stoichiometrically balanced combustion equation, the slope of the C_{st} line is 29.5 moles of O_2 per 5 moles of fuel. The estimated LOC would then be 1.91 x (29.5/5) = 11.27 vol % O_2.

An alternate approach would be to search the literature and find that the experimental values of the LOC for propane and butane are 11.5 and 12.0 vol % O_2, respectively (NFPA 69, 2002). If we use Le Châtelier's Rule, substituting LOC for LFL, we would get:

$$LOC_{mix} = \frac{1}{\frac{0.4}{11.5} + \frac{0.6}{12.0}} = 11.79 \text{ v/o}$$

A more conservative approach would be to establish the LOC based upon the lowest LOC for any fuel in the mixture.

<u>Adjusting the LOC for Operating Conditions</u>. Flammability properties such as LFL, UFL, and LOC are affected by the environmental variables of temperature and pressure. Most commonly, values reported in the literature were determined for "standard" conditions of 20 or 25 °C and 1 atmosphere. It is not unusual to be operating processes at other typically elevated temperatures and pressures. Increasing pressure and temperature raises the UFL and lowers the LFL (although the effect of pressure on the LFL is much less than on the UFL).

Bodurtha (1980) provides the following equations for adjusting LFL, UFL, and LOC for increases in temperature and pressure:

$$LFL_t = LFL_{25^\circ C} - (0.0008 * LFL_{25^\circ C})(t - 25)$$

$$UFL_t = UFL_{25^\circ C} + (0.0008 * UFL_{25^\circ C})(t - 25)$$

$$LOC_P = LOC - 1.5(\log P + 1)$$

$$UFL_P = UFL + 20.6(\log P + 1)$$

In using these equations, temperatures are °C, and pressures are megaPascals absolute.

Bodurtha does not provide an equation for adjusting the LOC with increasing temperature. However, keeping in mind both the graphical and algebraic approaches for finding the intersection of the LFL and C_{st} lines to estimate LOC, one might reason that the LOC should experience the same percentage decrease as the LFL. The first equation above shows an 8 % decrease in LFL for each 100 °C increase in temperature.

FM Global (2000) recommends that the LOC be decreased by 17% for each 100 °C increase in temperature up to 200 °C, with an additional decrease of 12% for each 100 °C increase above that. ESCIS (1994) suggests that the LOCs for gases and vapors decrease 0.5 to 1.0 vol % for each 100 °C rise in temperature and that the LOC for dusts decreases 1.6 vol % per 100 °C.

This is a good time to point out that estimates of flammability properties are convenient for use in preliminary studies, but that "back of the envelope" estimates should not serve as the basis for detailed design or to establish safe operating limits. At some point, a little time in the explosion hazards testing laboratory becomes both necessary and prudent.

<u>Effect of Different Inerts.</u> Recalling the description of why inerting works to suppress combustion, presented in Section 3, it is logical to infer that molecules with greater specific heats should be more effective inerts. Table 6 shows the relative specific heats for the more commonly used inerts. Figure 6 shows the impact that varying concentrations of four different inerts have on the flammability of methane.[17]

Table 6: Specific Heat of Common Inerts[(a)]

Inert	Specific Heat, (Btu/lbmole/°F)
N_2	7.0
CO_2	8.8
He	5.0
Ar	5.0
Water vapor	17

[(a)] At constant pressure; water vapor at 27 °C, other values at 0 °C

There appears to be a discrepancy between the relative heat capacities of water (H_2O) and CO_2 given in Table 6 and the relative effectiveness of the two inerts illustrated in Figure 6. Coward and Jones (1952) attribute a similar trend in another data set to the fact that the flammability tests using water vapor were run at elevated temperatures necessary to establish the desired concentration of water vapor. Tests for the other inerts were likely run near 25 °C.

---

[17] The flammability diagram is drawn on a rectangular, rather than triangular, coordinate system. The same information (i.e., the concentrations of the three components in any mix: fuel, air, and inert) can be obtained from the diagram. However, the air concentration must be obtained by subtracting the fuel and inert concentrations from 100 vol %.

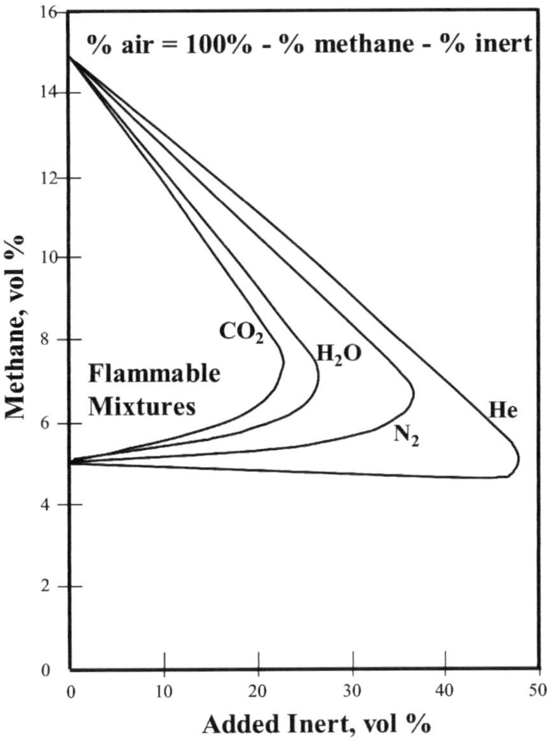

Figure 6: Effectiveness of Various Inerts

Sources of LOC Data. N_2 and CO_2 are by far the most commonly used inerts. NFPA 69 (2002) provides the most comprehensive list available of LOC data for gases, vapors, and combustible dusts. These data, compiled primarily from test work performed by the U.S. Bureau of Mines (BOM), are often repeated in other references.

Table 7 compares the N_2 and CO_2 LOCs for a number of common fuels. Ebadat (1994) suggests that the N_2 and CO_2 LOCs can be related according to the following equations:

The second equation appears to provide a better estimation of the N_2 LOC, yielding values that are typically 2 to 10% (relative) lower than the measure N_2 LOC, which is in the conservative direction.

Ebadat (2003) and Grossel and Zalosh (2004) describe the testing procedures for determining LOC.

Applying LOC Data. The methanol flammability diagram shown in Figure 5 would indicate that safe operations could be conducted below the 10.71 vol % O_2 LOC illustrated. However, this value would not be a suitable upper operating limit for oxygen concentration since there is naturally some uncertainty with respect to issues such as:

- Knowledge of what the true LOC is
- Changes in pressure and temperature that will affect both the flammability limits of the fuel and the LOC
- Variability of the inert gas purity (e.g., inerts from noncryogenic sources)
- Ability to monitor the O_2 concentration
- The stability of the process and how rapidly the O_2 concentration may change
- How rapidly the control system can sense and respond to the process change
- The desire for a workable operating margin to avoid false alarms and inadvertent process shutdowns

NFPA 69 (2002) requires the application of a safety margin between the determined LOC and the normal process operating O_2 concentration.

Table 7: Typical LOCs for N_2 and CO_2 as Inerts

Fuel	LOC N_2/Air (vol %)	LOC CO_2/Air (vol %)
Gases and Vapors		
Methane	12.0	14.5
Propane	11.5	14.5
Hydrogen	5.0	5.2
Carbon disulfide	5.0	7.5
Methanol	10.0	12.0
Dusts		
Coal	12 - 14	15 - 17
Corn starch	9	11
Polyethylene	-	12
Epoxy resins	-	12
Sulfur		12
Aluminum	5	2
Polyvinyl acetate	-	17

For systems where the "oxygen concentration is continually monitored," NFPA requires a safety margin of at least 2 vol % below "the worst credible case LOC." If the LOC is less than 5 vol % O_2, equipment is not permitted to be operated at greater than 60 % (relative) of the LOC.

For systems where the oxygen concentration is not continuously monitored, NFPA requires that the "oxygen concentration shall be designed to operate at no more than 60 percent of the LOC, or 40 percent of the LOC if the LOC is below 5 percent." NFPA further requires regularly scheduled checks of the O_2 concentration in the equipment. Such checks are waived for low pressure field storage tanks with an inert gas padding system (NFPA 69, 2002).

While NFPA uses the term "continually monitored," it is the intention of the NFPA 69 technical committee that this means "continuously measured" (Kirby, 2004).

Cunliff (2001) provides a very informative discussion of the management of inerting systems, pointing out that there are four levels of O_2 concentration that should be observed:

- LOC
- Max permissible oxygen concentration (MPOC)
- First alarm level
- Normal operating level

The MPOC corresponds to the "shall not be exceeded" O_2 concentrations established by NFPA. Many users would implement the MPOC as the trip point for an instrumented (interlocked) shutdown of the process. NFPA 654 (2000) requires that where oxygen monitoring is used (e.g., where a measured O_2 concentration trips an interlock), the installation must meet the integrity requirements established by ISA Standard S84.01 (ISA, 1996). NFPA 654 (2000) applies to systems handling combustible particulate solids (dusts), and NFPA has not yet made compliance with ISA S84.01 a more general requirement by adopting it as part of NFPA 69.

Partial Inerting. Situations may arise where it is feasible to inert a system, but not in full compliance with the NFPA requirements (i.e., to below the LOC, with application of the required safety margin). Zeeuwen et al. (1996) discusses the benefits to be derived from "partial inerting" of systems handling dust-air mixtures. The dust explosion severity, as reflected by the deflagration index (K_{st}), is reported to decrease almost linearly as the O_2 concentration is reduced from 21 vol % to the LOC; however, the maximum explosion pressure decreases significantly only near the LOC. Perhaps the most significant effect observed was the decrease in ignition sensitivity, as reflected by the minimum ignition energy (MIE). Decreasing the O_2 concentration from 21 to 17 vol % resulted in a tenfold increase in the MIE for a number of the dusts tested. It should be noted that this trend was observed only for spark ignition sources and that the ignition sensitive to continuous electric arcs and the minimum ignition temperatures of the dust layers and dust clouds were not so markedly affected.

Zeeuwen et al. (1996) cite partial inerting as one approach to reconciling the benefits and safety hazards of inerting. As noted, a tenfold increase in MIE was attained in a 17 vol % O_2 atmosphere which, as Table 3 shows, is not an environment in which one would choose to work but would be, nonetheless, life-supporting in an accidental exposure situation.

Partial inerting would decrease consumption of high purity inert or, alternatively, make inerting more feasible for those situations where only lower purity inert is available.

As noted, partial inerting does not fully comply with the requirements of NFPA 69 (2002) and it would not be appropriate to take credit for partial inerting as a primary explosion control. However, in a situation where another primary explosion control (e.g., explosion suppression per NFPA 69) was implemented, partial inerting could be a useful adjunct control aimed at reducing the probability of ignitions. Even when reliable protections are provided, it is still prudent to attempt to reduce the frequency with which they are challenged.

Use of Fuel Gases. In its guidance on oxidant concentration control, NFPA 69 (2002) includes the special case of systems operating above the UFL. As described before, excess fuel displaces oxidant, rendering the system noncombustible.[18] A special case of this approach to system safety is where additional fuel (e.g., natural gas) is intentionally added to keep the system fuel-rich. NFPA 69 specifies that the contents of vent headers operating near atmospheric pressure can be maintained noncombustible by the addition of at least 25 vol % natural gas or methane. This applies only where there is no oxygen enrichment of the gas stream and none of the gases or vapors have a UFL greater than 75 vol % (the LFL of hydrogen). Britton (1996) has documented the technical basis for this guidance.

When Inert Is Too Inert. There are at least two situations where the total exclusion of O_2 from a system actually compromises operational safety. Certain polymerization inhibitors added to stabilize monomers require a small concentration of dissolved O_2 to be effective. Unless there is a concentration of several vol % O_2 in the gases above the liquid, there will not be sufficient O_2 in solution to prevent polymerization (NFPA 69, 2002). For example, methyl acrylate and ethyl acrylate are commonly stabilized with hydroquinone monomethyl ether. Industry guidance recommends a minimum concentration of 5 vol % O_2 in the atmosphere in contact with the acrylate (ATOFINA et al., 2002).

Where inerting is provided for equipment processing aluminum, a low concentration of residual O_2 is required in the inert gas (NFPA 484, 2002). The concern is the production of fresh (bare) aluminum surface during operations such as machining or grinding, or as a result of particle size attrition during pneumatic conveying. Bare aluminum metal is pyrophoric, and sufficient O_2 must be present in the system as the surface is exposed to allow controlled oxidation and the formation of a passive oxide film. Removal of bare aluminum from a totally inerted environment and exposure to air could produce enough heat to initiate either a fire or an explosion.

9. Applications of Inerting

In the context of this paper, we are considering two aspects of inerting: purging and blanketing. As defined in Section 2, purging is the short-term supply of inert gas to a closed system in order to reduce the O_2 or fuel content of the system. Blanketing (or padding) is the long-term (perhaps continuous) maintenance of the inert atmosphere in the closed system.

---

[18] For example, the vapor in the gas tank of an automobile is normally too rich to burn because of the high vapor pressure of gasoline.

Purging. There are a variety of approaches to purging a closed system, and careful consideration is required to identify the most safe and cost-effective approach for a given situation. Considerations can include the following:

- What is the initial gas composition in the equipment?
- How inert must the equipment become?
- What inert is best suited from the standpoints of effectiveness and compatibility with process streams and materials of construction?
- What are the purity and availability of the inert?
- How quickly must the purging be accomplished?
- How complex is the equipment arrangement being purged?
- What are the pressure and vacuum ratings of the equipment being purged?
- What are the relative densities of the inert and the atmosphere initially present in the equipment?

Purging techniques can be broadly characterized as intended to either dilute the atmosphere inside the closed system or to displace the atmosphere. Dilution involves the intentional mixing of the inert with the atmosphere, and displacement involves careful avoidance of such mixing. Much of the guidance available to the process industries describes techniques relying upon dilution. However, CGA (2001a) provides detailed guidance on displacement, which will be discussed first.

Displacement Purging. Displacement is most easily achieved in piping where the inert can push the fuel ahead of it, with some back mixing at the interface. CGA (2001a) provides guidance for using this type of purging on long transmission lines and describes how it is possible to introduce a plug of inert into the line and follow it with a flow of air. Mixing between the fuel and the inert occurs at the front end of the plug, and mixing occurs between the inert and the air at the back end of the plug. The length of plug required to keep the air and fuel separated can be determined based upon the length and diameter of the pipe and the flow rate through the pipe.

Displacement purging of vessels largely relies on carefully controlled addition rates and the density difference between the initial atmosphere inside the equipment and the inert, to limit the mixing between the two. The relative gas densities govern whether the inert must be introduced at the top or bottom of the vessel (with the venting provided at the opposite end). For example, to purge a heavy hydrocarbon vapor from a tank with lower density N_2, the N_2 would be introduced at the top of the tank, with the tank vented near the bottom. Conversely, to purge natural gas from a tank using CO_2, the CO_2 would be introduced at the bottom with the tank vented at the top.

Some amount of mixing is unavoidable. CGA (2001a) cites the following factors as conducive to increased mixing:

- Large contact area (proportionally more mixing would be expected in a large-diameter, short vessel than in a tall, narrow vessel of equal volume, or in a pipe)
- Long purge times (provides longer time for mixing by diffusion)
- Mixing caused by a high velocity for the inert input

- Gravitational effects that would result if the inert was added at a location inconsistent with the relative densities (e.g., heavy inert added at the top of a vessel containing a lower density gas)
- Convection currents caused by temperature differentials within the vessel or between the inert and the initial vessel contents (another manifestation of density concerns)

Obviously, mixing will also be enhanced if there is only a slight difference in the densities of the inert and the initial vessel contents. Table 8 shows the densities (and specific gravities relative to air) of the common inerts, as well as methane, propane, and chlorine gas, and air saturated with benzene vapor.

Table 8: Density and Specific Gravity of Common Gases and Vapors

Gas or Vapor	Density, (lb./cu. ft.) [(a)]	Specific Gravity, Relative to Air
Air	0.0750	1.00
N_2	0.0725	0.97
CO_2	0.1150	1.53
Steam	0.0369	0.49
Ar	0.1034	1.38
He	0.0103	0.14
Methane	0.0412	0.55
Propane	0.1162	1.55
Cl_2	0.1867	2.49
Air saturated with benzene (Vapor pressure = 0.13 atm)	0.0915	1.22 (estimated)

[(a)] Saturated steam at 100 °C and 1 atm, others at 20 °C and 1 atm

Depending upon the size and complexity of the equipment arrangement, more than one inert inlet and/or vent may be required to adequately purge the equipment (e.g., piping connected to a vessel might have to be vented at one or more remote locations, or a large vessel might require several vents around its periphery).

If mixing were avoided entirely, only a single vessel volume of inert would be required to purge the vessel. A composition analyzer (e.g., a combustible gas analyzer or an O_2 analyzer) on the vent would show a constant reading until the last of the initial vessel contents were discharged, and would then show an abrupt composition change as inert begins to flow out. Perfect plug flow like this is not likely, however, and the time (and inert gas) required to purge the vessel will increase as the amount of mixing increases. Monitoring the composition changes and maintaining records of the efficiency of the purging will help in planning future purges of the same equipment.

The use of displacement purging generally implies an "all-or-nothing intent" (i.e., total replacement of the atmosphere in the vessel). The mechanism does not allow for reducing the concentration in the vessel to some controlled, uniform composition.

A special case of displacement purging involves filling the equipment with a liquid (most commonly water), then draining the liquid while backfilling the vessel with the inert. Such siphon purging, if properly configured and controlled, could approach 100% efficiency of purging (i.e., require only a single volume of purge). Siphon purging is not feasible unless the equipment and its supports can withstand the weight of the liquid, and it has the potential disadvantage of generating a large volume of liquid that could be contaminated and would require treatment before disposal.

The liquid selected for siphon purging must be compatible with system materials of construction and any process residues in the equipment. If the goal is to free the equipment of hydrocarbon vapors, a volatile hydrocarbon liquid would not be suitable for use since the liquid would assert its vapor pressure, contaminating the purge gas as it was introduced.

Dilution Purging. In dilution purging, the challenge is to design the purging system in a way to maximize the mixing occurring between the initial vessel contents and the added inert. Unlike displacement purging, dilution purging permits (given adequate mixing) establishing some composition endpoint other than the total replacement of the vessel atmosphere. There are four different mechanisms for dilution purging in common use:

- Sweep purging
- Pressure purging
- Vacuum purging
- Pressure/vacuum purging

Sweep purging involves providing a continuous supply of inert to the equipment being purged, with the gradual dilution of the vessel contents. The purge flow is continued for as long as necessary to affect the desired change in composition. Sweep purging, in contrast with the other dilution purging methods, has the advantage that it can be conducted at near-atmospheric pressures. No significant vacuum or pressure rating is required for the equipment being purged.

Equations for determining the purge flow rates and times required to inert a vessel are derived and discussed in a number of references (NFPA 69, 2002; Ebadat, 1994; Grossel and Zalosh, 2004; Crowl, 2003; Pilkington, 1999). For sweep purging, the rate and duration of the purge are often jointly expressed in terms of the number of volume turnovers required to reduce the concentration of one component in the vessel from its initial value to an intended value, as shown by the following equation:

$$\frac{Q_v t}{V} = \ln\left[\frac{C_o - C_g}{C_t - C_g}\right]$$

where:

Q_v = Volumetric flow rate for the purge
t = Required duration of the purge
V = Volume of the vessel being urged
C_o = Initial concentration of the undesired component in the vessel
C_g = Concentration of the undesired component in the purge gas stream (may be zero)
C_t = Intended concentration of the undesired component in the vessel at time t

Generally, the flow rate of the inlet gas and the flow path through the vessel determine the efficiency of the mixing of the inert with the vessel contents. This efficiency will normally be much less than 100%. NFPA 69 recommends the application of a safety factor of at least 4 (i.e., an efficiency no greater than 25%) to the calculated number of volume turnovers. See Section 11 for a discussion of issues related to purging vessels that are hard to mix because of their large height-to-diameter ratios.

Pressure purging can be used when the equipment (including any connected appurtenances) has an appreciable pressure rating and a source of high pressure inert gas is available. As implied by the name, pressure purging involves pressurizing the equipment with the inert, then depressurizing the equipment to discharge the undesired component that has been diluted by the inert. Multiple cycles may be required if (1) the equipment pressure rating or inert supply pressure is low, (2) the initial concentration in the vessel is high, or (3) the target concentration is low.

Vacuum purging operates in the reverse fashion. A partial vacuum is pulled on the vessel to remove the undesired component and the vacuum is subsequently broken by refilling the vessel with inert. The equipment and appurtenances must have an appreciable vacuum rating and a suitable vacuum source must be available. However, the backfilling of the vessel can be accomplished with relatively low pressure inert because the goal is only to return the vessel to near atmospheric pressure. Volatile liquids must be removed from the vessel to permit attaining the desired vacuum, and precautions must be taken to ensure that the vacuum source (e.g., vacuum pump) does not serve as an ignition source when combustible vapors are being removed from equipment. As with pressure purging, multiple cycles may be required, depending upon the circumstances.

The efficiency of pressure purging or vacuum purging is limited by the attainable pressures or vacuums. Where circumstances permit, vacuum and pressure purging can be combined to more effectively purge equipment through the combination of the available pressure differentials. The sequence can be either vacuum followed by pressure, or vice versa.

A variety of different sets of equations have been proposed for calculating the number of pressure, vacuum, pressure-vacuum, or vacuum-pressure cycles required for a given inerting sequence (Ebadat, 1994; Grossel and Zalosh, 2004; Crowl, 2003; Pilkington, 1999) Analysis has shown that the following single equation can be used to cover all four situations:

$$C_N = (C_o - C_g)\left(\frac{P_L}{P_H}\right)^N + C_g$$

where:

- N = The number of required cycles (discussed below)
- C_N = Concentration of the undesired component in the vessel after N cycles
- C_o = Initial concentration of the undesired component in the vessel
- C_g = Concentration of the undesired component in the purge gas stream (may be zero)
- P_L = Lowest pressure attained during the purge
- P_H = Highest pressure attained during the purge

This equation can be solved to directly provide the number of cycles, N:

$$N = \log\left(\frac{C_N - C_g}{C_o - C_g}\right) \Big/ \log\left(\frac{P_L}{P_H}\right)$$

The key to using a single equation for pressure, vacuum, pressure-vacuum, or vacuum-pressure purging is to properly define the values of P_L, P_H, and N to be used. This is illustrated for the four cases in Figure 7.

As indicated, when performing pressure purging, normal pressure is entered into the equations as P_L. When performing vacuum purging, normal pressure is P_H. The inputs are otherwise straightforward.

The literature does not provide explicit guidance with respect to mixing efficiencies for purging methods using pressure and/or vacuum. It is easy to envision, however, situations where complete mixing may not occur between the inert and the original system contents in portions of the system (e.g., at the ends of long pipelines remote from the point where the purge is introduced). Evaluation of the prudence of purge cycles, in addition to the theoretically calculated minimum, is required on a case-by-case basis. Multiple vent points (with pressure purging) or multiple purge inputs (with vacuum purging) will aid the distribution of the inert throughout the system and increase purging efficiency.

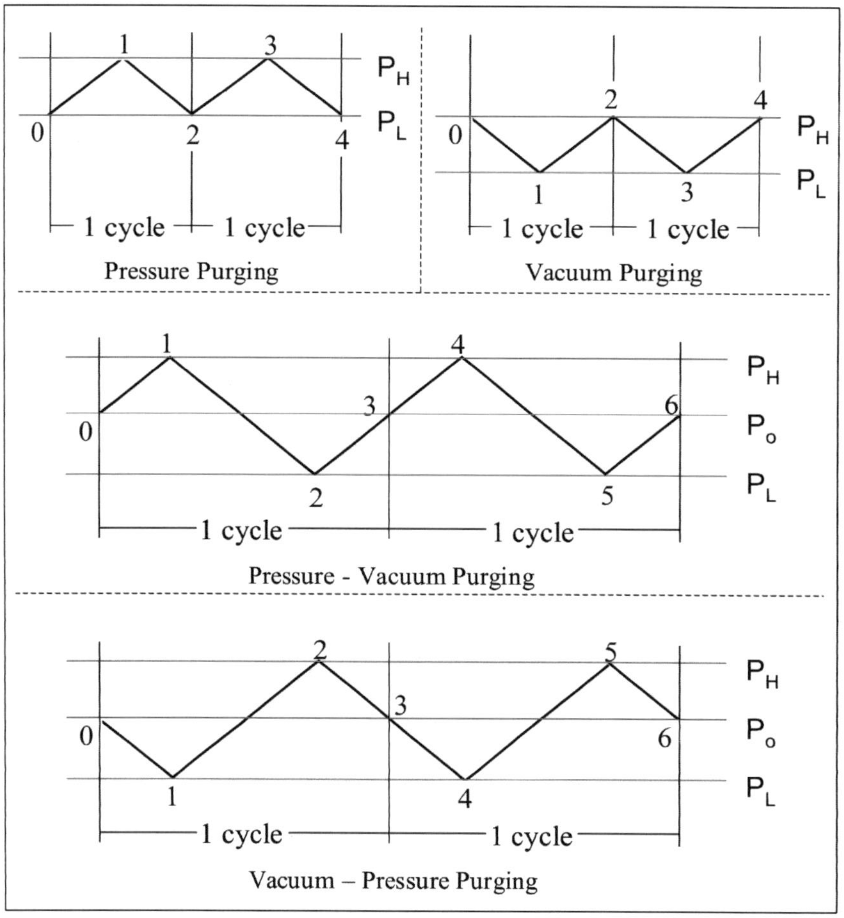

Figure 7: Various Purging Cycles Using Pressure and Vacuum

Vacuum purging takes less inert than pressure purging (e.g., pulling a ½ atmosphere vacuum in a vessel and refilling with inert removes half of the O_2 initially present and requires half of a vessel volume of inert, while establishing a 2 atmosphere pressure in the vessel and then venting back to atmospheric pressure also removes half of the O_2 but requires a full vessel volume of inert).

Vacuum-pressure purging can take less inert than pressure-vacuum purging if the ratio of P_O/P_L is greater than the ratio of P_H/P_O.

Using Flammability Diagrams to Plan Purging. Flammability diagrams can provide valuable assistance in planning the steps required to safely purge a vessel. The figures below show alternate means of establishing an air atmosphere in a vessel (e.g., in preparation for a maintenance activity). In both cases, the initial vessel atmosphere contains a mixture of 40 vol % fuel and 60 vol % air. Figure 8 shows what would happen if air were used to sweep the fuel vapor from the vessel. The composition in the vessel would pass directly through the widest portion of the flammability region.

The alternate approach, shown in Figure 9, calls for first adding N_2 to the vessel until the mix contains approximately 74 vol % N_2, 10 vol % fuel, and 16 vol % air. At that point, air can be added to the vessel to displace the remaining fuel.

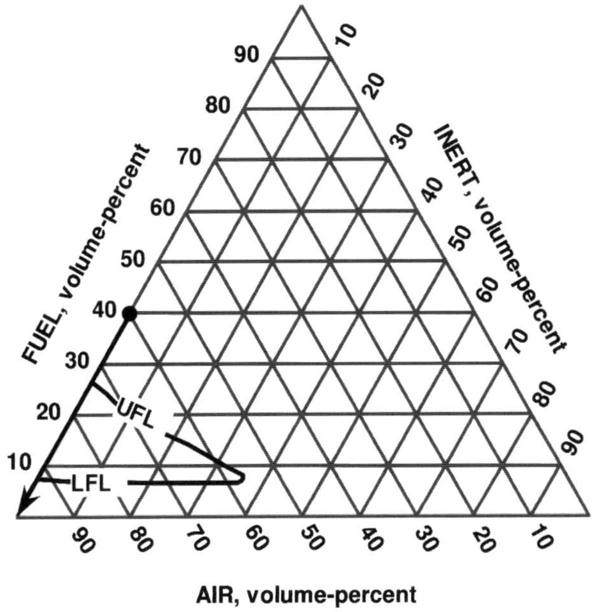

Figure 8: Hazardous Approach for Replacing Vessel Atmosphere

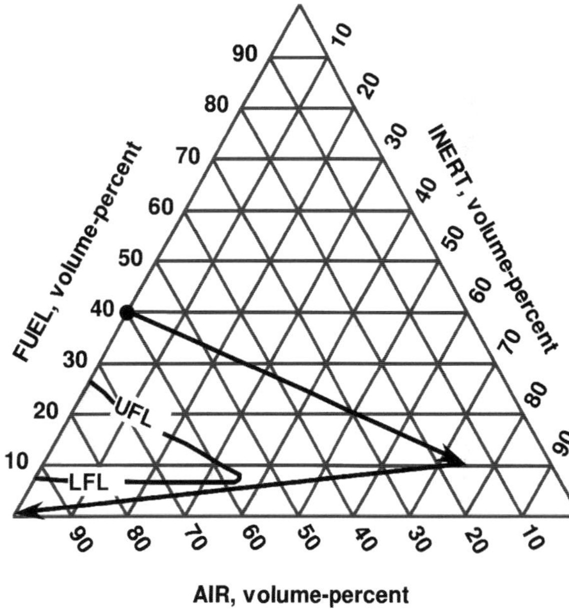

Figure 9: Safer Approach for Replacing Vessel Atmosphere

Figure 10 shows the purge pathway around the flammability diagram if the vessel initially contains 100 vol % fuel. An added inert concentration of about 85 vol % must be established before air can safely be added. This corresponds to a residual fuel concentration of 15 vol %.

After the flammability diagrams have been used to determine the reduction in O_2 concentration required from the N_2 addition, the equations provided previously can be used to determine the required rate and duration of a sweep purge or the required number of pressure or vacuum purge cycles. Do not forget the application of an appropriate safety factor for mixing efficiency.

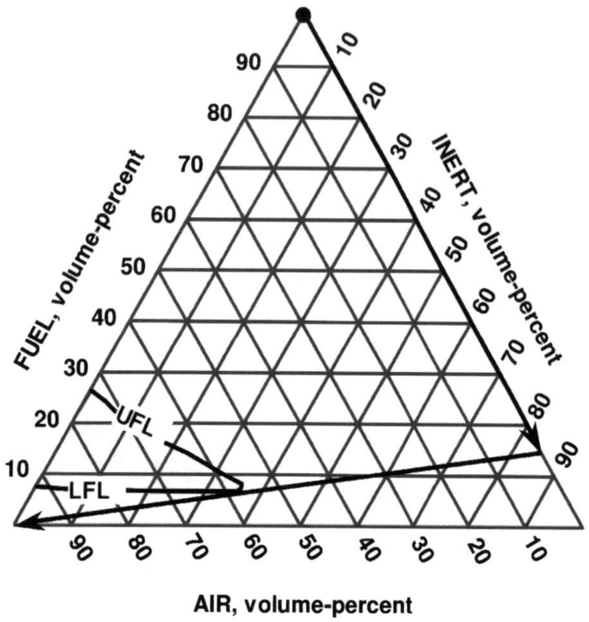

Figure 10: Purging a 100 vol % Fuel Atmosphere from a Vessel

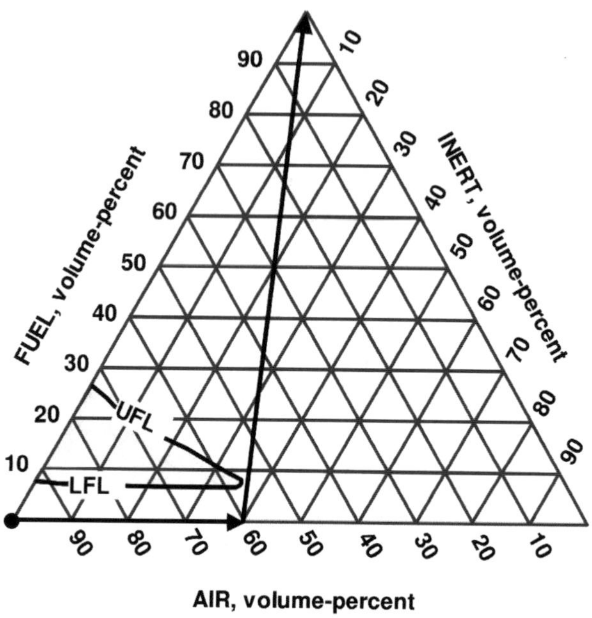

Figure 11: Purging Path for Returning Vessel to Service

Figure 11 shows the purge pathway for returning the vessel to service. N_2 must first be added to the vessel to reduce the air concentration to less than 60 vol %. At this point, fuel can be readmitted to the system without creating a combustible gas mixture in the vessel. While it might be simpler to take the N_2 concentration to nearly 100%, the plotted path is more efficient with respect to N_2 consumption and could be dictated by production schedule issues related to a timely return of the vessel to service. Consideration of mixing-related safety factors is also appropriate for return to service purging.

Figures 10 and 11 show, respectively, the operations of "purging out of service" (i.e., removing fuel before air can be introduced) and "purging into service" (i.e., removing air before fuel can be introduced). Crowl (2003) and CGA (2001a) describe approaches for determining the target fuel and air concentrations for purging out and into service. Note that the oxygen concentration corresponding to the "purge into service" air concentration is not equivalent to, and may be slightly greater than, the LOC for the fuel involved.

Keep in mind the relationship described previously with regard to composition changes occurring as we move along a straight line toward any of the vertices of the triangular diagram. As we add inert to the system, we move along a straight line toward the 100 vol % inert vertex of the diagram, keeping the fuel-to-air ratio constant. It should be noted that this relationship only holds true so long as there is no volatile liquid in the system. Figure 12 shows the same system addressed in Figure 9, but illustrates the composition change that would occur as N_2 is added to the vessel, if the fuel was present as vapor evolving from a heel of flammable liquid in the vessel. As pure nitrogen is added to the vessel, more of the fuel would evaporate to assert its vapor pressure in the vessel ullage, and the fuel vapor concentration would remain essentially constant until the last of the liquid evaporated.

A review of vapor pressure and flammability data for 15 common flammable liquids revealed two alcohols (ethanol and propanol) whose saturated vapor compositions at 25 °C and atmospheric pressure are above the LFL but below the stoichiometric concentration in air. Purging a vessel containing either of the liquids would, initially, worsen the flammability picture as the ullage composition (following a path such as shown in Figure 12) approached the C_{st} line. Fortunately, relatively little inerting would be required before the composition crossed the C_{st} line and became progressively richer as air was swept from the system.

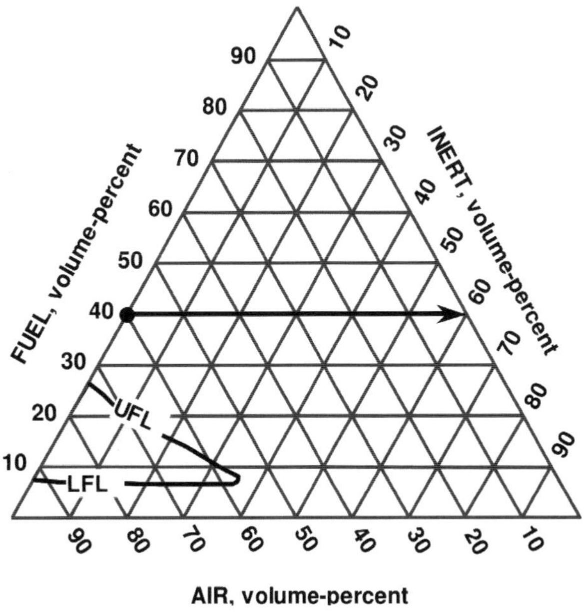

Figure 12: Effect of Volatile Liquid on Purging

Blanketing (or Padding). If a vessel is to be maintained partially or fully inerted, provisions must be made to compensate for air or oxygen that could enter the vessel as a consequence of:

- Backfilling the vessel as liquid is transferred out
- Daily cycling of ambient temperatures (day vs. night)
- Changes in atmospheric pressure
- Sudden changes in tank temperature due to rain storms
- Air infiltration due to minor leaks for systems operating under a partial vacuum
- Opening of vessels for sampling, inspection, or charging
- Air entrained or associated with solids or liquids charged to vessels
- Changes in the temperature of liquid flowing through the equipment (e.g., in a run tank)
- Operating errors and equipment failures

Other unique circumstances can defeat an inerting system; for example the *in situ* formation of O_2 from the decomposition of a peroxide in a process stream.

Part of the task of specifying and designing a blanketing supply system is the estimation of the demand put on the system by relevant challenges such as cited above, including consideration of credible simultaneous events.

The inert demand associated with the transfer of liquids from a system is easily quantified based upon either the peak anticipated transfer rate (for a maximum value) or the average throughput over a suitable period of time (for an average value), with one volume of inert required for each volume of liquid removed. An adequate supply of inert is not only required for oxidant exclusion, but also may be essential in preventing equipment damage if the equipment cannot withstand the vacuum created as liquid is removed. Consideration should be given to protecting equipment from such damage in the event of failure of the inert gas supply (e.g., through the use of vacuum breakers).

Diurnal temperature changes and the associated "breathing" of equipment must be considered for outdoor equipment. Blakey and Orlando (1984) provide a relationship for estimating the daily breathing loss upon warming, which must be made up for with inert upon cooling:

$$DBL = V_o\left[\left(\frac{T_s + T_{dc}/2}{T_s + \Delta - T_{dc}/2}\right) - 1.0\right]$$

where:

DBL = Daily breathing loss, standard ft^3
V_o = Average vapor volume (ullage) in vessel, ft^3
T_s = Average storage temperature, ^oF
T_{dc} = Average daily temperature change, ^oF
Δ = Adjustment for differential between the pressure setting on the purge supply regulator and the vessel pressure relief (e.g., backpressure valve), typically equivalent to 2 to 4 ^oF

Maps (Blakey and Orlando, 1984) are provided showing values of T_s and T_d for the United States. An example calculation for New Orleans showed that the monthly breathing loss for a hypothetical vessel was approximately equal to the ullage volume (i.e., indicating approximately one turnover of the vessel vapor space per month).[19]

Sudden, extreme changes in tank temperature resulting from weather changes such as thunderstorms can have a more pronounced effect on inert demands. Temperature changes will affect the density of the gases in the ullage and, more significantly, may cause condensation of vapors, resulting in marked pressure changes inside equipment. The American Petroleum Institute (API 2000, 1998) provides guidance for inbreathing requirements as a function of tank capacity. For tanks with capacities greater than 20,000 barrels (840,000 gallons), the inert requirements are approximately 2 scfh per ft^2 of total shell and roof area. For smaller tanks, the requirement is 1 scfh per barrel (42 gallons) of tank capacity. API notes that these inert supply requirements are substantially equivalent to a rate of temperature change in the tank ullage of 100 ^oF per hour. API recommends an engineering analysis to determine inert requirements for uninsulated tanks operating at temperatures in excess of 120 ^oF.

It should be noted that this guidance was developed for nonrefrigerated, aboveground tanks containing liquid petroleum or petroleum products suitable for storage in low pressure tanks operating at a pressure of less than 15 psig. As such, this guidance may not be suitable for all commodities. NFPA 69 reproduces this guidance in a nonmandatory appendix.

Inleakage for systems operating under a partial vacuum can be determined by testing. A vacuum, typical of normal operating conditions, is pulled on the equipment, which is then isolated from the vacuum source. By monitoring the rate of vacuum loss (pressure rise) and knowing the

[19] Note that this equation yields a volume, not a volumetric rate.

volume of the system and the temperature, an estimate can be made of the rate of air inleakage (a simple application of the ideal gas law).

Alternatively, as a first approximation, Gomez (1991) provides an empirical approach to estimating inleakage for "commercially tight systems." The following equation is based upon data from the Heat Exchange Institute:

$$MAL = A \; x \; V_s^{\,B}$$

where:

MAL = Maximum air leakage, lb/hr
A,B = Dimensionless coefficients, varying with P, see Table 8
P = Operating pressure, mm Hg absolute
V_s = Volume of system, ft^3

Air can be entrained in liquid transferred into an inerted system, particularly at the end of a transfer when the level in a feed tank or a drum is depleted and the pump starts to "gas out." An often overlooked source of air input comes with the charging of solids to vessels (e.g., paint mixers or resin reactors).

Table 8: Coefficients for Air Leakage Calculation

Pressure (mm Hg abs.)	A	B
90 – 760	0.1955	0.6630
21 – 89	0.1451	0.6617
3.1 – 20	0.1010	0.6579
1 – 3	0.05119	0.6568
<1	0.02521	0.6639

A void fraction of about 0.5 can typically be assumed for particulate solids. In other words, for every 1 ft^3 of "solids" charged to the vessel there are actually 0.5 ft^3 of particulates and 0.5 ft^3 of interstitial air. This can be a significant source of air into the system, particularly when products have a high solids content (i.e., little solvent, lots of solids). In addition, just opening the vessel to charge the solids can provide a pathway for air ingress into the system, particularly if the system is operating at a slight vacuum.

Numerous approaches have been developed for solving this problem. As an example, some facilities charge the solids to a hopper external to the mix vessel. The hopper is closed and purged with N_2 to displace the air from the solids bed. The solids are then charged to the vessel, perhaps through a rotary valve. Grossel and Zalosh (2004) and ESCIS (1994) provide additional solutions to this problem.

Ebadat (1994) provides an approach for calculating the increase in O_2 concentration in a continuously purged vessel to which multiple solids additions are made. A material balance on the ullage contents allows calculation of the increased O_2 concentration after the first solids addition (it is assumed that the vessel is not operating under a vacuum, which would draw additional air into the vessel):

$$C_1 = C_o\left(1 - \frac{V_{air}}{V}\right) + 21\left(\frac{V_{air}}{V}\right)$$

where:

C_1 = Concentration of air in the ullage after the solids addition
C_o = Initial concentration of air in the vessel
V = Volume of the ullage
V_{air} = Volume of added air, half of the bulk volume of solids added
21 = Concentration of O_2 in air

It is assumed that the time required for each solids addition is small in comparison to the intervals between additions. The decrease in O_2 concentration in the ullage between solids additions can be determined from the equation for continuous sweep purging:

$$C_2 = C_g + (C_1 - C_g)\, e^{\left(-Q_v t / V\right)}$$

where:

Q_v = Volumetric flow rate for the purge
t = Interval between end of solids addition and the start of the next
V = Volume of the ullage
C_1 = Oxygen concentration after solids addition (previous calculation)
C_g = Oxygen concentration in the purge gas stream (may be zero)
C_2 = Oxygen concentration after the purge and before the start of next solids addition

The value of C_2 can be inserted into the material balance equation in place of C_o, and the two equations can be used recursively to model the ullage O_2 concentration throughout the solids addition. If the model shows the O_2 concentration increasing to an undesired value, the interval between additions could be increased.

Analytical guidance is not available for quantifying all potential sources of air or other oxidant entry into inerted systems and, ultimately, engineering judgment must prevail. In the final analysis, there is no suitable substitute for a thorough process hazards analysis for evaluating the potential safety impact of equipment failure, operating error, and operational upsets.

<u>Conceptual Design Guidance for Blanketing Systems</u>. NFPA 69 suggests a number of approaches to controlling the flow of blanketing gas to a system and provides schematic diagrams for each approach (not reproduced here). Brief descriptions of the approaches follow.

If an ample supply of inert is available, cost is not an issue, and disposal of large quantities of potentially contaminated inert is not a problem, a continuous flow of inert to the system could be maintained. NFPA describes a system incorporating a restricting orifice in the inert supply line, with the orifice sized to meet the largest anticipated demand rate.

Variable-rate supply systems designed to respond to changes in demand should allow more affordable operation. NFPA suggests two approaches. The first addresses those situations where the liquid pumpout rate establishes the predominant demand for inert. To address this need, a solenoid valve in the purge supply line is sized to flow inert at a rate equal to the pumpout rate and is interlocked to the pump control circuit so that the solenoid valve is opened anytime the pump is energized. A smaller line, with a restricting orifice, is installed parallel to the solenoid valve in order to maintain a smaller continuous flow of inert through the system at all times.

The second variable-rate approach described by NFPA involves a flow (actually pressure) control valve, sized to meet the peak demand rate and modulated to control a certain pressure in the system. An alternative not described by NFPA would be to control inert flow to the system based upon measured O_2 concentration.

A blanketing system could be as simple as a pressure regulator in the supply line to the vessel and a conservation vent (or backpressure regulator) on the vessel vent. Relatively low pressures (on the order of 0.5 to 5 in. of water column) are required to maintain a satisfactory pad on a vessel. Pilot-operated regulators are often used to reliably achieve the simultaneous goals of high flow capacity and precise control of low discharge pressure. Giffin (1994) provides a short tutorial on basic considerations in the design of blanketing systems.

Grossel and Zalosh (2004) and ESCIS (1994) provide guidance on the design of padding systems for more complex equipment such as grinders, centrifuges, and dryers.

10. General Design Considerations for Inerting Systems

This section provides some general guidance on the design of inerting systems. Sources include FM Global (2000), Grossel and Zalosh (2004), and NFPA 69 (2002).

- Ensure the reliability of the inert gas supply, consistent with the severity of the consequences that could potentially result from failure of the supply. Consider cylinder or cryogenic liquid backup for onsite inert gas generators (i.e., membrane or PSA units). An alternative approach would be installing redundant generators. Evaluate the reliability of your pipeline supplier.

- Deal with reliable inert suppliers having effective quality control programs. The author is aware of one near incident when a truckload of liquid O_2, rather than liquid N_2, was delivered to a facility. Fortunately, the critical nature of the application required sampling and confirmation of the receipt prior to unloading, and the error was discovered. What assurances do you have of the purity of compressed gas receipts or of inerts provided by pipeline? Evaluate the potential variability in the purity of the inert supply, and monitor the O_2 concentration when appropriate (this is usually not a problem with cryogenic sources).

- Consider controls to ensure an adequately low dew point on inerts from combustion-based generators. Slope and drain or trap piping as appropriate.

- Carefully assess requirements for delivery rate and pressure at each consuming location, considering credible scenarios for coincident high demands in multiple locations. Low header pressure due to unanticipated demand could permit backflow from process vessels, resulting in contamination of the inert system. Provide backflow protection at points of use. A single check valve should not be considered reliable backflow protection. Do not increase the demand on an existing inert distribution system without first confirming that adequate flow rates and pressures can be maintained.

- Do not make <u>any</u> system changes without first evaluating the potential consequences. Apply management of change controls to the inerting system.

- Consider incompatibilities, both with process chemicals and materials of construction, when selecting inerts.

- Proper system design must include consideration of temperature and pressure ratings, especially with respect to cryogenic fluids. System design should be in accordance with recognized and generally accepted good engineering practice. Piping operating above 15 psig should be designed in accordance with ANSI/ASME B31.3, Process Piping (ANSI 2002).

- Properly clean the inert system after construction and maintenance activities. Consider the need for filters or screens to protect regulators, valves, and instrumentation from contamination with foreign material.

- Provide protection (such as backpressure control valves or excess flow valves) in branch headers to prevent failure of one line from depressurizing the entire system. Provide manual isolation valves on each major branch header.

- Do not permit cross-connections to other service systems. It is not unusual to see N_2 tied into the instrument air system as a clean, dry backup supply should the instrument air compressor fail. Such cross-connections provide the opportunity to either contaminate the purge N_2 system with air or to release N_2 into occupied areas where instrument air purges (that do not pose an asphyxiation hazard) are routinely made. CGA (2003d) cautions that instrument air can either accidentally or intentionally be used to supply breathing air or to power air tools in confined spaces.

- Provide adequate ventilation where purged equipment could leak inside occupied buildings. Air Products (1998a) recommends a minimum of six air changes per hour wherever liquid N_2 is used inside buildings. Remember that CO_2 and other dense inerts will accumulate in low spots. Consider installing O_2 monitors in high risk areas.

- Protect purged equipment against pressure extremes. For example, high pressure could result from failure of supply regulators. Low pressure could result if the inert gas supply is interrupted during liquid transfers out of vessels.

- Carefully consider the location of purge inlets and vents to ensure the desired distribution of inert within the equipment. Inlet velocities must be high enough to ensure proper mixing, but not excessive. Short-circuiting of inert gas flows between the inlet and the vent is a common cause of poor performance in vessel inerting. Properly consider the relative densities of the inert and the process gases when locating inlets and vents and in determining inlet flow rate requirements.

- Establish and adhere to inspection, testing, and preventive maintenance schedules for oxygen analyzers and other critical safety-related instrumentation to ensure attaining the required protective system reliability.

- Provide pressure, temperature, and flow instrumentation and alarms where warranted to allow troubleshooting system performance and to confirm required inerting flows.

- Where continuous measurement of O_2 concentration is not provided, establish and adhere to realistic schedules for periodic measurements to confirm proper inerting. The frequency of such checks should be consistent with the severity of potential consequences should the inerting fail.

- Discharge gases vented from equipment to a safe location and in compliance with environmental regulations. Consider asphyxiation hazards as well as the chemical, health, and flammability properties of the process constituents swept from equipment.

11. Evaluating Design Parameters for Vessel Inerting Supplies

As noted in Section 8, mixing efficiency is a critical issue in determining the duration of a purging operation and, as noted in the previous section, short-circuiting the inert flow between the inlet and the vessel vent is a common cause of inerting failure. For these reasons, it is important to have an understanding of the flow performance of the inlet inert stream, especially the depth of penetration into the vessel that can be expected. This is of special importance with respect to vessels having a large height-to-diameter ratio, such as storage silos for solids.

Turbulent gas jets have been well typified theoretically and experimentally and are described by Perry and Green (1999) and the American Society of Heating, Refrigeration and Air Conditioning Engineers (ASHRAE, 1989). The jets described are "free" jets, meaning they are discharging into open (unconstrained) environments or into enclosures that are very large relative to the dimension of the jet.

A gas jet exiting a nozzle expands as it entrains surrounding gas into its flow stream. The edges of the jet diverge at an angle on the order of 20° (relative to each other or 10° relative to the axis of the jet), as shown in Figure 13. The total momentum of the jet is conserved, and the velocity of the jet must decrease as it picks up mass due to entrainment. The centerline velocity of the jet can be estimated as (Perry and Green, 1999):

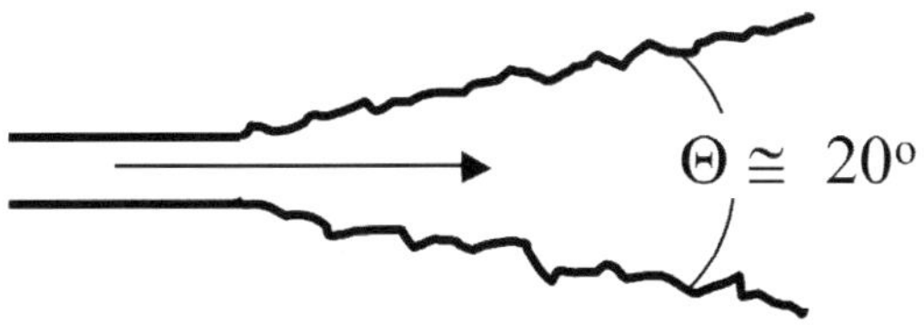

Figure 13: Expansion of Turbulent, Free Jet

$$\frac{v_c}{v_o} = K \frac{D_o}{x}$$

where:

v_c = Centerline velocity of jet, m/s[20]
v_o = Discharge velocity of jet, m/s
x = Distance along axis from point of jet discharge, m/s
D_0 = Diameter of the nozzle opening, m
K = 5 for $v_0 \approx 2.5$ to 5.0 m/s
= 6.2 for $v_0 \approx 10$ to 50 m/s

This relationship is valid for values of x/D_o ranging from 7 to 100. Other relationships are provided for the radial velocity distribution of the jet and the ratio of the total jet flow to the jet initial flow (which allows determination of the amount of entrained gas as a function of distance from the jet outlet).

According to the above equation, the jet will continue indefinitely, at an ever-decreasing velocity. In heating and ventilating parlance, the "throw" of a jet is defined as the axial distance from point of discharge at which the jet velocity drops below some arbitrary value. Perry and Green (1999) recommend a value of 0.3 m/s (1 ft/sec) to define the effective end of the jet. ASHRAE (1989) suggests the use of values ranging from 0.25 to 0.5 m/s (50 to 100 ft/min) for a terminal jet velocity that could not be distinguished from the velocity of random currents in "still air."

Jets discharging parallel and close to a wall will have a greater throw. Air entrainment rates for jets are a function of the area of the interface between the jet and the surrounding still gas. For a jet close to a wall the entrainment takes place over a surface having the shape of a half cone rather than a full cone, and less air is entrained at a given axial distance from the discharge. ASHRAE (1989) recommends that the values for K be multiplied by 1.4 (i.e., $2^{0.5}$) for jets in close proximity to, and flowing parallel to, a wall or similar surface.

The above discussion of free jets assumes that the jet and the surrounding gas have the same density. Lower density jets discharging horizontally into higher density surroundings will spread more rapidly, and higher density jets will spread less rapidly when discharged into lighter density surroundings (Perry and Green, 1999). This should affect the throw of the jets inversely (i.e., lighter jets will have a shorter throw in a denser surroundings and denser jets will have a greater throw in lighter surroundings). Horizontal jets are assumed above, so the gravitational effects of gas densities are not addressed either.

Unfortunately, purge jets often do not find a "free" environment in industrial applications (i.e., jets are discharged into enclosures [such as vessels] where the incoming gas must ultimately interact with the displaced gas that is moving in a counter flow direction, seeking an outlet). Perry and Green (1999) suggest that a jet can be considered a free jet when the cross-sectional area of the jet is less than one-fifth of the total cross-sectional area of the enclosure through which it is flowing (the area perpendicular to the axis of the jet). ASHRAE (1989) gives a more

---

[20] While Perry and Green (1999) are not specific, it is assumed that this velocity is measured at the exit plane of the nozzle at a pressure equal to the pressure of the surroundings and at a temperature equal to the bulk temperature of the gas. This is consistent with the BOM usage discussed below.

easily applied criterion, whereby a jet is assumed to be free if $x \leq 1.5(A_R)^{0.5}$, where A_R is the cross-sectional area of the enclosure perpendicular to the jet axis.[21]

Jets discharging into an enclosed environment have been studied by the BOM, with regard to applications for ventilation of tunnels in underground coal mines (BOM, 1943). Jets that are not free must contend with the flow resistance provided by the return air flow. At some point the jet stops diverging, leaving a return path between the jet and the enclosure walls, as shown in Figure 14. Jet momentum is lost to the pressure required to drive the return air past the incoming jet to the enclosure outlet. As a consequence, jet velocities decrease more rapidly and there is a point at which the jet is abruptly terminated. The BOM (1943) tests indicated that the jet did not extend beyond an axial distance of $4W_d$, where W_d is a characteristic dimension of the enclosure (such as the diameter of a vessel or the smaller of the height or width of a tunnel).

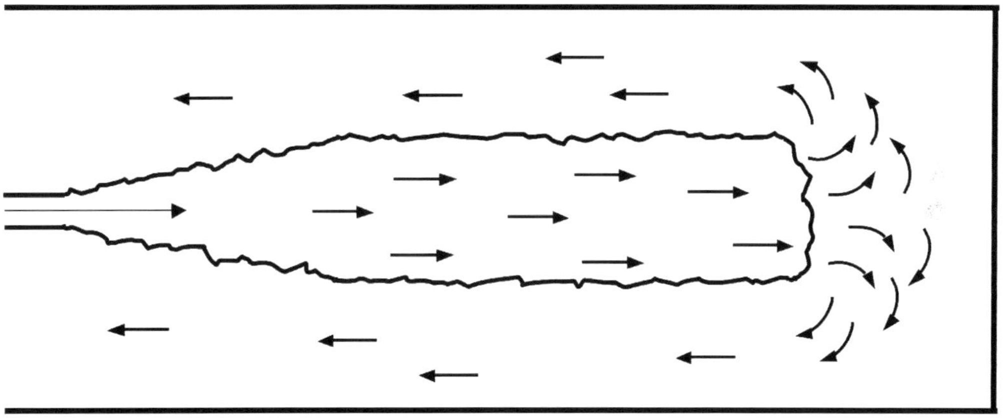

Figure 14: Flow Pattern of an Enclosed Jet

The BOM (1943) provides some very preliminary relationships (based upon testing of a limited number of equipment configurations in a tunnel that was 8 ft high and 8 ft wide) for determining the throw of a confined or constrained jet. The following equation calculates throw, assuming a 100 ft/min velocity for the end point of the jet:

$$X_{100} = W_d \left(\frac{\log(v_o) - 2 + \log(h) + \log(D_o) - \log(W_d)}{hk} \right)$$

where[22]:

- v_o = Discharge velocity of jet, ft/min
- X_{100} = Distance along the axis at which the jet velocity reaches 100 ft/min
- D_0 = Diameter of the nozzle opening, ft
- h = Constant, approximately 12 for the BOM test conditions

[21] This research work was performed for jets discharging parallel to the long axis of an enclosure. The concept of A_r (or W_d in the next paragraph) is not defined for jets entering the enclosure at an oblique angle or from the side.

[22] Note that the free jet equation from Perry and Green (1999) is in metric units and the confined jet equation from BOM (1943) is in engineering units.

k = Constant, approximately 0.027 for the BOM test conditions
W_d = Characteristic dimension for the enclosure, perpendicular to jet axis, ft

Figure 15 shows the calculated throw distance for a jet discharging into a vessel as a function of the vessel diameter. The values were calculated using the equation above, assuming an inlet diameter of 0.328 ft (0.1 m) and a jet discharge velocity of about 4900 ft/min (25 m/s). The volumetric flow rate for the jet discharge is approximately 420 cfm. The throw for a free jet, calculated using the equation from Perry and Green (1999), is approximately 100 ft (30.5 m), so the effect on the jet performance can be significant.

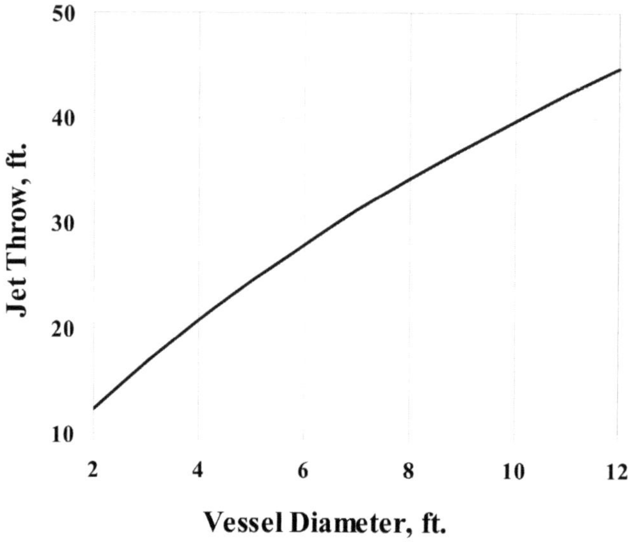

Figure 15: Estimated Throw for Confined Jet

The throw distance equation for the confined jet should be used with discretion because the underlying test work was limited to a narrow range of conditions. It should certainly not be used for values of W_d much in excess of 8 ft. In fact, at higher values the calculated throw distances start to decrease and eventually go negative.

Perhaps the greatest value for the equation is to help provide a qualitative appreciation of the role that the different variables provide in determining throw distance.

The free jet equation implies that the throw of the jet can be increased to any value desired by increasing the diameter and/or the velocity of the inlet gas stream. The BOM work helps demonstrate that, with real vessels, there is a limit to which we can increase throw. This is of particular importance with tall vessels with small diameters. Efficient mixing of the purge with the vessel contents cannot be expected if the purge cannot penetrate to the bottom of the vessel (or to the level of the liquid or solids contained in the vessel). Situations may arise where the vessel cannot be adequately purged from the top alone.

Acknowledgements

The author would like to acknowledge the assistance of Dr. Randy Freeman for his insightful technical review, Ms. Maureen Hafford, Ms. Leslie Adair, Ms. Tracy Johnson, and Ms. Julie Lowery for editorial assistance in making the manuscript comprehensible, and Ms. Angie Nicely for word processing support.

References

ABS (2004), American Bureau of Shipping, "Rules for Building and Classing Steel Vessels," Part 5, Chapter 1, Section 7, Paragraph 25, www.eagle.org/rules/downloads/svr2004/Part%205_1e.pdf

ANSI (2003), Z117.1 2003. *Safety Requirements for Confined Spaces*, American National Standards Institute, Washington, DC.

ANSI (2002), ANSI/ASME Standard B31.3, *Process Piping*.

Air Products (1998a), "Liquid Nitrogen," Safetygram 7.[23]

Air Products (1998b), "Gaseous Nitrogen," Safetygram 2.

Air Products (1998c), "Carbon Dioxide," Safetygram 18.

Air Products (1998d), "Dangers of Oxygen-Deficient Atmospheres," Safetygram 17.

Air Products, (1999), "Safe Handling of Cryogenic Liquids," Safetygram 16.

Air Products (2000), "Handling, Storage, and Use of Compressed Gas Cylinders," Safetygram 10.

Air Products (2003a), "The Hazards of Oxygen and Oxygen-Enriched Mixtures," Safetygram 33.

Air Products (2003b), "Product Migration of Liquefied Compressed Gases in Manifolded Systems," Safetygram 38.

Anon (2002), http://www.doesgodexist.org/JanFeb02/TrashingThePlanet.html[24]

API 2000 (1998), *Venting Atmospheric and Low-Pressure Storage Tanks*, Standard 2000, American Petroleum Institute.

ASHRAE (1989), *ASHRAE Handbook, Fundamentals*, I-P Edition, American Society of Heating, Refrigeration and Air Conditioning Engineers, Inc.

ATOFINA, et al. (2002), *Acrylate Esters, A Summary of Safety and Handling*, 3rd Edition, Intercompany Committee for the Safety and Handling of Acrylic Monomers, http://www.atofinachemicals.com/acrylics/literature.cfm

Blakey, P., and Orlando, G. (1984), "Using Inert Gases for Purging, Blanketing, and Transfer," *Chemical Engineering*, May 28.

Bodurtha, F.T. (1980), *Industrial Explosion Prevention and Protection*, McGraw-Hill.

---

[23] The Air Products publications cited here can be obtained at http://www.airproducts.com/Responsibility/EHS/ProductSafety/ProductSafetyInformation/Safetygrams.htm.

[24] I could not find my yellowed paper copy of the original version of this tongue-in-cheek article. Since this was attributed to ICI, I suspect that the author was Trevor Kletz.

BOM (1943), *Air Flow at Discharge of Fan-Pipes in Mines*, Report of Investigations R.I. 3730, U.S. Bureau of Mines.

BP (2003), *Hazards of Nitrogen and Catalyst Handling,* BP p.l.c.

Bretherick (1999), *Bretherick's Handbook of Reactive Chemical Hazards*, 6th edition, P.G. Urben, editor, Butterworth- Heinemann Ltd.

Britton, L.G. (1996), "Operating Atmospheric Vent Collection Headers Using Methane Gas Enrichment," *Process Safety Progress*, Volume 15(4).

CGA (2000), *Safe Handling of Compressed Gases in Containers,* Publication P-1, Compressed Gas Association.

CGA (2001a), *Purging Principles and Practices*, 3rd Edition, Compressed Gas Association.

CGA (2001b), *Oxygen-Deficient Atmospheres*, Publication SB-2, Compressed Gas Association.

CGA (2003a), *Precautions for Connecting Compressed Gas Containers to Systems,* Publication SB-10, Compressed Gas Association.

CGA (2003b), *Oxygen-Rich Atmospheres*, Publication P-39, Compressed Gas Association.

CGA (2003c), *Safe Installation and Operation of PSA and Membrane Oxygen and Nitrogen Generators*, Publication P-8.1, Compressed Gas Association.

CGA (2003d), *Safety of Instrument Air Systems Backed Up by Nitrogen*, Publication SB-28, Compressed Gas Association.

Coward, H.F., and Jones, J.W. (1952), "Limits of Flammability of Gases and Vapors," U.S. Bureau of Mines, Bulletin 503, NTIS AD-701 575.

Crowl, D.L. (2003), *Understanding Explosions*, Center for Chemical Process Safety.

CSB (1998), *Summary Report, Nitrogen Asphyxiation*, Report No. 98-05-I-LA, U.S. Chemical Safety and Hazard Investigation Board.

CSB (2003), "Hazards of Nitrogen Asphyxiation," Safety Bulletin No. 2003-10-B, U.S. Chemical Safety and Hazard Investigation Board.

CSB (2004), U.S. Chemical Safety and Hazard Investigation Board, incident log at http://www.csb.gov/circ/

Cunliff, B.J. (2001), "Avoiding Explosions by Means of Inerting Systems," IChemE Symposium Series No. 148.

Curtis, R. (1999), *Outdoor Action Guide to High Altitude: Acclimatization and Illnesses*, http://www.princeton.edu/%7Eoa/safety/altitude.html

De Graaf, J.D., et al., (2000), "The Bintulu ASU Incident," American Institute of Chemical Engineers, Spring National Meeting.

Dow Chemical, et al., (1999), *Ethylene Oxide User's Guide*, 2nd Edition, www.ethyleneoxide.com

Ebadat, V. (1994), "Preventing Explosion Hazards – Inert Gas Blanketing," *Solids Handling*, May/June.

Ebadat, V. (2003), "Is Your Dust Collection System an Explosion Hazard?" *Chemical Engineering Progress*, October.

ESCIS (1994), *Inerting, Methods and Measures for the Avoidance of Ignitable Substance-Air Mixtures in Chemical Production Equipment and Plants*, Expert Commission for Safety in the Swiss Chemical Industry, Booklet 3.

Ewing, C.T., et al. (1984), "The Extinguishing of Hydrocarbon Flames Based on the Heat-absorption Process Which Occur in Them," *Fire and Materials*, Vol. 8, No. 3.

FM Global (2000), Loss Prevention Data Sheet 7-59, "Inerting and Purging of Tanks, Process Vessels, and Equipment," Factory Mutual Insurance Company.

Giffin, G.L. (1994), "Keeping a Lid on Volatile Liquids," *Chemical Engineering*, February.

Gomez, J.V. (1991), "Calculate Air Leakage Values for Vacuum Systems," *Chemical Engineering*, June.

Grossel, S. and Zalosh, R. (2004), *Guidelines for Safe Handling of Powders and Bulk Solids*, Center for Chemical Process Safety.

Gullickson, R. (1996), "Reference Data Sheet on Inert Gases and Atmospheres," www.meridianeng.com/inertgas.html

Hardenburger, T. (1992), "Producing Nitrogen at the Point of Use," *Chemical Engineering*, October.

Hodson, L.L. (2001), "Safe Entry into Confined Spaces," *Handbook of Chemical Health and Safety,* American Chemical Society.

IFW (2004), Industrial Fire World Incident Log, http://www.fireworld.com/incidents/incidenttop.htm

ISA (1996), *Application of Safety Instrumented Systems for the Process Industries*, ISA Standard S84.01.

Kirby, D.C. (2004), personal communication.

McMullen, N., and Hojsak, M. (2003), "Reconsider Noncryogenic Systems for On-site Nitrogen Generation," *Chemical Engineering Progress*, September.

Meidl, J.H. (1970), *Explosive and Toxic Hazardous Materials*, Glencoe Press, Table 28, p. 293.

Michael, K.P. (1997), "Industrial Gas: Surveying Onsite Supply Option," *Chemical Engineering*, January.

NFPA 12 (2000), *Standard on Carbon Dioxide Extinguishing Systems*, National Fire Protection Association.

NFPA 30 (2003) *Flammable and Combustible Liquids Code*, National Fire Protection Association.

NFPA 53 (1999), *Recommended Practice on Materials, Equipment, and Systems Used in Oxygen-Enriched Atmospheres*, National Fire Protection Association.

NFPA 68 (2002), *Guide for Venting of Deflagrations*, National Fire Protection Association.

NFPA 69 (2002), *Standard on Explosion Prevention Systems,* National Fire Protection Association.

NFPA 70 (2000), *National Electrical Code*, National Fire Protection Association.

NFPA 77 (2002), *Recommended Practice on Static Electricity*, National Fire Protection Association.

NFPA 484 (2002), *Standard for Combustible Metals, Metal Powders, and Metal Dusts*, National Fire Protection Association.

NFPA 497 (1997), *Recommended Practice for the Classification of Flammable Liquids, Gases, or Vapors and of Hazardous (Classified) Locations for Electrical Installations in Chemical Process Areas*, National Fire Protection Association.

NFPA 499 (1997), *Recommended Practice for the Classification of Combustible Dusts and of Hazardous (Classified) Locations for Electrical Installations in Chemical Process Areas*, National Fire Protection Association.

NFPA 654 (2000), *Standard for the Prevention of Fire and Dust Explosions from the Manufacturing, Processing, and Handling of Combustible Particulate Solids*, National Fire Protection Association.

NIOSH (1976), *Criteria for a Recommended Standard — Occupational Exposure to Carbon Dioxide*, DHEW/NIOSH Pub. No. 76-194, National Institute for Occupational Safety and Health.

OSHA (2000), 29 CFR 1910.146, Confined Space Entry Standard, U.S. Occupational Safety and Health Administration.

OSHA (1998), 29 CFR 1910.134, Respiratory Protection Standard, U.S. Occupational Safety and Health Administration.

Perry, R.H. and Green, D.W. (1999), *Perry's Chemical Engineers' Handbook*, CD-ROM edition, McGraw-Hill

Pilkington, G. (1999), *Guidelines for Inert Gas Blanketing. Abbott Laboratories*, North Chicago, IL.

Pohanish, R.P., and Greene, S.A. (2003), *Wiley Guide to Chemical Incompatibilities*, 2nd Edition, John Wiley and Sons, Inc.

SAFTENG (2004), http://www.safteng.net/newsletter_alerts_subscription_page.htm

SOLAS (1974), Chapter II-2, Regulation 62, "Inert Gas Systems," International Convention for the Safety of Life at Sea (SOLAS).

USCG (1998), Title 46, CFR 32.53, "Inert Gas Systems," U.S. Coast Guard.

USCG (1999), Title 33, *Code of Federal Regulations*, Part 154, "Facilities Transferring Oil Or Hazardous Material In Bulk," U.S. Coast Guard.

USCG (2000), *Marine Safety Manual*, Vol. II, Section C, Chapter 5, "Inspection of Inert Gas Systems."

Ward, T.J. (1994), "Select the Best Nitrogen Supply," *Chemical Engineering Progress*, July.

Zeeuwen, P., et al. (1996), "Partial Inerting of Flammable Dust-Air Mixtures," 7th International Colloquium on Dust Explosions, Norway.

Additional Information Sources Not Referenced in This Paper:

Schofield, C., and Abbott, J.A., *Guide to Dust Explosion Prevention and Protection, Part 2 – Ignition Prevention, Containment, Inerting, Suppression, and Isolation*, Institute of Chemical Engineers, VCH Publishing, 1989.

VDI 2263 Part 2, *Dust Fires and Dust Explosions: Hazards – Assessment – Protective Measures,* 1992.

AUTHOR INDEX

AUTHOR INDEX

American Institute of Chemical Engineers
3 Park Avenue
New York, New York 10016-5991

ISBN: 978-1-60423-457-2

American Institute of Chemical Engineers

38th Loss Prevention Symposium 2004

Held at the 2004 AIChE Spring National Meeting

April 25-29, 2004
New Orleans, Louisiana, USA

Volume 2 of 2

Printed from e-media with permission by:

Curran Associates, Inc.
57 Morehouse Lane
Red Hook, NY 12571
www.proceedings.com

ISBN: 978-1-60423-457-2

American Institute of Chemical Engineers

38th Loss Prevention Symposium
2004

TABLE OF CONTENTS

Volume 1

SESSION LPS-7 CASE HISTORIES AND LESSONS LEARNED

Session Chair: John F. Murphy Co-Chair: Henry L. Febo

SIS Growing Pains:

How and why to make the switch from an internal to an industry standard

This paper was prepared for presentation at the AIChE 2004 Spring National Meeting 38th Annual Loss Prevention Symposium April 25-29, 2004 in New Orleans

SIS Growing Pains:
How and why to make the switch from an internal to an industry standard

J. Wayne Chastain and Kenneth Yount
Eastman Chemical Company

Abstract

Like many companies, Eastman Chemical Company is faced with a decision of moving from internal standards governing the design of safety instrumented systems (SIS) to one of the competing industry standards. This paper documents the thought process used by Eastman personnel in making the decision to move from an established internal standard for the specification and maintenance of SIS to an industry standard. Many of the issues which need to be considered in making such a move are documented including costs which should be considered, processes which may be impacted, corporate roles which need to be addressed, and options to be considered. Other organizations which are currently considering making such a change, or adopting an industry standard can hopefully draw from Eastman's experience in working through this issue.

Safety Instrumented Systems (SIS) have been in use in the process industries for many years. A Safety Instrumented System is defined as "System composed of sensors, logic solvers, and final control elements for the purpose of taking the process to a safe state when predetermined conditions are violated."[1]. Many other terms have been used for these devices in the past by different companies, industry groups, and governmental authorities including Emergency Shutdown Systems (ESD), Safety Interlocks, Safety Interlock System, and Safety Shutdown Systems (SSD).

In 1996 the Instrument Society of America released ANSI/ISA-S84.01 entitled "Application of Safety Instrumented Systems for the Process Industries." This was the first consensus standard issued on the design and use of safety instrumented systems for the general process industry. Response to this standard varied across the chemical process industry (CPI.) Some companies quickly moved to adopt the standard directly, likely finding that the methodology proposed in the standard did not vary tremendously from internal standards. Other companies are likely still not aware that this consensus standard exists.

Although the standard was issued in 1996, many companies' attention to this issue changed after March 23rd, 2000 when OSHA issued a letter of interpretation[2], stating, "As S84.01 is a national consensus standard, OSHA considers it to be a recognized and generally accepted good engineering practice for SIS." Various individuals and organizations in the industry have interpreted this letter by OSHA as an indication that in order to be compliant with the Process Safety Management regulation manufacturers would have to adopt the ANSI/ISA-S84.01 standard.

At the same time that the ISA was working on the SP84.01 standard, the International Electrochemical Committee was working on several standards addressing the design and use of SIS. The first of these was IEC 61508 "Functional safety of electrical/electronic/programmable electronic safety related systems", which was a general standard applicable to not only the CPI but any manufacturer using SIS in their plants[3]. The IEC followed this standard with the release of IEC 61511 "Functional Safety: Safety Instrumented Systems for the process industry sector"[4].

Indications are that the ISA Committee will revise the language in S84.01 to be identical to that found in IEC 61511 with few exceptions. Thus, it is very likely in the near future that one consensus industry standard for the design and use of SIS will exist. For those companies which are facing the decision of adopting this standard or some other standard or staying with their current internal standard, several issues must be addressed. Eastman Chemical Company has recently gone through this thought process of whether the company should stay with an internal standard, or adopt one of the consensus standards. The process used to make that decision was long and difficult and involved not only rationalizing what we hope is the correct answer, but also convincing all of the other players in the company who had a hand in making the decision that this was the right one for the company.

What did we have?

Tennessee Eastman Division of Eastman Chemical Company put in place an internal standard for the design and implementation of SIS in 1989. Unlike the current consensus standards, the Eastman standard was an architecturally based system. Four levels of SIS were specified in the Eastman standard with each level dictating a level of redundancy, varying requirements for use of "tried and true" technology, testing frequency, etc. Guidelines were present in the standard for when the various levels of SIS should be used, based on the consequence being protected against.

This standard was adopted to differing degrees by other Eastman Chemical Company sites, with some sites using only "good engineering practice" and a subjective selection and design criteria for SIS. Over the last several years Eastman Chemical Company has gone through a period of expansion through acquisition. Many of the acquired sites brought different standards for the design and implementation of SIS while some of the new sites had few SIS and fewer standards for their design and use.

Other companies may find themselves in a similar situation as they consider a switch to a consensus standard. Instead of one internal standard that may be in competition with the consensus standards, they may have several internal standards in use in different areas of the company. Certain divisions of the company may not have any standard that is being followed. This situation can create additional problems in obtaining buy in from the disparate groups who have very different backgrounds and levels of exposure to the ideas of risk management and SIS. It can also increase the resulting educational burden if the company chooses to standardize on a consensus standard or one of its own internal standards, due to the differences in background in the various corporate groups.

As a team within Eastman considered the key issues in standardizing the company on the internal standard that had been developed for the Tennessee site or adopting one of the consensus standards, the differences between the two had to be considered. The key differences between the internal standard that Eastman followed and the consensus standards which have been developed are:

- Basis (architectural vs. risk)
- Selection (consequence vs. required risk reduction)
- Documentation (little specified documentation vs. detailed safety requirements specification)
- SIS Reliability (architectural design vs. performance design)
- Verification (none vs. verification calculation)

Although other companies may have different types of internal standards, or no standard at all, the process of determining if adoption of the consensus standard makes sense, should address many of the same issues.

Why change?

Regulatory Issues

First, if a company has an internal standard, or several competing internal standards, or for that matter, no standard at all, why should that company choose to adopt one of these consensus standards for SIS design? One issue that will be raised by many is that after OSHAs letter of interpretation on the issue, if a company operates processes which are covered by the PSM standard, then the company must adopt S84.01 or it will not be in compliance with the PSM regulation. Each individual company will, of course, have to make an appropriate determination as to whether or not this is true, given the language of OSHA's letter. Eastman made the determination that this was not the case. OSHA never indicates that a company must adopt S84 in order to be in compliance, but simply indicates that as a consensus standard, S84 meets the criteria of following good engineering practice, as required by the standard. Indeed, OSHA specifies in this letter of interpretation that "With respect to SIS, OSHA does not specify or benchmark S84.01 as the only recognized and generally accepted good engineering practice." So, compliance with the PSM standard does not seem to be an adequate reason to adopt S84.01, if a company is following another standard which could be considered to meet the requirements of good engineering practice.

However, a company shouldn't ignore the implications of OSHA's letter of interpretation. OSHA has clearly indicated that compliance with S84.01 will meet the requirements of good engineering practice and the other requirements of PSM with regard to SIS with the following statement from the same letter, "If an employer documents per 1910.119(d)(3)(i)(F) that it will comply with S84.01 for SIS and it meets all S84.01 and other OSHA PSM requirements related to SIS, the employer will be considered in compliance with OSHA PSM requirements for SIS." Although following S84.01 is not required in order to be compliant with the PSM standard, by the language in OSHA's letter, following it does meet the requirements of good engineering practice. Without a letter to the effect that a company's internal standard also meets these requirements, it will be unclear if compliance has been achieved.

Although compliance with S84.01 or any other SIS standard is not a requirement of the PSM regulation in the United States, for those companies with operations overseas, particularly in the European Union, compliance with IEC 61511 may not be optional. There is an indication that governmental authorities in the EU have or will adopt IEC 61511 as a required standard for companies to follow. If this does occur, then at least in European operations, a company will have to make use of this consensus standard. If a company is required to adopt such a standard for a portion of its plants, it may be advantageous to adopt the same standard across the board for consistency.

Cost Savings

Another reason to switch to a consensus standard is the potential for cost savings. However, in conducting a cost benefit analysis of switching to a consensus standard, it is necessary to account for all of the factors which need to be taken into account, both favorable and unfavorable.

First, evaluating the favorable aspects of switching to one of the aforementioned consensus standards:

- Lowered costs associated with execution of contracts with engineering firms.

  One cost savings aspect of adopting a consensus standard is that engineering firms which are executing design work for your company will already be familiar with the details of the processes specified in these standards. This should result in a decrease in the required number of hours for the execution of the job due to the contract personnel already being high on the learning curve of the standard, as opposed to having to learn a company standard for SIS.

  Note that this argument is not as effective if your company always uses the same engineering firm who is already familiar with your internal SIS standards.

- Lowered cost due to elimination of over-design.

  The risk based methodology which can be used for the selection of the appropriate level of SIS for a particular event can result in a savings, if there is an increase in the precision of selection. Although a variety of methodologies can be used to select the Safety Integrity Level for SIS, the more sophisticated risk based methods such as Layers of Protection Analysis and Quantitative Risk Assessment should give more precise answers than simpler methods which might be used. In general, where less sophisticated methodologies are used, there should be a tendency toward over-design in order to ensure that processes are adequately protected. Increasing the precision of the selection, allows systems to be designed which adequately protect against the risks, but contain less error and are therefore less expensive.

Although these reasons are sound economic drivers for making the switch to one of the consensus standards, there are a variety of costs associated with making such a change that should also be accounted for:

- Development and deployment of new internal standards.

  In order to implement a consensus standard in a company with its own internal standards or with no standards, one requirement will be determining what processes of the company need to change and how those changes need to be implemented. Since this effort will generally be cross functional as well as impacting all of the operating areas of a company, in most cases a team will be required to make this evaluation. One requirement must be that the standard picked for implementation should be evaluated by an appropriate team to ensure that there are no changes to the standard that are recommended for adoption by the company. This team evaluation process can consume a significant number of man hours and is not without cost in and of itself.

- Alteration of internal processes.

  Once the processes that must be changed have been identified, a plan must be developed as to how these processes need to change in order to meet the requirements of the standard. Developing these required changes and implementing them will require time. Training employees on how the change of the systems will impact them and their work also may require a significant amount of work.

Both of these costs are one time costs associated with the development of the processes of implementing a consensus standard within a company. Although these do need to be accounted for, they are not as significant as potential on-going costs associated with making such a switch, such as:

- Costs of the design process.

  Although the precision of the selection techniques which may be used can result in lower installed costs for SIS using a consensus standard, both of the standards addressed in this article require a significant amount of engineering labor in order to meet the requirements for the design and documentation of the SIS. In many cases the requirements of the standards may not be any more severe than internal standards already being used by a company, but this potential cost needs to be evaluated and accounted for in the decision, if there are significant differences.

- Costs of maintenance processes.

  Because the consensus standards use a life-cycle approach to SIS, the costs which may be added by switching to a consensus standard do not end with the design and installation. There are testing, management of change, and decommissioning requirements which must be met in order to maintain the required integrity of the systems. These processes all have costs associated with them that a complete accounting needs to consider.

Depending on the how closely existing internal standards match the safety life cycle of the consensus standards, companies will have varying amounts of cost in switching to a consensus standard. For most, there will be some additional cost incurred, although this cost may be counter-balanced by the savings that can be found in other areas of the process. One factor which should be considered, however, is if a company finds that their systems are well below the requirements of S84.01 and switching to this standard or another consensus standard will result in a significant expenditure of funds, perhaps their current standards are not adequate for the task of ensuring the safe design of SIS.

Recent evaluations at Eastman Chemical Company have shown that the costs of switching to a consensus standard for several pilot projects have resulted in a slight savings. These evaluations only account for the differences in cost for engineering, design, and installation portions of the project. The ongoing maintenance costs, which may be higher under the consensus standard, have not been accounted for. However, these results are encouraging. Even if the maintenance costs result in a slightly higher life cycle cost for the SIS under the consensus standard, the risk is better understood and controlled under the new standard.

Which standard?

After evaluating the costs and benefits of switching to a consensus standard, the committee charged with determining a path forward on this issue for Eastman Chemical Company determined that switching was the desired option. Before presenting this as a proposal to management a decision had to be made as to which consensus standard would be followed. The two competing standards, S84.01 and IEC 61511, are mentioned above, as is the indication that it is very likely that these two standards will be harmonized in the near future. In order to be compliant with the future S84.01, it would seem reasonable to adopt IEC 61511. However, it is important to understand the differences between the two standards in making the determination of which one to adopt.

The standards are very similar in their implementation. Some of the few significant differences are:

- IEC 61511 is, in general, a broader and more prescriptive standard

  IEC 61511 is more prescriptive in its approach which can be seen through out the document when comparing it to the ISA S84.01 standard. As an example, ISA S84.01 specifies that a Process Hazards Analysis needs to be conducted but indicates that the details of such a process are outside of the bounds of the standard. IEC 61511, on the other hand, gives seven specific criteria which must be met by a "hazard and risk analysis" in order to be compliant with the standard. Note, that most if not all of these criteria would be met by a company covered by the PSM regulation in its PHA process due to the PSM requirements. But the IEC standard is more prescriptive than the PSM standard and might dictate a change to a company's PHA process and other procedures due to its prescriptive nature.

- System architectural requirements for SIS under IEC 61511

  The architectural requirements of IEC 61511 come in the form of "Requirements for hardware fault tolerance." Required fault tolerance is given in terms of the SIL of the SIS, the safe failure fraction of the sub-system, and the simplicity of the devices specifically being concerned with the presence of microprocessors or other "smart" devices as shown in Table 2. ISA S84.01 does not specify similar requirements for SIS. The inclusion of this requirement in IEC 61511 can increase the required redundancy above that otherwise required by the SIL.

- Additional documentation requirements under IEC 61511

  Continuing its tendency to be more prescriptive, IEC 61511 is much more detailed about the required elements of its safety requirements specification, requiring over 20 elements and a separate software safety requirements specification. The ISA S84.01 specifies 12 elements to be included. This is not to imply that the elements required by IEC 61511 will not be needed in the specification, but the language of ISA S84.01 leaves a great deal of additional latitude to the company to develop their own guidelines for SRS within the bounds set in the standard.

- Specific allowance for reliability of Basic Process Control System reliability in IEC 61511

  For layers of protection for an event which do not meet the requirements of an SIS as defined in the standard, but are implemented through the BPCS for the process in question, IEC 61511 specifies that a risk reduction factor of 10 is the maximum allowable value. ISA S84.01 is silent on this issue. While the value of 10 for an RRF is reasonable in the absence of other data or any additional study, this value may be too conservative in some cases and may dictate the use of a higher SIL SIS than would be required if a more in-depth analysis of the BPCS layers of protection had been conducted.

Based on the differences between the two consensus standards, a company has to determine which standard to use. Eastman Chemical Company decided to base our new SIS program on the current incarnation of S84.01 (1996). We did not feel that there was enough additional value in IEC 61511 to justify using it over the current edition of S84.01. But each company has to make this decision based on the merits of the two standards.

Who will the change effect?

If a decision is made to change a company's standard from an internal to one of the consensus standards, management will be the group that ultimately makes this decision. Above we have discussed the issues which must be accounted for in making a decision on this issue. However, before presenting a change that could be a large one for an organization to management, it is also important to understand the various groups and organizations that will be effected by this change. If possible buy in from each of the effected groups should be sought prior to requesting approval from management. Not only can this aid in obtaining approval for the change from management, but can ease the process of implementation.

Because of the life-cycle approach taken in each of the standards, a large number of groups can be affected to varying degrees.

- Engineering

  Engineering work processes with regard to SIS can be impacted more than any other group within an organization. A significant level of education is required in a company's engineering organization in order to allow effective usage of the consensus standard. This education will need to include expertise in conducting SIL calculations, development of a database to store SIL calculations and documentation, indications on how to deal with existing systems and grandfathering, development or acquisition of calculation tools, and ensuring that failure rate data needs have been met to support the required calculations. Training on appropriate documentation in Safety Requirements Specification will be necessary and will help to ensure appropriate communication of requirements to other engineers conducting detailed design of SIS. Note only does this effective communication ensure that systems are designed and installed as intended, but can reduce the cost of engineering projects.

- Safety

  Several processes which are typically owned by a company's safety organization will be affected by the change to a consensus standard. A company's Process Hazard Analysis procedure will need to be updated to include the steps required for identification of needed SIS. Since the risk management function is often maintained within a company's safety organization, the safety function will likely own the LOPA process for the organization or whatever process is used for the determination of required SIL for a given scenario. Implementation of this technology in the safety organization may require training of those personnel in LOPA or other risk evaluation tools that will be used to determine required SILs. The use of these risk evaluation tools to determine the required SILs may result in a significantly increased workload for this group. If this function will be maintained in some other organization, then these issues will impact that group instead.

- Maintenance

  Maintenance may be significantly impacted by the change to one of the consensus standards. Testing procedures are required for all SIS under the standards and there must a system for keeping track of the testing requirements and scheduling. After testing is conducted, results need to be documented and reported. Depending on the level of maintenance that is conducted under existing standards, this may be a significant increase in the required work load of the maintenance organization with regard to SIS. On the other hand, if a company already has in place a well organized preventative maintenance process, then only minimal changes may be required to the maintenance organizations work practices.

- Operations

  Operations needs to be aware of various factors that will impact them in the implementation of the one of the consensus standards. The role of the operations representative in any capital project will be impacted due to their need to understand the standard so they can participate in the appropriate selection of Safety Integrity Levels. After installation of a SIS under one of the consensus standards, operations may have additional labor needs to meet the functional testing requirements if the testing requirements in the consensus standard are more intensive than the testing requirements in the existing company standard. Another aspect which can impact operations from implementation of a consensus standard is that the requirements of the testing frequency can constrain run times for the plant.

- Quality Assurance

  The work processes of the organizations quality assurance organization may be impacted depending on how the consensus standard is implemented. If this organization is responsible for quality assurance on instruments or devices which are used as part of SIS then they will have to take into account the requirements of the standard in their work processes.

- Legal

  Several issues will have to be tackled by the company's legal group. First of all, the legal group should be responsible for making a recommendation to management on the risk to the company of not having consistent standards across a corporation. If some but not all of a company's sites are going to be transitioning to a consensus standard, there may be additional liability for those sites, and therefore for the corporation, which are operating under an old company standard or no standard at all. The legal group should also address any grandfathering issues as most companies will continue to have operating areas which have SIS designed under old corporate standards or which were not based on a standard but on the "good engineering practice" at the time of the installation.

Conclusion

Eastman Chemical Company has wrestled with the issues that need to be addressed in making the decision to adopt a consensus standard for the design, implementation, and operation of safety instrumented systems. This process of decision making involves issues of cost and benefit that include many intangibles that must be taken into account. This paper has attempted to document these issues so that other companies dealing with this decision might get a head start based on Eastman's findings as we have gone through this process. Eastman has arrived at the decision that an internal standard based on ISA

S84.01 (1996) meets the needs of our company. Other companies, depending on their independent evaluations, may very well arrive at different answers to this question.

Eastman Chemical Company is now in the early stages of implementation of this standard. New wrinkles may be uncovered and additional learnings made as we go through the process of integrating this standard into our work processes. If so, this paper may be the first in a series that can hopefully assist others as they attempt to make the decisions that will allow their own companies to better manage their risks.

References

1. ANSI/ISA-84.01-1996 - Application of Safety Instrumented Systems for the Process Industries. Research Triangle Park, NC: ISA, 1996.
2. U.S. Department of Labor Occupational Safety and Health Administration. OSHA Standards Interpretation and Compliance Letters, 3/23/2003 – "Compliance with PSM and ANSI/ISA-S84.01 for safety instrumented systems."
3. International Electrochemical Commission. "Functional Safety of Eletrical/Electronic/Programmable Electronic Safety-Related Systems," IEC 61508. Geneva: IEC, International Electrochemical Commission.
4. International Electrochemical Commission. IEC draft standard 61511, Part 1, "Functional Safety: Safety Instrumented Systems for the process industry sector," IEC 61511. Geneva: IEC, 1999.

Fault Tree and Layer of Protection Hybrid Risk Analysis

Marc Rothschild, P.E.
Rohm and Haas Company
Engineering Division
3100 State Road
Croydon, PA 19021
(215) 785-7327
mrothschild@rohmhaas.com

ABSTRACT

Layer of Protection Analysis (LOPA) is a relatively quick and straightforward method for quantifying risk. However, LOPA may be inadequate if used for compound failures when the required failure rate data is not available or when the failures are not independent. Fault Tree Analysis (FTA) can be used in these situations. FTA is designed to thoroughly and accurately evaluate compound failures and account for any dependencies between failures. If necessary, FTA can augment LOPA, combining the best qualities of both methods into one powerful hybrid tool for risk analysis.

1. INTRODUCTION

Layer of Protection Analysis (LOPA) is a simple and effective method for quantifying PHA's (such as HAZOP's and What-If's) by linking causes of failure with their safeguards. This quantification enables the analyst to prioritize the recommendations made by the PHA team. Furthermore, by associating cost with the various consequence levels, a LOPA can be used to perform a cost-benefit analysis. Many papers have been written on applications of the LOPA technique; examples are given in the references [1-5].

LOPA is a natural step following a HAZOP or What-If. These PHA studies identify the potential initiating causes and the safeguards of each identified consequence. A LOPA then evaluates the frequency of each consequence by quantifying the expected failure rate associated with the initiating cause and the probability of failure upon demand (PFD) of the safeguards. Also included in the LOPA equation are estimates of the conditional likelihood of exposure (presence factor) and the vulnerability of the exposed person to the consequence (likelihood of fatality given exposure). For flammable consequences, an additional term is used to estimate the conditional likelihood of ignition of the release. The LOPA equation for a given scenario takes the following form:

$$\text{Frequency}_{\text{consequence}} = \text{Frequency}_{\text{cause}} \times \text{PFD}_{\text{safeguard}\mathbf{1}} \times \text{PFD}_{\text{safeguard}\mathbf{2}} \times \ldots \times \text{PFD}_{\text{safeguard}\mathbf{N}} \times \text{Probability}_{\text{ignition}} \times \text{Probability}_{\text{presence factor}} \times \text{Probability}_{\text{personal vulnerability}}$$

While the uncertainty in the failure rate data precludes accurate analysis, LOPA is a powerful tool for prioritizing scenarios when order-of-magnitude accuracy is sufficient. Due to its ease and efficacy, LOPA is often the first approach taken in quantifying risk. However, one must be cautious when applying LOPA, because it has limitations.

The failure rate data required for a LOPA are generally only available for component failure and human error, but many failures are compound events that consist of combinations of these basic failures. The failure rates of these compound systems cannot be directly derived from the basic failure rate data. Another limitation is that the safeguards must be independent of each other and of the initiating event, and that the initiating events must be independent of each other. Otherwise, the LOPA will underestimate the consequence frequency. Interdependence may be apparent, such as failure of a shared component. At other times, the shared dependence may be less obvious, and common cause failure could result from conditions such as loss of instrument air, a process upset, or even from flawed maintenance practices.

Fault Tree Analysis (FTA) can be used when the above limitations restrict the applicability of a LOPA. Unlike LOPA, FTA can evaluate interdependent and compound failure events. FTA provides an additional benefit by producing a failure map that assists the analyst in identifying the strengths and weaknesses of the entire system.

While FTA offers many benefits over LOPA, FTA is less intuitive than LOPA and FTA requires specialized training and software. Therefore, it is desirable to apply both of these analytical techniques, combining the power of the fault tree with the simplicity of the LOPA.

Section 2 introduces a simple example and demonstrates the limitations of the LOPA approach. Section 3 introduces FTA and shows how fault trees can overcome those limitations. Finally, Section 4 shows how fault tree and LOPA methodologies can be combined into a hybrid analytical tool.

2. ILLUSTRATION OF LOPA LIMITS

Figure 1 presents a simple process flow diagram. This process involves unloading a toxic chemical from a tank truck into a storage tank. The operator monitors the level gauge (LG) and calls for a delivery when the tank level is low enough to sufficiently store the transfer. It is very important not to overfill the tank, since it would result in a toxic spill from the overfill line. To prevent overfill, there is an independent high-level transmitter (LT). Signals from both level instruments are sent to a logic solver. If either of these level instruments gives a high reading, then a signal is sent from the logic solver to an isolation valve, shutting off the charge to the tank.

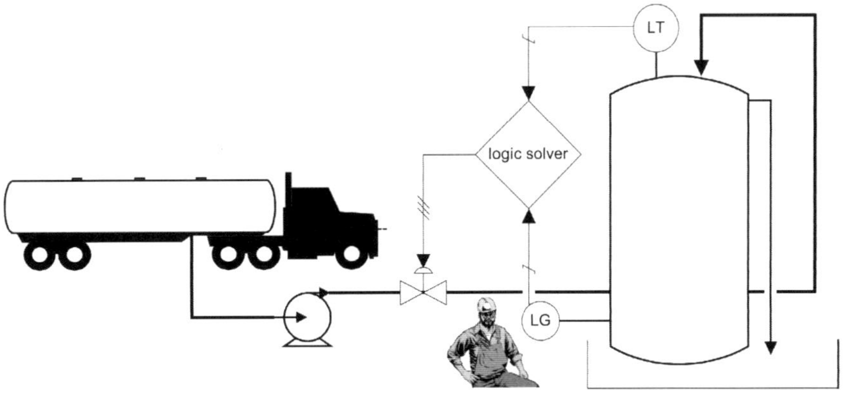

Figure 1 - Sample Problem Schematic

The challenge is to determine the frequency of overfilling the tank:

$$\text{Frequency}_{\text{overfill}} = \text{Frequency}_{\text{cause}} \times \text{PFD}_{\text{safeguards}}$$

The frequency of the initiating cause is the frequency at which the tank is filled. There are two safeguards against overfill. The primary safeguard is the tank operation. The operator is trained and instructed to only place an order when the tank is low (as indicated by the level gauge). The backup safeguard is the high-level control that shuts off the feed before the tank can be overfilled.

Expanding the above equation:

$$\text{Frequency}_{\text{overfill}} = \text{Frequency}_{\text{truck delivery}} \times [\text{PFD}_{\text{tank operation}} \times \text{PFD}_{\text{high level shutoff}}]$$

Although this equation is not complex, it nevertheless cannot be solved directly using LOPA. Both of the safeguards are compound failure events and therefore are unlikely to be found in failure rate databases. The PFD for tank operation is straightforward, and an experienced analyst would recognize that this compound PFD simply consists of the sum of the PFD of a false low reading from the level gauge plus the PFD of the operator ordering a truck transfer despite the level being too high. On the other hand, the PFD for the high-level protection is more complex and cannot be intuitively solved. Furthermore, there is interdependence between the two PFDs since they both rely on the level gauge. Ignoring this commonality would result in under-predicting the failure rate because the failure of the level gauge can undermine both safeguards.

3. FAULT TREE ANALYSIS

This section includes a *brief* overview of FTA. It is not possible to provide a comprehensive description in the space allotted. Fortunately, for those unfamiliar with this technique, there are texts available on the subject [6-9].

A fault tree is a logic diagram showing how systems can fail. The first step is to define the issue of concern, which is referred to as the "Top Event." The next step is to identify the immediate, necessary and sufficient causes for the Top Event. This process is repeated until the causes are dissected all the way to basic events.

Figure 2 presents a fault tree for the example given in Section 2 and Figure 3 gives a key to the fault tree symbols. This fault tree shows that the tank can be overfilled if the tank truck arrives to fill the tank AND both safeguards fail. The primary safeguard is the tank operation, which fails if the level gauge reads lower than the actual tank level OR if the operator neglects the level reading and accepts the transfer. The secondary safeguard - the high level shutoff - fails if BOTH the level gauge and the level transmitter measure lower than actual tank level OR if the logic solver fails OR if the shutoff valve fails to close upon demand. Note that the failure of the level gauge is common to both safeguards, thus making them interdependent.

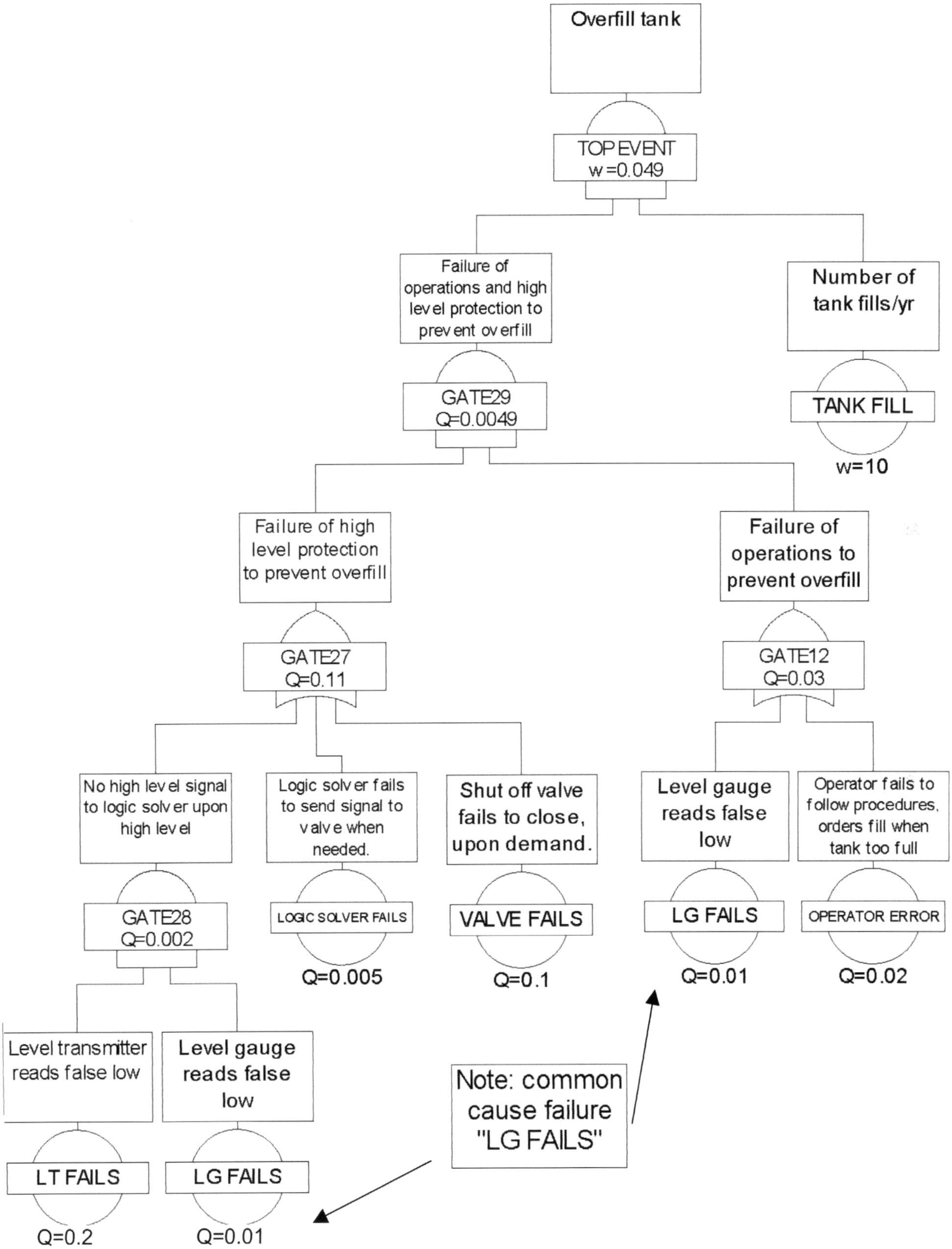

Figure 2 - Fault Tree for Sample Problem

Symbol	Description
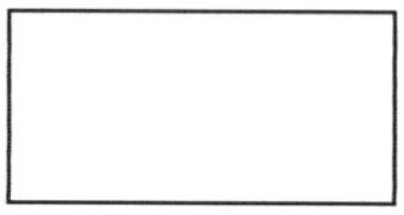	INTERMEDIATE EVENT - a fault that occurs because one or more antecedent causes occur through logic gates.
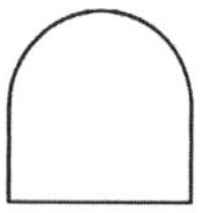	AND GATE – output occurs if *all* input faults occur.
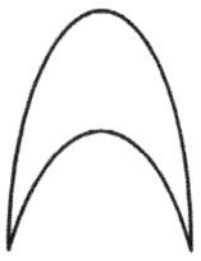	OR GATE – output occurs if *any* input faults occur.
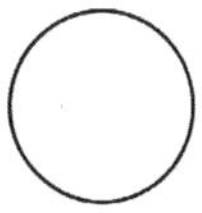	BASIC EVENT – an initiating fault requiring no further development since the appropriate resolution has been reached.

Figure 3 - Fault Tree Symbols

The following failure rate data[1] is applied to the fault tree to illustrate how fault trees evaluate compound failure events.

number of tank fills	10	per year
level gauge LG reads false low	0.01	per demand
level transmitter LT reads false low	0.2	per demand
operator loads tank in error	0.02	per demand
logic solver failure	0.005	per demand
shutoff valve fails to close	0.1	per demand

Applying the above failure data, the PFD for the safeguards can be determined. Note that the PFD of both safeguards failing (0.0049) is higher than the value calculated by the LOPA approach of multiplying together the probability of failure of the individual safeguards (0.03 x 0.11 = 0.0033). This higher failure rate is the result of the common cause failure of the level gauge, which would have been neglected in the LOPA[2].

[1] This failure rate data is fictional and is selected for illustrative purposes only.

[2] There are actually three likely LOPA approaches. In addition to ignoring the interdependency of the events, as given, the LOPA analyst, after recognizing the common failure between the safeguards, may elect to only count one of the safeguards. This approach results in an overestimation of the risk. Alternatively, the analyst may break the scenario into its component failures, which essentially is the approach taken by the FTA.

4. COMBINING FAULT TREES WITH LOPA'S

Despite the fact that some safeguards or initiating events are interdependent or are too complex to apply look-up failure rate data, one does not necessarily have to sacrifice the simplicity of a LOPA. Instead, the two techniques can be combined, conducting a FTA where necessary to augment a LOPA. The fault tree (Figure 2) gives the PFD for the combined safeguards of the example problem, which are compound events that share a common basic event. The resulting PFD for the combined safeguards can then be inserted into the LOPA formula as shown:

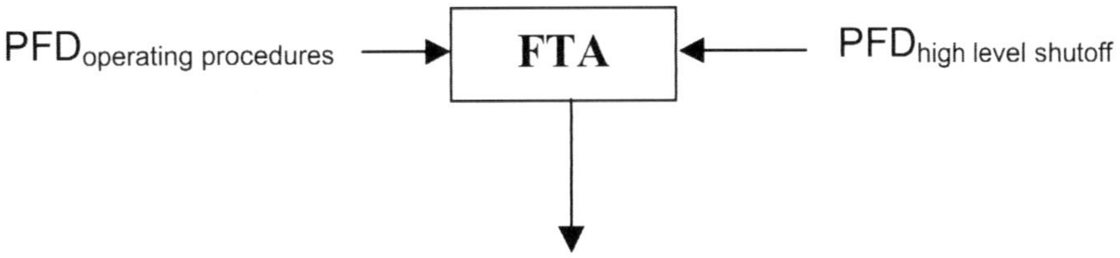

$$Frequency_{harm} = Frequency_{cause} \times PFD_{combined\ safeguards} \times Probability_{presence\ factor} \times Probabilty_{personal\ vulnerability}$$

The user can then combine the above scenario with the LOPA analysis for the other scenarios in the study. In this application, the fault tree analysis is used as a splint to combine two or more failure events that are interdependent.

Fault trees can also be developed in advance to solve for commonly-used architecture. Figure 4 shows examples of fault tree solutions for selected control schemes. These pre-solved fault trees could be used by the analyst to quickly evaluate the effectiveness of different control configurations as potential mitigation measures.

5. CONCLUSIONS

Layer of Protection Analysis (LOPA) is a very useful tool for prioritizing hazard scenarios and making risk-based decisions. However, it has limitations. LOPA requires that safeguards and initiating events be independent of each other. In addition, the failure rate data required by LOPA are generally only available for basic safeguards and initiating events and not for compound events. Fault Tree Analysis (FTA) is a very useful tool when LOPA does not apply. Unlike LOPA, FTA can evaluate interdependent and compound events. FTA also provides a visual failure map, assisting the analyst in better understanding the overall system. While FTA is very versatile, it requires specialized training and software and is not as intuitive as LOPA. Both methods have their respective limitations, but they can be synergistically combined to provide a powerful hybrid quantitative risk analysis tool.

Scenario: If the logic solver receives an input indicating an abnormal condition, it will send a signal to the control device so that it takes corrective action.

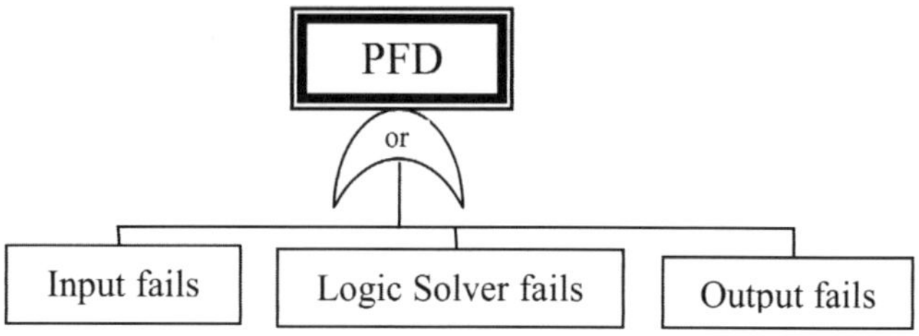

$PFD_{control} = PFD_{Input} + PFD_{Logic\ solver} + PFD_{Output}$[3]

Scenario: There are two inputs to the logic solver. If *either* input indicates an abnormal condition[4], then the logic solver will send a signal to the the control device so that it takes corrective action.

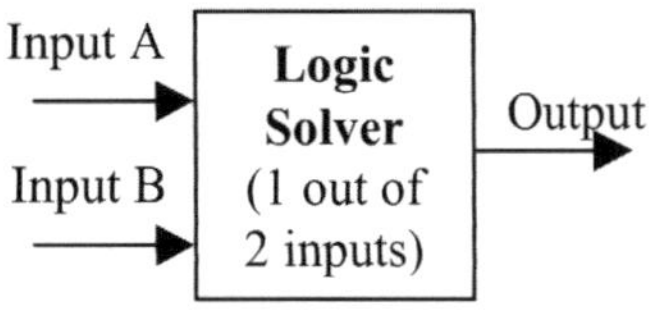

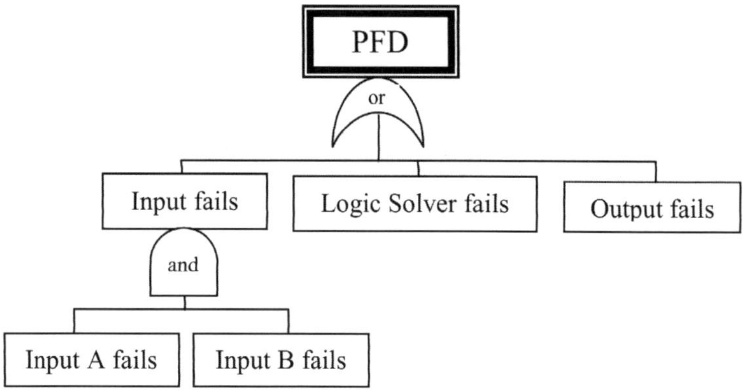

$PFD_{control} = (PFD_{InputA} \times PFD_{InputB}) + PFD_{Logic\ solver} + PFD_{Output}$

Figure 4 - Failure Analysis for Selected Control Configurations

[3] The equations in this figure use Boolean Algebra.

[4] i.e., input to logic solver fails if both Input A and Input B fails

Scenario: There are two inputs to the logic solver. If *both* inputs indicate an abnormal condition[5], then the logic solver will send a signal to the control device so that it takes corrective action.

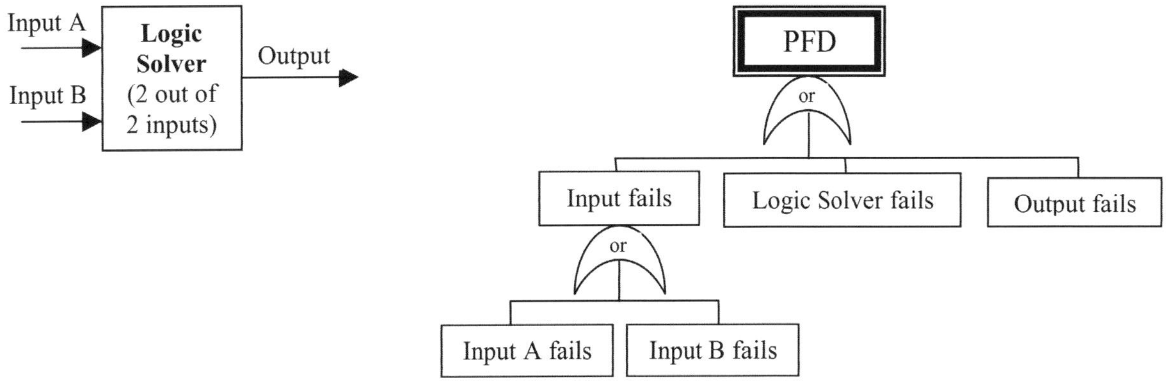

$$PFD_{control} = (PFD_{InputA} + PFD_{InputB}) + PFD_{Logic\ solver} + PFD_{Output}$$

Scenario: There are three inputs to the logic solver. If two out of three inputs indicate an abnormal condition, then the logic solver will send a signal to the the control device so that it takes corrective action.

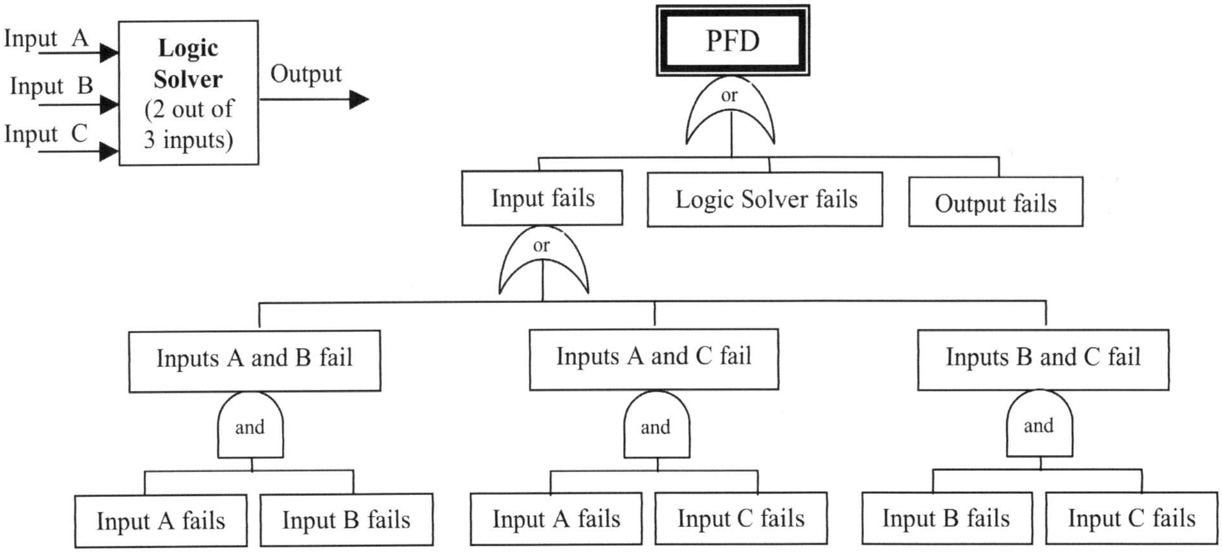

$$PFD_{control} = [(PFD_{InputA} \times PFD_{InputB}) + (PFD_{InputA} \times PFD_{InputC}) + (PFD_{InputB} \times PFD_{InputC})] + PFD_{Logic\ solver} + PFD_{Output}$$

Figure 4 - Failure Analysis for Selected Control Configurations (Cont.)

---

[5] i.e., input to logic solver fails if either Input A or Input B fails as both are required to signal the logic solver to take action.

REFERENCES

1. Dowell, A.M. III. "Layer of Protection Analysis: A New PHA Tool, After HAZOP, Before Fault Tree Analysis," in 1997 International Conference and Workshop on Risk Analysis in Process Safety. New York: American Institute of Chemical Engineers, 1997.

2. Center for Chemical Process Safety (CCPS), *Layer of Protection Analysis, Simplified Process Risk Assessment*. New York: American Institute of Chemical Engineers, 2001.

3. Dowell, A.M. III, "Layer of Protection Analysis and Inherently Safer Processes," Process Safety Progress, vol. 18 no. 4, pp 214 – 220, 1999.

4. Baybutt, P., "Human Factors Analysis for Process Safety – Apply LOPA-HF to a Fired Furnace," Hydrocarbon Processing, vol. 82 no. 4, pp 79 – 83, 2003.

5. Crowl, D.A. and Louvar, J.F., *Chemical Process Safety – Fundamentals with Applications*, 2nd ed., pp. 500 – 507. Upper Saddle River, NJ: Prentice Hall PTR, 2002.

6. U.S. Nuclear Regulatory Commission, *Fault Tree Handbook* (NUREG-0492). Washington, D.C., 1981.

7. Center for Chemical Process Safety (CCPS), *Guidelines for Chemical Process Quantitative Risk Analysis*, 2nd ed., pp 304 – 322. New York: American Institute of Chemical Engineers, New York, NY, 2001.

8. Lees, F.P., *Loss Prevention in the Process Industries*, 2nd ed., pp 9/13 – 9/31. Oxford: Butterworth-Heinemann, 1996.

9. Crowl, D.A. and Louvar, J.F. , *Chemical Process Safety – Fundamentals with Applications*, 2nd ed., pp. 491 – 499. Upper Saddle River, NJ: Prentice Hall PTR, 2002.

Oxidation Reaction Safeguarding with SIS

Edward M. Marszal, PE, CFSE (emarszal@exida.com)
Kevin J. Mitchell, PE, CFSE (kmitchell@exida.com)
Exida
2929 Kenny Suite 225
Columbus, OH 43221
(614) 451-7031

ABSTRACT

A variety of useful chemical compounds are economically produced using catalyzed oxidation reactions. These products include many common organic acids and anhydrides, industrial alcohols, and organic peroxides. Safely conducting catalyzed oxidation reactions on an industrial scale is a core competency of many chemical companies. However, there is a history of numerous incidents involving fire and explosion in oxidation reactors, and these accidents are compelling reminders of the hazards of oxidation reactions.

The primary hazard that is common to these technologies is the use of oxygen – either in air, enriched air, or pure form – as a reactant in contact with a combustible hydrocarbon, which is used either as a reactant or a solvent. Oxidation reactor design typically involves ensuring that residual oxygen levels in equipment are sufficiently low that they do not support combustion. This strategy safeguards against ignition of a flammable gas mixture within the reactor or downstream separation equipment. Normally, the basic process control system regulates the process chemistry and avoids potentially dangerous excursions involving high oxygen concentration. However, upset conditions often occur, and one of the commonly-employed safeguards to prevent an explosion is a Safety Instrumented System (SIS).

This paper explores some of the common risks that are encountered in oxidation process reactor sections. The paper also describes the instrumented safeguards that are typically used to prevent these risks from being realized and addresses some of the important details that should be considered during their design.

1.0 INTRODUCTION

Catalyzed oxidation reactions allow for a variety of useful chemical compounds to be economically produced. These products[1] include many common organic acids and anhydrides, industrial alcohols, and organic peroxides, as shown below.

- Terephthalic Acid (PTA)
- Isophthalic Acid (PIA)
- Phthalic Anhydride
- Adipic Acid
- Maleic Anhydride
- Phenol / Cumene Hydroperoxide (CHP)
- Butandiol, 1,4-
- Acrylonitrile
- Ethylene Oxide

Safely conducting catalyzed oxidation reactions on an industrial scale is a core competency of many chemical companies. However, there is a history of numerous incidents involving fire and explosion in oxidation reactors, and these accidents are compelling reminders of the hazards of oxidation reactions. Loss of control of an oxidation reaction can result a reactor explosion, with the potential for worker injury, significant environmental and property damage.

The primary hazard that is common to these technologies is the use of oxygen – either in air, enriched air, or pure form – as a reactant in contact with a combustible hydrocarbon, which is either as a reactant or a solvent. Oxidation reactor design typically involves ensuring that residual oxygen levels in equipment are sufficiently low that they do not support combustion. This strategy safeguards against ignition of a flammable gas mixture within the reactor or downstream separation equipment. Normally, the basic process control system regulates the process chemistry and avoids potentially dangerous excursions involving high oxygen concentration. However, upset conditions often occur, and one of the commonly-employed safeguards to prevent an explosion is a Safety Instrumented System (SIS).

The purpose of the SIS is to automatically return the process to a safe state when pre-determined safety conditions have been violated. They are often referred to as emergency shutdown systems, or safety interlock systems. ISA 84 defines a SIS as "a system composed of sensors, logic solvers, and final control

[1] This abbreviated list of substances was generated using catalyzed oxidation reactions was generated by reviewing the licensed processes shown in the Petrochemical Processes Special Report section of the March 2003 Edition of *Hydrocarbon Processing* Magazine. The list presented here is only intended to give a small example of the numerous catalyzed oxidation reactions utilized the in the chemical processing industries.

elements for the purpose of taking a process to a safe state when predetermined conditions are violated".

Design of SIS for oxidation reactor safety is governed, in part, by recent industry consensus standards from ISA and IEC.[2] These standards employ a performance-oriented approach in that they allow each individual company to define performance goals based on achieving a required amount of risk reduction rather than prescribing the hardware design of the SIS. Exida has performed numerous conceptual design projects involving SIS for oxidation reactors. This paper will illustrate some of the common Safety Instrumented Functions (SIF) used in oxidation reactor technology and illustrates practical application of the ISA and IEC standards.

2.0 GENERAL PROCESS DESCRIPTION

Commercial catalyzed oxidation reactions can take a number of forms. The primary difference between reaction types is the phase of the hydrocarbon reactant and the phase and type of catalyst used in the reaction. This white paper will focus on reactions where the hydrocarbon reactant (and reaction products) is in the liquid phase, and the catalyst for the reaction is also a liquid.

Figure 1 presents a typical process flow for the reaction section of a plant that employs a catalyzed oxidation reaction. The process mainly consists of a reactor vessel with an agitator. In some cases, the reactor may be jacketed to maintain the temperature of the reaction mixture.

Figure 1 – Typical Configuration of Process Section of Oxidation Process

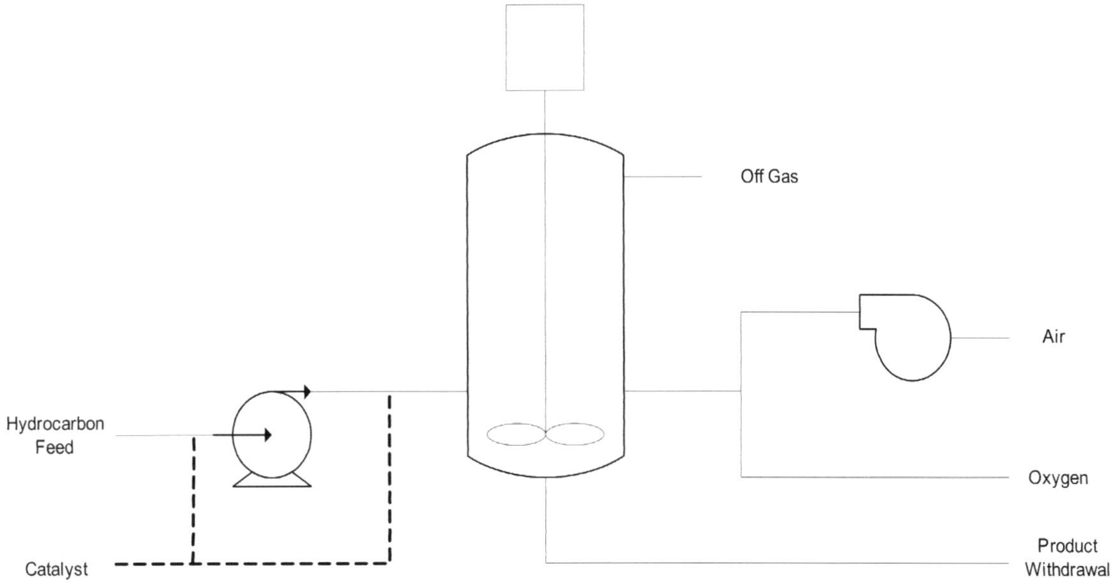

[2] Instrumentation, Systems, and Automation Society (ISA), ANSI/ISA S84.01, *Application of Safety Instrumented Systems for the Process Industry*, 1996. International Electrotechnical Commission (IEC), IEC 61508, *Functional Safety of electrical/electronic/programmable electronic safety-related systems*, First Edition, 1998. IEC 61511, *Functional Safety: Safety Instrumented Systems for the Process Sector*, FDIS, 2001.

The reactor is typically fed with two streams, the oxygen containing stream and the hydrocarbon stream. The oxygen containing stream varies, depending on process, from pure air to pure oxygen. When air is used, the air is typically compressed and fed into the reactor under pressure. In some cases, air is "spiked" with pure oxygen to make the reaction conditions more favorable. In other cases, the reaction is run using pure (chemical grade) oxygen.

The hydrocarbon feed is typically pumped into the reactor from a feed surge drum or feed mix tank. The liquid catalyst is either added directly to the feed mix in the feed mix tank, or continuously metered into the reaction vessel, sometimes through a separate process connection.

The reaction off gas is a combination of unreacted hydrocarbon feed, inert materials in the hydrocarbon feed, nitrogen (from the air feed), and a small amount of unreacted oxygen. The product is typically withdrawn in the liquid phase along with excess hydrocarbon feed materials and catalyst.

The reaction occurs in the liquid phase. The air and oxygen feed is injected into the liquid full portion of the reactor, which is agitated. The combination of agitation injection of gases acts to partially fluidize the reaction bed. The reaction is conducted using excess hydrocarbon feed in order to limit the amount of unreacted oxygen leaving with the off gas. The reaction can occur in either a continuous process or a batch operation.

3.0 PROCESS HAZARDS

The primary hazard involved in the reaction section (and downstream separation equipment) is the potential for the occurrence of flammable mixtures of hydrocarbon and oxygen occurring in the process equipment. If a source of ignition is put in contact with any flammable mixture that might be generated in the process, the result could be a fire or explosion.

The desired reaction in virtually all of the commercial oxidation processes are catalyzed to allow the creation of a valuable and desired product. In addition to use of the proper catalyst, the desired reaction path may also depend on appropriate temperature, pressure, and bed fluidization (mixing) in the reaction vessel. If all of these conditions are not present in the reactor system there is a potential for the desired reaction to fail to occur. This will then result in unreacted oxygen and hydrocarbon accumulating in a potentially flammable mixture in the reactor vapor space, and downstream equipment.

If a flammable mixture develops outside of the liquid reaction mixture inside the reaction vessel, ignition will lead to the uncatalyzed and undesired side reaction where oxygen and the hydrocarbon combust to form carbon dioxide, carbon monoxide, water, and various other reaction products. This undesired side reaction proceeds very rapidly and very exothermically given that a flammable mixture is present. The reaction will likely result in an explosion in the vessel where the reaction occurs, or loss of containment and a potential fireball if a fire occurs in the vessel instead of a sudden explosion.

4.0 INCIDENT CASE HISTORIES

1969 explosion in a Reactor Producing an Organic Acid

This operation involved a batch oxidation reaction. After the reactor was charged, air was introduced to begin the reaction. Because a grossly insufficient amount of hydrocarbon reactant was charged to the reactor, the reaction terminated unexpectedly after only 10 minutes, at a time when air flow had been ramped up to a maximum rate. As the reaction died off, oxygen concentration in the vent system began to rise rapidly. At the same time the temperature of the batch decreased because the reaction had stopped producing heat. Both conditions resulted in the vapors in the reactor vent entering the flammable operating region. The explosion caused extensive damage to the reactor and associated equipment.

1973 explosion in a Cumene Oxidation Reactor

Enriched air was being used to oxidize cumene to produce phenol. The plant had experienced plugging in the air distribution header to the oxidation reactor. These deposits were removed by flushing liquid back from the reactor through the header. During this procedure air flow had to be positively isolated. On the day of the incident two valves were left partially open in the air header. Enriched air entered the pipe and reacted with the hydrocarbon liquid. The pipe ruptured and ignited immediately, creating a massive fire that destroyed the entire plant.

1974 explosion in a Reactor Producing an Organic Acid

This oxidation reactor system involved continuous feed of catalyst to control the reaction. On the day of the incident, the reaction was proceeding normally, when it was discovered that the catalyst flow had been interrupted. An operator was sent to investigate and found that a manual block valve had been closed on the catalyst addition system. By the time the problem was corrected, the reaction had died off and oxygen levels were rapidly climbing in the reactor overhead system. The explosion blew off the vapor outlet line from the reactor and damaged associated piping.

1982 explosion in a Reactor Producing an Organic Acid

This batch oxidation reactor system used the concentration of oxygen in the overhead as read by analyzers as a key parameter in determining when a reaction had terminated. Increasing oxygen concentration in the overhead system indicated that the hydrocarbon reactant had been completely oxidized. On the day of the incident, operators were having problems with the oxygen analyzers causing them to periodically give a false high oxygen spike and subsequently cause the reactor to suddenly shutdown. During an attempt to restart the reaction, operators disabled the oxygen analyzers. Operators were being taxed with other operational problems in downstream separation equipment at the time the explosion occurred. It turned out that because the oxygen analyzers had been disabled, the control system did not terminate air flow to the reactor when the reaction had been completed. Minor damage to reactor equipment resulted from this explosion.

1983 explosion in a Reactor Producing an Organic Acid

During the day prior to the explosion, an operational problem resulted in the reactor being put in a hot hold condition. Air was isolated from the reactor and its contents were kept hot in anticipation of a reactor restart. Over many hours, air slowly leaked into the reactor past the air isolation valve, which was either leaking or left slightly open. Oxygen concentration built up in the reactor's vapor space, but this remained undetected. An explosion occurred, which was relieved through the reactor's emergency pressure relief device.

1987 explosion in a Reactor Producing an Organic Acid

During the initiation of the batch reaction, operating conditions (temperature and pressure) drifted into the flammable operating region. Operators activated an inert gas dilution system in an attempt to re-establish the reaction. They also disabled the oxygen analyzers which would have shutdown the reactor on high oxygen concentration. However, operators were unable to re-establish temperature and pressure control within normal operating limits. Less than 10 minutes after the start of the reaction, an explosion occurred, resulting in major damage to the reactor vessel, and its associated instrumentation.

1995 explosion in a Reactor Producing an Organic Acid

A leak in an air line internal to the oxidation reactor occurred. This allowed air to directly enter the reactor's vapor space and bypassed the air sparging system at the bottom of the reactor. A fire in the vapor space broke out immediately, and this actually depleted the concentration of oxygen in the overhead system. Reactor temperature measurements shot up rapidly. Within minutes overhead piping on the reactor failed due to overtemperature. The reactor contents were ejected under pressure and extensive fire damage resulted from this incident.

1999 explosion in an Air Line to an Oxidation Reactor

An operational upset occurred in a reactor producing an organic acid. Solvent from the reactor back-flowed into an air feed line due to problems maintaining the required differential pressure between the air line (normally higher pressure) and the reactor (normally lower pressure). On subsequent re-start of the reactor, enriched air was introduced into the feed line which started oxidation and combustion reactions with the solvent contained in the line. Field operators noticed the air feed line was glowing "cherry red". Within seconds, the line failed and a large fireball erupted.

2000 explosion in an Ethylene Oxide Manufacturing Plant

An explosion and fire occurred in the ethylene oxide manufacturing unit. Problems with an oxygen analyzer resulted in a decision to disable the device. This key safeguard normally monitored residual oxygen concentration in the process. Over a period of time oxygen concentrations increased above safe operating limits. Subsequently a detonation occurred resulting in extensive damage to the plant.

Lessons Learned

These case histories emphasize several key lessons which have been learned from oxidation reactor accidents, including:

- Ensure that residual oxygen concentrations are – at all times – well outside the flammable operating region.
- Ensure that the oxidation reactor is shutdown and air isolated upon detection of high residual oxygen concentration.
- Use pressure and temperature measurements to predict an approach to a flammable operating condition and initiate a reactor shutdown.
- Ensure operating pressures are maintained that do not allow flammable or combustible materials to backflow into air feed lines to the reactor.

These lessons bring the topic of Safety Instrumented Systems to the forefront of the discussion on how to design and operate oxidation reactors safely.

5.0 SAFETY FUNCTIONS

Safe operation of oxidation reactors is primarily achieved through careful control of the reactor operating conditions. The temperature and pressure of the reaction, along with the oxygen concentration in the vent system will determine whether the system is within the flammable operating region or outside (i.e., either fuel rich or fuel lean). At any given time, the operator must ensure that the process is not entering in the flammable region or even approaching it. This is often accomplished by either monitoring oxygen concentration in the process directly, through oxygen analyzers, or predicting a potentially flammable condition by using a combination of pressure and temperature measurements.

The basic process control system (BPCS) regulates normal process behavior. The normal operating conditions are set such that they are well outside the flammable region and they typically use a robust safety factor to ensure a wide margin of safety (See Figure 2). Safety critical alarms are set such that when process conditions deviate from normal operating ranges, operators have ample opportunity to intervene and correct the abnormal situation. Exida's experience in oxidation reaction technology shows that in most cases, operator intervention is successful in terminating a reaction before dangerous operating conditions develop.

Figure 2 – Layers of Protection

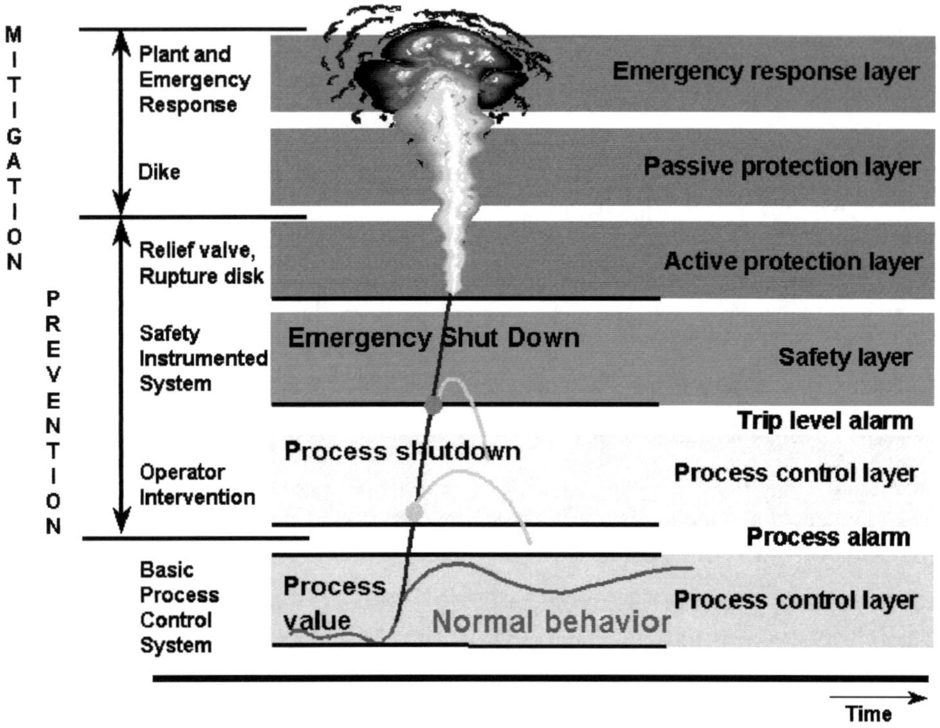

However, many companies who conduct oxidation reactions have adopted a philosophy that operator intervention alone is not a sufficient safeguard. This best-practice philosophy dictates that when certain pre-defined safety conditions are violated, control is taken away from the operator by the SIS, and the process is immediately brought to a safe state in an orderly manner. A logic solver, such as an industrial safety-rated Programmable Logic Controller (PLC), is used to continuously monitor process variables and initiate a safe shutdown. This logic solver is usually separate from the BPCS.

An oxidation reaction system will typically have between 5 and 15 Safety instrumented Functions. A safety instrumented function (SIF) is a set of specific actions to be taken when specific safety limits have been violated, and thereby will move the process from a potentially unsafe state to a safe state. On the other hand, a Safety Instrumented System (SIS) is the collection of equipment (sensors, logic solver, and final control elements) used to perform the SIF. Multiple SIF are often implemented in a single, complex SIS. Using the perspective of a Safety PLC, there can be many individual SIF that are executed in that single Safety PLC.

Some of the typical SIF found in oxidation reactors are shown in Table 1:

Table 1 – Safety Instrumented Functions

Item	Description	Inputs	Outputs
SIF-01	High oxygen concentration in oxidation reactor overhead causes reactor shutdown	O2 Analyzers	Air Isolation Valves Closed (Double Block and Seal)
SIF-02	Oxidation reactor low temperature or high pressure indicates approach to flammability limit and initiates a reactor shutdown	Reactor Temp. Reactor Press.	Air Isolation Valves Closed (Double Block and Seal)
SIF-03	Oxidation reactor fire (as indicated by high reactor temperature) initiates a reactor shutdown	Reactor Temp.	Air Isolation Valves Closed (Double Block and Seal)
SIF-04	Low differential pressure between air header and oxidation reactor initiates a reactor shutdown	Differential Press. Transmitter	Air Isolation Valves Closed (Double Block and Seal)

Additional SIF are typically used when enriched air is used in the reaction. The objective of these additional SIF is to ensure that oxygen concentration in the enriched air header does not exceed a certain pre-defined safe operating limit.

6.0 SIS IMPLEMENTATION

The ISA and IEC consensus standards guide the user in ensuring that high-availability safety systems are designed, installed, operated, and maintained in a manner that will promote ongoing integrity of plant operations. The performance-oriented nature of the standards allows for flexibility in implementing an approach that fits within a company's overall risk management framework, but it also requires a fundamental understanding of what SIS are required to do, and how well they need to perform to adequately manage risk.

An effective SIS design only begins with defining the Safety Instrumented Functions for the oxidation process. Establishing the key performance measurement for a SIF is the next step in the safety lifecycle shown in Figure 3. This is known as the Safety Integrity Level (SIL).

Figure 3 – Safety Lifecycle

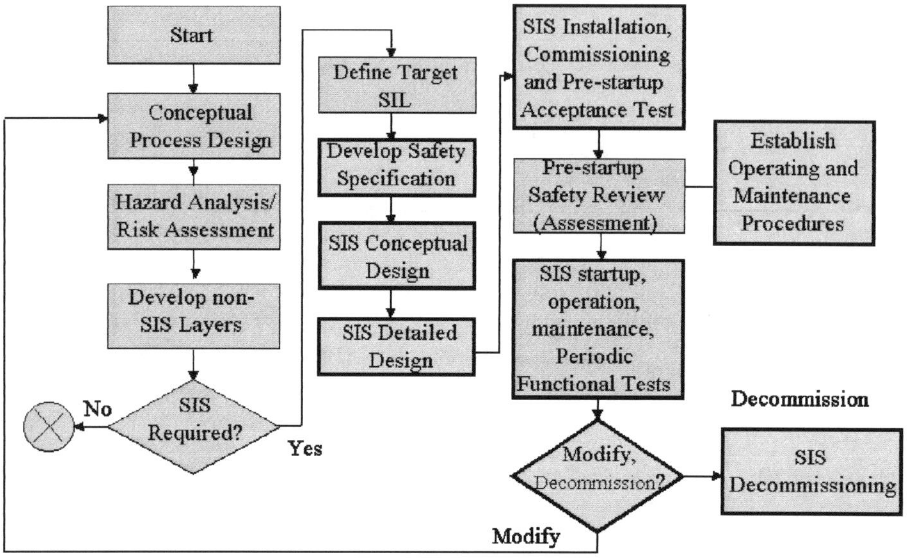

Companies are now specifying Safety Integrity Levels (SIL) based on the amount of risk reduction that is required to achieve a tolerable risk level. The SIS is then designed to meet or exceed this level of performance. The SIL represents the amount of risk reduction that is required from a Safety Instrumented Function (SIF), and it is categorized based on the average Probability of Failure on Demand (PFD_{avg}) as shown in Table 2.

Table 2 – Safety Integrity Levels

Safety Integrity Level	Probability of failure on demand (Demand mode of operation)	Risk Reduction Factor
SIL 4	0.001% to 0.01%	100,000 to 10,000
SIL 3	0.01% to 0.1%	10,000 to 1,000
SIL 2	0.1% to 1%	1,000 to 100
SIL 1	1% to 10%	100 to 10

The actual Safety Integrity Level that is selected for each SIF has an enormous impact on the design and testing requirements. Some of the most significant impacts are exhibited in the following areas:

- *Architecture* – A SIL1 design can usually be achieved using a single input / single output. However, if a SIL 2 or higher is required, a fault tolerant design may need to be employed, such as 1-out-of-2 (1oo2) voting on redundant oxygen analyzers. While this fault tolerance can result in significant improvements in Probability of Failure on Demand (and thus the achieved SIL level), it also can also significantly increase the frequency on nuisance trips due to instrumentation failures. A robust design must meet the required SIL as well as minimize the likelihood of a nuisance trip. Requiring a Safety Function to meet a SIL 3 requirement is possible, but often cost prohibitive.

- *Testing* – Air isolation valves in a SIS normally remain energized for very long periods of time before a demand is placed upon them, when they are required to quickly de-energize and to isolate the process. This results in an situation where an effective test of the SIS is the only way to ensure that a component has not failed in such a way that will defeat the entire system. More frequent testing decreases the probability that the system will fail when a demand is placed upon it. A higher Safety Integrity Level will often result in a requirement to test the system more often. Scheduling and completing this testing can be problematic for plants that have large on-stream times.

In addition to the quantitative requirements of ISA 84.01, the standard also lists a number of design criteria that must be considered and specified. These items include such considerations as:

- Tightness of shutoff
- Failure characteristics upon loss of utility (e.g., fail-open or fail-closed valves)
- Response time
- Required diagnostics, etc.

7.0 CONCLUSION

The historical record of accidents involving oxidation reactions is compelling evidence that process safety should not be taken lightly in such systems. Most of these accidents have occurred as the result of undesired high levels of oxygen in areas where oxygen is not desired, resulting in flammable atmospheres. In many cases, the risk posed by these hazards is reduced through the use of

Safety Instrumented Systems. Some typical safety instrumented functions of oxidation reactors might include:

- Isolation of air upon detection of high reactor vapor space oxygen concentration
- Isolation of air upon detection of potential reverse flow of reactant into the air system
- Isolation upon detection of a fire in the air system
- Isolation upon detection of unfavorable reaction conditions (e.g., low temperature and high pressure)

If you use Safety Instrumented Systems, you should ask if they have been designed, operated, and tested as per the requirements of ISA 84.01. If not, you should begin to carefully scrutinize your systems. They key questions you need to have answered are:

- How much risk reduction does our current SIS technology provide? Have we calculated a Probability of Failure on Demand for the system
- How much risk reduction do we need? Has this requirement been documented so we can justify our decisions?
- Does the existing design have sufficient amount of redundancy and fault tolerance to meet our risk reduction requirements?
- How often should our SIS be tested in order to ensure we meet our risk reduction requirements?
- Have all design requirements for the SIS been appropriately specified, including tightness of shutoff and process safety time?

If you don't have answers to these questions, or haven't started any of the steps in the safety lifecycle, you may need to take action.

THE ASSIGNMENT OF SIL TARGETS FOR COMPLEX SYSTEMS

Kimberly A. Dejmek, CFSE, PE and Ken A. Wehrman, Ph.D.

Baker Engineering and Risk Consultants, Inc.
4100 Greenbriar, Suite 130
Stafford, Texas 77477
United States of America
281-491-3881 ph., 281-491-3882 fax
Kdejmek@BakerRisk.com

Key Words: Safety Instrumented Systems (SIS), Safety Integrity Level (SIL), Quantitative Risk Analysis (QRA)

Prepared for presentation at the 38th Annual Loss Prevention Symposium, New Orleans, LA, April 26-29, 2004, Session T7003: Safety Instrumented Systems/Layers of Protection Analysis

December 2003

Unpublished

ABSTRACT

The assignment of performance targets, or target Safety Integrity Levels (SILs), is a critical step in the application of the Safety Instrumented System (SIS) standards, i.e., ANSI/ISA S84.0.01-1996, IEC 61508 and IEC61511. Although the SIL is a key concept in the implementation of the standards, the development and application of a method for determining the target SIL has been left to the owner/operator. However, the standards do provide guidance on this topic and present a number of techniques that can be considered, including risk matrix, risk graph, and Layers of Protection Analysis (LOPA). Generally, the methods for SIL assignment are qualitative or semi-quantitative risk assessment methods based upon the judgment of an assignment team. In most cases the methods based on expert opinion and limited historical data are adequate. However, in the case of reactive and other complex systems, the number and complexity of the scenarios often overwhelm these simplified approaches. There are warning signs that can aid in the identification of cases where the simplified methods may break down and provide non-conservative results. In cases such as this, a quantitative assessment should be conducted to evaluate the likelihood of Safety Instrumented Function (SIF) demands and the risk reduction supplied by the other protection layers, with the aim of determining the risk reduction required from the instrumented system.

INTRODUCTION

Although pressure relief devices are generally required for all pressure vessels, the difficulty in providing safe, adequate and reliable relief for many reactive systems may make over-pressure prevention rather than mitigation a more attractive option. Safety Instrumented Systems (SIS) can therefore be a critical protection layer for these systems. tend to be quite complex, which can complicate the selection of a performance target for SIS in this service.

In February of 1996, the ISA (The Instrumentation, Automation, and Systems Society) approved their standard entitled *Application of Safety Instrumented Systems for the Process Industries (ANSI/ISA 84.00.01-1996)[1],* and in 2003 the IEC (International Electrotechnical Commission) promulgated IEC 61511: *Functional safety – Safety instrumented systems for the process industry sector[2-4].* One basic concept of these standards is the definition of the Safety Integrity Level (SIL) for each Safety Instrumented Function (SIF) within the Safety Instrumented System (SIS). The SIL defines the safety performance criteria for the function and assigning the target requires an allocation of a portion of the overall required risk reduction to the safety instrumented function.

Figure 1 presents a graphical representation of the philosophy that serves as the basis for SIL target selection. Each undesirable outcome that will be influenced by the SIS is considered independently. The "inherent process risk", which is represented by the red arrow at the bottom of the figure, is evaluated based upon the expected likelihood and severity of the undesired event. Each of the available protection layers is identified and the associated risk reduction estimated. The green arrows represent the number and type of protection layers and provide an indication of the level of risk reduction associated with each. These layers include passive protection; basic process control, alarms, and operator intervention; the SIS; and event mitigation, which includes relief and flare systems, fire and gas detection and suppression, and emergency response. In this figure, each of the green arrows is the same size, however, for a specific event, the level of risk reduction provided by the various protection layers can vary. The concept of a SIL target is expressed in this figure through the size of the green arrow labeled "SIS". The SIL target directly expresses the risk reduction that is expected from the SIS for a particular scenario.

Although SIL is a key concept in the implementation of the standards, the development of a method for determining the target SIL has been left in the hands of the owner/operator. The standards do provide some guidance on this topic and include a number of example methods to consider. The available methods range from purely qualitative evaluation, up to and including quantitative risk analysis (QRA).

In general, these methods utilize a set of tools to guide a multi-disciplinary team in the selection of the target SIL. The assignment of a target SIL for a particular SIF typically requires the consideration of:

- Event likelihood,
- Event severity, and
- Effectiveness of the non-SIS protection layers.

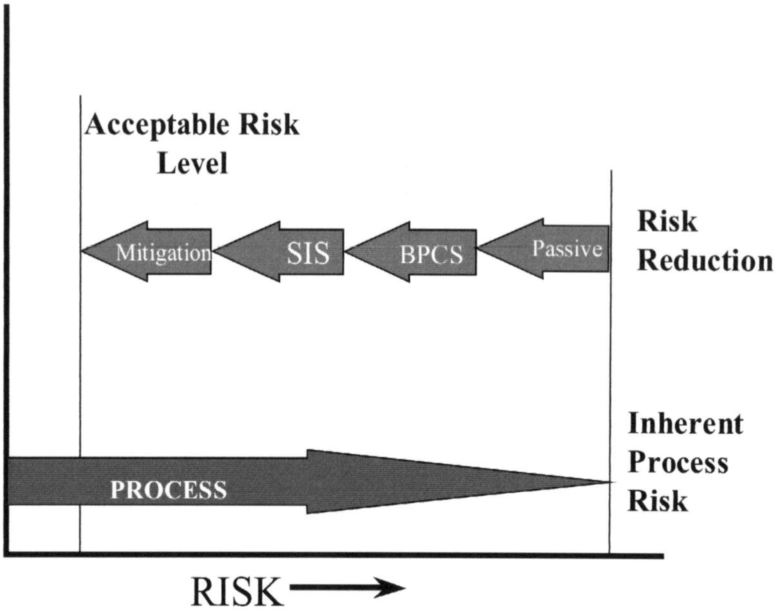

Figure 1 – Target SIL Assignment Philosophy

In the evaluation of most events, this is a straightforward process effectively addressed through a qualitative or semi-quantitative approach. However, there are scenarios that are too complex to be addressed by any of the less rigorous methods. Reactive systems, which generally have numerous potential initiators of high reaction rates and a reliance on a small set of actions to take the process to a safe state, typically fall into this category.

SIL SELECTION PROCESS

The Safety Integrity Level, or SIL, is specified as one of four discrete levels, as defined by Table 1. The SIL specifies "the safety integrity requirements of the Safety Instrumented Function to be allocated to the Safety Instrumented Systems [2]." Each SIL level is defined by a range of target average probability of failure on demand (PFD_{avg}).

Table 1 – Safety Integrity Levels for the Demand Mode of Operation

	Target Average PFD_{avg}	% Availability $(1-PFD_{avg}) \times 100$	Target Risk Reduction
SIL 1	$>10^{-1}$ to $\leq 10^{-2}$	>90 to ≤ 99	>10 to ≤ 100
SIL 2	$>10^{-2}$ to $\leq 10^{-3}$	>99 to ≤ 99.9	>100 to ≤ 1,000
SIL 3	$>10^{-3}$ to $\leq 10^{-4}$	>99.9 to ≤ 99.99	>1,000 to ≤ 10,000
SIL 4	$>10^{-4}$ to $\leq 10^{-5}$	>99.99 to ≤ 99.999	>10,000 to ≤ 100,000

In selecting the target SIL, there is a basic thought process that is addressed by most of the methods. The steps below outline the general process.

1. Describe the event that the SIF has been provided to prevent / mitigate.
2. Determine the consequence severity, without taking credit for any active safeguards, based upon potential safety, public, environmental, and financial impacts.
3. Identify *all* of the causes for which the SIF has been designed to detect and respond.
4. Evaluate the event likelihood of the event without taking credit for any active safeguards.
5. Determine the unmitigated risk by combining the consequence severity and event likelihood.
6. Identify any and all non-SIS independent protection layers (IPLs).
7. Evaluate the effectiveness of each IPL in the prevention or mitigation of the risk.
8. Determine the mitigated risk.
9. Compare the mitigated risk to the acceptability criteria.
10. Assign any remaining required risk reduction to the SIS with the target SIL.

SIL SELECTION METHODS

There are six methods that are generally considered for SIL assignment: modified HAZOP, consequence only, risk matrix (two and three-dimensional), risk graph, Layers of Protection Analysis (LOPA), and quantitative assessment. A number of articles, papers and books have been written describing the details of these methods [1-9]. Only a brief introduction to each has been provided here.

Modified HAZOP

The modified HAZOP method, which is introduced in Annex A of the ISA standard, is a purely qualitative SIL assignment technique [1]. As a part of the normal HAZOP process, the team identifies existing SIF or the need for additional SIF protection [4]. In cases where an SIF is deemed necessary, the team considers the severity of the consequences, the probability of event occurrence, and the expected risk reduction from non-SIS engineering and administrative controls. When all of these factors are taken together, the team comes to a consensus as to the appropriate safety integrity level.

Consequence Only SIL Assignment

This method is also purely qualitative, requiring only the consideration of the potential consequences of a process event. The consequences can be defined in terms of the potential safety, environmental, and economic impact associated with the event. An example set of definitions is provided in Table 2. Because the initiating event frequency, the availability of protection layers, and the probability that a dangerous event develops are not considered, this method can produce conservative results. However, this may be offset by the ease of application and the time saved in assigning SIL targets.

Table 2 - Example of the Consequence-Only Method

SIL 1	Potential for minor injuries. No environmental damage, release less than reportable quantity. Production and capital loss between \$10,000 and \$100,000.
SIL 2	Potential for serious injuries requiring hospitalization. Repairable environmental damage or release of reportable quantity. Production and capital losses between \$100,000 and \$1,000,000.
SIL 3	Potential for on-site fatality or multiple serious injuries. Environmental damage requiring long-term clean-up effort. Production and capital losses between \$1,000,000 and \$10,000,000.
SIL 4	Potential for multiple on-site fatalities or 1 or more off-site fatalities. Significant environmental impact that may threaten right to operate. Production and capital losses greater than \$10,000,000.

Risk Matrix

The risk matrix method is based upon the qualitative evaluation of the potential event consequences and likelihood. A team of process and operations specialists evaluates each safety function, first describing the potential consequences of the event and then developing a severity ranking. Typically, a risk matrix will have three to five severity ratings that span the range from minor impact to catastrophic event. As in the Consequence Only method, severity rankings can be defined for environmental and economic effects, as well as the on-site and off-site safety consequences.

Following the development of the safety severity, the team investigates the causes of the event, the safeguards (other than the SIS under consideration), and the event sequence in order to estimate the likelihood that the consequences will occur. A set of likelihood rankings is defined to represent the range of possible event frequencies. After listing the potential causes and protection layers, the team selects the appropriate likelihood ranking.

Using the severity and likelihood rankings together, the risk matrix is applied to determine the target Safety Integrity Level. Based upon their corporate risk tolerance, SIS philosophy, and ranking definitions, each corporation must determine the relationship between the rankings and the SIL. An example of a risk matrix has been provided in Figure 2.

The risk matrix is a widely accepted tool for the qualitative evaluation of risk. Many companies use a form of the risk matrix as a part of their Process Hazards Analysis (PHA) in order to determine if the existing engineering and administrative controls provide sufficient risk reduction, or if additional safeguards should be recommended. Although familiarity with the method could be considered a benefit, the application of risk matrix to the selection of a target safety integrity level is somewhat different to that used in a PHA, which may contribute to confusion.

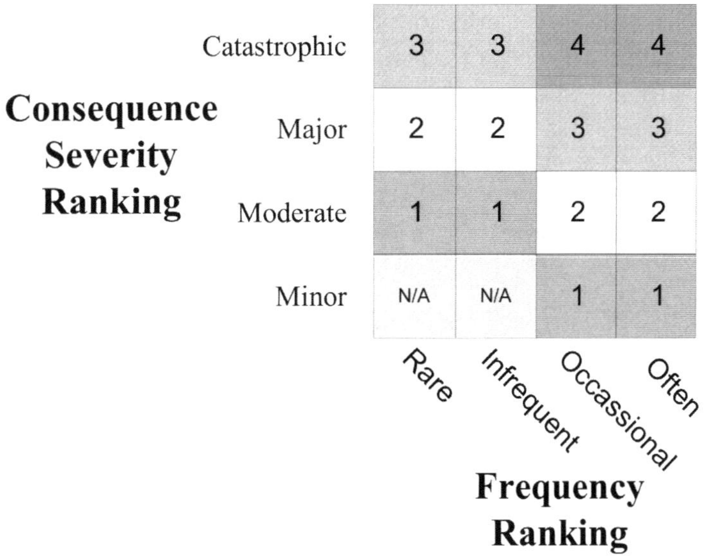

Figure 2 - Example of the Risk Matrix Method

Another variation of the risk matrix method is also used in the determination of target SIL. This method, called the Safety Layer Matrix in Annex A of the ISA standard, incorporates a third variable, Protection Layers, in addition to the Severity Ranking and the Frequency Ranking [1]. When applying this method, the evaluation of the Frequency Ranking does not include any protection layers. The Frequency Ranking only considers likelihood of the event causes and external events that occur for the consequences to be realized. Any risk reduction is addressed through the application of the Protection Layer Ranking. In order to be considered in this ranking, the systems must meet the definition of a protection layer [7]. Typically, three levels of Protection Layer Rankings are considered. These are based upon either the amount of risk reduction provided by the protection layers (e.g., low, medium, or high) or on the number of protection layers that exist other than the SIS under consideration.

Risk Graph

The IEC standards 61511 and 61508 provide an alternative method to the risk matrix called risk graph. This method provides a SIL correlation based on four factors:

- Consequence (C),
- Occupancy (probability that the exposed area is occupied) (F),
- Possibility of avoiding the hazardous event (P), and
- Frequency of the unwanted occurrence (W).

This method is a qualitative technique requiring development of definitions for each parameter to ensure that the four parameters are properly chosen by the assignment team. In addition to the consequence and event likelihood that are part of the basic risk matrix, the risk graph focuses attention on the evaluation of the risk to an exposed individual. The four factors are evaluated from the perspective of a theoretical person being in the incident impact zone at the time of the incident.

Once these factors are determined, the risk graph is utilized to determine the associated SIL. This method requires a multi-disciplinary team to ensure that the four parameters listed above are properly chosen. An example risk graph is shown in Figure 3 [3].

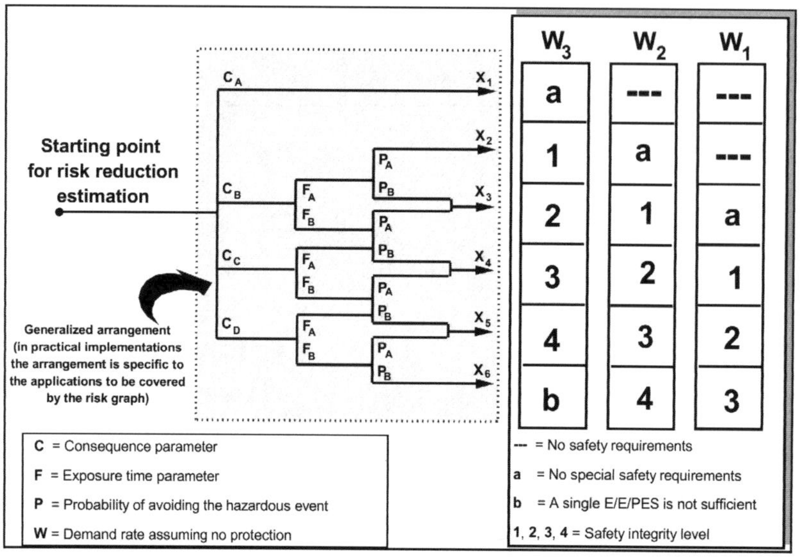

Figure 3. Example of the Risk Graph Method

Layers of Protection Analysis

Layers of Protection Analysis (LOPA) is a semi-quantitative method of risk analysis. The consequence evaluation remains qualitative, as in the previously described methods. The evaluation of the event likelihood is quantitative, based upon "order of magnitude" estimates of the initiating event frequency and the availability of the protection layers. The basic concept of LOPA is summarized in the following steps [6]:

- Identify impact events, determine the types of impact, and classify event severity.
- List all of the causes for each impact event.
- Estimate the frequency of each initiating cause.
- List the Independent Protection Layers (IPLs) for each cause-consequence pair.
- Determine the PFD for each IPL.
- Calculate the mitigated event frequency for each cause-consequence pair.
- Sum the frequencies for each cause-consequence pair that will place a demand on the SIS.
- Compare the total mitigated event frequency to the acceptability criteria for the associated event severity classification.
- Determine SIL based upon required risk reduction or identify other risk reduction measures, if required to meet the risk acceptability criteria.

Quantitative Risk Assessment (QRA)

The quantitative approach to SIL assignment is the most rigorous technique. Quantitative evaluation of both the consequences and the event likelihood are possible; however, it is typical for the consequence evaluation to remain qualitative. The event likelihood is calculated by modeling the combined influence of the potential causes, protective systems, and any external events that are required for the consequence to be realized. These external events allow all of the steps in the event sequence to be quantified. This includes probabilities of events such as immediate or delayed ignition, individuals present in the hazard zone, atmospheric conditions that will support the development of an explosive mixture, release size sufficient to lead to undesired consequences, etc. This information can be used to construct and quantify a fault tree or event tree.

Quantitative analysis is sometimes the only method that will allow complex events with multiple causes, protection layers and interdependencies to be evaluated. The method does require a thorough understanding of the event sequences and failure data for each basic event.

REACTIVE AND OTHER COMPLEX SYSTEMS

Most corporate guidelines on SIS address only a single method; however, there is no single SIL assignment method suitable for all situations. Since most of the methods represent a semi-quantitative form of risk assessment, a makeshift QRA, it might be thought that full QRA is a panacea. However, the effort and expertise required to perform such a study is unwarranted for the majority of SIL assignments. There are cases, however, where it provides the only means of obtaining a valid assessment of the overall risk, the risk reduction provided by the non-SIS projection layers and the risk reduction required from the SIS.

In order to use QRA tools effectively in the SIL assignment process, it is important to recognize the appropriate situations. The warning signs include:

- Numerous scenarios relying on the same SIF and/or non-SIS safeguards,
- None of the initiating events is significantly more likely than the others,
- Safeguards are not equally applicable to all initiating events,
- Multiple SIFs work to prevent a single scenario,
- Interdependencies exist between safeguards, i.e., common equipment,
- Interdependencies between initiating events and safeguards,
- Safeguards are not independently capable of preventing the undesirable outcome,
- Safeguards fail to meet the minimum criteria required for consideration by the less rigorous technique,
- Multiple event outcomes of concern, e.g., fire, explosion, toxic consequences can arise from a single event,
- Passive safeguarding, such as elevated relief/depressure discharge piping, hardened buildings, blast walls, etc. make consequences unclear.

In the case of overpressure protection for reactive systems, a number of these characteristics are often present. Because the actions that stop the reaction are often the only available actions that prevent the accident scenario, numerous SIFs, as well as the manual intervention, may all be acting upon the same equipment items, e.g., the isolation of feedstocks, the injection of a kill fluid, depressuring etc. It is also likely that there will be some safeguards that, although not completely effective in preventing the event in all cases, may reduce the likelihood or the potential consequences. This may include actions such as introducing additional cooling. It is also likely that there may be numerous initiators of the overpressure scenario including overfeed of reactants, overfeed of catalyst, loss of mixing or circulation, loss of cooling, ineffective cooling, failed temperature control, blocked/plugged reactor outlets, exposure to external fire, etc. These overpressure scenarios may share some common safeguards and, in addition, there may be safeguards that are designed for only one specific cause.

It may be tempting to assume that the simplified approaches have a degree of built-in conservatism, and that the resulting SIL target insures sufficient risk reduction. This however, can be an invalid and dangerous assumption. When simplified methods are applied to complex systems, the tendency is to break the complex system into a number of smaller pieces and evaluate each one independently. However, by failing to evaluate the combined likelihood of a demand on the SIF, the assigned SIL target may be too low.

Another common mistake is the integration of the SIL selection directly into the PHA process, which also introduces the potential to underestimate the required SIL target. HAZOP and other PHA methods are inductive techniques, beginning with a specific cause and evaluating all of the potential subsequent consequences. Target SIL assignment requires the simultaneous consideration of all of the initiating events in order to evaluate the likelihood of a demand on the SIF. Even in somewhat simple systems, this can introduce an error into the assessment that may underestimate the required SIL. Although this problem is inherent in the Modified HAZOP method, any of the other methods can also be applied with the same fatal flaw.

QRA IN SIL SELECTION

Quantitative studies can support the SIL selection process by providing greater insight and understanding of the potential consequences and the event likelihood. Depending upon the specific scenario under investigation, either or both types of analyses may be considered.

Consequence modeling can be used by the SIL assignment team to better estimate the potential for on-site and off-site exposures to toxic and flammable effects. Dispersion modeling provides an estimate of the boundaries of a toxic or flammable cloud. It can also be applied to specific questions such as whether an elevated atmospheric release from relief or depressure discharge piping has the potential for ground level effects, as illustrated in Figure 4. Fire modeling can be used to estimate the radiation effects from pool fires, jet fires and fireballs. The team can then use these results to evaluate the on-site and off-site safety impacts and the potential for event escalation due to thermal exposures. Over-pressure scenarios also require the consideration of blast effects either from a vessel rupture or a vapor cloud explosion (VCE) following the loss of containment. Often passive protection layers are employed to address potential explosion effects and their risk reduction should be evaluated. By determining the explosion energies, flame speeds and associated blast contours and through the analysis of the structural response to these blast loads, the vulnerability of personnel in occupied buildings can be estimated. This allows teams to consider the risk reduction provided by blast walls around reactors and explosion hardened control buildings. Quantitative consequence analysis can be extremely helpful to SIL assignment teams attempting to refine and confirm their initial qualitative evaluations.

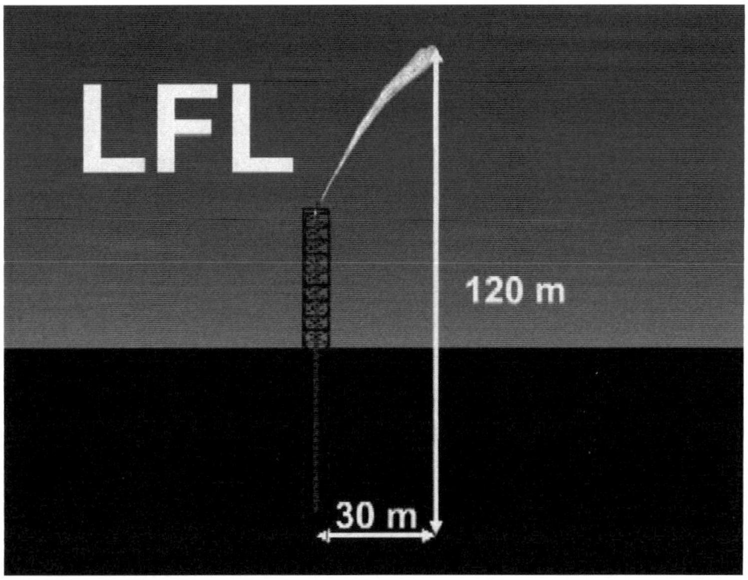

Figure 4 – Example Dispersion Model Result

Quantitative tools are more often used by SIL assignment teams to refine the event likelihood. Fault tree analysis (FTA) and event tree analysis (ETA) can provide teams with much needed information and understanding. Event tree analysis can be applied to scenarios with numerous outcomes based upon the availability of the protection layers and the presence of various conditions, e.g., wind direction, ignition sources, time of day, etc.

Figure 5 provides an example of an event tree and illustrates how ETA can be used to estimate the frequency of scenarios with multiple outcomes. Although this type of evaluation can also be accomplished through the LOPA technique, the graphical nature of ETA makes the analysis easier to understand, communicate and validate.

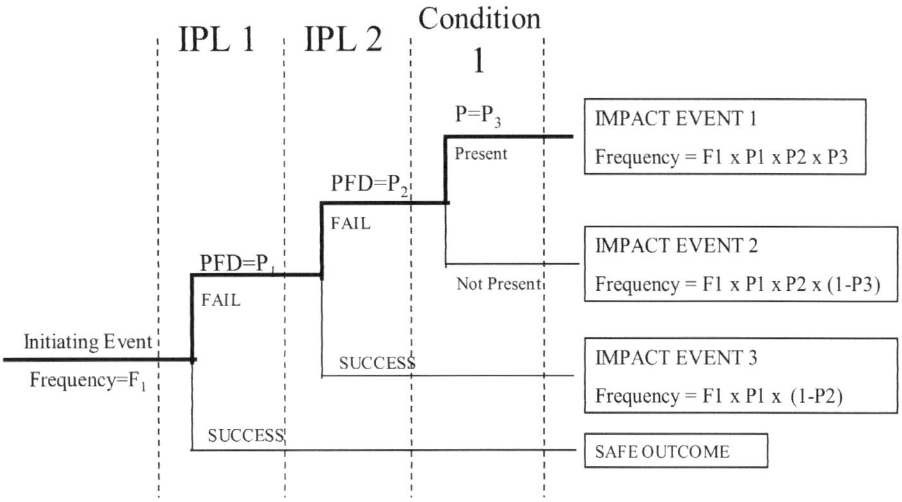

Figure 5 – Example Event Tree

Fault Tree Analysisis a deductive technique for calculating the likelihood of a final event, based upon an analysis of all of the possible paths and sequences that lead to it. This tool is perfect for evaluating complex events with numerous paths and interdependencies. When solved using cut-sets, rather than a simple gate-by-gate approach, any duplications or common components are detected and correctly assessed.

After first identifying an "undesirable" event, it is necessary to determine all precursor events or causes which can lead to the undesirable event. The inter-relationship between all events and causes is diagrammatically represented in the form of an inverse tree. The single undesirable event is placed at the top and all precursors and causes are placed on branches coming out of the top event via logic gates defining the causality, i.e. AND and OR. The tree is drawn sufficiently downward to reach *basic* or *undeveloped* events for which failure rates or probabilities of occurrence can be estimated.

Figure 6, presented on the last page of this paper, represents a summary fault tree for the evaluation of a reactor over-pressure scenario. The top event is the "frequency of reactor overpressure." Each of the types of initiating events and protection layers is defined by an "undeveloped event," indicating that further development of the logic for that branch is necessary before the tree could be solved. This summary tree is provided to illustrate how the evaluation of a complex event such as reactor overpressure would be started. As each branch is developed, the specific initiating events due to human error, equipment failure, and/or human error would be identified. Additionally, any safeguards specific to that branch would be addressed. The summary tree also shows that non-independent protection layers, such as manual and automatic initiation of the kill system, can be taken into account.

It is worthy of note that the simple rule of adding at the OR gates and multiplying at the AND gates is inappropriate when solving complex trees where a particular event may be found on more than one branch. In these cases, it is necessary to perform some Boolean simplification or rearrange the logic in some other manner. The most common and useful means for accomplishing this is *cut set analysis.*

CONCLUSIONS

As corporate and regulatory pressures to eliminate atmospheric discharge continue, the need increases to prevent relief demands and reduce the dependence upon relief devices as the principal protection layer. The role of SIS in the risk management of over-pressure protection is increasing. Therefore, it is critical that these instrumented systems are properly specified to insure that they deliver adequate risk reduction.

Simplified approaches such as risk matrix, risk graph, and LOPA are effective tools for target SIL assignment in most situations. However, these methods are unacceptable for the analysis of complex systems, producing results that either over-specify, or more dangerously, under-specify the SIS integrity requirements. Quantitative analysis should be used in support of the simpler tools, in these situations. The results of any quantitative study can usually be integrated back into the qualitative or semi-quantitative method to provide consistency with the other SIL selection decisions.

NOMENCLATURE

ETA – Event tree analysis
FTA – Fault tree analysis
HAZOP – Hazard and operability study
IEC – International Electrotechnical Commission
IPL – Independent protection layer
ISA – Instrumentation, Systems, and Automation Society
LOPA – Layers of protection analysis
PFD_{avg} – Average probability of failure on demand
QRA – Quantitative risk assessment
SIF – Safety instrumented function
SIL – Safety integrity level
SIS – Safety instrumented system

REFERENCES

[1] "Application of Safety Instrumented Systems for the Process Industries," ANSI/ISA 84.00.01-1996, 1996, ISA, Research Triangle Park, NC.

[2] "IEC 61511-1: Functional safety: Safety instrumented systems for the process industry sector- Part 1: Framework, definitions, system, hardware and software requirements," 2003, IEC.

[3] "IEC 61511-3: Functional safety: Safety instrumented systems for the process industry sector- Part 3: Guidance for the determination of safety integrity levels," 2003, IEC.

[4] Summers, A.E. and Ford, K.A., 1998, "Assigning Safety Integrity Levels," 53rd Annual Symposium on Instrumentation for the Process Industries, College Station, Texas.

[5] Dowell, A. M., III, 1999, "Layers of Protection Analysis and Inherently Safer Processes," Process Safety Progress, Volume 18 No. 4, pp. 214-220.

[6] Dowell, A.M., III, 1997, "Layers of Protection Analysis: A New PHA Tool, After Hazop, Before Fault Tree," International Conference and Workshop on Risk Analysis in Process Safety, October 21-24, 1997, Atlanta, GA, American Institute of Chemical Engineers, New York, pp. 13-28.

[7] CCPS, 2001, Layers of Protection Analysis: Simplified Process Risk Assessment, New York: American Institute of Chemical Engineers, Center for Chemical Process Safety. *Guidelines for Safe Automation of Chemical Processes*, Center for Chemical Process Safety, American Institute of Chemical Engineers, New York, 1993.

[8] *Guidelines for Hazard Evaluation Procedures*, Center for Chemical Process Safety, American Institute of Chemical Engineers, New York, 1985.

[9] Dejmek, K.A., "Key Factors in the Selection of a Safety Integrity Level Assignment Method," ISA EXPO 2000, New Orleans, LA, August 2000.

[10] Marszal, E.M., and Scharpf, E.W., 2002, Safety Integrity Level Selection – Systematic Methods including Layer of Protection Analysis, First Edition, Instrumentation, Systems, and Automation Society (ISA), Research Triangle Park, NC.

[11] Marszal, E.M., *"Hydrocracker SIL Selection Case Study,"* ISA 2002 Technical Conference Series, Safety Instrumented Systems for the Process Industries, Baltimore, MD, May 2002.

[12] Windhorst, J.C.A., *"Over-pressure Protection by Means of a Designed System Rather Than Pressure Relief Devices,"* International Conference and Workshop on Risk Analysis in Process Safety, Atlanta, GA, October 1997.

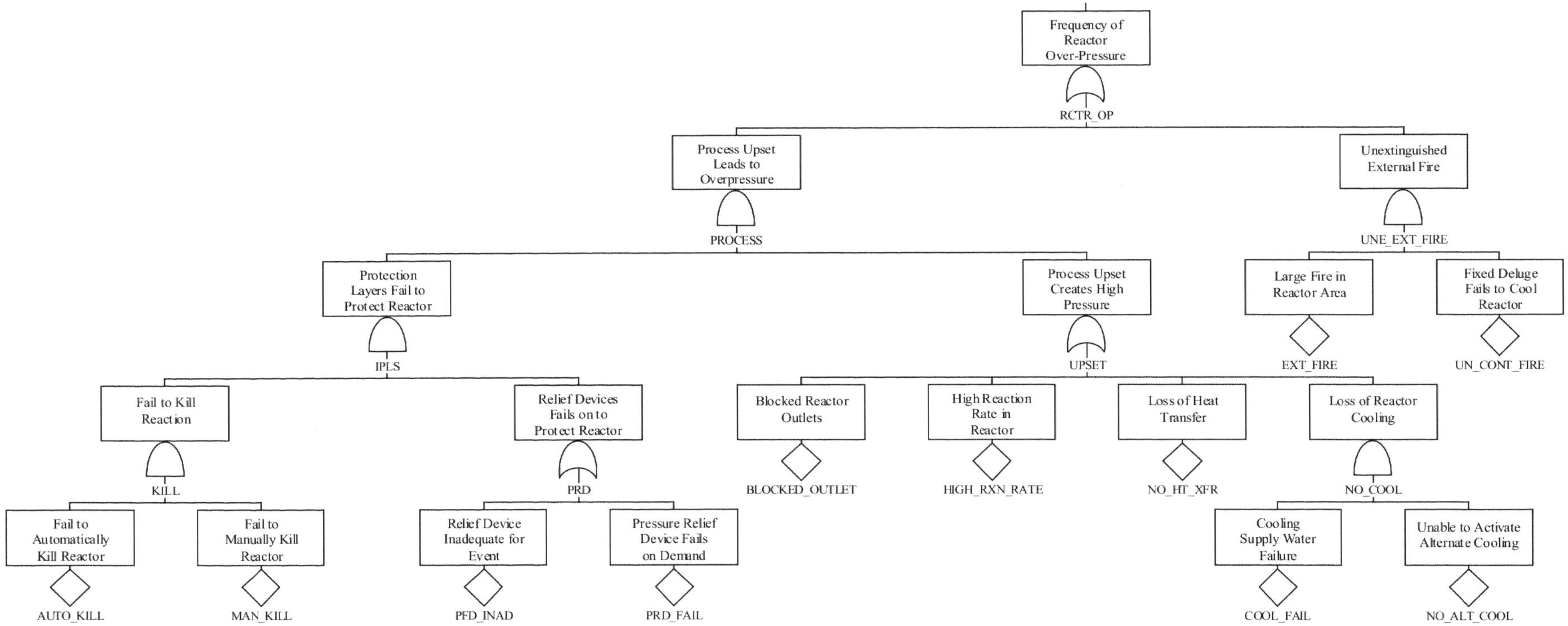

Figure 6 – Sample FTA (Top of Tree Only)

SAFETY MEASURE EVALUATION AND DESIGN FOR AN OFFSHORE PROCESSING FACILITY USING RECENTLY PROPOSED METHODOLOGY SCAP

Faisal I. Khan
Faculty of Engineering & Applied Science
Memorial University of Newfoundland
St. John's, NF
Canada A1B 3X5
<fkhan@engr.mun.ca>

and

Paul R. Amyotte
Department of Chemical Engineering
Dalhousie University
Halifax, NS
Canada B3J 2X4
<paul.amyotte@dal.ca.>

Prepared for Presentation at
American Institute of Chemical Engineers
38th Annual Loss Prevention Symposium
New Orleans, LA
April 26 – 29, 2004
Session T7003: Safety Instrumented Systems/Layer of Protection Analysis

November 2003

UNPUBLISHED

ABSTRACT

Risk assessment associated with engineering studies is aimed at providing decision support to achieve a cost effective and safe design of offshore process facilities. This is achieved by reducing the platform process risk through selection and design of cost effective safety measures. The main benefit from this approach is the design of optimal safety measures. In addition to optimization, significant savings are made possible by making the right decisions at the right time.

In this paper, a methodology for risk-based process safety decision making is described for offshore oil and gas (OOG) process activities. The methodology is applied to various offshore process units – compressor, separators, flash drum and driers of an OOG platform. Based on the risk potential, appropriate safety measures are designed for each unit. The paper also illustrates that implementation of the designed safety measures reduces the high risk values to an acceptable level.

Keywords: Offshore risk assessment, accidents, risk analysis, risk-based design, safety measures design

INTRODUCTION

The use of Quantitative Risk Analysis (QRA) in the offshore industry dates back to the mid 1970s. Further development of offshore QRA came in the early 1980s when the Norwegian Petroleum Directorate issued its guidelines for safety evaluation of platform conceptual designs. These guidelines required that QRA be carried out for all new offshore installations in Norway at the conceptual design phase. Since 1992 in the UK, safety case legislation has required, not formally but in practice, the industry to use offshore risk analysis as a component in the development of a safety case (Vinnem, 1998).

The concept of a 'life cycle' for an offshore installation is gaining wide acceptance within the offshore industry. The idea is to consider a development in totality from concept through design, construction, operation and to abandonment. A subset of this approach can be thought of as a 'safety life cycle' where only safety is considered. Figure 1 outlines one such safety life cycle. Most regulatory agencies require the submission of a safety case for offshore installations for each major phase, i.e. design, operation, and abandonment. The operational safety case is the one which companies with existing installations have been working on for the last two years or so (Finucane, 1994). A safety case must contain: an executive summary, details of the offshore installation, a description of the company safety management system, analysis of hazards affecting the installation, and identification and implementation of safety improvements.

A large component in many of these operational safety cases is a quantified risk analysis of the major hazards. This includes quantification of risk to individuals and/or groups of workers. The offshore fully manned production facility typically involves a number of stages of oil, gas and water separation, gas compression, and dehydration. The risk present on a typical offshore installation may be categorized as: process risk, dropped object risk, structural failure risk, helicopter accident risk, and ship collision risk. Depending upon the details of a specific installation, the percentage contribution of each risk category could vary significantly; for example, fire and explosion risk may be between 20 to 80% of individual risk (Finucane, 1994).

The inherent safety level of an offshore process facility can be increased by selecting the optimum design in terms of the installation/field configuration and the layout, which reduces the risk to a level that is As Low As Reasonably Practicable (ALARP). This requires the identification of major risk contributors and their assessment early in the project life cycle (HSE, 1993). The biggest challenge operators face is that of demonstrating the implementation of ALARP. Currently, most operators use a cost-benefit analysis method to demonstrate ALARP with respect to safety improvements (Finucane, 1994). Figure 2 shows one such structured approach (modified after Finucane, 1994). It is well recognized that if the structured approach of identification and assessment is not carried out early in the project, it is possible that the engineering judgment approach will fail to identify all of the major risks and thus loss prevention expenditures will be targeted in areas where there is little benefit. This would result in expensive remedial actions later during the life of the project.

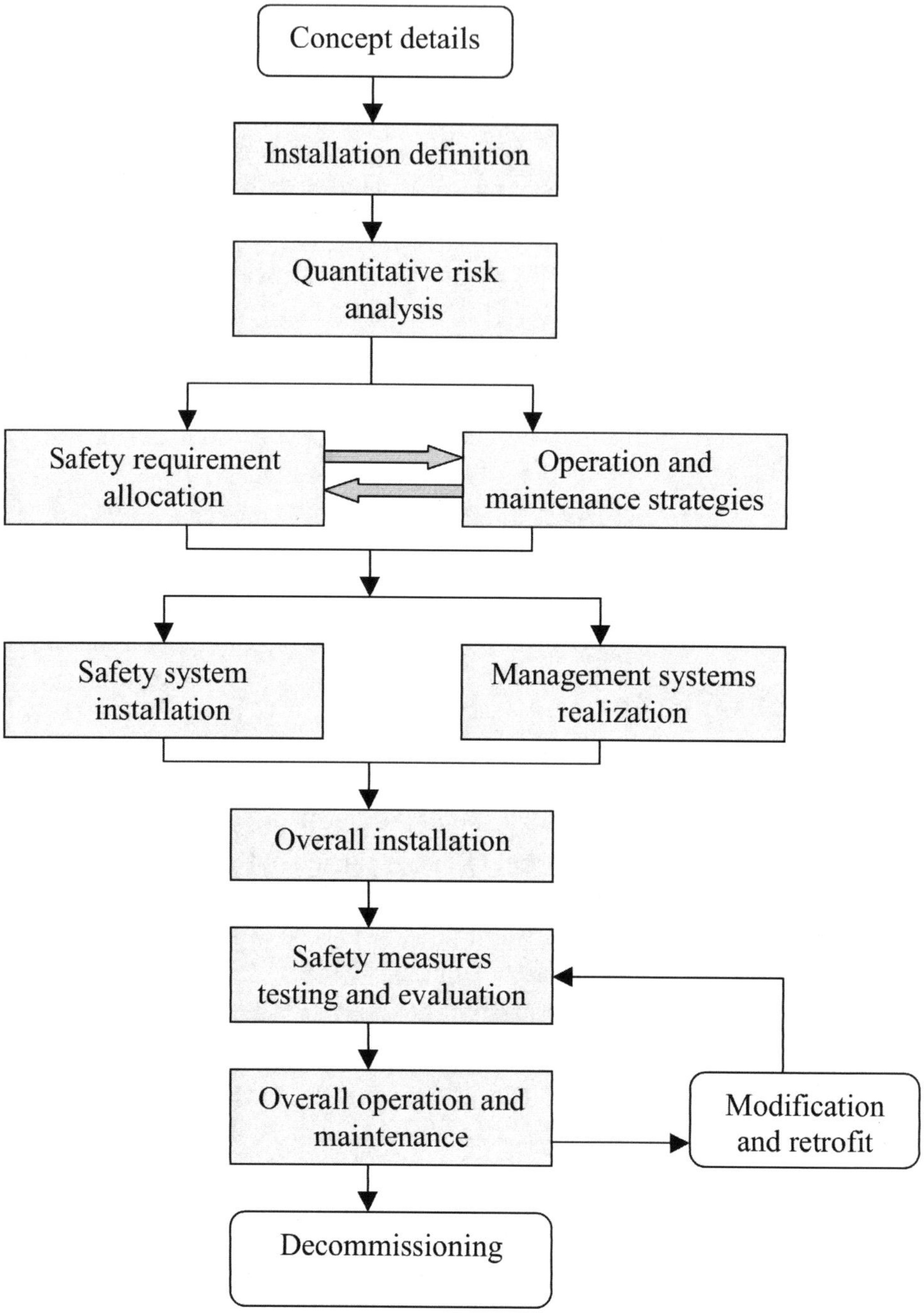

Figure 1 Safety life cycle for an offshore installation.

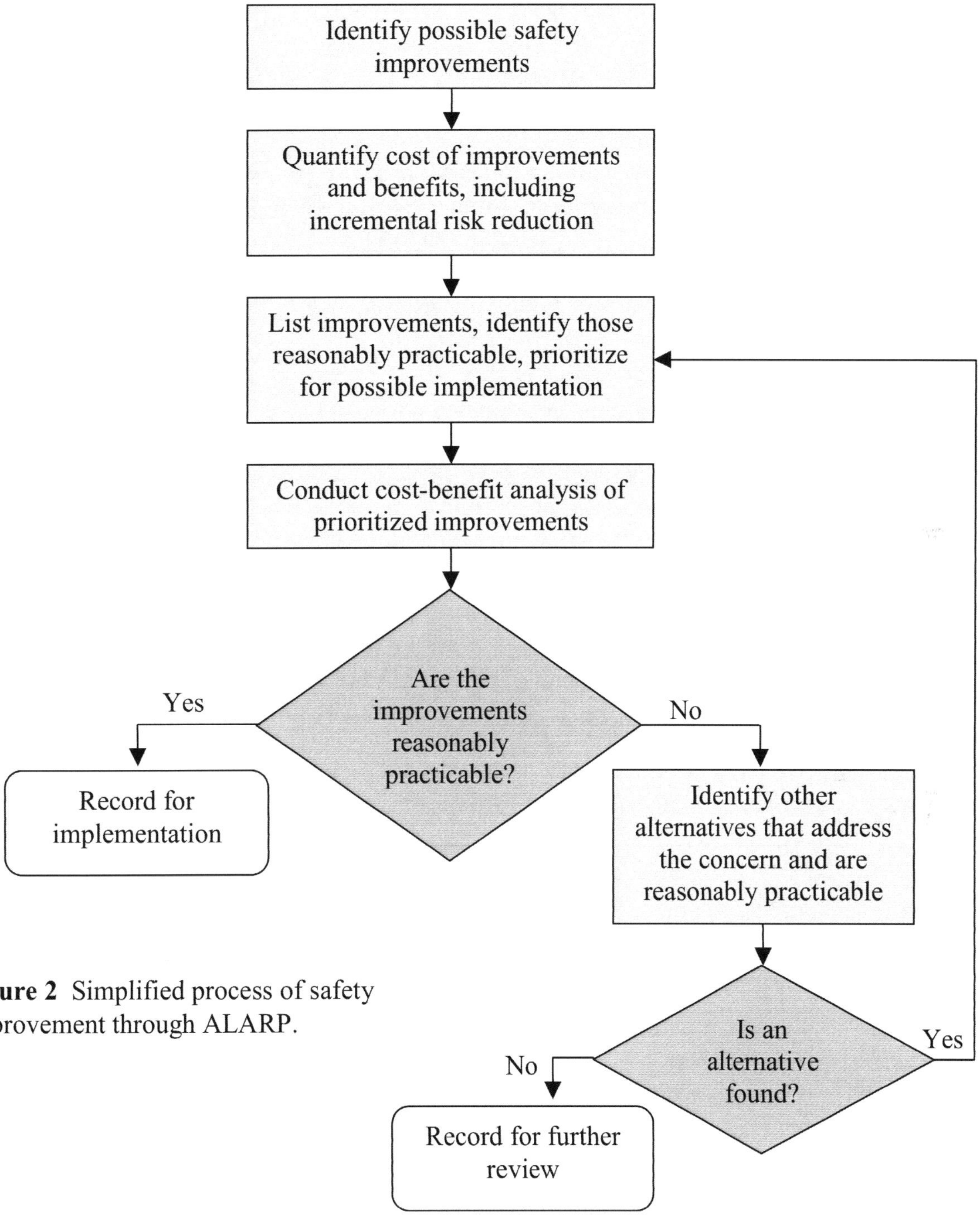

Figure 2 Simplified process of safety improvement through ALARP.

QUANTITATIVE RISK ANALYSIS OF OFFSHORE PROCESS OPERATIONS

The four main objectives in using QRA are: i) estimating risk (in absolute or relative terms), ii) determining design loads and conditions, iii) understanding hazard causation and potential escalation pathways, and iv) risk ranking. Most of these objectives usually imply quantification of risk to personnel, and often to the environment and assets. While there are many consequence models available for determination of personnel fatalities and injuries, considerable modeling work is still needed for determination of consequences to the environment and assets (Vinnem, 1998). The time and effort required to complete a full QRA for an offshore installation is a function of the complexity and size of the installation, which in the extreme case could require many months of effort spread over a prolonged period (Crawley, 1999; Crawley and Grant, 1997). This time scale precludes the support of the rapidly changing design, which is a feature of the concept development phase.

The United Kingdom Offshore Operators Association (UKOOA) has developed guidelines for an instrument-based protective system for application to OOG installations (UKOOA, 1995). Safety Integrity Level (SIL) determination is the key element of these guidelines. In its simplest form, the allocation of a SIL for a safety system is a way of specifying the appropriate level of reliability to match both the hazard and the tolerable risk. Therefore, to determine a SIL, one needs to consider both the severity and the likelihood of an incident.

Crawley and Grant (1997) proposed a screening tool for offshore risk assessment. This tool permits the risk assessment of many design options in a methodical, consistent and auditable manner. It is aimed at reducing front-end design costs and targeting design efforts in a cost-effective and safety-oriented manner. Vinnem (1998) presented a good overview of QRA use in offshore industries and emphasizes that QRA is an important tool in regulation development in various jurisdictions (e.g. UK, Norway, US and Canada). He has nicely emphasized that probabilities used in a QRA, whether in an absolute or relative sense, should be considered as 'notional' values, i.e. that they cannot directly be considered estimates of 'true' values.

Recently, Falck et al. (2000) and Brandsater (2002) have discussed the use of QRA in the design of an oil production system. They have detailed the use of QRA in safety and emergency preparedness analysis during the engineering and construction phase of the project. Though they have emphasized the use of QRA in the conceptual design stage, no such guidelines or methodology have been discussed. Further, they have emphasized the use of QRA as a decision support system for offshore process facilities.

The effective linkage between the QRA and the design team is essential and is key to QRA success. It is essential to ensure a proper understanding of the design problems so that these can be effectively addressed in the QRA; similarly QRA results must be understood by the design team and decision-makers. The QRA process should be synchronized with the engineering activity (Falck et al., 2000). It is necessary to aim for a living QRA, i.e. a risk model of the platform that is updated and refined in details as required. Assumptions made at an early stage to compensate for missing information need to be followed up and eventually replaced by factual information when available. Finally, the QRA results need to be 'translated' into engineering terms (Brandsater, 2002).

To meet these challenges, Khan and co-workers have proposed a quantitative methodology for safety measure design based on a feedback system of fault tree and credible accident scenarios.

The methodology, named SCAP (***S***afety/***C***redible/***A***ccident/***P***robabilistic fault tree analysis), has been applied to many onshore process industries (Khan et al., 2001a, b) and an offshore platform (Khan et al., 2002). It is effective in deciding and designing safety measures based on the risk potential. SCAP has been closely integrated with other engineering studies for scenario-based design in areas such as detector layout, passive protective measures, ventilation, and location of fire walls, blast walls, flame detectors, etc.

SCAP: A METHODOLOGY FOR RISK-BASED SAFETY ASSESSMENT AND CONTROL MEASURES DESIGN

The SCAP methodology has been revised for its application to offshore process facilities. The revised methodology encompasses all the characteristics of the original SCAP methodology. As previously mentioned, to demonstrate the effectiveness of SCAP it has been applied to an offshore process facility (Khan et al., 2002). Here, we recapitulate the methodology and its application with more detailed discussion on safety systems evaluation.

The architecture of the methodology is presented in Figure 3. As mentioned in Khan et al. (2002) the methodology is applicable at any stage of the design, and is particularly useful at the early design stage when the designer is free to adopt the suggested safety measures or modifications. Application of this methodology at the early design stage is possible as it requires data that are readily available at this stage. The reliability of a study conducted with such early stage data (involving uncertainties) is debatable, but can be argued on the basis of the following points:

i) The design and evaluation of safety measures based on the risk potential considers risk in relative terms. If there are uncertainties in risk computation, they are present in all units, and in relative terms will not have much effect.

ii) The techniques used in SCAP, such as analytical simulations with fuzzy set theory and MCAA (Maximum Credible Accident Analysis), are robust and less susceptible to input data uncertainty (Khan et al., 2001a, b). Therefore, early stage data (involving uncertainties) will not significantly affect the final outcome of the study.

This methodology tries to make the concept of a risk-based safer design a reality. It involves risk assessment steps which are interactively linked with the implementation of safety measures. The resultant system reveals the extent of risk reduction by each successive safety measure. Again, this methodology has been applied to many onshore process industries and has proven to be efficient and easy to use, as well as requiring limited data (Khan et al., 2001a, b).

The major steps of the revised SCAP methodology remain the same with modifications in their sub-steps (Figure 3). This paper provides a brief recounting of the revised SCAP methodology and its application to an OOG facility, with a detailed discussion on its applicability and effectiveness. The details of SCAP and other tools used in this paper are discussed in Khan et al. (2001a, b).

Hazard Identification Step

This step utilizes the revised Hazard Identification and Ranking (HIRA) system, as it is flexible and able to consider the vulnerability of offshore operations (Khan et al., 2002). The revised HIRA is comprised of two indices: fire and explosion damage index (B1) and toxic damage index (B2).

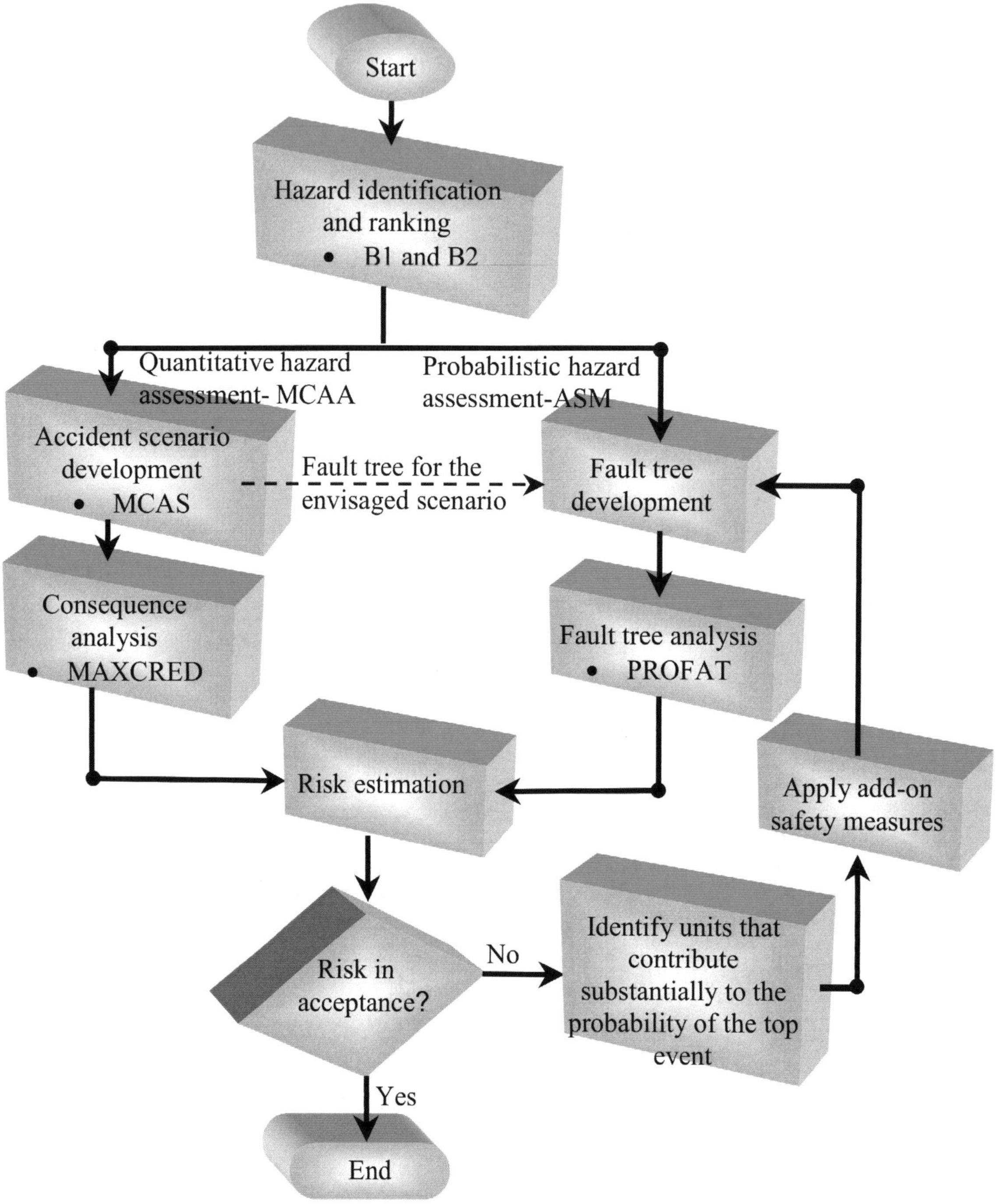

Figure 3 The SCAP algorithm.

Fire and explosion damage index (B1)

B1 is a representation of lethal heat and overpressure loads over an area. It is measured in terms of the radius of the area (in meters) affected lethally by overpressure and heat loads (50% probability of causing fatality). Estimation of B1 involves the following steps:

i) classification of the various units in an industry into the five categories mentioned below:
   - storage units
   - units involving physical operations such as heat transfer, mass transfer, phase change, pumping and compression
   - units involving chemical reactions
   - transportation units
   - other hazardous units such as furnaces, boilers, direct-fired heat exchangers, etc.,

ii) evaluation of energy factors,

iii) assignment of penalties, and

iv) estimation of damage potential.

Toxic damage index (B2)

B2 quantifies the toxic load over an area in terms of the radius (in meters) affected by a toxic load of 50% probability of causing a fatality. It is derived by using transport phenomena and empirical models based on the quantity of chemical(s) involved in the unit, the physical state of the chemical(s), the toxicity of the chemical(s), the operating conditions, and the site characteristics. The dispersion is assumed to occur under slightly stable atmospheric conditions. We have opted for 'slightly stable atmospheric conditions' as these represent a median of high instability and high stability. We believe that this assumption for dispersion may also hold in an offshore process facility where partial confinement may lead to a low likelihood of dilution.

The estimation of B2 is done with one core factor named the 'G factor' and seven penalties. The details of the G factor and penalty calculation may be found in Khan and Abbasi (1998a).

Why is revised HIRA appropriate here?

Revised HIRA is appropriate for the present application because:

i) It considers the impact of various process operations and associated parameters for hazard identification.

ii) It accounts for vulnerability due to the degree of unit congestion, characteristics of the surrounding unit, and site characteristics.

iii) It considers several operating conditions generally encountered in an offshore process operation.

iv) It provides quantitative results of good reliability.

v) It does not require a case-to-case calibration as its magnitude directly signifies the hazard level.

Quantitative Hazard Assessment Step

This step aims to quantify hazards; MCAA (Maximum Credible Accident Analysis) is the preferred approach. MCAA is comprised of two steps (Khan and Abbasi, 1997, 1998b): i) accident scenario forecasting, and ii) consequence analysis (damage estimation for the envisaged accident scenario).

Accident scenario forecasting

Forecasting likely accident scenarios is the most important step in this exercise. A number of accident scenarios can be envisaged in a unit; however, it may not be possible to analyze all these scenarios, particularly at an early design stage. A system which short-lists the important scenarios is needed. The screening or short-listing of accident scenarios has been debated since it was originally proposed by CCPS (1989). Subsequently, a modified 'worst-case accident scenario' approach has been practiced (Hirst and Carter, 2000). Although the CCPS and worst-case approaches are effective and easy to use, they focus only on one accident parameter – consequence. Recently, Khan (2001) proposed a 'maximum credible accident scenario' (MCAS) approach which considers both consequences and the likelihood of accident occurrence. Khan (2001) demonstrated that although certain accidents may not be the worst in terms of consequences, their high probability of occurrence is a major concern. These accidents often escalate and cause a catastrophe which is not even modeled by a worst-case accident scenario.

The MCAS approach centers on the theme of credibility, which is defined as a combination of impact area and the probability of occurrence, and is estimated as:

$$C = (L1^2 + L2^2)^{1/2}$$

where L1 and L2 represent the credibility factors estimated for fire and explosion hazard and toxic hazard, respectively.

Consequence analysis

Many computer-automated tools are available for a detailed consequence assessment for offshore facilities. COMEX, VENTEX, CLICHÉ, SCOPE2, ARAMAS, OHART, and PLATO are the most frequently used. Gardner et al. (1995) reviewed these hazard assessment tools. Complex computer models are also available for fire and explosion characteristics estimation, e.g. FLACS, μFLACS, REAGAS, EXSIM, and EXPSIM (Finucane, 1994; van Wingerden, 1994; Jones and Irvine, 1997).

These models are frequently used for a detailed QRA. However, their application at the early design stage is not an easy task due to the large data requirement and lengthy processing time. Though these models yield reliable detailed results, they may not be helpful at the early design stage. A computer-automated tool, MAXCRED (Khan and Abbasi, 1999), and its latest version, MAXCRED-III (Khan and Abbasi, 2000), perform MCAA. This tool enables the simulation of accidents and an estimation of their damage potential. MAXCRED-III also incorporates the domino/cascading effect (see Khan and Abbasi, 2000 for details).

MAXCRED-III has five main modules (options): scenario generation, consequence analysis, domino considerations, documentation, and graphics. In the scenario generation module, accident scenarios are generated for the unit under study. This step is based on the MCAS approach. The consequence analysis module involves the assessment of likely consequences if an accident scenario materializes. The consequences are quantified in terms of damage radii of different propensities. The domino module analyzes the damage potential of the primary event at the point of location of the secondary unit, and checks for the likelihood of occurrence of the secondary accident. If the probability of the secondary accident is sufficiently high, then appropriate accident scenarios are developed and analyzed for consequences. The graphics module enables the visualization of risk contours in the context of accident sites. This option has two facilities: (i) site drawing, and (ii) contour drawing. The documentation module of

MAXCRED-III deals mainly with the handling of different files: data file, scenario file, output file, and flow of information.

Probabilistic Hazard Assessment (Fault Tree Analysis) Step

The objective of this step is to quantify the probability of occurrence of the earlier envisaged accident scenario. Fault tree analysis (FTA), the most appropriate technique for this application, uses deductive reasoning to determine the occurrence of an undesired event. FTA along with component failure and human reliability data can help in determining the frequency of occurrence of an accidental event.

Methods for FTA include the analytical method (Yllera, 1988), the Monte Carlo simulation method (Papazoglou et al., 1992; Rauzy, 1993), and the Markov simulation method (Hauptmanns and Werner, 1990; James et al., 1993). Khan and Abbasi (2001) have proposed a methodology for probabilistic fault tree analysis: Analytical Simulation Methodology (ASM). ASM combines analytical methods with fuzzy mathematics, Monte Carlo simulations, and structure modeling. The ASM is easier, faster and involves less uncertainty in its predictions (Khan and Abbasi, 2001). A computer-automated tool, PROFAT (PRObabilistic FAult Tree analysis) was developed to perform ASM. ASM involves five main steps: i)fault tree development, ii) Boolean matrix creation, iii) finding of minimum cutsets and optimization, iv) probability analysis, and v) improvement index estimation.

The methodology is resolved into the computer software PROFAT, which is written in C++ and consists of five main modules: data, minimum cutsets analysis, probability analysis, improvement factor analysis, and general purpose modules, each of which performs a specific task, and is linked with the other modules (Khan and Abbasi, 2001).

Risk Quantification and Design of Safety Measures Step

Using the results of the hazard assessment and probabilistic hazard assessment steps, the individual risk and/or fatality accident rate (FAR) is computed and then compared with the regulatory standards. If they exceed the acceptance criteria, extra safety measures need to be implemented on the unit. After deciding the necessary safety options to be implemented, the probabilistic hazard assessment and hazard quantification steps are repeated and the latest individual risk and/or FAR is again computed and compared with the regulatory standards. This is repeated until the risk and/or FAR fall within the acceptable range.

RISK-BASED SAFETY MEASURES DESIGN FOR AN OFFSHORE PROCESS UNIT

The above methodology has been applied to decide the safety measures for various process units on an offshore platform; the detailed results of this study are given in Khan et al. (2002). Here, we are briefly recapitulating the important points of this case study with a detailed discussion on safety methods and their evaluation. The process plant on an offshore platform generally has three main parts: i) the wellhead, ii) separators, and iii) gas compression. The simplified layout of the process plant of a typical platform is depicted in Figure 4. Production lines from individual wells terminate at the wellhead, with each line being topped by a 'Christmas tree'. The well fluid passes through a manifold and is withdrawn at a production separator through a wing valve. The well fluid passing through the separators is separated into four major components. Oil is pumped through the main oil line to the onshore facility. Part of the condensate is pumped along with the oil. Gas is compressed using centrifugal compressors; it is subsequently passed through the flash drum where the temperature is reduced and condensate is formed and separated out. The gas, is subsequently dried and purified. It is then further compressed to high pressure through reciprocating compressors. Part of the gas is used at the wells and for power generation on the platform; the remaining gas is pumped to the onshore facility with a small amount being flared. The study does not include wellhead hazards but focuses on the major parts of the process plant (separation and compressors).

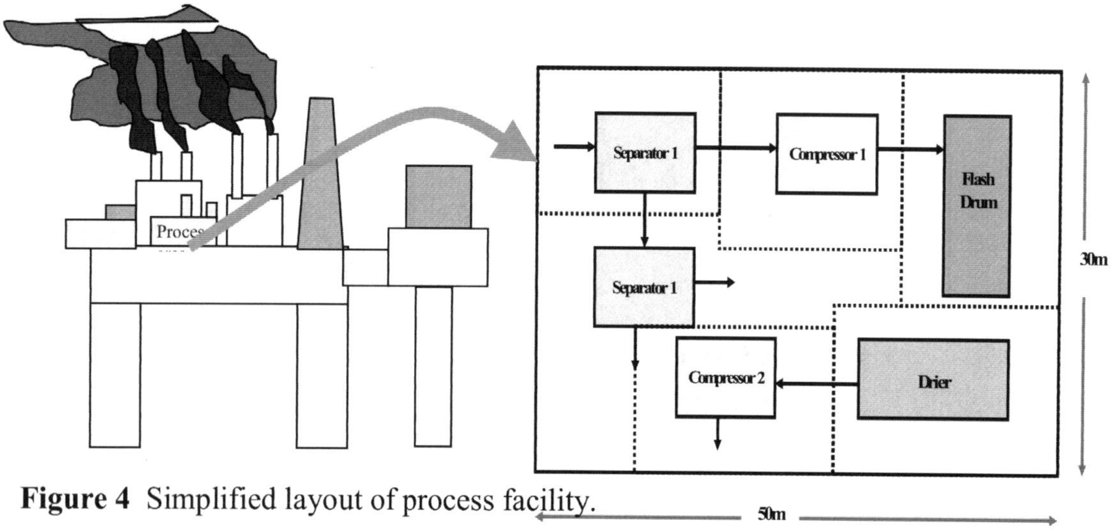

Figure 4 Simplified layout of process facility.

Hazard Identification

The complete process facility (separators, compressors, and pipelines) was subjected to a detailed study. Safety measures were designed and implemented on each process unit; however, in order to prioritize by importance, a hazard identification study was first conducted. The results of the study are plotted in Figure 5. It is evident from Figure 5 that the separators, compressors, drier, and flash drum are highly hazardous, whereas the oil and gas pipeline and pumps are moderately hazardous. To illustrate the SCAP methodology, a detailed study is presented on two highly hazardous units – condensate separator and compressors (1 and 2).

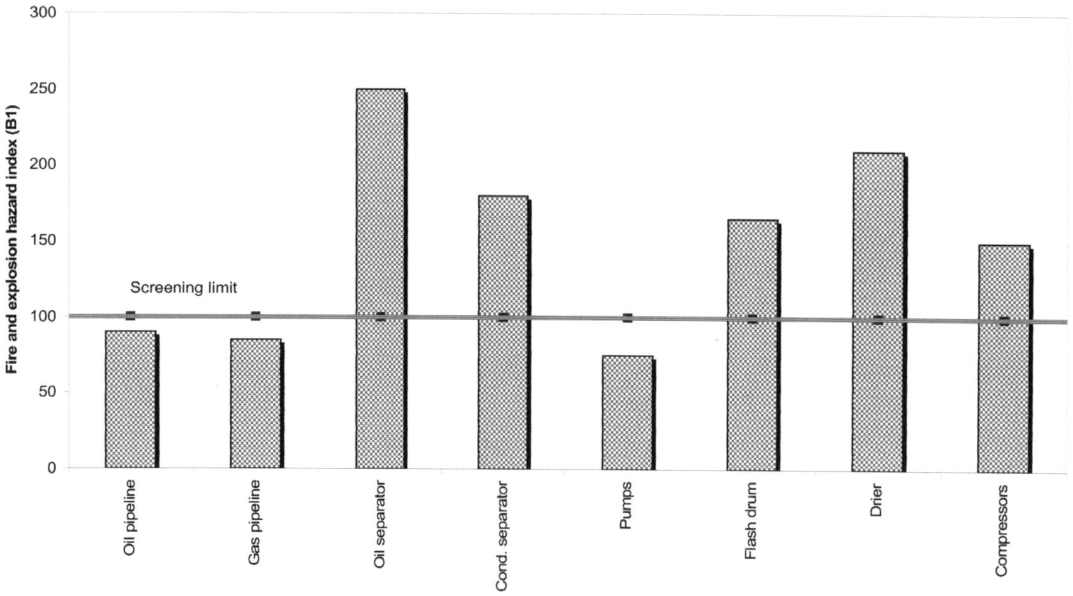

Figure 5 Results of hazard identification and ranking step.

Quantitative Hazard Assessment

Maximum Credible Accident Scenario Development

Many accident scenarios have been envisaged for each unit. The list of the most credible scenarios for the units presently under investigation is presented in Table 1. The credibility of an accident scenario was assessed considering the MCAS procedure as described earlier.

Table 1 Maximum Credible Accident Scenarios (MCAS) for different units.

Process Unit	Most Credible Accident Scenario	Description
Condensate separator	Vapor Cloud Explosion (VCE) followed by pool fire	Release of a chemical from the condensate separator would form a vapor cloud which on ignition would cause a VCE. Unreleased liquid in the unit would burn as a pool fire.
Compressors 1 and 2	Jet fire	Continuous release of flammable gas from compressors 1 and 2 on ignition would cause a jet fire.

Consequence Analysis

MAXCRED was used for detailed consequence analysis. In the following sections, the results for the separator and compressor units are discussed; results for other units are available in Khan et al. (2002).

Condensate separator

The results for MCAS in the condensate separator – Vapor Cloud Explosion (VCE) followed by pool fire – are presented in Table 2. A VCE followed by fire would cause considerable damage. It is evident from Table 2 that damage of a high degree of severity due to overpressure and shock wave would be operative over an area of ~50 m radius, while moderate damage (50% probability of lethality) would occur over an area of ~75 m radius. The unburned chemical in the unit would burn as a pool fire. The heat load generated due to the pool fire would be lethal over an area of 55 m radius. The heat load and shock wave generated by this unit may initiate secondary and a higher order of accidents in the units within close proximity, such as condensate and gas pipelines.

Compressors 1 and 2

The results of damage calculations for the compressor units (1 and 2) are presented in Table 3. It is evident from these results that this scenario would cause moderate damage. There is no likelihood of overpressure development; however, a fire jet of ~5 m in length would be operative. The lethal heat load of 50% probability of causing fatality and damage would be operative over an area of 35 m radius. It is likely that the jet flame would cause serious damage in the neighboring unit either through direct impingement or by external heat load. The units that would be affected by this accident are the flash drum and the drier.

Table 2 Results of consequence analysis for accident in condensate separator.

Parameters		Values
Unit: ***Condensate separator***		
Scenario: *VCE followed by pool fire*		
Explosion: VCE		
Total energy released by explosion	(kJ)	: 1.23e+07
Peak overpressure	(kPa)	: 320.00
Variation of overpressure in air	(kPa/s)	: 345.00
Shock velocity of air	(m/s)	: 353.00
Duration of shock wave	(ms)	: 8.0
Damage Radii (DR) for various degrees of damage due to overpressure		
DR for 100% complete damage	(m)	: 53
DR for 100% fatality or 50% complete damage	(m)	: 74
DR for 50% fatality or 25% complete damage	(m)	: 86
Fire: Pool fire		
Burning area	(sq.m)	: 265.00
Burning rate	(kg/s)	: 10.00
Heat flux	(kJ/sq.m)	: 2654.00
Damage Radii (DR) due to thermal load		
DR for 100% fatality/damage	(m)	: 34
DR for 50% fatality/damage	(m)	: 55
DR for 100% third degree of burn	(m)	: 69
DR for 50% third degree of burn	(m)	: 78

Table 3 Results of consequence analysis for accident in compressor units.

Parameters		Values
Unit: ***Compressor units***		
Scenario: *Jet fire*		
Fire: Jet fire		
Flame length	(m)	: 5.45
Burning area	(sq.m)	: 792.00
Burning rate	(kg/s)	: 10.00
Heat flux	(kJ/sq.m)	: 1493.00
Damage Radii (DR) due to thermal load		
DR for 100% fatality/damage	(m)	: 24
DR for 50% fatality/damage	(m)	: 35
DR for 100% third degree of burn	(m)	: 44
DR for 50% third degree of burn	(m)	: 57

Probabilistic Hazard Assessment (PHA)

Probabilistic hazard assessment was conducted for all units; however, only the results for the condensate separator and compressors are presented here. The failure frequency data used in the PHA (presented in Tables 4 and 6) were obtained from Worldwide Offshore Accident Databases (WOAD, 1998), HSE reports (HSE 1994, 1996), and offshore data from E&P Forum (1995). Using these data, fault tree analysis has been conducted to estimate the failure probability of each accident scenario with the results given below.

Condensate separator

The most credible accident scenario for this unit is envisaged as a VCE followed by a pool fire. There are 21 basic events that contribute directly and indirectly to the occurrence of this accident (Table 4). The likely logical sequences of events that lead to this accident are depicted in Figure 6.

The developed fault tree (Figure 6) was analyzed using PROFAT, and the results are presented in Table 5. The overall occurrence rate of this accident scenario is computed as 9.474E-04 per year. Table 5 indicates that events 18, 20, 12, and 17 contribute 17%, 17%, 12%, and 10%, respectively, in the occurrence of this accident. Controlling these events would considerably reduce the overall probability of occurrence of this accident scenario.

Table 4 Elements of the fault tree developed for a probable accident in condensate separator.

Number in Figure 6	Elements	Failure Frequency (/yr)
1	Leak from joints	0.045
2	Leak from main pipeline	0.003
3	Leak from joints	0.045
4	Leak from main pipeline	0.003
5	Leak from vessel	0.0015
6	Leak from fracture, joints or crack	0.0004
7	Leak from pipe connections	0.0065
8	Leak from safety valve	0.0055
9	Leak from pressure release valve	0.015
10	Leak from control valves	0.025
11	Outlet pipe chocked	0.0035
12	High pressure upstream line	0.17
13	Sudden phase change	0.017
14	External heat absorption causing increase in pressure	0.016
15	Ignition due to explosion energy	0.15
16	Ignition due to external heat from surroundings	0.20
17	Ignition due to electric spark	0.25
18	Release from pipe after explosion	0.10
19	Release from vessel aftermath of explosion	0.05
20	Ignition due to external explosion energy	0.20
21	Ignition due to fire heat load	0.25

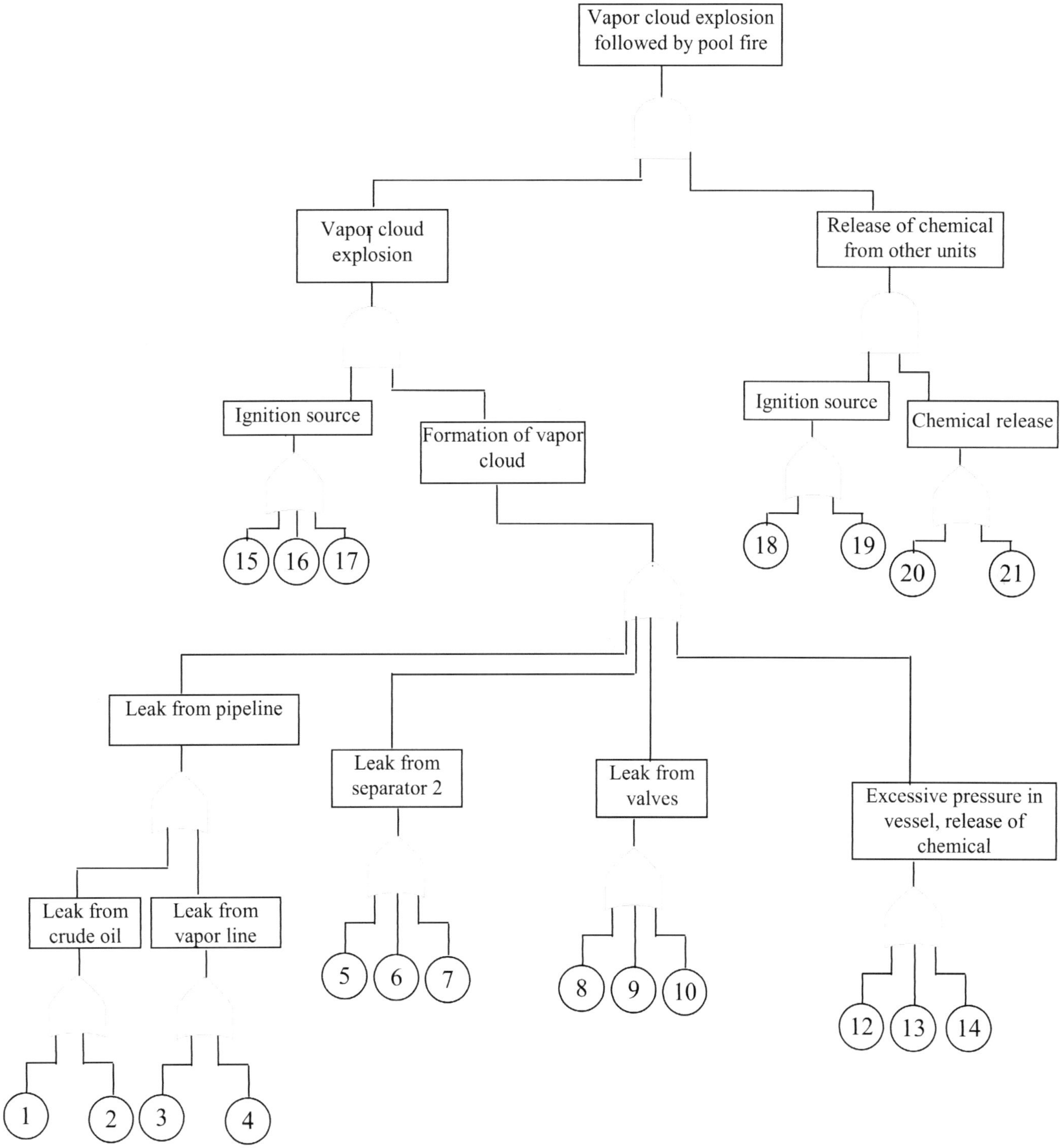

Figure 6 Fault tree diagram for condensate separator; details of basic events are presented in Table 4.

Table 5 Fault tree analysis results (output of PROFAT) for condensate separator.

Event not-occurring	Probability	Improvement	Improvement Index
0	9.474457e-04	0.000000e+00	0.000000
1	8.279830e-04	4.778510e-04	3.155792
2	9.397716e-04	3.069656e-05	0.202724
3	8.279830e-04	4.778510e-04	3.155792
4	9.397716e-04	3.069656e-05	0.202724
5	9.436756e-04	1.508045e-05	0.099593
6	9.465664e-04	3.517309e-06	0.023229
7	9.302496e-04	6.878450e-05	0.454262
8	9.329916e-04	5.781649e-05	0.381828
9	9.077042e-04	1.589659e-04	1.049832
10	8.810459e-04	2.655993e-04	1.754053
11	9.383557e-04	3.635992e-05	0.240126
12	4.958510e-04	1.806379e-03	11.92956
13	9.023399e-04	1.804231e-04	1.191538
14	9.050069e-04	1.697551e-04	1.121085
15	7.109045e-04	9.461649e-04	6.248599
16	6.318837e-04	1.262248e-03	8.336055
17	5.529077e-04	1.578152e-03	10.42232
18	3.161132e-04	2.525330e-03	16.67761
19	6.318094e-04	1.262546e-03	8.338019
20	3.161281e-04	2.525270e-03	16.67722
21	6.318094e-04	1.262546e-03	8.338019

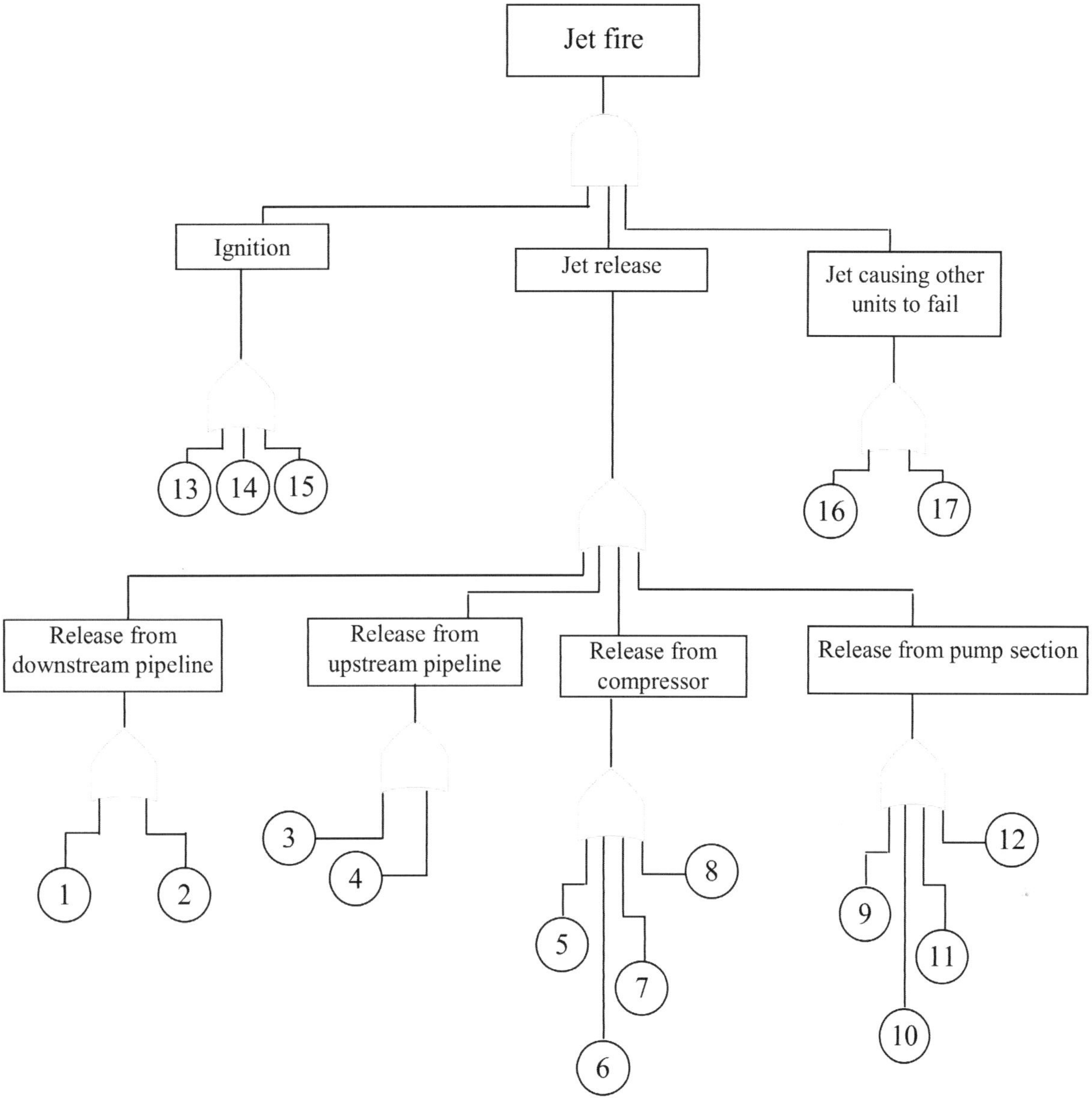

Figure 7 Fault tree diagram for compressor unit; details of basic events are presented in Table 6.

Table 6 Elements of the fault tree developed for a probable accident in compressor units.

Number in Figure 7	Elements	Failure Frequency (/yr)
1	Leak from compressor downstream pipeline	0.0065
2	Leak from compressor downstream pipeline joints	0.090
3	Leak from compressor upstream pipeline	0.003
4	Leak from joints of compressor upstream pipeline	0.045
5	Release from casing of compressor	0.050
6	Leaking of seal	0.120
7	Release from impeller	0.100
8	Compressor completely failed causing release of chemical	0.070
9	Leak from junction of pump and pipeline	0.010
10	Leak from rotor	0.060
11	Pump failed to operate and caused release of chemical	0.150
12	Leak from casing	0.200
13	Ignition due to explosion energy	0.150
14	Ignition due to external heat from surrounding	0.200
15	Ignition due to electric spark	0.250
16	Fire caused failure of pipeline leading to chemical release	0.010
17	Fire caused vessel to fail and release of chemical from vessel	0.005

Table 7 Fault tree analysis results (output of PROFAT) for compressors units.

Event not-occurring	Probability	Improvement	Improvement Index
0	1.364250e-02	0.000000e+00	0.000000
1	1.355903e-02	3.339117e-04	0.205645
2	1.248035e-02	4.648631e-03	2.862933
3	1.360403e-02	1.539034e-04	0.094784
4	1.306202e-02	2.321958e-03	1.430014
5	1.299739e-02	2.580464e-03	1.589220
6	1.209246e-02	6.200195e-03	3.818488
7	1.235117e-02	5.165338e-03	3.181155
8	1.273893e-02	3.614304e-03	2.225926
9	1.286812e-02	3.097529e-03	1.907662
10	1.170394e-02	7.754267e-03	4.775589
11	1.170394e-02	7.754267e-03	4.775589
12	1.105588e-02	1.034648e-02	6.372044
13	7.998807e-03	2.257479e-02	13.90304
14	9.132371e-03	1.804053e-02	11.11054
15	1.026367e-02	1.351535e-02	8.323643
16	9.132714e-03	1.803916e-02	11.10970
17	4.584522e-03	3.623193e-02	22.31400

Compressors 1 and 2

The fault tree comprising of 17 basic events has been developed for the MCAS in the compressor units (Figure 7). The probabilities of the occurrence of these basic events are presented in Table 6.

The developed fault tree was analyzed using the PROFAT algorithm, which computed the total occurrence rate of the top event as 1.364E-02 per year. Results reveal that events 17, 13 and 14 are the most crucial ones and contribute about 47% in initiating the accident (Table 7). Controlling these basic events would drastically reduce the probability of occurrence of this scenario.

Risk Quantification

Using the results of the previous steps, risks were computed for all the units identified as hazardous and moderately hazardous. Interesting results are observed. Though the compressor units are moderate in damage-causing capabilities, they were found to pose the greatest risk. This is because of their high probability of failure. The unit observed to be the most disastrous as a result of damage calculations – the oil separator– was found to pose comparatively less risk, due to its low probability of failure. Figure 8 presents a summary of the average individual risk factors caused by the two main units along with the ALARP criteria. Analysis of these results reveals that the compressor units followed by the condensate separator pose high individual risk. Their risk and FAR values exceed the ALARP acceptance criteria. These units need attention in order to bring these high risks to an acceptable level.

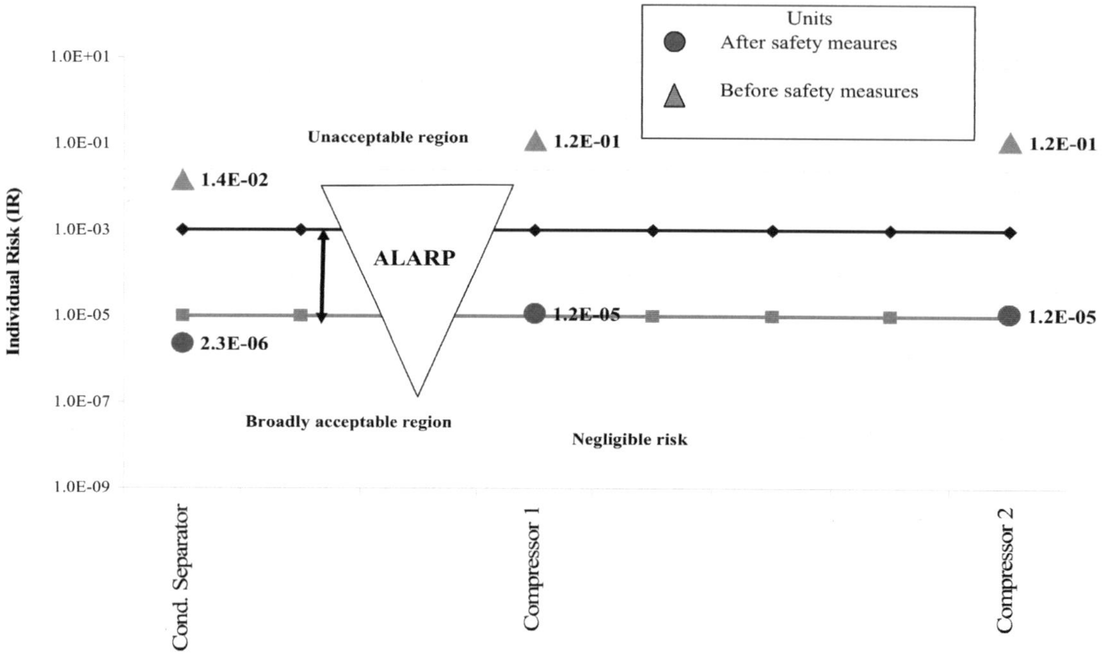

Figure 8 Individual risk with ALARP criteria.

Risk reduction through safety measures – MCAA-PFTA controller system

Safety measures evaluation

Explosion venting

The most important method of mitigating the effects of gas explosions is venting. A vent opening is introduced to limit pressure build-up (van Wingerden, 1994). The size of this opening is chosen in such a way that pressure build-up due to the explosion is sufficiently compensated by outflow of burned and unburned gas. However, it is not just the size of the vent opening which is important. The location of the vent opening and the choice of vent cover (vent panels) are also important.

When there is sufficient venting close to the ignition point, burned gases will be vented quickly. As a result, expansion flow in the direction of obstacles and the turbulence generated behind the obstacles will be limited (van Wingerden, 1994). Hence, the overpressure will be low. However, when the venting is less effective in the early phase of the explosion, the free unburned gas will be pushed ahead of the flame, a strong turbulent flow field will be generated and a positive feedback mechanism will accelerate the flame and cause high overpressure (van Wingerden, 1994). This emphasizes that venting hot combustion products at an early stage of the explosion is a very effective means of reducing flame acceleration. The most important factors for effective venting are: the size of the vent area, how the vent area is distributed, the direction of explosion relief, and how quickly explosion relief panels are activated. The vent area should simply be as large as possible. There are many methods available nowadays to determine the correct size for a vent opening. It is important that the gas explosion venting be directed into open areas with a minimum of obstructions.

Compartment shape

The shape of the compartment and the location of the vent areas are closely linked. There are two main principles applied in optimizing the shape of a compartment: an ignition point anywhere in the compartment should be as close as possible to the major vent areas, and strong turbulence in the unburned gas ahead of the flame and long flame travel distances should be avoided. For a compartment with explosion venting on two end walls, the ideal shape is a cubic box. A relatively low explosion pressure can be expected in such a configuration (van Wingerden, 1994). Most explosion scenarios will give high pressures if the module is elongated and vent openings are only located on the two ends. It is even more important to avoid an elongated shape if the compartment only has a vent opening in one of the side walls. In the case of ignition at a closed end wall, the flame can accelerate over a long distance and venting has no beneficial effect since it only leads to flow past obstacles and hence to turbulence generation.

Water deluge

Water deluge is another important safety measure to mitigate fires and explosions. It is important to understand that water deluge should not be considered as an alternative to gas explosion venting, but as an additional method under certain conditions thus making explosion venting more effective. Bjerketvedt and Bjokhaug (1991) undertook a pilot experimental investigation addressing the effect of water sprays on gas explosions. They used a 1:5 scale model of an offshore module. The main beneficial effect identified was the effect of water vapor on the burning rate. Extraction of heat from the flame front played a minor role. In strong explosions, the large droplets break up due to the flow speeds generated in the unburned mixture ahead of the flame front. These flow speeds cause drag forces which act on the droplets and deform them. It was observed that water deluge is only effective for explosions in which high flame accelerations occur.

Unfortunately, there are disadvantages related to the use of a water deluge system for explosion suppression. Since the activation time for an ordinary deluge system is much longer than the duration of the explosion, the deluge system has to be activated on gas detection. On activation, the water pumps are started and water will flow to the deluge system in the relevant module. This means that there is a period during which there is a possibility of ignition without water deluge being in full operation. In addition, there have been reports of accidents where the probable ignition source was a discharge in electrical equipment due to moisture from the deluge system (van Wingerden, 1994; Wighus, 1994). Water sprays and deluge should therefore be activated only in compartments with waterproof electrical equipment.

For fire protection, fine water sprays have been introduced and tested for offshore applications (Shetty et al., 1998). The experiments so far have shown that they are a good alternative for protection of enclosed or even partly enclosed spaces where hydrocarbon fuels may create large fires. Their efficiency in extinguishing smaller fires is less as compared to Halon protection. Even if the fine water sprays do not extinguish the smaller fires, a reduction of fire intensity is obtained, and the potential damage can be greatly reduced. The fine water spray may be produced in different ways, and three different types of system are available. A twin-fluid system where water is forced through narrow holes in a nozzle by air or other gases is self-contained with a water reservoir and a pressurized gas bottle. Another system supplies high-pressure water (100 bar) to a nozzle with narrow holes, and a third type of system utilizes a rotating nozzle to produce fine droplets. The rotating nozzle system operates at low water-supply pressures (5 – 10 bar) (Wighus, 1994).

Obstruction management

A compartment will contain process equipment, pipe-work, rooms etc. During a fire and explosion, these objects will obstruct the flow and cause turbulence. These objects will also interfere with venting. The main principle of the guidance here is to arrange the obstructing objects so that: (i) minimum turbulence is generated, and ii) venting is not blocked.

Ignition source management

Both the strength and location of the ignition source play an important role in determining the course of fire and explosion events offshore. In general, effective mitigation of fires/explosions is obtained if the ignition point is close to the vent area. However, if the venting of combustion products is not sufficient to keep the flame speed at a low level, ignition at the edge of a congested area may cause high explosion pressures. It is imperative that attempts are made to avoid ignition sources. Known, potential ignition sources should be located such that worst-case scenarios are avoided and flame arrestors/suppressors provided at all vulnerable locations.

Blast walls

When aiming to add barrier walls in order to reduce the global overpressure or heat load, one should consider (Berg et al., 2000):

- The location of the critical equipment with respect to the wall position. The wall should be used to protect the equipment from the explosion/fire heat load and not to confine the area around the critical equipment.
- The location of the possible ignition source. The wall should be located between the potential existing ignition sources and the critical equipment.
- The ability of the wall to contain a possible gas leakage. Therefore, both dispersion and explosion simulations should be run for any wall partitioning assessment.

Weak and strong walls can have the same gas-cloud containing capacities. However, in the case of a large leakage, with almost full gas coverage, the resulting pressure increase from an accidental explosion is expected to be generally smaller with weak walls than with strong walls (Berg et al., 2000). If the wall is able to contain a possible gas leakage then:

- The pressure increase on the side containing the gas cloud will be higher with a strong blast wall than with a weak wall.
- The pressure decrease on the opposite side will be higher with a strong blast wall than with a weak wall.

One should then choose between a weak or strong wall depending on where the critical equipment is located. The cost and weight of a strong wall will be very different from a weak wall.

Effect of separation

Adding a gap between the process modules can reduce the potential overpressure level, especially in the area located just after the gap. This is due to a significant drop in flame acceleration. Furthermore, there seems to be a maximum critical gap size beyond which the pressure level is not reduced even by increasing the gap width. The study conducted by Berg et al. (2000) found that there was no further significant reduction in overpressure by increasing the gap beyond 15 m. As space is always a limiting factor and concern for an offshore platform, this option is not very promising.

Risk re-evaluation

A risk reduction exercise was conducted by incorporating various safety measures and add-on control measures as described above. The possible safety and hazard control options considered to reduce the risk are given in Table 8 (Santos-Reyes and Beard, 2000; Lees, 1996). From these, various combinations of control measures were selected to reduce the risk potential of a unit. When these measures are taken into account, the unit fault tree is modified as shown in Figure 9 (compressor unit). On analyzing the new fault tree (Figure 9), the frequency of occurrence of the top event (envisaged accident) is reduced to 1.311E-06, which is 10,373 times (i.e. several orders of magnitude) lower than the previous value. The individual risk and FAR values after the implementation of control measures for this unit lie well within the acceptable range (Figure 8). The FAR value was reduced from 11,127 to 1.0.

Table 8 Control measures implemented over different units to reduce the risk.

Control Measures	Frequency of Failure (/yr)
Flame arrester	0.040
Water sprinkling system	0.045
Flammable gas detector	0.065
Advanced control mechanism, i.e. feed forward, cascade control, neural network based control, DDC	0.005
Advanced final control element (digital controller)	0.002
Installation of emergency relief system against over-pressurization of separators, flash drum, and drier	0.050
Check valve with relief provision to flare	0.030
Installation of bypass line	0.004
Preventive maintenance of pumps	0.100
Preventive maintenance of compressors	0.150
Preventive maintenance of pipeline	0.070
Leak detector in compressor and pumping section	0.057
Installation of safe venting system on pipeline	0.010
Installation of blast barriers	0.030
Installation of external cooling system for separators and drier	0.045
Installation of inert gas purging system to prevent flammable gas cloud formation	0.065

Similarly, after deciding the safety measures (Table 8), the fault tree for the condensate separator is modified as shown in Figure 10 and processed through PROFAT for probability estimation. The results reveal that after implementing the safety measures, the probability of occurrence decreases to 1.555E-08. Using the revised value of the probability of occurrence, the average individual risk decreases to 1.55E-07 and the FAR value reduces from an original value of 1291 to 0.01. These values fall within the acceptable zone of the ALARP criteria.

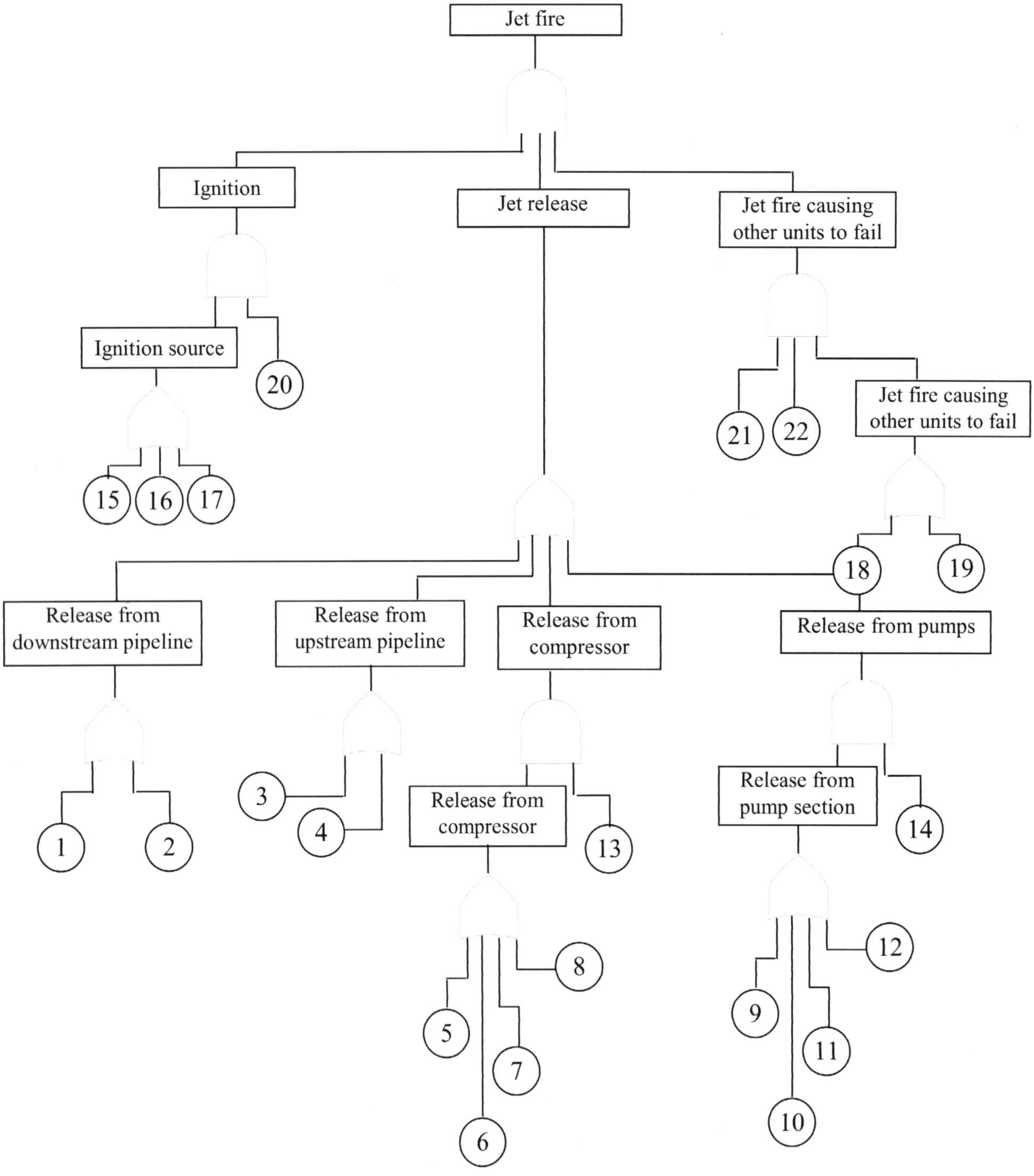

Figure 9 Modified fault tree diagram for compressor unit after implementing safety measures.

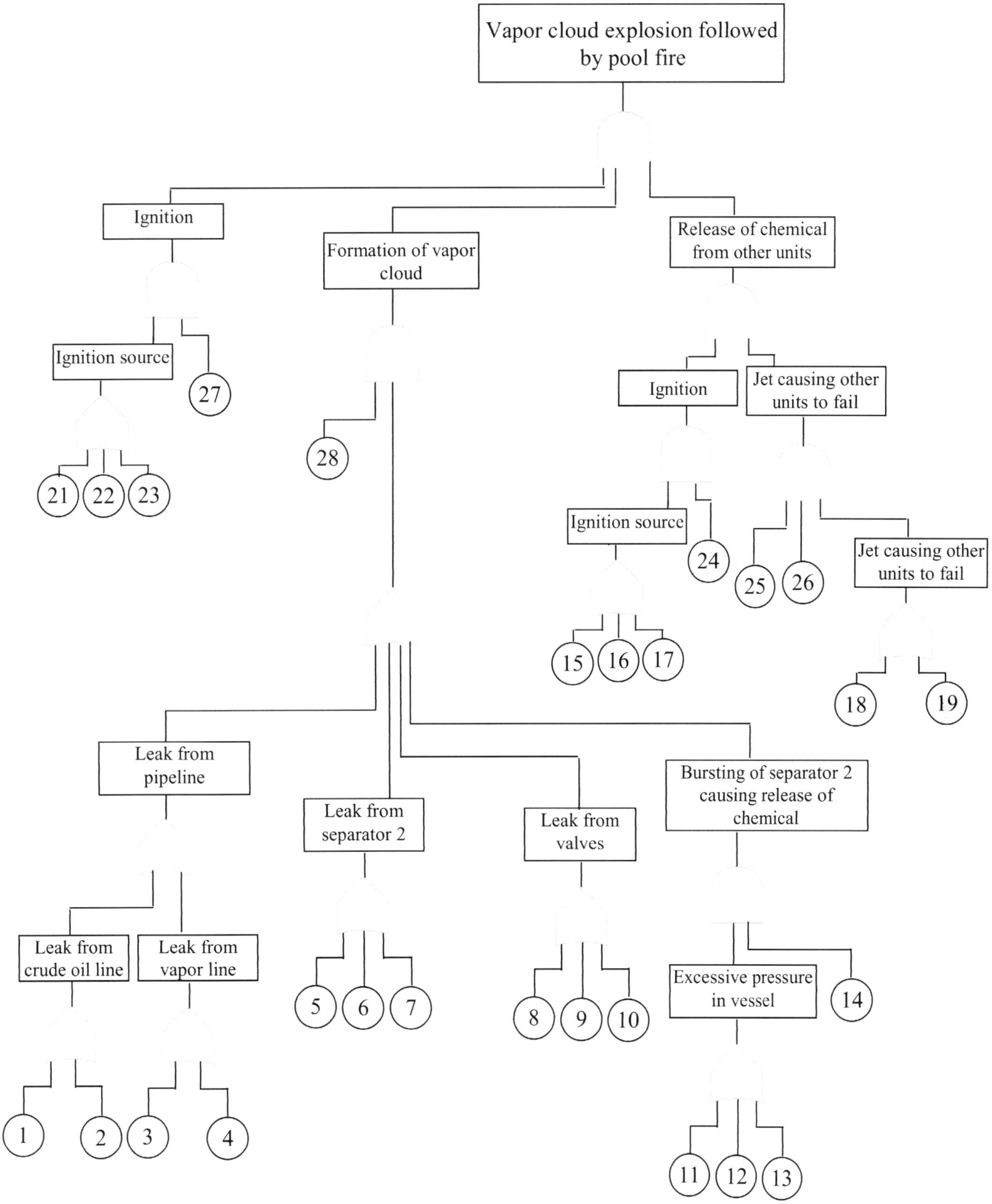

Figure 10 Modified fault tree diagram for condensate separator after implementing safety measures.

Further (and as shown in Khan et al., 2002), upon incorporation of safety measures for the oil separator, the flash drum and the drier, the frequency of occurrence reduces to a range of 1.0E-06 to 1.0E-08. The average individual risk and FAR values for these units after implementing the safety measures fall well within the ALARP acceptable region.

CONCLUDING REMARKS

The main objective of risk assessment and the associated engineering studies is to provide decision support to achieve a cost effective and safe design. In practice, this is accomplished through the design requirements and a reduction of the platform risk level with cost effective safety measures.

The main benefit from this approach is that of obtaining optimal safety measures with the end result being a safer platform design. In addition to optimization, it is realistic to assume that significant savings may be made by making the right decisions at the right time. A wealth of experience shows that risk assessments carried out too late (on existing or frozen designs) result in excessive costs for modifications and changes, or reveal solutions where unsafe designs cannot be satisfactorily resolved or mitigated.

In this paper, a strong case is made for risk analysis to be considered a design tool, much more so than as a tool for verification of a safe design, as emphasized by Falck et al. (2000). Although serious concerns were raised when risk assessment was introduced in the early 1980s, the usefulness of the technique was realized by many within the first couple of years. Presently, the approach is considered effective and successful. The important role of risk considerations as a design tool builds on comparative risk assessment, not the assessment of risk in an absolute sense. A risk of 10^{-5} has no other meaning than expressing that this hazard is much less significant than those other hazards at a risk level of 10^{-4} or 10^{-3}. The design tool risk assessment always needs to be quantitative – in terms of consequence calculations and also probability analysis. In the US as well as Canadian offshore areas, developments are being taken into deeper waters where larger installations and more significant economic exposure are prevalent. Both the economic exposure as well as the exposure of personnel to hazards will be expected to call for dedicated detailed engineering risk assessments.

This paper presents a revised version of the recently proposed SCAP methodology for risk-based safety management for offshore process activities through a quantitative feedback system of probabilistic risk assessment. It illustrates the application of the discussed methodology to a typical offshore process facility. The methodology is a combination of five quantitative steps; each requires an independent technique and computer-aided tools.

The effectiveness of this methodology has been demonstrated by applying it to a typical offshore process facility. From the initial phase of the case study, it was observed that compressor units inherit maximum risk due to their higher probability of failure. However, after implementing safety measures, the probability of occurrence was reduced drastically, thereby causing a substantial reduction in risk.

ACKNOWLEDGEMENTS

The authors gratefully acknowledge the financial support provided by the Natural Sciences and Engineering Research Council of Canada (NSERC) and the Canada Foundation for Innovation (CFI).

REFERENCES

Berg, J.R., Bakke, J.R., Feranley, P. and Brewerton, R.B. (2000). *CFD Layout Sensitivity Study to Identify Optimum Safe Design of a FPSO*, In Proceedings of Offshore Technology Conference, May 1-4, 2000, Houston, TX, USA.

Bjerketvedt, D. and Bjokhaug, M. (1991). *Experimental Investigation – Effect of Water Sprays on Gas Explosions*, Prepared by Christian Michelsen Institute for UK Department of Energy, OTH 90 316, HMSO, London, UK.

Brandsater, A. (2002). Risk Assessment in the Offshore Industry, *Safety Science*, **40**, 231.

CCPS (1989). *Guidelines for Chemical Process Quantitative Risk Analysis*, American Institute of Chemical Engineers, NY, USA.

Crawley, F.K. (1999). The Change in Safety Management for Offshore Oil and Gas Production Systems, *Process Safety and Environmental Protection*, **77**, 143.

Crawley, F.K. and Grant, M.M. (1997). Concept Risk Assessment of Offshore Hydrocarbon Production Installations, *Process Safety and Environmental Protection*, **75**, 157.

E&P Forum (1995). *Leak and Ignition Database*, Available from DNV, Hovik, Norway.

Falck, A., Skramstad, E. and Berg, M. (2000). Use of QRA for Decision Support in the Design of an Offshore Oil Production Installation, *Journal of Hazardous Materials*, **71**, 179.

Finucane, M. (1994). The Adoption of Performance Standards in Offshore Fire and Explosion Hazard Management, *Fire Safety Journal*, **23**, 171.

Gardner, D.J., Hulme, G., Hughes, D.J., Evans, R.F. and Brington, P. (1995). In IChemE Symposium Series No. 134, Institution of Chemical Engineers, Rugby, UK.

Hauptmanns, V. and Werner, W. (1990). *Engineering Risks*, Springer-Verlag, Berlin, Germany.

Hirst, I.L. and Carter, D.A. (2000). A "Worst Case" Methodology for Risk Assessment of Major Accident Installations, *Process Safety Progress,* **19**, 78.

HSE (1993). *Draft Offshore Installations (Fire and Explosion, and Emergency Response) Regulations and Approved Code of Practice*, Consultative Document 64, Health and Safety Executive, Sheffield, UK.

HSE (1994). *Statistical Report from Leak and Ignition Database*, Health and Safety Executive, London, UK.

HSE (1996). *Offshore Accident/Incident Statistics Reports*, OTO 96.954. Health and Safety Executive, London, UK.

James, H., Harris, M. J. and Hall, S. F. (1993). Comparison of Event Tree, Fault Tree and Markov Methods for Probabilistic Safety Assessment and Application to Accident Mitigation, *Major Hazards Onshore and Offshore*, **IChemE Series # 130**, 59.

Jones, J.C. and Irvine, P. (1997). PLATO Software for Offshore Risk Assessment: A Critique of the Combustion Features Incorporated, *Journal of Loss Prevention in the Process Industries*, **10**, 259.

Khan, F.I. (2001). Development of Maximum Credible Accident Scenarios for Realistic and Reliable Risk Assessment, *Chemical Engineering Progress*, November 2001, 56.

Khan, F.I. and Abbasi, S.A. (1997). A Maximum Credible Accident Analysis Based Quantitative Risk Assessment Study of Chemical Process Industry, *Indian Chemical Engineer*, **A39**, 92.

Khan, F.I. and Abbasi, S.A. (1998a). Multivariate Hazard Identification and Ranking System, *Process Safety Progress,* **17**, 157.

Khan, F.I. and Abbasi, S.A. (1998b). Techniques and Methodologies for Risk Analysis in Chemical Process Industries, *Journal of Loss Prevention in the Process Industries*, **11**, 261.

Khan, F.I. and Abbasi, S.A. (1999). MAXCRED – A New Software Package for Rapid Risk Assessment in Chemical Process Industries, *Environment Modeling and Software*, **14**, 11.

Khan, F.I. and Abbasi, S.A. (2000). Assessment of Risks Posed by Chemical Industries – Application of a New Computer Automated Tool MAXCRED-III, *Journal of Loss Prevention in the Process Industries*, **13**, 12.

Khan, F.I. and Abbasi, S.A. (2001). Analytical Simulation and PROFAT II: A New Methodology and a Computer Automated Tool for Fault Tree Analysis in Chemical Process Industries, *Journal of Hazardous Materials*, **A75**, 1.

Khan, F.I., Husain, T. and Abbasi, S.A. (2001a). Design and Evaluation of Safety Measures Using a Newly Proposed Methodology "SCAP", *Journal of Loss Prevention in the Process Industries*, **15**, 129

Khan, F.I., Iqbal, A., Ramesh, N. and Abbasi, S.A. (2001b). SCAP: A New Methodology for Safety Management Based on Feedback From Credible Accident – Probabilistic Fault Tree Analysis System, *Journal of Hazardous Materials,* **A87**, 23.

Khan, F.I., Sadiq, R. and Husain, T. (2002). Risk-Based Process Safety Assessment and Control Measures for Offshore Process Facilities, *Journal of Hazardous Materials*, **A94**, 1.

Lees, F.P. (1996). *Loss Prevention in the Process Industries*, Butterworths, London, UK.

Papazoglou, I.A., Nivoliantiou, A.O. and Christou, M. (1992). Probabilistic Safety Analysis in Chemical Installation, *Journal of Loss Prevention in the Process Industries*, **5**, 181.

Rauzy, A. (1993). New Algorithms for Fault Tree Analysis, *Reliability Engineering and System Safety*, **40**, 203.

Santos-Reyes, J. and Beard, A.N. (2000). *A Systematic Approach to Managing Fire Safety on Offshore Installations*, In Proceedings of Offshore Technology Conference, May 1-4, 2000, Houston, TX, USA.

Shetty, N.K., Soares, C.G., Christensen, P.T. and Jensen, F.M. (1998). Fire Safety Assessment and Optimal Design of Passive Fire Protection for Offshore Structures, *Reliability Engineering and System Safety*, **61**, 139.

UKOOA (1995). *Guidelines for Fire and Explosion Hazard Management*, United Kingdom Offshore Operators Association, London, UK.

van Wingerden, K. (1994). Course and Strength of Accidental Explosions on Offshore Installations, *Journal of Loss Prevention in the Process Industries*, **7**, 295.

Vinnem, J.E. (1998). Evaluation of Methodologies for QRA in Offshore Operations, *Reliability Engineering and System Safety*, **61**, 39.

Wighus, R. (1994). Fires on Offshore Process Installations, *Journal of Loss Prevention in the Process Industries*, **7**, 305.

WOAD (1998). *Worldwide Offshore Accident Databank*, Veritec, Oslo, Norway.

Yllera, J. (1988). *Modularization Methods for Evaluating Fault Tree of Complex Technical System*, In Kandel, A. and Avni, V. (Eds), Engineering Risk and Hazard Assessment, Vol 2, CRC Press, Inc, FL, USA.

Layer of Protection Analysis: Generating Scenarios Automatically from HAZOP Data

Arthur M. (Art) Dowell, III, P.E.
Senior Technical Fellow
Hazard Analysis
Rohm and Haas Company
6519 State Highway 225
Deer Park, TX 77536 USA

Thomas R. (Tom) Williams
Senior Risk/Reliability Engineer
ABS Consulting
Risk Consulting Division
10301 Technology Dr
Knoxville, TN 37932-3392 USA

Prepared for Presentation at the
American Institute of Chemical Engineers
2004 Loss Prevention Symposium
Session T7003, Safety Instrumented Systems/Layer of Protection Analysis
New Orleans, LA
April 26-29, 2004

UNPUBLISHED

Layer of Protection Analysis: Generating Scenarios Automatically from HAZOP Data

Arthur M. (Art) Dowell, III, P.E.
Rohm and Haas Company
Deer Park, TX USA

Thomas R. (Tom) Williams
ABS Consulting
Knoxville, TN USA

ABSTRACT

This paper details the concept of automatically generating LOPA scenarios from a process hazard analysis (PHA) conducted using a methodology such as HAZOP. Specialized software selects consequences that meet severity criteria or risk criteria. It then takes each end consequence, follows each link path to an initiating cause, and presents each rolled up link path as a single LOPA scenario, complete with all the safeguards (i.e., candidate protection layers) found along the link path. The scenarios can be presented in database or spreadsheet format. The rolled-up LOPA spreadsheet allows the analyst(s) to identify safeguards that are independent protection layers and assign appropriate values to each independent protection layer. The spreadsheet calculates the resultant mitigated risk (or mitigated likelihood) in real time. This makes it easy for the analyst(s) to determine which independent protection layer or group of independent protection layers provide the most effective means for reaching or maintaining a target risk threshold.

The concept (demonstrated using ABS Consulting's HazardReview LEADER™ software) makes the process of going from PHA results to LOPA results a lot less time consuming. It avoids retyping and reduces the risk of overlooking scenarios. The paper will present lessons learned from applying the tools in real PHA/LOPA applications.

INTRODUCTION

Layer of protection analysis (LOPA) helps companies understand, in a rational and consistent manner, how many safeguards are enough for a particular accident scenario. LOPA takes a pre-defined cause-consequence pairing (typically identified during a qualitative hazard evaluation), determines how many independent protection layers (IPLs) are provided by existing and/or recommended safeguards, and evaluates whether this number of IPLs provides adequate risk mitigation. LOPA goes beyond the typical use of a risk matrix but is less detailed than a fault tree analysis. It basically separates the question of "How likely is it?" into two issues:

(1) likelihood (frequency) of the initiating event, and
(2) probability of failure on demand (PFD) of the IPLs.

LOPA can provide a company with the following information for a scenario on a consistent basis:

(1) Worst-case unmitigated risk (assuming all safeguards fail),
(2) As-is mitigated risk (with existing safeguards in place), and
(3) The improvements necessary to reach a target risk threshold, as described in Dowell (1, 3, 4, 5, 6) and CCPS (2).

The general format of a LOPA table is shown in Table 1 from Dowell (1).

Table 1: General Format of LOPA Table

Consequence & Severity	Initiating Event (Cause)	Initiating Event Challenge Frequency /yr.	Preventive Independent Protection Layers Probability of Failure on Demand (PFD)				Mitigation Independent Protection Layers (PFD)	Mitigated Consequence Frequency /yr.
			Process Design	BPCS (DCS)	Operator Response to Alarms	SIF (PLC relay)		

The severity of the consequence is estimated using appropriate techniques, which may range from simple "look-up" tables to sophisticated consequence modeling software tools. One or more initiating events (causes) may lead to the consequence; each cause-consequence pair is called a scenario. LOPA focuses on one scenario at a time. The frequency of the initiating event is estimated (usually from look-up tables or historical data). Each identified safeguard is evaluated for two key characteristics:

- Is the safeguard effective in preventing the scenario from reaching the consequence? AND,
- Is the safeguard independent of the initiating event and the other IPLs?

If the safeguard meets BOTH of these tests, it is an IPL. LOPA estimates the likelihood of the undesired consequence by multiplying the frequency of the initiating event by the product of the PFDs for the applicable IPLs using Equation 1 from CCPS (2).

$$f_i^C = f_i^I \times \prod_{j=1}^{J} PFD_{ij} \tag{1}$$

$$= f_i^I \times PFD_{i1} \times PFD_{i2} \times \ldots \times PFD_{iJ}$$

Where f_i^C = frequency for consequence C for initiating event i

f_i^I = initiating event frequency for initiating event i

PFD_{ij} = probability of failure on demand of the jth IPL that protects against consequence C for initiating event i.

Typical initiating event frequencies and IPL PFDs are given by Dowell (1, 3) and CCPS (2). Figure 1 illustrates the concept of LOPA – that each IPL acts as a barrier to reduce the frequency of the consequence. Figure 1 also shows how LOPA compares to event tree analysis. A LOPA analysis describes a single path through an event tree, as shown by the heavy line in Figure 1.

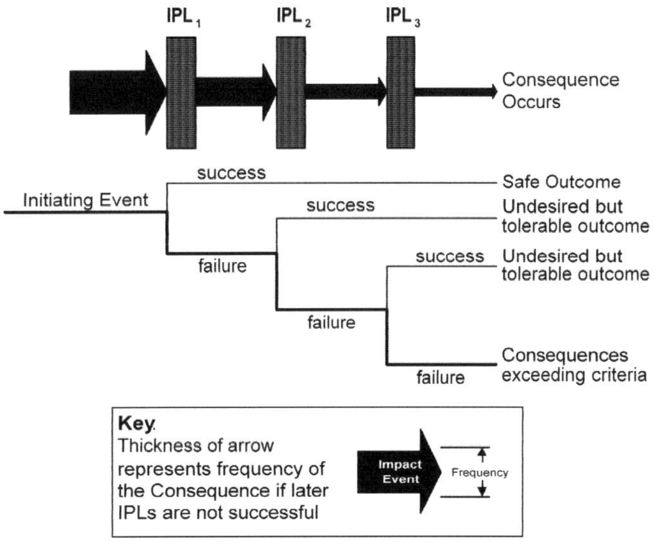

Figure 1: Comparison between LOPA and event tree analysis
[Copyright ©2001, AIChE, reproduced by permission]

DEVELOPING LOPA SCENARIOS

One approach to developing LOPA scenarios is to use a simple screening risk matrix in the HAZOP or other process hazard analysis methodology. Each consequence is ranked for its severity, and the associated causes for the consequence are placed into categories for their unmitigated frequencies, that is, the frequency before application of safeguards. The risk associated with a scenario – a cause-consequence pair – is estimated by the intersection of the consequence severity and the cause frequency on the risk matrix. Many companies have established guidance criteria to select higher risk scenarios for additional analysis. For example, the "Red" zone on the risk matrix may represent consequence severities of one or more fatalities with a frequency above a given threshold. A company's guidance criteria may require LOPA or more complex quantitative analysis for all scenarios in the "Red" zone.

Translation of HAZOP information into LOPA scenarios is shown graphically in Figure 2 (Dowell, 3). Note that not all the information from the HAZOP is included in the LOPA. Consequences that do not meet the risk matrix criteria are omitted. Very low-frequency causes may be omitted. Safeguards that do not provide risk reduction will not be given credit as IPLs in the LOPA (but they may be noted in the LOPA documentation). Additional IPLs may be added as a result of the LOPA study.

The user can manually review the PHA documentation, identify consequences that meet the risk matrix criteria for additional analysis, and develop LOPA scenarios for those consequences, including the associated causes and safeguards. Such activities are tedious, and information can be overlooked or left out, particularly if the PHA is not documented logically, thoroughly, and consistently.

To help ensure logical, thorough, and consistent PHA documentation for processes involving interrelated process parameters and interconnected equipment, interrelated HAZOP deviations are often linked electronically such that the consequence of one deviation is shown as a cause of another deviation, and vice versa. While this is the most efficient, logical, and thorough way to document a PHA in many cases, linking makes manual extraction of LOPA scenarios more difficult.

Alternatives to this approach include the following:

(1) Avoiding the use of logical cause-effect linking in favor of documenting complete HAZOP scenarios within single deviations. While this approach often results in erroneous, misleading, or incomplete HAZOP results, it can help minimize the effort of porting HAZOP results to LOPA.

(2) Using specialized software that queries linked HAZOP scenarios and assembles the causes and safeguards along the entire cause-effect link path to create a complete LOPA scenario for each cause-consequence pair. This approach is equally effective in minimizing the HAZOP-to-LOPA effort while also allowing the HAZOP to be conducted and documented in a logical, thorough, and consistent manner.

HazardReview LEADER includes a LOPA module that implements the second, more thorough approach. It rolls up individual cause-consequence LOPA scenarios from more broad PHA scenarios recorded in HazardReview LEADER. It is particularly powerful when used in conjunction with the LEADER Links feature.

LEADER Links are used to show cause-effect relationships between multiple HAZOP deviations. When used correctly, LEADER Links help prevent duplication or multiple crediting of safeguards as well as helping to ensure that safeguards are listed only at deviations where they are directly applicable. This is the first step in moving a PHA toward LOPA. For example, a relief valve is not a direct safeguard against high level, but high level may lead to high pressure and cause the relief valve to open. Thus, the LEADER Links methodology credits the relief valve as a direct safeguard at the high-pressure deviation but not at the high-level deviation. A LEADER Link then shows the cause-effect relationship between the high-level deviation and the high-pressure deviation.

This concept is important for understanding the power of LEADER's LOPA module. The LOPA module rolls up linked scenarios into Excel spreadsheets. That is, it takes each end consequence, follows each link path to an initiating cause, collects all the existing and recommended safeguards for each link path, and presents each rolled up link path as a single LOPA scenario, complete with all the safeguards (i.e., candidate protection layers) found along the link path.

Each rolled-up LOPA spreadsheet allows the analyst(s) to assign appropriate values (or credits) to each safeguard, and the spreadsheet calculates the resultant mitigated risk (or mitigated frequency) in real time. This makes it easy for the analyst(s) to play the "what-if" game to determine which safeguard or group of safeguards provides the most effective means for reaching or maintaining a target risk threshold.

The LOPA module does not provide the answers, but it makes the process of going from PHA results to LOPA results a lot less time consuming. A trained LOPA analyst is needed in order to apply the LOPA rules appropriately and consistently.

There are some pitfalls to avoid when using the LEADER Links methodology. If linking is done inappropriately, the user may find a multitude of essentially duplicate scenarios that must be screened by hand.

Key points for successful linking and generation of LOPA spreadsheets include the following:

(1) Avoid assigning risk matrix severities to intermediate consequences. In the example above, where high level leads to high pressure, if there is no safety consequence for high level by itself, high level should not be assigned a risk matrix severity. Assigning a safety severity to high pressure is sufficient to ensure that high level and its preceding causes will be captured in the rolled up LOPA scenario.

(2) Assign safeguards only to the specific deviations where they apply; see the relief valve example above. This will avoid having a particular safeguard appear multiple times in a particular LOPA scenario.

(3) Exercise discipline and consistency in linking. For example, similar analysis nodes should have similar link paths.

(4) Minimize parallel link paths having the same ultimate cause and the same ultimate consequence. Where parallel link paths are appropriate, use explanatory text to differentiate the two paths.

(5) Avoid circular links. For each ultimate consequence, the software needs to be able to work back through each link path to an ultimate cause without revisiting the deviation having the ultimate consequence.

(6) Use the same text to describe the same consequence, cause, or safeguard, wherever each item occurs. This will help in eliminating duplicate items in the LOPA scenarios.

There are also some important things to keep in mind when developing a LOPA protocol for your company. These items have a direct impact on the software you choose and how, when, and by whom LOPA studies will be conducted. The following are two of the most critical items to consider:

(1) LOPA is an objective engineering study, not a subjective brainstorming exercise. It does not have to be quantitative or even semi-quantitative, but it does need to be objective. With this said, it is nearly impossible to develop objective LOPA results during a PHA team meeting. LOPA needs to be conducted outside of the influences of the various interests and biases of a typical PHA meeting. (However, some organizations do report successful use of LOPA protocols during the PHA meeting. Note that the risk tolerance criteria used for LOPA decisions must be based on a per-scenario frequency. If the risk tolerance criteria involves summation of multiple scenarios, it is much better to do the LOPA analysis after the PHA is complete [Dowell, 6].) The experienced opinion of the authors is to do LOPA after the PHA.

(2) Most risk matrixes being used for risk ranking in PHA meetings are not appropriate for use in LOPA. The reasoning is not obvious or easily understood without practice. However, the following is a brief attempt to explain this issue.

Typical risk matrix severity categories used in PHAs are based on personnel, public, and environmental impacts rather than quantity, type, and conditions of material or energy released. It requires a subjective judgment to determine what type of protection layer reduces the expected frequency of a fatal injury from once per year to once in 10 or 100 years, or what types of conditions make the potential for a fatal injury "not credible" (i.e., the perceived frequency of a severe consequence is so low that the analyst(s) assigns a lower severity category).

It is much more objective and defensible to state what type of protection layer reduces the frequency of a 1000-lb release of flammable material above its boiling point from once per year to once in 10 or 100 years, or what types of conditions make this type of release "not credible" such that a 100-lb release becomes the assigned severity. However, very few risk matrixes used in PHAs have these types of severity categories.

Typical PHA risk matrix categories can be used for LOPA, but the analyst(s) must be very careful and understand the assumed conditions that are built into each category. Most PHA teams do not have this level of understanding. To help bridge this knowledge gap, companies have taken two basic tacks, including (A) providing specialized LOPA training to select engineers/analysts and (B) developing more specific or advanced LOPA rules and tables to help minimize subjectivity when estimating the frequency of fatality or injury from a given release event, described in Dowell (1, 3, 6) and CCPS (2). Again, we believe LOPA and particularly the more advanced LOPA rules are best applied by trained engineers/analysts outside of PHA meetings.

CONCLUSION

LOPA has proven to be an effective tool to determine if there are enough safeguards and sufficient risk reduction to meet the risk tolerance criteria for scenarios developed from PHA information. However, preparing for LOPA can require tedious efforts in pilfering through complex and duplicative PHA information to develop meaningful LOPA scenarios. These efforts can be minimized by applying risk matrix rules consistently, carefully documenting PHA information in a logical manner, and using specialized software that automates the rollup of LOPA scenarios from interrelated HAZOP deviations.

HazardReview LEADER™ is a trademark of ABS Consulting

REFERENCES

1. Dowell, A. M., III, "Layer of Protection Analysis for Determining Safety Integrity Level," *ISA Transactions* **37** 155-165, 1998.
2. Center for Chemical Process Safety, *Layer of Protection Analysis: Simplified Process Risk Assessment*, American Institute of Chemical Engineers, New York, NY, 2001
3. Dowell, A. M., III, "Layer of Protection Analysis: A New PHA Tool, After HAZOP, Before Fault Tree Analysis," Presented at Center for Chemical Process Safety *International Conference and Workshop on Risk Analysis in Process Safety*, Atlanta, GA, October 21, 1997, American Institute of Chemical Engineers, New York, NY, 1997.

4. Dowell, A. M., III "Layer of Protection Analysis and Inherently Safer Processes," *Process Safety Progress*, **18**, 4, 214-220, 1999.
5. Dowell, A. M., III "Layer of Protection Analysis: Lessons Learned." *ISA Technical Conference Series: Safety Instrumented Systems for the Process Industry*, May 14-16, 2002, Baltimore, MD.
6. Dowell, A. M., III, and D. C. Hendershot, "Simplified Risk Analysis – Layer of Protection Analysis (LOPA)", American Institute of Chemical Engineers 2002 National Meeting, Indianapolis, IN, November 3-8, 2002

NOMENCLATURE

AIChE	American Institute of Chemical Engineers
BPCS	Basic Process Control System
CCPS	Center for Chemical Process Safety
DCS	Distributed Control System
HAZOP	Hazard and Operability Analysis
IEC	International Electrotechnical Commission
IPLs	Independent Protection Layer
ISA	The Instrumentation, Systems, and Automation Society
LOPA	Layer of Protection Analysis
PFD	Probability of Failure on Demand
PHA	Process Hazard Analysis
PLC	Programmable Logic Controller
SIF	Safety Instrumented Function

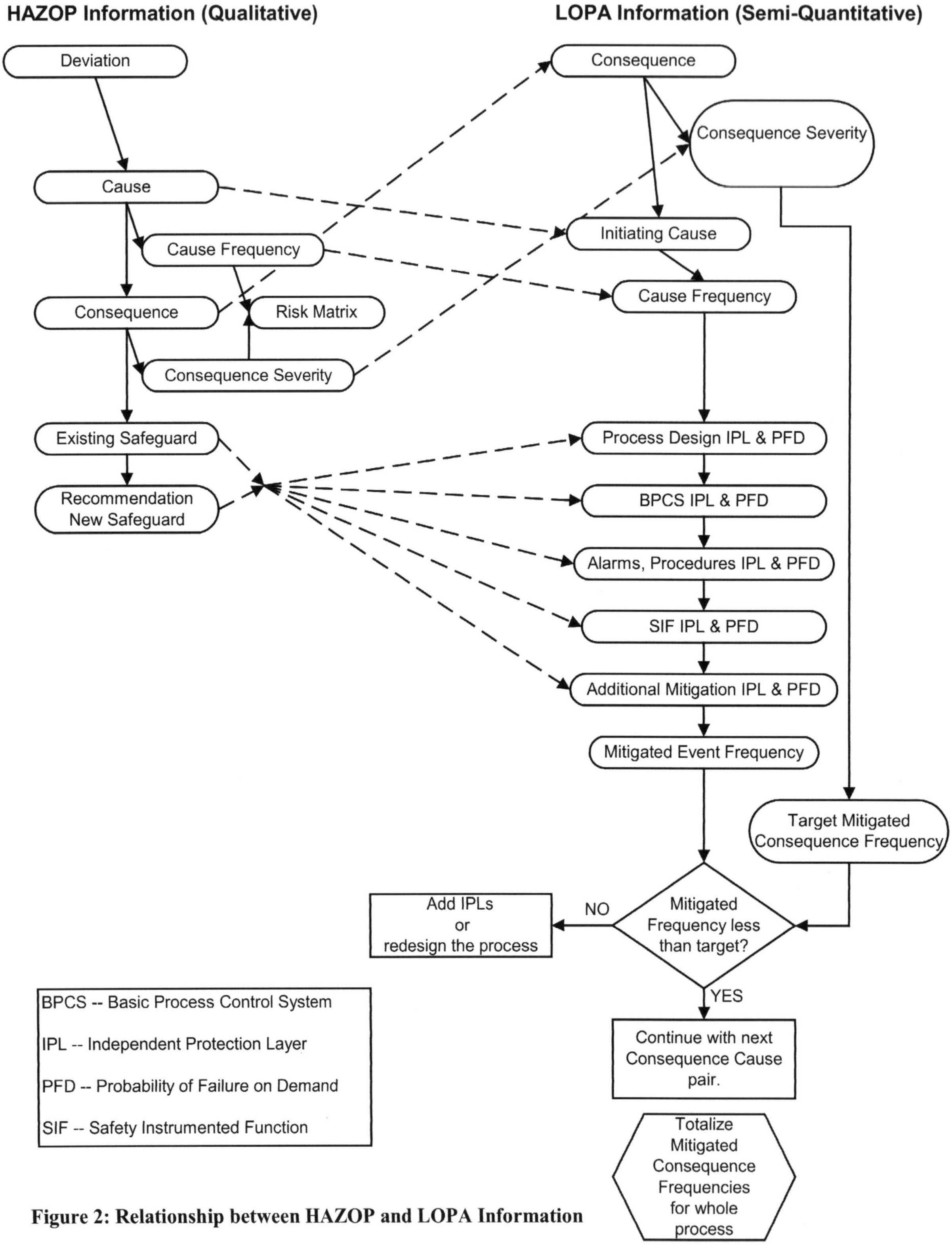

Figure 2: Relationship between HAZOP and LOPA Information

Fire Protection for Large Storage Tanks – Where do we stand?

- **Akhil K. Das & Ajay A. Ambhorkar**
*(Freelance consultants)**

Introduction

Tank storage forms an essential component of any chemical process industry (CPI) and more so in the hydrocarbon processing industry (HPI). Tank storage also represents large portions of overall inventory of any installation – a point sometimes realised only after a major loss event. Yet, tanks and tank farms are often faced with the "*erect, use and forget*" syndrome – management visits to these areas are generally few and far between; technical personnel assigned work there feel that professional challenges are absent; budget allocations for upgrading are hard to come by as it is often considered a "deferrable" item of expenditure. One may not have to scratch deep to uncover such instances. It is in this environment that the issue of fire protection of large storage tanks has to be considered. The discussion in this paper will be limited to fire protection of atmospheric over ground storage tanks. (Figure 1)

Tank Fire Loss Experience

Incident Log[1] reported 2 refinery fires in Kenya and Australia in the month of June 2003. In August 2003, out of 4 refinery fires, one was a tank fire. American Petroleum Institute (API) data for the year 1999[2] reports explosions, but not fires, in atmospheric storage. Data for a 10-year period (1992-2002)[3] for a multinational oil major indicate that there were 6 tank fires in tank storage in different sites over the world. Based on data from API and other sources, Barry[4] gives a frequency of 0.0012 tank fire per year. The statistics assign 91% of tank fires to fixed roof tanks. Of the remaining 9% of the cases (floating roof tanks), 83% were rim fires. Statistics may help recognise the origin, fire growth, and fire control process of tank fires but, to be really beneficial, they should aim to collect information over time and from different parts of the globe.

Tank Fire Scenarios

The term "tank fire" may convey the immediate impression of a fully-involved, full surface flaming fire in a tank. The situation may, however, be quite varied and Table 1 gives a list of potential scenarios.

Fixed roof tank	Fire at vent
	Generation of large quantity of vapours due to fire exposure
	Roof blown out without fire
	Roof blown out and full surface fire
	Boilover
Floating roof tank	Rim fire on roof
	Full surface fire with sunken / stuck roof
Fixed-cum-floating roof	Rim fire on inner (floating)roof
	Full surface fire with sunken / stuck roof
Common	Loss of containment due to overfilling
	Rupture of tank shell and loss of containment
	Flooding of dyke with fire fighting water causing tank to "float"

Table 1: *Potential incident scenarios*

Fire Loss Prevention Practices

In the early years, storage tanks primarily comprised shells of riveted steel plates and roofs of wood and tar coated paper[(5)]. Frequent fire experience with such tanks led to improved design, storage, and handling practices and fire protection approaches. These developments were reflected in the formulation of codes and guidelines. One of the foremost improvements that came about was the use of floating roof tanks for storing high vapour pressure, low flash hydrocarbons. Other early developments included the shift from riveted shells to welded shells and the use of weak roof-to-shell welds in fixed roof tanks.

It was probably more of learning through loss experiences than planned research that led to improved tank protection provisions. Many of these were "rules of thumb" proposed and adopted. Some of the protection provisions practised over the years are listed below and a brief discussion on some follow:

- Locating tanks downwind of potential sources of ignition
- Use of floating roof tanks for high vapour pressure hydrocarbon liquids
- Separation distances between tanks and between tanks and other facilities
- Segregation
- Lightning "protection"
- Impounding (remote or by dyking)
- Exposure protection with water (deluge and/or medium velocity water spray)
- Exposure protection through insulation
- Foam protection (fixed or otherwise)
- Protection for seals of floating roof tanks

As loss experience improved, there was also a trend to increase tank capacities primarily with economics in mind and today we have what we may term as *millionaire* (million-barrel) tanks. As a corollary, while the frequency of fires has reduced, their severity (the other component of risk) has multiplied.

Fixed vs. Floating Roof Tanks

Fixed roof tanks are commonly used to store low vapour pressure liquids (true vapour pressure below 100 kPa) and are considered the minimum acceptable type(6). The roof may be flat, conical, or dome-shaped. The roof-to-shell joint in fixed roof tanks are kept weak to provide a vent for high pressure build up in the vapour space of the tank and which cannot be safely relieved through the venting devices. Grossel(7) states that this is a design feature in the petroleum sector but not in other CPIs.

Floating roof tanks may be

- External floating roof where the roof simply floats on the liquid surface and there is a seal attached to the roof perimeter to cover the annular space between the roof and the shell.

- Internal floating roof (either directly floating or on pontoon) with a fixed roof intended either for weather protection or for quality assurance.

In the latter, vapour accumulations over the floating roof may affect the operating buoyancy of the roof(5). The vapour space in such tanks may also constitute a hazard due to accumulation of vapours and may require either inerting (if the hazard is assessed as serious) or the provision of louvers. Vervalin(8) awards the highest fire safety rating to plain floating roof tanks and this is in agreement with Barry's report.

The subject of emissions from storage tanks has assumed importance in view of the increasingly stringent environmental protection requirements, e.g., volatile organic compounds (VOCs) in air emissions. To control such emissions, tank vents are being manifolded to control provisions such as flares and scrubbers. While this has addressed the environmental issue, safety problems have increased(7) such as

- Propagation of fires from one vessel to another and
- Conversion of flame propagation from deflagration to detonation during passage through a manifold.

Additional design considerations have accordingly been recommended to minimise these potential effects. When implemented, these features can render the system more complex to design, build and maintain.

Lightning Protection and Electrical Grounding

The National Lightning Safety Institute[9], quoting from the *Journal of Hazardous Materials, 40 (1995)*, states that at storage and processing activities in North America, 16 out of 20 accidents involving petroleum products storage tanks were due to lightning strikes. While lightning spikes cannot be prevented, it is the subject of protection that has generated much debate and activity over the years. Until less than 3 decades ago, antenna mast type lightning arresters mounted on tanks was a common sight. Other designs were also developed and used including tall structures meant to provide coverage within a defined cone. Today it is recognised that fires may be caused by lightning[10]

- Either due to direct strikes (flammables exposed either to the heat effect or to the stroke channel) or
- Or due to secondary effects viz. bound charge, electromagnetic pulse, electrostatic pulse and earth currents.

Carpenter[10] states that lightning protection standards for the petroleum industry provide no guarantee of protection as statistics point to secondary effects (rather than direct strikes) being the cause of most petroleum fires attributed to lightning. Apparently, conventional protection systems may only increase the risk of impact by the secondary effects and dissipation array systems (DAS) probably provides the answer to this problem.

Ideas on grounding of tanks have remained an unclear area with protagonists of different perceptions prescribing provisions in keeping with their respective schools of thought. One school of thought suggests that grounding connections are not required if the tank is in contact with the earth or its connecting piping is grounded. Grossel[7] also suggests that grounding connections are required if tanks are not in contact with earth (e.g. on concrete foundations). Carpenter[10], on the other hand, states that the concrete foundation of a tank augments the connection to earth as the effective resistivity of concrete (2.5 ohm-metres) is much lower than most soils.

Layout and Spacing

Layout and spacing are important features of tank farm design. Besides the safety distances mandated by statutes or recommended in codes of design and practice, access is a vital feature as proposed in API 2001[11].

Traditionally, separation distances between tanks have been a mandatory provision for protection of tanks. Much research has been done in the last 2 or 3 decades to determine bases for separation distances based on thermal radiation due to fire in one tank. The Institute of Petroleum (IP)[12] prescribes spacing of tanks based on the type of tank (fixed or floating roof) and irrespective of the class of contents. The basis of tank spacing stipulated by the National Fire Protection Association (NFPA)[13], on the other hand, takes into consideration both the type of tank and the material class. NFPA also makes a special provision by differentiating for crude tanks located in isolated production facilities and those located elsewhere between by permitting a spacing of as low as 0.9m for tanks less than 3000 barrel capacity. GE GAP Guidelines, representing an insurance industry basis, recommends tank spacing based on tank type and capacities, but does not consider the class of material contained[14]. The GE guidelines also recommend that, in case vapour cloud explosion is expected, the minimum recommended spacing requirement may need to be reviewed and the greater value used. These codes and standards are prescriptive and are wanting in performance-based aspects.

Taking this into view, spacing between tanks based on allowable thermal radiation due to fire in an adjacent tank has been studied for over two decades now. It should be borne in mind that, in addition to the level of incident flux, factors such as the level of contents in a tank, the duration of exposure, and protection provided in terms of insulation and water spray provisions should also be considered. The limiting radiant flux figures of 37.8 and 37.5 kW/m^2 have been quoted and used by Das[15] in his study on Safe Separation Distances for Gasoline Storage Tanks. Codes of Practice have tried to incorporate such concepts. For instance, IP[16] recommends a maximum of 32 kW/m^2 for protected tanks (8 kW/m^2 for unprotected tanks). Clark and Escalante[17] have in their model for radiation and exposure protection quoted the AIChE recommended value of 13.5 kW/m^2. It may be noted that recommended limiting radiant thermal flux values are spread over a wide range. Most of these limiting incident flux values were arrived at based on flame temperatures and known proportions of radiant heat release in large scale pool fires. Baum and McGrattan[18] have proposed a radiative transport model based on the emission of Lagrangian elements. R2a[19] has developed dynamic models of radiant emissions of different tank fire scenarios and these can offer good insight on the subject.

NFPA also recommends an allowable heat input of 63100 W/m^2 of exposed surface area for vessels wetted by the contents while, for unwetted portions, it is 18930 W/m^2 [20]. It is, however, not clear if this is intended to cover process vessels only or it caters to storage vessels as well. API 2001 also recommends the provision of space for manoeuvring of fire fighting appliances.

Despite the large volumes of work done by researchers in this area, however, their findings have not found translation to the practicing world and the spacing distances recommended in various codes remain as they were years ago.

Impounding

Loss of containment from a storage tank may either be the effect of a fire or the release may itself provide fuel for a fire. In either case, environmental damage may ensue due to emissions. In any case, the release has to be brought under control and impounding is a means to such control. Impounding can be designed in two ways:

- **Remote impounding** where, through natural topography or by design, the release is diverted to a location where, even if it gets ignited, adjacent property including tanks may not be affected.
- **Impounding by dyking** in which the contents of the largest tank within the dyke may be contained.

While a tank dyke, by design, provides adequate holding capacity, the effective pool diameter of a dyke with hydrocarbon spread across its length and breadth is much greater than that of the tank itself. A dyke may help prevent the spread of burning liquid but, in case this "dyke pool" gets ignited, the flame height and thermal radiation effects could be more damaging. Different codes of practice also permit locating more than one tank in a common dyke in some cases and in which loss of content with fire in one tank can affect the safety of the other tanks. Besides, in the extreme case of a *boilover* (where this potential exists), a dyke can provide very little protection. At the risk of sounding blasphemous, it may even be asked if a tank dyke may at all be considered a mode of fire protection.

Further, considering the general preference for earthen dykes, the potential for soil contamination and ground water pollution in case of loss of containment remain high. This is in addition to the common experience of erosion of earthen dykes.

Exposure Protection

Experience shows that the portion of tank shell over the liquid surface can get damaged in fire scenarios either due to fire in the same tank or by radiation from fire in an adjacent tank. In the case of a fixed roof tank, this is more likely after a sustained fire has resulted in separation of its roof; for a floating roof tank, this and consequent complications involving the floating roof is always a potential threat. Fortunately, as historical data bear out, only a small proportion of floating roof tank fires go beyond the rim seal fire stage. Such damage to the shell may lead to structural failure and release of contents from the tank.

For protection against thermal radiation, insulation on the tank shell may provide some degree of protection; the same protection against tank shell damage will not be available for a tank on fire. In the case of insulated tanks, application of cooling water will be without any benefit. Water cooling is a very effective means of maintaining structural integrity of the tank. Hawthorne[21] provides some rules of thumb for fire-fighters to estimate water requirements from hose streams. At the earlier stages of designing for exposure protection through cooling water application, the concept of deluge valves was considered adequate for fixed roof tanks until it was realised that the water outflow was not uniformly distributed due to imperfections of the roof surface; this led to the design of multiple rings of water spray nozzles for the roof. Appreciation of the phenomenon of *kick off* of water from hot tank surfaces also led to the development of shell protection through multiple ring systems in place of a single top ring of water spray nozzles. Guidelines cannot of course provide for separate cooling water requirements for the portions of the shell above and below the liquid level.

Codes of practice also recommend different cooling water application rates,[22] depending on factors such as closeness and exposure to radiation from fire, prevailing wind direction, etc. These factors will vary in different fire scenarios and even during the same fire scenario. System design, therefore, will essentially need to be generic and based on worst case scenarios. It is observed during drills and real fires that it may not be practicable to apply the theoretically recommended rates of water and water supply may then appear inadequate. Impromptu scale-up or scale-down is also not practicable out there in a real fire scene. In such an environment, it appears inevitable that water inventories and application will be over-designed.

Tank Fire Combating Provisions

Despite other developments, foam and water are the major materials use in tank fire combat. The associated **foam** making and delivery hardware could be of different types such as:

- **Fixed** system where all components of the hardware from foam station to delivery outlets are fixed. Activation of the system may be manual, automatic or in-between.
- **Semi-fixed** system where the delivery portion including piping is fixed but the foam concentrate and foam making and injection aids have to be brought to site.
- **Mobile** system are wheeled facilities that carry all components and material.
- **Portable** system provides basic protection for small hazards.

It may also be mentioned that the above nomenclature is not universal.

Foam delivery into the tank may be through

- Sub-surface injection (possible only for fixed roof tanks) or

- Top delivery in all types of tanks through monitors, foam towers, foam chambers or foam makers.

All of the above systems and the delivery methods have their own pros and cons which need to be considered while selecting a system. Depending on the type of fuel, type and size of tank, location of a tank vis-à-vis other facilities etc., different provisions may be required in different tanks even at the same location. While codes may provide the basics for selection and design, engineering judgement should be exercised and expert advice taken to facilitate optimisation. It should also be remembered that, in case of a sustained major fire, fixed hardware may suffer damage and monitors or portable foam towers may need to be used.

Selecting a system can also be difficult at times due to different view points or due to differences between codes and experience. For instance, NFPA does not recognise the use of monitors on tanks of diameter greater than 18 m (60 ft) whereas fires on tanks up to 58 m (190 ft) have been successfully put out with monitors[(24)]. Apparently, the relevant code in its current form does not take into account the technological advancements which have been made. Similarly, NFPA 11 suggests that foam will travel about 30 m (100 ft) across a liquid surface[(24)] while retaining its blanketing property. What, then, should one do for tanks of diameter greater than 60m (200 ft)? Managements should be conversant with such inherent limitations to be able to take decisions about selection of systems.

In respect to designing, codes recommend foam application over a certain period. This also forms a basis for calculating required foam inventories. However, the actual time required to put out a fire cannot be predicted. In addition, experience shows[(25)] that foam application should be continued for around 30 minutes after the fire has gone out.

With respect to nozzles, greater foam throw is achieved with non-aspirating nozzles than with the aspirating type[(25)]. They can also be used as water monitors with flexibility to quickly modify discharge patterns from jets to wide angle sprays. On the other hand, aspirating nozzles help obtain optimum foam expansion.

Once a tank surface is fully burning, foam applied through fixed tank-mounted hardware may prove insufficient and monitors may need to be used. If HVLR (high volume long range) monitors are used, very large water flows are required and this may affect availability of water for cooling. The delivery point of the monitor discharge should also be carefully selected to be at the region where updraft is low and where air intake into the fire plume provides a low pressure zone. This is vital as attempts to deliver foam into other portions of the plume may be a waste due to the very high updraft (may be 50 km/hr or more) of the plume.

The experience of the fire in the Sandoz warehouse in Switzerland and the consequent pollution in the Rhine in the 1980's highlighted once again the environmental issues related to fire fighting in general and to the use of foam in particular. There was also an incident in 1991, when the effluent treatment plant in a US refinery was affected following the use of alcohol resistant AFFF. Research[26] on the subject is on-going and environmental stipulations are evolving but there are, as yet, no international standard for *eco friendly* foams.

Water supply is perhaps the most critical factor in tank fire control. The recommended requirements may be designed in, but to control flows (particularly for exposure protection) may not be practicable on site.

Some codes recommend that to design a fire protection system and compute material inventory, the simultaneous occurrence of one fire in the process units and another fire in a tank are considered. However, if fire in one tank spreads to another or there are 2 simultaneous tank fires, requirements will be very different and managements must have contingency plans to handle such events including emergency procurement of foam and other essential items.

Statistics referred to earlier show that a small proportion of tank fire incidents have occurred in floating roof tanks and that most of them have been **rim fires**. It is not clear whether the fact that rim fires did not become full surface fires was due to Halon protection. However, as the use of Halon for seal protection was never universal, apparently it was foam application on the seal that checked fire growth. Therefore, the non-availability of Halon following its phase-out may not be a major cause for concern in matters of tank fire protection.

An important point with respect to all fire protection provisions (both hardware and software) is maintenance and preparedness testing. History records numerous instances when actual fire combat operations have failed because of inattention towards the health of hardware and inadequate pre-planning and preparedness. Contingency plans must include as many modes of failure as can possibly be envisaged and means of overcoming all of them if shocks and surprises are to be avoided. Some such points to be considered are

- Regular periodic inspection and maintenance of hardware
- Inventory checks of provisions like foam and water supply
- Additional manpower requirements
- Additional requirement of appliances
- Varying weather conditions
- Presence of dangerous gases like hydrogen sulphide

- Personal protective equipment for emergency personnel
- Regular drills and improvement actions

Fire Combat Operations

Although the mention of a "tank fire" could generally give the impression of a fully involved fire with the entire surface of an open roof tank on fire, actual scenarios could be diverse as given in Table 1. API 2021[27], in Table 2, gives a bird's eye view of the fire potential in different types of tanks, the required pre-planning and also the various approaches to combating tank fires. From an overall point of view, tank fires have to be fought with the aim to do one (and possibly all four) of the following:

- fight the fire in the affected tank
- prevent its spread and
- protect adjacent tanks or other facilities from the effects of the fire
- provide life safety.

Fighting tank fires calls for preplanning and preparedness to take decisions based on ground reality. These require that the manpower have

- knowledge of the fire characteristics of the tank contents
- understanding of behaviour of fires in tanks
- thorough familiarity with fire fighting hardware, fire fighting media and their limitations
- hard core experience and discipline
- (may be) daredevilry

Although preplanning is essential, the plan must have built into it scope for modification to suit conditions on site. The decision to pump out contents from a burning tank, for instance, has proven to be right or wrong in different cases. Experts have also tried to push in water and float the oil to prevent outflow of oil through a rupture on the shell; this might lead to situations similar to *boilover* in some cases.

API 2021[28] describes the general approaches for the above as

- Passive
- Defensive
- Offensive

each meant to be adopted under different situations of the fire itself, the perceived growth (or dying) pattern, endangered people and property, available fire control resources, management philosophy for loss control, etc. However, adopting a passive approach has resulted in some major disasters when the fire spread and did not stay confined to the expected affected area.

Possible Management Strategy

It has rightly been said that matters related to safety in today's (and more so in tomorrow's) technological scenario cannot be approached through the *fly-fix-fly* process. Inherent safety may be the panacea for tank protection and Khan *et al*[29] provide a useful approach in this respect. Hazards need to be identified, risks estimated and preplanning done in line with statutory, management, and societal concerns. It is no different in the arena of tank fire protection and facility managements need to know the many aspects which need to be addressed and considered. API 2021A[30] provides some suggestion in this regard. Figure 2 of the present paper in the form of a decision tree can form the basis of this aid for managements while considering the fire protection for over ground atmospheric storage tanks. The logic proposed in the tree cannot be considered comprehensive; it is aimed at providing a basis for design and pre-planning and is wide open for improvement.

Potential losses resulting from scenarios in Table 1	Personal safety
	Material lost by fire
	Loss of containment
	Fire damage of tank and auxiliaries
	Damage to adjacent tanks & other facilities
	Damage to ambient atmosphere
	Damage to soil
	Damage to ground water
	Boilover
Risk assessment	Likelihood of any of the above
	Potential severity
	Public liability
	Tolerability of risk
Economics	Cost-benefit analysis
	Tolerable deductible
	Sole or multiple facility

Table 2: *Factors to be considered while assessing risk of tank fire incident*

The left half of the tree refers to various aspects that may contribute to the tank fire hazard. The right half deals with specifics of the aspects and involves management decision-making factors. This includes deciding on such factors as risk tolerance, resource allocation, financial, physical or technological constraints, potential public liability, environmental requirements, etc.

The bottom box of aspects (left half) deals with restoration and goes beyond control and containment of the incident. This aspect addresses the issue of returning to business which is important because that (not only putting out the fire) indeed is the crux of emergency management.

With the wide variety of protection alternatives recommended or available, management may be faced with the daunting task of selecting the provisions appropriate for their perceived needs.

Conclusion

Like in all other occupancies, it is best if fires involving tanks are totally eliminated. However, in spite of best efforts at all stages, human failures, natural events and technological sources may lead to fires in atmospheric storage tanks. Hence, there is a definite need to be prepared to handle fires in tanks, dykes, or associated facilities. Managements must appreciate the large inventories that may be held in tanks and which are susceptible to literally be burnt up.

For this, managements need to assess the risks of their tanks and plan for protection. However, wide variations in guidelines and codes of practice, mismatch between codes and field observations, difference of experience between experienced fire fighters, and a variety of other reasons may make the task of risk assessment uncomfortable. Further, despite all designed provisions and pre-planning, it is often seen that experienced heads and hands are the most vital component of tank fire management.

The decision tree proposed in this paper may provide a basis for management decision-making. It may also provide fire professionals with a checklist and help to justify their recommendations to management.

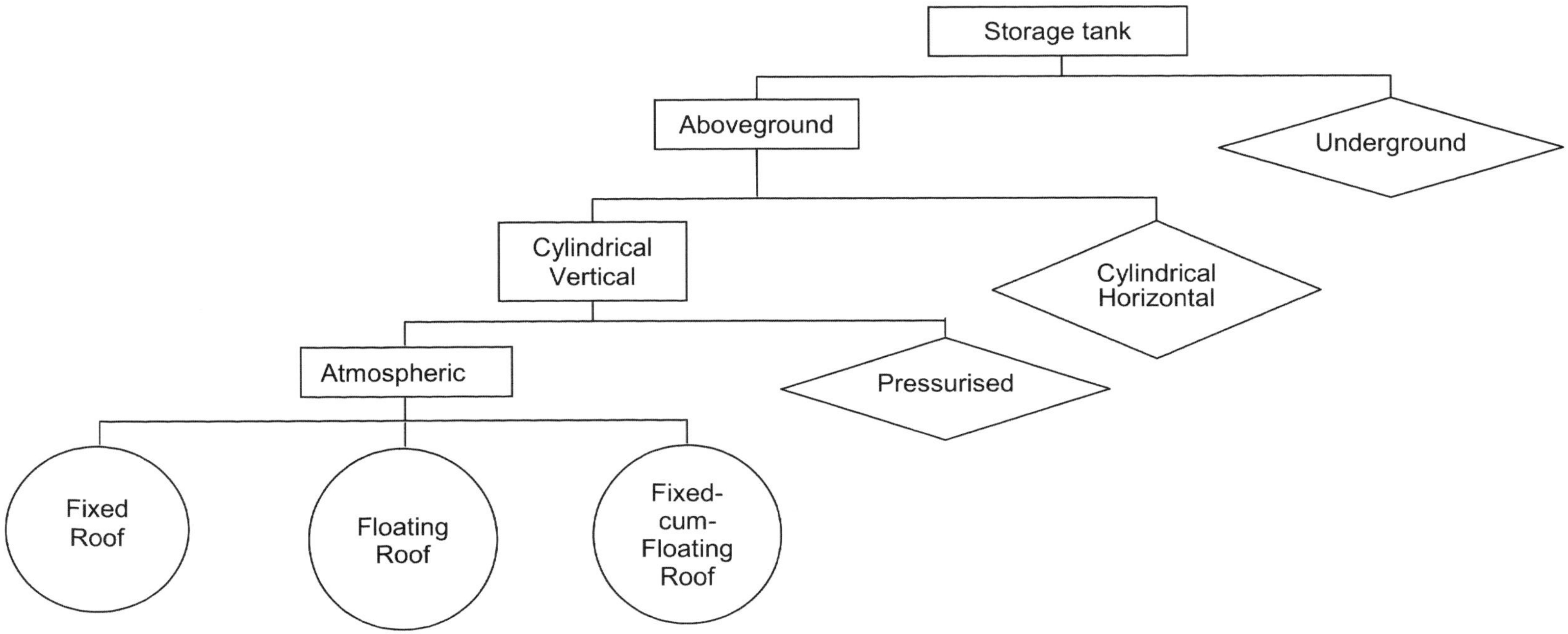

Figure 1: *Types of Storage Tanks – Symbols based on Fault Tree logic*

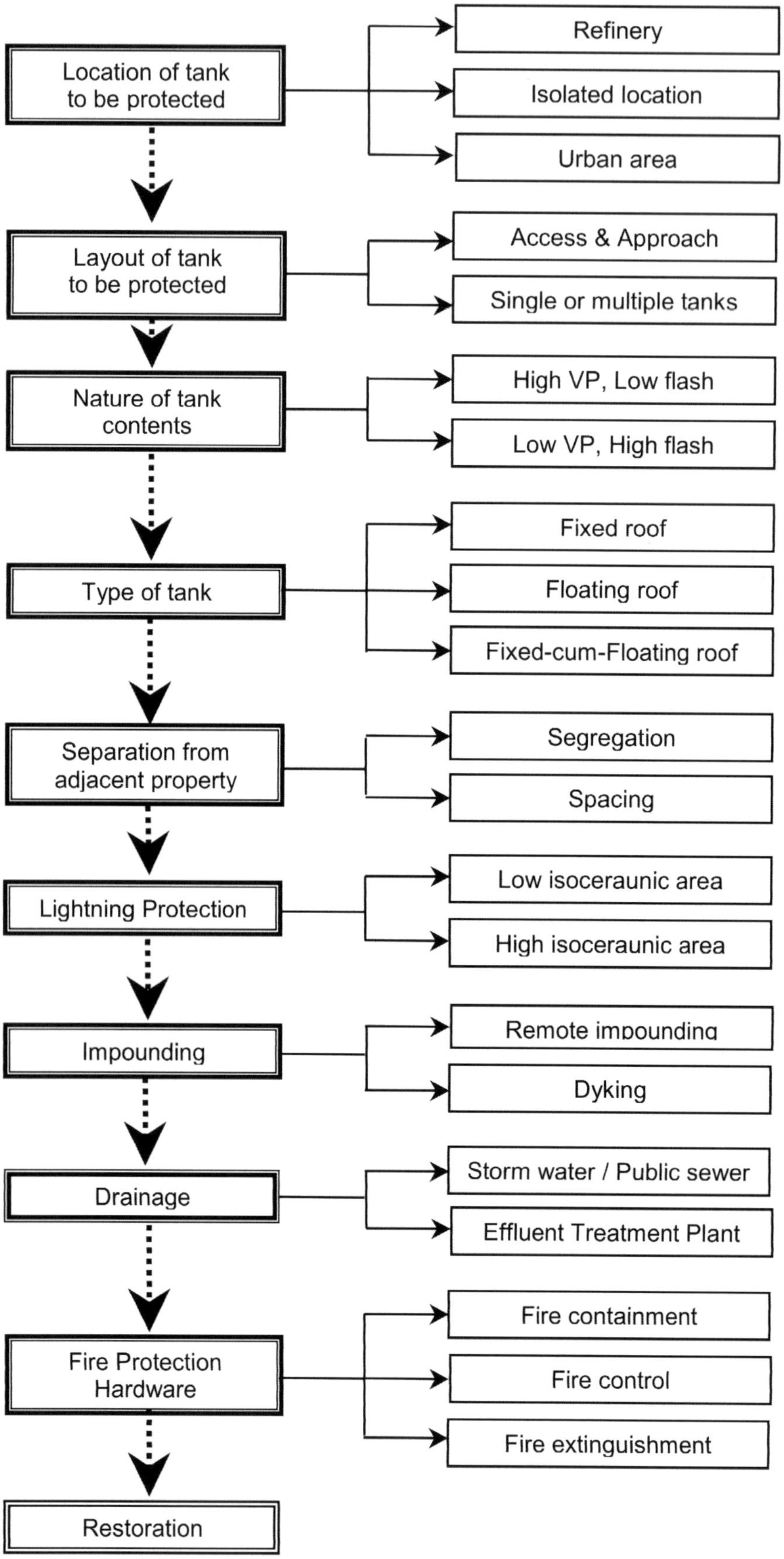

Figure 2: *Management Considerations for Tank Fire Protection Planning*

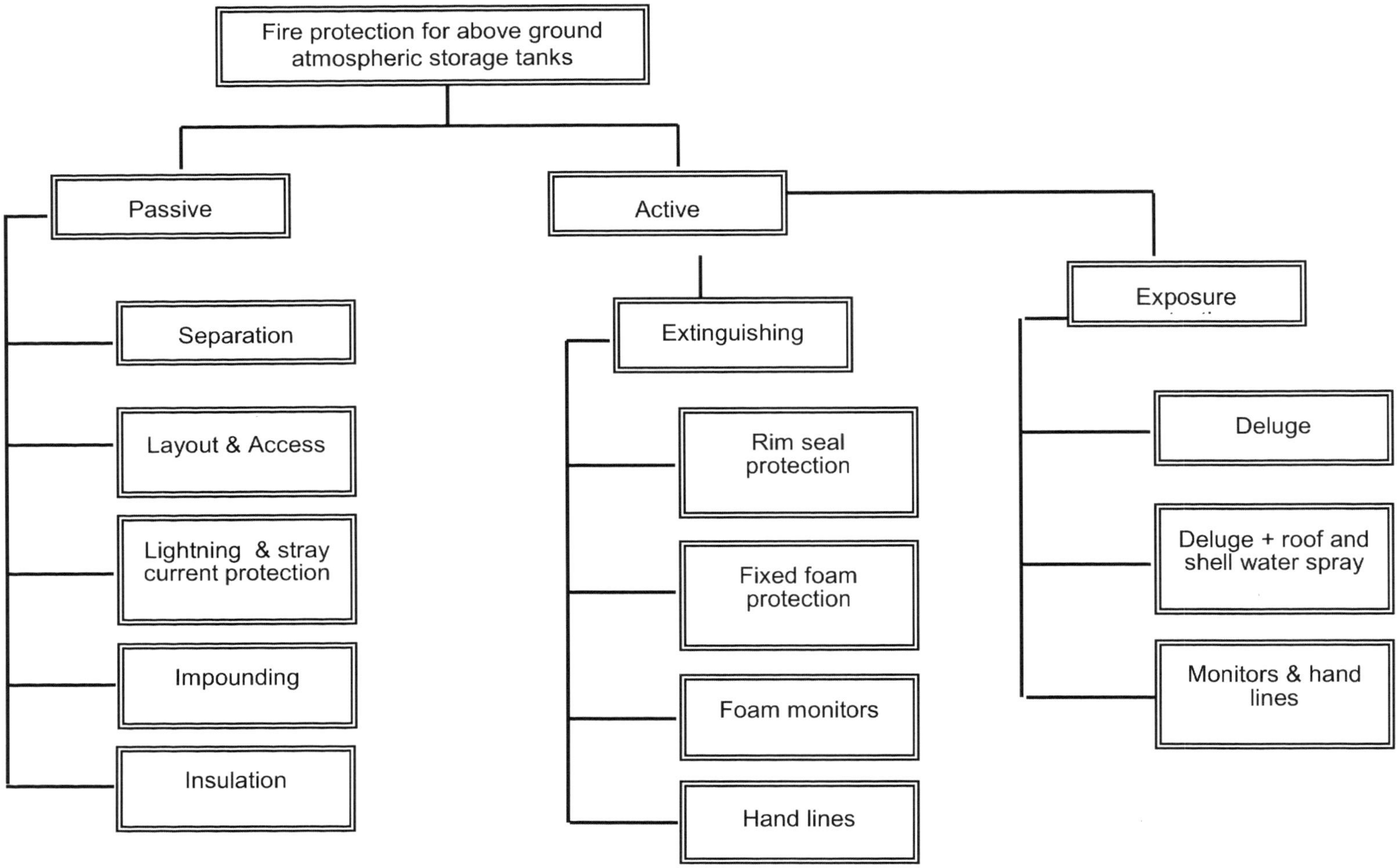

Figure 3: *Some Modes of Fire Protection for Atmospheric StorageTanks*

References

1. Monthly Incident Log, Industrial Fire World, Industrial Fire World Incident Log.htm, www.fireworld.com/incidents/incidenttop.htm

2. Flowers, M., 1999 Process Safety Performance Measurement Report, American Petroleum Institute, April 2000.

3. Doyle, J., Explosions & Fires: 1992-2002, Addendum #1 to Riding the Dragon, Environment Health Fund, Boston, MA, USA, November 2002. www.shellfacts.com

4. Barry, T.F., Historical Fire Incident Data Use & Sources, www.fireriskforum.com, June 2003

5. Fixed or Semi-Fixed Foam Fire Protection Systems for Storage Tanks *CHEMGUARD INC.*, Mansfield, Texas, USA 76063 Rev: 04012001

6. Amrouche Yacine et al, General Rules for Aboveground Storage Tank Design and Operation, pp 54-58, Chemical Engineering Progress, AIChE, December 2002

7. Grossel S., Approaches to Chemical Safety Management in the Chemical Industry, tis.eh.doe.gov/web/chem._safety/ws2000/proceedings/Grossel/20Nov,2003

8. Vervalin, C.H., Ed., Fire Protection Manual for Hydrocarbon Processing Plants, Gulf Publishing Company, Houston, TX, USA, 1981.

9. Kithil, R., Annual USA Lightning Costa & Losses, National Lightning Safety Institute, www.lightningsafety.com, October 2003

10. Carpenter, R.B., Lightning Protection for Flammable Storage Facilities, Lightning Eliminators & Consultants, Inc., Boulder, CO, USA, 1996

11. Section 3.6.2 and 3.6.4, pp 12, API Recommended Practice 2001, Fire Protection in Refineries, American Petroleum Institute, Washington, DC, USA., 1998.

12. Table 5.1, the Institute of Petroleum Model Code of Safe Practice in the Petroleum Industry, Part 3, Refining Safety Code, the Institute of Petroleum, London, UK, 1993.

13. NFPA 30, Flammable and Combustible Liquids Code, National Fire Codes, National Fire Protection Association, Quincy, MA, USA, 1996.

14. Oil and Chemical Plant Layout and Spacing, GE GAP Guidelines, GAP.2.5.2, GE Global Asset Protection Services, September 3, 2001.

15. Das, A.K., Safe Separation Distances Between Hydrocarbon Storage Tanks due to Radiation from Fire in One Tank, M.S.(Fire Protection Engg.) Thesis, Worcester Polytechnic Institute, 1983

16. Table 5.1, the Institute of Petroleum Model Code of Safe Practice in the Petroleum Industry, Part 19, Fire Precautions at Petroleum Refineries and Bulk Storage Installations, the Institute of Petroleum, London, UK, 1993.

17. Clark, R. & Escalante, F., Model Establishes the Radiation and Exposure Protection Required for Refinery Storage Tanks, NFPAOnline Exclusive, November 2002, www.nfpa.org/JournalArchive/onlinefire.

18. Baum, H.R., and McGrattan, K.B., Simulatuion of Oil Tank Fires, Interflam '99, pp 1117-1128, Proceedings 18thInternational Interflam Conference, June-July 1999.

19. R2a (Risk & Reliability Associates), Melbourne, Australia, www.r2a.com.au

20. NFPA 15, Water Spray Fixed Systems, National Fire Codes, National Fire Protection Association, Quincy, MA, USA, 1996.

21. Hawthorne, E., pp 46, Petroleum Liquids – Fire Emergency Control, Prentice Hall, Englewood Cliffs, NJ, USA, 1987.

22. A.2.4.3.1 and A.2.4.3.2, Appendix 2, Application Rates for Fire Water and Foam, Model Code of Safe Practice, 19, Fire Precautions at Petroleum Refineries and Bulk Storage Installations, the Institute of Petroleum, London, UK, 1993.

23. Appendix 9, Water Requirement for Protecting Adjacent Tanks Against Heat Radiation, the Institute of Petroleum Model Code of Safe Practice in the Petroleum Industry, Part 3, Refining Safety Code, the Institute of Petroleum, London, UK,1993.

24. Storage Tanks, Buckeye Fire Equipment, www.buckeyef.com

25. Murray, D., Tank Fire Fighting, The Catalyst, The Official Newsletter of JOIFF, Dublin, Ireland, December, 2001

26. Eco-Friendly Foam, angusfire.co.uk, Reprinted from Industrial Fire Journal, September 1999

27. Table 2, pp 11, API Recommended Practice 2021, Management of Atmospheric Storage Tank Fires, American Petroleum Institute, Washington, DC, USA, 2001

28. Section 6.6, pp 13-14, API Recommended Practice 2021, Management of Atmospheric Storage Tank Fires, American Petroleum Institute, Washington, DC, USA, 2001

29. Khan, F. et al, Evaluation of Available Indices for Inherently Safer Design Options, Process Safety Progress (AIChE), 22, June 2003.

30. API Publication 2021A, Interim Study – Prevention and Suppression of Fires in Large Aboveground Atmospheric Storage Tanks, American Petroleum Institute, Washington, DC, USA, 1998

❋ Freelance consultants presently working (through M/s Heston, a UK-based HR company) on assignment of developing and delivering training at the Safety, Health & Environment Training Centre of Kuwait National Petroleum Company.

Spent Sulfuric Acid Storage Tank Explosion Incident

David M. Heller
US Chemical Safety and Hazard Investigation Board
Washington, D.C.

Introduction

On July 17, 2001, an explosion occurred at the Motiva Enterprises LLC Delaware City Refinery (DCR) in Delaware City, Delaware. Jeffrey Davis, a boilermaker with The Washington Group International, Inc. (WGI), the primary maintenance contractor at DCR, was killed; eight others were injured (Figure 1).

Figure 1. Tank 393, collapsed at northeast corner of storage tank dike. (Note floor of tank 393 to the right of tank shell.)

Because of the serious nature of this incident, the U.S. Chemical Safety and Hazard Investigation Board (CSB) launched an investigation to determine the root and contributing causes and to issue recommendations to help prevent similar occurrences. The CSB released its investigation report on August 28, 2002, at a public meeting in Wilmington, Delaware.

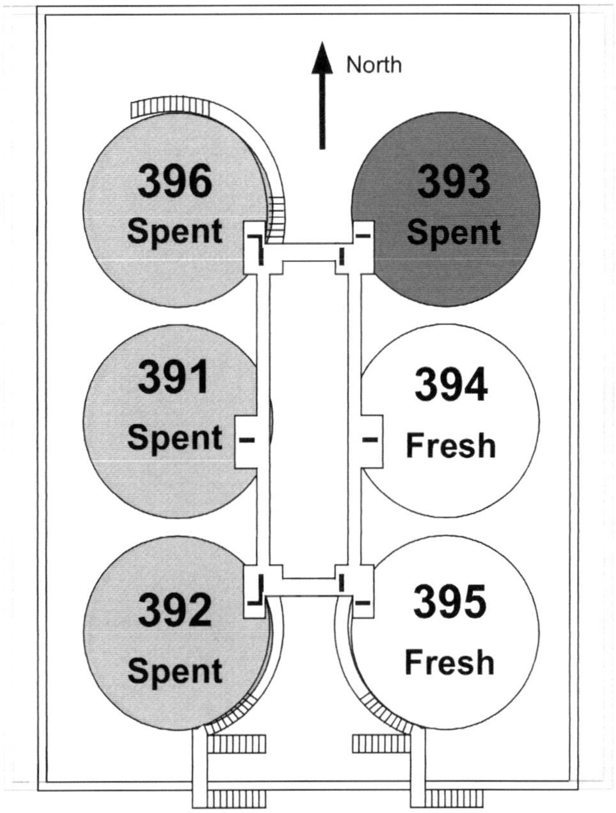

Figure 2. Layout of acid storage tanks and catwalks.

Background

Tank 393 was one of six 415,000-gallon carbon steel (CS) tanks originally built in 1979 and located in a common diked area (Figure 2). The tanks stored fresh[1] and spent[2] H_2SO_4 used in the refinery's sulfuric acid alkylation process. Over the years, the tanks had experienced significant localized corrosion. Leaks were found on the shell of tank 393 annually from 1998 through May 2001; all of the reported leaks were repaired, except for one discovered in May 2001. However, at the time of the incident, several additional holes in the roof and shell of tank 393 were unreported.

---

[1] Fresh H_2SO_4 typically contains 99 percent acid and 1 percent water.

[2] Spent H_2SO_4 typically contains 88 to 95 percent acid and up to 5 percent water, with the balance being hydrocarbons, including some light hydrocarbons that can vaporize.

As leaks occurred and were repaired, the Motiva tank inspectors repeatedly recommended an internal inspection of tank 393. However, despite the imminent hazard presented by this particular tank, Motiva repeatedly postponed its inspection, originally scheduled for 1996. Between 1996 and 2000, even though tank 393 was emptied three times to change its service between fresh and spent acid, Motiva maintained that inventory constraints prevented taking the tank out of service. In 2000, after the U.S. Environmental Protection Agency (EPA) identified 18 tanks that required internal inspection and after Motiva reduced its maintenance budget, the tank inspection program was reprioritized–which further deferred the inspection of tank 393 until January 2002.

Tank 393 was one of four tanks originally designed for fresh H_2SO_4 that had been converted to store spent acid. Spent H_2SO_4 normally contains small amounts of flammable materials. Light hydrocarbons in the acid can vaporize and create a flammable atmosphere above the liquid surface if a sufficient concentration of oxygen is present. To guard against this hazard, Motiva installed a carbon dioxide (CO_2) inerting system[3] and a conservation vent[4] with flame arrester.[5] However, the system was poorly designed and did not provide enough flow of CO_2 to prevent the formation of a flammable atmosphere in the vapor space of tank 393.

The Incident

In June and July 2001, Motiva directed WGI to repair the catwalk that provided access to instrumentation on the roofs of the acid storage tanks. The catwalk had deteriorated due to acid vapors in the atmosphere around the tanks. The contractor employees worked this job on several days, cutting out damaged sections of the grating and welding down new sections (Figure 3).

---

[3] Inerting refers to the use of a nonflammable, nonreactive gas to render the combustible material in a system incapable of supporting combustion.
[4] A conservation vent (also referred to as a PV valve) is a weight-loaded, pilot-operated, or spring-loaded valve used to relieve excess pressure or vacuum in a tank.
[5] A flame arrester is a device intended to prevent a flame from propagating through an open vent into a vessel.

Figure 3. Tank 394 in foreground and remains of catwalk, from the north.

On July 17, as the contractors were working, a spark from the hot work most likely either entered the vapor space of tank 393 or contacted flammable vapors escaping from one of the holes in the tank and flashed back into it. An explosion[6] in the vapor space of the tank generated sufficient pressure to separate the tank's floor-to-shell joint, causing the shell to lift up and come down across the dike wall, and instantaneously release its contents of 264,000 gallons of spent H_2SO_4. The adjacent tank 396 also lost its contents; acid continued to leak from several of the remaining storage tanks for several weeks.

Flammable and combustible material in the spent acid ignited, and a fire burned for approximately one-half hour. The total mass of spent H_2SO_4 released–estimated by EPA at 1.1 million gallons–overwhelmed the area diking and plant containment and wastewater systems. Approximately 99,000 gallons of acid reached the Delaware River, killing fish and other aquatic life.

Mr. Davis, who had been working on the catwalk, was fatally injured. Eight other workers suffered acid burns, burning eyes and lungs, and nausea.

History of Leaks

Tank 393 and other acid tanks had a history of leaks. Major repairs to tank 393 were made in 1994, and the tank had at least one leak in each of the years 1998, 1999, 2000, and 2001(Figure 4). The leak reported in May 2001 was not repaired. Significant leaks also occurred in other H_2SO_4 tanks at Motiva, two of which were condemned and replaced due to corrosion.

Figure 4. Side of tank 393 showing series of patches.

In 1994, Motiva tank inspectors recommended that thickness measurements of tank 393 be conducted in 1996. In inspection reports written in 1999, 2000, and 2001, they recommended an internal inspection "as soon as possible." Motiva took no action on these recommendations (Table 1).

---

[6] The explosion was modeled by CSB as a weak deflagration that generated a maximum pressure of approximately 5.0 pounds per square inch gage (psig).

Table 1

History of Motiva API 653 Tank Inspector Findings and Recommendations on Tank 393

Date	Comments From Inspection Reports
1992	Approximately half of the corrosion allowance is used up in large sections of the entire exposed area.
1994	At the present rate of corrosion, this tank will have used up all of the corrosion allowance for the bottom two shell courses in approximately 4 years. Inspection dept. to sonoray in 2 years to monitor corrosion and calculate for continued service.
June 1998	This tank is overdue for an external sonoray survey as recommended in [1995].
July 1998	Remove this insulation again in 2 years for inspection department to sonoray shell.
September 1999	Due to recent tank failure, it is recommended that this tank be taken out of service as soon as possible for an internal inspection and sonoray survey.
April 2000	Due to the recent leaks on this tank, it is recommended to remove it from service for an internal inspection as soon as possible.
July 2000	This tank to be taken out of service as soon as possible for an internal inspection due to corrosion on the shell.
May 2001	This tank has a history of leaks and internal corrosion. This tank should be taken out of service as soon as possible for an internal inspection and permanent repair.

Tank 393 was emptied three times (April and October 2000, and April 2001) as service was alternated between fresh and spent acid. Each of these occasions presented an opportunity to prepare tank 393 and conduct an internal inspection.

The design and implementation of the Motiva H_2SO_4 tank inspection program was inadequate. Motiva's plan was to inspect its tanks at intervals prescribed by American Petroleum Institute (API) Standard 653[7] (i.e., every 5 years for external inspections and 10 years for internal inspections). However, API 653 notes that inspection frequencies must be modified based on the corrosivity of the stored material. Motiva inspectors recommended revised frequencies, but the inspections did not occur.

[7] API Standard 653, Tank Inspection, Repair, Alteration, and Reconstruction, December 1995.

NACE International (National Association of Corrosion Engineers) RP 0294-94[8] recommends an internal inspection every 5 years and an external in-service inspection every 2 years for tanks in concentrated H_2SO_4 service. H_2SO_4 regeneration contractors[9] contacted by CSB follow these NACE guidelines.

After an internal inspection in 1994, Motiva inspectors recommended that thickness measurements be conducted for tank 393 in 1996; this did not occur. In addition, after 1994, no complete external inspections were conducted (i.e., inspectors only spot-checked shell thickness in the vicinity of leaks). It is likely that an external inspection would have identified the holes in the roof of the tank and the need for a thorough internal inspection. API 653 requires the evaluation of flaws, deterioration, or other conditions that might affect the performance of a tank and the determination of its suitability for the intended service.

Management did not consider the leaks in tank 393 to constitute an imminent hazard to safety or the environment. They stated in interviews their belief that patching the leaks allowed the tank to operate safely, even though inspectors noted that the repairs were temporary and required an internal inspection to ensure vessel integrity. They also believed that lowering the liquid level in the tank below the leak point addressed the hazard.

---

[8] NACE Recommended Practice (RP) 0294-94, Design, Fabrication, and Inspection of Tanks for the Storage of Concentrated Sulfuric Acid and Oleum at Ambient Temperatures, 1994.

[9] H_2SO_4 regeneration contractors take spent acid from refiners, convert it back to fresh acid, and return it for reuse.

Management of Change

Tank 393 was converted from fresh to spent acid service in March 2000 with minimal engineering support. Conversion involved the addition of CO_2 inerting and a conservation vent/vacuum breaker with a flame arrester.

- In its work order for the conversion, Motiva did no engineering and did not issue a request for engineering support from WGI.
- Sizing of inerting flow requirements for normal and upset conditions was inadequate, and the CO_2 gas flow was insufficient to maintain a nonflammable atmosphere in the tank.
- The tank lacked emergency venting provisions, such as weighted vent covers or a weak seam roof, which would likely have prevented it from separating at the floor and catastrophically releasing its contents.

Motiva did not use its MOC system for the conversion of tank 393 from fresh to spent acid service. As a result, the conversion did ***not*** benefit from the following good practices:

- Review and sign-off on the proposed changes by subject area experts (e.g., corrosion, tank design) and higher level management.
- Process hazard review.
- Prestartup safety review.

Inadequate Inerting System

The vapor space of tank 393 was not adequately inerted. Design deficiencies provided an insufficient flow of CO_2 to keep the internal atmosphere below the flammable range.

- CO_2 was not hard-piped into tank 393, but rather was supplied via a temporary rubber hose running off the inerting system of an adjacent tank (396). Because of the long length and small diameter of the hose, an insufficient amount of CO_2 reached tank 393 (Figure 6).

- The hose supplying CO_2 was not attached to tank 393. It was dropped into a hole in the roof, where a nozzle had been located before it corroded (Figure 5). The temporary hose setup was in use for over a year; it was not converted to conventional hard piping, as on the other acid tanks.

- A common overflow line connected the vapor space of tank 393 to two tanks open to the atmosphere.

Figure 5. Hole in roof of tank 393 through which temporary hose was inserted.

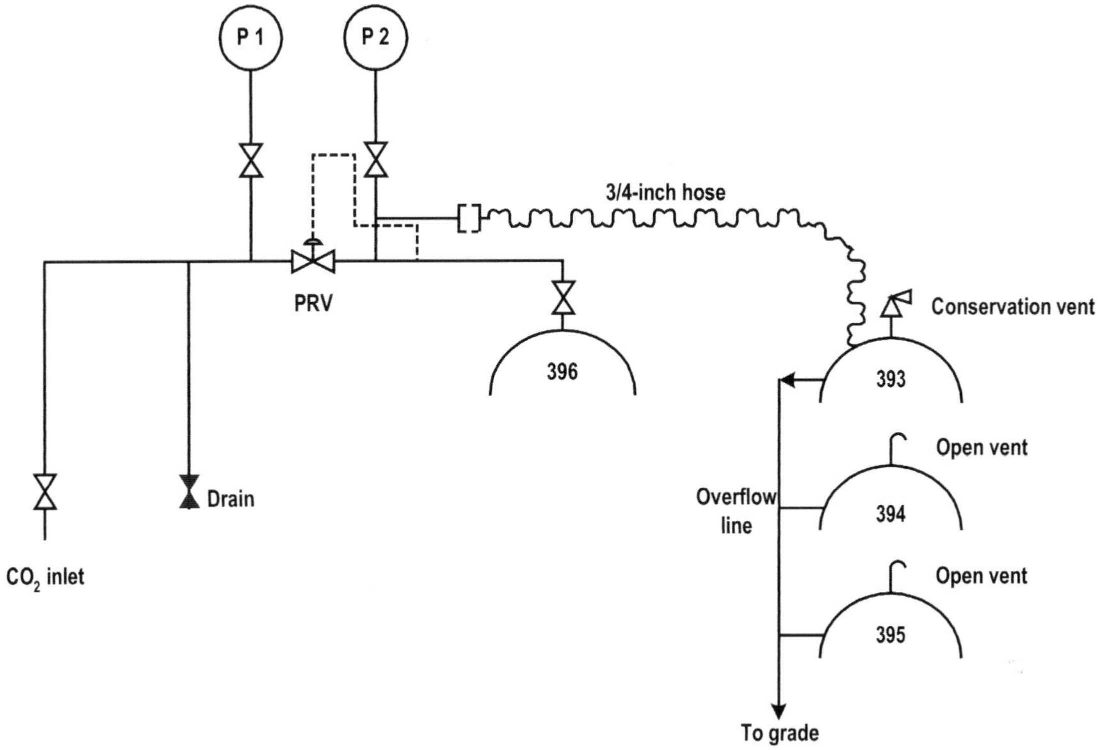

Figure 6. Inerting system schematic (CO_2 flow to tank 393 and overflow connection to 394/395).

The vapor space above the liquid surface in tank 393 contained flammable vapor. Spent H_2SO_4 from alkylation units normally contains sufficient flammable hydrocarbons to generate a flammable atmosphere given the presence of oxygen. As the tank contents settle, some of the hydrocarbons evaporate into the vapor space and may reach flammable concentrations.

Because of the holes in the tank and an ineffective inerting system, tank 393 exhibited severe localized corrosion beyond that considered normal in concentrated H_2SO_4 service. Both rainwater and water moisture in the ambient air entered the tank through holes caused by the corrosion. H_2SO_4 becomes more corrosive as its concentration decreases with the addition of water.

Unsafe Condition Reporting System

An operator submitted an Unsafe Condition Report[10] on June 27, 2001, after rejecting a hot work permit on the acid tank catwalk because of high flammable vapor readings. This report cited the holes in tanks 393 and 396, and noted that a hose inserted into an open hole in the tank roof supplied the inerting gas for tank 393. In the 3 weeks between submission of the Unsafe Condition Report and the day of the incident, Motiva investigated this report but did not correct the deficiencies noted or implement temporary safeguards, such as banning hot work in the vicinity of the holes.

Inadequate Hot Work Permitting Procedures

Motiva allowed hot work to be performed in the vicinity of a tank with holes in its roof and shell. Hot work should not have been authorized; Motiva was aware of the condition of the tanks and the fact that they contained flammables. In addition, once the work was authorized, inadequate precautions were taken to prevent an ignition of flammable vapors. On July 17, and on several occasions in the days prior to the incident, contract workers used burning and welding equipment to cut out and replace sections of the catwalk above the acid storage tanks.

- Contractors, permit writers, and approvers were unaware of the hazards posed by conducting hot work in the vicinity of the holes on the tops of tanks 393 and 396.
- On at least two previous occasions in the weeks prior to the incident, hot work permits to work on the catwalk grating had been denied–once because of excessive amounts of sulfur dioxide (SO_2) gas, and a second time because of a high reading for flammable vapors. However, despite this

[10] Unsafe Condition Reports can be written by plant workers to bring management and union attention to safety and health issues that are not adequately addressed by the immediate supervisor.

history, the July 17 hot work permit did not specify the need for periodic retesting or continuous monitoring of the atmosphere around the work area for flammables.

OSHA PSM Coverage

Motiva did not consider the acid tank farm to be covered by requirements of the Occupational Safety and Health Administration (OSHA) Process Safety Management (PSM) Standard (1910.119). The incident would likely not have occurred if good process safety management practices had been adequately implemented, such as mechanical integrity, management of change (MOC), and prestartup safety review.

The OSHA PSM Standard is a systematic approach to safety and the prevention of catastrophic incidents. Enacted in 1992, the standard details 14 elements of good safety management. Processes containing any of a specific list of hazardous substances, including flammables, must adhere to these elements. However:

- The PSM Standard does not cover H_2SO_4.
- Although flammables above a threshold quantity are covered, the Meer decision[11] determined that coverage does not extend to stored flammables in atmospheric tanks even if connected to a process. OSHA has not challenged this decision.[12]

OSHA did not cite Motiva for violations of the PSM Standard following the July 17 incident.

[11] U.S. Secretary of Labor v. Meer Corporation, OSHRC Docket No. 95-0341, 1995.
[12] Memorandum from John B. Miles, Jr., Director, Directorate of Compliance Programs, OSHA, to Regional Administrators, "Standard Interpretations, Coverage of Stored Flammables Under the Process Safety Management Standard," May 12, 1997.

Root and Contributing Causes

Based on the evidence and findings, the CSB determined the following root and contributing causes for the incident:

Root Causes

Motiva did not have an adequate mechanical integrity management system to prevent and address safety and environmental hazards from the deterioration of H_2SO_4 storage tanks.

- The repeated recommendations of the tank inspectors that tank 393 be taken out of service "as soon as possible" for an internal inspection were unheeded.
- A leak in the shell of tank 393, observed in May 2001, was not repaired. Instead, the tank liquid level was lowered below the leak point, and the tank remained in service.
- Management failed to recognize the imminent hazard posed by the holes in tank 393 and did not promptly initiate repairs or take the tank out of service

Motiva engineering management and MOC systems inadequately addressed conversion of the tanks from fresh to spent acid service.

- The CO_2 inerting supply to tank 393, installed in 2000, was incapable of maintaining a non-flammable atmosphere.
- CO_2 was supplied to tank 393 via a temporary hose run off the inerting system of an adjacent tank. The hose was dropped into a hole in the roof of tank 393.
- No engineering calculations were made to determine proper sizing for the inerting system.

- The tank conversion was completed without review of changes by technical experts, process hazard analyses, or prestartup safety reviews–all elements of a proper MOC program.

The Motiva hot work program was inadequate.

- Motiva scheduled and permitted hot work to be conducted above and around tanks that contained flammable vapors and had known holes; tank 393 had a leak in its shell and open holes in its roof, and tank 396 also had an open hole in its roof.
- After authorizing the hot work, Motiva management did not institute adequate precautions to ensure worker safety, such as continuous monitoring.

Contributing Causes

The Motiva refinery system for investigating Unsafe Condition Reports, informing workers about such reports, and tracking the satisfactory resolution of issues was inadequate.

- In the 3 weeks between submittal of the Unsafe Condition Report on June 27 and the day of the incident, management did not correct the reported deficiencies or implement temporary safeguards.
- Motiva operators would likely not have authorized hot work in the vicinity of the tank if they had understood the hazards, nor would contract employees have conducted the work.

The Motiva Enterprises LLC management oversight system failed to detect and hold Motiva refinery management accountable for deficiencies in the refinery's mechanical integrity, engineering management, and MOC systems.

Recommendations

The CSB issued the following substantial recommendations as a result of the incident investigation:

To the Occupational Safety and Health Administration

Ensure coverage under the Process Safety Management Standard (29 CFR 1910.119) of atmospheric storage tanks that could be involved in a potential catastrophic release as a result of being interconnected to a covered process with 10,000 pounds of a flammable substance.

To the Delaware Department of Natural Resources and Environmental Control

Ensure that regulations developed for the recently enacted Jeffrey Davis Aboveground Storage Tank Act require that facility management take prompt action in response to evidence of tank corrosion that presents hazards to people or the environment.

To Motiva Enterprises–Delaware City Refinery

1. Implement a system to ensure accountability for mechanical integrity decision making. Include the following specific items:

    - Review of inspection reports by subject area experts, such as metallurgists or equipment design engineers, to ensure adequate analysis of failure trends and suitability for intended service.

    - Establishment of a planning system to ensure the timely repair of equipment.

    The Center for Chemical Process Safety (CCPS) publication, *Plant Guidelines for Technical Management of Chemical Process Safety*, Chapter 3, "Accountability Objectives and Goals," presents a model for such a system.

2. Review the design of existing tankage that contains or has the potential to contain flammables to ensure that, at a minimum:

- Inerting systems are installed where appropriate and are adequately sized and constructed.
- Emergency venting is provided.

3. Ensure that management of change reviews are conducted for changes to tank equipment and operating conditions, such as:

- Tank service and contents.
- Tank peripherals, such as inerting and venting systems.

4. Revise the refinery hot work program to address the circumstances that require use of continuous or periodic monitoring for flammables.

5. Upgrade the refinery Unsafe Condition Report system to include the following:

- Designation of a manager with decision-making authority to resolve issues.
- Establishment of a mechanism to elevate attention to higher levels of management if issues are not resolved in a timely manner.
- Identification of a means to ensure communication of hazards to all potentially affected personnel.

Work with the Paper, Allied-Industrial, Chemical & Energy Workers International Union (PACE) Local 2-898 to design and implement the improved system.

To Motiva Enterprises LLC

Conduct periodic audits of storage tank mechanical integrity and design, Unsafe Condition Reports, hot work, management of change, and accountability for these systems at Motiva oil refineries. Ensure that the audit recommendations are tracked and implemented. Share the findings with the workforce.

To the American Petroleum Institute (API)

1. Work with NACE International (National Association of Corrosion Engineers) to develop API guidelines to inspect storage tanks containing fresh or spent H_2SO_4 at frequencies at least as often as those recommended in the latest edition of NACE Standard RP 0294-94, Design, Fabrication, and Inspection of Tanks for the Storage of Concentrated Sulfuric Acid and Oleum at Ambient Temperatures.

2. Revise API tank inspection standards to emphasize that storage tanks with wall or roof holes or thinning beyond minimum acceptable thickness that may contain a flammable vapor are an imminent hazard and require immediate repair or removal from service.

3. Ensure that API recommended practices address the inerting of flammable storage tanks, such as spent H_2SO_4 tanks. Include the following:

    - Circumstances when inerting is recommended.

    - Design of inerting systems, such as proper sizing of inerting equipment, appropriate inerting medium, and instrumentation, including alarms.

NACE International (National Association of Corrosion Engineers)

1. Work with the American Petroleum Institute to develop API guidelines to ensure that storage tanks containing fresh or spent H_2SO_4 are inspected at frequencies at least as often as those

recommended in the latest edition of NACE Standard RP 0294-94, Design, Fabrication, and Inspection of Tanks for the Storage of Concentrated Sulfuric Acid and Oleum at Ambient Temperatures.

Paper, Allied-Industrial, Chemical & Energy Workers International Union (PACE) Local 2-898

Work with Motiva management on the design and implementation of an improved Unsafe Condition Report program.

Jeffrey Davis Aboveground Storage Tank Act

As a result of the Motiva incident and other incidents that were not covered by State or Federal process safety regulations, the State of Delaware enacted legislation in July 2002 to control the installation, operation, maintenance, and repair of aboveground storage tanks. The purpose of the Jeffrey Davis Aboveground Storage Tank Act–named in memory of the WGI worker killed in the Motiva incident–is to provide for the safe containment of petroleum and other regulated substances in aboveground storage tanks exceeding 12,499 gallons in capacity.

The legislation applies to substances covered under the Comprehensive Environmental Response, Compensation, and Liability Act (CERCLA , 42 USC 103), liquid petroleum products, and carcinogens, as well as any other substances that DNREC may add during its rulemaking. H_2SO_4 is a CERCLA hazardous substance subject to 40 CFR 302. Tanks that are subject to the Delaware Extremely Hazardous Substances Risk Management Act requirements are exempt from this legislation.

The Jeffrey Davis Aboveground Storage Tank Act directed DNREC to enact regulatory requirements for spill prevention and control, as summarized below:

- Product inventory or similar control system to adequately identify tank releases.
- Procedures to follow when systems indicate abnormal loss or gain of a substance that is not explainable by spillage, temperature variations, or other known causes.
- Corrective action in response to a tank release.
- Record of actions taken in accordance with the above items.
- Enforcement program.
- Standards to ensure against any future release from a tank being taken out of service or reintroduced into service.

There is also a requirement for regulations to be developed that provide for "appropriate inspection, maintenance, monitoring, and repair of aboveground storage tanks . . ." The legislation further specifies that, at a minimum, an inspection report be developed whenever a tank is emptied for maintenance or repair or removed from service. Inspection reports must describe the thickness of tanks and the repairs needed; a followup completion report is required. Newly constructed tanks require a report on welding certification, and nondestructive testing (NDT) prior to placing the tank in-service. In addition to submitting a copy to DNREC, all reports are to be kept on file for the life of the tank.

DNREC was given the authority to promulgate additional regulatory requirements consistent with the directives of the legislation. The legislature directed DNREC to "consider" the standards and recommended practices of NFPA, API, National Association of Corrosion Engineers (NACE) International, American Society for Testing and Materials (ASTM), Underwriters Laboratories (UL), Petroleum Equipment Institute (PEI), and Steel Tank Institute (STI).

DNREC has issued draft regulations and the regulations are currently under public review.

Selected References

American Petroleum Institute (API), 2002. *Safe Welding, Cutting, and Hot Work Practices in the Petroleum and Petrochemical Industries*, API Publication 2009, Seventh Edition, February 2002.

API, 1998. *Venting Atmospheric and Low-Pressure Storage Tanks*, API Standard 2000, Fifth Edition, April 1998.

API, 1998. *Welded Steel Tanks for Oil Storage*, API Standard 650, Tenth Edition.

API, 1995. Tan*k Inspection, Repair, Alteration, and Reconstruction*, API Standard 653, Second Edition, December 1995.

Center for Chemical Process Safety (CCPS), 1995. *Plant Guidelines for Technical Management of Chemical Process Safety*, American Institute of Chemical Engineers (AIChE).

Dean, S., and G. Grab, 1986. "Corrosion of Carbon Steel Tanks in Concentrated Sulfuric Acid Service," *Materials Performance*, July 1986, pp. 48-52.

Dillon, C.P., 1977. *Materials Selector for Hazardous Chemicals, Vol. 1: Concentrated Sulfuric Acid and Oleum*, Materials Technology Institute of the Chemical Process Industries, Inc.

Fyfe, D., R. Vanderland, and J. Rodda, 1977. "Corrosion in Sulfuric Acid Storage Tanks," *Chemical Engineering Progress*, March 1977, pp. 65-69.

NACE International (National Association of Corrosion Engineers), 1994. *Design, Fabrication, and Inspection of Tanks for the Storage of Concentrated Sulfuric Acid and Oleum at Ambient Temperatures*, NACE Standard RP 0294-94.

NACE International, 1991. *Materials for the Handling and Storage of Commercial Concentrated (90 to 100%) Sulfuric Acid at Ambient Temperatures*, NACE Standard RP 0391-01.

INVESTIGATION OF A NAPHTHA STORAGE TANK FIRE

Prepared by:

Thomas V. Rodante

Baker Engineering and Risk Consultants, Inc.
3330 Oakwell Court, Suite 100
San Antonio, TX 78218-3024
www.BakerRisk.com

Prepared for presentation at:

AIChE 2004 Spring Annual Meeting
Hyatt Regency, New Orleans, LA
April 25-29, 2004

ABSTRACT

In October 1988, one of the world's largest combined loss storage tank fires occurred at a refinery located on the island of Pulau Merlimau, Singapore. At the height of the incident, the blaze involved three floating roof naphtha storage tanks, each approximately 134 feet in diameter and containing a total of 294,500 barrels of product. The resultant property loss was estimated at over U.S.$6.6 million in 1988 dollars.

At several stages, the fire threatened to involve tankage in adjacent dikes containing kerosene, reformate, motor gasoline, and diesel product. Since the refinery was located on an island, equipment and manpower were ferried to the site. Despite the size of the fire and adverse logistics, firefighting efforts were successful in containing the incident to the primary dike tankage.

This paper investigates the incidents leading up to the fire, operational and design engineering considerations, and analyzes the basic fire fighting strategy, pre-fire and emergency response plans, fire water management, and fixed / semi-fixed foam systems.

INTRODUCTION

The subject refinery is located on the small island of Pulau Merlimau approximately 10 miles southwest of downtown Singapore. The main island of Singapore is approximately one mile to the east (see Appendix I). At the time of this incident, access to the refinery was only by boat, and all equipment, supplies, vehicles, etc. had to be transported by barge and ferry (see Appendix II). Currently there is a new causeway providing a direct land route to this refinery and others on similar islands. The Singapore Government has been in the process of reclaiming the land between several islands to eventually form a much larger industrial complex on a single land mass connected to the main island of Singapore.

The refinery plants were originally constructed in 1973 and have been modified and expanded over the years to its current crude throughput capacity of approximately 285,000 barrels per stream day (BPSD). Due to land limitations, many older tanks were built in common dikes and spacing between them was minimized but still in compliance with industry practice and local regulatory requirements.

During the afternoon of October 25, 1988 after two days of heavy rainfall, a serious fire occurred in Tank No. 1, a 41m (134ft) diameter by 20m (65ft) high, 160,000 bbls external floating roof naphtha storage tank. The subject tank shared a common diked area with two other identical tanks (Nos. 2 & 3). At the height of the incident, the blaze involved all three floating roof naphtha tanks located in the common diked area, and had jumped to the rim seal of a naphtha tank in an adjacent dike. Damage had been contained to the primary dike resulting in the loss of all three tanks, 294,500 bbls of product, and four minor injuries. At several stages the fire threatened to involve kerosene, reformate, gasoline, and diesel tankage in adjacent dikes.

The refinery fire brigade fought the fire with assistance from the Singapore Fire Service, industrial mutual aid partners, the Civil Defense, Port of Singapore Authority, and Armed Forces. The fire was brought under control by the fifth day and final extinguishment was declared on the following day. At the time, equipment damage was estimated at U.S.$3 millionwith $3.6 million in product losses. The refinery was shut down for 15 days resulting in additional consequential losses. In 2003 dollars, total insurance losses are estimated at $18.8 million.

The following paper was prepared from a review of data contained in published reports, newspaper articles, subsequent industry investigations, and related association technical committee activities. The author of this paper was employed at the time of this incident by one of the joint venture refinery partners in the position of Corporate Fire and Safety Manager and was a primary member of the investigating team.

This 15-year-old incident is being re-examined now to review the lessons learned and hopefully prevent a new incident from taking place. Historical reviews of incidents reinforce the lessons learned and help dispel erroneous assumptions or theories that may have evolved.

EVENTS LEADING TO THE IGNITION

During the period of 08:00 on October 24th to 08:00 on the 25th, Singapore experienced a heavy monsoon rainfall of 91mm (3.6in). During the previous night, there was a rainfall that one person described as the worst he had seen in 23 years.

Since the day shift on October 23rd, Tank No. 1 had been filling with naphtha rundown directly from two crude distillation units. A review of tank gauge readings (height of product in the tank – see Appendix IV) showed that the level had been rising at an average of 300mm per shift. During the day shift on the 24th (one day prior to the incident), the recorded level had risen 490mm, which was 190mm more than the average rise per shift. This discovery had gone unnoticed. The next shift's reading revealed a 100mm rise, which was only one third of what should have been an expected rate. This anomaly also went unnoticed. The next reading indicated that there was actually a fall in level of about 190mm. This fall in tank level was not noticed by the operators prior to recording the gauge reading in the daily log and going off shift.

After the shift change, the October 25th morning shift personnel were reviewing the previous night's log records. It was then noticed that the last gauge reading was lower then the previous entry, and an operator was sent to double check the tank gauge. By 08:30 the indicated tank level had fallen by further 585mm and by 09:40 hours the level dropped another 3,800mm. At this point a visible check at the Tank No. 1 platform was made. The roof was found to be almost totally submerged with two small sections on either side of the anti-rotation pole still above the liquid level. Bubbling was noticed in a few areas directly above the submerged pontoons. Some witnesses observed a slight shell deformation at the anti-rotational pole attachment. An attempt to close the roof drain was made, but the valve was stuck in the open position. There was a strong odor of vapors in parts of the tank dike but curiously little flow was seen coming from the drain outlet.

Figure 1 Floating Roof Partially Sunk

At 10:00 hours the naphtha rundown streams were diverted from Tank No. 1 to tankage in another diked area. By 10:15 hours a decision was made to gravity transfer from Tank No. 1 to Tank No. 2, the adjacent tank in the common diked area. Flammable vapor readings taken downwind on the dike wall showed a vapor concentration of 8 percent of the lower explosive limit. At this point the decision was also made to use firefighting foam application on Tank No. 1 to cover the surface of the exposed naphtha.

At approximately 11:30 hours a meeting was held to discuss the foaming operations. It was decided to begin the attempt after the lunch break. Around 12:30 hours it became evident that there was a problem with the anti-rotational pole. At the point where the pole attaches to the top of the tank shell, the shell had become seriously deformed. (Appendix III shows the relationship of the tank platform and anti-rotation pole.) It appeared that the tank shell had been pulled inward to the extent that the top plate was distorted by about 45 degrees. The decision was then made to stop transferring out of Tank No. 1.

The area around the tank farm had been closed off, and at 12:50 hours a foam proportioning fire truck was positioned and made ready to deliver foam through the semi-fixed foam system. At approximately 13:15 hours foam application through the tank's fixed foam pourers was started, and at approximately 13:25 hours the entire tank surface area erupted in flames. Three hours elapsed from notification of the problem to the start of foam application. During these three hours the level in the tank was lowered by approximately 1,300mm by transfers to other tanks. It is estimated that the tank held approximately 100,000 bbls of naphtha at the time of ignition.

The fire alarm was urgently sounded and the highest response level was immediately declared. This emergency response level initiated the call-out of the Singapore Fire Service and the Mutual Aid plan participants.

FIRE FIGHTING EFFORTS

The refinery fire service immediately began to establish cooling streams onto the shell of Tank No. 1 by using portable ground level monitors positioned on the dike wall. At approximately 13:35 hours, the first member of the Mutual Aid plan arrived on scene bringing several pieces of portable fire fighting equipment with them. During the initial two hours, the principal fire fighting activities were directed at establishing additional streams from portable ground level water monitors both on and within the dike to cool the shell of the burning tank. The assumption was that cooling the shell would reduce the burning rate. Additional portable water monitors were positioned in-between the burning tank and the tank immediately adjacent to it to serve as a water curtain in an attempt to prevent ignition of the adjacent tank. No cooling water streams were directed onto the shell or roof of the adjacent tank. Fixed cooling water deluge systems on three kerosene storage tanks in the adjacent dike were also activated.

Portable foam nozzles, one on the west wall of the dike and a second on the east side, were positioned to attempt extinguishment of the burning tank. However, the foam streams could not reach the top of the tank. Foam concentrate was delivered to these monitors from a foam tender, and water was supplied directly from the refinery's hydrant system.

Product transfer from Tank No. 1 to any available tankage space was once again started. Initially, Tank No. 3 within the common diked area received most of the product from the burning tank. Tanks with available space outside of the primary dike were not available. A tanker ship was ordered to act as floating storage and arrived at the refinery dock around 20:00 hours on the 25th. Three hours later, naphtha from the burning tanks was being loaded into the ship. Product temperature was monitored to ensure that it was not too close to its boiling point.

About an hour and a half into the fire, foam was applied to the floating roof seal area of Tank No. 2 through the semi-fixed foam system. The hope was that the foam blanket would prevent the seal area from igniting. At this point additional support from the Singapore Fire Service, the Port of Singapore Authority fireboats, and other surrounding refineries in the Mutual Aid scheme began to arrive with heavy equipment and manpower.

Two hours after ignition of Tank No. 1, a rim seal fire ignited on the adjacent naphtha Tank No. 2 that was, by this time, almost full. The fire started in the seal area closest to Tank No. 1 (i.e. most exposed to radiant heat from the burning tank). Two hours later (now approximately 17:30 hours on the 25th), the rim seal fire became a full surface fire after the roof sunk. By this time the Mutual Aid scheme had been fully implemented. Further assistance was on standby from Changi airport, the Civil Defense, and the Armed Services. It was vitally important that this scheme worked well because after the second tank became fully involved, the situation was critical.

Figure 2 – Second Tank Ignites

By 18:00 hours, foam was applied via the semi-fixed system to the rim seal area of Tank No. 3, once again in an attempt to prevent ignition of this tank as well. The potential effectiveness of this strategy was seriously hampered from the beginning since some of the foam discharge pourers were blocked. Shortly thereafter, the man-way cover on Tank No 2. began leaking, resulting in a ground level dike fire. Around midnight on the 25th, the third (Tank No. 3) caught fire on its rim seal and once again, approximately two hours later (02:00 hours on the 26th), also became a full surface inferno. Thereafter, it was a major struggle to contain the fire within the primary diked area.

During the day of the 26th, intense radiant heat and changing wind conditions dictated that fire fighting efforts be re-directed to preventing the fire's spread onto the tanks in the adjacent dike to the east. This dike contained four smaller cone roof tanks in kerosene service and one smaller open-topped floating roof tank in naphtha service. All tanks were approximately 23m (75ft.) in diameter. The kerosene tanks were full and no attempt was made to transfer product out of these tanks. It was felt that the product would act as a heat sink protecting the integrity of the tank shell. Later, when the threat to the most seriously exposed tanks appeared critical, the refinery began de-inventorying operations on the naphtha tank and one of the kerosene tanks. Water curtains were established and cooling water streams directed onto heat-affected zones. Despite the fact that there was no evidence of direct heat impingement, cooling water deluge systems were activated for LPG storage vessels located two dikes away. The decision was made that should the kerosene tanks ignite and jeopardize the LPG area, all nonessential personnel would be evacuated from the refinery. Fortunately this did not occur.

A cone roof kerosene tank and an open top floating roof naphtha tank (No. 4), directly adjacent to burning tanks No. 2 and No. 3, received direct flame impingement for almost one hour. Eventually at approximately 07:30 on the 27th, a seal fire started in the naphtha Tank No. 4. Attempts to extinguish this seal fire with the semi-fixed foam system appeared useless due to the obvious leaks in the system piping. Firefighters ascended the tank stairwell with foam hand lines and quickly extinguished the seal fire.

Due to the accumulation of firewater within the original tank dike area on the morning of the 26th, portable sump pumps were placed in the dike to supplement the drains. An additional drainage system was to be established by cutting a small section of the dike wall above the water level and then installing two pipes on top of the dike. The removed dike section was to be back-filled and valves in the pipe used to control the flow. After the initial cutting had started, officials asked to have the wall resealed before the pipes could be laid. Eventually, the weakened section of the dike wall collapsed against the rising accumulated liquid. Firefighters directed water streams at the flowing burning liquid while others closed the breach with sand bags.

The situation started to ease progressively after the 27th as tanks burned themselves out, and by the 30th, all external assistance came to an end. The Mutual Aid scheme was a success and reflects the dedication and credit owed to those involved. Even the combined resources of somewhere between 300 to 500 men were strained by the scope and duration of the fire containment activities.

No serious injuries occurred during the incident; however, four firefighters required medical attention and were hospitalized for a short period.

All three naphtha tanks were totally destroyed. Some surrounding tanks received minor heat damage, but no significant problems were discovered. The amount of naphtha burned during the incident totaled 294,500 bbls. Approximately 40,000 gallons of foam concentrate were airlifted to Singapore during this incident.

TANK DESIGN and HISTORY

- Design: API 650 Appendix C
- Constructed: 1973
- Height: 19.8m (65ft.)
- Diameter: 40m (131ft.)
- Capacity: 25,400 M^3 (160,000 BBLS)
- No. Pontoons: 24
- Last inspection (external shell and roof pontoon topsides) carried out in 1985
- Common dike with two identical tanks
- Dike capacity: 110% (176,000 BBLS)
- Shell to Shell Separation: 21m (70ft.)
- Product service: Straight Run Naphtha

OPERATING PROCEDURES

At the time of this incident, remote reading level gauges were not installed in the tank farm. Level readings of every tank were to be taken at the beginning and end of each shift. Tanks with product flowing into or out of the tank were required to be gauged every four hours. This practice was not adhered to during the days leading up to the incident. Tank level (i.e. product height) was recorded on an operator's log sheet, and observations of moving tanks recorded in the operator turnover logbook by the outgoing shift. Oncoming shift operators are required to read the turnover logbook at the beginning of their shift and verify tankage levels.

While the oncoming outside operator began his rounds taking tank gauge readings, the inside operator was reviewing the logbook. It was the inside operator beginning his shift on the 25th of October who first noticed the final tank gauge entry was lower than the previous entry. When the outside operator came into the control room with his current readings, an even larger drop in tank level was confirmed.

ANALYSIS

Roof Sinking

Meteorological records indicated that during the morning of October 24th, there was an extremely heavy rainfall over a 3-hour period. Again on the 25th, at 04:00 and 07:00 hours there was another downpour. The recorded rainfall between 08:00 on the 24th to 08:00 on the 25th was 91mm (3.6inches). This amount of rain alone equates to 112mm (4.4inches) of water depth at the center well on the roof of Tank No. 1.

Inspection records indicated that Tank No. 1 never had a detailed turnaround and inspection outage, while both sister tanks No. 2 and No. 3 had. Repair records for both sister tanks suggested there was a pattern of heavy corrosion on the anti-rotation / gauge poles, roof plate welds, and holes in the pontoon sections. Previous external inspections of Tank No. 1 revealed problems with the center well drain screen becoming blocked, causing water to accumulate towards the gauge pole side of the roof. At the beginning of the incident, an attempt was made to close the roof drain. However, little water and/or naphtha flow was observed even though the roof was submerged at this time. During the post incident demolition of Tank No. 1, the center well mesh screen was found partially blocked. It is reasonable to expect that given the same service conditions, Tank No. 1 would have experienced the same deterioration and maintenance requirements as did the sister tanks, but had not been refurbished to the same extent.

Figure 3 Roof Drain Totally Submerged

The roof design was such that it would remain afloat if either two pontoons were flooded or all pontoons were dry and the center well had 254mm (10in.) of accumulated water. Examination of other floating roofs revealed that the pontoon man-way covers were not bolted shut or in any other way secured. Past incidents suggest that unsecured man-way covers can float or be blown off as liquid displaces air in a leaking pontoon. Also, during demolition it was observed that the pontoon sleeves through which the poles pass were totally corroded through at a height of 400mm (16in.) above the pontoon floor. Normal product level in the sleeve chamber would be 260mm (10in.). Lastly, a split was found between the wall and deck seam of one pontoon. Such a split, along with the other previously mentioned defects, would provide multiple entry points for rainwater and / or product to gain entry into various pontoons.

Another suspect design was the emergency roof drains. These drains are simple overflows located around the center well, intended as a last-resort path to flow rainwater off the center well into the stored product before sufficient weight of the water causes the roof to sink. On the underside of the emergency drain is a siphon-type seal filled with water. The higher density of water ensures that hydrocarbon product will not be able to displace the water seal and flow onto the roof. If the water seal is impaired or the differentials mitigated by the additional weight of the roof as a result of water pooling on a hung-up or partially sunk roof, it is conceivable that the product could backflow onto the roof.

Given the high rainfall, blockage in the roof drain, mechanical breeches in the pontoons, and the possibility of product backflow, it is highly probable that the combined effect was rainwater filling the pontoons adjacent to the gauge pole causing the roof to partially submerge, tilt, and eventually hang. As a combination of product and rainwater spilled over the top of the un-welded bulkhead seams, the roof progressively sank. As product continued into the tank with the roof now pinned by the gauge and anti-rotation poles, the remaining buoyancy of the last dry pontoons approximately 90 degrees to either side of the poles caused the roof to buckle along an axis from one pole to the other, like a folded taco shell. (Also refer to tank diagram in Appendix III.)

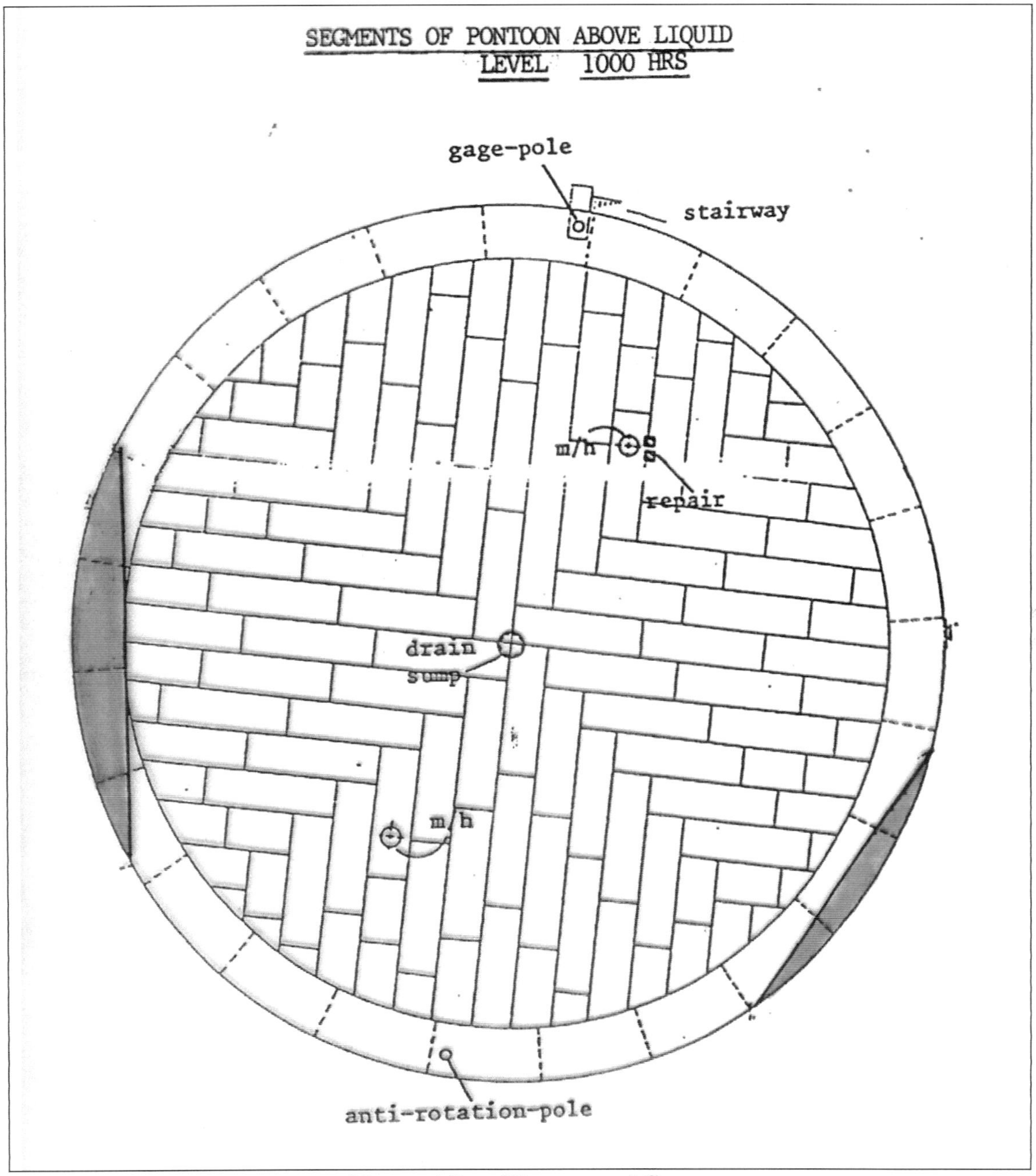

Figure 4 Roof Pontoons Afloat

Operations

In reviewing a graph of tank level vs. time (refer to Appendix IV), it becomes clear at what point the problems with the roof began. The graph shows an average filling rate of approximately 300mm per shift until between 07:00 and 15:00 hours on the 24th, when the slope of the line clearly increases. The increased fill rate while production rundown remained constant can be explained by the following hypothesis: As rainwater accumulated on the roof increasing its weight, more product had to be displaced to keep the roof afloat. The additional product displacement due to the increasing roof weight causes a rise in the level float assembly commensurate to the combination of incoming product plus rainwater accumulated on the roof.

Between 15:00 and 23:00 hours on the 24th, the slope of the line decreases to almost a flat rise — again, while product rundown remains constant. This can be explained as the period when the now-tilted roof begins to hang up on either or both gauge / anti-rotation poles, and thus cannot rise as fast as the incoming liquid level. Incoming product now must find its way onto the roof or push the remaining buoyant sections upward, resulting in a creased roof.

The dramatic drop off after 06:30 on the 25th suggests that somewhere between then and the last reading at 01:00 hours, the roof began sinking at an increasing rate. The 190mm drop recorded by the night shift was not recognized by these operators prior to their shift change. Recording individual data points on a log sheet wasn't sufficient to show the significance of the one outstanding datum point, nor is it clear enough to show a change in trends.

Foaming the tank was given second priority to transferring product out. As product was drawn out from the tank in an attempt to level the roof and reduce inventory should an ignition occur, the more severely sunken gauge pole side of the roof was drawn downward. More and more of the roof's weight and that of the flooded pontoons was held up by the anti-rotation pole. By the time foam was applied some three hours later, the stress was such that extensive shell deformation had occurred. It is suggested that eventually the pole or its connection to the tank shell abruptly failed. During tank demolition, evidence of mechanical failure of the anti-rotational pole top supports was found.

Ignition

Initial theories regarding ignition source called into question the wisdom of foaming the exposed liquid surface in light of the possibility of static discharge. While laboratory experiments have shown that flowing foam through a nozzle can indeed induce a static electric potential, it was concluded that this mechanism was not a very probable ignition source. Firstly, the laboratory experiments showed that for an electric potential to exist, the discharge stream had to be projected through the air onto an electrically isolated liquid. In this incident, foam discharge was a relatively gentle flow through a fixed foam pourer and onto a deflector plate, both of which were attached to the tank shell. Also note that naphtha is not considered a low conductivity fuel. Lastly, the humidity level was very high. None of the actual conditions during foam application to Tank No. 1 were conducive to static buildup and subsequent discharge.

No outside ignition source could be confirmed, such as traffic in the area, or witnessed flash-back from ground level. Given the suspected corrosion problems on the gauge and anti-rotational poles as evidenced by the histories on the sister tanks and the post fire inspection evidence, it was concluded that the most likely ignition source was friction sparks subsequent to the mechanical failure of the pole support.

Fire Fighting

Ground Level Monitors

Two Skum ground-level monitors, each with a rated capacity of 1,000 USGPM of foam solution at 150psi inlet pressure, were initially used to attack Tank No. 1. Water supply was taken directly from the refinery hydrant system supplied by four 2.5-inch hose lines. Foam streams from neither monitor were able to reach the tank top.

The calculated minimum foam application rate for this size tank is 2,256 USGPM. Allowing for stream losses, it can be seen that at least three Skum monitors of the six available were required. Further, since the refinery firewater pumps were rated at 150psi discharge pressure, friction plus elevation losses made it impossible to have supplied the monitors at their required inlet pressure directly from the hydrants. Large flow monitors (and by today's standards, these are no longer considered high flow) must be supplied via fire truck to boost inlet pressure.

It is concluded that with effective use of available resources, a credible attempt could have been made to extinguish naphtha Tank No. 1.

Tank Cooling

A computer model of radiant heat levels received by Tank No. 2 revealed that between 7,800 to 9,500 BTU/Hr/sq.ft. flux levels would have been incident on the top shell in front of the rim seal area at a point closest to the burning tank. Since these conservatively calculated heat levels are considered manageable by direct application of cooling water streams, it is postulated that with proper techniques, a credible attempt could have been made to prevent the ignition of the second floating roof tank.

Semi-Fixed Foam Systems

The semi-fixed foam systems on three separate occasions and on three separate tanks (naphtha tank Nos. 2, 3, and 4) failed to meet anticipated performance criteria of preventing or extinguishing a rim seal fire. Operational, performance, and design failure modes were investigated with deficiencies discovered in each area.

On the operational side, interviews were held with refinery and mutual aid personnel. Additionally, a test drill was conducted and observed. There was confusion over the required operating pressure and a lack of understanding regarding friction and elevation pressure losses.

Numerous accounts of foam flowing out of the air inlets at the base of foam pourers and out of piping elbows, were reported by eye-witnesses. Stories of foam not flowing from some fixed pourers, and foam coming around splash plates rather than down the sides of the tank were also reported. Lastly, the main suspicion pointing to improper maintenance and performance testing was the lack of a refinery procedure.

The design of the installed semi-fixed foam system for floating roof tanks did not conform to the 1988 edition of NFPA code 11. The foam pourers on the subject tanks were placed on 105-foot centers with a foam dam height below the weather shield. The application rate (assuming proper inlet pressure) was 0.5 USGPM/sq ft., and thus is considered more than adequate. However, given the foam dam height, the observed very large drainage slots on other tanks, and distance to the next pourer, it is postulated that the generated foam never traveled to form a continuous blanket around the rim seal area.

Figure 5 Drainage Slots

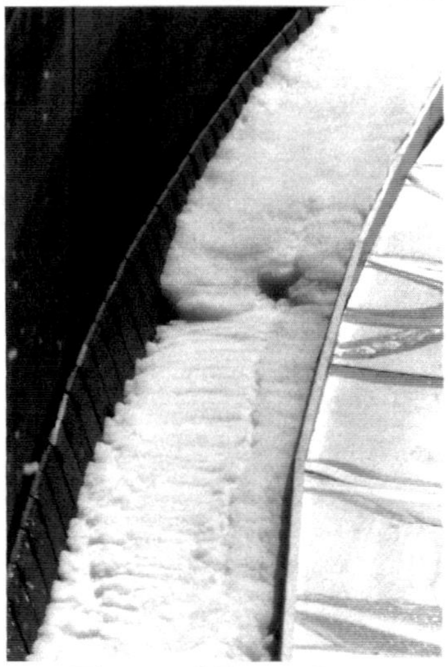

Figure 6 Foam Dam

Figure 7 Foam Travel

CONCLUSIONS

Review of the sequence of events, equipment design, and operating practices related to this incident suggest that the sinking of the floating roof was both foreseeable, and preventable. At the time of this incident, successful extinguishment of a full surface fire for such a large diameter tank had not previously been achieved. Regardless, had the fire protection systems been installed, maintained, and thus functioning as designed, and firefighting techniques been more exacting, it is anticipated that the fire could have at least been held to the original tank and possibly extinguished.

- The investigating team concluded that a combination of heavy rain, maintenance problems, and inherent design features caused the floating roof on naphtha Tank No. 1 to start sinking.
- Operator procedures combined with human behavioral factors, common around shift changes, resulted in a delayed recognition of the situation.
- Slow actions to blanket the vapor surface and attempts to re-float the roof and/or draw off product exacerbated the problem.
- The most likely cause of the ignition was a mechanical failure as a consequence of the sinking of the tank roof and subsequent mechanical sparking of the support for the anti-rotational pole where it is attached to the rim of the tank shell.
- Lack of adequate performance testing and maintenance on the tank semi-fixed foam systems resulted in their ineffectual operation.
- Lack of adequate performance testing and the resulting knowledge of the firewater distribution system led to ineffectual manual fire-fighting tactics.

FINDINGS

Prevention

The basic Prevention principle is to ensure that floating roofs do not sink, and if they do, be prepared to control the potential consequences.

Maintenance:

- Major internal and external inspections must be performed at their prescribed frequencies. Complete and clear documentation, e.g. findings, next inspection due date, required repairs, etc. must be communicated to all interested parties.
- Maintenance information from other equipment should be evaluated when considering alterations to inspection frequencies. Heavier than anticipated corrosion on similar equipment warrants increased monitoring, different inspection techniques, and/or design changes.
- Leaking pontoon sections and corroded gauge/guide poles must be repaired in a timely manner.
- Floating roof center well drain filter meshes should be checked regularly for blockage.
- Pontoon man-way covers should be tightened down on their fastening lugs or otherwise secured.
- Fixed and mobile fire fighting equipment must be maintained in good repair. This means conducting actual performance flow tests of all emergency equipment and systems.
- Firewater distribution systems should be full flow performance tested. Water supply curves showing both quantity and residual pressure should be prepared for all major areas. Identified system bottlenecks should be fixed and new supply curves developed. The resultant supply curves should be compared to the anticipated fire demand.

- All foam system designs should be reviewed against the requirements of NFPA 11, "Low Expansion Foam and Combined Agent Systems." Foam pourers on floating roof tanks should be spaced at a maximum of 24.4m (80ft.) centers. Foam dams should be at least 50mm (2in.) higher than the weather shield or secondary shield, but not less than 610mm (24in.) high.

Design:

- All edges of pontoon bulkheads should be provided with a continuous fillet weld as per recommendations in API 650 Appendices.
- Water sealed emergency roof drains should be avoided or additional inspection be required to ensure the seal is functioning properly.
- Pontoon area as specified in API 650 is a minimum requirement. Consideration should be given to increase the natural buoyancy of external floating roofs in areas subject to monsoon-type rainfalls.
- The installation of intermediate dikes within multi-tank diked areas will limit the spread of a spill and a consequent fire.
- Safe separation distances as recommended in industry publications are usually based upon still-to-calm wind conditions and may not be sufficient to prevent adjacent tankage from ignition when exposed to a full surface fire.
- Prepare pre-fire plans for critical areas. These should contain information on predetermined minimum required application rates, required foam inventories, available water supply rates and pressures, etc., taking into account actual performance data. Signs should be provided at tank semi-fixed foam inlets explaining the operating conditions at the fire truck required for making foam.
- High flow portable foam monitors require an inlet pressure of 150psig or more. "Steamer" connections at selected hydrants should be available to reduce the reliance on small diameter supply hoses and associated friction losses. Large diameter 127mm (5in.) soft supply hoses should be used to connect fire apparatus to the hydrant systems so as to minimize pressure loss and the clutter of multiple long lengths of fire hose.
- Fixed cooling water spray rings should be sectionalized into two or three sections each supplied by an independent riser.

Operations:
The basic principle in tank farm operation is to adopt practices that promote the recognition of potential upset conditions in a timely manner.

- Visual inspection of roofs should be conducted during periods of unusually heavy rainfall.
- Tankage movement log sheets should be designed to provide a quick means of status review by both supervises and operators. Unusual movements must be recognized and investigated promptly.
- Promote good communications between operations at shift change to review the status of oil movements.

- Consider integrating rate of flow, rate of change logic into Programmable Logic Control (PLC) or Digital Control Systems (DCS) warnings.
- Floating roof tanks fitted with emergency roof drains utilizing a water seal require an operating procedure to ensure the integrity of the water seal.

Incident Control:
The basic control principal is to contain the fire to its place of origin.

- Properly declare the appropriate alarm condition when a potentially serious incident occurs. Due to the potentially serious consequences resulting from a sunken floating roof, quick decisive action is a must.
- When a roof sinks, immediately apply firefighting foam onto the flammable liquid surface to reduce vapor generation and the risk of ignition.
- Stabilize a sunken roof before attempting to draw off product or re-float the roof. Moving product out of the tank with an unstable floating roof, particularly if it is on fire, is inherently risky.
- For cooling water streams to be effective they must be applied directly onto the heat affected areas of storage tanks. Unwarranted water usage depletes water pressures on the firewater distribution mains and exacerbates the drainage problem.
- Extinguishment should only be attempted after all required resources are in place to apply at least the minimum required application rate for the recommended minimum duration. As proven by countless fire incidents, foam application to a burning tank should not be attempted until all required resources have been assembled and are ready to be used.
- Annual performance tests should be conducted on all mobile fire apparatus.

APPENDIX I - MAP OF ISLANDS OFF SINGAPORE

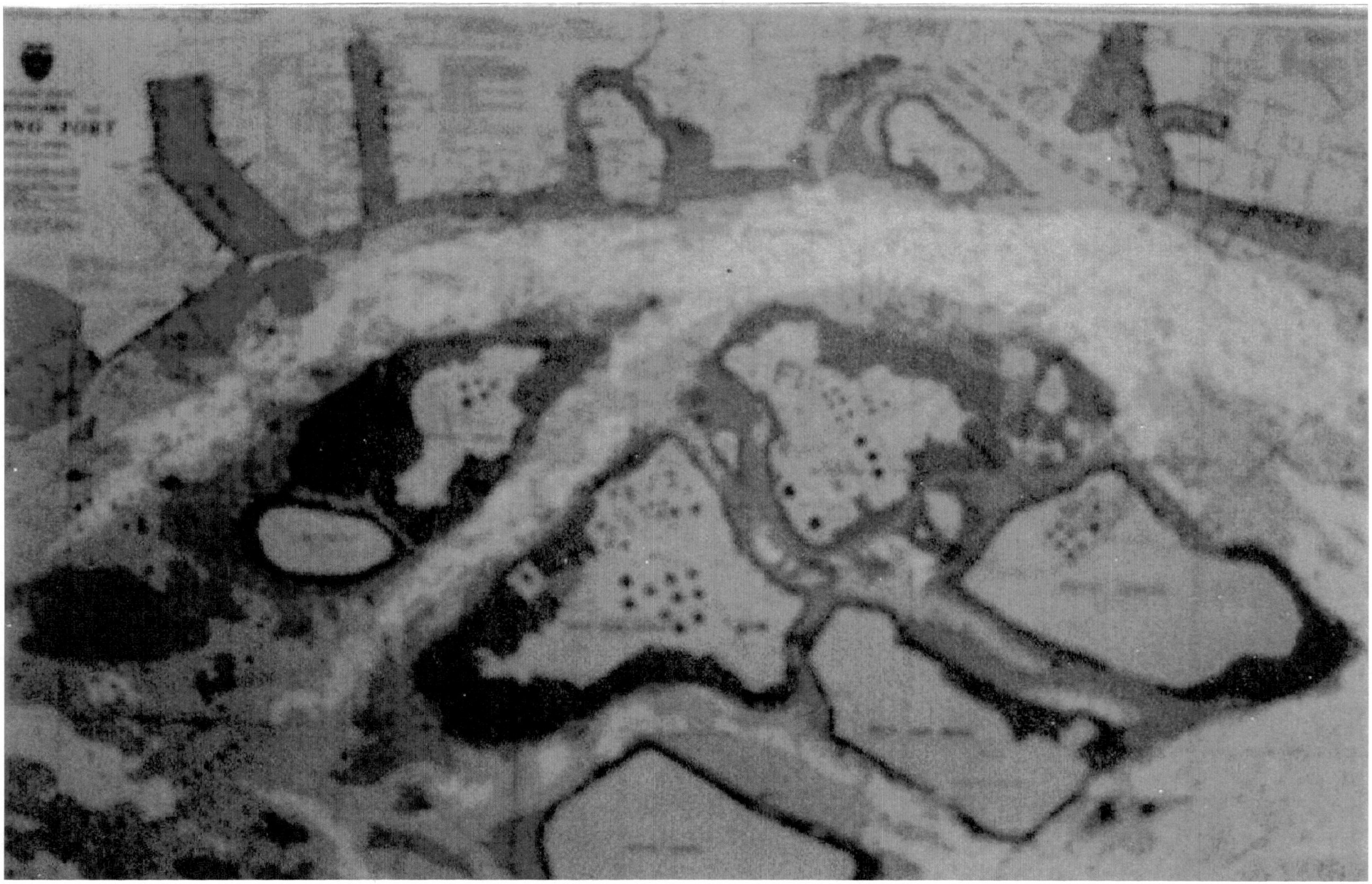

APPENDIX II – PULAU MERLIMAU ISLAND

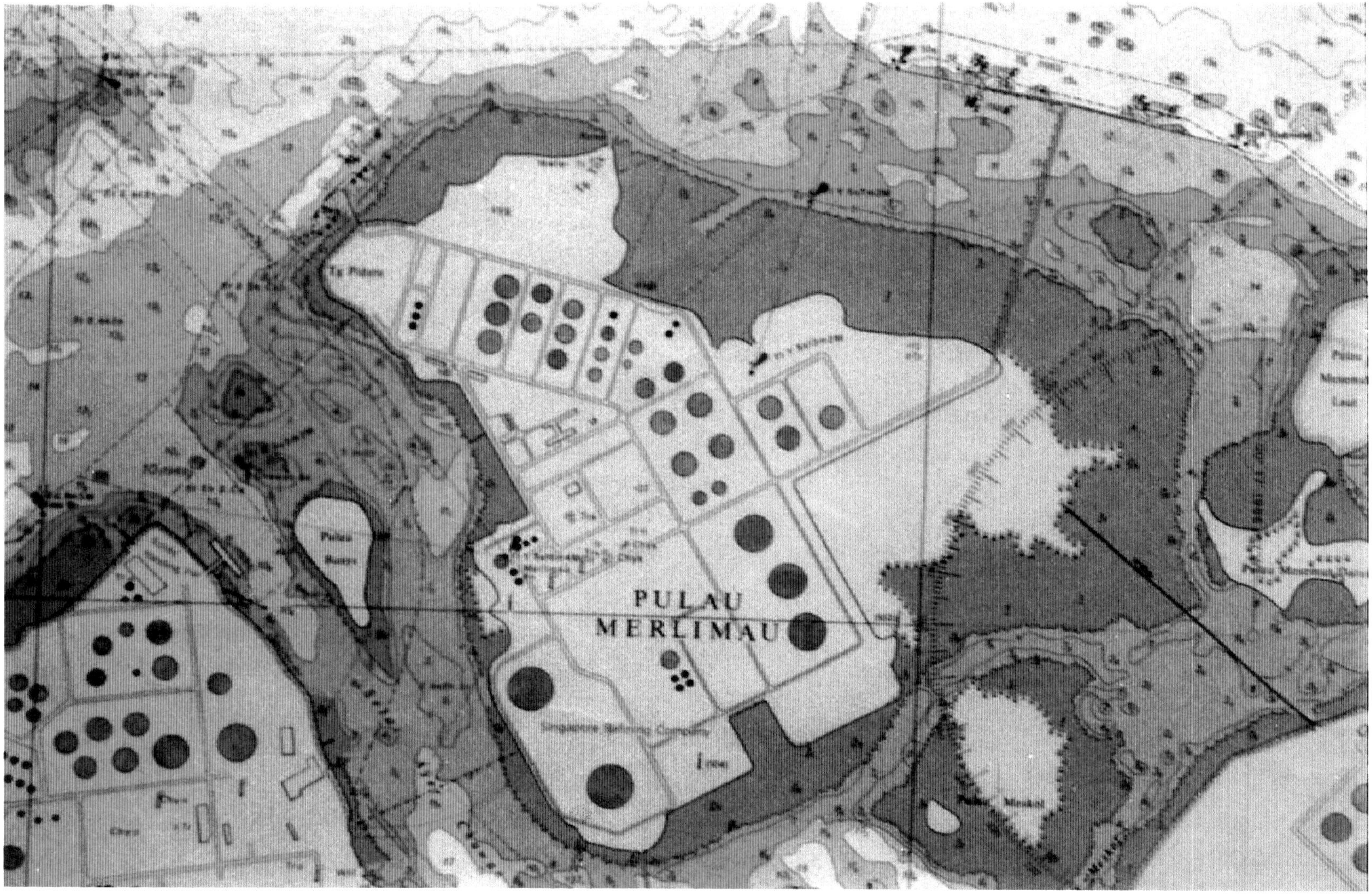

APPENDIX III - TANK DIAGRAM

NAPHTHA TANKS

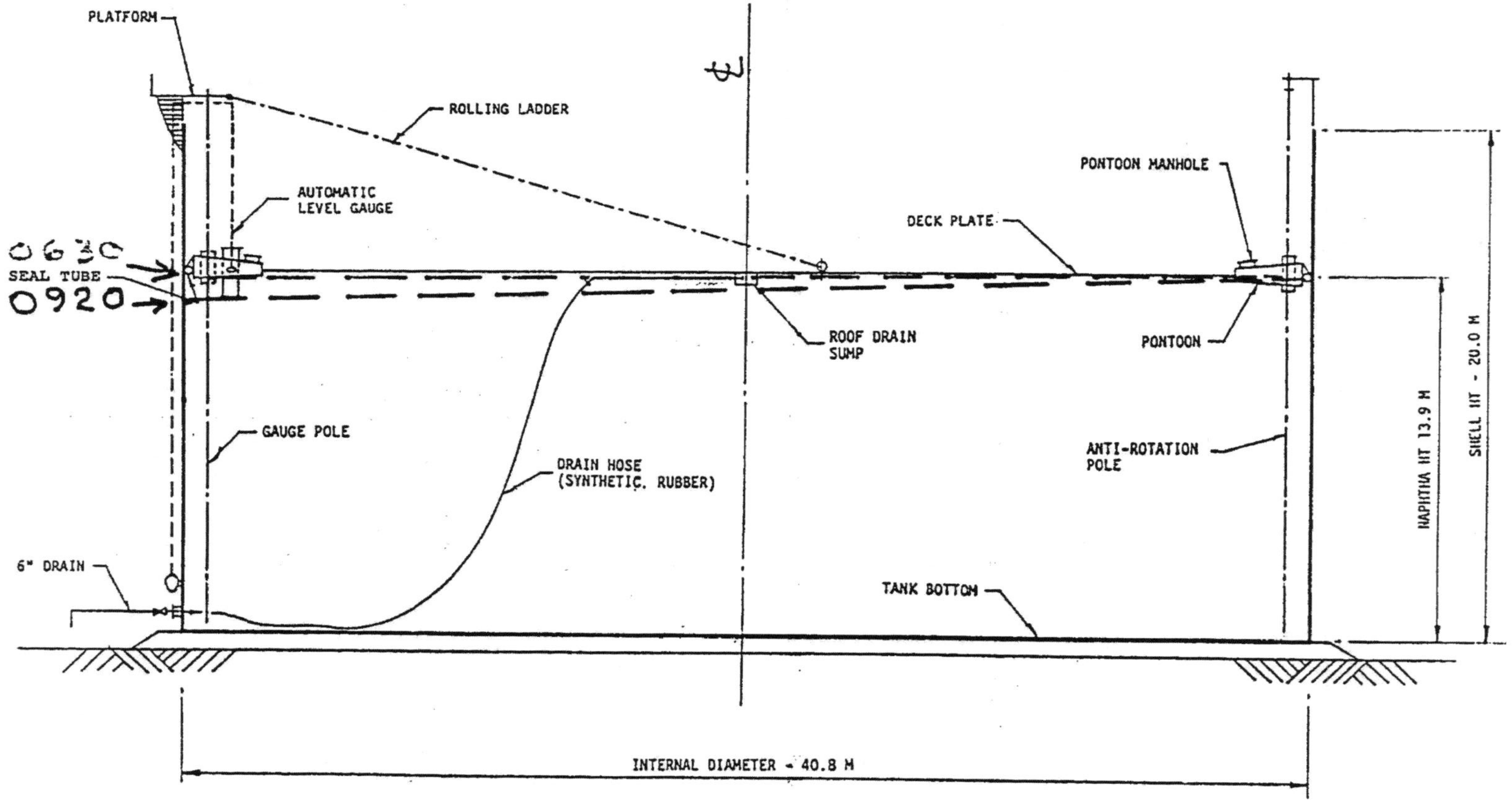

APPENDIX IV

ROOF POSITION

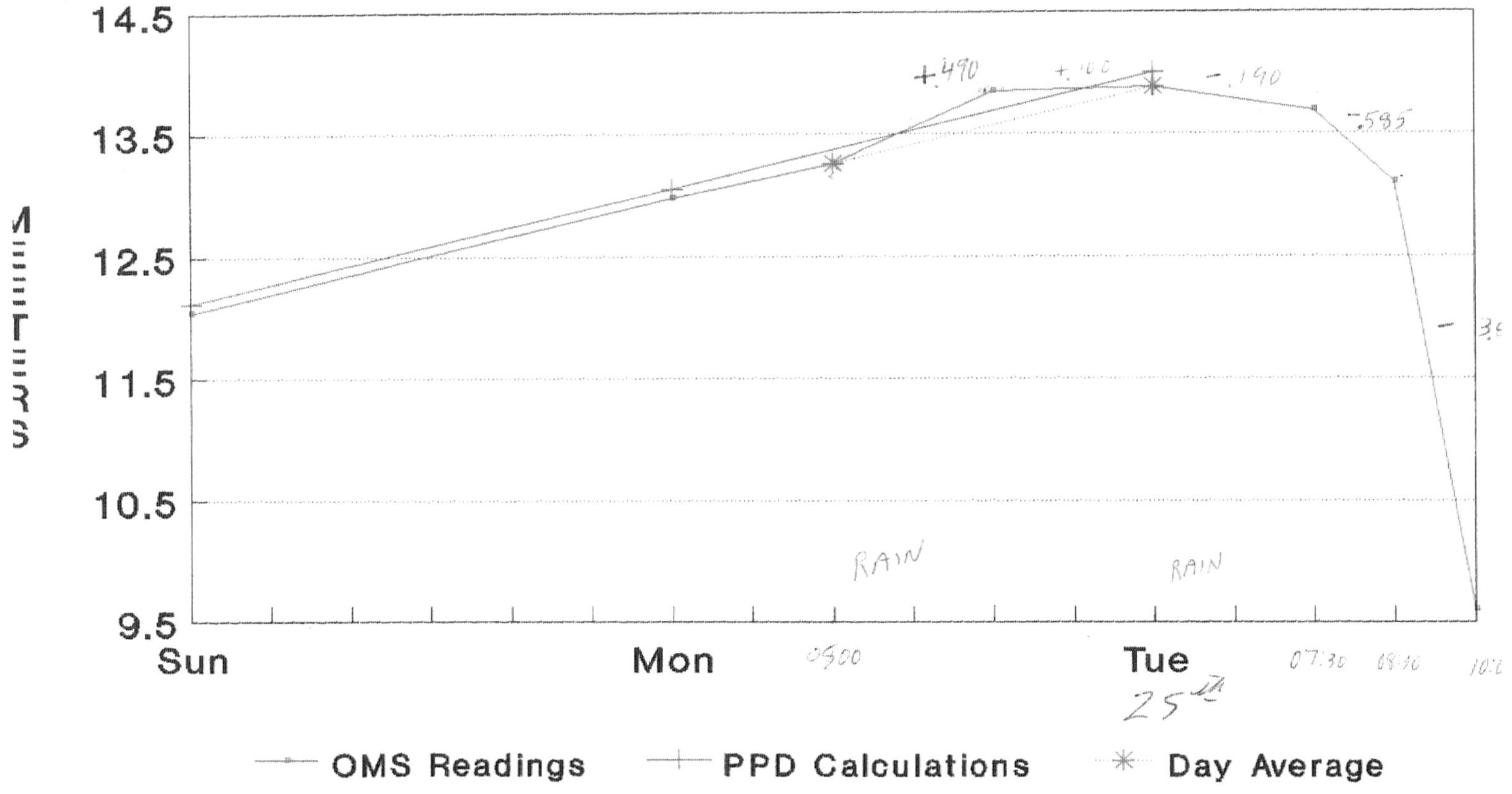

Presented to: **American Institute of Chemical Engineers**
Loss Prevention Symposium, New Orleans, LA - April, 2004

Fire Protection Considerations for an LNG/CNG Fueling Facility

by
Jack Woycheese, PE, FSFPE
Senior Engineer and Director,
International Business Development
Hughes Associates, Inc.

Fire Protection Considerations for an LNG/CNG Fueling Facility

Abstract

This presentation will address fire protection engineering design considerations incorporated into an enclosed Liquefied Natural Gas (LNG) storage building housing two 30,000 gal cryogenic LNG tanks. LNG is pumped to vaporizers to Compressed Natural Gas (CNG) fueling facilities for a large metropolitan bus fleet. Due to a NIMBY attitude and code siting issues the storage building and related equipment was modified to incorporate findings of a FMEA and to meet the intent of applicable provisions of NFPA 57, NFPA 59A and a proposed CalOSHA pressure vessel regulation, Title 8. The presentation will discuss siting, drainage, ESD and isolation, pressure relief, detection and suppression design criteria, as well as involvement of the critical stakeholders including the local authority having jurisdiction(s) to achieve a cost effective design within the project scope, budget and timeline.

Introduction

The alternative fuels discussed in this paper are LNG, CNG, and L/CNG. The latter is an LNG storage facility that converts the LNG to CNG for fuel. [1]

For the past several years, the South Coast Air Quality Management District (SCAQMD) in Southern California has encouraged bus companies to develop an Alternative Fuels program. As a result, LGN or CNG is increasingly common in California.

SCAQMD has subsidized the

- Purchase of buses running on alternative fuels

---

[1] CNG is favored when vehicles have short daily mileage; LNG, for longer mileage.

- Construction of fuel storage and refueling facilities for the alternative fuel buses.

Since the start of the program, several major bus operators have implemented programs to introduce alternative-fuel buses and have included essential storage and refueling facilities. These operators include:

- Omnitrans facilities in San Bernardino, CA
- Big Blue Bus in Santa Monica, CA
- Orange County Transit Authority in Anaheim and Garden Grove, CA

In most cases, these companies store LNG in aboveground storage vessels, which comply with design and siting requirements of applicable NFPA standards.

Focus of Paper

As the focus of this paper, we will review the fire protection issues and requirements involved in LNG and CNG as an alternative fuel. In particular, we will examine the application of fire protection principles to an unusual installation. in a location, which had gas compressors, which occasionally released odorized natural gas toward a nearby schoolyard.

While the concentrations were well below unacceptable levels for fire or explosion, the odor raised concerns for the school officials, teachers, and students. As a result, neighborhood sensitivities focus on the potential release of natural gas.

The company intended to address these concerns through a project to install LNG as a source of fuel, thus enabling the removal of the gas compressors..

Background

LNG/CNG-specific information is essential to understand the issues and resolution. These elements in this section include:

- Alternative Fuels
- Facilities
- Representative Codes and Standards

Alternative Fuels

Environmental goals drive the demand for alternative fuels. In California and many other states, local environmental agencies are encouraging fleet operators to develop refueling infrastructures for automobile drivers. They do so by underwriting part of the cost of refueling facilities and fleet operation vehicles.

Fuels for Refueling Facilities & Fleet Operation Vehicles

Alternative fuels for such programs include:

- Methanol and ethanol
- Electrical power (batteries)
- Natural gas (methane) in the form of liquefied natural gas (LNG) and compressed natural gas (CNG)
- Hydrogen (hydrogen fuel cells and hydrogen mixtures)

The choice of fuel is a function of many factors, such as supplier capabilities, cost, vehicle use, subsidies, fleet size, and political factors.

In some cases, fuel comes from utility gas pipelines. Compressed gas from the pipeline is stored in cascade storage tanks (explained later in this paper). Generally variations in the percentage and makeup of non-methane components in the pipeline natural gas affect the quality of pipeline sources, although some utility systems have greater quality (i.e., higher percentages and more consistent components) than others. Variations in gas quality affect vehicle performance and engine maintenance. Such effects have led to increasing LNG refueling facility storage. Fuel stored as LNG can be used either in LNG form, or converted to CNG.

Properties of LNG and CNG

Generally natural gas is a mixture of methane and heavier hydrocarbons, such as ethane, propane, butane. LNG is natural gas in a cryogenic state. Cryogenics are materials colder than minus 150°F. Examples include nitrogen (-320 °F), Oxygen (-297 °F), Argon (-302 °F), Hydrogen (-420 °F), Helium (-454 °F) and Methane (natural gas, -260 °F).

Some health and safety issues with cryogens include the potential for cold burns upon skin contact, and anoxia if released to the atmosphere because cryogens can displace oxygen.

Properties of LNG

LNG, methane cooled to cryogenic temperatures and liquefied, is

- Odorless
- Colorless
- Non-toxic
- Flammable when within the range of 5% to 15% by volume with air.

When released to atmosphere, the liquid heats in contact with the ground and with air, converting to gas. Initially, the evolved gas is heavier than air. Upon acquiring additional heat, the gas approaches the characteristics of methane, which are lighter than air at temperatures above -180 °F. LNG vapors behave similarly to CNG when the temperature of the gas reaches -30 °F.

Properties of CNG

CNG, compressed methane, is

- Odorless
- Colorless
- Non-toxic
- Flammable when within the range of 5% to 15% by volume with air.

When released to atmosphere, CNG expands as a lighter than air gas.

Facilities

Any designer of LNG/CNG facilities should identify the legally adopted codes and standards for the facilities' locations and should confirm the editions of those codes and standards with the enforcing agencies, otherwise known as the Authority Having Jurisdiction (AHJ).

Generally the applicable standards include:

- The local Building and Fire Codes (including associated Mechanical, Electrical and Plumbing Codes)
- NFPA 57, Liquefied Natural Gas (LNG) Vehicular Fuel Systems Code
- NFPA 52, Compressed Natural Gas (CNG) Vehicular Fuel Systems Code
- In some jurisdictions portions may be used of NFPA 59A, Standard for the Production, Storage, and Handling of Liquefied Natural Gas
- In addition, in California the Department of Occupational Safety and Health (DOSH) pressure vessel code (Title 8) has been revised to incorporate CNG and LNG regulations. Other state Occupational Safety & Health or Pressure Vessel Code administrators may have similar regulations
- NFPA 88A, Standard for Parking Structures

Basic Design of a CNG Station

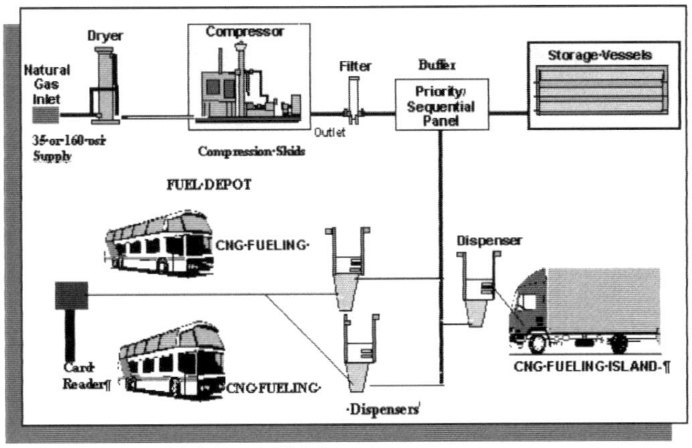

The basic CNG station is designed for the following processes:

1. Receives natural gas from a natural gas inlet, such as a pipeline.
2. Dries and sends gas to a compressor.
3. Filters and sends to storage CNG at 4000 to 5000 psi.
4. Accepts Card-Lock or credit cards at vehicle-fueling dispensers
5. Fills the vehicle fuel tanks from one of three storage tanks.
6. Tops off the vehicle fuel tanks from a second of the three storage tanks.

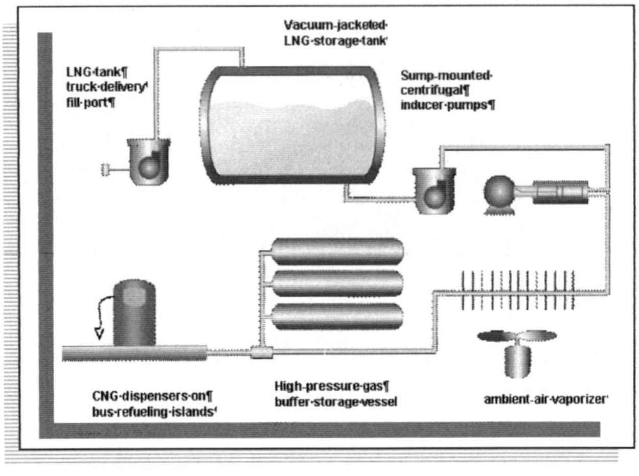

Basic Design of an L/CNG Station

Design considerations for an L/CNG station include these processes:

1. LNG is delivered by truck, or produced from gas pipeline through liquefiers locally.
2. LNG is transferred to the storage tanks through a facility transfer pump. Storage is in one or more cryogenic tanks, which may be above or below ground.
3. From the storage tanks, LNG is transferred to reciprocal pumps, which increase pressure from about 60 psi to 5000 psi.
4. From the reciprocal pumps, the high-pressure LNG is vaporized in ambient heat exchangers creating CNG.
5. For fuel in LNG form, the LNG is transferred from the storage tanks to LNG dispensers. The LNG dispenser may have a local heat exchanger to control the delivered fuel temperature.
6. CNG is stored in Cascade storage tanks and dispensed, as per the CNG facility discussion above.

Temporary Facilities

Temporary facilities are a stopgap measure to enable refueling operations either during construction of permanent facilities or during a facility's shut down.

In all cases, the proposed installation should be subject to the applicable requirements of NFPA 52 or 57, as appropriate, with fire protection considerations applied by a qualified fire protection engineer.

Alternatives

A common design for temporary facilities is an LNG trailer or bobtail truck left in position with refueling either being sent to a temporary vaporizer and on-site or temporary compressor for CNG refueling, or pumped as LNG to an LNG dispenser.

Some designs incorporate a skid-mounted LNG storage tank with associated equipment as needed for CNG or LNG refueling.

Representative Codes & Standards for Site Considerations:

Site spacing considerations are provided in the applicable codes and standards. As the following list is representative and not all-inclusive, a designer should confirm all site-requirement standards and codes, which are applicable to a given project:

NFPA 52:

1. Compression, storage, and dispensing equipment located outdoors shall be
    a. Above ground,
    b. Located away from electric power lines or where exposed by their failure,
    c. A minimum of 10 ft (3.0 m) from the nearest important building or line of adjoining property that can be built upon or from any source of ignition
    d. Not less than 10 ft (3.0 m) from the nearest public street or sidewalk line,
    e. At least 50 ft (15 m) from the nearest rail of any railroad main track.
    f. Designed with a minimum separation of 20 ft (6.1 m) between containers and aboveground tanks containing flammable or combustible liquids
2. During outdoor fueling operations, the point of transfer shall be located at least 10 ft (3.0 m) from any important building, mobile home, public sidewalk, highway, street, or road and at least 3 ft (1 m) from storage containers.

NFPA 57

1. LNG tanks and their associated equipment shall not be located where exposed to failure of overhead electric power lines operating over 600 V.

2. The minimum separation distance between LNG containers and exposures, and between fired equipment shall be located as follows from any impounding area or container drainage system:

Container Water Capacity		Minimum Distance from Edge*	
gal	m^3	Ft	m
<125	0.5	0	0
125-500	0.5-1.9	10	3
501-2,000	1.9-7.6	15	4.6
2,001-15,000	7.656.8	25	7.6
15,001-30,000	56.8114	50	15
30,001-70,000	114-265	75	23
*Of Impoundment or Container Drainage System to Buildings, Property Lines			

3. Points of transfer are located not less than 25 ft (7.6 m) from the nearest important building not associated with the LNG facility, from the line of adjoining property that can be built upon, or from fixed sources of ignition.

4. Drainage should be arranged away from buildings and equipment. Transfer point drainage should be within an impound area, or to a safe area. If within an impound, the impound must hold the contents of a 10 minute spill, or less if based on observable conditions and approved by the AHJ.

5. Tanks are required to be located within dikes sufficiently sized to contain the largest tank contents.

Project Details

Fire Protection Considerations for an Aboveground Storage Vault

In one project location, two 30,000 gal storage tanks (12 ft diam by 60 ft horizontal) adjacent to the refueling dispensers were unacceptable to neighbors. The owner requested solutions from the designer.

Rejected Proposals

Proposal	Reason for Rejection
Install tanks on other side of site, adjacent to property line.	Could not comply with spacing requirements from the tanks to the property line.
Install tanks below ground	Ground water and soil conditions
Bury tanks aboveground	Complications in accessing the tanks for routine inspection and maintenance

Accepted Proposals

The accepted proposal located the tanks within a 28,400 sq.ft. vault and followed, as guidelines:

- Fire Code concepts for the design of underground vaults for flammable tank storage
- NFPA 30 vault requirements.

The design basis was as follows:

1. The vault is a one-story facility built of concrete masonry units. Approximately one third of the floor is four feet below grade to accommodate requirements for a minimum amount of suction pressure (known as LNG head) leading to the pumps. The roof is 13 feet above grade. The pumps are in the below-grade section; and the tanks, in the grade-level section.
2. A high-expansion-foam, fire-extinguishing system was selected. The manufacturer's tests demonstrated that HiEx foam would be effective in an LNG release in
   - Mitigating vapors (with a minimum 3 ft depth of foam)
   - Extinguishing fire (with a minimum 6 ft depth of foam)

   The design basis for the foam system was based on a concentration of foam at 2.75% for 60 minutes applied through a 40gpm nozzle (6000 cfm) and a 220-gpm nozzle (14,000 cfm). The resulting foam depth is sufficient either to mitigate vaporization or to control or extinguish a fire.
3. Activation is at one of two manual pull stations: one at the main exit from the vault, and the other at a safe location.
4. A facility operator stands by at the manual pull station during any tank-truck-transfer operations.
5. Automatic Emergency Shut Down (ESD) actuation occurs with fire detector activation. Gas detection provides alarm at 20% LFL, and ESD at 60% LFL.
6. Ventilation (6 ACF/Hr) is provided, actuated at 20%LFL.

7. Explosion relief is provided per NFPA 68, Guide for Venting of Deflagrations. Only roof vents are needed.
8. The two doors to the facility are blast resistant to assure pressure relief through the explosion vents.
9. Electrical Area Classification is Class I, Div.2 throughout the interior of the vault.
10. Tanks are separated inside by intermediate dike walls, and arranged to drain toward the pump section.
11. Personnel entry is controlled through a confined entry permit system.
12. During any filling operation, doors to the nearby maintenance shop are closed, and bus traffic in prohibited within 50 ft.
13. ESD activation shuts off power to the facility, closes valves at tank discharge nozzle, and shuts down pumps.

The AHJ considered these features as sufficient mitigation for an installation of cryogenic storage tanks in closer proximity to the property line than permitted for aboveground tanks.

Conclusion: Tips for Successful LNG/CNG Projects

1. Conducting a process safety analysis (HAZOP, What If Checklist, etc,) – **highly recommended** We strongly recommend conducting a process safety analysis (PSA) early during the design phase and following up to assure resolution of all items. While it is tempting to apply the results of PSAs conducted on similar designs, our experience has shown that, even with similar designs, local conditions and slight design changes will introduce fire-safety issues, which need mitigating or resolving.
   An important outcome of a PSA is the basis not only for training and but also for sharing basic knowledge of the operation of the facility and the purpose of the instrumentation and controls.
2. Implementing appropriate elements of Process Safety Management, such as per API RP 750, AIChemE/CCPS Guidelines or OSHA —**highly recommended**.
3. Assuring that the Piping and Instrumentation Diagram (P&ID) and Process Flow Diagram reflect the as-built (project working drawing) condition.
4. Testing the emergency shut down after completion.

- Assure documentation of a comprehensive matrix, detailing the expected actuation and outputs.
- Test each actuation device and assure the desired effect is produced.
- Validate full-scale testing of, in particular, programmable logic controllers (PLCs or similar control systems).
- Field-test all gas, fire and manual actuation devices.

5. Reviewing site plan, spacing and drainage issues as early as possible.
6. Assigning responsibility for resolution of any identified deficiencies and following up to assure resolution.
7. Documenting any equivalencies granted and including signed approvals in file for future reference.
8. Providing initial and periodic training of operators, maintenance and inspection personnel. Consider web-based training to facilitate the training and to track the results.
9. Applying fixed facility requirements to temporary facilities.
10. Conducting a fire protection review of the effect of maintaining vehicles using CNG and LNG in existing or planned maintenance garages. In particular, electrical area classification may be inappropriate and precautions established for work which could lead to a release of LNG or CNG.

Additionally, in California, all facilities require certification for engineering integrity and compliance with regulations by a California Registered Engineer(s) competent in the design, fabrication and construction of LNG facilities, i.e., knowledge of LNG and CNG containers, cryogenic operations and equipment, piping systems, fire detection, fire protection equipment, methane gas detection, and other components of the facility.

AVOIDING ELECTROSTATIC HAZARDS IN STORAGE TANKS

Thomas H. Pratt

Burgoyne Incorporated
2864 Johnson Ferry Road
Marietta, GA 30062

ABSTRACT

The petroleum industry has a lot of experience with electrostatic ignitions in the storage of liquids. This experience is reflected in their standards and guidelines and many of these are applicable to chemical operations; however, there are important differences to be kept in mind with using them. Additionally, there are many quantitative guidelines set forth in the standards and recommended practices which should be taken advisedly before they are implemented. This paper discusses their origin so that they can be taken in the proper context and modified to meet their intent while designing operating chemical facilities.

INTRODUCTION

In the design of facilities having large storage tanks the engineer is faced with many challenges, not the least of which is the electrostatic ignition of flammable vapors. The engineer must take care not to become complacent about the electrostatic aspects of the design because the consequences of having an ignition may be tremendous. So what is the engineer to do? Fortunately, we have a wealth of experience with large storage tanks which can be drawn on to offer sound guidance for the design, and this guidance has been distilled and set forth in the recommended practices.

When designing and operating chemical systems which have large storage tanks the engineer should look to (at least) the American Petroleum Institute *Protection Against Ignitions Arising Out of Static, Lightning, and Stray Currents, API Recommended Practice 2003*, Sixth Edition 1998 and the National Fire Protection Association *NFPA 77, Recommended Practice on Static Electricity 2000 Edition.* There are of course other resources which have excellent guidance, but for the present discussion we will constrain ourselves to these two since they are the primary Recommended Practices in the USA.

The first and foremost design consideration is that of providing and maintaining proper bonding and grounding of all conductive components of the equipment involved in the operation. An electrically isolated conductive object around flammable and combustible materials is the same sort of hazard as having an open flame!

LIQUID CONDUCTIVITY

In the design and operation of a chemical plant there are many criteria to be considered, especially where one is dealing with liquids of low conductivity. Here is where the API practices come in handy for the chemical operation since the refined hydrocarbons have low conductivities and there is a wealth of experience. It follows then that the API practices can be transposed directly to practices in a chemical operation. This is usually the case but there is one outstanding exception.

There are some operations which must be considered in the design stage where the relaxation time of a liquid is important; i.e., where one must wait for an electrostatic charge to relax or dissipate before performing some subsequent operation.

The relaxation time, τ, of a liquid is an inverse function of its conductivity; therefore those liquids having low conductivities will have long relaxation times.

$$\tau = \varepsilon \varepsilon_0 / k$$

$\tau \equiv$ Relaxation time, seconds
$\varepsilon \equiv$ Dielectric constant
$\varepsilon_0 \equiv$ Permittivity of free space (8.85×10^{-12} secS/m)
$k \equiv$ Conductivity, S/m

When relaxation times are a consideration, consider the case of hydrocarbons which have dielectric constants around 2.3. Since the dielectric constants are within a narrow range it makes it easy to write a practice around a property of the hydrocarbon rather than around its relaxation time; i.e., write a practice around the conductivity of the liquid. This is done extensively in API RP2003 where the demarcation of 50 pS/m is used to separate conductive liquids from nonconductive liquids for the purposes of the practice. The demarcation point of 50 pS/m is cited in API RP2003 some 16 times.

This demarcation point has stood the test of time for petroleum products but when observed in the cold light of day it may not serve well for other liquids found in chemical operations. As an example ethyl ether with a dielectric constant of 4.6 would have a relaxation time at a conductivity of 100 pS/m the same as a hydrocarbon at 50 pS/m. It has therefore been suggested [Britton, 1999, p. 103] that the demarcation point for chemical systems be set at 100 pS/m. So in the absence of other compelling data, one should use 100 pS/m for chemical operations.

ELECTRIC FIELDS

The casual observer may consider that proper bonding and grounding is all that is required to keep electrostatic ignition sources at bay. Such is not the case for large storage tanks because there can be problems with electric fields in large volumes where there are none under the same conditions in small volumes. That is, where there may be a small or negligible problem with an operation in small vessels, there may be a significant problem in larger ones. (Remember the problems we had when we went to super tankers, the same procedures led to some big explosions. - But that is another story.) These sorts of electrostatic problems arise because of the more intense electric fields in the larger volumes. For a given charge density (Coulombs per cubic meter) there is a direct relationship between vessel size and electric field strength. Thus, the larger the tank the stronger the field in the center of the tank and the greater the potential problem from ignition by brush discharge.

The common spark discharge occurs between two electrodes of different potential (voltage) and can be eliminated by bonding or grounding. But brush discharge is a one electrode discharge where the discharge occurs between a blunt, grounded electrode and an intense electric field [q.v., Britton, 1999 and Pratt 2000]. Brush discharges can be incendive to flammable vapors and can occur when a metal object is inserted into an electric field or when an electric field is created around a metal object.

Electric fields inside of storage tanks arise as a result of the following conditions:

• When the body of the liquid contains a charge, Figure 1.

• When there is a charged mist in the outage, Figure 2.

In either case there will be an electric field in the outage and if the outage contains a flammable vapor, precautions must be in place to avoid a source of ignition.

When filling a storage tank with a liquid of low conductivity there will be an accompanying streaming current which will result in having a charged liquid in the tank and its concomitant electric field in the outage. When the filling stops, the charge will dissipate and the electric field will go away in a time frame depending on the conductivity of the liquid. However, while filling there will be a steady state condition where the rate of charging by the streaming current will equal the rate of dissipation by charge relaxation and the liquid will be charged. The amount of charge on the liquid will be determined by the relative rates of charging and dissipation. Obviously, the lower the conductivity - the larger the charge, for a given streaming current.

The charge in the liquid will create an electric field in the outage. Therefore one does not want to have an electrode (a metal object of the proper geometry) in such an outage. The classic examples of incidents where this mechanism has resulted in ignitions are (1) that of sampling or gauging a tank during the filling operation; i.e., inserting a metal object into an electric field, Figure 3; and (2) that of filling a tank which has a metal object such as a thermal well protruding into the outage; i.e., an electric field is created around a metal object.

A question can now be posed as to how long after the filling operation has been completed can one sample or gauge a tank? API RP2003 says 30 minutes for tanks larger than 10,000 gallons, five minutes for tanks between 5,000 and 10,000 gallons, and one minute for tanks less than 5,000 gallons. These waiting periods are much longer than what would be expected from charge relaxation but are due to the slow settling of water or other foreign debris. Again this practice is derived from experience and measurements in petroleum operations and may or not be transferrable to chemical operations; eg., where the chemical is miscible with water.

Another way to get an electric field into a tank is by creating a charged mist in the outage. When a small droplet of a liquid is formed there are mechanisms by which the droplet will have an electric charge. when a number of them are formed under similar conditions, a charged mist results. The sign and magnitude can vary from day to day and batch to batch in any operation in which such mists are formed. If a flammable atmosphere is also present, one does not want to have an electrode of the proper geometry in the charged mist.

This is the geneses of the admonitions against splash loading. The practices stipulate that the inlet should be at the bottom of the tank and there are restrictions to the linear flow velocities in order to minimize the amount of misting which may occur.

A common way to inadvertently get a mist into an outage is by air entrainment into the liquid line coming into the tank, and one certainly should not clear the incoming lines with air.

The insidious problem with charged mists is that they can persist for significant periods. If one performs an operation where a mist is suspended in an outage along with a flammable vapor (eg., washing) an electrostatic charge can remain on the mist. The safety guides recommend that a period of five hours be given before any metallic object is inserted into the outage (ISGOTT, 1988, ¶ 19.5.4).

The presence of a mist can be detected by the Tyndall effect by shining a flashlight into the outage or even better by using a field test meter to detect the presence of an electric field.

FLOATING ROOF TANKS

Floating roof tanks are a good way to eliminate the outage of a storage tank. If you don't have an outage, it cannot have an electric field or contain a flammable vapor. One must take care however to insure that the floating roof is electrically bonded to the tank shell. This is usually done by the shunt for lightning protection but one should insure that proper bonding of the roof to the tank shell is indeed in place.

When the floating roof has landed on its legs it becomes a fixed roof tank and the same precautions should be taken. Also, in the case of an internal floater the outage may contain flammable atmospheres if the seals leak and the outage should be treated accordingly.

NONCONDUCTIVE TANKS

When it comes to above ground nonconductive tanks, there are some differences between NFPA 77 and API RP2003. The former recommended practice states that "Tanks constructed of nonconductive materials are not allowed for storing Classes I, II, and IIIA liquids...."; i.e., liquids having a flash point below 200°F. API RP2003 states that "It is not recommended to store flammable liquids in nonconductive (eg., plastic, fiberglass) aboveground tanks."; i.e., liquids having a flash point below 100°F. This leaves liquids having flash points between 100°F and 200°F in limbo. The API recommended practice goes on to say that if a liquid has a conductivity less than 50 pS/m then the tank should have "an enclosing grounded conductive shield" and should have a grounded "metal plate with a surface area no less than 30 in^2 per 100 gallons located at the tank bottom". The former requirement is imposed to shield the concomitant electric field from the exterior of the tank and the latter requirement is imposed to provide a place where accumulated charges in the liquid can find their way to ground. (Note that a 10,000 gallon fiberglass tank would require a chain link fence wrapped around it and a square metal plate 4.5 feet on an edge on the bottom of the inside; i.e., nonconductive tanks may not be economical.)

These practices are something to be kept in mind when changing equipment from one service to another. As an example consider a case where a fiberglass tank was installed and used in service of corrosive materials of high conductivity. The shields or plates would not have been installed. If that process is abandoned and another is put in its place where nonconductive liquids are used, the equipment would not be directly transferrable to the new service.

CONCLUSIONS

The first line of defense against the ignition of flammable and combustible materials is to keep all conductive (metal) objects bonded and grounded. In large storage tank operations there is the additional concern of having a significant electric field in the outage where brush discharge ignition can take place at a grounded conductor. The defense against such ignitions is to have measures in place to eliminate or reduce the strength of such electric fields. Failing that, one must have operational procedures in place such that metallic objects will not be inserted into outages where electric fields may be present.

REFERENCES

American Petroleum Institute, "Protection Against Ignitions Arising Out of Static, Lightning, and Stray Currents" *API Recommended Practice RP2003*, Sixth Edition, September 1998.

National Fire Protection Association, *NFPA 77 Recommended Practice on Static Electricity,* 1993 Edition, National Fire Protection Association, Quincy, MA 02269.

International Safety Guide for Oil Tankers and Terminals, Third Edition, International Chamber of Shipping, Oil Companies Marine Forum, International Association of Ports and Harbors, Printed in England by Witherby & Co. Ltd., London, EC1, 1988.

Britton L. G., *Avoiding Static Ignition Hazards in Chemical Operations, a CCPS Concept Book*, American Institute of Chemical Engineers, New York, 1999, ISBN 0-8169-0800-1

Pratt T. H., *Electrostatic Ignitions of Fires and Explosions*, American Institute of Chemical Engineers, Center for Chemical Process Safety, New York, 2000, ISBN 0-8169-9948-1.

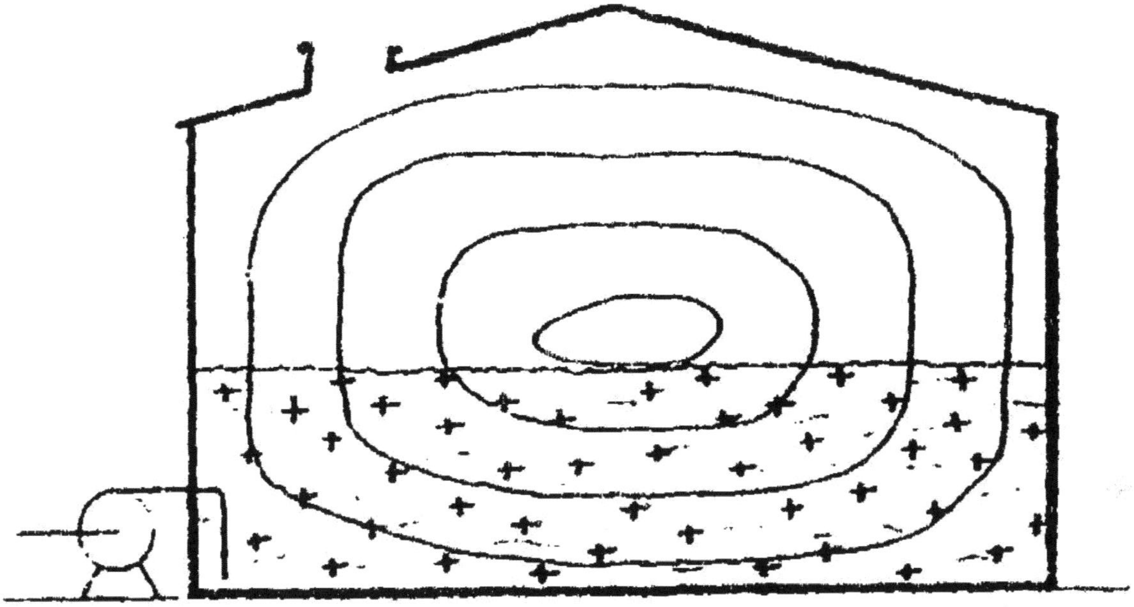

Figure 1: Storage tank with a charged liquid showing equapotential field lines

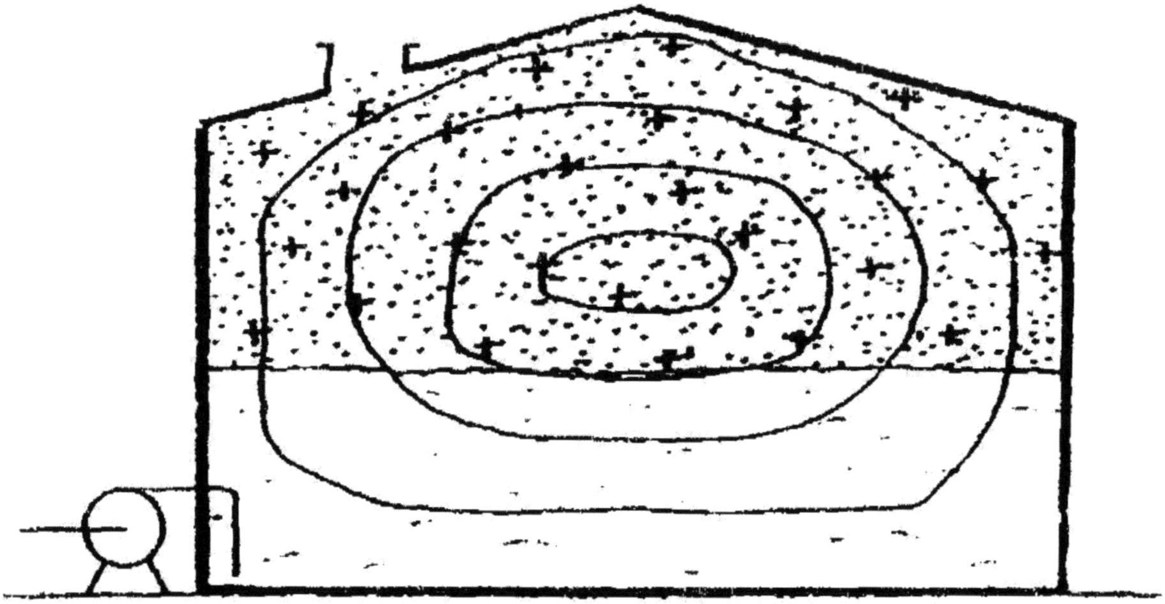

Figure 2: Storage tank with a charged mist showing equapotential field lines

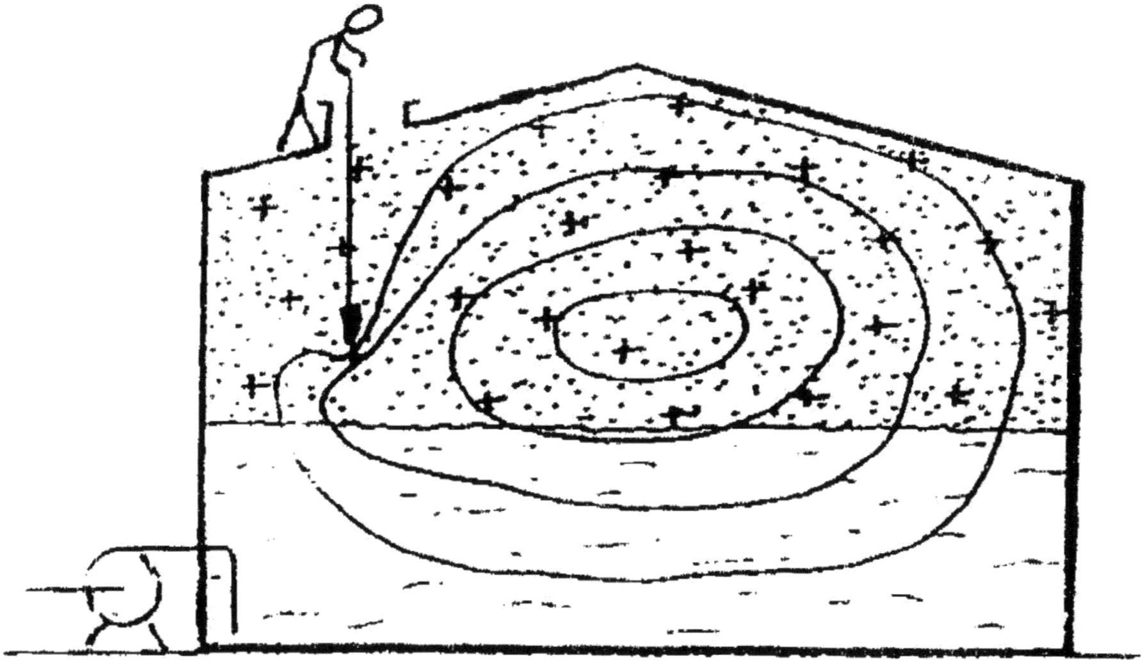

Figure 3: Storage tank being guaged showing compression of field lines.

Aboveground Storage Tank Entry Certification Program

Guy R. Colonna, P.E.
National Fire Protection Association
1 Batterymarch Park
Quincy, MA 02169

Prepared for presentation at

American Institute of Chemical Engineers
38th Annual Loss Prevention Symposium
New Orleans, LA

April 28, 2004

Unpublished

Aboveground Storage Tank Entry Certification Program

Guy R. Colonna, P.E, NFPA, 1 Batterymarch Park, Quincy, MA 02169

Hundreds of thousands of aboveground petroleum and petrochemical liquid storage tanks provide storage for the country's oil and related products. Operators and facility workers routinely work in, near or on these structures. To do that safely requires specific procedures. The nature of the work varies – routine maintenance on pipelines, pumps, valves and other appurtenances; routine repairs to that equipment or to the tank itself; cleaning for internal inspection or change of product; significant repairs to the tank.

The petroleum industry has long had procedures directing the processes for decommissioning, repairing and recommissioning aboveground petroleum and petrochemical liquid storage tanks. Complementary to federal regulations and other industry consensus standards, the petroleum industry procedures address the various tank designs, repair situations and unique procedures developed and adopted by the industry to maintain these important storage containers. As new regulations evolve and third-party services become the norm, the petroleum industry has again looked ahead to establish standard practices and ensure safety.

Work in or on these tanks requires that fundamental procedures for confined space safety be applied. An industry certification program has nearly been completed. When implemented it makes the industry standards the industry benchmark by relying on individuals who will be trained and certified in their ability to consistently apply those benchmark procedures found in the industry standards. This paper offers a review of the hazards encountered in this workplace and the safe work practices found in the industry standards, and introduces the training and certification program ready for implementation.

ON APRIL 6, 2000, TANK T-1001 at the Williams Petroleum tank farm at the Port of Anchorage, AK, erupted in flames. Though no one was seriously injured or killed, the results could have been more devastating based upon the procedures followed and the lack of awareness on the part of "experienced" workers.

At the time of the incident, "work" involved tank cleaning by a contractor specializing in this operation. The previous product in the tank was jet fuel (flash point: 100°F (37.8°C), but since its construction in 1993, it more normally contained naphtha (flash point: 0°F (-17.8°C). The cleaning job included three basic steps: remove the log seal around the perimeter of the floating roof by use of a razor knife to cut the bottom of the neoprene flap, pull out the contaminated foam core, and then continue to remove the rubber from the roof.

The seal, a hollow rubber tube wrapped around the roof perimeter, should be recognized as a type of "hollow structure" that is likely to contain raw product trapped inside (in this case, naphtha or jet fuel or a combination of the two).

The experienced cleaning contractor used one worker inside the tank that was equipped with appropriate protective clothing and supplied-air respirator. In addition to the protection for the entrant, the contractor provided an attendant at one of two open manholes located at the base of the tank and mechanical ventilation at the other open manhole. Since the operation involved a flammable/combustible liquid product, the contractor used explosion-proof equipment and ensured that the ventilation fan was bonded to the tank. However, neither the contractor nor anyone else at the site perceived any danger associated with the razor knife as it was pulled through the rubber seal. The knife generates static that can accumulate in the naphtha trapped within. This static or additional static caused by the subsequent spillage of raw product from the seal down seven feet to the floor of the tank could have caused the fire (1).

Maintaining aboveground storage tanks involves numerous tasks that bring workers into close contact with a variety of hazards. The petroleum industry recognized that its tanks often posed unique hazards above those typically found in other industries. The industry developed its own industry standard and modeled if after the OSHA General Industry standard on *Permit-Required Confined Spaces (PRCS), 29 CFR 1910.146* (2). The petroleum industry standard, developed by the American Petroleum Institute (API), is now an ANSI accredited standard, *ANSI/API Standard 2015, Requirements for Safe Entry and Cleaning of Petroleum Storage Tanks* (3).

API 2015 covers mandatory requirements for isolating tanks and appurtenances, gas and vapor freeing, testing, classifying, entering, cleaning and responding to emergencies associated with petroleum and petrochemical aboveground, atmospheric and low pressure storage tanks and provides for compliance with OSHA and NFPA standards. The approach taken in API 2015 matches that found in OSHA's confined space standard – one based on three principles.

An effective confined space program can be built on the three industrial hygiene principles of Recognition, Evaluation and Control. OSHA and API have both incorporated these principles into their basic requirements. Recognition represents the hazard assessment phase of the operation and requires all spaces to be identified. OSHA PRCS requires all "confined" spaces to be identified and "classified."

So what is a confined space? The current standardized definition includes three elements (2):

- Space is large enough to permit entry and work
- Space is not intended for continuous human occupancy
- Space has limited or restricted means for entry and exit

A few things to keep in mind – entry means any time any part of a worker's body crosses the "plane" of the space opening. So, it means that entry occurs even when not entirely inside a space. Another point about the entry/exit; some spaces might have numerous access points, but would still qualify as a confined space because the access to them is restricted. Restricted could include those instances where the opening is placed in the side of a space above a worker's head requiring some maneuvering to climb to the opening and enter "horizontally" followed by a vertical drop once inside. So, keep in mind the two parts of this element – limited and restricted.

Recognition and the industry standards also require an additional assessment – what hazards are present in these spaces? OSHA and API have separated the more hazardous confined spaces from other spaces through the concept of "permit-required." The OSHA standard becomes more effective once it has been determined that one of several conditions exists – conditions that make a confined space a "permit-required confined space (2)."

A tank (space) is classified as a PERMIT REQUIRED CONFINED SPACE if it meets ALL of the confined space requirements and also has ONE or MORE of the following characteristics:

- Contains or has the potential to contain a hazardous atmosphere.

- Contains a material with the potential to engulf an entrant.

- Has an internal configuration such that an entrant could become trapped or asphyxiated by inwardly converging walls or by floors that slope downward, tapering to smaller cross-sections.

- Contains any other recognized serious safety or health hazard.

A hazardous atmosphere exposes workers to risk of death or incapacitation, impairs ability to self-rescue, or causes injury or acute illness. Hazardous atmospheres include (2):

- Flammable gas, vapor or mist in excess of 10% of the lower flammable limit (LFL)
- Airborne combustible dust in excess of 10% of the lower flammable limit (LFL)
- Oxygen content above 23.5 % by volume or below 19.5% by volume in air
- Toxic contaminant levels above the Permissible Exposure Limit (PEL)
- Any concentrations immediately dangerous to life or health (IDLH)

The assessment completed at this stage flags those spaces that pose a more significant hazard if entry and work is planned and prevents any work from proceeding until the next step is completed – evaluation.

Since atmospheric hazards constitute a large fraction of the hazardous conditions encountered during entry and work on aboveground petroleum and petrochemical liquid storage tanks, evaluation will at a minimum require atmospheric monitoring of all entry and work areas within a tank. The atmospheres pose a threat to the workers and contribute to increased fire and explosion dangers. A report issued by J & H Marsh & McLennan Consulting Services on large property damage losses in the hydrocarbon-chemical industries indicates that fires and explosions represent a significant portion of the losses over a thirty-year period (1968 – 1997)(4). Even though the specific nature of these incidents is not directly related to the storage tank operations within those facilities, it demonstrates the importance of properly recognizing, evaluating, and controlling these atmospheres when performing maintenance or repairs where the atmospheres may be present.

API 2015 provides much more detail on the "other" hazards that OSHA includes in its PRCS definition. In addition to API 2015, another industry document, *ANSI/API RP 2016, Guidelines and Procedures for Entering and Cleaning Petroleum Storage Tanks*, provides information and alternative methods to assist tank-cleaning personnel (5). This document, a companion to API 2015, provides more practical "how to" guidance for the supervisors and workers engaged in tank cleaning and entry.

API developed a system found in API 2015 that incorporates the various atmospheric thresholds and applies a designation that indicates the status for potential entry situations. These designations are similar in concept to the long-standing standard safety designations found in *NFPA 306, Standard for the Control of Gas Hazards on Vessels* (6). This standard, since 1922, has formed the basis for confined space entry and work conditions within maritime construction and repair operations. NFPA Certificated Marine Chemists apply this standard in conjunction with OSHA and US Coast Guard regulations for shipyard employment and maritime vessel safety (7 – NFPA article).

The API 2015 designations include (3):

- Entry with Special Precautions
- Entry with Restrictions
- Entry without Restrictions
- Entry when Inerting

The designations enable a facility to easily put into action the Recognition and Evaluation findings and thus become part of a Control measure. The designations are included on the permit, which is the key Control measure imbedded in both OSHA and API procedures.

The designations are only effective, however, if the evaluation performed on the various atmospheric hazards is conducted by a qualified individual that understands all aspects of atmospheric monitoring within confined spaces. The OSHA and API requirements include the designation of several key personnel, who must be trained to perform their assigned duties. These individuals need not all be unique in their designation; however, there are specific limitations on their actions in some cases. OSHA and API require an entrant, an attendant, an entry supervisor and rescue/emergency response personnel. It is possible in some instances for these designated individuals to share some assignments – for example, the entry supervisor (who establishes that safe conditions exist for entry and work and signs the permit) could in some instances also be the entrant or the attendant. Because some of the training overlaps, it is common practice for workers to be trained to serve in several roles within the confined space team.

One individual that OSHA and API do not explicitly include is the atmospheric tester – for use in this discussion, qualified atmospheric tester. There is actually no such individual, but it is important to understand what knowledge and skills such an individual might possess. The qualified atmospheric tester should:

- Be able to identify typical confined spaces and hazards
- Be familiar with structure and configuration of typical tanks
- Be able to use, calibrate and maintain instrumentation
- Be able to interpret results from atmospheric tests
- Be familiar with regulations and standards applicable to entry and work activities
- Be able to perform atmospheric tests and record results

The quantitative evaluation obtained from these tests coupled with the qualitative information about any additional hazards (such as potential for entrapment or engulfment or other physical or mechanical hazards) forms the basis for the determination whether a safe condition exists for entry and work. Using the API designations, the tank entry supervisor completes the requisite permit and the process shifts into the third and final stage – Control.

Control represents that element used both at the initial or pre-entry stage as well as during the actual work to maintain safe conditions. It includes engineering controls (ventilation, isolation (lock-out/tag-out), and substitution); administrative controls (permits, safe work practices (SOP), warning signs, training); and personal protective controls (clothing, respiratory protection, emergency response equipment). The permit enables the entry and work activity that has been requested to proceed. The process followed to obtain that permit – recognition, evaluation and control – is repeated as necessary any time conditions change or additional work is considered. API and its members have highlighted the role of the entry supervisor in a proposed new program currently under development.

Under the API Individual Certification Program (ICP), industry members representing owner/operators as well as contractors have supported the newest of these programs – Tank Entry Supervisor certification. API launched its first ICP in 1989; these programs were developed to provide the following (8):

- Maintenance of safety and technical standards
- Management control of inspections
- Specially trained inspectors for petroleum and related processing equipment
- Reduced potential for downtime
- Increased regulatory compliance

The objective for the Tank Entry Supervisor ICP program is to provide a baseline of professional competency to enhance safe work performance in and around petroleum and petrochemical liquid aboveground storage tanks. The benefits to industry operations and safety are (8):

- Improved safety and quality of tank operations
- Provide owners and operators a benchmark for assessing and selecting tank entry personnel
- Better trained facility personnel
- Professional recognition for tank entry personnel

Before embarking on this initiative the API Safety Committee examined the landscape to determine whether tank entry safety was an area warranting this level of effort. OSHA and NIOSH published several reports in the mid-to-late 80's documenting an alarming number of confined space incidents and fatalities (involving many sectors of industry including the petroleum industry) (9, 10). One source identified the following factors as contributing to the fatalities presented in the various reports; the data from these reports indicate that workers (11 – other NFPA):

- Fail to recognize confined spaces and the attendant dangers
- Underestimate the dangers
- Trust their senses
- Become complacent
- Attempt to rescue their buddy

OSHA targeted these factors in its PRCS regulatory program; however, as shown in Table 1, incidents involving entry and work in confined spaces (including aboveground petroleum and petrochemical liquid storage tanks) occurred before the regulations came into effect and continue even 10 years after publication of the OSHA requirements. API's response to this unacceptable safety record includes the development and update of key industry consensus standards, sponsored training at industry forums, and the development of a new Tank Entry Supervisor ICP. This program is an experience and exam-based certification program for the supervisor-level experts in the industry.

The certification process includes an assessment of work/field experience (balanced as requested against educational credentials), a written examination (150 multiple choice questions), and recertification every 3 years. Tank Entry Supervisor certification indicates proficiency in understanding and applying API 2015 and 2016; *NFPA 326, Safeguarding Tanks and Containers for Entry, Cleaning or Repair*; and 29 CFR 1910.146, Permit-Required Confined Spaces to entry and work situations. A task group of industry safety and industrial hygiene experts working with tank maintenance contractors and other subject matter experts identified seven activity areas for knowledge and proficiency. The seven areas identified by the team are (8, 12):

1. Project Planning (not including field work)
2. Tank Preparation
3. Ventilation and Atmospheric Testing
4. Entry for inspection and work planning
5. Tank cleaning
6. Entry for Repairs/Modifications
7. Return to Service

These areas have been expanded into knowledge-based elements that then represent specific aspects to be examined as part of the certification examination. The task group has also worked on "item-writing" activities where specific exam questions are developed to correspond to these knowledge areas. API anticipates offering its first certification exam during March 2004.

As noted throughout this paper, a thorough understanding and proficiency in confined space safe practices makes up a significant component of a Tank Entry Supervisor's knowledge base. For that reason, API has collaborated with NFPA to promote training in safe tank entry fundamentals. The training course is independent of API's certification initiative but complements the knowledge base that a prospective certification candidate might require. The course focuses on the recognition, evaluation and control approach to confined space safe practices and relies on the API 2015/2016 documents and OSHA PRCS to lay the procedural groundwork for a permit-based entry and work system for confined spaces. The course offers, through hands-on practical activities, a complete review of atmospheric hazards, sensor design and function, testing "do's and don'ts", calibration and maintenance instruction, and one-on-one actual testing using a tank simulator. The API certification is based upon a written examination; however, proficiency in atmospheric monitoring can only be assessed in a hands-on, demonstrated performance that examines skill as well as knowledge.

API and its industry members intend this ICP to become a benchmark. As further evidence of the API and industry commitment to this effort, OSHA's Office of Alliance Programs has approved an alliance between OSHA, API and NFPA to "promote safe and healthy working conditions within the petroleum and petrochemical liquid storage industries." The Alliance is scheduled to be signed into effect in the 1st Quarter of 2004. It remains in effect for 2 years and has several goals that target the industry and rely on both organizations and their memberships to promote safety and health through training, sharing information and industry outreach activities.

Confined space entry and work remains a serious threat to workers and the workplace. The OSHA reports cited previously document confined space fatalities occurring due to both toxic and asphyxiating atmospheres and to fires and explosions. OSHA regulations and API standards on safe practices have been in place for some time now, yet incidents continue to occur. Every one is one too many. Experience in the maritime industry where a certificated individual – the Marine Chemist – is required as part of safe procedures for entry and work (particularly hot work) stimulates a level of awareness that has improved conditions in this workplace. The Marine Chemist, because of their unique training and experience with the subject of confined space safe practices, are included in the OSHA PRCS standard cited as a source of technical and practical expertise.

Perhaps the API Tank Entry Supervisor certification program will experience similar results within the petroleum and petrochemical liquid storage industry. A benchmark is in the making. Now the industry needs to step up and be measured against it.

References

(1) Finkel, Martin H., *Anatomy of a Confined Space Fire*, **Professional Engineer**, May 2002, p. 18 – 20.
(2) Title 29, Code of Federal Regulations, Permit-Required Confined Spaces, 29 CFR 1910.146.
(3) ANSI/API Standard 2015, *Requirements for Safe Entry and Cleaning of Petroleum Storage Tanks*, API, Washington, DC, 2001.
(4) *Large Property Damage Losses in the Hydrocarbon-Chemical Industries – A Thirty-year Review*, J&H Marsh & McLennan Consulting Services, 18th Edition, 1998.
(5) ANSI/API RP 2016, *Guidelines and Procedures for Entering and Cleaning Petroleum Storage Tanks*, API, Washington, DC, 2001.
(6) NFPA 306, *Standard for the Control of Gas Hazards on Vessels*, NFPA, Quincy, MA, 2003.
(7) Colonna, Guy R., *75 Years of Maritime Fire Protection*, **NFPA Journal**, March/April 1997, p. 90 – 94.
(8) Briskin, Tina, *API Individual Certification Programs*, Marine Chemist Association Seminar, Mobile, AL, July 2003.

(9) *Selected Occupational Fatalities Related to Toxic and Asphyxiating Atmospheres in Confined Work Spaces as found in Reports of OSHA Fatality/Catastrophe Investigation*, US Department of Labor - Occupational Safety & Health Administration (OSHA), July 1985.
(10) *Worker Deaths in Confined Spaces*, NIOSH, January 1994.
(11) Colonna, Guy R., How to Work Safely in Confined Spaces, NFPA Journal, September/October 1994, p. 80 – 87.
(12) William Erny, API, personal communication, January 2004.

Table 1 – Confined Space Incidents**(8)

Date	Location	Type of Incident	Fatality/Injury	Work being performed
March 2003	United Kingdom	Atmospheric – lack of O_2	2 deaths	Work – not specified; out of service fuel tank
June 2002	Newfoundland, Canada	Explosion	1 burn	Cleaning gasoline tank
April 2002	California	Atmospheric – lack of O_2	2 deaths	Cleaning, argon inerted tank
April 2002	Texas	Fire	No injuries	Cleaning out of service tank
December 2001	Texas	Explosions	1 burn	Product removal for vacuum cleaning gasoline tank
August 2001	Texas	Fall	1 death	Roof repair
February 2001	California	Explosion	1 death 1 burn	Cleaning sludge, gasoline storage tank, improper use of vacuum equipment
April 2000	Alaska	Fire	No injuries	Repair tank seal for jet fuel tank

***Source: Tina Briskin, Marine Chemist Association Annual Seminar, Mobile, AL, July 2003.*

PROTECTION OF FACILITIES AGAINST MALEVOLENT USE OF VEHICLES

Walter Sawruk, P.E.

Senior Consultant
ABS Consulting
Shillington, PA

Prepared for Presentation at

American Institute of Chemical Engineers
38th Annual Loss Prevention Symposium
Engineering Solutions to Facility Security Challenges

New Orleans, LA

April 26 – 29, 2004

ABSTRACT

It is widely recognized that a more urgent need exists since 9/11 to protect vital assets at chemical facilities and refineries from terrorist threats. Industry voluntary initiatives addressing these heightened security concerns have been undertaken with the primary emphasis thus far on performance of security vulnerability assessments (SVAs). Some facilities having completed the SVA have now begun implementation of engineered security enhancements, including continuous vehicle barrier systems for protection against malevolent use of vehicles. The perceived threat is that the terrorist vehicle is used to transport terrorists to the desired target within the facility or, it is transporting a large bomb that detonates following impact with the vehicle barrier. Design and qualification of active vehicle security barriers has long been standardized through the implementation of government technical specifications [5] that include full scale testing requirements. However, similar standards and design criteria currently do not exist for passive barrier systems. Furthermore, the range of passive barriers designed and tested for military applications either do not meet the challenges posed by today's terrorist threat, or these tested military designs are plagued by high fabrication and installation costs and undesirable construction features.

This paper describes the design process used to develop reliable, relatively inexpensive passive vehicle barrier systems for large facilities and kinetic energy demands on par with the highest rated active barriers currently available. Sensitive data are not revealed, however, sufficient information is provided to guide a competent engineer through the design process. The design process described begins with development of barrier performance objectives that include obtaining a reliable prediction of vehicle penetration distance and a high degree of confidence that vehicle incapacitation and entry denial is achieved. Discussed next is a means to obtain a quantitative definition of the terrorist threat in terms of vehicle type, size, weight and speed, which translates to kinetic energy demand for the barrier. A discussion of how natural and man-made terrain features can be used to effectively reduce or eliminate the vehicle demand is included. Discussion continues with a detailed description of the analytical process for vehicle impact with the barrier from which vehicle penetration distance, barrier performance, and vehicle damage estimates are obtained. Cost data are also presented. By following the engineering approach described herein, a reliable, cost effective passive vehicle barrier system design can be achieved for large facilities and high kinetic energy demands.

INTRODUCTION

Faced with the task of providing protection of a facility against malevolent use of vehicles, one must consider the threat characteristics, asset vulnerabilities, consequences of failure to neutralize the threat, and available resources. This overall process, or security vulnerability assessment, should lay the groundwork for an engineered approach to a defensive strategy employing tactics for the deterrence, delay, and defeat of terrorist attack by vehicle. The malevolent vehicle threat is that the vehicle rapidly transports terrorists to the desired target within the facility or, it is transporting a large bomb that detonates following impact with the vehicle barrier. Specific data required to commence an engineered solution includes:

- Design basis vehicle (DBV) type, size, weight (GVW including payload), maneuverability, and acceleration characteristics
- Design basis explosive (DBX) size (equivalent TNT weight) transported by the DBV
- Critical asset vulnerability to external overpressures in terms of standoff distance requirements as a function of explosive charge size
- Vehicle barrier system (VBS) configuration for the facility

One can readily find access to numerous publications offering a wide assortment of active and passive security barriers. Much of this information, especially that pertaining to passive barriers, does not provide adequate technical data to enable the performance of the barriers against specific vehicle threats to be determined. The engineer is thus faced with a design challenge to ensure satisfaction of the vehicle barrier performance objectives. Unfortunately, a lack of economical designs based on full-scale crash testing and standard design methods for passive barriers currently exists. Hence, the focus of this paper is the development of an engineering approach to the design of passive vehicle barrier systems with an emphasis on inertia type barriers.

VEHICLE BARRIER PERFORMANCE OBJECTIVES

Performance objectives of vehicle barrier systems are summarized as follows.

- Vehicle drive system must be disabled upon high-speed impact with the barrier
- Barriers must have sufficient weight and sliding resistance to prevent the vehicle from slowly pushing through individual barriers
- Barriers must have sufficient strength to absorb a high-speed impact without failure (breaking apart) prior to disabling the vehicle
- Vehicle penetration distance must be limited to some predetermined value based on the threat and assets being protected (e.g. blast effects from a vehicle bomb detonated at the point where the vehicle stops)

In addition to these performance objectives, deterrence value, resistance to tampering, aesthetics, maintenance, and cost considerations often have a major influence on the VBS design.

VEHICLE THREAT DEFINITION

Vehicle Characteristics

Vehicle characteristics include type, size, weight (GVW including payload), maneuverability, and acceleration data. Vehicle threats can range from compact sedans carrying 500 lbs TNT to semi-trailers with 60,000 lbs TNT as depicted by the *TSWG Terrorist Bomb Threat Stand-Off* card reproduced in Figure 1. The passenger cargo van (or mid-size truck) having a payload rating of approximately 4000 lbs is representative of the vehicle used in the 1995 bombing of the A. P. Murrah Federal Building [6]. In addition, the heaviest vehicle historically used for crash testing of active barriers per the US Department of State's specification SD-STD-02.01 [5] is a mid-size truck having a GVW of 15,000 lbs and maximum speed of 50 mph.

Terrorist Bomb Threat Stand-Off

THREAT	THREAT DESCRIPTION	EXPLOSIVES CAPACITY[1] (TNT EQUIVALENT)	BUILDING EVACUATION DISTANCE[2]	OUTDOOR EVACUATION DISTANCE[3]
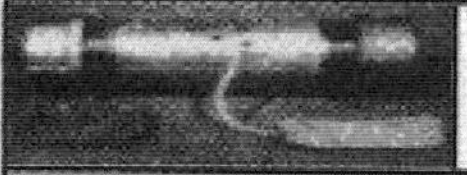	PIPE BOMB	5 LBS/ 2.3 KG	70 FT/ 21 M	850 FT/ 259 M
	BRIEFCASE/ SUITCASE BOMB	50 LBS/ 23 KG	150 FT/ 46 M	1,850 FT/ 564 M
	COMPACT SEDAN	500 LBS/ 227 KG	320 FT/ 98 M	1,500 FT/ 457 M
	SEDAN	1,000 LBS/ 454 KG	400 FT/ 122 M	1,750 FT/ 534 M
	PASSENGER/ CARGO VAN	4,000 LBS/ 1,814 KG	640 FT/ 195 M	2,750 FT/ 838 M
	SMALL MOVING VAN/DELIVERY TRUCK	10,000 LBS/ 4,536 KG	860 FT/ 263 M	3,750 FT/ 1,143 M

This card supersedes any previous undated versions 11/99

THREAT	THREAT DESCRIPTION	EXPLOSIVES CAPACITY[1] (TNT EQUIVALENT)	BUILDING EVACUATION DISTANCE[2]	OUTDOOR EVACUATION DISTANCE[3]
	MOVING VAN/ WATER TRUCK	30,000 LBS/ 13,608 KG	1,240 FT/ 375M	6,500 FT/ 1,982 M
	SEMI-TRAILER	60,000 LBS/ 27,216 KG	1,570 FT/ 475 M	7,000 FT/ 2,134 M

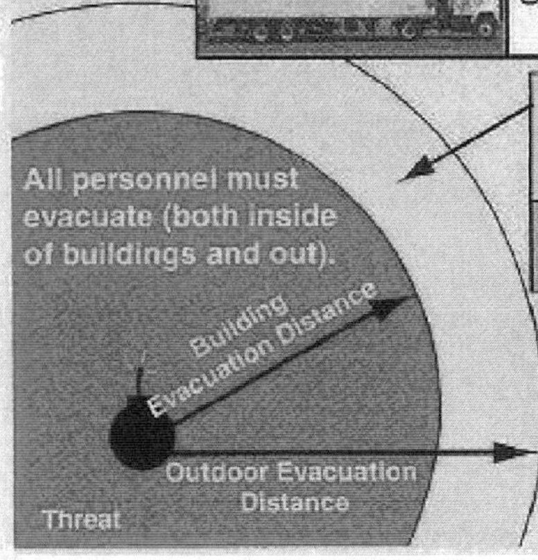

All personnel must either seek shelter inside a building (with some risk) away from windows and exterior walls, or move beyond the Outdoor Evacuation Distance.

Preferred area (beyond this line) for evacuation of people in buildings and mandatory for people outdoors.

[1] Based on maximum volume or weight of explosive (TNT equivalent) that could reasonably fit in a suitcase or vehicle.
[2] Governed by the ability of an unstrengthened building to withstand severe damage or collapse.
[3] Governed by the greater of fragment throw distance or glass breakage/ falling glass hazard distance. Note that pipe and briefcase bombs assume cased charges which throw fragments farther than vehicle bombs.

Figure 1 - TSWG Vehicle Bomb Threat Stand-Off Card[1]

[1] Available for public purchase at: http://bookstore.gpo.gov/eresponse.html.

A mid-size truck (GVW = 15,000 lbs) is used in SD-STD-02.01 to establish the basis for three standard test conditions designated as K4, K8 and K12 with corresponding vehicle speeds and kinetic energies of 30 mph/450 ft*kips, 40 mph/800 ft*kips and 50 mph/1250 ft*kips, respectively. Acceleration versus speed data for such a vehicle on a straight, level surface is depicted in Figure 2 using data from Fricke [7]. Speed and kinetic energy versus distance data can then be generated as shown in Figure 3.

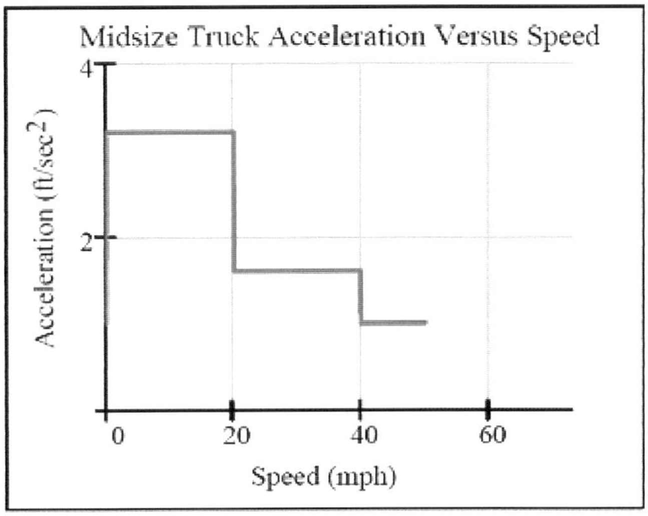

Figure 2 - Typical Speed Versus Acceleration Function for a Mid-Size Truck

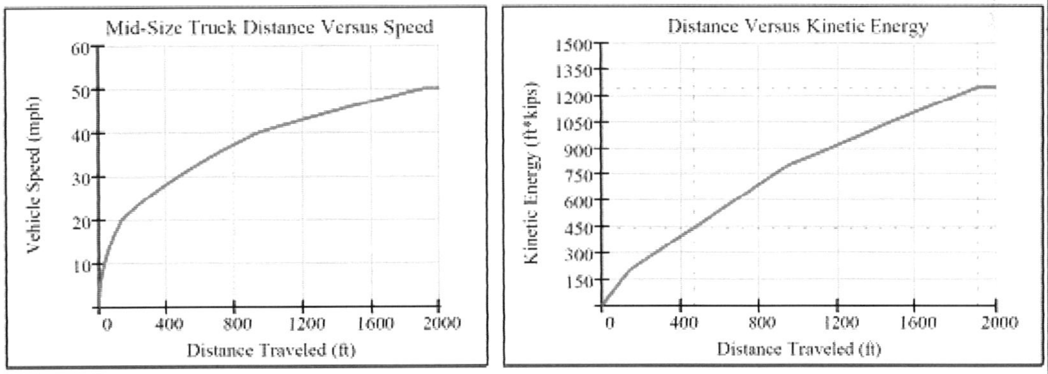

W_{veh} = 15000.0 lbf	**Kinetic Energy**	**Distance**	**Speed**
	K4 = 450.0 ft·kips	S_{K4} = 469.0 ft	$V_1(S_{K4})$ = 30.0 mph
	K8 = 800.0 ft·kips	S_{K8} = 939.0 ft	$V_1(S_{K8})$ = 40.0 mph
	K12 = 1250.0 ft·kips	S_{K12} = 1909.1 ft	$V_1(S_{K12})$ = 50.0 mph

Figure 3 - Distance Versus Speed and Kinetic Energy for a Representative Vehicle

Prior to developing an energy demand profile for the VBS, additional speed and kinetic energy data are required for the vehicle on curved and sloped paths. A simplified, conservative approach for curved paths, neglecting vehicle dynamics, is to determine the maximum speed as a function of curve radius based on the maximum centrifugal force which can be balanced through friction [4]. Friction coefficients for a wide range of surface conditions are available from Fricke [7]. Figure 4 depicts such results for a mid-size truck on a paved surface in good condition. Gravity effects the vehicle acceleration versus speed function as depicted in free body diagram of a vehicle on a slope in Figure 5. These basic 'tools' can used to determine the kinetic energy demand profile of the VBS.

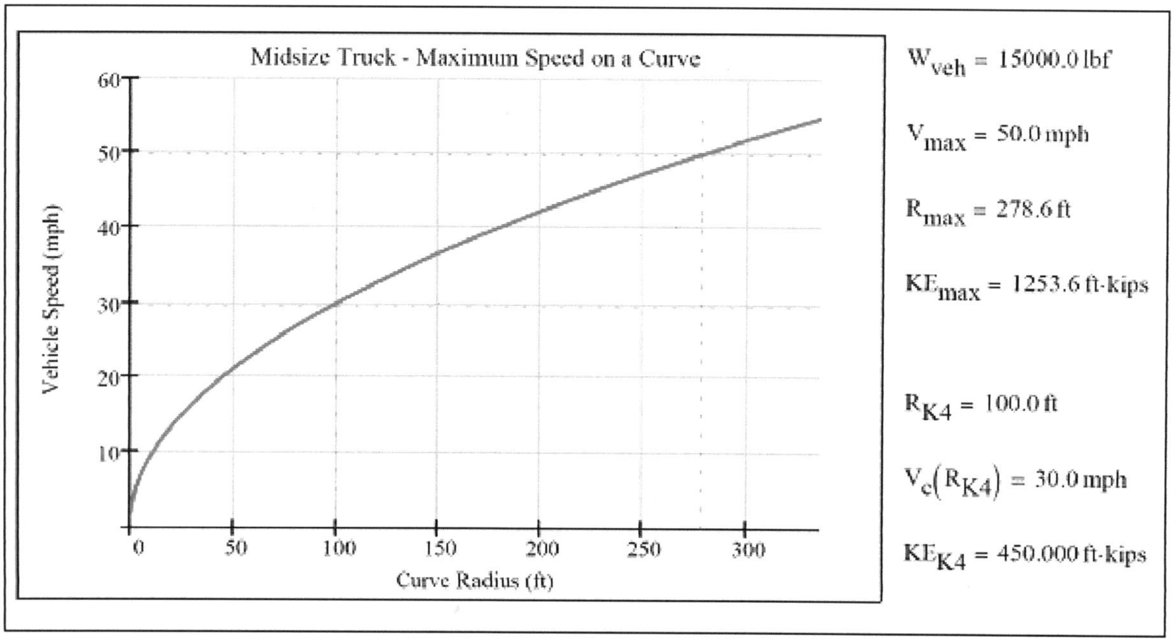

Figure 4 - Maximum Controllable Speed Around a Curve for a Mid-Size Truck

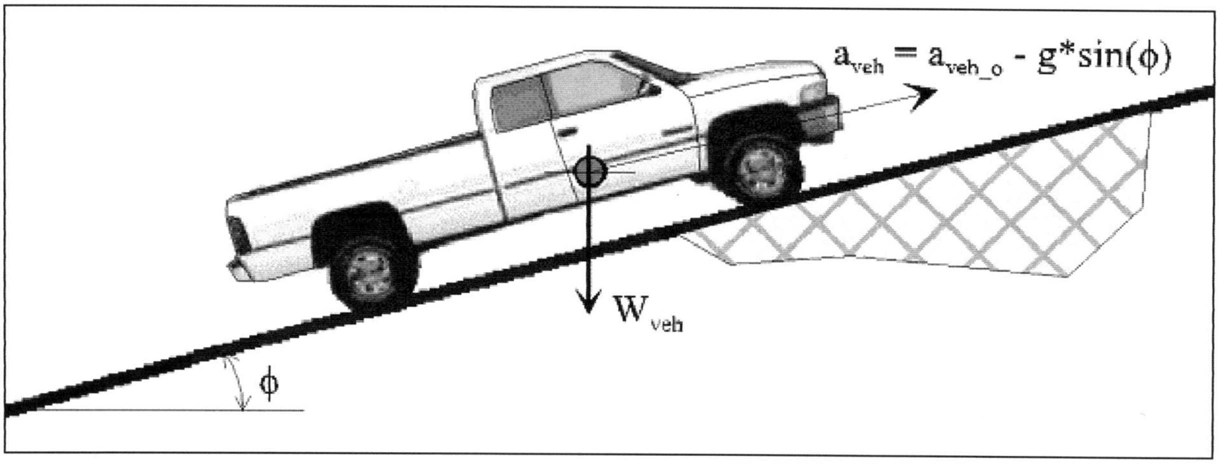

Figure 5 - Free Body Diagram of Vehicle on a Slope

Kinetic Energy Demand Profile for the VBS

Potential vehicle paths are established by a site inspection. Potential vehicle paths often originate far off-site on neighboring public and private properties. Range finders, GPS units and aerial photos facilitate the inspection.

Figure 6 depicts a representative multi-segment vehicle path with the end point being the vehicle barrier. Vehicle speeds are calculated for each segment sequentially. Figure 7 depicts vehicle paths superimposed on a sample VBS general arrangement drawing. Kinetic energy demand on the barrier is determined using the normal component of the vehicle speed at impact. Speed reduction measures such as installation of obstacles in the vehicle path are generally most useful in local areas. Speed bumps offer no advantage. This information is used to develop the kinetic energy demand profile for the entire VBS.

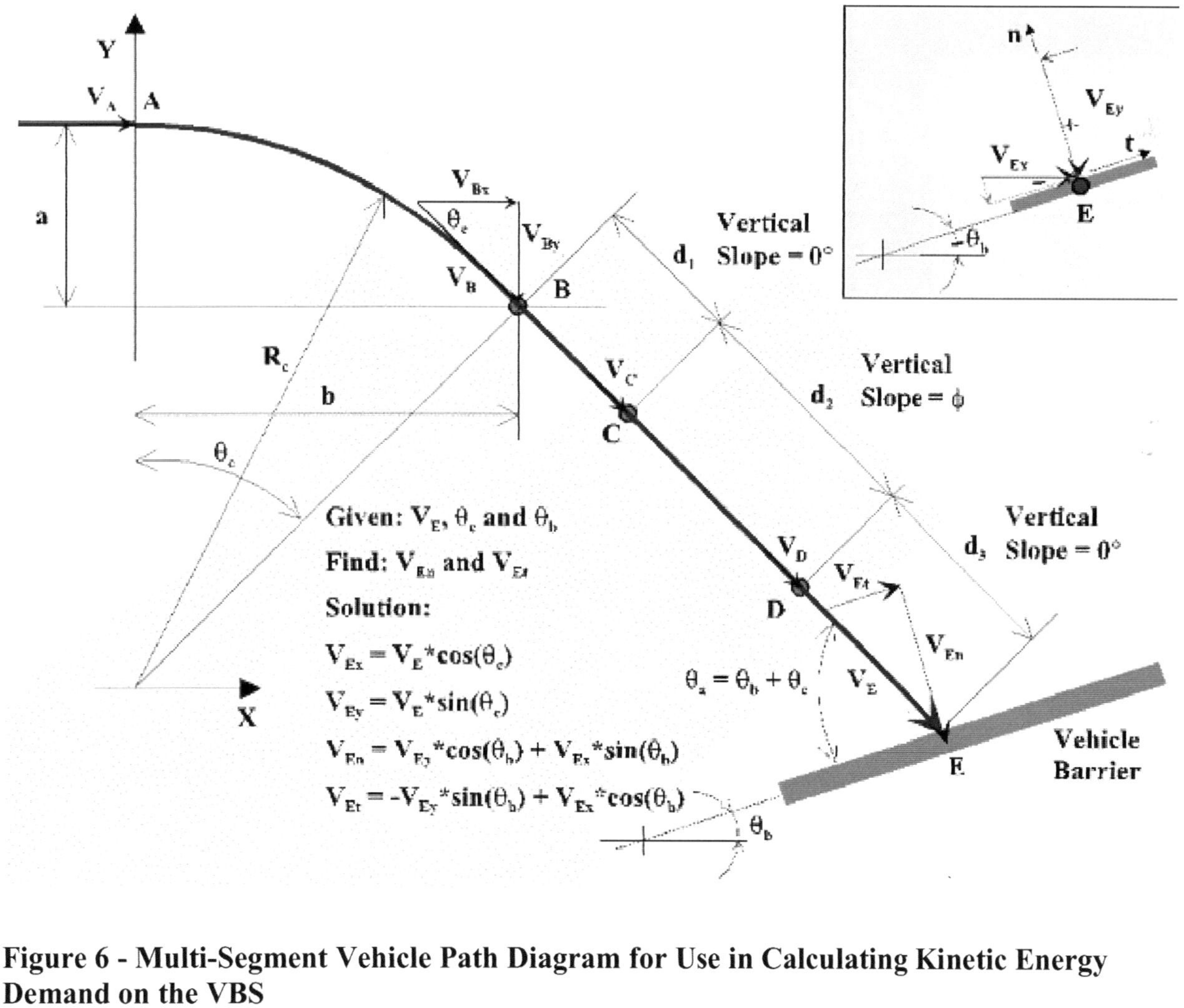

Figure 6 - Multi-Segment Vehicle Path Diagram for Use in Calculating Kinetic Energy Demand on the VBS

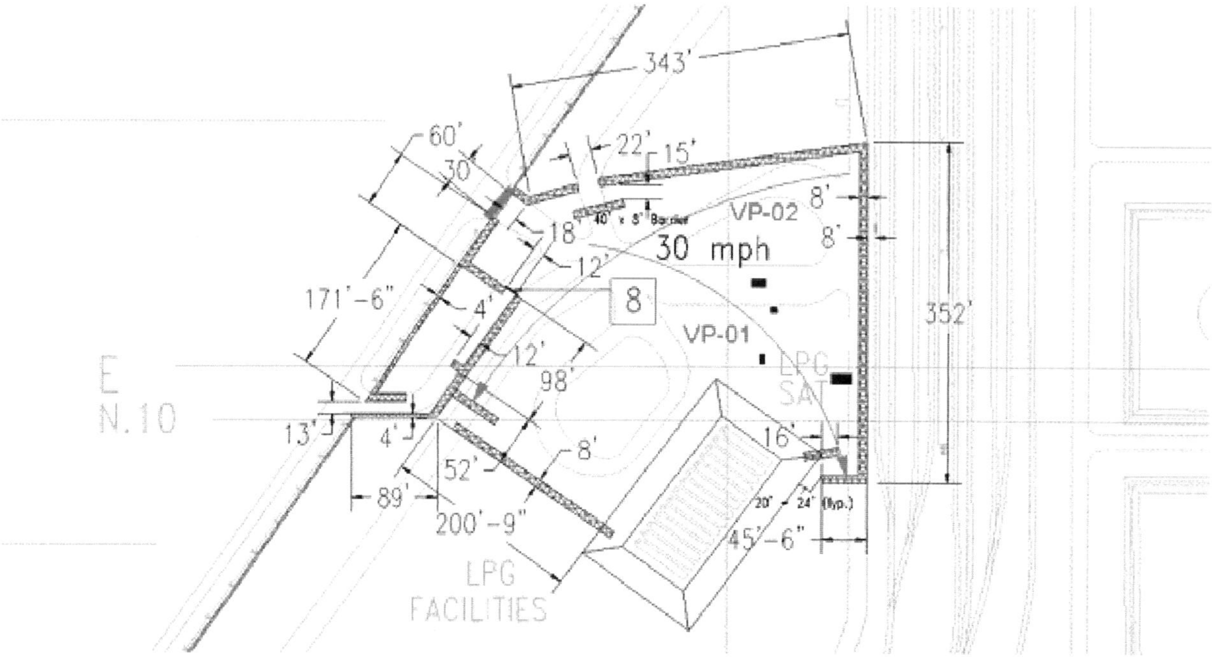

Figure 7 - Sample Vehicle Path Design Requirements Superimposed on the VBS General Arrangement Drawing

PASSIVE BARRIER CATEGORIES

Resistance Mechanisms

Passive barriers can be categorized according to the mechanism used to resist vehicle penetration. Passive barriers such cables, beams and bollards rely primarily on their strength and internal strain energy absorption capacity to halt the forward motion of the vehicle. These barriers may be categorized as strength based designs. Large surface mounted boulders and precast concrete blocks resist vehicle penetration through their inertia effects for high-speed impacts. Such barriers may be categorized as inertia barriers.

Strength Based Designs

Strength based designs rely on the strength but more specifically, energy absorption capacity. Design practice is to balance the total kinetic energy of the vehicle with the internal strain energy of the barrier. High strength steel cables offer an efficient choice (high energy capacity with low volume of material). Energy absorption capacity is proportional to the cable length at the sacrifice of increased vehicle penetration distance. Steel beams (i.e., flexural members) are a viable option for relatively short spans such as across roadways. Bollards (e.g., concrete filled steel pipes) fall into this category.

Inertia Barriers

Inertia barriers rely upon their high mass to inflict damage to the vehicle. Cost savings can be achieved by placing such barriers on the surface as opposed to partial embedment in the earth. Figure 8 shows such a barrier designed for high-speed vehicle impact. Further discussion on the design approach for such barriers is given below.

Figure 8 - Representative Inertia Barrier Designed for High-Speed Vehicle Impact

Ditches and Berms

Figure 9, obtained from MIL-HDBK-1013/14 [4], provides guidelines for the use of ditches and berms as vehicle barriers. These guidelines can be applied to natural features such as creek beds and rock outcroppings. Erosion and maintenance costs are important considerations for such barriers.

Kinetic Energy Ratings of Published Passive Barrier Designs

Figure 10 depicts the kinetic energy demand for three representative vehicles over a range of vehicle speeds. Also shown in the figure are kinetic energy ratings of several passive barriers established by full scale vehicle crash tests as reported in MIL-HDBK-1013/14 [4]. Note that none of the tested passive barrier designs achieve the so-called "K12" energy rating of 1250 ft*kips. Hence the demand for qualified, high-energy rated passive barriers remains high.

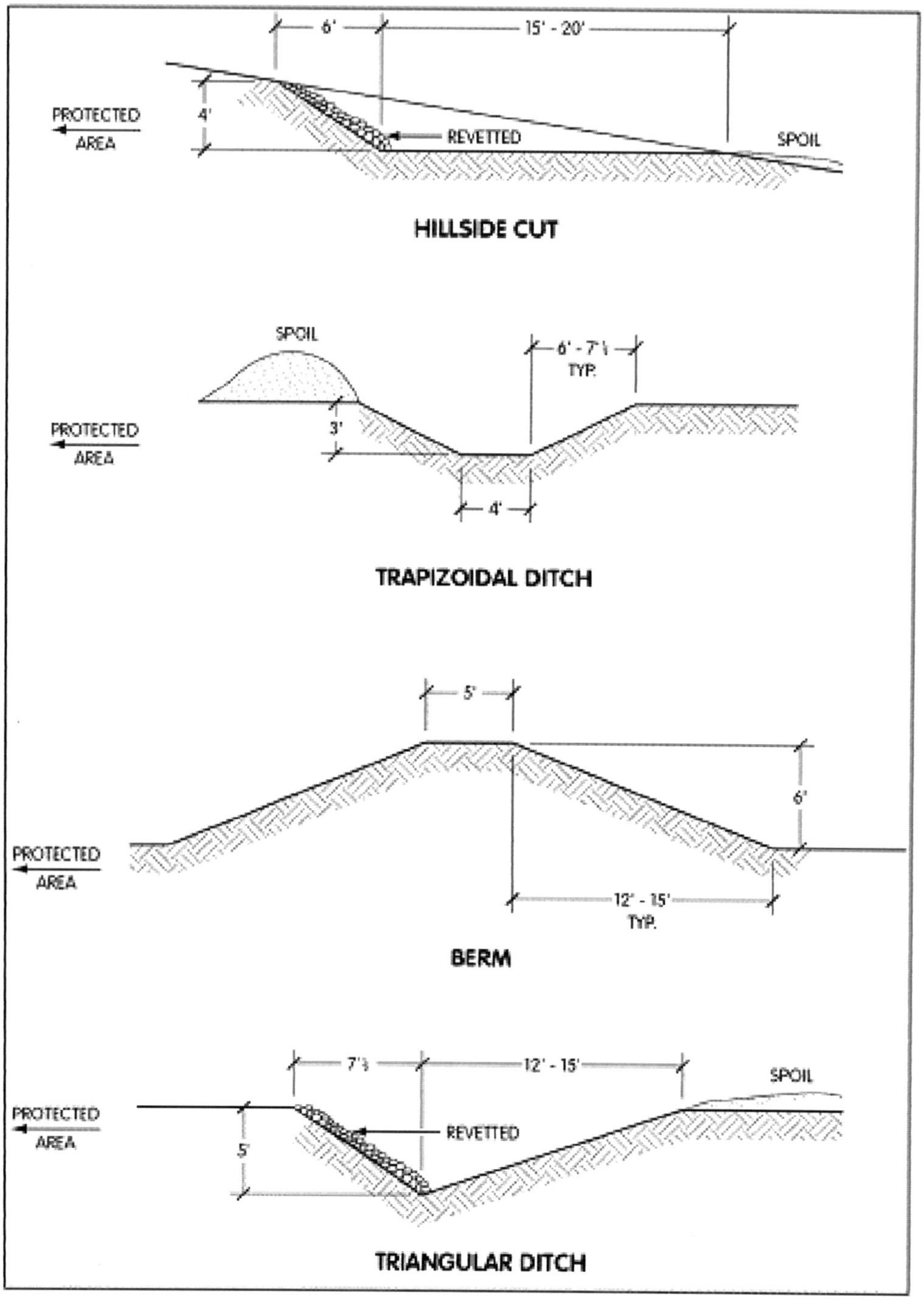

Figure 9 - Guidelines for Ditches and Berms Used as Vehicle Barriers

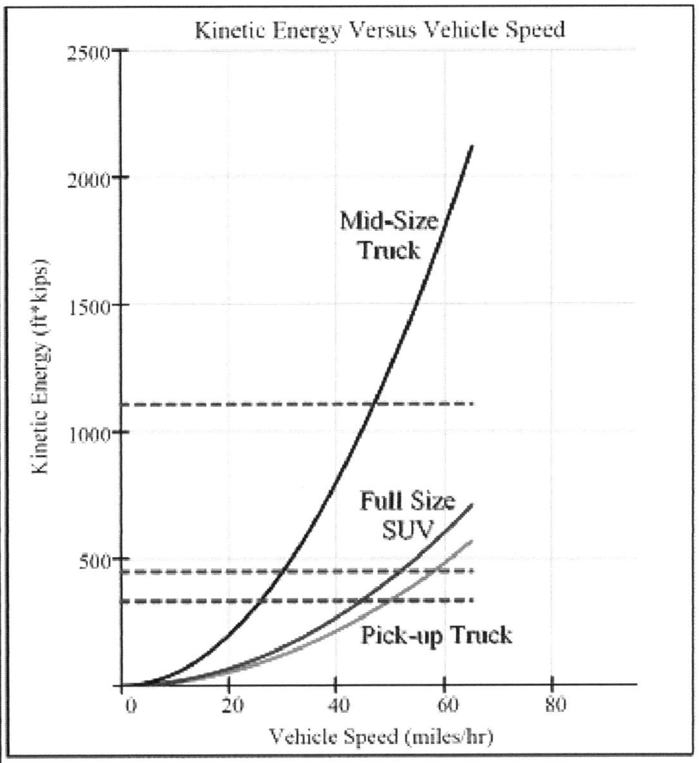

Representative Vehicle Weights:

Pick-up truck	$W_{PU} = 4000\,lbf$
Full size SUV	$W_{SUV} = 5000\,lbf$
Mid-size truck	$W_{MST} = 15000\,lbf$

Vehicle Kinetic Energy:

$$KE = \frac{1}{2} \cdot \frac{W}{g} \cdot V^2$$

Representative Passive Barrier KE Ratings Validated by Test (MIL-HDBK-1013/14):

Steel bollards	$KE_{BOL} = 346\,ft\cdot kips$
Concrete planters	$KE_{PLNTR} = 1107\,ft\cdot kips$
Anchored Jersey barriers,	$KE_{JB} = 334\,ft\cdot kips$
R/C Wall (19 inch),	$KE_{RCW} = 450\,ft\cdot kips$

Figure 10 – Representative Vehicle Kinetic Energies and Passive Barrier Ratings

INERTIA BARRIER DESIGN PROCEDURE

Design Objectives

A barrier system meeting the performance objectives stated above at the lowest installed cost, with little or no maintenance, is the primary design objective for most cases. However, aesthetics sometimes overrides cost considerations. For inertia barriers, it is generally less expensive to simply install more massive barrier units on grade with minimal surface preparation than it is to achieve higher vehicle impact resistance through partial embedment of a lighter weight barrier. The design approach for such a surface mounted, inertia vehicle barrier is discussed in more detail.

Design Approach

Initial impact of the vehicle with the inertia vehicle barrier can be analyzed using classical mechanics methods. Likewise, subsequent motion of the vehicle and the barrier can be predicted using energy methods. However, it is a very difficult task to demonstrate analytically that the vehicle drive system is disabled or that the driver has sustained serious injuries upon impact with the barrier. These aspects are typically addressed through full scale testing. Hence, the US Department of State's specification SD-STD-02.01 [5] is used for qualification testing of active vehicle barriers (gates, pop-up bollards and pop-up wedges). Therefore, a conservative energy-based approach based on vehicle crash studies performed for the government, the automobile insurance industry, and crash gate manufacturers is taken here to address vehicle damage.

The planar impact of two rigid bodies is illustrated in Figure 11. The brief explanation of the dynamic impact analysis formulation provided here is based on the works by Brachs [2, 3]. In this application, Body No. 1 represents the vehicle and Body No. 2 represents the inertia barrier. Consider these bodies to be lying in a horizontal plane. Each body possess translational as well as rotary inertia. Eccentricities are measured from the point of impact to the center of gravity for each body. Since these eccentricities are generally nonzero coupled with the inclusion of tangential friction forces acting at the point of impact, both bodies will translate as well as rotate about their respective vertical axes.

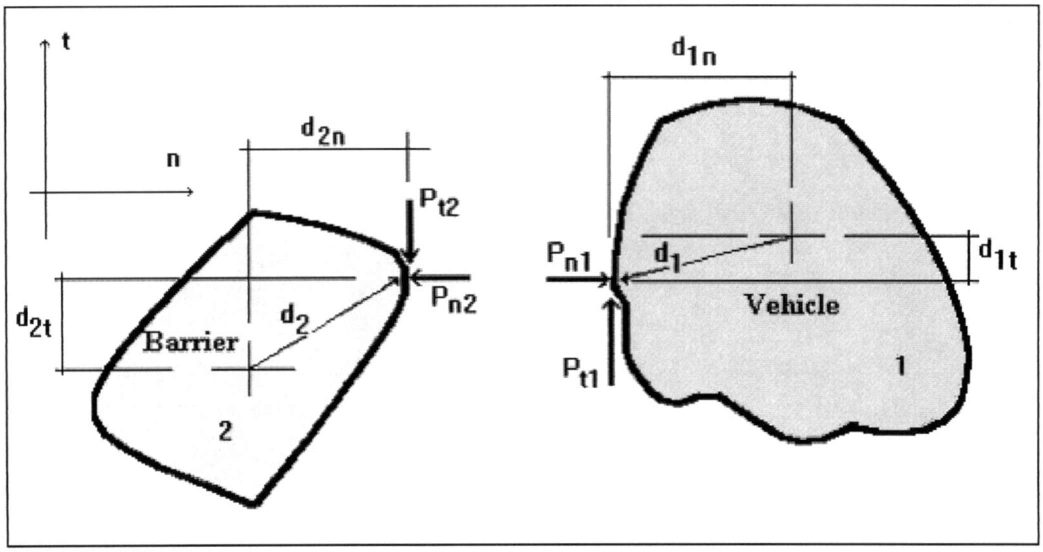

Figure 11 - Free-Body Diagram of Vehicle Impact With Inertia Barrier

Conservation of linear momentum in the normal, "n", and tangential, "t", directions is written as follows. Initial velocities are written using lowercase variable names and final velocities are written in uppercase.

$$m_1V_{1n} + m_2V_{2n} = m_1v_{1n} + m_2v_{2n}$$

$$m_1V_{1t} + m_2V_{2t} = m_1v_{1t} + m_2v_{2t}$$

For this application, v_{1n} equals the vehicle design speed and, $v_{1t} = v_{2n} = v_{2t} = 0$.

The coefficient of restitution, e, is used to relate the initial and final velocity components at the point of impact. Note that $\omega_1 = \omega_2 = 0$ for this application.

$$V_{1n} + d_{1t}\Omega_1 - V_{2n} + d_{2t}\Omega_2 = -e(v_{1n} + d_{1t}\omega_1 - v_{2n} + d_{2t}\omega_2)$$

A coefficient of friction, μ, at the point of impact relates the tangential impulse, P_t, to the normal impulse, P_n, occurring during the impact,

$\mu = P_t/P_n$

where,

$P_n = m_1(V_{1n} - v_{1n}) = -m_2(V_{2n} - v_{2n})$

$P_t = m_1(V_{1t} - v_{1t}) = -m_2(V_{2t} - v_{2t})$

The friction coefficient is selected such that sliding will exist between the two bodies during the impact, thereby ensuring that excessive tangential velocities are not calculated. This approach also ensures that kinetic energy is not artificially created. Each body's angular momentum is related to the resultant impulse, P_r, as follows;

$I_{o1}(\Omega_1 - \omega_1) = d_1 P_r$

$I_{o2}(\Omega_2 - \omega_2) = d_2 P_r$

where, I_{o1}, I_{o2} = rotary inertia for Body Nos. 1 and 2 about the center of gravity of each body.

A sufficient number of equations have now been defined to enable a solution to be obtained for the six unknown quantities; V_{1n}, V_{1t}, Ω_1 for Body No. 1 and V_{2n}, V_{2t}, Ω_2 for Body No. 2. Once these quantities are known, the instantaneous kinetic energies of each body can be determined.

Vehicle penetration distances and barrier sliding distances are calculated using the normal component of the instantaneous velocities after impact, V_{1n} and V_{2n}, in an energy balance equation. All kinetic energy is dissipated through sliding friction between the body and the ground.

Vehicle damage estimates are performed on the basis of energy dissipation and to a lesser extent on the vehicle crush and the change in vehicle speed immediately after the initial impact. These quantities are results of the impact analysis. Occurrence of energy loss through deformation of the vehicle is implied through the use of a coefficients of restitution less than unity. Estimates of the coefficient of restitution for vehicle impacts against rigid barriers over a range of vehicle speeds are available from AiTS [1]. Vehicle crush estimates are performed using crush coefficients available from sources such as AiTS [1], Fricke [7], McHenry [9] and Siddall [11]. Examination of the data presented by these sources in addition to energy absorbing highway safety barrier studies by Hirsch [8] and published active security barrier test results, leads to two observations for a head-on collision of a mid-size truck collision with a barrier. These observations form the acceptance criteria regarding vehicle drive incapacitation for qualification of inertia barriers by analysis as opposed to full scale testing.

- Kinetic energy loss of 500 ft*kips or more through vehicle deformation provides a high degree of confidence of drive system incapacitation
- An instantaneous change in vehicle speed of 30 mph or more provides a high degree of confidence of drive system incapacitation

In addition to the high-speed impact, the barrier system must provide sufficient resistance to prevent a vehicle from slowly pushing its way through. A simplified static analysis procedure based on the maximum traction that can be developed at the drive wheels is performed to demonstrate satisfaction of this requirement. A slightly more involved dynamic (i.e., equivalent static) approach as outlined by Taborek [12] can be used to account for inertia forces and weight shifting between the vehicle axles.

Sample Design Results

Results of a dynamic impact analysis performed for a set of six surface-mounted, precast concrete barriers are depicted in Figure 12. Barrier height and width were constant at 3 ft and 4 ft, respectively. However, the barrier length was varied from 6 ft to 11 ft in 1 ft increments for this example. Barrier placement allowed for 3 ft gaps end-to-end for drainage and pedestrian access. It was found that a minimum length of 8 ft is required to ensure sufficient damage to the vehicle. The amount of energy absorbed for the 8 ft long barrier is approximately 505 ft*kips with a uniform frontal crush of approximately 9 inches.

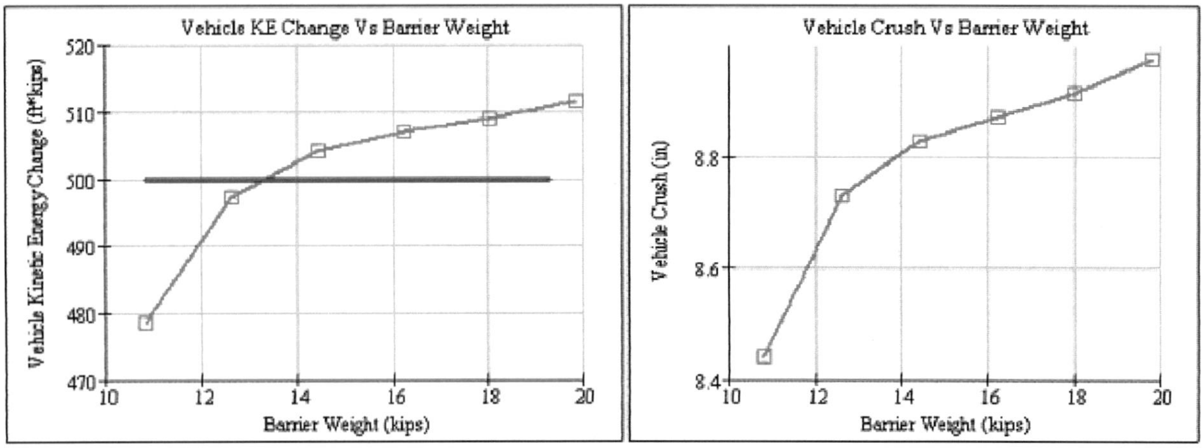

Figure 12 - Inertia Barrier Impact Analysis Results

Figure 13 depicts the calculated vehicle penetration distances for a 50 mph impact of the mid-size truck on the same set of precast concrete barriers. A penetration distance of approximately 43 ft is predicted for the minimum acceptable weight barrier. The vehicle penetration distance is taken into account for locating the VBS with respect to vital assets for blast effects.

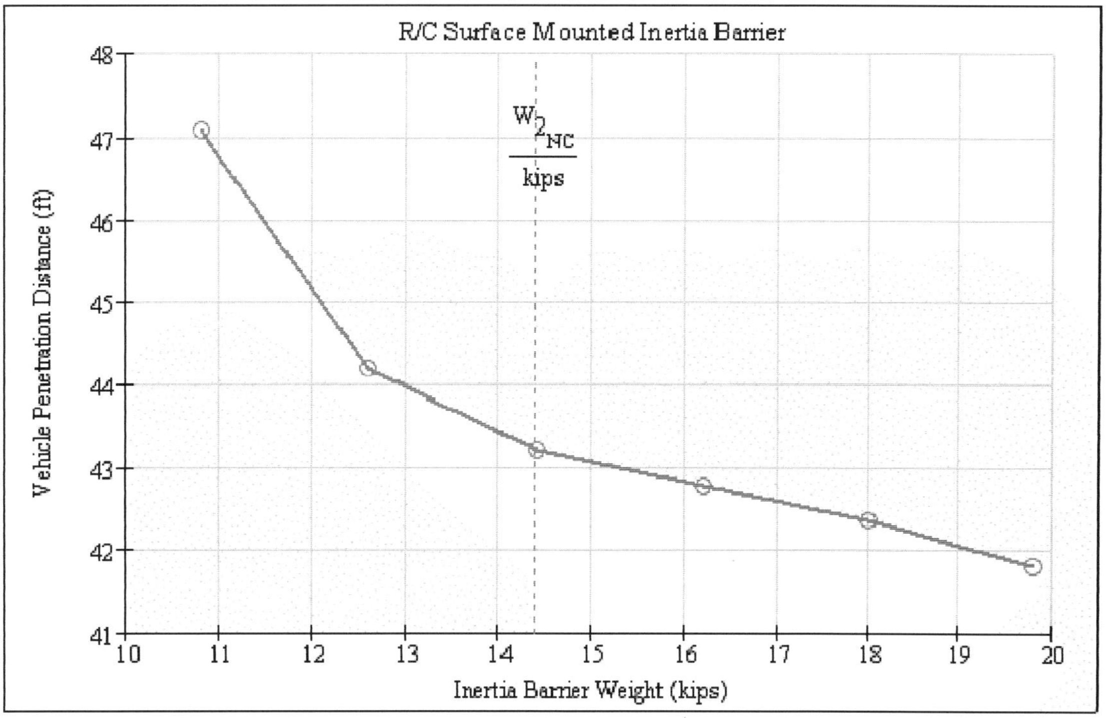

Vehicle Specifications:	Weight,	W_1 = 15.0 kips	Impact speed,	v_1 = 50.0 mph	
Barrier Sizes/Weights:	Height,	H_2 = 3.0 ft	Width,	B_2 = 4.0 ft	GAP = 3.0 ft
	Length,	L_2^T = (6.0 7.0 8.0 9.0 10.0 11.0) ft			
	Weight,	W_2^T = (10.8 12.6 14.4 16.2 18.0 19.8) kips			
Minimum Acceptable Barrier Weight,		$W_{2_{NC}}$ = 14.4 kips			

Figure 13 - Inertia Barrier Weight Versus Vehicle Penetration Distance

Sample Designs

Samples of representative inertia barriers designed for high kinetic energy demands are depicted in the following photos. Surface mounted inertia barriers consisting of large limestone boulders and precast concrete rectangular blocks are shown in Figure 14. The precast concrete blocks are reinforced and fitted with lifting devices to facilitate handling.

Figure 15 depicts a decorative variation of the concrete blocks that were cast-in-place on grade. The decorative finish was achieved through the use of form liners.

A hybrid design that relies upon a combination of strength and inertia effects is the double "Jersey" barrier system shown in Figure 16. The standard Department of Transportation (DOT) precast concrete barriers are rated at approximately 100 ft*kips of energy as stand-alone highway traffic safety devices. However, by incorporating a large mass of in-fill material (e.g., sand, earth or gravel) between two rows of these barriers spaced at approximately 6 ft to 8 ft apart, achieving high kinetic energy ratings is feasible. The precast concrete units are connected end-to-end using standard hardware. No know vehicle impact tests exist for this design.

Figure 14 - Large Boulders and Precast Concrete Blocks Used as Inertia Barriers

Figure 15 - Decorative Surface Mounted Precast Concrete Blocks Used as Inertia Barriers

The decorative landscape stone wall shown in Figure 17 serves as a vehicle barrier in an area where aesthetics are a primary design consideration. Behind the wall is an 8 ft wide earth berm landscaped with shrubs which provides the resistance to vehicle penetration. Pedestrian walkways are incorporated into the vehicle barrier design.

Figure 16 - Double Jersey Barriers with Gravel Infill Material (Under Construction)

Figure 17 - Decorative Landscaped Stone Wall with Earth Berm

Natural features are sometimes effective as vehicle barriers, such as the creek and embankments shown in Figure 18. Consideration must be given to long term erosion of the embankments which can render natural features useless as a vehicle barrier. Embankment configurations are evaluated for effectiveness as vehicle barriers using the guidelines provided in MIL-HDBK-1013/14 [4] which are summarized in Figure 9. Use of natural features as part of the VBS is attractive not only due to aesthetics but also for as a significant cost savings.

Figure 18 - Natural Features (Stream and Embankments) Used as a Vehicle Barrier

COST DATA

Installed cost data for several passive barrier designs are provided in Table 1. These data are compiled from several recent US projects. Installed costs may vary considerably for other sites depending on availability of standard DOT "Jersey" barriers, suitable (consistent size and integrity) boulders and other local conditions. Additional cost data may be obtained from MIL-HDBK-1013/14 [4].

Table 1 - Installed Costs of Selected Passive Barriers

Barrier Type	Description	Installed Cost per Linear Foot
Double Jersey Barriers	DOT Precast concrete units, 6 ft spacing, interlocked end-to-end, surface mounted	$150
Limestone Boulders	Single row 5 ton boulders, surface mounted	$35
	Double row 5 ton boulders, surface mounted	$70
Precast Concrete Wall	Precast concrete wall, 3 ft high by 4 ft wide in 10 ft (minimum) lengths, surface mounted, 3 ft gaps end-to-end	$144
Landscape Wall/Soil Berm	CMU wall, 3 ft high backed by 8 ft wide soil berm	$100

CONCLUSIONS

Following performance of a security vulnerability assessment, one may be faced with the task of providing protection of a facility against malevolent use of a vehicle by terrorists, presumably for the purpose of delivering a large bomb as close as possible to critical assets. A significant challenge in such a case will be to obtain qualified designs of passive barrier systems meeting the performance objectives delineated herein. Additional considerations often include deterrence value, resistance to tampering, aesthetics, maintenance, and installed costs. A design process for a reliable, relatively inexpensive passive barrier system for large facilities and meeting the performance objectives has been presented. Sample passive barrier designs and recent cost data were also provided.

REFERENCES

1. AiTS, "AiDamage3", Shab Hill/Birdlip, Glos UK, (http://www.aitsuk.com/index.htm), 31-Oct-02.
2. R. M. Brach, "Rigid Body Collisions", ASME Journal of Applied Mechanics, March 1989, Vol. 56, pp. 133-138.
3. R. M. Brach, "Friction, Restitution, and Energy Loss in Planar Collisions", ASME Journal of Applied Mechanics, March 1984, Vol. 51, pp. 164-170.
4. DOD, **SELECTION AND APPLICATION OF VEHICLE BARRIERS**, MIL-HDBK-1013/14, US Department of Defense, Washington, D.C., 01-Feb-99.
5. DOS, "Specification for Vehicle Crash Testing of Perimeter Barriers and Gates", SD-STD-02.01, US Department of State, Washington, D.C., 1985.
6. FEMA/BPAT, **THE OKLAHOMA CITY BOMBING: IMPROVING BUILDING PERFORMANCE THROUGH MULTI-HAZARD MITIGATION**, FEMA 277, Federal Emergency Management Agency/Building Performance Assessment Team, Washington, D.C., August 1996.
7. L. B. Fricke, **TRAFFIC ACCIDENT RECONSTRUCTION - VOLUME 2 OF THE TRAFFIC ACCIDENT INVESTIGATION MANUAL**, Northwestern University Traffic Institute, Evanston, IL, 1990.
8. T. J. Hirsch, D. L. Ivey, et. al., **TEST AND EVALUATION OF VEHICLE ARRESTING, ENERGY ABSORBING AND IMPACT ATTENUATION SYSTEMS**, Final Report Contract No. CPR-11-585 for US DOT FHA, Texas Transportation Institute/Texas A&M Research Foundation, College Station, TX, Nov. 30, 1971.
9. R. R. McHenry and B. G. McHenry, "Effects of Restitution in the Application of Crush Coefficients", SAE Paper 970960, 1997.
10. J. A. Neptune, "Crush Stiffness Coefficients, Restitution Constants, and a Revision of CRASH3 and SMAC", SAE Paper 980029, 1998.
11. D. E. Siddall and T. D. Day, "Updating the Vehicle Class Categories", SAE Paper 960897 from SP-1150, 1996.

12. J. J. Taborek, **MECHANICS OF VEHICLES**, Penton Publishing Co., Cleveland, OH, 1957.

NOMENCLATURE

e = coefficient of restitution, $0 \leq e \leq 1$

μ = coefficient of friction at contact point between Body Nos. 1 and 2

μ_m = value of μ which maximizes total energy loss

μ_o = value of μ which corresponds to zero relative tangential velocity at contact point

μ_{1s}, μ_{2s} = coefficients of sliding friction between Body Nos. 1 and 2 and the soil

M_1, M_2 = translational masses of Body Nos. 1 and 2

I_1, I_2 = mass moments of inertia of Body Nos. 1 and 2

d_1, d_{1n}, d_{1t} = distances between the center of mass and point of impact for Body No. 1

d_2, d_{2n}, d_{2t} = distances between the center of mass and point of impact for Body No. 2

r_{g1}, r_{g2} = radii of gyration for Body Nos. 1 and 2

v_{1n} = initial translational velocity of Body No. 1 in the normal direction

v_{1t} = initial translational velocity of Body No. 1 in the tangential direction

v_{2n} = initial translational velocity of Body No. 2 in the normal direction

v_{2t} = initial translational velocity of Body No. 2 in the tangential direction

ω_1, ω_2 = initial rotational velocities of Body Nos. 1 and 2

V_{1n} = final translational velocity of Body No. 1 in the normal direction

V_{1t} = final translational velocity of Body No. 1 in the tangential direction

V_{2n} = final translational velocity of Body No. 2 in the normal direction

V_{2t} = final translational velocity of Body No. 2 in the tangential direction

Ω_1, Ω_2 = final rotational velocities of Body Nos. 1 and 2

ke_1, KE_1 = initial and final kinetic energies of Body No. 1, respectively

ke_2, KE_2 = initial and final kinetic energies of Body No. 2, respectively

D_{1sn}, D_{2sn} = sliding distances of Body Nos. 1 and 2 in normal direction

Φ_1, Φ_2 = sliding angle of Body Nos. 1 and 2 about axis normal to the plane

GAP = gap distance between barriers; used to determine eccentricities of impact locations

Applying Inherently Safer Technologies For Security Of Chemical Facilities

David A. Moore, PE, CSP

President and CEO
AcuTech Consulting Group
Chemetica, Inc.
88 Kearny Street
San Francisco, CA 94108
dmoore@acutech-consulting.com

Prepared for Presentation at

American Institute of Chemical Engineers
38th Annual Loss Prevention Symposium
Engineering Solutions to Facility Security Challenges

New Orleans, LA

April 26 – 29, 2004

Abstract:

Inherent Safety is a recognized process risk management strategy, and has value for risk reduction of intentional chemical releases as well as accidental releases. Most of industry is not formally using inherent safety as a strategy for security risk management, although proposed security regulations and industry guidelines are focusing on the method as a first choice.

This paper will address the concepts of inherent safety, and will review how they can be used to effectively reduce security risk. Examples will be presented against typical design basis threats. For example, companies possibly can provide standoff distance vs. adding active security measures, or reduce or distribute inventories instead of paying excessive costs for security countermeasures. It will also review the limitations of inherent safety and business constraints it may pose.

Given that inherent safety is a rather subjective concept, it makes the matter a difficult one to understand, implement, and regulate. Key to its adoption is the need for practical tools to illustrate how to evaluate a risk and apply the concepts. This paper will describe a step-wise approach that can be used for evaluating the opportunities for inherently safer technologies during security vulnerability analyses.

Companies should be knowledgeable of inherent safety, actively document attempts to use the concepts, and should prepare themselves for the challenges of regulation of the issue should it come to that requirement.

Introduction

Inherent Safety is a recognized process risk management strategy, and has value for risk reduction of intentional chemical releases as well as accidental releases. At issue is that most of industry is not formally using inherent safety as a strategy for process risk management for managing accidental releases. Now, with the terrorist events of September 11, 2001, U. S. companies are recognizing that inherent safety can be applied to address security risk management.

In fact, proposed Federal security regulations and industry guidelines are focusing on the method as a first choice. As a result, there is a greater need to understand inherent safety principles and to formally develop methods for analyzing both accidental and intentional loss. In particular, the interest in regulating chemical facility security by inherent safety requires industry to be very familiar with principles and practices. Regardless of regulatory drivers, the use of inherent safety principles for security risk reduction is advisable and a formal methodology to ensure completeness is recommended.

Inherent Safety

Background on Inherent Safety

Inherently Safer Systems (ISS) is a philosophy that is encouraged in the industry to focus engineers and managers on reducing or eliminating a hazard within a chemical process, as a goal, to the greatest extent feasible. No absolute ground rules exist for accomplishing this – it is a philosophy. The Center for Chemical Process Safety of the American Institute of Chemical Engineers (CCPS) published a book in 1996 to promote the concept.[1]

The ISS method employs four key inherently safer strategies:

1. Minimizing;
2. Substituting;
3. Moderating;
4. Simplifying.

Each of these strategies can accomplish hazard reduction or elimination. "Hazard is defined as a physical or chemical characteristic that has the potential for causing harm to people, the environment, or property."[2] The concept is based on the belief that if one can eliminate or moderate the hazard, not only is the risk reduced, it may be possible to remove the risk altogether from consideration. Alternatively, an inherently safer system would make the hazard less likely to be realized and less intense if there is an accident.

Lack of Formality of Use of the Method

The history of inherent safety as a documented strategy for loss prevention is rather recent, but application of the concept is a very old practice in the industry. Safety professionals agree that this is a good 'way of thinking', and is a best practice in process safety management. Most engineers give informal consideration to inherent safety since it may be integral to other issues such as minimizing costs, limiting equipment design, minimizing inventory, or other engineering and business optimization strategies. If the objective of inherent safety (and security) is not given formal consideration, however, it may be overlooked or underutilized. To encourage and structure the application of the method, more formal protocols are recommended.

Proposed Security Regulations Requiring Inherent Safety Concepts

At this time in industry, while ISS is appreciated, it is not prescribed in any U.S. chemical process security regulation we are aware of nor is it widely practiced in a formal, documented way by industry throughout the world. Most companies would recognize it as a philosophy, and would encourage its use where possible.

---

[1] Center for Chemical Process Safety, *Inherently Safer Chemical Processes: A Life Cycle Approach, Center for Chemical Process Safety*, AIChE, 1996.

[2] Ibid, CCPS

The "Chemical Facilities Security Act of 2003" as proposed in the U.S. Senate in 2003[3], is an example of proposed legislation that may require the consideration of inherent security. This bill had a requirement for facilities to give consideration to "Alternative Approaches". In this bill the definitions of alternative approaches and consideration are as follows:

> ALTERNATIVE APPROACHES.—The term "alternative approaches" means ways of reducing the threat of a terrorist release, as well as reducing the consequences of a terrorist release from a chemical source, including approaches that:
> (A) use smaller quantities of substances of concern;
> (B) replace a substance of concern with a less hazardous substance; or
> (C) use less hazardous processes.
>
> CONSIDERATION.—The term "consideration" includes:
> (A) an analysis of alternative approaches, including the benefits and risks of such approaches;
> (B) the potential of the alternative approaches to prevent or reduce the threat or consequences of a terrorist release;
> (C) the cost and technical feasibility of alternative approaches; and;
> (D) the effect of alternative approaches on product quality, product cost, and employee safety.

The regulation of inherent security or safety is not without some difficulties, particularly with the subjective nature of the use of inherent safety and the lack of formal guidance. Key to the potential complications of regulation of inherent safety is several issues:

1. the need to judge the degree of inherent safety to be applied as opposed to other process risk management strategies;
2. judging the effectiveness of inherent safety;
3. evaluating and accepting alternative approaches.[4, 5]

Layers of Protection and Overall Process Risk Management Strategy

CCPS has published several publications that sanction the use of several process risk management strategies of Layers of Protection and the use of all four process risk management strategies:

1. Inherent
2. Passive
3. Active
4. Procedural

---

[3] Proposed Senate Bill 'Chemical Facilities Security Act of 2003', Senator Inhofe, October, 2003.

[4] Section 450-8.016(D)(3) of the County Ordinance Codes as amended by Ordinance No. 2000-20, Contra Costa County, California.

[5] Moore, David, "Experiences in the Regulation of Inherent Safety", Mary Kay O'Connor Process Safety Center, Texas A&M University System, 2002 Annual Symposium, "Beyond Regulatory Compliance, Making Safety Second Nature", October 29 - 30, 2002, Texas A&M University, College Station, Texas.

In fact, these two concepts are described in the same textbook on inherent safety.[6] These strategies are illustrated in Figures 1 and 2.

The layers of protection concept is that every hazard needs to have a series of layers of prevention, detection, and mitigation systems to either assist in prevention of the incident or in reducing the potential impacts of any event. Key to the concept is the idea of reducing the risk of the chain of events from initiation of an incident to prevention of or reduction of the hazardous outcome.

Figure 1 – Process Risk Management Strategies

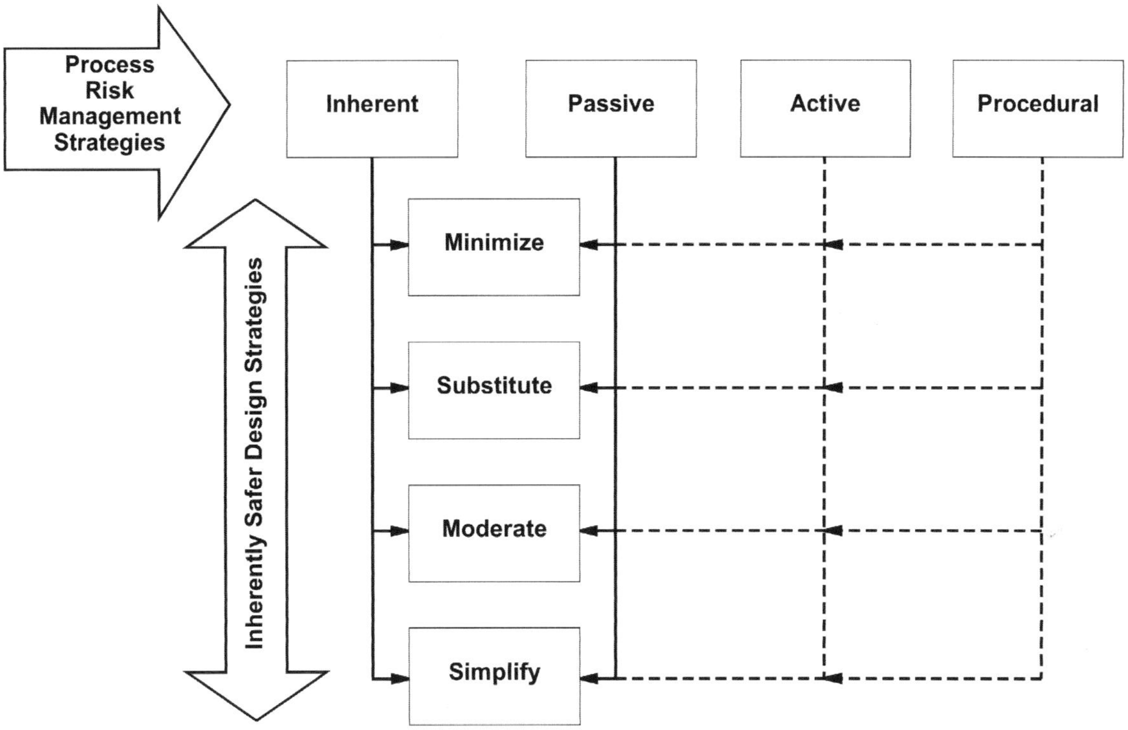

Source: Fig 2.2, CCPS Concept Book *"Inherently Safer Chemical Processes"*[5]

[6] Ibid, CCPS

Figure 2 – Layers of Protection Concept

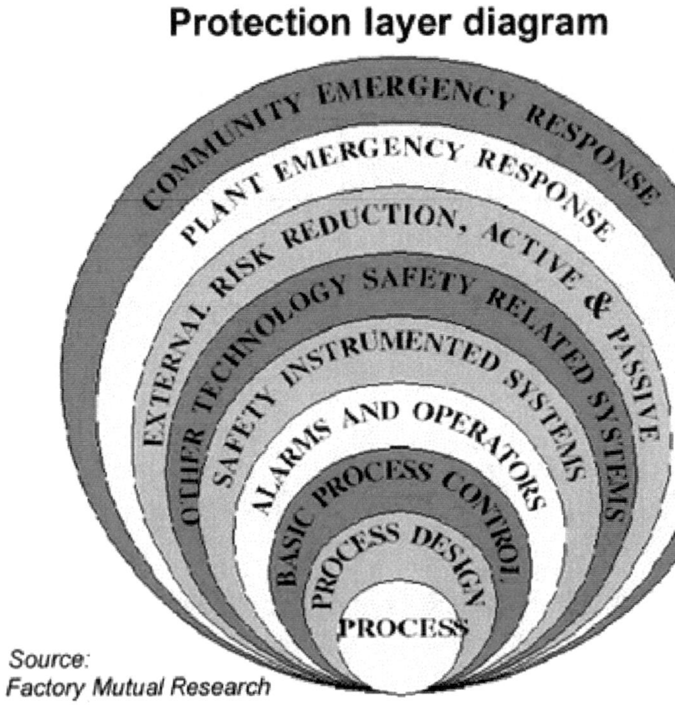

Opportunities for Inherent Safety

CCPS states that there are three stages of an accident sequence, the initiation, the propagation, and the termination stages, as shown in Figure 3. "Inherently Safer Strategies (and Layers of Protection) can impact the accident process at any of the three stages.

The most effective strategies will prevent initiation of the accident. Inherently safer design can also reduce the potential for propagating an accident, or provide an early termination of the accident sequence before there are major impacts on people, property, or the environment".[7] This description by CCPS says that safety professionals believe that there are many opportunities to employ inherent safety – not only to eliminate a hazard but also to moderate an accident sequence in different ways that, in effect, lowers the risk. Inherent safety and layers of protection/process risk management strategies are not opposing strategies – they are complementary strategies in an overall scheme to reduce risk. In fact, the strategies cannot be taken out of context and be effective.

---

[7] Ibid, CCPS

Figure 3 – CCPS Accident Sequence Concept

Differences in Inherent Security

Accidental risk and intentional risk have some similarities, but there are important differences. Depending on the degree of exposure potential of the company or the public from an intentional release, the attractiveness of a target, and the ease of attack, companies may face entirely different risks than the accidental risk the facilities were designed to manage.

The risk that is being analyzed for chemical process security is defined as an expression of the likelihood that a defined threat will target and successfully attack a specific security vulnerability of a particular target or combination of targets to cause a given set of consequences. This is contrasted with the usual accidental risk definitions. The risk variables are defined as shown in Figure 4.

Figure 4
Intentional Release vs. Accidental Release
Risk Definitions

Intentional Release Risk is a function of: • Consequences of a successful attack against an asset and • Likelihood of a successful attack against an asset.	Accidental Release Risk is a function of: • Consequences of an accidental event and • Likelihood of the occurrence of the event.
Likelihood is a function of: • the Attractiveness to the adversary of the asset, • the degree of Threat posed by the adversary, and • The degree of Vulnerability of the asset.	Likelihood is a function of: • The probability of an event cascading from initiating event to the consequences of interest and the frequency of the events over a given period.

The need for security enhancements will be determined partly by factors such as the degree of the threat, the degree of vulnerability, the possible consequences of an incident, and the attractiveness of the asset to adversaries. In the case of terrorist threats, higher risk sites are those that have critical importance, are attractive targets to the adversary, have a high level of consequences, and where the level of vulnerability and threat is high.

Security Vulnerability Assessments (SVAs) are conducted to evaluate vulnerabilities, SVAs are not a quantitative risk assessment, but are performed qualitatively using the best judgment of the SVA Team. The expected outcome is a qualitative determination of risk to provide a sound basis for rank ordering of the security-related risks and thus establishing priorities for the application of countermeasures.[8]

A basic premise is that all security risks cannot be completely prevented. The security objectives are to employ four basic strategies to manage the risk including Deter, Detect, Delay, and Respond. Appropriate strategies for managing security can vary widely depending on the circumstances including the type of facility and the threats facing the facility. As a result, it is difficult to prescribe security measures that apply to all facilities in all industries, but instead it is suggested to use SVA as a means of identifying, analyzing, and reducing vulnerabilities.

The intentional release risk includes possible attacks by outsiders or insiders, or a combination of the two adversaries. It may be perpetrated by a number of different adversaries with varying intents, motivations, weapons, tactics, and capabilities. These issues need to be sorted out in a threat assessment, which is, in effect, a risk-based assessment that forms the basis of the design basis threat assumption the facility designs and operates to.

The selection of the threats should include reasonable local, regional, or national intelligence information, where available. This step also includes determining the target attractiveness of each asset from each adversary's perspective.

A responsible company has to give thought to the possible threats and attempt to organize the many combinations and permutations into a threat matrix such as Figure 5. Key to this matrix is the first variable – who is the adversary and what is the target?

Threat to a particular asset varies with several factors including the degree of interest that an adversary may have in the asset, as well as the degree of impact possible if the asset was attacked, disabled, copied, compromised, or stolen. For this reason, the threat assessment includes a step whereby each asset is analyzed from the perspective of each potential adversary to determine the degree of attractiveness of the asset to the adversary. Attractiveness is therefore another factor in the determination of risk. Assumptions can be made on the primary vs. secondary threat to adjust the emphasis of the analysis.

---

[8] *Guidelines for Managing and Analyzing the Security Vulnerabilities of Fixed Chemical Sites*, American Institute of Chemical Engineers, August 2002.

Figure 5 - Threat Matrix

Threat Matrix	Disposition		
Type of Adversary	Internal	External	Colluded
Terrorist	Secondary Threat	Primary Threat	Secondary Threat
Employee/Contractor	Primary Threat	Secondary Threat	Secondary Threat
Activist	Secondary Threat	Primary Threat	Secondary Threat

■ (solid)	Primary Threat
▩ (checkered)	Secondary Threat

For inherent security, the primary intent is to reduce the hazard (which is the first principle of inherent safety as well as for security). The second intent is applying the concept of layers of protection (which is the second principle of inherent safety as well as for security). This is intended to limit the effects of an event along the chain of events from initiation to impact of the event. Thirdly, a facility should employ the concepts of layers of security or defense in depth as shown in Figure 6. This concept is similar to layers of protection however the emphasis of this strategy is for the limitation of the adversary to access to address the target asset. In this concept, the adversary is assumed to be interested to attack, steal, degrade or contaminate the target asset. The layers of security are intended to deter, detect, delay, and respond to the actions of the adversary and the subsequent events caused by the attack.

Figure 6 – Defense in Depth Chemical Process Security Concept

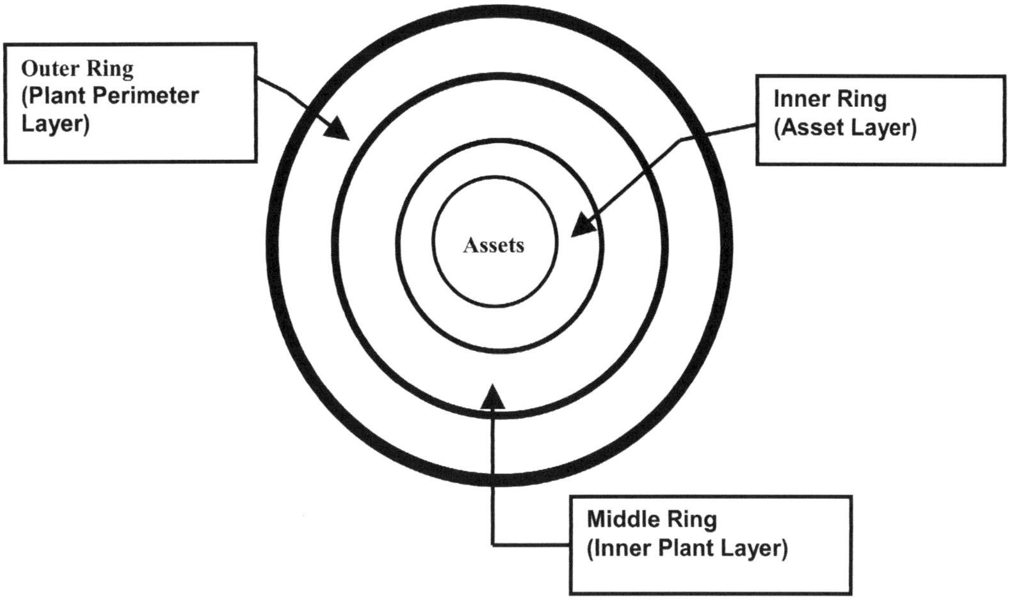

The facility has the opportunity to detect, deter, delay, or respond to the event at any point in the layers of security as shown in Figure 6, and at any time along the process security incident sequence, as shown in Figure 7. Optimized security would deter and detect at the earliest stages of the event and suppress the adversary prior to initiation of the event.

Figure 7 –Process Security Incident Sequence Concept

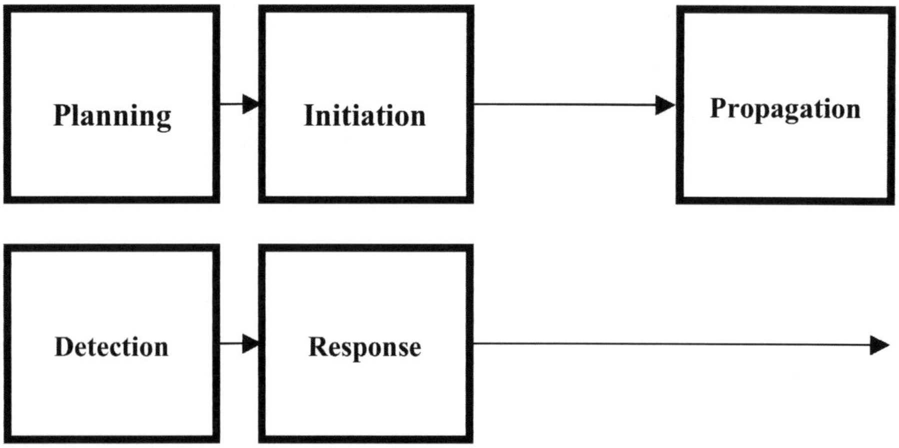

The stages of the chemical process security incident are shown as two simultaneous activities of both the adversary (upper flow diagrams) and the facility (lower flow diagrams) include the following considerations:

1. **Planning** – the adversary stages the incident including planning, surveillance, data gathering, initial positioning for attack;
2. **Initiation** – the adversary begins the attack (physical or cyber action against asset) and encounters the first layer of security at the perimeter (counter surveillance, physical barriers, lighting, intrusion detection);
3. **Propagation** – the successful attack on the asset results in the propagation of the incident including release from containment, contamination, degradation of assets, and subsequent damages;
4. **Detection** – the facility identifies the adversary's actions including physical detection of the presence of an unauthorized person, identification of illegal actions; or initial detection of the event caused by a successful attack;
5. **Response** – the facility begins a sequence of actions to communicate the incident has occurred or to control or mitigate the event.

Inherent Security Methodology

Application of inherent security principles is done during the SVA process. During the analysis of each asset, the inherent safety considerations previously mentioned are reviewed. Inherent security should be considered as a first strategy, and opportunities for inherent security may exist either fundamentally (removing or reducing the hazard altogether), or may be possible to apply along the process security incident sequence. The inherent safety/security principles must act in concert and should address each stage of the sequence of events with consideration for each of the strategies of deter, detect, delay, and respond. For example, some applications of the concept for the different strategies of security are shown in Figure 8.

Figure 8 – Examples of Inherent Security Principles

Example Inherent Security Strategies – Deter	Example of Inherent Security Strategies – Detect	Example of Inherent Security Strategies – Delay	Example of Inherent Security Strategies - Respond
1. Information security practices 2. Buffer zones 3. Reduction of inventory	1. Clear zones around facility 2. High contrast surfaces 3. Location of guard facilities	1. Natural barriers 2. Target assets away from perimeter – standoff distance 3. Assets are within multiple layers of barriers	1. Limited inventory 2. Increased buffer zone 3. Diking, grading, sloping

Benefits and Drawbacks of Inherent Principles for Chemical Process Security

The benefits of employing the inherently safer strategy is that, hopefully, every decision that is made to design, build, operate, and maintain a process attempts to employ an inherently safer system than the original concept or existing process operation. This could result in lower overall risk, since the original hazards may be eliminated or greatly reduced. But it could be that the mere application of an inherently safer solution does not guarantee a safer or more secure process. In fact, it is recognized that any change to a process may result in other undesirable effects, even if the change was in the best interest of safety or security. Safety and security often have conflicts.

As an example, a refinery decided to consolidate a number of control rooms that had been located throughout various process units in order to reduce the risk of exposure of workers to accidental events. They proposed to move the control room to a more remote area of the refinery that wasn't within the exposed zone of the effects of any explosion scenario. As a result, the move was seen as inherently safer.

As a Security Vulnerability Analysis was conducted on the proposed change, the SVA team identified that the consolidation of the refinery control room meant that the criticality of the control rooms had increased since all of the control functions were moved into one building. Secondly, they determined that the proposed location of the control room added several new security vulnerabilities. This included that the proposed location was along an unsecured, poorly lit perimeter that may have made the accessibility of the target to the adversary more attractive and easier to attack without detection. The location was surrounded by contractor and visitor areas that had little restrictions for access so the potential for undetected access to the area was increased. Also, the parking area near the proposed location allowed exposure to rogue vehicles.

Also, an inherently safer/secure change alone does not necessarily constitute a safer/secure plant. If by employing layers of protection the risk can be sufficiently reduced or if it can be reduced to a greater degree than if one or more inherently safer strategies are employed, layers of protection may be a preferred strategy. Even if an inherently safer/secure strategy is employed, the hazard still may be present, and so it may be desirable or necessary to employ layers of protection in addition.

Security Vulnerability Assessment in the Chemical Industry

Brian R. Dunbobbin, Thomas J. Medovich, Marc C. Murphy, Annie L. Ramsey

Air Products and Chemicals Inc.
7201 Hamilton Blvd
Allentown, PA 18195
(610) 481-6736
dunbobbr@apci.com

Prepared for Presentation at

American Institute of Chemical Engineers
38th Annual Loss Prevention Symposium
Engineering Solutions to Facility Security Challenges

New Orleans, LA

April 26 – 29, 2004

ABSTRACT

Following the events of September 11, 2001, Air Products and Chemicals Inc. (APCI) developed a Security Vulnerability Assessment (SVA) methodology, consistent with the Center for Chemical Process Safety (CCPS) guidelines. This methodology is designed for efficient and thorough evaluation of a large number of facilities ranging from small industrial gas sites to large chemical plants. This methodology evaluates the potential consequences and attack scenarios at a facility and the attractiveness of the facility as a terrorist target. The team then provides recommendations for engineering and security improvements. Participation in early SVA development exercises with industry and governmental agencies made it clear that it is critical to have a team approach that includes process safety, security, and site operations functional expertise.

This paper presents an overview of the APCI SVA methodology and summarizes major findings and lessons learned. Findings from an SVA provide multiple levels of protection for our assets and the public. The engineering and security solutions from the evaluation are intended to deter, detect, delay, and respond to an attacker. They include:

1. Inherently safer alternatives such as reducing inventory, designing fail-safe systems, and improving plant layout.
2. Enhanced physical security systems such as fences, access control, and monitored intrusion detection.

INTRODUCTION

The events of September 11, 2001 motivated the American Chemistry Council (ACC) member companies, including Air Products and Chemicals Inc. (APCI), to address counter terrorism issues at chemical facilities in order to evaluate and improve, where appropriate, the facility design and protective measures available in the possible event of a terrorist attack. We participated in developing the Responsible Care code and followed the implementation schedule set by the code.

APCI is implementing the Security Vulnerability Assessment program globally to all of our facilities through regional teams following the same methodology. Teams have been established for North America, Europe, and Asia. The security effort at APCI is addressed to People, Product (and business) transactions, Information, and Process plant and equipment operations. The Security Vulnerability Assessment (SVA) program focuses on the security of our process plants and equipment operations while keeping protection of the employees, contractors, customers, and the communities as a primary concern.

Because APCI has a large number of facilities ranging from small industrial gas sites to large chemical plants, a Security Vulnerability Assessment (SVA) methodology, consistent with the Center for Chemical Process Safety (CCPS) guidelines, was developed by APCI in order to perform efficient and thorough evaluation of these various types of facilities. This APCI SVA methodology followed the CCPS methodology and adapted the technique to be more appropriate for our specific facilities. This methodology uses a team approach to evaluate the potential attack scenarios and consequences at a facility and the attractiveness of the facility as a terrorist target. SVA is a structured methodology used to facilitate a discussion about security and itself will not prevent a terrorist attack.

The sites are ranked from the highest risk (Tier 1) to the lowest risk (Tier 4). This paper provides an overview of the APCI SVA methodology. We found this methodology is robust enough to be applied to transportation issues also, although transportation is outside the scope of this paper. We also utilized the learning from the higher tier SVAs to develop a checklist that is used for performing SVAs at lowest risk sites.

SVA METHODOLOGY

Facility Security Prioritization Process

The initial security screening, which assigns priority tier levels for all APCI facilities was completed following the ACC Prioritization Process. The team completing the screening had detailed knowledge of the facilities and included representatives from Process Safety, Security, and Line Management. The screening process evaluated facilities based on the following:

- Types of attack
- Offsite impact
- Chemical hazard
- Potential for use as a weapon of mass destruction (WMD) or precursor of weapon of mass destruction
- Attractiveness of target

The highest risk facilities were assigned to Tier 1 and the lowest to Tier 4. Tier 4 facilities represent sites where chemicals are handled but they represent the lowest risk and/or consequence of the four tier levels.

An APCI Chemical of Concern (COC) list was developed for use in the screening process. This list was developed by reference to various safety, health and environmental as well as import/export, and other regulatory lists and guidance, and includes chemicals that have potential for offsite impact and chemicals that may be used as potential WMD or precursor of WMD. It also includes certain chemicals that may be potential theft target for drug use or manufacture.

APCI SVA Team

APCI utilized a project team approach in completing the SVAs. The SVA team possesses, or have access to, sufficient knowledge, experience, and perspective to adequately analyze the security vulnerability related hazards. The team leader should have knowledge and/or prior experience of SVAs. The team leader should also be impartial so facilitation of the findings during the meeting is objective and consistent throughout the company. SVA team leaders had been either a SVA project manager or a Process Safety engineer who is not strongly focused on physical security systems and does not have a direct reporting relationship to the facilities.

The SVA team at a minimum consisted of people who collectively possessed the following knowledge and/or skills:

- Security vulnerability analysis procedures and methods.
- Security procedures, methods, standards, and systems.
- Process safety.
- Knowledge of the site under study including:
    - Potential hazards associated with the process chemistry, raw materials, finished goods, and the physical location of each.
    - Process and equipment design basis.

  The engineering and operation staff associated with each facility was used to meet this requirement.
- Project management skills for organization and execution of the SVAs and effective implementation of the recommendations.

Other skills that may be considered as appropriate are:

- Military doctrine, especially in terrorism, weapons, targeting and insurgency / guerilla warfare and knowledge of weapons of mass destruction (WMD). Police bomb squad, and other appropriate consultants have been used to provide input to the SVAs.
- Adversary characteristics and capabilities knowledge, especially of trans-national terrorist groups. We maintained weekly contact with the FBI and other Federal and Local Authorities and participate in several industrial and governmental groups related to terrorist activities.
- Safety, Industrial hygiene, Environmental issues. These skills are from our in-house EH&S organization in combination with outside industrial and governmental groups.

The SVA team membership may be a minimum of three people with each member providing multiple skills described above.

APCI SVA Procedure

Tier 1 through Tier 3 facilities were assessed using the SVA procedure described in this section. Tier 4 facilities used a "check-list" approach that will be described in a later section.

Our work process started with the SVA project manager scheduling the SVAs, beginning with the Tier 1 facilities. The SVA project manager and the facility manager selected the members of the SVA Teams. The facility manager or that manager's representative is a member of the SVA team. A pre-meeting was held about two weeks before the on-site visit. The facility manager gathered the facility information, and completed the initial pages of the SVA document up to and including the facility plot plan at the pre-meeting. The date for the on-site visit and the agenda for the meeting is also agreed to at the pre-meeting. A thorough pre-meeting has been shown to minimize the time spent on site by the entire SVA team.

During the on-site visit, the SVA Team reviewed the pre-meeting work of the SVA document prepared by the facility, and completed the remainder of the SVA document using the facility information gathered by facility personnel. The SVA Team and the facility management team hold a closing conference at the conclusion of the site visit to ensure agreement of the recommendations. The SVA Team, the facility manager, and the SVA project manager prioritized the SVA recommendations, taking into account the risks and timing of implementing the recommendations.

APCI SVA Documentation

The SVA methodology and documentation developed by APCI is available on the ACC public web site (www.responsiblecaretoolkit.com). This is used for Tier 1 through 3 facilities. The work process for the SVA is shown in the flowchart in Figure 1 attached. Each step is described below:

- SVA Team Composition – The people on the SVA team for the facility are identified along with their functional responsibility.

- Facility Documentation – All information required for the SVA team to properly assess the security vulnerabilities of the site is identified in this section and made available as required for the SVA.

- Facility Staffing – Information on the total number of employees and contractors on site, the number of people present per shift, and the number of people present by location is identified in this section. This provides the SVA team an understanding of the dynamics of the people flow on the site.

- Existing Security Systems – A checklist is used to assess the existing security at the facility. The checklist covers Site Perimeter, Access Control, Lighting, Alarm/Intrusion Detection Monitoring, Uniformed Security Officers, Closed-circuit Television Systems, Product Storage and Security, and Previous Security Concerns or Activities.

- Hazard Identification and Consequence Evaluation – This section documents the potential hazards for the chemicals of concern that are present at a facility and covers an evaluation of the potential consequences of terrorist attack. All four key security events: off-site impact, theft, product contamination, and business interruption are considered.

- Plot Plan – The areas containing Chemicals of Concern and critical control/utility areas are identified on the plot plan to provide a visual representation of these "Areas of Concern." Orientation of roads and perimeters is also identified.

- Potential Attack Scenarios – For each "Area of Concern," the potential methods that terrorists (insiders or outsiders) might use to cause problems and existing security countermeasures are identified. To aid in the consistency of the evaluation, a list of potential attack scenarios was developed. The SVA team would decide which potential attack scenarios from this list apply to the facility and then brainstorm additional attack scenarios that should be considered.

- Assessment of Attractiveness – This takes into account the attractiveness of attacking a particular area of concern based on the resulting publicity, off-site population impact, severity of impact from product contamination, an dease of transportation for theft of chemicals that are potential WMD or precursor of WMD.

- Gap Analysis – After assessing what security measures are in place to counteract the scenarios developed in previous steps and the attractiveness of attack for each "Area of Concern," additional potential security, engineering or operational measures that would help to reduce the risk of terrorist attack are evaluated and documented in this section.

- Recommendations – After reviewing the gap analysis, the SVA team develops a list of recommendations for reducing the vulnerability of the facility. The SVA team, the facility manager, and the SVA project manager prioritize the SVA recommendations.

Tier 4 Security Vulnerability Assessment Report

A Tier 4 SVA checklist was developed based on the ACC Modified SVA Methodology for Tier 4 facilities. By comparing the existing layers of protection to those required in the checklist, additional layers of protection may be determined to be necessary.

The facility manager completes the checklist by marking "Yes, No, or Not Applicable (N/A)" to each question. Explanation of the current system or practice should be provided for each "No" and "N/A" answer. The facility manager returns a copy of the completed checklist to the SVA project manager and a copy to the appropriate supervisory management.

The SVA project manager reviews any checklists with "No" answers and contacts Corporate Security for an evaluation on the possible need for security upgrades. The SVA project manager documents the appropriate recommendations in this report. Recommendations for improved countermeasures made during this initial review should be considered preliminary ideas that may be modified upon further review by the core team, including the facility manager, the operation/manufacturing manager, project manager, Process Safety, and Corporate Security.

The SVA project manager and the appropriate facility management review the recommendations and assign responsibility for the implementation of any security upgrades.

LESSONS LEARNED

Training

To share information on the on-going activities for security concerns and to help raise the employees' awareness, training packages were developed. A general awareness training package was required to be taken by all APCI employees. An overview SVA training package was used as an introduction to SVA for all SVA team members prior to performing the site SVA. A more in-depth training package on the SVA procedure was required for people who lead or facilitate SVAs. This ensured a consistent approach when conducting SVAs at all of our facilities.

Generic Forms, Guidelines, and Standards

To aid in performing consistent and efficient SVAs among similar processes, generic templates of the SVA form were developed for use during SVAs at the various sites. In addition, general guidelines and standards for required security measures for the different tier levels and facility types were developed and reviewed with business management to ensure consistency in application of the security measures.

Engineering and Security Solutions

The engineering and security solutions from the SVA evaluation are intended to deter, detect, delay, and respond to an attacker. They include inherently safer alternatives and physical security systems.

Inherently Safer Alternatives

As a general practice, plants were designed to operate safely with built-in layers of protection to fail-safe normally with independence of control and protective systems. Based on the history of process design and safety, which includes quantitative risk assessment and risk and consequence modeling, the SVA teams evaluated whether existing protective systems are adequate and where necessary, additional layers of protection were implemented.

In several locations, the SVA teams looked at the feasibility of reducing the chemical inventory or the consolidation of two or more sites. Reducing the chemical inventory reduces the potential off-site consequences. By consolidating sites, the high concern chemicals can be consolidated into a smaller number of facilities that can be more easily secured and monitored. Teams also looked at plant layout by considering potential domino effect of the neighboring process equipment. These solutions result in minimizing the potential consequence and attractiveness of a site as a potential target.

Physical Security Systems

Standards have been developed to set the minimum security measures required for the different tier level facilities for new projects. This ensured future SVAs are performed with consistent guidance with what has been done with the existing facilities. These standards cover topics such as fence, access control, plant layout, lighting, and etc.

Facility Design and Layout:

The overall design of a facility can improve or weaken the security posture of the site. As a result, some of the following items are now being considered during the design of each new facility and are being retrofitted were applicable to existing facilities:

- Providing a separation between personnel parking areas and the production environment.
- Personnel and vehicles entering and exiting the facility do so only through designated security controlled points.
- Consideration of increased measures of security around critical, toxic, hazardous, and flammable process and product storage areas. These areas should incorporate appropriate means to adequately protect against deliberate and willful attempts from unauthorized persons to disrupt these areas. Wherever practical, appropriate means of protection include the following:
    - Limiting entrance points into structures housing critical process and product storage areas. In some cases, compartmentalized security perimeters have been erected around critical process and product storage areas that cannot otherwise be secured in closed or confined spaces.
    - Consideration of equipment location, leaving old equipment in place as obstruction, as well as positioning of bund walls, dikes, retention basins, and interior roadways as additional security measures.
    - Positioning tank labels and signage in areas that are less visible to outsiders.

Facility Perimeter Barriers:

As a result of the ongoing work of the APCI SVA teams, a greater emphasis has been placed on the design of perimeter barriers around the property or fixed asset. New designs of perimeter barriers seek to incorporate the following three concepts of prevention:

- Prevent people or vehicles from entering company property except in those places managed or controlled by some level of security intervention.
- Prevent people and materials from leaving the company property except through these same control points.
- Prevent people and vehicles from having access to certain buildings, places, and areas within the company property.

When designing these systems, it is important to give adequate consideration for emergency escape routes. The perimeter barrier construction should be of suitable materials to effectively deter, detect, and delay an unauthorized attempt to gain access through, over, or under the barrier by people with limited resources. At a minimum, the SVA team seeks to maintain the following standards for perimeter barriers:

- Acceptable materials for perimeter barriers are building surfaces, metal, including drawn wire mesh (chain-link) with barbed wire or razor wire, masonry walls (concrete, brick, and stone) that are not easily scaled.
- Established "clear zones" on either side of the protective perimeter barrier that is kept free of any objects (saplings, weeds, overhanging tree branches, stored materials, etc.) that could possibly damage the perimeter and allow for an unauthorized entry. See Picture 1 below.
- Appropriate signage placed along the perimeter per local ordinance, e.g. "PRIVATE PROPERTY" or "NO TRESPASSING".

- Routine inspections for damage or signs of forced entry and any noticed damages reported to site management and/or person(s) with security responsibility.
- Where buildings form part of the site perimeter, all doors and windows are secured from unauthorized entry and have contact switches to an alarm or access control system.
- Active entrance points are limited in number and consistent with safe operation.
- Perimeter gates/doors that provide immediate access to critical process and/or product storage areas have a means to control access through a lockable secure point.

Picture 1. Facility perimeter.

Facility Security Lighting:

During the SVA, a review of the requirements for establishing appropriate security lighting around critical areas of the facility are addressed. The lighting review will work to better understand if the site lighting design provides for optimal visual and system surveillance capabilities. In addition, site security lighting should provide safety for employees as well as a visible security deterrent.

Facility Internal Barriers:

Improved measures for existing facilities and new design standards require increased security around certain high risk, highly attractive, critical, toxic, and hazardous areas shall take steps to ensure the area are secured against unauthorized entry. Hardening of these areas could generally be accomplished in one or more of the following manners:

- Construction of fencing and gates across open dock areas.
- Construction of a secured fenced area inside an existing structure.
- Construction of a secondary, secured fenced perimeter around the entire product storage area.
- Whenever possible, access gates are limited in number with consideration for operational efficiency and safety concerns.

Picture 2. below shows installed hardened fencing products such as expanded metal mesh or "wire wall" to provide hardened physical security protection. Expanded metal mesh or "wire wall" products are designed to withstand attempts to breach the barrier with wire cutters and scaling.

Picture 2. Internal Barriers

Facility Closed-Circuit Television Systems:

The increased awareness of security concerns as a result of the SVA process has led to a greater application of security systems and technologies at APCI facilities. For example, many locations are now utilizing closed-circuit television systems to monitor and record activity of critical areas, hazardous product storage areas, site perimeters, and entrance points, etc. Certain locations are now installing closed circuit television (CCTV) systems for process monitoring or unattended operations monitoring. All video, regardless of function or monitoring capacity, should be archived to digital video recorders. The use of digital video recorders allows multiple persons in multiple locations to monitor the video simultaneously. For instance, many high risk sites in North America are currently being monitored remotely via the APCI Security Operations Center (SOC) in Allentown, PA as well as locally by site operations. See Picture 3.

Picture 3. APCI Security Operations Center

Facility Intrusion Monitoring Systems:

Intrusion detection systems provide a site with true active monitoring and are designed to report a change in state, such as a door being forced open or motion detection on a dock or fence line, to a centralized area for acknowledgement of receipt of alarm. Centralized areas can include onsite control rooms, site security centers, remote security operations centers, contracted central stations, and even local police departments. APCI has been developing advanced intrusion detection platforms at many high risk sites around the world that combine sophisticated passive infra-red (PIR) sensors with video motion detection (VMD) technologies. In combination, these two technologies provide an integrated system that provides for a high detection probability when actively monitored.

Facility Access Control Systems:

Regardless of security risk, each facility is required to develop an access control program. Facility access control programs range from standard lock and key systems to card access systems. The primary purpose for establishing access control programs or systems is to:

- Actively monitor and control movement of personnel and vehicles through the perimeter and/or structures.
- Secure access to critical or hazardous structures.

Access control systems can also be used to provide personnel accountability functions in an emergency on certain locations, further increasing the coordination between safety and security. APCI has adopted a global standard for card access technology for each of our facilities around the world.

Facility Entrances and Exits:

During a site SVA, a great deal of time is spent understanding the traffic flow of persons and vehicles entering, on-site, and exiting the property. The aim is to better understand how to secure the facility entrance and exit points appropriately to maintain a desired level of security without interfering with safety or plant operability. Securing vehicle entrance points will likely be by manual, lockable gates, or entrance and exit gates secured with an automated gate controlled with a monitored access control system.

Facilities that require strict personnel accountability are required to have all personnel enter and exit through a full-size turnstile controlled with a monitored card access control system on both inbound and outbound sides. See Picture 4. Emergency exits that bypass the turnstile and sound an alarm are available for personnel to use in the event of an onsite emergency.

Picture 4. Full-height Turnstile

Uniform Security and Protective Services:

The use of uniform security services at some locations is determined after a clear understanding of need versus cost. In general in the United States, the use of systems such as card access, CCTV, and automated gates normally is more cost effective than uniform security services for security monitoring and site access control. However, some larger locations with increased traffic, large numbers of employees, or operations that may require manned activity, dictate the need for an on-site security presence. In most cases, uniform security will be combined with security technology for greater capabilities. During the SVA process, the team will work with the site management and operations to clearly define the uniform services responsibilities at the location (gate access, emergency response, administrative, patrols, safety inspections, etc.).

CONCLUSION

APCI developed a SVA methodology, using a team approach that includes process safety, security, and site operations functional expertise, to carry out efficient and thorough evaluation of a large number of facilities ranging from small industrial gas sites to large chemical plants. APCI is applying the SVA program globally by utilizing regional teams while following the same methodology. The engineering and security solutions are intended to deter, detect, delay, and respond to an attacker. The solutions include inherently safer alternatives, which were accomplished by designing fail-safe systems, reducing inventory, and improving plant layout. The solutions also include enhanced physical security systems and procedures. Work processes and standards have also been developed to address future installations.

Figure 1. Security Vulnerability Assessment Work Process

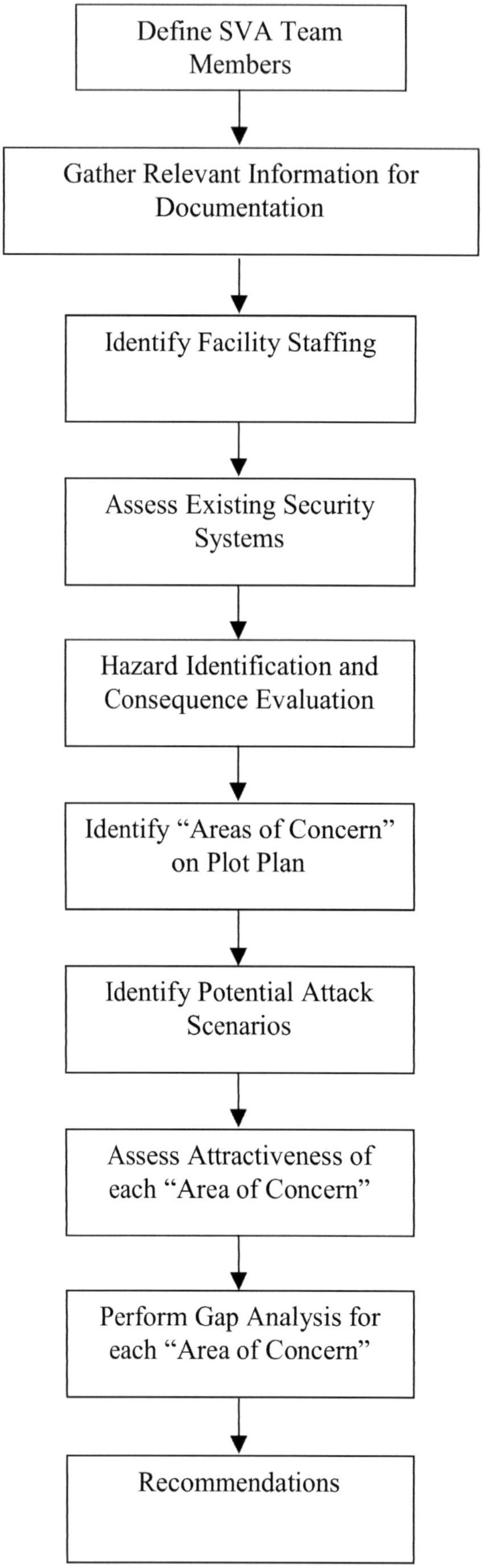

Human Challenges In Facility Security Engineering

Dennis Attwood, Ph.D, CPE and Bill Effron, CSP

RRS Engineering
League City, TX 77573

Prepared for Presentation at

American Institute of Chemical Engineers
38th Annual Loss Prevention Symposium
Engineering Solutions to Facility Security Challenges

New Orleans, LA

April 26 – 29, 2004

Introduction

With the growing concern for the security of process operations from sabotage and terrorist attack, most integrated oil and chemical companies are examining their vulnerabilities to attack and modifying their approaches to security. Most of these modifications involve acquiring high-tech detection equipment, increasing site surveillance activities, and upgrading security procedures and practices. In many cases, the new high-tech equipment has not been designed to take into account the capabilities and limitations of the operators or physically located to allow effective monitoring by the security staff. Increased site surveillance does not consider the vigilance and alertness of security personnel. Typically, procedures and processes are not designed so that people can comprehend and act on them quickly and without error.

The following paragraphs examine four areas of a facility security system to determine whether they have been designed with the capabilities and limitations of the users in mind.

1. Equipment that has been designed to visually detect security breaches.

2. Processes that have been put in place to increase site alertness.

3. How alertness and fatigue of security personnel may be affected by shift schedules.

4. The procedures that have been developed to respond to security violations.

The paper will recommend changes in the equipment, processes and procedures to take advantage of what people can and cannot do.

Visual Monitoring

In many installations, the first line of detection of a loss, a threat or an intrusion comes from visual detection using CCTV cameras and monitors. The ability of security personnel to detect abnormal events relies on several human and equipment characteristics:

1. The Visual Ability of the Human

It is generally known that visual capability decreases as a person ages. In fact, the most rapid decline in visual acuity (the ability to resolve detail) occurs, on average, when a person reaches 40 years old. It is estimated that visual acuity is reduced as much as 30% between the ages of 20 and 60 years (Attwood et al, 2003). Other changes in the eye occurring with age that could affect the visual detection ability of security personnel include:

a. Yellowing of the lens which significantly reduces the amount of light reaching the retina (the part of the eye that converts light to visual signals to the brain)

b. Decrease in the focusing power of the eye (accommodation) due to hardening of the lens on the front of the eye that changes shape as we focus on near and far objects and the deterioration of the muscles in the eye that change the shape of the lens

c. Changes, around the age of 60, in the sensors in the eye that create the visual image and the nerves that transport the information to the brain. These changes affect the size of the visual field and sensitivity to low light levels

For older security personnel, special steps must be taken to ensure that they can perform as well as younger personnel. These might include:

a. Higher levels of lighting and larger targets
b. More viewing time to absorb the information
c. Higher levels of contrast between the target and its background. White on black is much better than shades of grey
d. Good corrected vision
e. Locating displays at the same distance from the eye to avoid the need for rapid accommodation

2. The Visual Resolution of the CCTV System

The resolution of a CCTV system is largely dependent on how many parts the visual scene is divided into, the light sensitivity of the system and the quality of the optics that focus the image. Higher resolution cameras and monitors are better at detecting small images or fine detail than low, though they also cost more. The resolution required is dependent on the task that the human is being asked to perform. If the task is merely detecting the presence of a large object, like a truck at a plant gate, the resolution can be low. But, if the task is to detect the name of the trucking company or a license plate number, the resolution must be much higher. The point here is that the equipment must support the user's task. So, it's essential that a full analysis of the security person's job be completed before the equipment is specified. Task analysis techniques are available in many books and can be performed by any competent human factors specialist.

3. The Visual Environment in Which Both the Camera and Monitor are Located

For those readers who have at one time or another pointed their camcorder into the sun, they know the problems that the wrong visual environment can create. As a rule of thumb, the human should not be exposed to a visual task where the background is more than 10 times brighter than the object. If the camera is pointed into the sun or at bright lights, the background is many hundreds of times brighter than the object that he or she is trying to detect or follow. The object becomes poorly defined (such as a silhouette or a blur). On the other hand, the human should not be asked to perform a visual task where the contrast between the object and the background is very low (where contrast is defined as the ratio of the brightness of the object to that of the background). In the first example, the camera should be positioned with the bright lights (sun or luminaires) behind the camera so they light the object. In the contrast condition, painting the background so the contrast to the object is maximized, or lighting the scene can improve detection.

Finally, the security person cannot detect what he doesn't see. The optimal visual field for the eyes is within a cone of 15 degrees all around the eye. So, CCTV monitors that contain the most important information should be inside this cone of optimum vision. A monitor mounted far to the side of or behind a seated user will not be monitored, no matter how important it is.

Alertness and Vigilance

Human performance is optimal when we are neither too busy nor too bored. Although the potential for work overload is a common concern with security personnel, work under-load should also be considered important because it can result in a loss of attention and alertness potentially reducing performance. Jobs that consist of monotonous, boring work are not only non-motivating but may be associated with unreliable performance.

Perceptual performance is determined by our ability to focus and maintain attention on the incoming events. The ability to sustain attention is called vigilance. Vigilance tasks require the operator to detect small changes in the information being presented over long unbroken periods of time. Vigilance is often defined as "waiting around for nothing to happen". In the process industries, many tasks can be classified as vigilance tasks including: security monitoring, control room operations, and long-haul trucking.

Several factors can contribute to the vigilance decrement including alcohol, stress, drugs, and environmental factors such as heat and noise. In a similar manner, several factors can attenuate the vigilance decrement and make detection of the faint signal more possible. They include:

1. Time on task: The shorter the time on task, the better is the vigilance performance.
2. Intensity of the signal: The more the signal differs from the background in terms of contrast, color, size or movement, the more likely it will be detected.
3. Importance: Those signals that are most important to the observer or are of most interest will likely be detected sooner than those that are not as important or interesting.
4. Exercise: Research suggests that exercise generally improves monitoring performance (Ray, 1991 and Norin and Wyon, 1992). So, exercising while monitoring could be worthwhile.

Shift Schedules and Fatigue

Fatigue has been a major factor in significant accidents in the process industry (e.g. Three Mile Island, Bhopal). Fatigue can be caused by factors that define the schedules that are implemented for shift workers. It is generally agreed, for example, that time of day is the most important predictor of fatigue in shift workers. Night work is related with increased fatigue among plant workers and with increased accident frequency. It is also well known that the body clock of a worker who is on a daytime-sleep/nighttime-wake cycle affects both nighttime drowsiness and daytime sleep. The worker is not only tired at night, but he or she finds it difficult to sleep during the day. Data from a recent US/Federal Highway Administration (FHWA) study, for example, indicate that drivers on long night runs spent only 4.37 hours in bed which was 40-85 minutes less than the time in bed for all other conditions.

Work schedules can also affect the fatigue and alertness of security personnel. Circadian shifts and lack of sleep can reduce alertness and detection performance, decision making and judgment, information processing and memory as well as self-initiated cognitive activity, like maintaining situation awareness and planning. Moreover, people over 40 years of age (Barnes, 1949; Grandjean, 1988) are less tolerant of changes in schedule, take longer to adapt to shift changes, tire more easily, and are more prone to disturbed sleep than those under 40 years.

Ensuring that security personnel are tolerant of the effects of shift schedules on their performance will require attention from the company, the worker, and his or her family. Studies have shown, for example, that an individual shift worker can have a profound effect on his or her own health and safety by following a few commonsense lifestyle principles such as:

1. Paying attention to eating schedules and to the foods eaten at night.
2. Giving sleep the attention that it deserves. Not letting social or domestic pressures take away sleep time.
3. Advising family and friends of your schedule so they respect the need for privacy.
4. Maintaining physical fitness.
5. Napping prior to the first night shift.

The company can also prevent fatigue by promoting awareness of the personal and family coping strategies and by optimizing work schedules and facilities. For example, companies can:

1. Design night jobs to maintain alertness by ensuring that there are frequent changes in the job tasks.
2. Provide rest facilities for "on-call" or extended hours.
3. Ensuring that food and beverage availability is the same as that during the day.
4. Modifying or eliminating tasks that require continuous bouts of high vigilance.

Security Procedures

Procedures are important because they provide the rules to be followed in established situations and the process to be used when situations are not defined. Procedures also provide an information base for training new personnel.

Procedures should only be written for activities that require them. Too many procedures can be a detriment to any job. Simple activities, that are done frequently and whose consequence of error is low, should not have procedures. On the other hand, tasks that are not performed often (e.g. emergencies), or are complex, requiring mental calculations and quick decisions, or tasks whose consequence of error are high require procedures.

Good procedures are designed to provide only the information that the user needs in a manner that he or she can use immediately. The procedure should not contain information or instructions that are part of the user's basic skill set. The procedure should be written using a consistent format, often called a style guide, so the user is familiar with it. Important aspects of a procedures' format include:

1. Beginning each sequential step with an active verb. Narrative style should not be used.
2. Distinguishing warnings, cautions, and notes from an active step.
3. Choosing a sans serif font that is large enough to be read under adverse conditions (13 or 14 point type).
4. Using hierarchical numbering schemes. For example:

   1. …….

      1.1 ……..

      1.2 …….

   Two levels are usually sufficient

5. Inserting enough information on each page of the procedure so the user understands what he or she is reading and where they are in the process. For example, each page should contain the procedure title, page number, total number of pages, version number and date published

In many instances, procedures are not used or are mis-used resulting in poor outcomes. Many times the non- or mis-use of procedures can be traced to a problem within the system that creates and manages them. For example:

1. Conditions in the workplace do not support the use of procedures.
2. The equipment necessary to carry out a procedure is not be available.
3. Workers are not instructed on the use of the procedure.
4. The procedure is out of date.
5. The employee does not understand the consequences of deviating from a procedure.
6. Not enough time has been set aside to follow a procedure.

Checklists are available to evaluate the effectiveness of written procedures (Attwood et al, 2003).

Summary

In summary, the increased complexity of security systems may come at the expense of the user. While many of the systems currently available automate a number of the functions that the human security person used to perform, every system requires human intervention and must be designed to match their capabilities and limitations.

References

Attwood, D.A., Deeb, J.M. and Danz-Reece, M.E. "Ergonomic Solutions for the Process Industries." Elsevier/Gulf Professional Publishing, Boston MA, 2003.

Barnes, R.M. "Motion and Time Study: Third Edition." John Wiley and Sons, New York, 1949.

Grandjean, E., "Fitting the Task to the Man: Fourth Edition." Taylor and Francis, London, 1988.

Norin, F., and Wyon, D.P. "Driver Vigilance – The Effects of Compartment Temperature." Report No. 920168, Society of Automotive Engineers, Warrendale, PA.

Ray, P.S. "Effects of Fatigue and Heat Stress on Vigilance of Workers in Protective Clothing." Proceedings, 35th Annual Meeting of the Human Factors Society, 1991, pp 885-889.

Generalized Findings from a Process Threat Management Case Study

James R. (Rob) Whiteley
405-744-9117
whitele@ceat.okstate.edu

Jan Wagner
405-744-4077
jwagner@ceat.okstate.edu

School of Chemical Engineering
Oklahoma State University
423 Engineering North
Stillwater, OK 74078-5021

Prepared for Presentation at

American Institute of Chemical Engineers
38th Annual Loss Prevention Symposium
Engineering Solutions to Facility Security Challenges

New Orleans, LA

April 26 – 29, 2004

1. Introduction – The Process Threat Management Problem

After the tragic events of 9/11/01, terrorist and criminal acts are now considered credible incidents in the process industries. Deliberate attacks on the nation's petroleum refineries and chemical plants pose a significant threat to employee and public welfare, national security, and the U.S. economy. The primary response of government and industry has been to focus on improved physical and cyber security to prevent attacks and the associated consequences.

While prevention is clearly preferred, the potential for successful attacks must be recognized and addressed [*1*]. If a facility is attacked, the extent of the inflicted damage is determined by the available plant safety systems and procedures. We refer to this "inside the gate" response as *Process Threat Management* [*2*].

This paper discusses the terrorist or criminal threat from a process, rather than security, point of view. All process plants are designed to deal with unintentional events such as equipment failure, loss of utilities, fire exposure from spills, etc., that threaten safe operation of the facility. However, the installed safeguards were designed without considering acts of sabotage or attack by a thinking adversary. There is a need to determine how existing process safety systems and practices will perform in response to this new threat.

This paper presents generalized findings from a collaborative industry/academia case study for a new unit in a domestic U.S. refinery. The goals of the study were:

1. If no changes are made in existing safety systems, what type of performance can be expected in response to a terrorist attack? Is a reduction in impact possible? If yes, how?

2. How can existing Process Hazards Analysis (PHA) methods be exploited and/or modified to address process threats? Are new tools needed? If yes, what are the required capabilities?

3. What should be the operating strategy during an attack? Can existing plant automation (regulatory control system, advanced control systems, safety instrumented systems) be employed in new ways to minimize damage from terrorist attacks? Is new or additional automation needed?

This paper focuses on the first two issues.

2. Background

Site Security and Vulnerability Assessment

Industry responded quickly to the new threat of domestic terrorism. *Site Security Guidelines for the U.S. Chemical Industry* were developed jointly by the American Chemistry Council (ACC), the Chlorine Institute, and the Synthetic Organic Chemical Manufacturers Association (SOCMA) [*3*] and issued before the end of 2001. Similar guidelines for the petroleum industry were published by the American Petroleum Institute (API) in 2003 [*4*]. The American Chemistry Council made enhanced security activities mandatory for its members in January 2002. The most recent additions to the ACC Responsible Care® security code require independent third-party verification.

Development of security vulnerability assessment methodologies was initiated by several industry groups with results published in 2002 and 2003.

- The American Institute of Chemical Engineers (AIChE) Center for Chemical Process Safety (CCPS), *Guidelines for Managing and Analyzing the Security Vulnerabilities of Fixed Chemical Sites*, August 2002.

- Synthetic Organic Chemical Manufacturers Association, *SOCMA Manual on Chemical Site Security Vulnerability Analysis Methodology and Model*, November 2002.

- American Petroleum Institute / National Petrochemical and Refiner's Association, *Security Vulnerability Assessment Methodology for the Petroleum and Petrochemical Industries*, May 2003.

The federal government moved quickly along a parallel path. The National Institute of Justice, working with Sandia National Laboratory, developed the following methodology:

- National Institute of Justice / Sandia National Laboratory, *A Method to Assess the Vulnerability of U.S. Chemical Facilities*, November 2002.

Each of the methodologies listed above employ traditional risk assessment techniques and provide a well-defined, systematic framework to identify security threats, risks, and vulnerabilities. It is important to note that all of the techniques are "asset-based" rather than "process-based."

The vulnerability assessment methods all require analysis "of a manufacturing process's response to a terrorist attack." However, the methodologies do not include guidance on how to perform this assessment. There is a wide body of literature on how to perform traditional process hazard analyses [*5-10*]. However, when developed, these techniques did not consider deliberate acts and the impact on plant equipment and safety systems. Consequently, there is uncertainty regarding the level of protection provided against this new type of destabilizing event.

News Media and Public Awareness

The potential impact of a terrorist attack due to release of toxic material in a densely populated area has been broadly publicized by the news media. A frequently quoted study by the U.S. Army Surgeon General concludes that as many as 2.4 million people could be killed or injured in an attack against a U.S. toxic chemical plant [*11, 12*]. Another frequently quoted study is a 1999 report by the Agency for Toxic Substances and Disease Registry [*13*]. News releases from environmental activist organizations frequently reference EPA records indicating that 123 chemical facilities in the U.S. threaten a million or more nearby residents and each of over 700 plants put at least 100,000 people at risk. Most of these potential incidents are associated with uncontrolled vapor release of chlorine, sulfuric acid, hydrogen fluoride, or sulfur dioxide.

Public concern regarding the potential for a catastrophic release is undoubtedly raised by reports from government agencies citing inadequate facility security. Two recent examples include a March 2003 report from the U.S. General Accounting Office (GAO) [*14*] and a January 2003 report from the Congressional Research Service [*15*]. During the same period, the Department of Homeland Security issued at least one warning [*16*] that "Al Qa'ida operatives may attempt to launch conventional attacks against the U.S. nuclear/chemical-industrial infrastructure to cause contamination, disruption, and terror. Based on information, nuclear power plants and industrial chemical plants remain viable targets." A February 2003 article by Weinstock [*17*] described the ease with which reporters or activists were able to enter chemical facilities.

The public has high expectations regarding safety in plant operations. This is due in part to long-standing industry initiatives and programs such as Responsible Care®. When considering the threat of deliberate acts against process plants, it is likely that societal expectations exceed the capability of existing plant safety systems. The first step in setting reasonable public expectations is to evaluate the expected performance of these systems.

Summary of Work Performed to Date

Improving physical site security has been the main response to the 9/11 tragedy. Much attention has been paid to the three G's - guards, gates, and guns.

The need to provide cyber security to protect plant data acquisition and control systems has been recognized [*18, 19*] and is being addressed. A summary of the government and chemical sector initiatives in cyber security is presented in an electronic newsletter prepared by the Critical Infrastructure Protection Project [*20*].

There has been little published in journals and magazines associated with the process industries. Most of the published articles address security or vulnerability assessment [*21-25*]. Baybutt [*26, 27*] makes a clear case that traditional risk assessment methods used for OHSA mandated process safety management must be augmented to address deliberate acts. In [*1*], Baybutt highlights the distinction between plant security (secureguards) and plant safety (safeguards) and introduces inherent security principles analogous to the well-established principles of inherent safety.

To the authors' knowledge, there have been no published reports on the expected performance of plant safety systems to a terrorist or criminal attack.

3. Case Study

The authors teamed with a major U.S. refiner to explore the expected performance of existing plant safety systems in response to a terrorist attack. A new refinery gas processing unit was selected for study. The team had access to all available documentation (P&IDs, HAZOP, etc.) for the unit. The process and safety design for the unit reflects current practice.

The unit selected for study is generic in the sense that all refineries include one or more units that provide the same service. The primary functions of the unit are gas compression, H_2S removal, and component separation via fractionation. The unit was installed adjacent to other operating units within an integrated refinery. Vents from the overpressure protection system for the new unit were connected to an existing flare system. The unit is operated by a modern computer control system from a centralized control room.

From a risk management standpoint, there is nothing exceptional about the gas processing unit. No toxic chemicals such as hydrogen fluoride or sulfuric acid are involved. The unit was not selected for study based on special vulnerability or threat to the community under catastrophic conditions, but rather because of its typical nature. With the exception of a reactor, the gas processing unit incorporates all of the major unit operations associated with refineries and chemical plants. The representative nature of the unit is consistent with the goal of the study to identify generalized rather than specific findings.

Analysis Methods

The starting point for the analysis was the traditional PHA. The intent was to leverage existing results and methodologies as much as possible with minimum additional effort. In this case, the PHA was a HAZOP.

The case study was based on terrorist acts; the threat presented by activists was not considered. We differentiate terrorists and activists based on the assumed objectives of the two groups. We characterize a terrorist as a thinking adversary whose goal is maximum damage and loss of life. We assume activists are primarily interested in business disruption without intentional loss of life.

Traditional safety systems are designed to deal with one unpredictable (unintentional) event followed by a sequence of predictable consequences. In a terrorist or criminal act, multiple events of unpredictable nature over an unknown time frame must be considered. Consequently, safety events previously deemed implausible must now be considered. Furthermore, the number of possible event combinations and sequences has increased dramatically. The problem can appear overwhelming as the finite number of scenarios to be analyzed pre-9/11 has been uncapped [*2*]. This was the most challenging aspect of the problem.

Evaluation of safety system performance was performed from two directions. The first involved following the HAZOP and considering relevant terrorist-initiated acts at each node. This represents a highly structured, deviation-driven approach. The second approach was consequence-driven. In this approach, catastrophic scenarios were identified. The expected performance of existing safety systems was noted for each scenario. The potential to mitigate the impact of a scenario by modifying or adding safety systems was also noted. This represents a much less structured approach.

Deviation-Driven (HAZOP) Analysis

A review of the process hazard analysis for the gas processing unit was undertaken to gain some insight into ways hazard analysis methods might be leveraged to address process threats. The process safety information available at the time is summarized in Table 1.

The original HAZOP analysis sessions required several hundred man-hours. The study was well documented. A review to the original HAZOP study to include intentional causes required approximately 12 man-hours.

In a node-by-node review of the hazard and operability study, the original guidewords and deviations were followed. However, possible causes were extended to include intentional attempts to release material or energy from the process or initiate deviations outside of safe operating limits. Scenarios included misalignment of the process by opening or closing block valves, venting material through bleed or sample valves, and compromising operating procedures.

The original HAZOP team considered misalignment processes, either explicitly or implicitly, for all nodes where this was possible. As a result, adequate safeguards were in place to mitigate the consequences of intentional misalignment of units. However, there is an additional area of vulnerability under administrative controls. Lock-out and tag-out systems are an integral part of energy control procedures during construction and maintenance. However, these systems are also used to prevent the operation of valves in a process or utility system. Sometimes these are on block valves on vital services such as flares and firewater service. The security of locked valves should be included in process hazard analysis.

Safety integrity systems or emergency shutdown systems could also be taken off-line in the case of a well-planned assault on a process. In most modern chemical processes, these safety systems are independent of the process control system, but still accessible to operators. The process hazard analysis should address the vulnerability of these systems.

The HAZOP for the gas processing plant identified operating procedures as safeguards in a small number of situations. However, the scenarios were irrelevant with respect to terrorist initiated acts. Although not identified as a concern in this case, operating procedures may be negated as safeguards under conditions of a terrorist attack. Additional discussion of this point is presented later.

Safe operation of the gas processing unit relies heavily on passive shutdown and relief systems. Review of the HAZOP and other documents in Table 1 suggested that the existing systems to detect and respond to release, overpressure, and other emergency events may be adequate to address the consequences of a terrorist attack assuming the safeguard systems are unaffected by the attack. In cases of intentional or unintentional release of material, both H_2S and hydrocarbon detectors were provided to alarm operators. The firewater system included sprinkler systems on all pumps, heat exchangers, and vessels with inventories of flammable materials, in addition to monitors and hydrants around the perimeter. The relief valve sizing scenarios considered external exposure to fire for vessels (including heat exchangers) that contained significant inventories of flammable liquids. However, consistent with current practice, consequences with partially or totally compromised safety systems were not considered in the original HAZOP.

Facility layout and equipment siting were part of the PHA for the gas processing unit. American Petroleum Institute Recommended Practices, National Fire Protection Association Codes, and Construction Industry Institute Process Industry Practices were applied in arranging major process blocks, utilities, off-sites, and buildings. Industry guidelines were followed in establishing separation distances for major equipment to minimize the potential damage from fires or explosions in adjacent areas. If a saboteur were successful in causing a fire or explosion in one unit, the unit layout would mitigate the escalation of the incident into other units or plants to the extent determined by layout practices.

Consequence-Driven Analysis

The starting point for this analysis was a worst-case scenario involving maximum release of flammable and toxic materials from the gas-processing unit due to a terrorist attack. Ignition was assumed to occur immediately resulting in a large hydrocarbon fire (no vapor cloud explosion). Consideration of the impact of such an event on existing safeguards led to three major areas of concern.

1. In the case of an organized assault on a plant, emergency responders and operators might be prevented from carrying out procedures in the field, such as manning fire-water monitors or operating manual valves.

2. Facility siting and layout should minimize the escalation of major incidents to other processes within the plant.

3. Vital services must be reliable.

Additional discussion regarding each concern follows.

<u>Inability to Implement Procedural Safeguards</u>

Operating procedures can be viewed as the innermost protection layer for a chemical process [*28*]. There are several aspects of a terrorist attack that may reduce or completely eliminate the efficacy of operating procedures as a safeguard. In a case of insider collusion, essential procedures may not be implemented in an emergency due to sabotage or intentional negligence. In the case of the gas processing unit, intentional failure to shut off the feed from an upstream unit would clearly increase the impact of a fire or material release.

The threat to field personnel and damage to infrastructure are more likely examples of how operating procedures may be rendered ineffective during a terrorist attack. Emergency procedures requiring operators to shut down or isolate equipment in the field may not be completed as personnel seek safe haven. This concern has been encountered in the private sector, i.e., incidents of firemen coming under attack when responding to emergencies during riots. Emergency response plans must address plant and personnel security, and security vulnerability analysis must address emergency response. Plant security systems are outside the scope of this presentation.

Likewise, the loss of radio communication between personnel, damaged valves and equipment, or severed instrument wiring may prevent planned response actions. These same factors can also compromise the next layer of protection, critical alarms and supervision. For intentional incidents, operating procedures may not count as a layer of protection, or safeguard.

Plant Siting and Layout

In our case study, plant siting and layout information was not available for areas outside the gas processing plant. Had the information been available, one approach may have been to conduct a "plant safety analysis" in addition to the process safety analysis. The plant safety analysis would consider worst-case attack scenarios for each unit and assess the impact on adjacent units within the plant, as well as impacts on other plants at the site. The plant safety analysis could use the same methods as a PHA, but with a broader scope. The analysis would consider interactions between units and plants at a site. Facility siting, emergency response systems, and other vital services need to be assessed on a plant or facility level – not just the process unit level typical of PHAs.

A question was also raised whether the emergency conditions used as the basis for existing siting and layout practices need to be made more severe. Such a change may be justified if attack scenarios are deemed capable of producing larger, more damaging material releases and fires.

Vital Services

Risk management at the facility level focuses attention on the reliability of vital services. Firewater systems are recognized as vital services that must be extremely reliable. However, other services may be unique to a unit or site. In the gas processing plant, the flare system is a vital service, not only for the gas plant but also for other plants tied into the flare header. Capacity limitations are addressed in the design of the pressure relief system for a unit and in the HAZOP. However, the capacity limitations are based on relief rates that exclude the possibility of a criminal attack. This raises the question whether the flare system is adequately sized to accommodate the maximum potential vent rate during a terrorist attack. We did not have sufficient information to make this evaluation for the gas processing unit. Unexpected equipment failure in an area not subject to attack may occur due to excessive back-pressure if the flare system is overloaded due to an attack on the facility.

Furthermore, the reliability of the flare system was not directly addressed at the unit level. Could this service be intentionally compromised? If so, what are the consequences? Are additional safeguards required or justified? These questions are raised in the context of the flare system but apply to all vital services.

A related concern with plant wide implications on vital services is the increased possibility of losing all instrument and control links with a centralized control room.

4. General Findings

Evaluation of the results from the two analysis approaches leads to the following initial conclusions.

4.1 Process hazard analysis has evolved into a fairly robust process for documenting the process safety information, identifying hazards and safeguards, and managing risk. Current practices can be extended to include additional scenarios related to intentional initiation of events. For example, possible causes for deviations can augmented to include "intentional."

4.2 Procedural layers of protection may not be reliable, if they can be intentionally ignored or compromised.

4.3 Facility layout and siting studies should not be limited to new facilities. For existing facilities, these studies may reveal opportunities to minimize risk of injury, environmental and property damage, and business interruptions. Siting and layout reviews may also reveal hazards that may not be identified by other elements of process safety management.

4.4 Process hazard analysis that considers the broader interactions between units in a plant and between plants at a site may lead to inherently safer facilities. Site security systems will play a larger role in process and facility safety.

Acknowledgments

The authors gratefully acknowledge the assistance and information provided by their industrial collaborators.

References

1. Baybutt, P. "Inherent Security: Protecting Process Plants Against Threats." *Chemical Engineering Progress*, Vol. 99, No. 12, pp. 35-38, 2003.

2. Whiteley, J. R., M. S. Mannan, and S. J. Brouillard. "Process Threat Management." *Proceedings of Mary Kay O'Connor Process Safety Center 2003 Symposium.* College Station, TX pp. 107-115. 2003.

3. American Chemistry Council, Chlorine Institute, and Synthetic Organic Chemical Manufacturers Association. "Site Security Guidelines for the U.S. Chemical Industry." 2001.

4. American Petroleum Institute. "Security Guidelines for the Petroleum Industry." 2003.

5. Mannan, M. S., D. Hendershot, and T. A. Kletz, "Fundamentals of Process Safety and Risk Management," in *Encyclopedia of Chemical Processing and Design*, R. G. Anthony, Editor, Marcel Dekker, Inc.: New York. pp. 49-94, 2002.

6. Crowl, D. A. and J. F. Louvar. *Chemical Process Safety: Fundamentals with Applications*. Prentice Hall, Englewood Cliffs, NJ, 1990.

7. Center for Chemical Process Safety. *Guidelines for Chemical Process Quantitative Risk Analysis*. American Institute of Chemical Engineers, New York, 1989.

8. Center for Chemical Process Safety. *Guidelines for Hazard Evaluation Procedures: 2nd Edition with Worked Examples*. American Institute of Chemical Engineers, New York, 1992.

9. Center for Chemical Process Safety. *Guidelines for Consequence Analysis of Chemical Releases*. American Institute of Chemical Engineers, New York, 1999.

10. Wells, G. *Hazard Identification and Risk Assessment*. Institution of Chemical Engineers, Rugby, Warwickshire, UK, 1996.

11. Pianin, E. "Study Assesses Risk of Attack on Chemical Plant." Washington Post, p. A08, http://www.washingtonpost.com/ac2/wp-dyn/A10616-2002Mar11, March 12, 2002.

12. Kostant, A. "Terrorist Attack on Chemical Plants Could Endanger Millions." Environmental Media Services, http://www.ems.org/chemical_plants/zz.ems.02.06.20.html, June 20, 2002.

13. Department of Health & Human Services - Agency for Toxic Substances and Disease Registry. "Industrial Chemicals and Terrorism: Human Health Threat Analysis, Mitigation and Prevention." 1999.

14. Stephenson, J. B. "Homeland Security: Voluntary Initiatives Are under Way at Chemical Facilities, but the Extent of Security Preparedness Is Unknown." GAO-03-439, U.S. General Accounting Office, 2003.

15. Schierow, L.-J. "Chemical Plant Security." Order Code RL31530, Congressional Research Service, The Library of Congress, 2003.

16. National Infrastructure Protection Center. "Homeland Security Information Update." in *Information Bulletin 03-003,* 2003.

17. Weinstock, M. "Easy Targets." *Government Executive*, No. February, 2003.

18. Teumim, D. J. "Are Your Plants and Pipelines Safe from Cyber Attack?" *Chemical Engineering Progress*, Vol. 98, No. 2, pp. 69-73, 2002.

19. Dacey, R. F. "Critical Infrastructure Protection: Significant Challenges in Developing Analysis, Warning, and Response Capabilities." in *Subcommittee on Technology, Terrorism, and Government Information, Committee on the Judiciary*, United States General Accounting Office: Washington 2001.

20. Critical Infrastructure Protection Project. "The Cip Report." p. 1-16, 2002.

21. Marszal, E. M. "Process Security Management: Assess the Threats, Control the Risk." *Chemical Engineering*, 2003.

22. Heinold, D. and D. Smith. "Site Vulnerability Assessments for Facilities That Use, Store, or Transport Chemicals." *ENSR Insight*, Vol. 3, 2002.

23. Schwartz, J. "Hit or Miss?" *Adhesives Age*, Vol. 45, No. 7, p. 12, 2002.

24. Baybutt, P. "Process Security Management Systems: Protecting Plants against Threats." *Chemical Engineering*, Vol. in press, 2003.

25. Ragan, P. T., M. E. Kilburn, S. H. Roberts, and N. A. Kimmerle. "Chemical Plant Safety - Applying the Tools of the Trade to a New Risk." *Chemical Engineering Progress*, Vol. 98, No. 2, pp. 62-68, 2002.

26. Baybutt, P. "Assessing Risks from Threats to Process Plants: Threat and Vulnerability Analysis." *Process Safety Progress*, Vol. 21, No. 4, pp. 269-275, 2002.

27. Baybutt, P. and V. Ready. "Strategies for Protecting Process Plants against Terrorism, Sabotage and Other Criminal Acts." *Homeland Defense Journal*, Vol. 2, No. 2, p. 1, 2003.

28. Center for Chemical Process Safety. *Guidelines for Engineering Design for Process Safety*. American Institute of Chemical Engineers, New York, 1993.

Table 1. Available Process Safety Information

Process Description
- Process Steps
- Process Flow Diagrams
- Safe Operating Limits
- Maximum Intended Inventory

Piping and Instrument Diagrams

Hazards of Materials
- Inadvertent Mixing Table
- Reaction Explanation
- MSDS Reference

Design Basis Information
- Heat and Material Balance
- Relief System Design
- Area Electrical Classification
- Equipment List
- Piping Schedule

Protective Safety Systems
- Safety Plot Plan and Fire Protection Equipment
- Ventilation Systems
- Mechanical Safety Systems
- Instrument Safety Systems

Electronic Management of Change Process

This paper was prepared for presentation at the AIChE
2004 Spring National Meeting
38th Annual Loss Prevention Symposium
April 25-29, 2004 in New Orleans

Electronic Management of Change Process

R. Wayne Garland
Eastman Chemical Company

Abstract

Protection safeguards must reliably and effectively detect, diagnose and control process deviations before the deviations can result in loss events such as fires and explosions. Any process change can potentially affect protection safeguards. Therefore it is critical to identify process changes and assess their impact on layers of protection, safety instrumented systems (SIS), operating procedures and alarms. Process change designs should be reviewed and any needed modifications to safeguards should be implemented as part of the change.

Existing management of change processes rely primarily on human knowledge and interaction among various functional groups, making them highly susceptible to human error. In an effort to reduce the risk involved with process changes, we developed a standardized, electronic management of change (MOC) system and process. It facilitates reviews of process changes, update of company Process Safety Information (PSI) and notification of affected personnel. It also enhances the ability to track and audit changes.

Implementation included all necessary software and development/selection of training. The system is web based and includes a standardized electronic form with checklists (including some flexibility for site or division variation) and an automatic workflow routing. All system functions were automated including routing of documents & drawings, record keeping, approvals, notification, tracking, reporting, printing, searching and audit trails. The process has now been in use for more than 1.5 years and has been utilized for over 3500 process changes with good results. The major advantages of the system identified by users are tracking, documentation and compliance, automatic routing and automatic reminders.

Problem Statement

In order to insure that an appropriate low level of risk is maintained, the entire life cycle of a manufacturing process must be considered. Process safety should be incorporated into design and should be reviewed using Process Hazard Analysis (PHA) methodology. However a design may only be done once and processes may undergo a PHA only every 5 years. During the intervening time periods many changes can be made to the process that appear minor but could potentially have major impacts on both safety and regulatory compliance. Changes are made to improve safety, quality, cost and capacity. To assure that the desired low level of risk is maintained and that layers of protection do not degrade, changes need to be flagged and reviewed appropriately for impact on process safeguards. Also changes should be reviewed for potential regulatory impact. Therefore a good management of change process is a critical part of the life cycle of a manufacturing process. This is one of the 4 points in the manufacturing process life cycle where layers of protection, SIS, alarms and operating procedures should be reviewed as illustrated in Figure 1.

To ensure that the safety level of the manufacturing process is maintained and not degraded or compromised, it is vitally important to follow a systematic management of change process that involves the correct persons with the appropriate expertise.
However management of change is inherently complex because:

1. There are many process safety regulations, codes, standards and good engineering practices. Some examples are:

- OSHA 1910.119 Process Safety Management
- OSHA 1910.106 Flammable & Combustible Liquids
- Industrial Chemical Control Laws such as TSCA
- Hazard Communication regulations (HAZCOM)
- NFPA - National Electric Code, Life Safety Code, Fire Prevention Code
- ASME Boiler & Pressure Vessel Code

2. Manufacturing processes can be highly technical and specialized.

3. Safety systems can be complex with multiple layers. This includes Safety Instrumented Systems (SIS), procedures and alarms.

4. There can be company specific requirements.

5. Multiple hand-offs between groups across organizational boundaries are usually involved.

6. Communication with all affected parties is needed.

It is extremely difficult for all potential originators of manufacturing process changes to be kept fully informed and knowledgeable in all of these areas, even if there are administrative procedures for management of change. Therefore the originator of a manufacturing process change often faces a daunting task.

Management of change (MOC) processes are highly susceptible to human error because they rely primarily on human knowledge and interaction among various functional groups. For humans the probability of failure on demand generally used in Layers of Protection analysis (LOPA) ranges from 0.1 to 0.01. For a typical chemical manufacturing company there can be thousands of process changes per year with multiple steps required to execute each change. This results in hundreds or thousands of potential failure points. Most of these will have negligible consequences, but a small percentage could have serious consequences. This is a typical safety pyramid concept as illustrated in Figure 2.

At Eastman Chemical Company, the situation was further complicated because we did not just have one management of change (MOC) process. Rather, each site and division had its own process. These different processes were confusing for centralized groups such as engineering and safety. This complicated not only the changes themselves but also auditing and reporting.

In order to improve our MOC process we wanted a system that was less subject to human error. Even though it would still be largely an administrative system, we felt that we could automate it and incorporate interlocks, flags and multiple layers of protection. Please note that we are now talking about layers of protection in the MOC process itself to assure that critical steps are not overlooked. This led us to an electronic solution. A standardized MOC process should be less susceptible to error for the following reasons: (1) centralized groups would not be confused by multiple processes (2) interactions between areas would not be confused by different processes for each area (3) MOC process enhancements could be incorporated once rather than trying to propagate them to multiple processes. Therefore our object was to design and implement a standardized electronic MOC process.

We wanted not only an electronic form or an electronic approval process but rather a complete management of change process from start to finish including approval, execution, startup, and completion (see Appendix 1 for a block flow diagram of these four phases). Also we wanted a process that would help us comply with all HSE regulations – not just process safety or PSM. We wanted one process that could be used for all changes (even ISO 9000 quality standards) so we would not have multiple processes with the associated maintenance and training costs. After some study we were convinced that management of change is sufficiently generic to allow this to be done. We did some research to determine if there was any commercially available software that would meet these needs but at that time (mid 1999) there was nothing available that we thought would meet our needs. Therefore we decided to develop a process.

Figure 1 – Manufacturing Process Life Cycle

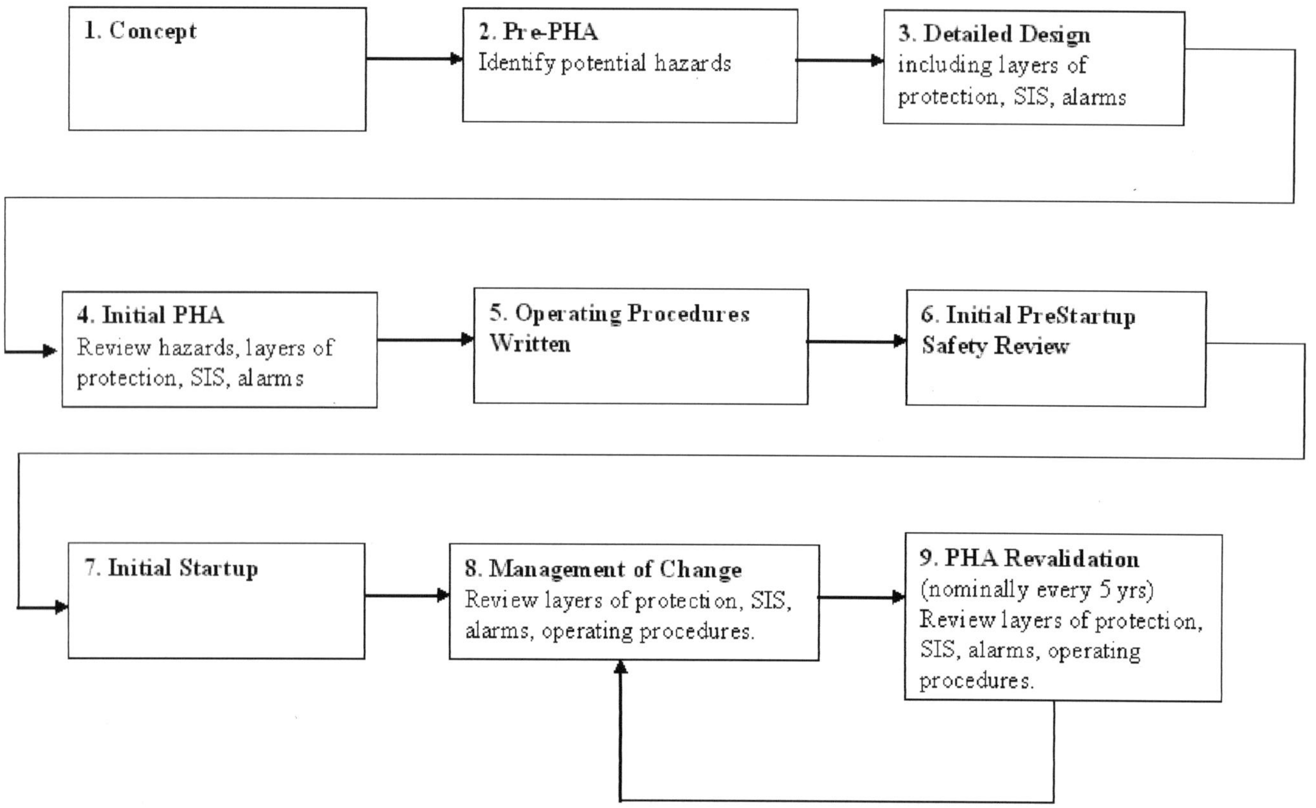

Figure 2 - A Matter of Probabilities

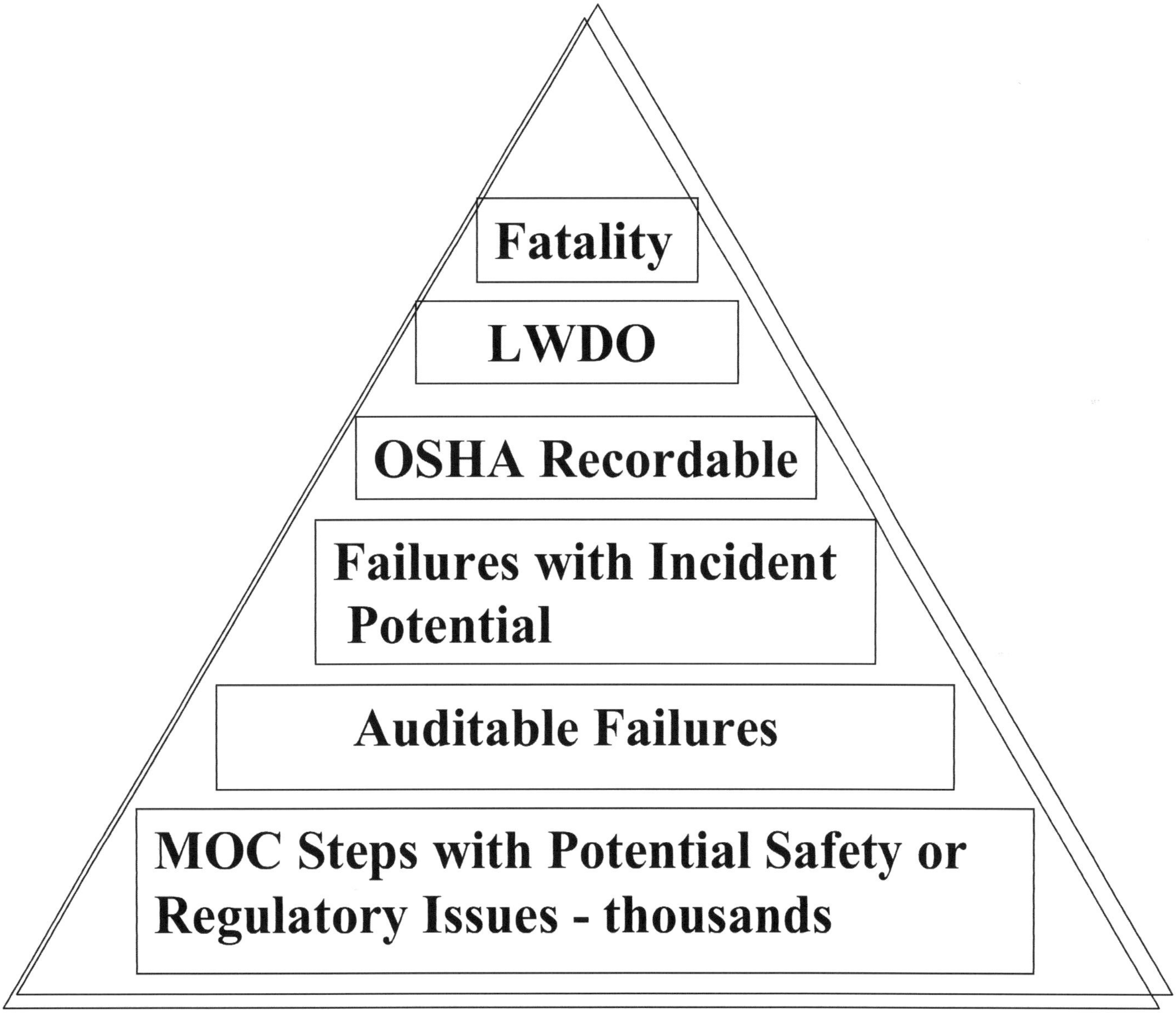

Design of eMOC

In an effort to reduce the risk involved with manufacturing process changes, we developed a standardized, electronic management of change process tool that we call eMOC. It facilitates reviews of manufacturing process changes, update of company Process Safety Information (PSI) and notification of affected personnel. It also enhances the ability to track and audit change. We utilized best practices from all areas to develop a standard system.

To design the eMOC process tool we used a multifunctional team consisting of representatives from manufacturing, engineering, IT and HSE groups. We piloted it in 4 areas at 3 different plant sites starting in September 2000. Based on pilot data we made extensive modifications and went "live" with the production system on line in mid-2002.

Implementation included all necessary software and development of training materials. The system is web based and includes a standardized electronic form with checklists (including some flexibility for site or division variation) and an automatic workflow routing. All system functions are automated including routing of documents & drawings, record keeping, approvals, notification, tracking, reporting, searching and audit trails. We wanted a totally electronic system but we had a few areas still using hard copy drawings and also found the electronic drawing viewer to be somewhat cumbersome. Therefore we designed the system to allow the usage of electronic drawings to be optional. In the future better electronic drawing viewer technology should hopefully be available.

There were several hurdles that the design team had to overcome. The first was organizational – achieving agreement among all the organizations on a standard MOC process. There were also technical hurdles on selection of hardware and software. In order to minimize costs, we wanted to utilize software that was already available and already in use at our company if possible. This would minimize not only purchase costs, but also training and maintenance costs. Since we needed to route and store documents, we utilized Electronic Document Management (EDM) software, Livelink from Open Text Corporation. This software allows the creation of forms and also workflow maps for routing documents. There is a linked Oracle database. Other software we utilized included Adobe PDF forms, Microsoft Office software and Myriad electronic drawing viewer from Informative Graphics Corporation. Making all these different software packages work together smoothly was challenging. We also had to overcome some technical hurdles on hardware. During the pilot we tried to share a server with some other applications and found that eMOC ran too slowly. In the end we set up a completely separate system with its own server, database and Livelink instance in order to achieve acceptable performance in terms of system speed and reliability.

We found during design of eMOC that there are several trade-offs which require some judgment by the design team.

- Standardization versus flexibility. We wanted to achieve a sufficient degree of standardization to achieve the benefits, yet allow some flexibility for individual areas to incorporate unique considerations due to specific technology, organizational differences, etc. We also needed flexibility for different types of changes. We designed a standard MOC form with the most commonly used information and allow optional attachments that can be area or change specific.

- Layers of protection versus ease of use – we wanted a system that had multiple layers of protection such as interlocks, checklists, specific instructions, flags and reminders but was not too burdensome to use. Specific interlocks included are as follows:
  - Approvals must be obtained before the execution steps are sent
  - Safety reviews must be verified as completed before the startup step is sent
  - Training/notification must be verified as completed before the startup step is sent
  - PSI (Process Safety Information) updates must be verified as completed before the startup step is sent
- On-map versus off-map – we wanted to include a workflow step for all the critical elements but not to make the workflow so complicated that it could not be understood by the users. Elements which were not critical could be part of a combination workflow step or handled outside the workflow routing. In the beginning we made the workflow too complicated (too many steps) and had to go back later and make some simplifications.
- Checklist detail – we wanted checklists of sufficient detail that the originator could fill them out without in-depth knowledge of process safety and HSE regulations. But we did not want the checklists to be so long that they would be burdensome and time consuming. We wanted them to be "tickler" lists to prompt the originator when to get other people involved such as professional safety personnel. We actually ended up with some of the checklist questions automatically initiating workflow steps; however most of them are intended to be information for the originator and approvers. A Safety checklist and an Environmental checklist are built into the MOC form.
- Simple versus complex changes. Making the system flexible enough to handle the many different types of changes was challenging. We made the workflow "smart" by building in logic blocks and alternate paths. This is all behind the scenes and is not seen by the user. They see only the MOC form and options are included on the form. The system reads information from the MOC form and determines the exact routing.
- Simplicity versus functionality. We wanted a system that was as "user friendly" as possible. Our ideal was to have a "fill in the blank" form. We wanted people to be able to fill in the form and then press one button to initiate a change. On the other hand we wanted functionality to be available such as editing and adding attachments to the MOC package. We found (at least with the software we had available) that a tool friendly enough for the occasional user may not have sufficient functionality. In the end we decided to focus on the frequent users and advise the occasional users to request help from their area MOC Coordinator or other expert when they need to initiate a change.

The major parts of the eMOC process tool are:

1. Web interface. See Figure 3. Our objective was to put links to everything the user would need on one screen.
2. Forms – this includes the MOC form with built in safety and environmental checklists. See Figure 4. There are four types of changes that can be selected on the MOC form – facility, control system software, procedures and information systems. The MOC type selected will affect the workflow routing. More than one type can be selected if applicable. In addition to the

MOC form itself, there is also some other optional forms and checklists that can be attached if desired.

3. The workflows. These use information from the MOC form to automatically route the MOC package. Logic blocks in the workflows make the necessary routing decisions. Each workflow step has a performer (responsible person) who is assigned either from the MOC form or the predetermined cost center roles. The performer automatically gets an e-mail note with a link that takes him to the correct location. When he clicks this link he sees specific instructions for that workflow step, some links and a disposition button(s). This is shown in Figure 4. Some steps are mandatory, some are optional and some are triggered by specific questions on the safety and environmental checklists. There are specific workflow steps for procedure updates, engineering design, concept safety review and PreStartup safety review. It is at these steps that Safety Instrumented Systems (SIS), layers of protection, alarms and operating procedures are reviewed. The workflows also contain some administrative interlocks. There are 2 different types:
    a. The routing itself prevents certain workflow steps from being sent until others are completed. In particular, the Startup step is not sent until all pre-startup requirements are met – approvals, safety reviews, training and notification.
    b. Some workflow steps have mandatory questions that must be answered before the workflow will proceed. Generally these are verifications that certain critical tasks have been performed.

   A block flow diagram showing more details about the workflow is attached as Appendix I.
4. Cost Center roles – there is a lookup table with names for frequently used roles listed by cost center. Therefore the originator does not have to enter these names for every change. He just enters the cost center and the system automatically knows the names. See Figure 6 for a list of roles. There are 7 mandatory roles, meaning that they are mandatory for certain types of changes (not necessarily for every type). There are also 12 optional roles, meaning that they are never mandatory but are available to be used as appropriate and desired.
5. Reports – there are several automatically generated reports that aid in managing and tracking MOC packages. Most of these utilize the built in Live Report functionality in Livelink. See Figure 7 for examples. For a few reports we had to export the data from Oracle to some other reporting software.
6. The interactive training area or "sandbox". In this area people can generate training MOC packages and learn about how the system operates.
7. The User Manual. This is a reference document. Help is available in the User Manual on almost any topic related to eMOC.

Figure 3 – eMOC Home Page

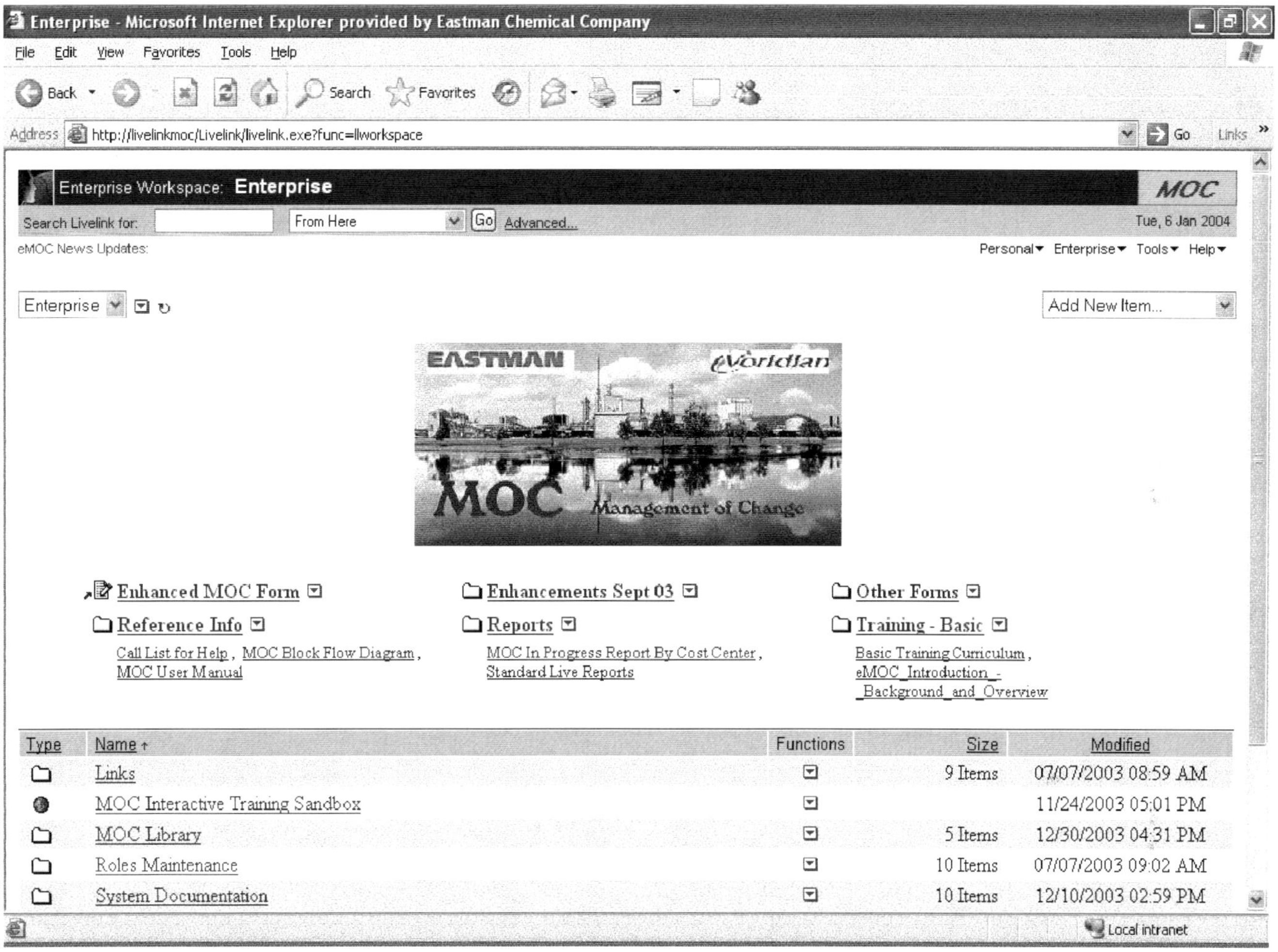

Figure 4 – eMOC Form Selected Parts

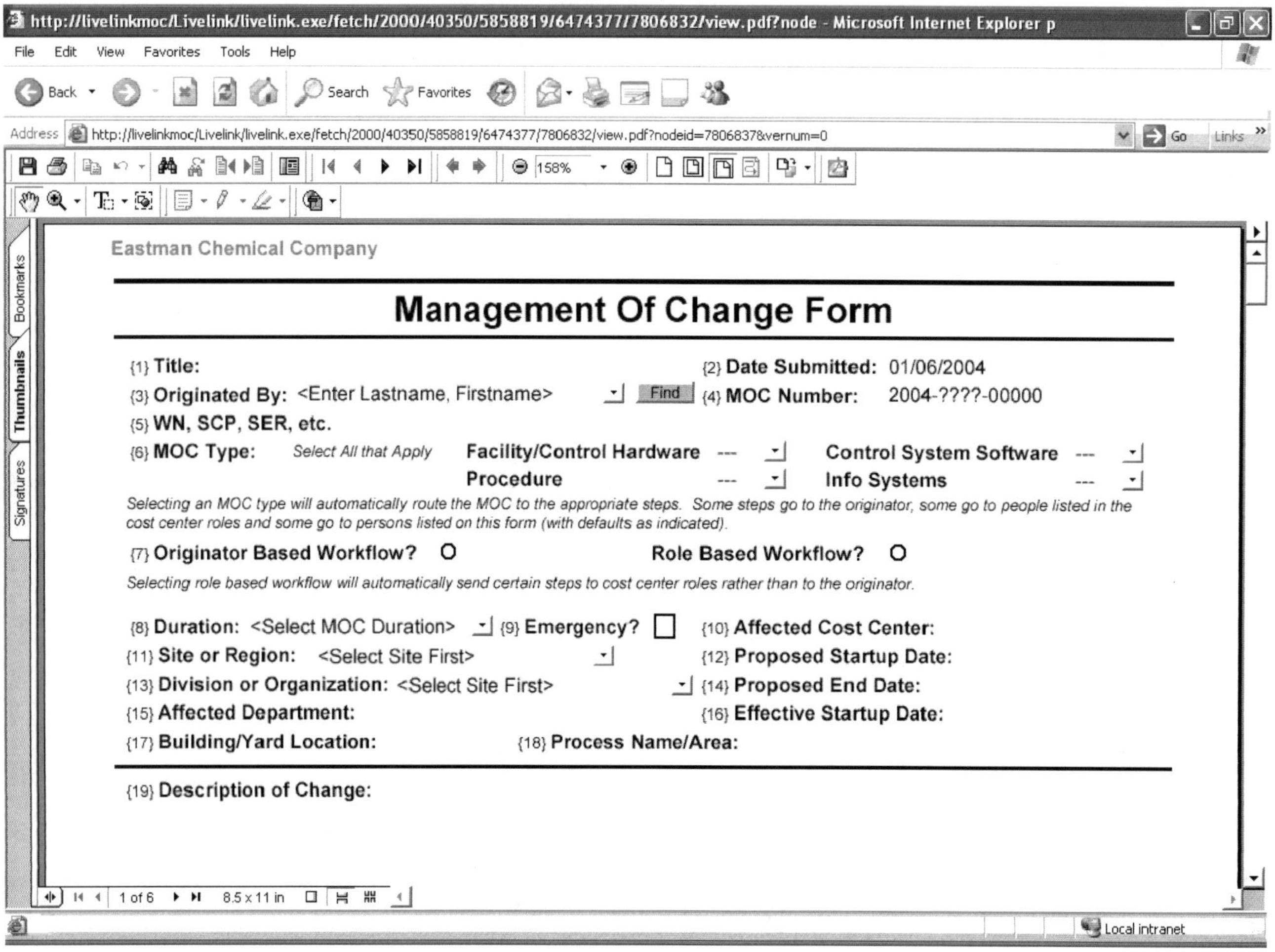

http://livelinkmoc/Livelink/livelink.exe/fetch/2000/40350/5858819/6474377/7806832/view.pdf?node - Microsoft Internet Explorer p

Eastman Chemical Company

Management Of Change Form

{1} **Title:** {2} **Date Submitted:** 01/06/2004

{3} **Originated By:** <Enter Lastname, Firstname> Find {4} **MOC Number:** 2004-????-00000

{5} **WN, SCP, SER, etc.**

{6} **MOC Type:** *Select All that Apply* **Facility/Control Hardware** --- **Control System Software** --- **Procedure** --- **Info Systems** ---

Selecting an MOC type will automatically route the MOC to the appropriate steps. Some steps go to the originator, some go to people listed in the cost center roles and some go to persons listed on this form (with defaults as indicated).

{7} **Originator Based Workflow?** O **Role Based Workflow?** O

Selecting role based workflow will automatically send certain steps to cost center roles rather than to the originator.

{8} **Duration:** <Select MOC Duration> {9} **Emergency?** ☐ {10} **Affected Cost Center:**

{11} **Site or Region:** <Select Site First> {12} **Proposed Startup Date:**

{13} **Division or Organization:** <Select Site First> {14} **Proposed End Date:**

{15} **Affected Department:** {16} **Effective Startup Date:**

{17} **Building/Yard Location:** {18} **Process Name/Area:**

{19} **Description of Change:**

Figure 4 – eMOC Form Selected Parts (continued)

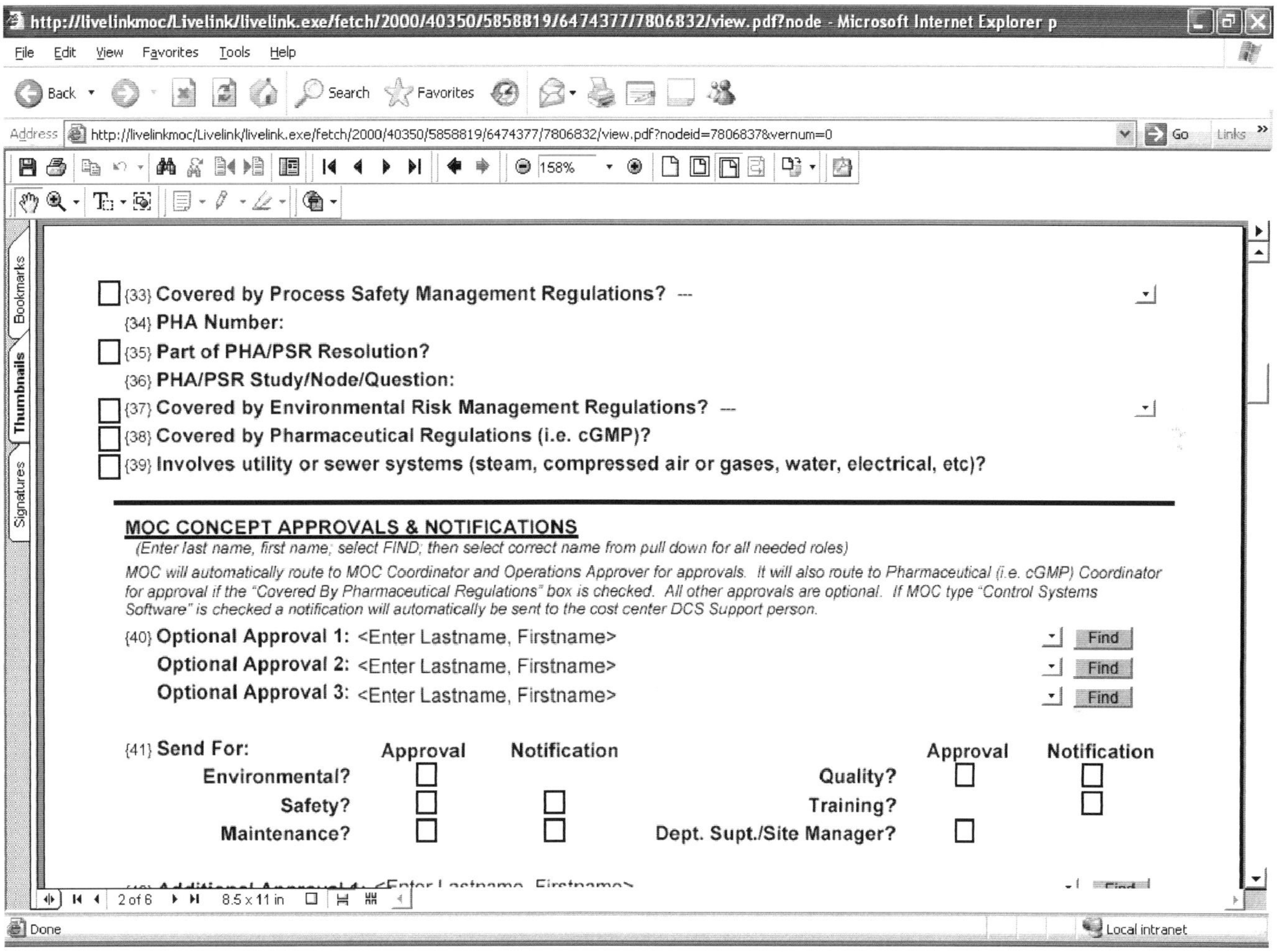

http://livelinkmoc/Livelink/livelink.exe/fetch/2000/40350/5858819/6474377/7806832/view.pdf?node - Microsoft Internet Explorer p

{33} Covered by Process Safety Management Regulations? ---

{34} PHA Number:

{35} Part of PHA/PSR Resolution?

{36} PHA/PSR Study/Node/Question:

{37} Covered by Environmental Risk Management Regulations? ---

{38} Covered by Pharmaceutical Regulations (i.e. cGMP)?

{39} Involves utility or sewer systems (steam, compressed air or gases, water, electrical, etc)?

MOC CONCEPT APPROVALS & NOTIFICATIONS

(Enter last name, first name; select FIND; then select correct name from pull down for all needed roles)

MOC will automatically route to MOC Coordinator and Operations Approver for approvals. It will also route to Pharmaceutical (i.e. cGMP) Coordinator for approval if the "Covered By Pharmaceutical Regulations" box is checked. All other approvals are optional. If MOC type "Control Systems Software" is checked a notification will automatically be sent to the cost center DCS Support person.

{40} Optional Approval 1: <Enter Lastname, Firstname> Find

Optional Approval 2: <Enter Lastname, Firstname> Find

Optional Approval 3: <Enter Lastname, Firstname> Find

{41} Send For:

	Approval	Notification		Approval	Notification
Environmental?	☐		Quality?	☐	☐
Safety?	☐	☐	Training?		☐
Maintenance?	☐	☐	Dept. Supt./Site Manager?	☐	

Figure 4 – eMOC Form Selected Parts (continued)

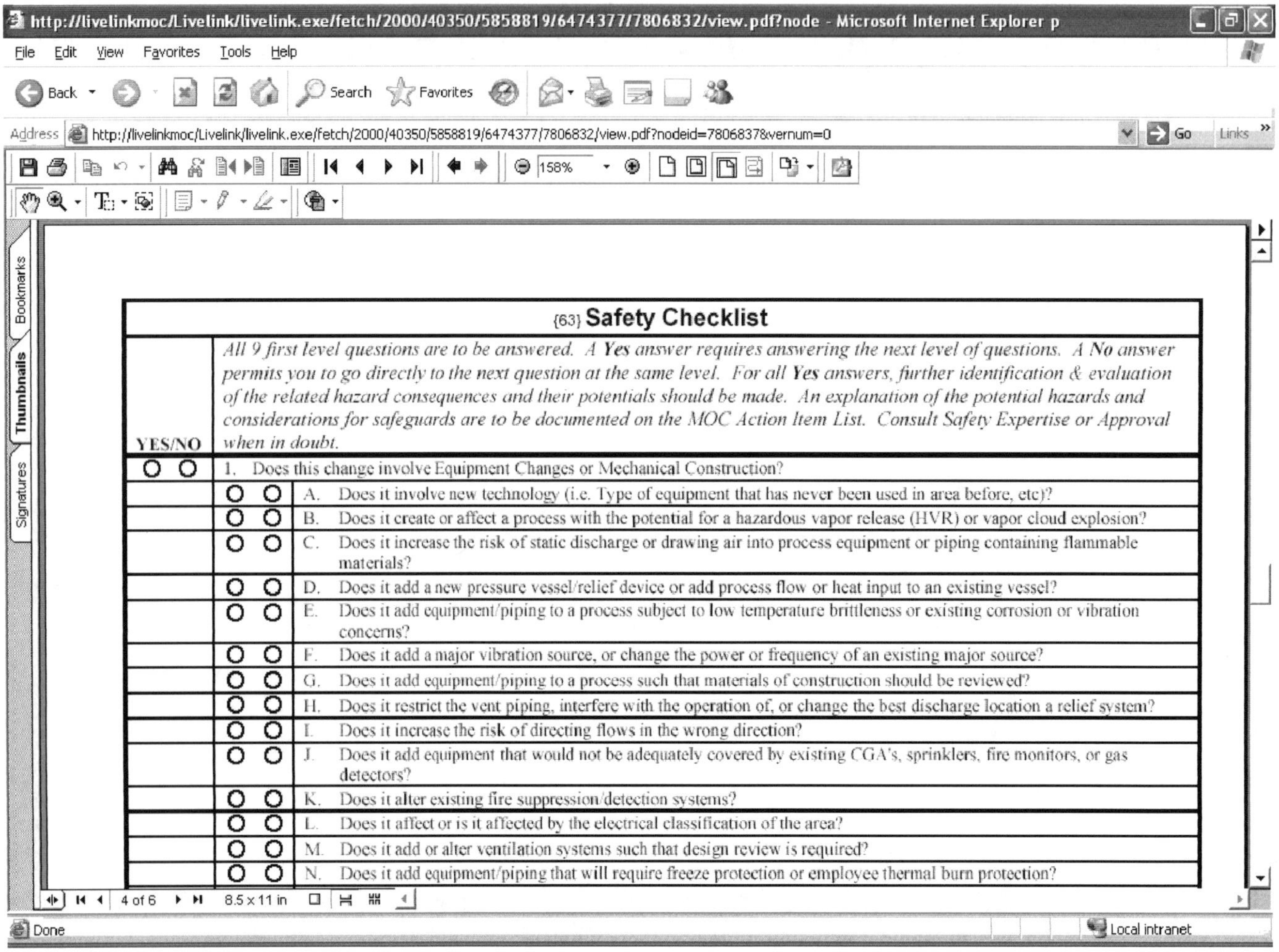

{63} **Safety Checklist**

YES/NO		*All 9 first level questions are to be answered. A **Yes** answer requires answering the next level of questions. A **No** answer permits you to go directly to the next question at the same level. For all **Yes** answers, further identification & evaluation of the related hazard consequences and their potentials should be made. An explanation of the potential hazards and considerations for safeguards are to be documented on the MOC Action Item List. Consult Safety Expertise or Approval when in doubt.*
O O		1. Does this change involve Equipment Changes or Mechanical Construction?
	O O	A. Does it involve new technology (i.e. Type of equipment that has never been used in area before, etc)?
	O O	B. Does it create or affect a process with the potential for a hazardous vapor release (HVR) or vapor cloud explosion?
	O O	C. Does it increase the risk of static discharge or drawing air into process equipment or piping containing flammable materials?
	O O	D. Does it add a new pressure vessel/relief device or add process flow or heat input to an existing vessel?
	O O	E. Does it add equipment/piping to a process subject to low temperature brittleness or existing corrosion or vibration concerns?
	O O	F. Does it add a major vibration source, or change the power or frequency of an existing major source?
	O O	G. Does it add equipment/piping to a process such that materials of construction should be reviewed?
	O O	H. Does it restrict the vent piping, interfere with the operation of, or change the best discharge location a relief system?
	O O	I. Does it increase the risk of directing flows in the wrong direction?
	O O	J. Does it add equipment that would not be adequately covered by existing CGA's, sprinklers, fire monitors, or gas detectors?
	O O	K. Does it alter existing fire suppression/detection systems?
	O O	L. Does it affect or is it affected by the electrical classification of the area?
	O O	M. Does it add or alter ventilation systems such that design review is required?
	O O	N. Does it add equipment/piping that will require freeze protection or employee thermal burn protection?

Figure 5 – Workflow Step

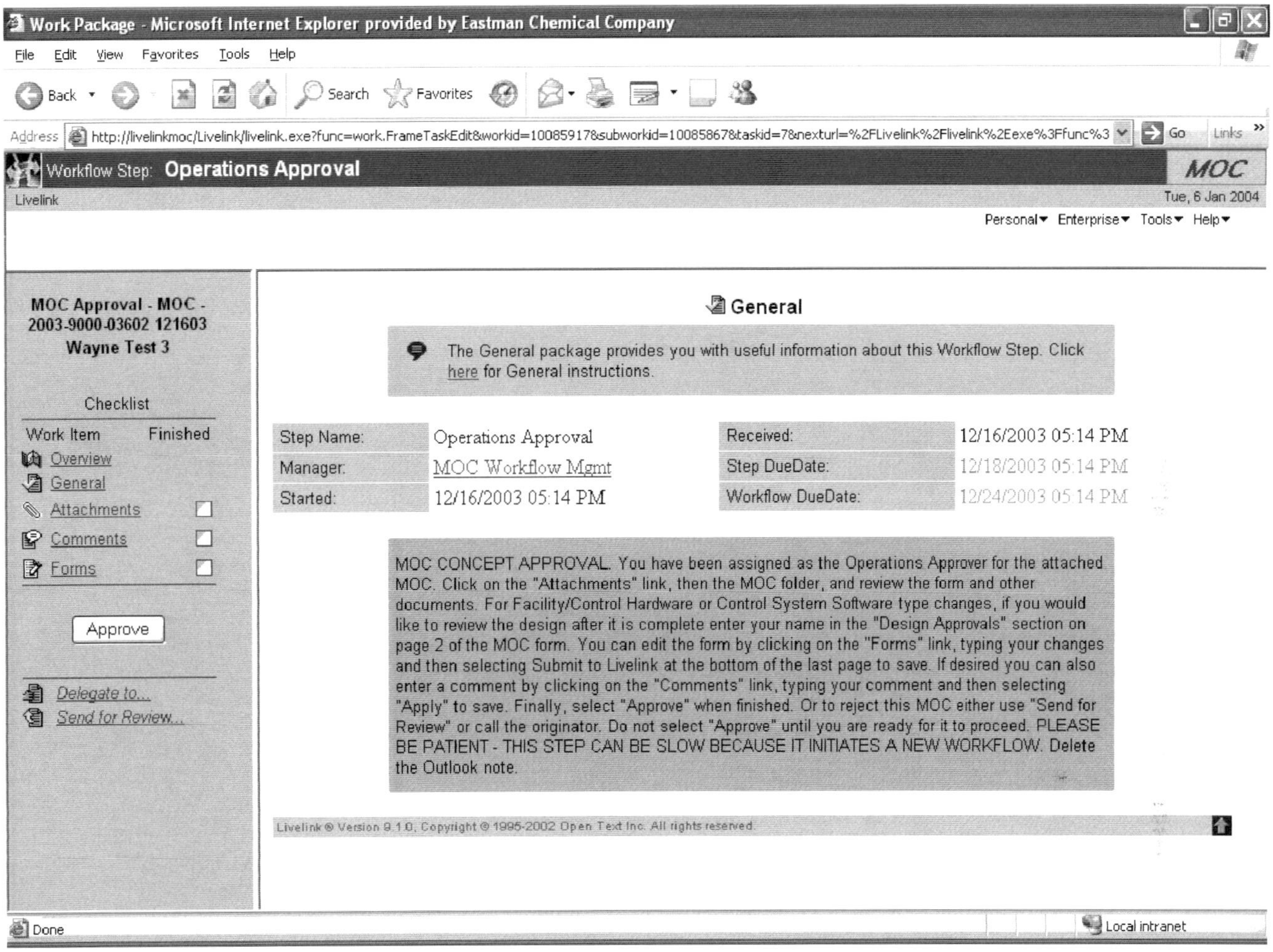

Figure 6 – Roles

Mandatory means that this role is mandatory for certain types of changes. There are 4 types – facility, control system software, procedure and information systems. Optional means that it is optional for all changes.

Roles

DCS Support	mandatory
Department Head	optional
Dept Environmental Coordinator	optional
Electrical MI Coordinator	optional
Engineering Contact	optional
Hazard Communication (HAZCOM) Coordinator	mandatory
Industrial Chemical Control Law (ICCL) Coordinator	mandatory
Information System Coordinator	optional
MI Coordinator	optional
MOC Coordinator	mandatory
Maintenance Approval	optional
Operations Approval	mandatory
Operations PSIC	optional
Procedures Editor	optional
Quality Approval	optional
Spare Parts Updater	mandatory
Safety Approval	optional
Training Approval	optional
cGMP Coordinator	mandatory
TOTALS	mandatory = 7 optional = 12

Figure 7 – Example Reports

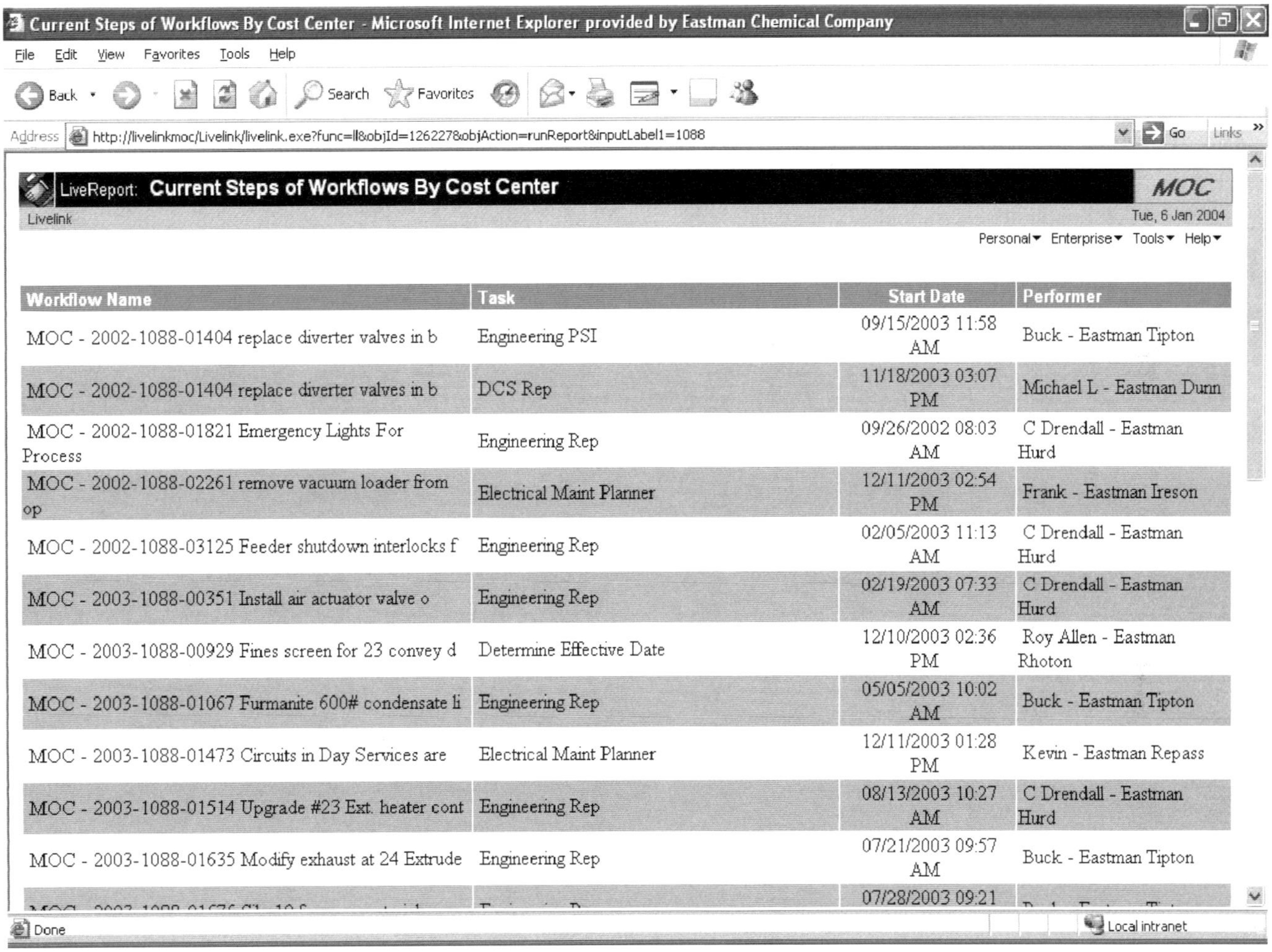

Workflow Name	Task	Start Date	Performer
MOC - 2002-1088-01404 replace diverter valves in b	Engineering PSI	09/15/2003 11:58 AM	Buck - Eastman Tipton
MOC - 2002-1088-01404 replace diverter valves in b	DCS Rep	11/18/2003 03:07 PM	Michael L - Eastman Dunn
MOC - 2002-1088-01821 Emergency Lights For Process	Engineering Rep	09/26/2002 08:03 AM	C Drendall - Eastman Hurd
MOC - 2002-1088-02261 remove vacuum loader from op	Electrical Maint Planner	12/11/2003 02:54 PM	Frank - Eastman Ireson
MOC - 2002-1088-03125 Feeder shutdown interlocks f	Engineering Rep	02/05/2003 11:13 AM	C Drendall - Eastman Hurd
MOC - 2003-1088-00351 Install air actuator valve o	Engineering Rep	02/19/2003 07:33 AM	C Drendall - Eastman Hurd
MOC - 2003-1088-00929 Fines screen for 23 convey d	Determine Effective Date	12/10/2003 02:36 PM	Roy Allen - Eastman Rhoton
MOC - 2003-1088-01067 Furmanite 600# condensate li	Engineering Rep	05/05/2003 10:02 AM	Buck - Eastman Tipton
MOC - 2003-1088-01473 Circuits in Day Services are	Electrical Maint Planner	12/11/2003 01:28 PM	Kevin - Eastman Repass
MOC - 2003-1088-01514 Upgrade #23 Ext. heater cont	Engineering Rep	08/13/2003 10:27 AM	C Drendall - Eastman Hurd
MOC - 2003-1088-01635 Modify exhaust at 24 Extrude	Engineering Rep	07/21/2003 09:57 AM	Buck - Eastman Tipton
[illegible]	[illegible]	07/28/2003 09:21	[illegible]

Figure 7 – Example Reports (continued)

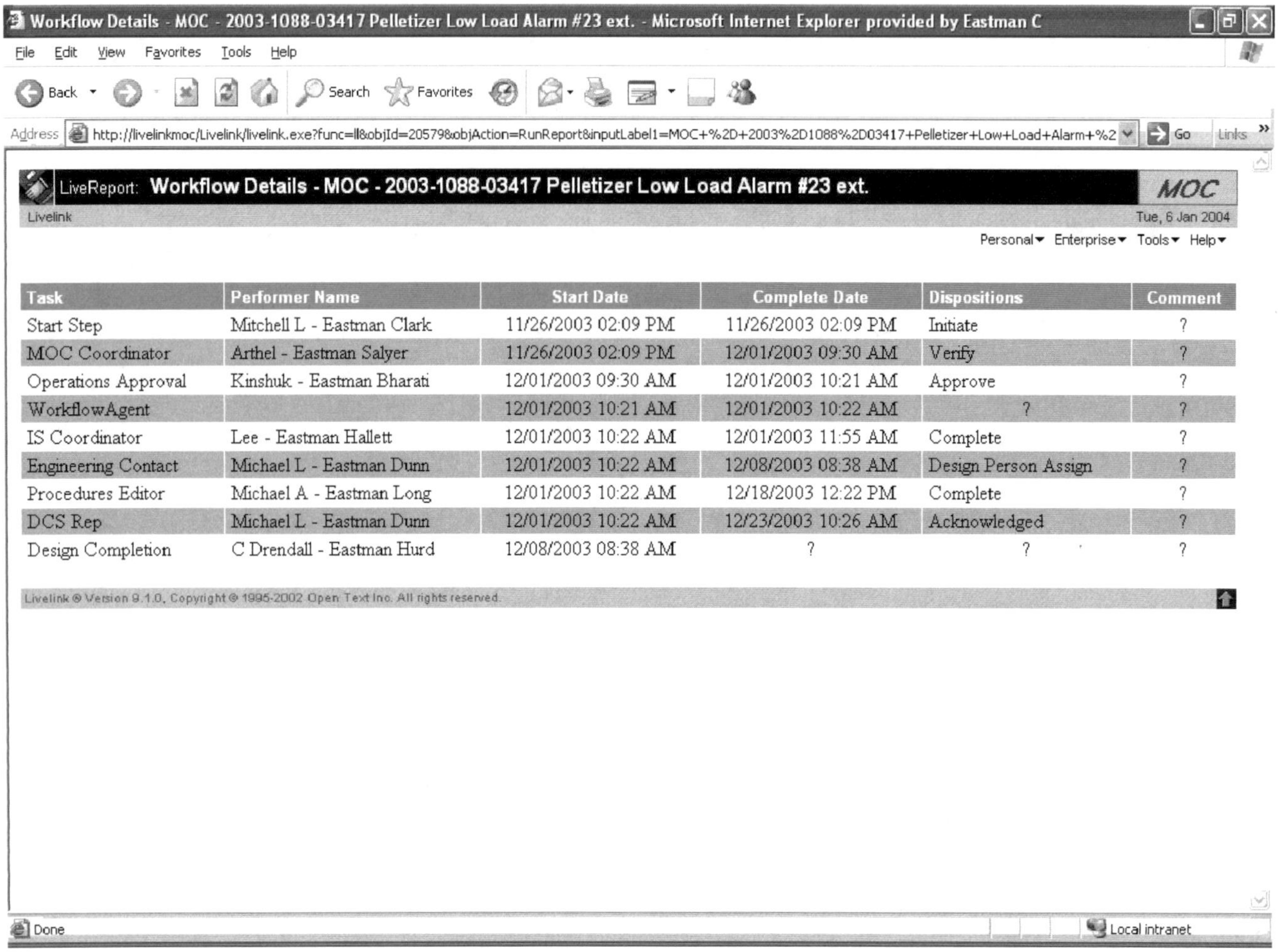

Task	Performer Name	Start Date	Complete Date	Dispositions	Comment
Start Step	Mitchell L - Eastman Clark	11/26/2003 02:09 PM	11/26/2003 02:09 PM	Initiate	?
MOC Coordinator	Arthel - Eastman Salyer	11/26/2003 02:09 PM	12/01/2003 09:30 AM	Verify	?
Operations Approval	Kinshuk - Eastman Bharati	12/01/2003 09:30 AM	12/01/2003 10:21 AM	Approve	?
WorkflowAgent		12/01/2003 10:21 AM	12/01/2003 10:22 AM	?	?
IS Coordinator	Lee - Eastman Hallett	12/01/2003 10:22 AM	12/01/2003 11:55 AM	Complete	?
Engineering Contact	Michael L - Eastman Dunn	12/01/2003 10:22 AM	12/08/2003 08:38 AM	Design Person Assign	?
Procedures Editor	Michael A - Eastman Long	12/01/2003 10:22 AM	12/18/2003 12:22 PM	Complete	?
DCS Rep	Michael L - Eastman Dunn	12/01/2003 10:22 AM	12/23/2003 10:26 AM	Acknowledged	?
Design Completion	C Drendall - Eastman Hurd	12/08/2003 08:38 AM	?	?	?

Results

eMOC adds the following improvements over our manual (paper-based) management of change processes:

- Automatic routing
- Automatic reminders for each task
- Detailed instructions for each task
- Administrative interlocks
- Automatic tracking
- Automatic flagging
- Automatic reports
- Electronic files with automatic backup
- Additional training
- Improved checklists

Another advantage of eMOC is that it is easily transportable to other plant sites or subsidiaries. It is basically "drop in" with the only setup requirement being training.

The eMOC production system has been running for over 18 months since July 2002. During this time period 3768 MOCs have been initiated. Of these 1377 have been completed, 2349 are still active and 42 have been cancelled. We recently conducted an evaluation of results. The users themselves evaluated the system based on 2 major criteria: (1) safety and regulatory compliance (2) functionality & time required.

Using a standard form, each of the 10 major sites/divisions evaluated eMOC compared to their old paper process. Average results show:
(1) Both safety and documentation of regulatory compliance have been improved
(2) Functionality of eMOC is satisfactory
(3) Overall time required is slightly more but is justified by the benefits obtained.

The major positives of eMOC listed by users were:
(1) Tracking
(2) Documentation and compliance
(3) Automatic routing
(4) Automatic reminders.

The major negatives listed were:
(1) Technology issues, primarily software
(2) Complexity of the system and amount of training required.

Eastman Chemical Company has applied for and received a provisional patent for eMOC.

Conclusions

1) eMOC has achieved its major objective of improving safety and helping to assure regulatory compliance.
2) eMOC is a tool, not a complete management system. Auditing of the process is still needed to close the loop, although a standardized electronic system can make auditing simpler and faster.
3) There are several keys to success for this type of project, which include:

    a) A cross functional team for design including representatives from all key areas.
    b) Management support including line management and a steering team at the appropriate level of senior management.
    c) Objectives and performance measures agreed upon before the design process starts.
    d) Appropriate decisions on the design tradeoffs listed in the "Design of eMOC" section above.
    e) Piloting. Data from the pilot should be used to make any appropriate. modifications before the tool is implemented across the board in manufacturing processes.
    f) Training and startup support. There should be a core group of knowledgeable trainers. In addition, an Implementation Coordinator should be named for each major area (site/division). Their duties include involving all appropriate area personnel and scheduling training.
    g) Up front investigation of technical considerations for both software and hardware to identify the most appropriate platforms and any potential watch outs.
    h) Assignment of sufficient IT resources.
    i) User friendliness and a minimum workload on users.

Appendix I – Workflow Block Diagram

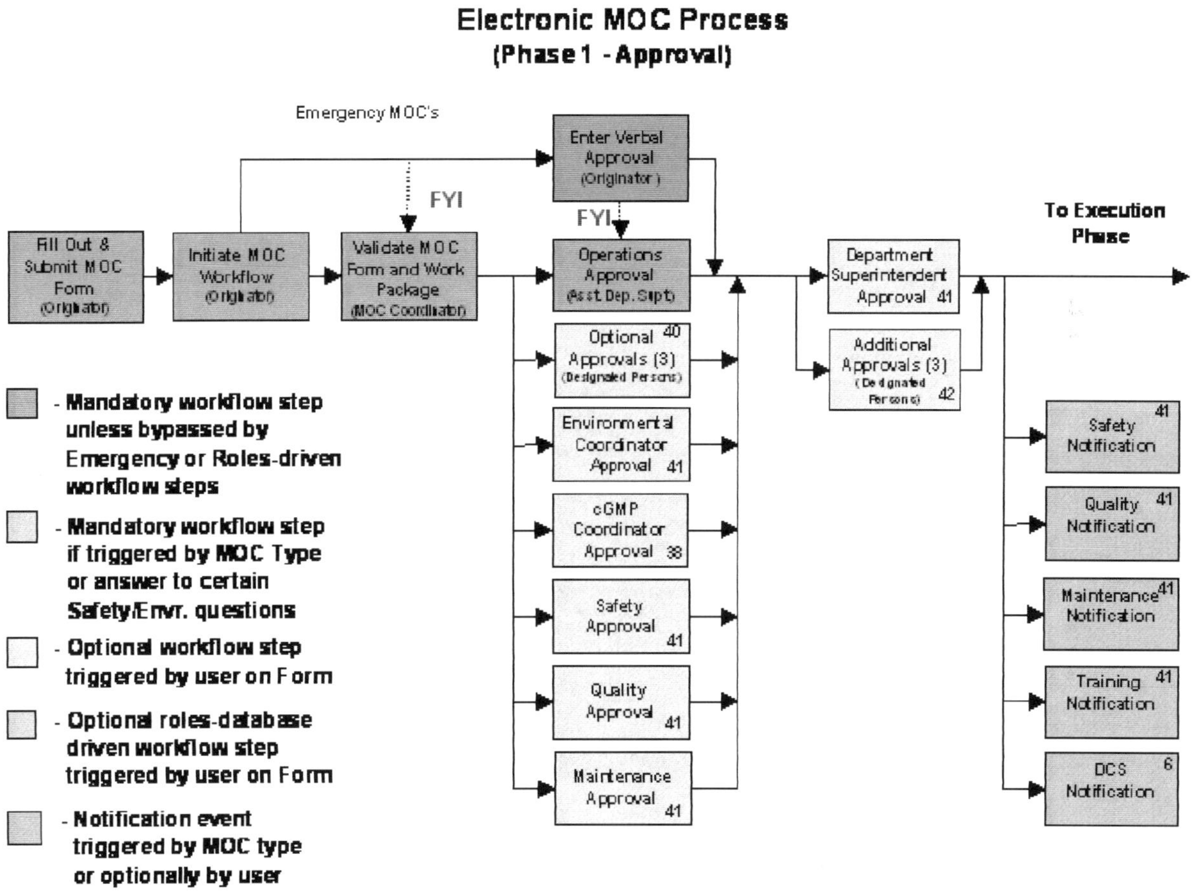

Electronic MOC Process
(Phase 2 - Execution)

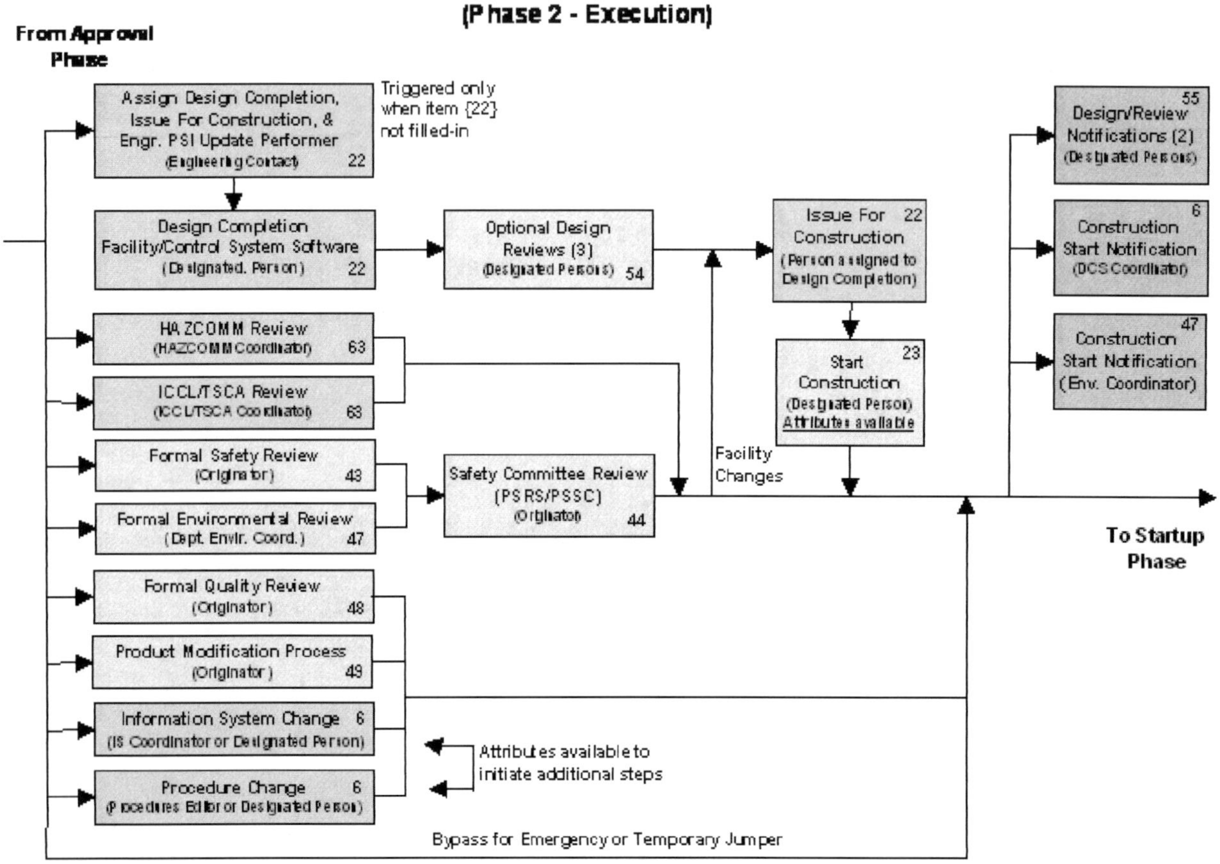

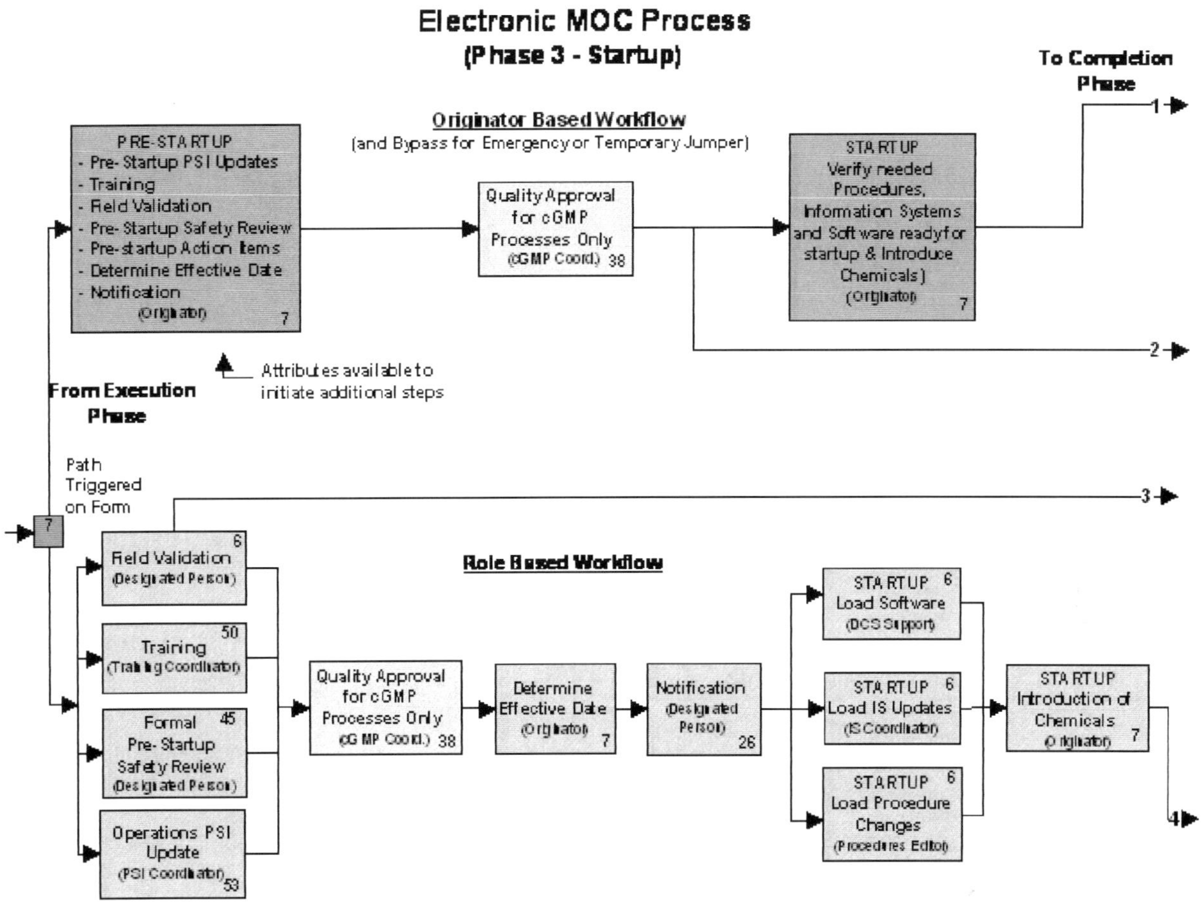
Electronic MOC Process
(Phase 3 - Startup)
To Completion Phase
Originator Based Workflow
(and Bypass for Emergency or Temporary Jumper)
PRE-STARTUP
- Pre-Startup PSI Updates
- Training
- Field Validation
- Pre-Startup Safety Review
- Pre-startup Action Items
- Determine Effective Date
- Notification
(Originator) 7
Quality Approval for cGMP Processes Only (cGMP Coord.) 38
STARTUP Verify needed Procedures, Information Systems and Software ready for startup & Introduce Chemicals) (Originator) 7
1
2
Attributes available to initiate additional steps
From Execution Phase
Path Triggered on Form
7
3
Role Based Workflow
Field Validation (Designated Person) 6
Training (Training Coordinator) 50
Formal Pre-Startup Safety Review (Designated Person) 45
Operations PSI Update (PSI Coordinator) 53
Quality Approval for cGMP Processes Only (cGMP Coord.) 38
Determine Effective Date (Originator) 7
Notification (Designated Person) 26
STARTUP Load Software (DCS Support) 6
STARTUP Load IS Updates (IS Coordinator) 6
STARTUP Load Procedure Changes (Procedures Editor) 6
STARTUP Introduction of Chemicals (Originator) 7
4

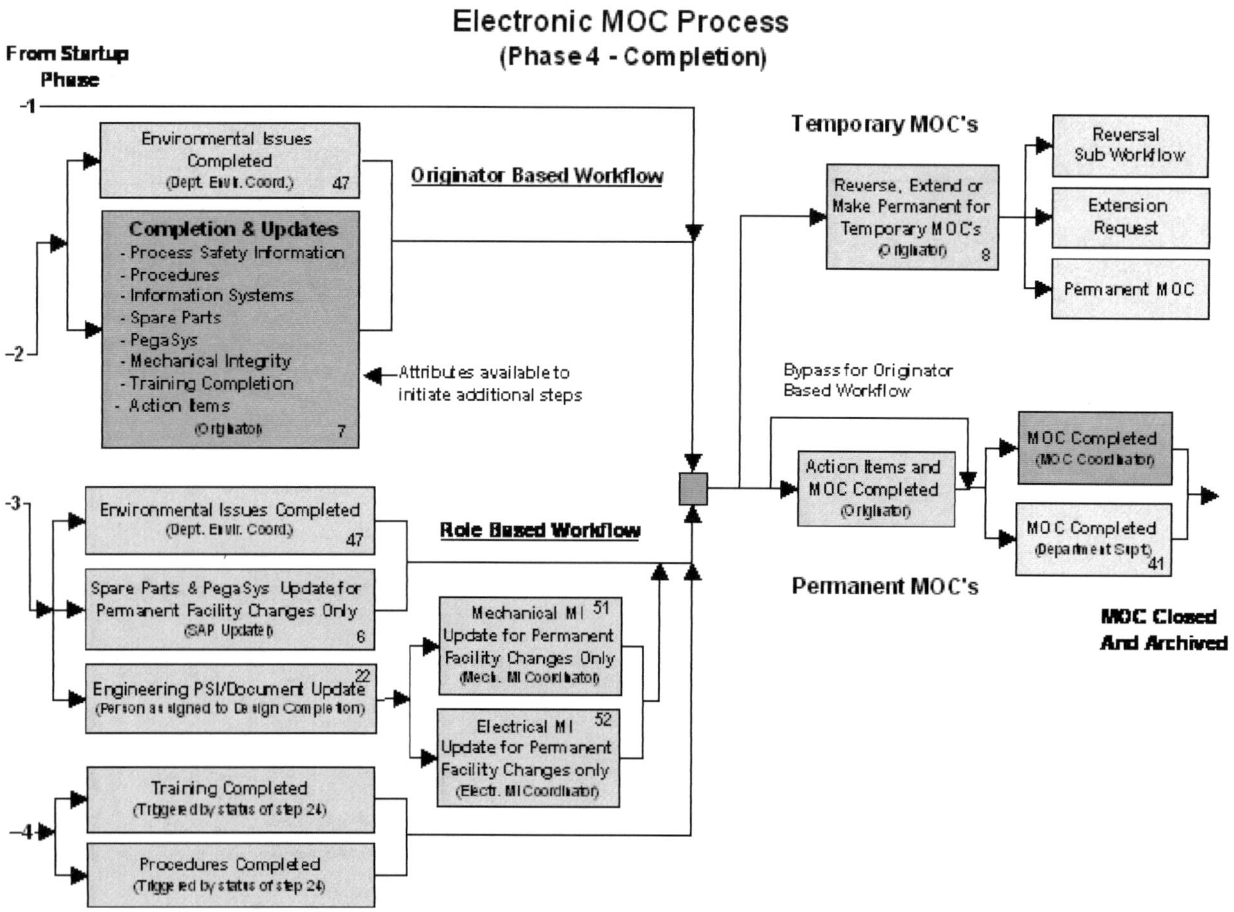

Electronic MOC Process
(Phase 4 - Completion)
From Startup Phase
-1
-2
-3
-4
Environmental Issues Completed (Dept. Envir. Coord.) 47
Completion & Updates
- Process Safety Information
- Procedures
- Information Systems
- Spare Parts
- PegaSys
- Mechanical Integrity
- Training Completion
- Action Items
(Originator) 7
Originator Based Workflow
Attributes available to initiate additional steps
Environmental Issues Completed (Dept. Envir. Coord.) 47
Spare Parts & PegaSys Update for Permanent Facility Changes Only (SAP Updater) 6
Engineering PSI/Document Update 22 (Person assigned to Design Completion)
Role Based Workflow
Mechanical MI 51 Update for Permanent Facility Changes Only (Mech. MI Coordinator)
Electrical MI 52 Update for Permanent Facility Changes only (Electr. MI Coordinator)
Training Completed (Triggered by status of step 24)
Procedures Completed (Triggered by status of step 24)
Temporary MOC's
Reverse, Extend or Make Permanent for Temporary MOC's (Originator) 8
Reversal Sub Workflow
Extension Request
Permanent MOC
Bypass for Originator Based Workflow
Action Items and MOC Completed (Originator)
MOC Completed (MOC Coordinator)
MOC Completed (Department Supt) 41
Permanent MOC's
MOC Closed And Archived

ABSTRACT

Lessons for Management:

Management System Failures Identified in Incidents Investigated by the US Chemical Safety and Hazard Investigation Board

By

Angela S. Blair, P.E.
Chemical Incident Investigator
US Chemical Safety & Hazard Investigation Board

A key mission of the US Chemical Safety and Hazard Investigation Board (CSB) is to determine the root causes of incidents, report the findings and issue recommendations to prevent similar incidents from occurring. CSB investigators respond to a variety of events occurring in a wide range of workplaces, from chemical plants and refineries to food-flavoring factories and steel mills. The facilities are all sizes, from single owner-operated concerns with a few employees to major manufacturing plants for multi-national corporations.

The details of each investigation are unique and the root causes are pertinent to each specific case. However, a common thread that emerges in CSB investigations is the inadequacy of management systems that might have prevented the incident from occurring. Examples of the systemic issues identified in CSB reports are:

- Lack of hazard review to predict and prevent incidents
- Insufficient investigation and follow-up after previous incidents
- Inadequate training of staff
- Failure to implement effective mechanical integrity programs

These issues are well recognized as elements of a process safety management (PSM) program, although many incidents investigated by the CSB occurred at facilities that are not regulated by OSHA's process safety management rule[1]. Indeed, a number of these incidents occurred at facilities that are well outside society's definition of "chemical plants."

This paper contains a summary and discussion of the management system failures identified as causally related in CSB investigation reports. The analysis and conclusions are those of this investigator, and do not necessarily reflect the opinions of the Chemical Safety & Hazard Investigation Board.

---

[1] 29 CFR 1910.119 *Process Safety Management of Highly Hazardous Chemicals*

Lessons for Management:

Management System Failures Identified in Incidents Investigated by the US Chemical Safety and Hazard Investigation Board

By

Angela S. Blair, P.E.
Chemical Incident Investigator
US Chemical Safety & Hazard Investigation Board

Introduction

This paper contains a review of the investigational products of the U.S. Chemical Safety and Hazard Investigation Board (CSB). Since becoming operational in 1998, the CSB has completed 19 case studies and incident investigations, as of this writing. Although the incidents occurred in industries over a diverse range that includes farming, foundries, refineries and chemical plants, at least one trend has emerged from these investigations. In *every CSB investigation, management system failures were identified as causally related to the incident.*

Please note that the opinions and conclusions expressed in this paper are those of the author, and do not necessarily represent the opinions of the Chemical Safety and Hazard Investigation Board.

In recent presentations to industry and professional groups, CSB Chairman Carolyn Merritt lists five success factors to preventing unintended events:

- Hazard Recognition
- Engineering and Design
- Management Systems
- Human Factors
- Culture

One could successfully argue that hazard recognition, engineering, design, human factors and culture are, or are driven by, management systems. Therefore, management systems are the keys to prevention. Management system failures lead to unintended events...typically called accidents or incidents. The purpose of this paper is to highlight the most prevalent types of management system failures implicated in CSB investigations. The investigation reports further the CSB mission of incident prevention by promoting improvement in these critical management systems.

Definitions

The term "management systems" has taken various meanings over the past decades. The International Standards Organization defines a management system as

> "the organization's structure for managing its processes - or activities - that transform inputs of resources into a product or service which meet the organization's objectives, such as satisfying the customer's quality requirements, complying to regulations, or meeting environmental objectives."[2]

The Center for Chemical Process Safety states that

> "Management systems consist of explicit sets of arrangements for planning, organizing, implementing, and controlling work within complex organizations. They are established by management to assist in achieving their organizations' goals and objectives in an efficient and effective manner."[3]

These two definitions, while somewhat wordy and potentially confusing, do not contradict each other. To keep things simple, for the purposes of this paper we'll define management systems as *the collection of policies, procedures and work instructions through which management accomplishes its goals.* This paper deals with investigations of serious industrial incidents, therefore we're interested primarily in those management systems that serve to promote safety and reduce incidents.

Among safety professionals it is tempting to equate safety management systems with *process safety management (PSM.)* Certainly the OSHA and EPA regulatory requirements for chemical process safety provide a framework for

---

[2] ISO 9000/14000 website, http://www.iso.ch/iso/en/iso9000-14000/index.html

[3] *Technical Guidelines for Technical Management of Chemical Process Safety,* CCPS, New York, NY 1989, p. 12.

safety management systems. Similarly, the Center for Chemical Process Safety (CCPS) promotes process safety through integrated management system elements. It is nevertheless critical to keep in mind the large number of companies and facilities not covered by the existing PSM regulations. Within these organizations, management systems may not adhere to the established process safety framework, yet they are no less critical.

The use of the term "management systems" in this paper also implies those activities and conditions that are within the control of the management of the facilities and companies involved in the incidents. By this definition one would not regard community emergency planning as a management system entirely within the control of the company or facility.

The term "root cause" has also been a subject of lively debate within the investigating communities. For the purposes of this paper, the definition of root cause is generalized according to the definition given by CCPS:

> [Root causes are] prime reasons, such as failures of some management systems, that allow faulty design, inadequate training, or improper changes, which lead to an unsafe act or condition and result in an incident. Root causes are also known as underlying causes. If root causes were removed, the particular incident would not have occurred.[4]

Contributing causes, as the term is used in CSB investigation reports, are key causal factors that contributed to the severity of the incident consequences or increased the likelihood of the incident.

The smaller products of CSB investigations, case studies and safety bulletins, do not refer to root and contributing causes, per se. However

---

[4] *Guidelines for Investigating Chemical Process Incidents,* CCPS, New York, NY 1992, p. 8.

factors that contributed to the incident, or systems that may have prevented the incidents are identified. These causal factors are also counted as root causes for the purposes of this paper.

Analysis

As stated in the introductory paragraph, CSB has completed 19 investigations[5] to date. Of the 85 root and contributing causes identified in the investigation reports, 73 can be classified as relating to management systems. Put another way, *improvements in management systems could have prevented or lessened the severity of nearly three-fourths of these incidents.*

A spreadsheet summary of the root and contributing causes identified in these 19 CSB reports is included in the Appendix.

In fairness, it is important to admit that adherence to the CSB prevention mission skews the investigation process toward a search for management system causes for incidents. In order to gain the maximum preventative effect, CSB strives to identify and address the underlying management deficiencies and systemic solutions pertinent to each incident. CSB investigative teams are *required* to pursue the causal path well beyond the immediate physical cause of the event.

The 73 management systems causes identified in CSB investigations can be further classified into specific areas of management involvement. The categories used for this paper are similar to the familiar PSM elements, however a few of the categories are somewhat broader. The breakdown of causes by category is given in Figure 1.

---

[5] Updated data are available by contacting the author at angela.blair@csb.gov

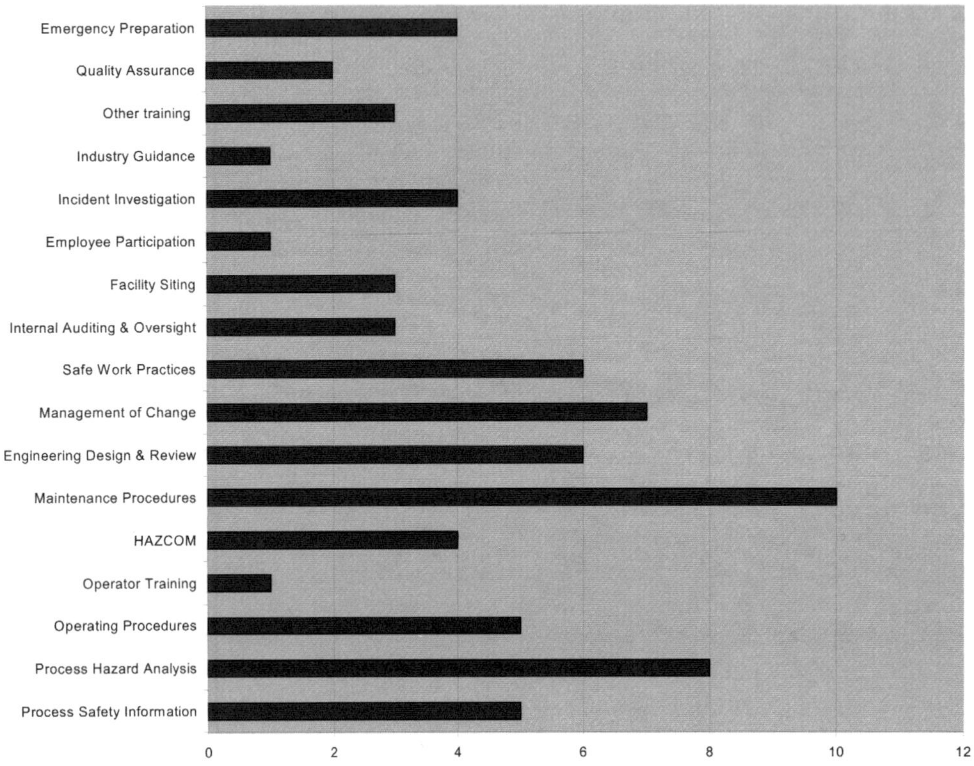

Figure 1. Breakdown of Management System Categories Identified as Root or Contributing Causes in Incidents Investigated by the CSB

Starting with a simple statistical analysis, the following categories of management systems were most often identified by CSB as root and contributing causes of incidents:

- Maintenance Procedures (10*)
- Process Hazard Analysis (8)
- Management of Change (7)
- Safe Work Practices (6)
- Engineering Design & Review (6)

*the number of times this management system appears as a root or contributing cause

These five categories account for half of the management system causes as root and contributing causes in CSB investigations. The prevalence of these five types of causes also suggests a general pattern, especially if the reader were to take a closer look at the reports in which these causes were listed (see the matrix in the appendix.)

Failure to predict hazards and implement preventive measures, insufficient planning and inadequate engineering involvement in the daily and non-routine plant operations resulted in injuries, property damage and, unfortunately, loss of life. The repeated appearance of process hazard analysis, management of change and engineering design and review as causes of incidents is significant. *The need for engineering involvement and technical support is incident prevention is clear.* The incidents at Concept Sciences and BP Amoco both illustrate the consequences of technical shortcomings.

The category labeled "maintenance procedures" encompasses traditional mechanical integrity elements including inspection and testing, preventive maintenance and maintenance instructions. Several incident investigations revealed deficiencies in managing maintenance activities.

Maintenance work often represents a departure from routine operations and must be approached with careful planning, performed by properly trained employees on adequately prepared equipment, using the appropriate tools. The CSB investigations into the incidents at Motiva Enterprises and Tosco's Avon Refinery both revealed failures in the maintenance management systems.

Hazard recognition (for example, process hazard analysis) or some type of predictive hazard review should occur at all phases of facility planning, design, construction, operation and dismantlement. Furthermore, a hazard recognized must become a hazard *addressed and communicated* to effectively prevent an incident. The incident at BP Amoco, like the tragedy of the Space Shuttle Columbia, was an illustration of the consequences when information about a hazard (in the BP Amoco case, thermal decomposition) is not in the hands of the right people. The gap must be closed between the research chemists and the hazard analysts with bi-directional communication. Effective hazard evaluation and incident prevention requires a team approach involving all aspects of project design, implementation and operation.

There is at least one other way to gain value from the types of causes identified in CSB investigations. The following analysis moves beyond mere statistical analysis, and attempts a sorting of the causes based on value judgment.

By grouping the management systems into even broader categories, then re-examining the CSB cause data, one can see even more clearly the powerful influence of management oversight and proper engineering on incident prevention. This is one conclusion drawn from the data. Certainly, the reader can apply different filters, and reach other conclusions. The vital necessity of effective management systems to prevent incidents remains obvious, regardless of the method one uses to sort the data.

The following table shows the basis for the groupings shown in Figure 2.

Management System Cause Types	Category	% of Management System Causes	Number of Appearances
Process Safety Information	Engineering	6.85	5
Process Hazard Analysis	Engineering	10.96	8
Operating Procedures	Procedures	6.85	5
Operator Training	Training	1.37	1
HAZCOM	Training	5.48	4
Maintenance Procedures	Procedures	13.70	10
Engineering Design & Review	Engineering	8.22	6
Management of Change	Oversight	9.59	7
Safe Work Practices	Procedures	8.22	6
Internal Auditing & Oversight	Oversight	4.11	3
Facility Siting	Engineering	4.11	3
Employee Participation	Oversight	1.37	1
Incident Investigation	Oversight	5.48	4
Industry Guidance	Oversight	1.37	1
Other training	Training	4.11	3
Quality Assurance	Oversight	2.74	2
Emergency Preparation	Oversight	5.48	4
Totals		**100.00**	**73**

Table 1. Management System Causes by Broad Category

The logic of these category groupings begs a bit of explanation. Even though management systems like process hazard analysis involve input from operations, maintenance and other groups, this is typically seen as an engineering or technical function. Likewise, engineering is typically responsible for developing and maintaining process safety information.

When this new grouping is expressed graphically, one can see that technical involvement (engineering) is equally valued in accident prevention with management oversight and procedures.

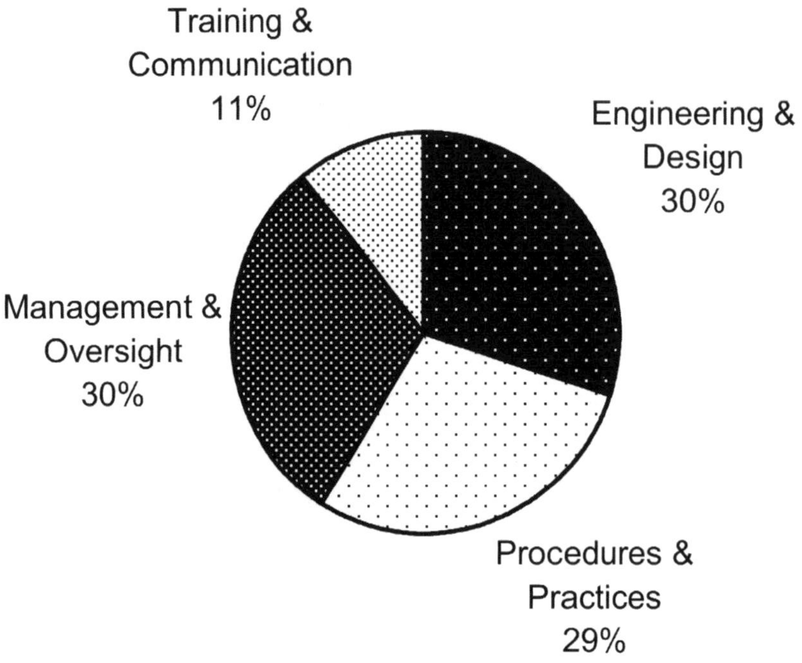

Figure 2. Incident Root and Contributing Causes Expressed in Broad Management System Categories

Examples

The following few examples illustrate the range of management system issues identified as causal factors in CSB investigations, and cover the broad spectrum of facilities.

Morton International
Paterson, New Jersey
Runaway Reaction

On April 8, 1998, a runaway reaction in the Yellow 96 dye plant at Morton International in Paterson, New Jersey resulted in an explosion and fire. Nine people were injured when a 2000-gallon reactor overpressured.

This incident occurred at a large facility, with significant available technical resources within a sizeable corporation. This represents the upper ("big company") end of the CSB investigative spectrum.

CSB found that this runaway reaction incident stemmed from the failure to identify and address the reactive hazards inherent in the process. Although laboratory data was available that indicated the potential for runaway reactions, this information did not reach the teams responsible for hazard analysis and scale-up design of this new process. Two root causes and two contributing causes were identified. These causes addressed:

- Insufficient hazard analysis;
- Inadequate follow-through on hazards addressed;
- Inadequate process information available for hazard analysis, and
- Inadequate management of change review.

DPC Enterprises

Festus, Missouri

Liquid Chlorine Release

On the morning of August 14, 2002, a liquid chlorine unloading hose ruptured during a railroad tank car unloading operation at DPC Enterprises, L.P., near Festus, Missouri. By the time the leak was isolated three hours later, 48,000 pounds of chlorine was released over a 3-hour period. Good fortune and prevailing wind directions prevented significant injuries, although 66 people in the surrounding community sought medical evaluation after exposure to chlorine vapor.

CSB found that the chlorine unloading hose failed because it was made from the wrong material. Stainless steel braided hose had been inadvertently substituted for it's look-alike Hastelloy-C in a supply chain mistake. Inadequate testing and maintenance of the emergency shutoff system and insufficient emergency planning contributed to the release of half a rail car of poisonous chlorine.

This was a small facility operating on behalf of a medium-sized corporation with holdings in several states. The company had implemented management systems for mechanical integrity and emergency planning, but these systems were inadequate to prevent the incident. The three root causes and five contributing causes identified by CSB addressed:

- Quality assurance problems in the supply chain and on the part of DPC
- Inadequate testing and inspection of emergency shutoff systems
- Inherent difficulties in visually identifying Hastelloy-C chlorine transfer hoses

- Inadequate community notification
- Insufficient emergency planning on the part of the company and the community responders

Environmental Enterprises, Inc.

Cincinnati, Ohio

Hydrogen Sulfide Exposure

On December 11, 2002, a mechanic was overcome by hydrogen sulfide gas while walking through a wastewater treatment plant to retrieve a tool. He was fortunate to be discovered very quickly by fellow employees and dragged to safety, where he recovered.

This accidental exposure resulted from the use of the wrong vessel for chemical treatment. CSB investigators found that, although a properly enclosed and vented treatment tank was available, an operator had used an open-top clarifier to perform final polishing of the wastewater. The hydrogen sulfide monitor that should have warned the victim of the presence of this poisonous gas had not been recently calibrated, and was not working at the time of the incident.

This event occurred in a small facility that is the only one of its kind for this privately held company. This represents the other end (the "small company" end) of the CSB incident spectrum from BP Amoco. Again, CSB found management system failures that contributed to the incident. In the case study report, the following management systems were listed as having potential preventive value:

- Process operation training
- Hazard communication
- Mechanical integrity
- Incident investigation
- Management oversight

Conclusions

This paper presents data on19 investigation reports published by the US Chemical Safety and Hazard Investigation Board. The root and contributing causes identified in these reports present a strong and repeated argument for improved safety management systems. The incidents demonstrate that no facility, large or small, and no process, simple or complex, is immune from the need for management systems to prevent incidents.

Simple statistical analysis and value-based groupings of the types of causally associated management systems provides a means of understanding these issues. The areas for broadest impact and preventive value are identified.

References

1. Adams, Edward E. 1995. *Total Quality Safety Management – An Introduction*, American Society of Safety Engineers (ASSE), June 1995.

2. Center for Chemical Process Safety (CCPS), 1992. *Guidelines for Investigating Chemical Process Incidents,* American Institute of Chemical Engineers (AIChE).

3. CCPS, 1994. *Guidelines for Preventing Human Error in Process Safety,* AIChE.

4. CCPS, 1989, *Guidelines for Technical Management of Chemical Process Safety,* AIChE.

5. United States Chemical Safety and Hazard Investigation Board (CSB), 1998. *Investigation Report: Explosive Manufacturing Incident, Sierra Chemical Company, Mustang, Nevada, January 7, 1998.*

6. CSB, 1999. *Summary Report, Nitrogen Asphyxiation, Union Carbide Corporation, Hahnville, Louisiana, March 27, 1998.*

7. CSB, 1999. *Investigation Report, Propane Tank Explosion, Herrig Brothers Feather Creek Farm, Albert City, Iowa, April 9, 1998.*

8. CSB, 2000. *Investigation Report, Chemical Manufacturing Incident, Morton International, Inc., Paterson, New Jersey, April 8, 1998.*

9. CSB, 2000. *Investigation Report, Catastrophic Vessel Overpressurization, Sonat Exploration Company, Temple 22-1 Common Point Separation Facility, Pitkin, Louisiana, March 4, 1998.*

10. CSB, 2001. *Investigation Report, Refinery Fire Incident, Tosco Avon Refinery, Martinez, California, February 23, 1999.*

11. CSB , 2001. *Investigation Report, Steel Manufacturing Incident, Bethlehem Steel Corporation, Burns Harbor Division, Chesterton, Indiana, February 2, 2001.*

12. CSB, 2001. *Safety Bulletin, Management of Change, August 2001.*

13. CSB, 2002. *Case Study, Explosion at Concept Sciences: Hazards of Hydroxylamine, March 2002.*

14. CSB, 2002. *Investigation Report, Thermal Decomposition Incident, BP Amoco Polymers, Inc., Augusta, Georgia, March 13, 2001.*

15. CSB, 2002. *Investigation Report, Refinery Incident, Motiva Enterprises LLC, Delaware City Refinery, Delaware City, DE, July 17, 2001.*

16. CSB, 2002. *Investigation Report, Hydrogen Sulfide Poisoning, Georgia-Pacific, Naheola Mill, Pennington, Alabama, January 16, 2002.*

17. CSB, 2003. *Investigation Report, Petroleum Products Facility Incident, Third Coast Industries, Friendswood, Texas, May 1, 2002.*

18. CSB, 2003. *Investigation Report, Chlorine Release, DPC Enterprises, L.P., Festus, Missouri, August 14, 2002.*

19. CSB, 2003. *Investigation Report, Vapor Cloud Deflagration and Fire, BLSR Operating, LTD., Rosharon, Texas, January 13, 2003.*

20. CSB, 2003. *Case Study, Hydrogen Sulfide Exposure, Environmental Enterprises, Inc., Cincinnati, OH, December 11, 2002.*

21. CSB, 2003. *Investigation Report, Chemical Waste-Mixing Incident, Kaltech Industries Group, Inc., Borough of Manhattan, New York, NY, April 25, 2002.*

22. CSB, 2003. Investigation Report, Explosion and Fire, First Chemical Corp., Pascagoula, Mississippi, October 13, 2002.

Appendix

CSB Investigations

Root and Contributing Cause Data

Appendix
CSB Investigation Reports Root and Contributing Causes Categorized by Management Systems

	Totals	**Sierra Chemical Co**	**Sonat Exploration**	**Union Carbide**	**Morton International**
Date		January 7, 1998	March 4, 1998	March 27, 1998	April 8, 1998
Location		Mustang, NV	Pitkin, LA	Hahnville, LA	Paterson, NJ
Killed	37	4	4	1	0
Injured	172	6	0	1	9
Description		Explosion	Vessel overpressurization & rupture	Nitrogen asphyxiation	Runaway reaction, explosion
Root Causes	51	4	2	1	2
Contributing Causes	34	1	1	2	2
Total causes	**85**	**5**	**3**	**3**	**4**
Management System Causes	**73**	**4**	**3**	**2**	**4**
Non-Management System Causes	12	1		1	
Management Systems		**Sierra Chemical Co**	**Sonat Exploration**	**Union Carbide**	**Morton International**
Process Safety Information	5				1
Process Hazard Analysis	8	1	1		1
Operating Procedures	5		1		
Operator Training	1				
HAZCOM	4				
Maintenance Procedures	10				
Engineering Design & Review	6		1		
Management of Change	7				1
Safe Work Practices	6			2	
Internal Auditing & Oversight	3	1			
Facility Siting	3	1			
Employee Participation	1	1			
Incident Investigation	4				1
Industry Guidance	1				
Other training	3				
Quality Assurance	2				
Emergency Preparation	4				
Totals	**73**	**4**	**3**	**2**	**4**

Appendix
CSB Investigation Reports Root and Contributing Causes Categorized by Management Systems

	Herrig Brothers	Condea Vista	Equilon Enterprises	Concept Sciences	Tosco Avon Refinery
Date	April 9, 1998	October 13, 1998	November 25, 1998	February 19, 1999	February 23, 1999
Location	Albert City, IA	Baltimore, MD	Anacortes, WA	Hanover Township, PA	Martinez, CA
Killed	2	0	6	5	4
Injured	7	4	0	14	1
Description	Propane tank BLEVE	Uncontrolled reaction & explosion	Fire following loss of utilities	Thermal decomposition, explosion	Fire during maintenance
Root Causes	3	1	1	4	2
Contributing Causes	1				2
Total causes	**4**	**1**	**1**	**4**	**4**
Management System Causes	**1**	**1**	**1**	**4**	**4**
Non-Management System Causes	3				
Management Systems	**Herrig Brothers**	**Condea Vista**	**Equilon Enterprises**	**Concept Sciences**	**Tosco Avon Refinery**
Process Safety Information				1	
Process Hazard Analysis				1	
Operating Procedures					
Operator Training					
HAZCOM					
Maintenance Procedures					2
Engineering Design & Review				1	
Management of Change		1	1		1
Safe Work Practices					
Internal Auditing & Oversight					1
Facility Siting				1	
Employee Participation					
Incident Investigation					
Industry Guidance					
Other training	1				
Quality Assurance					
Emergency Preparation					
Totals	**1**	**1**	**1**	**4**	**4**

Appendix
CSB Investigation Reports Root and Contributing Causes Categorized by Management Systems

	Bethlehem Steel	BP Amoco	Motiva Enterprises	Third Coast Industries
Date	February 2, 2001	March 13, 2001	July 17, 2001	May 1, 2002
Location	Chesterton, IN	Augusta, GA	Delaware City, DE	Brazoria County, TX
Killed	2	3	1	0
Injured	4	0	8	0
Description	Fire During Maintenance	Thermal decomposition, release	Explosion, fire & acid spill	Fire
Root Causes	3	3	3	2
Contributing Causes	2	4	2	1
Total causes	**5**	**7**	**5**	**3**
Management System Causes	**5**	**7**	**5**	**1**
Non-Management System Causes				2
Management Systems	**Bethlehem Steel**	**BP Amoco**	**Motiva Enterprises**	**Third Coast Industries**
Process Safety Information		1		
Process Hazard Analysis		1		1
Operating Procedures	1			
Operator Training				
HAZCOM	1			
Maintenance Procedures	2		1	
Engineering Design & Review		2		
Management of Change	1	1	1	
Safe Work Practices		1	1	
Internal Auditing & Oversight			1	
Facility Siting				
Employee Participation				
Incident Investigation		1	1	
Industry Guidance				
Other training				
Quality Assurance				
Emergency Preparation				
Totals	**5**	**7**	**5**	**1**

Appendix
CSB Investigation Reports Root and Contributing Causes Categorized by Management Systems

	Georgia-Pacific	Kaltech Industries Group	DPC Enterprises	First Chemical Corporation	Environmental Enterprises	BLSR Operating, LTD
Date	**January 16, 2002**	**April 25, 2002**	August 14, 2002	October 13, 2002	December 11, 2002	January 13, 2003
Location	**Pennington, AL**	**New York, NY**	Festus, MO	Pascagoula, MS	Cincinnati, OH	Rosharon, TX
Killed	**2**	**0**	0	0	0	3
Injured	**8**	**36**	66	3	1	4
Description	**Inadvertent mixing, toxic gas release**	**Unintended reaction, explosion**	Hose failure, chlorine release	Thermal decomposition, explosion	H2S Release	Vapor cloud deflagration
Root Causes	**2**	**2**	3	4	7	2
Contributing Causes	**2**	**2**	5	2		5
Total causes	**4**	**4**	**8**	**6**	**7**	**7**
Management System Causes	**4**	**2**	**6**	**6**	**6**	**7**
Non-Management System Causes		**2**	2		1	
	4	**4**				
Management Systems	**Georgia-Pacific**	**Kaltech Industries Group**	**DPC Enterprises**	**First Chemical Corporation**	**Environmental Enterprises**	**BLSR Operating, LTD**
Process Safety Information	**1**					1
Process Hazard Analysis	**1**			1		
Operating Procedures		**1**			1	1
Operator Training					1	
HAZCOM		**1**			1	1
Maintenance Procedures			2	2	1	
Engineering Design & Review	**1**			1		
Management of Change						
Safe Work Practices					1	1
Internal Auditing & Oversight						
Facility Siting				1		
Employee Participation						
Incident Investigation					1	
Industry Guidance						1
Other training	**1**					1
Quality Assurance			2			
Emergency Preparation			2	1		1
Totals	**4**	**2**	**6**	**6**	**6**	**7**

AN INHERENT SAFETY BASED INCIDENT INVESTIGATION METHODOLOGY

Attiq Goraya, Paul R. Amyotte
Department of Chemical Engineering
Dalhousie University
Halifax, NS
Canada B3J 2X4
<agoraya@dal.ca>
<paul.amyotte@dal.ca.>

and

Faisal I. Khan
Faculty of Engineering & Applied Science
Memorial University of Newfoundland
St. John's, NF
Canada A1B 3X5
<fkhan@engr.mun.ca>

Prepared for Presentation at
American Institute of Chemical Engineers
38th Annual Loss Prevention Symposium
New Orleans, LA
April 26 – 29, 2004
Session T7006: Case Histories and Lessons Learned

November 2003

UNPUBLISHED

ABSTRACT

A methodology has been developed to enable the explicit use of the principles of inherent safety in an incident investigation protocol. The usefulness of this approach is demonstrated by application to the Westray coal mine explosion that occurred in Nova Scotia in 1992. This process-related disaster resulted in the deaths of 26 workers, destruction of the underground workings, and bankruptcy of the parent company. The purpose in presenting this case study is twofold – to validate the methodology, and to identify the inherent safety considerations that could have prevented the incident. These findings have application beyond the realm of coal mining and extending well into the world of the chemical process industries.
Keywords: Incident investigation, inherent safety, domino loss causation model, case study

INTRODUCTION

Process related incidents can cause production interruption, asset damage, environmental harm, injuries and fatalities. A timely and effective incident investigation is an opportunity to gain new insights into the root causes of a loss-producing event. Within the spirit of continuous improvement, investigations of incidents (both loss-producing and near misses) provide opportunities to learn how to prevent similar events in the future. The reward of a successful investigation is therefore the prevention of incidents and better protection of employees, equipment and other business assets, and the environment. Incident investigation reports, including detailed incident descriptions and recommendations, are commonly shared within the chemical industry. Many professionals believe that this sharing of incident reports has played a major role in the steady improvement of industrial safety.

The Relation of Incident Investigation to Management of Process Safety

The Center for Chemical Process Safety (CCPS) was formed by the American Institute of Chemical Engineers (AIChE) in 1985 to promote the improvement of process safety among those who handle, use, process and store hazardous materials. CCPS has identified multiple components that are involved in the successful management of process hazards and has grouped these components into 12 elements; incident investigation is listed as element 9. These 12 elements are regarded as the components of a process safety management system. Process Safety Management (PSM) is the application of management systems to the identification, understanding and control of process hazards to prevent process related injuries and incidents (CCPS, 1989). In Canada, the approach to PSM is based largely on the CCPS model and is described on the website of the Canadian Society for Chemical Engineering (www.chemeng.ca).

To be able to address, improve and control process safety, analysts and the engineers working on process hazard analysis need to have full access to detailed information about the process, materials involved, design of plant, operating procedures, maintenance records, and industry wide operating experience. Essential operating experience includes equipment reliability data and detailed incident investigation reports from similar plants. These data and reports form a feedback loop in the process safety management system. This feedback loop thus makes it possible to improve safety through better identification of process hazards.

The Need for Incident Investigation

An effective Process Safety Management system should be retrospective and prospective simultaneously. This means that the PSM system needs to ensure that the lessons learned from past incidents can be utilized in normal day-to-day operations and when planning for new facilities. Incident investigation provides the important linkage between the lessons of past incidents and safer designs and operations in the future (CCPS, 1992a).

Among the many advantages of incident investigation are the reports and case studies that investigations yield; these help in preventing the recurrence of incidents. The importance of incident investigation reports and case studies is also reported by Crowl and Louvar (2002). In loss prevention, case studies are considered as one of the most effective learning tools.

Inherent Safety

Loss prevention in the Chemical Process Industries (CPI) is generally considered in three ways (Etowa et al., 2002): i) engineered safety, ii) procedural safety, and iii) inherent safety. Engineered, or add-on, safety involves the addition of safety devices at the end of the design. These safety devices do not perform any fundamental operation, but are designed to act when a process upset occurs. Procedural safety measures, or administrative controls, utilize safe work practices and procedures to reduce risk. On the other hand, inherent safety uses the inherent properties of a material or process to eliminate or reduce the hazard. The fundamental difference between inherent safety and the other two categories is that inherent safety seeks to remove the hazard at the source as opposed to accepting the hazard and looking to mitigate the effects.

The concept of inherent safety was first proposed in the late 1970's by Trevor Kletz in his Jubilee Lecture to the Society of Chemical Industry in Widnes, England (published as Kletz, 1978). Inherent safety is making some inroads into the CPI, with several companies attempting to justify the cost/safety benefit ratio of inherent safety (Zwetsloot and Ashford, 2003). A recent review by two of the current authors provides further details, examples and status reports on inherent safety (Khan and Amyotte, 2003). Figure 1 (after Khan et al., 2003 and Amyotte et al., 2003) presents a systematic approach to loss prevention which illustrates a desired hierarchical relationship among engineered, procedural and inherent safety considerations.

Principles of Inherent Safety

An inherently safer plant, by virtue of its design, generates little or no damage in the event of an incident. The principles of inherent safety describe the different ways to achieve an inherently safer plant. The four most general and widely applicable principles are as follows:

- **Minimize**,
- **Substitute**,
- **Moderate**, and
- **Simplify**.

The idea behind the minimization principle is to minimize the hazard, be it through using smaller quantities of hazardous material or performing a hazardous procedure fewer times. With the substitution principle, one looks to replace a hazardous material, process route, or procedure with one that is less hazardous, thus eliminating or reducing the hazard. The principle of moderation seeks to use hazardous materials under less hazardous conditions when the use of those materials cannot be eliminated or minimized. The final principle is simplification, in which the goal is to design processes and equipment to reduce the opportunities for errors by eliminating excessive use of add-on safety features and protective devices.

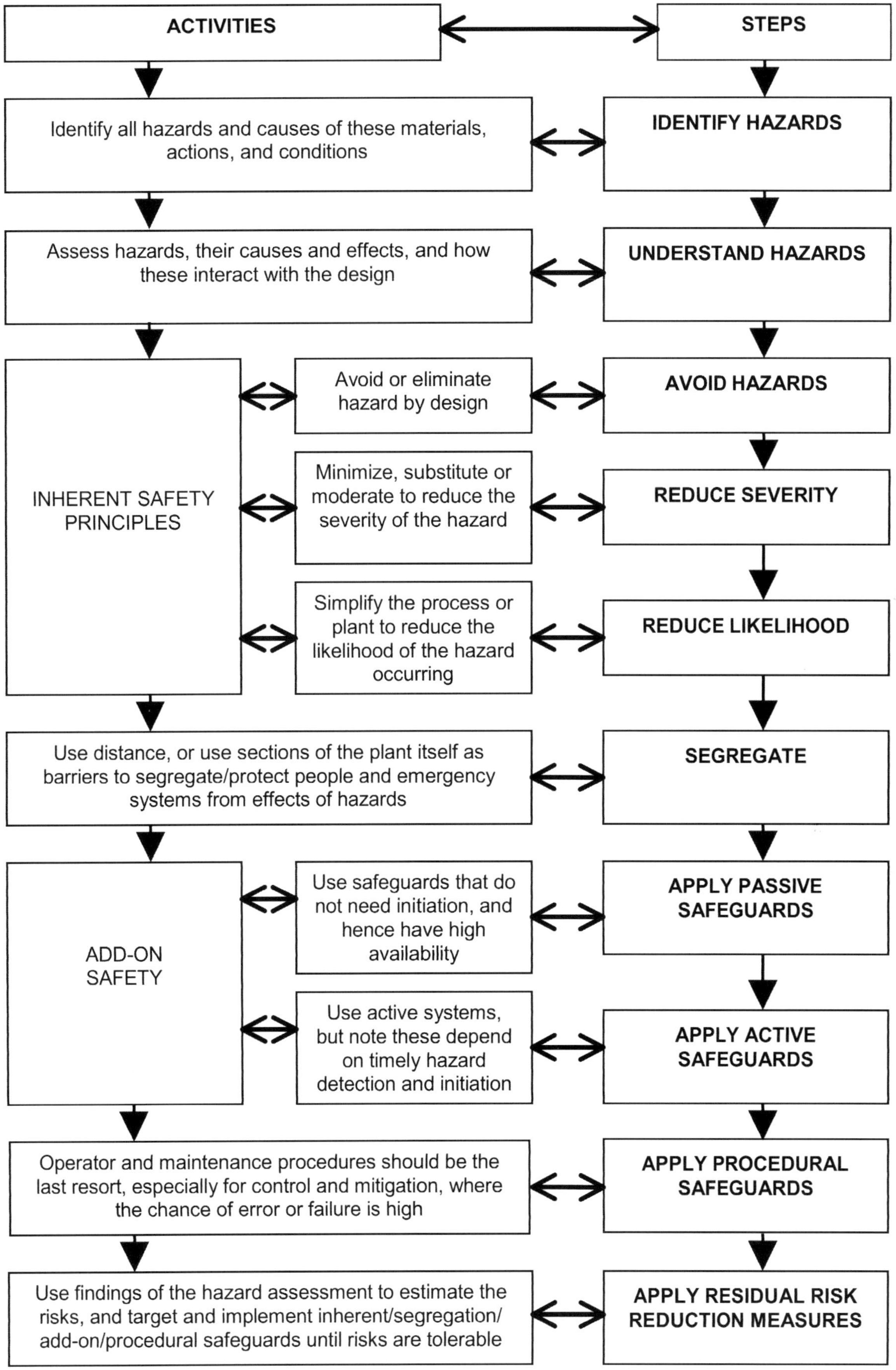

Figure 1 A systematic approach to loss prevention.

Objective of Current Work

Process chemists and engineers of today are increasingly considering inherently safer options in their design decisions; it is our belief that the authors of incident reports do so less often. It is our further belief that explicit attention to inherent safety considerations during incident investigation can, however, be effective in preventing incidents with similar root causes in the future.

Thus, the objective of this paper is to incorporate inherent safety considerations in an incident investigation protocol. This should result in investigation reports and case studies which have more of a fundamental basis in the principles of inherent safety. As previously commented, the study of case histories is a valuable and powerful tool in loss prevention. A recently reported example by Mahnken (2001) illustrates how case histories can be incorporated into process hazard analyses.

INCIDENT INVESTIGATION PROTOCOL

The inherent safety based incident investigation methodology developed in the current work is shown in Figure 2. The framework for the protocol was adapted from that presented by Rosser (2000), with additional input from CCPS (1992a), and is composed of four main parts (moving from top to bottom down the centre of Figure 2, with the numbers 1 – 4 as described below):

1. ***Preparation and Initial Response***

   Team Planning, Incident, Loss, and Notification to Activate Team

2. ***Data Gathering***

   On-Site Investigation

3. ***Data Analysis***

   Identification of Immediate Causes, Identification of Basic Causes, and Identification of Lack of Management Control Factors

4. ***Recommendation Reporting, Implementation and Follow-up***

   Recommendations, Implement Recommendations, and Follow-up

The protocol as shown in Figure 2 clearly displays evidence of the essential management functions of *plan, do, check, act.* It is expected that there will be overlap between steps as the continual improvement concept of PSM is applied within a given investigation. Further, other steps would likely be employed by different organizations based on both the scale of their operations and the magnitude of the incident being investigated. Our intent here is to provide a simple, basic framework which clearly indicates that:

- planning for investigation before an incident is essential,
- loss can occur in many different forms,
- some form of root cause analysis is required to ultimately identify management system deficiencies, and
- recommendations must address different levels of causation and must be acted upon with verification of results to ensure the intended improvements have been achieved.

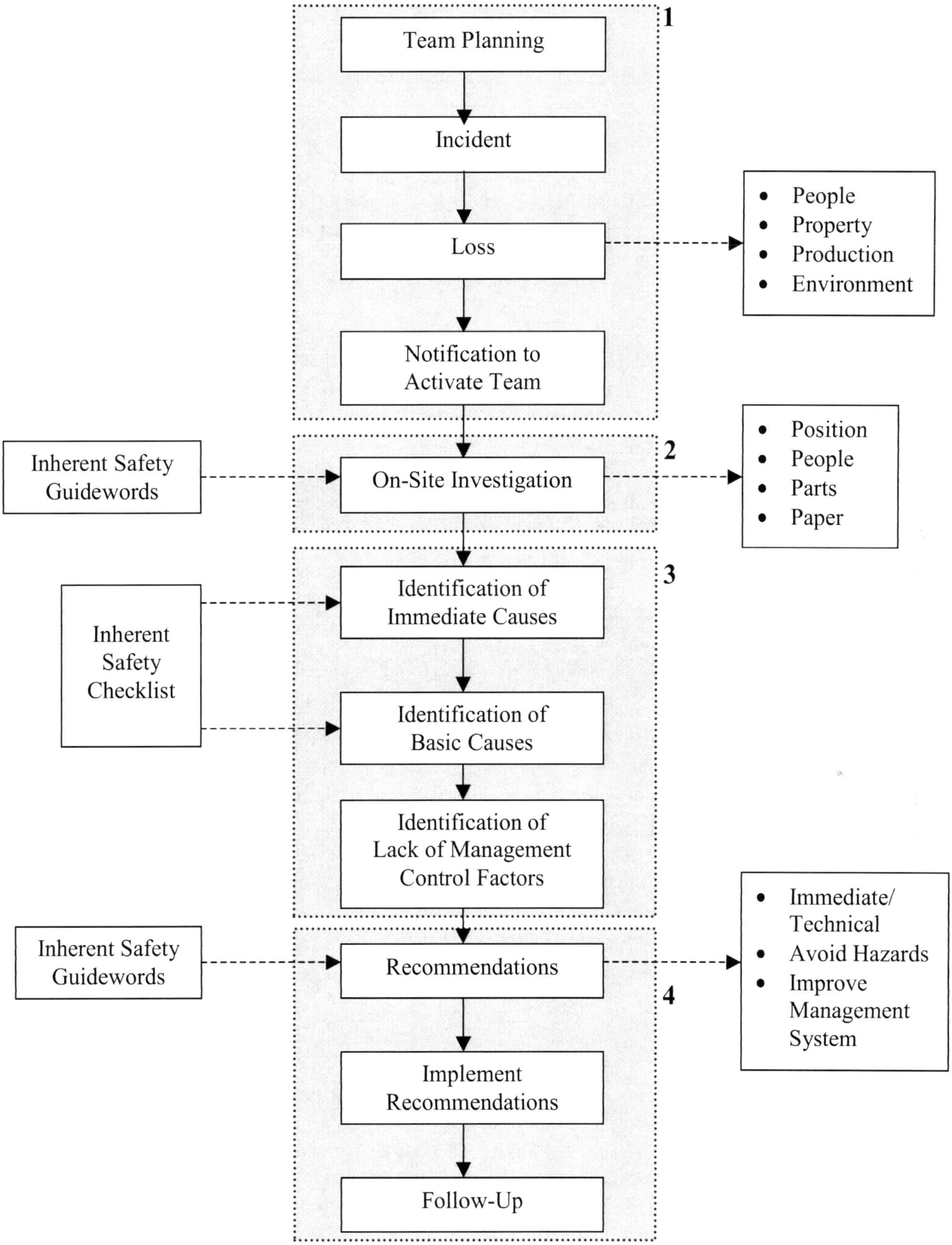

Figure 2 An inherent safety based incident investigation methodology.

The following sections highlight the key features of the investigation protocol given in Figure 2.

Categories of Loss

Wilson and McCutcheon (2003) advocate an *integrated* approach to industrial safety through management of the continuous reduction of risks to people, property (assets), production and the environment. Accordingly, a loss-producing event in any one of these categories should be investigated. Further, there is increasing recognition that investigation of near-misses affords significant opportunities to learn and prevent future events that could actually result in some form of loss.

Categories of Data (Evidence)

Classification of investigation data (evidence) into the categories of position, people, parts and paper has been recommended by numerous practitioners; see, for example, Keller (1996), Wilson (1998), Rosser (2000) and Wilson and McCutcheon (2003). Table 1 (after Keller, 1996) and Figure 3 (after Wilson, 1998) provide details on evidence nature and fragility, respectively.

Table 1 Classification of investigation data (evidence).

Category	Example
Position	• physical location of injured personnel, equipment, spills, skid marks, etc. • position of electrical and mechanical switches and controls • location and arrangement of scattered product • proximity of ladders and other tools
People	• witnesses and those directly involved in incident (potential interviewees) • equipment designers, manufacturers, purchasers • experts on maintaining and operating equipment
Parts	• parts of any equipment that may have failed • parts of any equipment that may have interfered with or interrupted normal operation
Paper	• personnel training records • equipment maintenance records • safety compliance records • procedure manuals

Domino Loss Causation Model

The 'heart' of an incident investigation is the data analysis stage; it is here that the search for root causes is conducted. There are several loss causation models available in the open literature and many companies have their own proprietary models. The current work employs the domino loss causation model developed by Bird and Germain (1996) and shown in Figure 4. Wilson (1998) and Wilson and McCutcheon (2003) give further details on the model and its use within an incident investigation methodology. As described by Amyotte and Oehmen (2002), the domino model is a structured method of identifying and categorizing the multiple events that cause an incident. The remainder of this section describes the model components as they appear in both Figures 2 and 4.

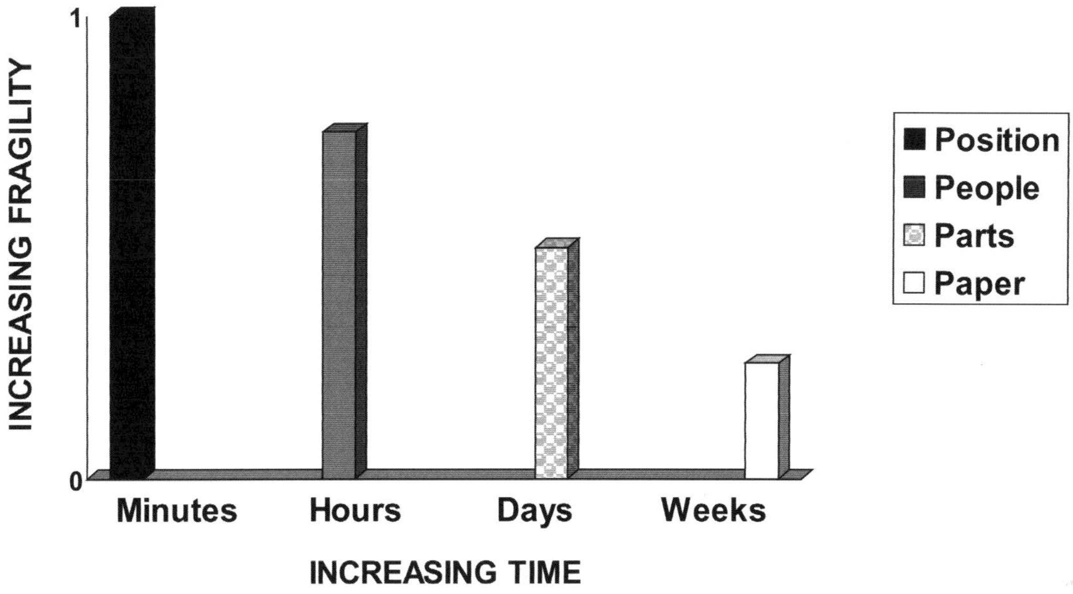

Figure 3 Illustration of the need to preserve evidence in an incident investigation. (The height of the bar representing each category illustrates the fragility of that category, and is therefore an indication of the approximate relative timeframe for data preservation.)

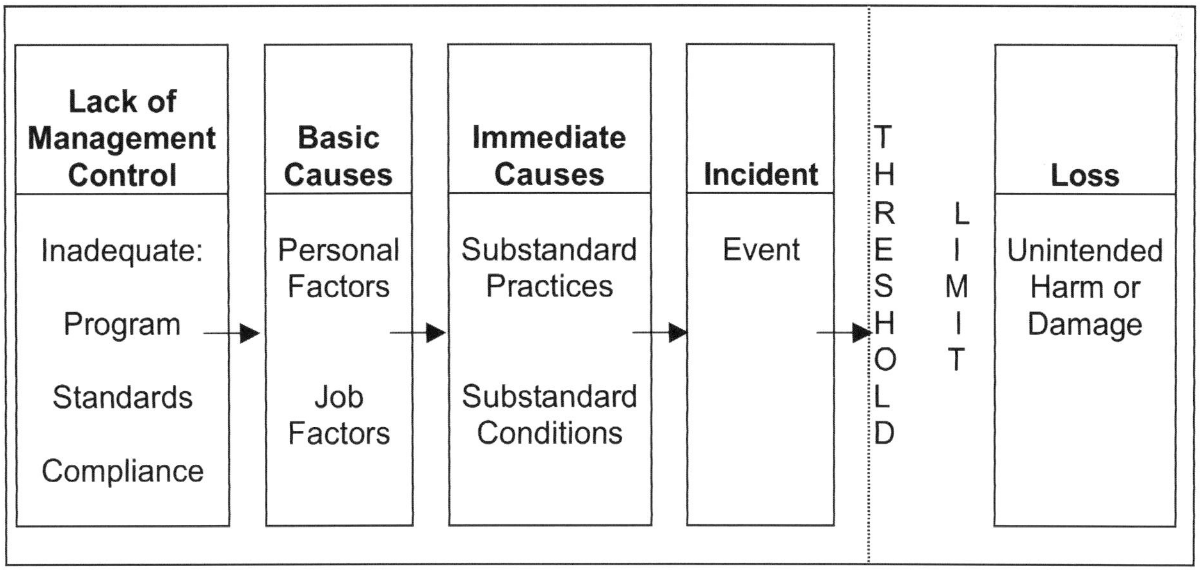

Figure 4 Domino loss causation model.

First, the loss that occurred is described in terms of the following:

- Injuries and/or deaths to the people involved,
- Damage to property on-site or in the surrounding area,
- Damage to the process and/or equipment involved, taking into account revenue lost due to inactive operation of the process, and
- Environmental damage.

Second, the incident that occurred is described in detail, in terms of the substances (e.g. raw materials) and events that transpired (e.g. fire, explosion) and resulted in physiological and/or structural threshold limits being exceeded. Then, the immediate causes of the incident are identified. Substandard practices could include failure to use personal protective equipment, removing safety devices, and the failure to follow procedures. Substandard conditions include inadequate ventilation, excessive exposure to harmful materials, defective equipment, and inadequate personal protective equipment.

After the immediate causes of the incident are analyzed, the basic causes are investigated. Some personal factors that could bear responsibility are inadequate employee capability, lack of knowledge by the employee about possible dangers, a high stress level in the work environment, and improper job motivation imposed by management. Job-related factors could also be responsible, such as inadequate leadership, inadequate maintenance, inadequate work standards, and the abuse or misuse of equipment.

Lack of control of the workplace by management leads to most of the basic causes previously mentioned. An inadequate loss management program can lead to incidents and shows a fundamental lack of management control. Inadequate program standards or inadequate compliance with existing standards also illustrate a need for changes that should be incorporated by management. This level of causation is often referred to as management system deficiencies.

The basic premise of the domino loss causation model is thus one of multiple causes – incidents are seldom, if ever, the result of a single event. On the contrary, there are typically several levels of causation. Immediate causes (essentially symptoms) exist because of deeper, more systemic basic causes, which in turn find their origin in factors that lie primarily within the control of management.

Kletz's Layered Investigation Model

The layered investigation approach of Trevor Kletz (Kletz, 1979, and as described in Crowl and Louvar, 2002) was adopted for the formulation of recommendations arising from the investigation. This means that three levels of recommendations are generated:

- immediate technical recommendations (first layer),
- recommendations to avoid the hazards (second layer), and
- recommendations to improve the management system (third layer).

With reference to the domino model, these layers generally correspond to the causation levels of immediate causes, basic causes, and lack of management control factors, respectively (albeit with some overlap among the three layers and the three causation levels).

Inherent Safety Guidewords and Checklist

Guidewords and checklists are well-established, proven tools that are widely used in hazard identification. Incident investigation is in many respects hazard identification 'after the fact'; it is therefore reasonable to also use the same tools for investigation purposes. Thus, we have attempted to explicitly incorporate inherent safety considerations into the proposed incident investigation protocol via the use of guidewords and a checklist.

Recommended guidewords and their description are shown in Table 2. These guidewords are simply the four most general and widely applicable principles of inherent safety (as previously mentioned). The description for each is purposely brief and is focused on materials, process routes, equipment and procedures. Use of these guidewords as 'mind triggers' during on-site investigation and formulation of recommendations, as shown in Figure 2, will help ensure that the concepts of inherent safety are visible at both the front and back ends of the protocol. These guidewords are intended as a supplement to existing tools that may already be in use for gathering and assessing evidence and formulating recommendations.

Table 2 Inherent safety guidewords.

Guideword	Description
Minimize	Use smaller quantities of hazardous materials when the use of such materials cannot be avoided. Perform a hazardous procedure as few times as possible when the procedure is unavoidable.
Substitute	Replace a substance with a less hazardous material or processing route with one that does not involve hazardous material. Replace a hazardous procedure with one that is less hazardous.
Moderate	Use hazardous materials in their least hazardous forms or identify processing options that involve less severe processing conditions.
Simplify	Design processes, processing equipment, and procedures to eliminate opportunities for errors by eliminating excessive use of add-on (engineered) safety features and protective devices.

Four series of checklist questions built around these guidewords are given in Table 3. The intention here is to use these more detailed inherent safety indicators during the identification of both immediate causes and basic causes (as illustrated in Figure 2). As with the guidewords themselves, the intention is not to replace existing checklists, but rather to supplement these more traditional tools with ones that explicitly incorporate inherent safety considerations.

The checklist in Table 3 is not all-inclusive, and is in fact a mix of both process safety and occupational safety issues. It is intended as an illustrative example of the type of thinking required to move beyond the usual engineered/procedural form of checklist questions. Some companies will have their own inherent safety based checklists; additionally, the Appendix in Bollinger et al. (1996) contains a sample inherently safer process checklist. These authors also comment that Appendix B in CCPS (1992b) contains an extensive checklist with many questions related to inherent safety. (For example – *Can the supply pressure of raw materials be kept below the working pressure of vessels receiving them?* – could easily fall under *moderation*).

Table 3 Inherent safety checklist.

Guideword	Checklist Question
Minimize	• Is the storage of all hazardous gases, liquids and solids minimized? • Are just in time deliveries used when dealing with hazardous materials? • Are all hazardous materials removed or properly disposed of when they are no longer needed or not needed in the next X days? • Are all tasks involving working at heights done in as few trips as possible? • Are all tripping hazards minimized and all walking surfaces tractional during all weather conditions? • Are low noise equipment and machinery taken into consideration when making new purchases? • Is shift rotation optimized to avoid fatigue? • Are awkward positions and repetitive motions minimized? • Are attempts made to completely eliminate raw materials, process intermediates or by-products? • Are elbows, bends, and joints in piping minimized?
Substitute	• Can a less toxic, flammable or reactive material be substituted for use? • Is there an alternate way of moving product or equipment as to eliminate human strain? • Can a water based product be used in place of a solvent or oil based product? • Are all allergenic materials, products and equipment replaced with non allergenic materials, products and equipment when possible?
Moderate	• Can potential releases be reduced via lower temperatures or pressures, or elimination of equipment? • Are all hazardous gases, liquids and solids stored as far away as possible to eliminate disruption to people, property, production and environment in the event of an incident? • When purchasing new equipment, are acceptable models available that operate at lower speeds, pressures, temperatures or volumes? • Are workplaces designed such that employee seclusion is minimized? • Are food and beverages offered at the workplace appropriate to minimize negative impact and maximize positive impact? • Are all power tools de-energized when not in use for extended periods? • Are all lights, sprinklers, accesses/exits, ventilation ducts, electrical installations and machines clear of piled materials?
Simplify	• Are all manuals, guides and instructional material clear and easy to understand, especially those that are used in an emergency situation? • Are equipment and procedures designed such that they cannot be operated incorrectly or carried out incorrectly? • Are machine controls located to prevent unintentional activation while allowing easy access for stopping the machine? • Are all machines, equipment and electrical installations easily isolated of all sources of power? • Is the workplace designed for consideration of human factors (i.e. an ergonomically designed workplace)?

CASE STUDY: WESTRAY COAL MINE EXPLOSION

A case study illustrating the investigation protocol in Figure 2 was developed for the Westray coal mine explosion, which occurred in Plymouth, Nova Scotia (Pictou County), Canada, on May 9, 1992, killing 26 miners. (Figure 5 from Richard (1997) shows an aerial view of the mine site, pre-explosion.) The methane levels in the mine were consistently higher than regulations, which was caused by inadequate ventilation in the mine. Flammable dust concentrations were also higher than permissible levels due to inadequate cleanup of coal dust and the fact that there was no crew in charge of rock dusting (inerting the coal dust with limestone or dolomite). These and many other factors contributed to the poor work conditions that continually existed in the Westray mine and made it an incident waiting to happen. All of these substandard conditions and practices could be attributed to the lack of concern that management had towards safety issues in the mine, which was one of the primary root causes of the problem at Westray.

Other authors, such as Munro (2000), have commented on the roles of the inspectorate and other government officials in the Westray disaster (roles which are outside the scope of the current work). The analysis that follows relies exclusively upon public domain references such as the book by Jobb (1994) and, primarily, the Report of the Westray Mine Public Inquiry – Justice K. Peter Richard, Commissioner (Richard, 1997). The current paper builds upon a previous publication from our research group (Amyotte and Oehmen, 2002). Emphasis in this previous work was on the domino loss causation model itself and its application in understanding root causes. Here we have extended the analysis to clearly identify how a lack of inherent safety considerations played a key role in the incident initiation and propagation. The sections which follow focus on the following components shown in Figure 2: Incident, Loss, On-Site Investigation, Identification of Immediate Causes; Identification of Basic Causes; Identification of Lack of Management Control Factors, and Recommendations.

Incident

There was an explosion in the Westray mine on the morning of May 9, 1992. This incontrovertible fact means that the fire triangle criteria (fuel, oxidant and ignition source) were satisfied. The Inquiry Report (Richard, 1997) gives the fuel source as initially an accumulation of methane (due to methane layering at the mine roof) which, once ignited, triggered a full-scale coal dust explosion. The oxidant was the oxygen in the mine atmosphere. The methane ignition source is postulated to have been sparks from the continuous miner striking pyrites or sandstone at the coal face being worked in the Southwest section of the mine (Richard, 1997). The existence of other potential ignition sources has been recognized by Richard (1997) and several other authors (e.g. Munro, 2000).

This sequence of a primary gas explosion leading to a secondary coal dust explosion is well-documented in the literature (e.g. Hertzberg et al., 1982) and is a well-understood phenomenon in industry. The term *secondary explosion* is somewhat misleading, however, because of the devastating effects of a coal dust explosion due to the overpressures and rates of pressure rise generated. Figure 6 (Richard, 1997) shows the damage to the No. 1 Main portal, which was the point of entry to the underground workings. (The No. 1 Main was located behind the two concrete storage silos shown in Figure 5). Lifting and removal of the metal panels covering the portal is an indication of the destructive overpressures generated hundreds of meters beneath the surface.

Figure 5 Aerial view of the Westray mine site in Plymouth, Nova Scotia (pre-explosion).

Figure 6 Damage to portal at No. 1 Main.

Loss

Losses (in all categories of people, property, production and environment) resulting from the Westray explosion include:

- Deaths of 26 human beings.
- Extensive damage to the mining equipment and the mine itself (which was eventually purposely flooded).
- Bankruptcy of the parent company, Curragh Resources. The mine had to be closed because it was rendered inoperable, this removed an additional company revenue source.
- Default of millions of dollars in provincial and federal government loans.
- Millions of dollars of severance, worker's compensation, pension plan, and unemployment insurance payments.
- Cost of holding a public inquiry and running an extensive criminal investigation and trial involving the mine and underground managers.
- Loss to Nova Scotia Power (electrical utility company) of approximately 700,000 tonnes of Westray coal.
- Future wages of the 26 miners killed in the explosion.
- Future income of other residents of Pictou County as a result of the mine becoming inoperable.
- Enormous personal losses faced by the 26 miners' families and friends, on which it is impossible to put a price tag.

On-Site Investigation

The objective of on-site activities is to gather information to be used in determining root causes. The initial activities following the explosion were obviously related to rescue and recovery attempts; gathering of evidence came later. The evidence acquired is well-documented by Richard (1997) and covers every facet of the categories of position, people, parts and paper. Application of the inherent safety guidewords (Table 2) to this evidence illustrates some of the hazardous materials, equipment and procedures that might have been minimized, substituted, moderated, or simplified. It can be helpful to also view these guidewords in their negative sense – e.g. maximization (negation of minimization) of fuels, ignition sources, etc. Examples of the Westray evidence follow.

Methane

The co-presence of methane is a fact of life in most coal mining operations, and Westray was no exception. In this application, methane is a hazard primarily because it is a flammable gas. Methane has lower and upper flammability limits of 5 % and 15 % by volume in air, respectively, and reaches its maximum explosibility at near-stoichiometric concentrations of about 9.6 % by volume in air. Methane is lighter than air and tends to rise to the roof of the mine once liberated by the mining process, thus forming a layer. This process, known as *methane layering*, provides a rich source of fuel which needs only an ignition source for a methane fire to occur. Ventilation is the primary defense against the hazards of methane (as discussed shortly).

Coal Dust

Coal dust, both airborne and deposited on surfaces, is a major health and safety hazard in coal mines. Coal dust, like most other organic materials, is explosible when suspended in air at concentrations above its minimum explosible concentration (which is typically in the range of 50 – 100 g/m^3). Coal dust deposits present a smoldering/fire hazard, as well as offering a ready fuel source for a dust explosion in the event of an aerodynamic disturbance that raises some of the layered dust into a cloud.

Richard (1997) concluded that coal dust at least ankle-deep had accumulated on many of the underground roadways in the Westray mine, making foot travel difficult. When floor dust gave rise to difficult traveling conditions for vehicles, attempts were made to alleviate the situation by pushing the dust to the sides.

Non-Flameproof Vehicles

There were instances in which workers, supervisors, and the management at Westray employed non-flameproof equipment underground (Richard, 1997). The presence of the non-flameproof equipment underground naturally led to the presence of active ignition sources in a highly hazardous atmosphere.

Fuel Storage

According to Richard (1997), records show that fuel for mining machinery was improperly stored underground. Management had instructed that the fuel be stored underground and that the vehicles be refueled underground. This storing of fuel and re-fueling of vehicles underground added to the already existing hazardous materials and hazardous practices.

Auxiliary Ventilation System

When work is being conducted in a coal mine beyond the main airflows, ventilating air must be brought to the working area by an auxiliary ventilation system. Westray mainly used auxiliary fans in the exhaust mode (Richard, 1997). This procedure has several inherent difficulties which are discussed later in the paper.

Identification of Immediate Causes

Immediate causes are, as the name implies, the factors that immediately precede and therefore lead to an incident. They can be broken out into the categories of substandard practices and conditions.

Substandard Practices

Substandard practices are behaviors that could permit the occurrence of an incident. Application of an inherent safety checklist (e.g. Table 3) to the Westray evidence (Richard, 1997) helps to identify the substandard practices given in Table 4.

Substandard Conditions

Substandard conditions are circumstances that could permit the occurrence of an incident. Again, application of an inherent safety checklist (e.g. Table 3) to the Westray evidence (Richard, 1997) helps to identify the substandard conditions given in Table 5.

Table 4 Substandard practices categorized by inherent safety principle (guideword).

Guideword	Substandard Practice
Minimize	• poor housekeeping with respect to clean-up and removal of coal dust • storing fuel and re-fuelling vehicles in non-flame-proof areas underground
Substitute	• improper use of ventilation system e.g., re-routing without provision for miners. Munro (2000) also refers to the use of contaminated air to ventilate working areas and the practice of allowing intake air to pass by unventilated abandoned workings.
Moderate	• inadequate rock dusting. It is well-established that significant amounts of rock dust, a thermal explosion inhibitor, are required to reduce the explosion pressure and rate of pressure rise during a coal dust explosion to acceptable levels (i.e. to render mine dust non-explosible). The co-presence of methane results in an increase in the amount of rock dust required for explosion suppression.
Simplify	• continuation of mining in spite of inoperable methane detection devices

Table 5 Substandard conditions categorized by inherent safety principle (guideword).

Guideword	Substandard Condition
Minimize	• high methane concentrations • thick layers of coal dust on the mine floor, having an unacceptably high level of combustible matter. The Coal Mines Regulation Act (CMRA, 1989) of the province of Nova Scotia (which applied to the Westray mine) stipulates an upper limit of 35 weight % combustible matter in mine dust. Alternatively, this limit may be expressed as a minimum of 65 weight % incombustible matter (taken to be the inherent ash and moisture in the coal dust plus admixed rock dust). The Act (CMRA, 1989) further specifies that the percentage of incombustible dust is to be increased by 10 % for each 1 % of methane present in the mine atmosphere. The scientific basis for these regulations rests on well-understood phenomena.
Substitute	• inadequate ventilation system design and capacity. Munro (2000) includes the following points in his list of ventilation system deficiencies: inadequate airflow for clearing methane from the working face during mining, inadequate airflow to prevent the layering of methane at the roof, and inadequately constructed airflow devices for controlling airflow underground.
Moderate	• high concentrations of airborne coal dust
Simplify	• inadequate system to warn of high methane concentrations in the mine atmosphere

Identification of Basic Causes

Basic causes are the underlying or root causes for the existence of immediate causes. As with the immediate level of causation (substandard practices and conditions), it is helpful to separate basic causes into two broad categories – personal and job factors.

Personal Factors

Personal factors are basic causes that relate to the individual worker. As with the immediate causation level, application of an inherent safety checklist (e.g. Table 3) to the Westray evidence (Richard, 1997) helps to identify the personal factors shown in Table 6. For example, the minimization question in Table 3 dealing with shift optimization could trigger the first personal factor listed in Table 6.

Additionally, the factors in Table 6 can be identified by the more traditional approach of examining each immediate cause and simply asking why that cause existed. For example, the substandard practice of continuation of mining in spite of inoperable methane detection devices (Table 4) has a deeper cause (at least partially) of improper motivation by which production proceeded at the expense of safety (Table 6).

Table 6 Personal factors categorized by inherent safety principle (guideword).

Guideword	Personal Factor
Minimize	• physiological stress caused by methane overexposure and fatigue due to 12-hour shifts
Substitute	• lack of mining experience of personnel working in the mine
Moderate	• lack of knowledge of safe underground work practices
Simplify	• improper motivation by which production proceeded at the expense of safety • psychological stress caused by fear of reprisal for reporting safety concerns

Job Factors

Job factors are basic causes that relate more to the work system than the individual worker. Using the same approach as for personal causes, the job factors shown in Table 7 can be identified.

Identification of Lack of Management Control Factors

Lack of control in managing loss is generally due to deficiencies in one or more of three areas – the loss management program or system itself, the standards identified and set for the loss management program, and the degree of compliance with such standards (Bird and Germain, 1996).

Table 7 Job factors categorized by inherent safety principle (guideword).

Guideword	Job Factor
Minimize	• inadequate follow-through on recommendations from mine inspectorate personnel
Substitute	• inadequate engineering during mine design and planning with respect to potential loss exposures (e.g. mine ventilation; intersecting geological fault lines leading to frequent roof-falls)
Moderate	• lack of safe work practices and procedures • inadequate purchasing and maintenance of rock dust inventory
Simplify	• inadequate leadership in terms of assignment of responsibility (e.g. mine examiner and production supervisor responsibilities undertaken by the same person) • lack of proper orientation and training for new employees • poor communication of standards (e.g. concerning the roles and responsibilities of occupational health and safety committees)

Inadequate Program Elements

Inadequate program elements that contributed to the Westray explosion include:

- Management commitment and accountability to safety matters.
- Management of change procedures.
- Incident investigation (including near miss reporting and investigation).
- Training (orientation, safety, task-related, etc.).
- Task definition and safe work practices and procedures.
- Workplace inspections and more detailed hazard identification methodologies.
- Program evaluation and audits.

Inadequate Program Standards

Inadequate program standards that contributed to the Westray explosion include standards relating to virtually all the program elements listed above, including inadequate:

- Concern expressed by management toward safety matters.
- Follow-through on inspections for substandard practices and conditions.
- Action on hazard reports submitted by employees.
- Job instructions for employees.
- Equipment maintenance.
- Scheduling of management/employee meetings to discuss safety concerns.

Inadequate Compliance Factors

Inadequate compliance factors that contributed to the Westray explosion include:

- Poor correlation between management actions and official company policy concerning the relationship between safety and production.
- Compliance to industry practice and legislated standards concerning numerous aspects of coal mining: methane concentrations, rock dusting, control of ignition sources underground, etc. The previously mentioned Coal Mines Regulation Act (CMRA, 1989) of the province of Nova Scotia represented the primary external standard for Westray.

Recommendations

Using Kletz's layered approach (Kletz, 1979), recommendations arising from an incident investigation are made according to the level of causation: i) immediate technical recommendations, ii) recommendations to avoid the hazards, and iii) recommendations to improve the management system. Here it is felt that that use of the inherent safety guidewords in Table 2 will serve as a reminder of the usefulness of these principles in both incident prevention and mitigation.

Immediate Technical Recommendations (First Layer)

Had the damage been localized to the extent that resumption of mining operations was feasible, this layer of recommendations would have been aimed at addressing the substandard practices and conditions indicated in Tables 4 and 5. Clearly, items such as removal of coal dust accumulations would have immediate benefits. Relocation to the surface of vehicles and fuel underground at the time would also have an immediate positive result. While important, these actions would have limited long-term impact without further consideration of the hazards underground and the safety management system itself.

Recommendations to Avoid the Hazards (Second Layer)

Avoidance of hazards is fundamentally different than simply removing the hazard once. Hazard avoidance requires forethought and can only be achieved with a thorough system of hazard identification. Possible recommendations to avoid some of the hazards present in the Westray mine are presented in Table 8, using the inherent safety guidewords from Table 2. These recommendations deal with the immediate causes in a more permanent manner and also begin to address the basic causes of the incident.

Recommendations to Improve the Management System (Third Layer)

Recommendations aimed at improving the management system must address the numerous deficiencies identified in the previous section on identification of lack of management control factors. These would include: i) program elements such as hiring procedures, training, safe work procedures, and hazard identification, ii) program standards such as appropriate follow-through on inspection results, and iii) compliance issues relating to both internal (i.e. company) and external (e.g. legislated) standards.

Table 8 Hazard avoidance possibilities categorized by inherent safety principle (guideword).

Guideword	Recommendation to Avoid Hazard
Minimize	• *Methane:* degasification (extraction of methane prior to mining) • *Methane:* adequate ventilation design and system • *Coal dust:* housekeeping program for continuous removal • *Fuel storage:* elimination of underground storage or minimization of fuel inventories underground • *Ignition sources:* elimination of non-flameproof equipment underground • *Shift length:* reduction from 12 hours
Substitute	• *Auxiliary ventilation system*: forcing, rather than exhaust, system for adequate airflow to clear methane from the working face of the mine • *Main ventilation fan*: alternate design and location so as not to pick up dust and other debris from the coal return conveyor belt.
Moderate	• *Coal dust:* purchase and maintenance of an adequate inventory of rock dust; implementation of a program for rock dusting underground • *Coal dust:* roadway consolidation (a process to control the formation and dispersal of dust on mine roadways by application of a combination of rock dust, moisture-absorbing material and a binding agent)
Simplify	• *Methane:* installation of a reliable, robust mine air monitoring system

There are clearly lessons to be heeded from Westray for the coal mining industry – reminders of the importance of, for example, rock dusting and underground ventilation. These reminders are technical in nature, and would seem to find little application in the CPI, other than perhaps by extension into the realm of explosion suppression and mitigation techniques such as relief venting. The most critical lessons from Westray, however, are the ones that transcend industrial boundaries and are related to the basic principles of loss management.

These lessons are found in the third layer recommendations listed previously. These recommendations indicate the need for attention to both *system* and *attitude* perspectives. A loss management *system* – implemented, supported and enforced by management personnel – is absolutely essential. As mentioned in the introduction to this paper, the process safety management approach widely used in the Canadian chemical industries is based largely on the CCPS model. The suite of PSM elements recommended by the Canadian Society for Chemical Engineering is given in Table 9. The parallel is striking between the PSM elements in Table 9 and the lack of management control factors identified in this work as being largely responsible for the Westray disaster.

Table 9 Elements of Process Safety Management (PSM).

1. Accountability: Objectives and Goals	7. Human Factors
2. Process Knowledge and Documentation	8. Training & Performance
3. Capital Project Review & Design Procedures	9. Incident Investigation
4. Process Risk Management	10. Company Standards, Codes & Regulations
5. Management of Change	11. Audits & Corrective Actions
6. Process and Equipment Integrity	12. Enhancement of Process Safety Knowledge

The only acceptable *attitude* toward industrial safety – both morally and, in the case of present-day Nova Scotia, legally (OHSA, 1996 and Dehaas, 2001) – is expressed by the *Internal Responsibility System*, or *IRS*. This concept, which is the foundation of occupational health and safety legislation in Canada, states that every individual in an organization is responsible for health and safety. Primary responsibility lies with each person (manager, supervisor, employee, contractor, etc.) to the extent of their authority and ability to ensure a safe and healthy workplace. While the Internal Responsibility System is in one sense a management system for preventing incidents, it is perhaps more appropriately viewed as a system of personal beliefs shared universally (Strahlendorf, 2001). Expressed in this way, it becomes clear that attitude toward industrial safety at all levels in a company hierarchy is at least as important as the actual safety management system itself.

CONCLUDING REMARKS

In this paper we have attempted to demonstrate that the principles of inherent safety can be an effective addition to an incident investigation protocol. We have recommended the use of inherent safety guidewords and checklists to explicitly incorporate these principles. The proposed approach has been applied to the Westray coal mine disaster. This process-related incident clearly illustrates that such events can be prevented with proper leadership by management on safety issues. The importance of implementing an appropriate loss management program in industrial organizations cannot be overstated.

ACKNOWLEDGEMENTS

The authors gratefully acknowledge the financial support provided by the Natural Sciences and Engineering Research Council of Canada.

This paper is dedicated to the memory of the twenty-six miners who were killed in the Westray mine explosion.

REFERENCES

Amyotte, P.R. and Oehmen, A.M. (2002). Application of a Loss Causation Model to the Westray Mine Explosion, *Process Safety and Environmental Protection*, **80**, 55-59

Amyotte, P.R., Khan, F.I. and Dastidar, A.G. (2003). Reduce Dust Explosions the Inherently Safer Way, *Chemical Engineering Progress*, **99**, 36-43.

Bird, F. E. and Germain, G. L. (1996). *Practical Loss Control Leadership*, Det Norske Veritas, Loganville, GA, USA.

Bollinger, R.E., Clark, D.G., Dowell III, A.M., Ewbank, R.M., Hendershot, D.C., Lutz, W.K., Meszaros, S.I., Park, D.E. and Wixom, E.E. (1996). *Inherently Safer Chemical Processes: A Life Cycle Approach*, D.A. Crowl (editor), American Institute of Chemical Engineers, New York, NY

CCPS (1989). *Guidelines for Technical Management of Chemical Process Safety*, American Institute of Chemical Engineers, New York, NY, USA.

CCPS (1992a). *Guidelines for Investigation of Chemical Process Incidents*, American Institute of Chemical Engineers, New York, NY, USA.

CCPS (1992b). *Guidelines for Hazard Evaluation Procedures*, 2nd edition with worked examples, American Institute of Chemical Engineers, New York, NY, USA.

CMRA (Coal Mines Regulation Act) (1989). *Chapter 73 of the Revised Statutes*, Province of Nova Scotia, Queen's Printer, Halifax, NS, Canada.

Crowl, D.A. and Louvar, J.F. (2002). *Chemical Process Safety Fundamentals with Applications*, 2nd edition, Prentice Hall, NJ, USA.

Dehaas, D. (2001). *Editorial: Hats Off to Nova Scotia*, OHS Canada, **17**, 4.

Etowa, C.B., Amyotte, P.R., Pegg, M.J. and Khan, F.I. (2002). Quantification of Inherent Safety Aspects of the Dow Indices, *Journal of Loss Prevention in the Process Industries*, **15**, 477-787.

Hertzberg, M., Cashdollar, K. L., Lazarra, C. P. and Smith, A. C. (1982). *Inhibition and Extinction of Coal Dust and Methane Explosions*, RI 8708, United States Bureau of Mines, Pittsburgh, PA, USA.

Jobb, D. (1994). *Calculated Risk – Greed, Politics and the Westray Tragedy*, Nimbus Publishing Ltd., Halifax, NS, Canada.

Keller, J.J. (1996). *Accident Investigation in the Workplace*, Video Training Package, J.J. Keller & Associates , Inc., Neenah, WI, USA.

Khan, F.I. and Amyotte, P.R. (2003). How to Make Inherent Safety Practice a Reality, *Canadian Journal of Chemical Engineering*, **81**, 2-16.

Khan, F.I., Sadiq, R. and Amyotte, P.R. (2003). Evaluation of Available Indices for Inherently Safer Design Options, *Process Safety Progress*, **22**, 83-97.

Kletz, T.A. (1978). What You Don't Have, Can't Leak, *Chemistry and Industry*, **May 6**, 287-292.

Kletz, T.A. (1979). Layered Accident Investigations, *Hydrocarbon Processing,* **Nov**, 373-382.

Mahnken, G.E. (2001). Use Case Histories to Energize Your HAZOP, *Chemical Engineering Progress*, **97**, 73-78.

Munro, B. (2000). Westray Mine – Managing Risk in the Public Sector, In *Proceedings of 4th Biennial Process Safety and Loss Management Conference (PSLM 2000)*, November 5-8, 2000, Calgary, AB, Canada.

OHSA (Occupational Health and Safety Act) (1996). *Chapter 7 of the Acts of 1996*, Province of Nova Scotia, Queen's Printer, Halifax, NS, Canada.

Richard, Justice K.P. (1997). *The Westray Story – A Predictable Path to Disaster*, Report of the Westray Mine Public Inquiry, Province of Nova Scotia, Canada.

Rosser, B.W.J. (2000). Incident Investigation – The Most Important Element in any H.S.E. Management System, In *Proceedings of 4th Biennial Process Safety and Loss Management Conference (PSLM 2000)*, November 5-8, 2000, Calgary, AB, Canada.

Strahlendorf, P. (2001). *The Internal Responsibility System*, OHS Canada, **17**, 30.

Wilson, L. (1998). *Basic Learnings in Industrial Safety and Loss Management*, Association of Professional Engineers, Geologists and Geophysicists of Alberta, Edmonton, AB, Canada.

Wilson, L. and McCutcheon, D. *Industrial Safety and Risk Management*, University of Alberta Press, Edmonton, AB, Canada.

Zwetsloot, G.I J.M. and Ashford, N. A. (2003). Inherently Safer Design Based on Rapid Risk Analysis, *Safety Science*, **41**, pp. 219-240.

Distant Replay

What Can Re-Investigation of a 40-year-old Incident Tell You?

A look at Eastman Chemical's 1960 Aniline Plant Explosion

Peter N. Lodal
Technical Fellow
Plant Protection Technical Services
Eastman Chemical Company, Kingsport Operations
Kingsport, TN

On October 4, 1960, Eastman Chemical Company suffered the worst accident in its 83-year history, when an aniline manufacturing facility exploded. 16 people were killed, and more than 400 injured as a result of the blast. This paper analyzes the incident and its aftermath using both historical records and modern analytical techniques. The results provide useful insight into both the technical and cultural safety issues raised, as well as valuable information that can be applied to current processes.

Introduction

Tuesday, October 4, 1960. Most of the nation was preparing for the opening of the World Series, which was to begin the following afternoon in Pittsburgh (1). The second Nixon-Kennedy debate was to be televised that Friday (2), and with exactly a month to go before the election, interest was extremely high.

For Kingsport, Tennessee, though, all attention was focused locally, and a small manufacturing community was left in shock and mourning. At 4:43 PM local time, Tennessee Eastman's aniline manufacturing plant exploded, ultimately leaving 16 dead, 48 hospitalized, and more that 250 people requiring medical treatment of some type.

As I began my career in chemical equipment design, the "old-timers" would occasionally make reference to the explosion obliquely, with comments like "we don't do that any longer because of the aniline plant", or "so-and-so would have been promoted had it not been for the aniline plant." However, few details were available, and so we were left to wonder just exactly what happened.

About 12 years after I started work, one of our vice-presidents, who had been injured in the explosion, gave our work group the most thorough presentation of the effects of the explosion I had seen, and it piqued my interest to learn more. Many other people felt the same way, and so we began to gather as much information as we could about the incident, recognizing that, even 35+ years after the fact, this was still a sensitive issue in the community.

As we began to gather information, photographs, contemporary eyewitness accounts, and did interviews with employees and retirees, a number of people asked us, "Why are you bothering to look at an incident which is ancient history?", or "What do you really expect to learn so long after the fact, especially since we don't make this material any more?" Frankly, we didn't know what to expect, but the aniline plant explosion was such a seminal event in our company's culture, we felt that it was worth a try.

The Aniline Plant

Figure 1 shows a photo of the aniline facility taken on December 26, 1957, about the time the plant started up. The chemistry practiced is a simple two-step process (Figure 2) using mixed nitric/sulfuric acid to nitrate benzene to nitrobenzene, and then hydrogenating the nitro group over a catalyst to create aniline. The physical setup to accomplish this (Figure 3) is a two-reactor, three distillation column unit. The first column separated the nitrobenzene/water low boiling azeotrope from the recycled aqueous nitric/sulfuric mixture. The second and third columns performed a low-boiler/high-boiler cleanup of the aniline itself, resulting in a clean, overhead aniline stream as product.

The October 4 explosion occurred in the nitrator, the base of the nitrobenzene column, and the associated thermosiphon reboiler and sludge evaporator. Figures 4 through 14 show the extent of the devastation, which completely destroyed not only the aniline plant, but several adjacent warehouses filled with drums of combustible plasticizer and severely damaged a number of other buildings on site. Reports of broken windows offsite were numerous; fortunately, there were no serious injuries in the community. Estimates of property damage ran in the $4-5 million range, with significant (but undocumented) business interruption losses as well.

By any measure, this explosion was a catastrophe, both for the company and its employees, and the surrounding community. A full incident investigation team was put together, with both Tennessee Eastman and Eastman Kodak personnel. Outside help was secured from Factory Mutual, DuPont and Hercules. The final incident report, dated January 31, 1961, concluded that "... the cause of the nitrobenzene column failure and initiation of detonation is not known." This is where we began our inquiry.

Surviving Evidence

Although no official copy of the incident report is still in existence, the surviving notes and photographs are, along with one key piece of analytical data--the underflow composition of the nitrobenzene column less than an hour prior to the explosion. Because none of the physical evidence is still existent, all of our conclusions are speculative. However, using newer analysis techniques, and obtaining additional physical property data that did not exist in 1960 have allowed us to piece together a possible scenario that satisfactorily explains the surviving data.

Analytical Sample

The starting point for our "investigation" was the surviving sample from the base of the nitrobenzene column. By design, this column was intended to make a complete separation of the nitrobenzene/water azeotrope from the nitric/sulfuric/water solution, recycling the acid stream to the nitrator, and taking the nitrobenzene forward to the hydrogenator. Normally, there would have been no nitrobenzene in the column underflow. However, the surviving sample showed 18 % (wt) nitrobenzene (Figure 15), indicating a significant deviation from design. Our first task was to theorize how this might have happened.

This system is shown on a series of residue curve maps (Figures 16-22).

Residue curve maps (3) are a graphical illustration of liquid composition changes under simple distillation conditions. This type of diagram is very useful in illustrating system separation boundaries as defined by azeotropes. For a three-component system the illustration is depicted in triangular form, similar to a liquid-liquid phase diagram.

The three main components of the system as shown in Figure 16 (nitrobenzene, water and nitric acid) are at the apexes of the triangle. Figure 17 shows the addition of the normal operating line. Since water is generated in the reaction, there is an excess of water, allowing for the complete removal of all nitrobenzene under normal circumstances. Thus, the bottom of the operating line approaches the water/nitric acid azeotrope, with all nitrobenzene driven overhead.

It is important to note that the residue curve map for this system contains a distillation boundary between nitrobenzene and the nitric acid-water high-boiling azeotrope. The implication of this distillation boundary is that if for some reason the column feed should become somewhat enriched in nitrobenzene, the column underflow composition, being limited to not cross the boundary, cannot remain nitrobenzene-free. The starburst on Figure 18 indicates the composition of the surviving column underflow sample taken just before the detonation. The composition is now known to be coincident with the residue curve distillation boundary. Figure 19 illustrated the change in the system operation line, with corresponding feed concentration change, necessary to effect the underflow composition indicated by the sample.

Distillation Staging Analysis

Using a simplified three-component system model, the separation capability of the nitrobenzene column was reconstructed. This column was a 30-sieve tray unit, with the feed at tray 8 (as numbered from the bottom up). Expected tray 10 (just above the feed) temperatures were in the 102-104 °C range, which was to be expected given the 99 °C atmospheric boiling point of the nitrobenzene/water azeotrope. However, temperatures were running significantly higher on the day of the incident, in the 121-125 °C range instead. By modeling this column with significantly deteriorated tray efficiencies below the feed, modeling predicts both temperatures in this range, and significant concentrations of nitrobenzene in the underflow. With insufficient separation ability to drive the low boiling azeotrope overhead, the nitrobenzene (as a pure component the heaviest component in the system) preferentially drops to the base of the column. Once in the column base, this material was then recycled to the nitrator, resulting in the shift of nitrobenzene concentration in the column feed and further buildup of nitrobenzene in the column underflow, as shown in Figure 19.

While there are several scenarios that might explain why the column separation capability was compromised, the most likely (in our opinion) one involves the fact that this plant was in a restart mode after a significant maintenance turnaround.

Throughout the 2.5 years this plant operated, there were significant corrosion issues in the nitration section of the plant. This is not surprising, given the aggressive nature of an aqueous mineral acid/water system. It was also the first plant Eastman had run using titanium as a material of construction. During the summer and early fall of 1960, the nitrator, the base of the nitrobenzene column, and the nitrobenzene column reboiler were all replaced with new units. It is theorized that upon restart, the significantly larger heat transfer capability of a new, clean reboiler may have caused a surge of vapor into the column upon restart, either damaging the trays, or dislodging the manways. If this were the case, the separational capabilities of the damaged trays would be significantly compromised, and could explain the abnormal temperature and composition data that survived the incident.

Cause of the Detonation

Although tray damage caused by excessive vapor in the column can explain the compositional changes, by itself it still does not explain why the material detonated. For this explanation, we return to the residue curve maps. Figure 20 shows a dotted line connecting the water/nitric acid azeotrope and pure nitrobenzene. The project team had run shock sensitivity tests on several synthetic mixes of this material along the nitrobenzene/nitric-water azeo line (dotted line on the residue curve diagram) and had seen no shock sensitive behavior.

Figure 21 shows this same residue curve map overlaid with a liquid-liquid phase diagram of the underflow material. The shaded area shows the area of known shock sensitivity for the nitrobenzene/nitric acid/water system, available in the late 1950s to the plant designers. According to this information, the column underflow composition, even in its deteriorated condition, is still significantly out of the shock-sensitive zone.

Part of the boundary of the shock-sensitive composition region corresponds to the binodal curve demarking one-phase and two-phase liquid behavior. Apparently, in this system two-phase liquid mixtures are not shock sensitive, or are not found to be so by the drop weight impact test. Figure 21 shows the shock sensitive region at 25 °C; Figure 22 shows the region at 100 °C, much closer to the actual column operating temperature.

The data available to the plant designers was all obtained at ambient temperature (corresponding to Figure 21). Elevated temperature data was not available in 1960. However, when more recent investigations are studied, the single phase region expands greatly with increasing temperature (Figure 22), such that the column bottoms composition is now within the single phase, shock sensitive region.

Conclusions

As stated earlier, our conclusions are, and must remain speculative, since none of the physical evidence survives. However, the analysis of the available data suggests the following sequence of events:

1. Plant started during day shift on October 4, 1960.
2. Initial startup of new, clean reboiler results in undetected damage to column trays.
3. Compromised separation ability of the nitrobenzene column results in unexpected nitrobenzene concentrations in the column base. Because this material is recycled to the nitrator, the overall system concentration of nitrobenzene continues to increase throughout the day.

4. Because of difficulties in getting the plant lined out, the foreman decides to put the nitrobenzene column on total reflux at about 3:15 PM.
5. By 4:43 PM, the concentration in the column underflow has increased to the point where it penetrates the single-phase, shock sensitive boundary (Figure 23).
6. An unknown disturbance (valve closing, pump kicking on, etc.) causes the mixture to detonate, resulting in a fireball of about 150 ft in diameter, with a TNT-equivalency estimated at 6 tons.
7. Neighboring warehouses are set afire by the fireball. Injuries are due to the fireball, the blast pressure wave, and significant shrapnel generated by the explosion.

Learnings

There are many learnings that can (and have been) extracted from an analysis of this incident. Many are specific to Eastman's Kingsport Site and our current reactive chemistry management practices. However, there are some general learnings that apply far beyond the scope of this particular incident, and I have listed what I feel are the most significant.

1. *The explosion did not occur because of disregard for the collection of any safety data.*

I had heard anecdotal stories of insufficient testing or attention to safety on the part of personnel, and found them to be completely untrue. Not only did the design team obtain a full set of shock sensitivity tests on what they thought would be the most hazardous material, they also changed their procedures sometime during operation of the plant to obtain shift samples of the column underflow to avoid concentrating polynitrated byproducts in the column underflow. Because of this change, we have the surviving underflow sample that provided a key to this analysis.

2. *Careful analysis, even years after the fact, can indeed reveal new information, as well as reinforce that which is already known.*

The use of residue curve mapping techniques combined with the ternary phase diagram revealed a combination of circumstances that was not anticipated in 1960. This has resulted in changes to our current chemical safety testing protocols.

3. *Safety culture, though hard to define precisely, is strongly influenced, and affected long term by significant events.*

Eastman instituted its Process Safety Review Committee (PSRC) structure immediately after this event, and it has continued to this day. Each site has a committee, chaired by the highest ranking site manufacturing official, which meets monthly to review potential process safety and reactive chemistry issues that affect the site. The result of this structure has been an order of magnitude reduction in the rate of serious incidents, sustained over a four decade period.

Another significant development from this incident was the development of a computer algorithm to estimate potential shock sensitivity and maximum energy release potential via the rearrangement of any molecular composition. This program, CHETAH, is now an ASTM standard (4).

4. *Review of those events can have a very positive, reinvigorating influence on an already established safety culture.*

I have given variations of this presentation to approximately 500 Eastman people, none of whom were on the active payroll in 1960. I often give it after lunch, which is sometimes difficult to maintain audience attention. I have NO problems with attention for this subject. People see buildings they walk by every day, and what they looked like after the explosion, and it has a profound effect on their thinking. You no longer have to answer the questions "Can it really happen here?" or "How bad can it really be?" After presenting people with the facts, they know that, without continued vigilance, the answers are "It already has" and "Really, REALLY bad." I have yet to see an Eastman audience leave a presentation without a change of attitude.

Acknowledgements

The author would like to dedicate this paper to the memory of the 16 Eastman employees who died as a result of this event. I would also like to thank the dozens of people who shared first-hand experiences and recollections, anecdotes and memorabilia, which helped make this effort worthwhile. While many contributed to this effort, a special note of thanks goes to Dr. Jeffrey Siirola, for his technical analysis of the incident presented here.

References

1. Reichler, J.H. (ed.), "The Baseball Encyclopedia," 6th edition, p. 2639 Macmillan Publishing, New York (1985)
2. White, Theodore H., "The Making of the President—1960", p 283 Athenaeum Press (1961)
3. Perry, J.H, and D.W. Green (ed.) "Chemical Engineering Handbook" 7th edition, pp 13-56 through 59, McGraw-Hill, New York (1997)
4. ASTM DS51D "Computer Program for Chemical Thermodynamic and Energy Release Evaluation—CHETAH® Version 7.3 ISBN 0-8031-2094-X

Figure 1

Aniline Chemistry

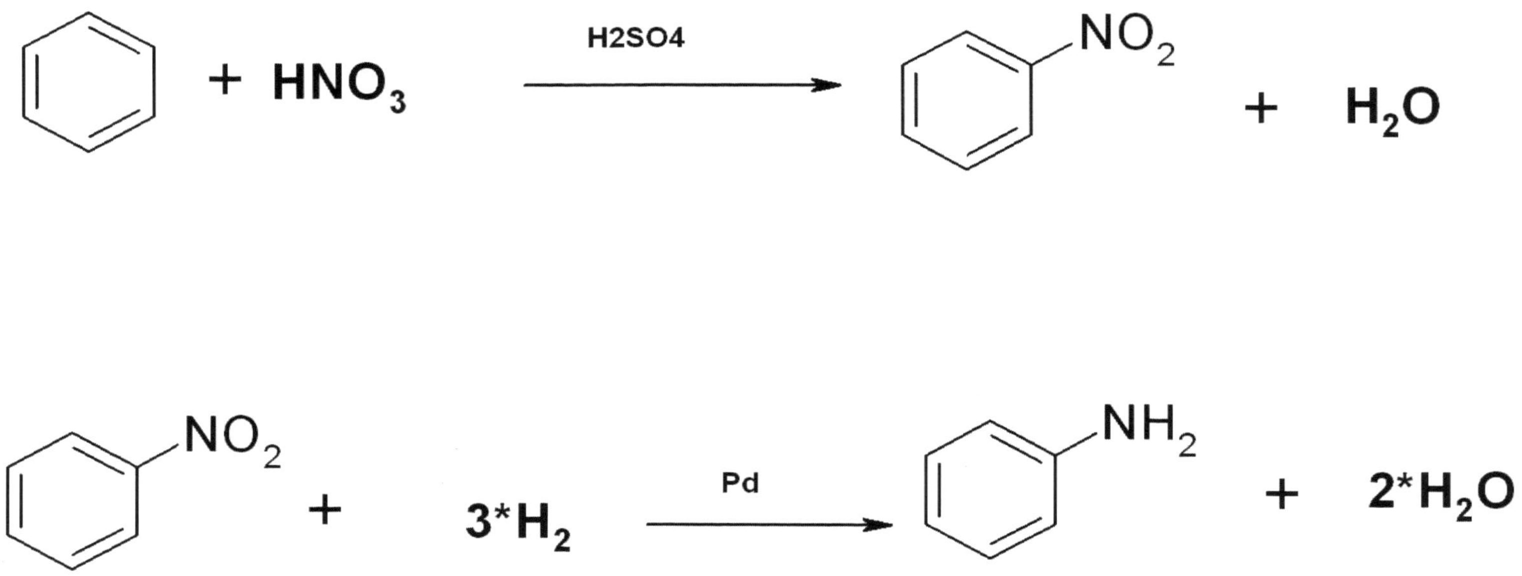

Figure 2

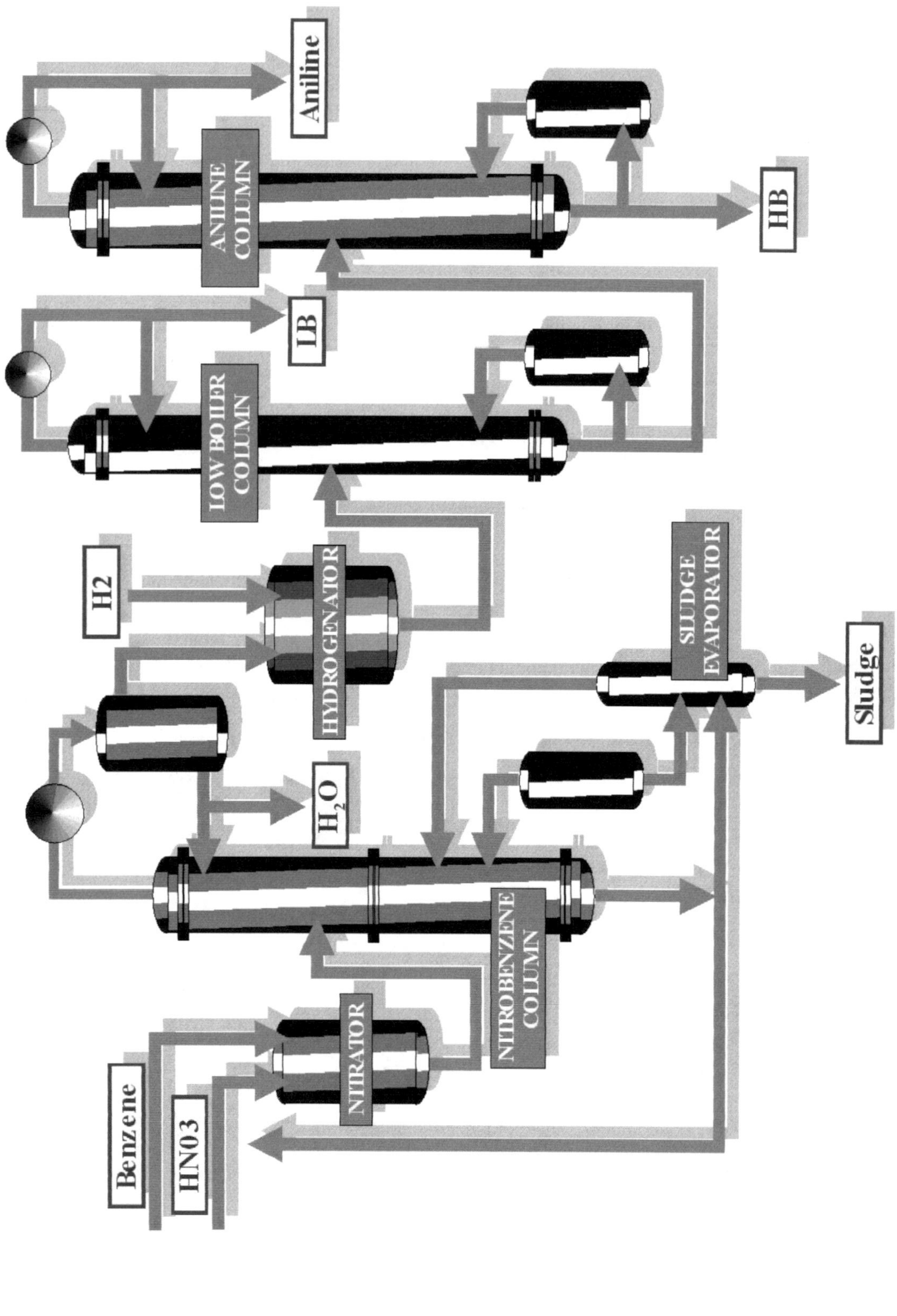
Benzene
HN03
NITRATOR
NITROBENZENE COLUMN
H2O
H2
HYDROGENATOR
SLUDGE EVAPORATOR
Sludge
LOW BOILER COLUMN
LB
ANILINE COLUMN
Aniline
HB

Figure 3

Figure 4

Figure 5

Figure 6

B-156 AT UPPER LEFT

Figure 7

Figure 8

Looking southwest from Building 207
Building 122 in background

Figure 9

East and north sides Bldg. 27

Figure 10

Figure 11

View from Bldg. 12- Bldgs. 122,118 & 143 in background

Figure 12

Figure 13

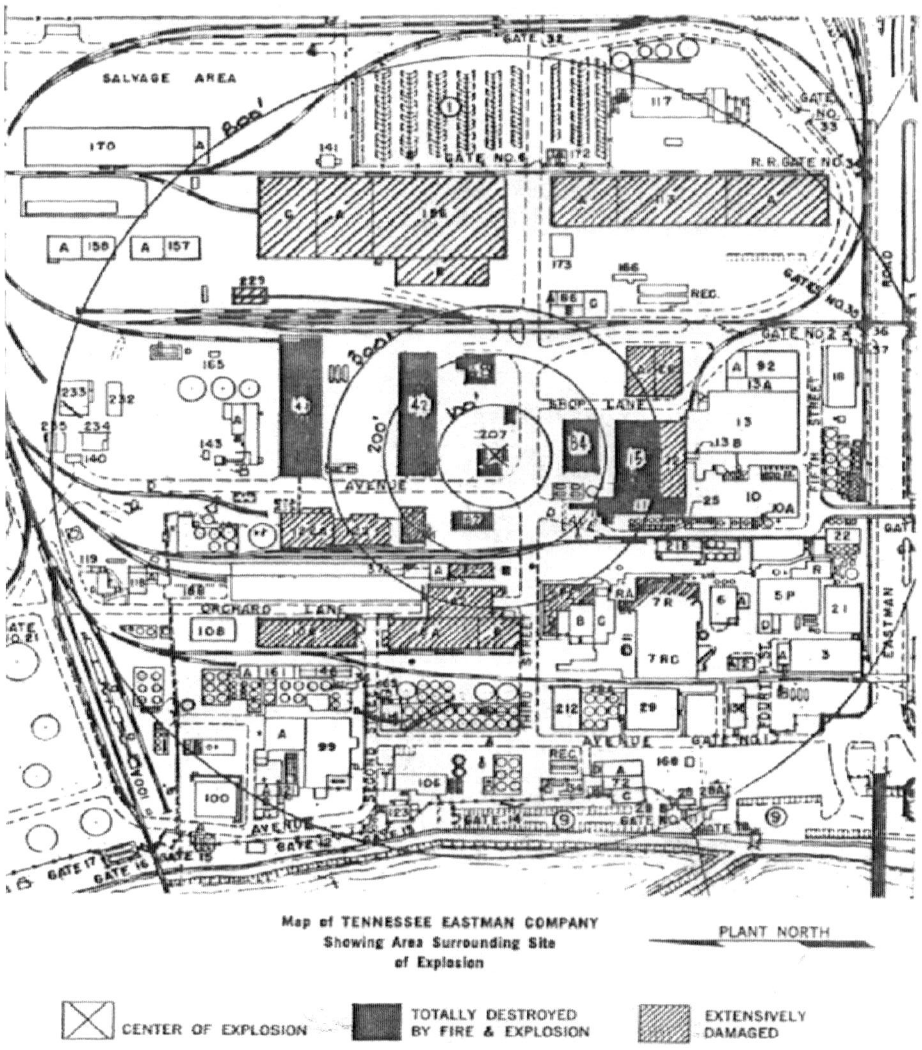
SALVAGE AREA
GATE 32
GATE NO. 6
GATE NO. 33
R.R. GATE NO. 34
GATE NO. 35
GATE NO. 2
EASTMAN ROAD
FIFTH STREET
FOURTH ST.
THIRD STREET
SECOND
AVENUE
ORCHARD LANE
200'
100'
207
Map of TENNESSEE EASTMAN COMPANY
Showing Area Surrounding Site
of Explosion
PLANT NORTH
CENTER OF EXPLOSION
TOTALLY DESTROYED BY FIRE & EXPLOSION
EXTENSIVELY DAMAGED

Figure 14

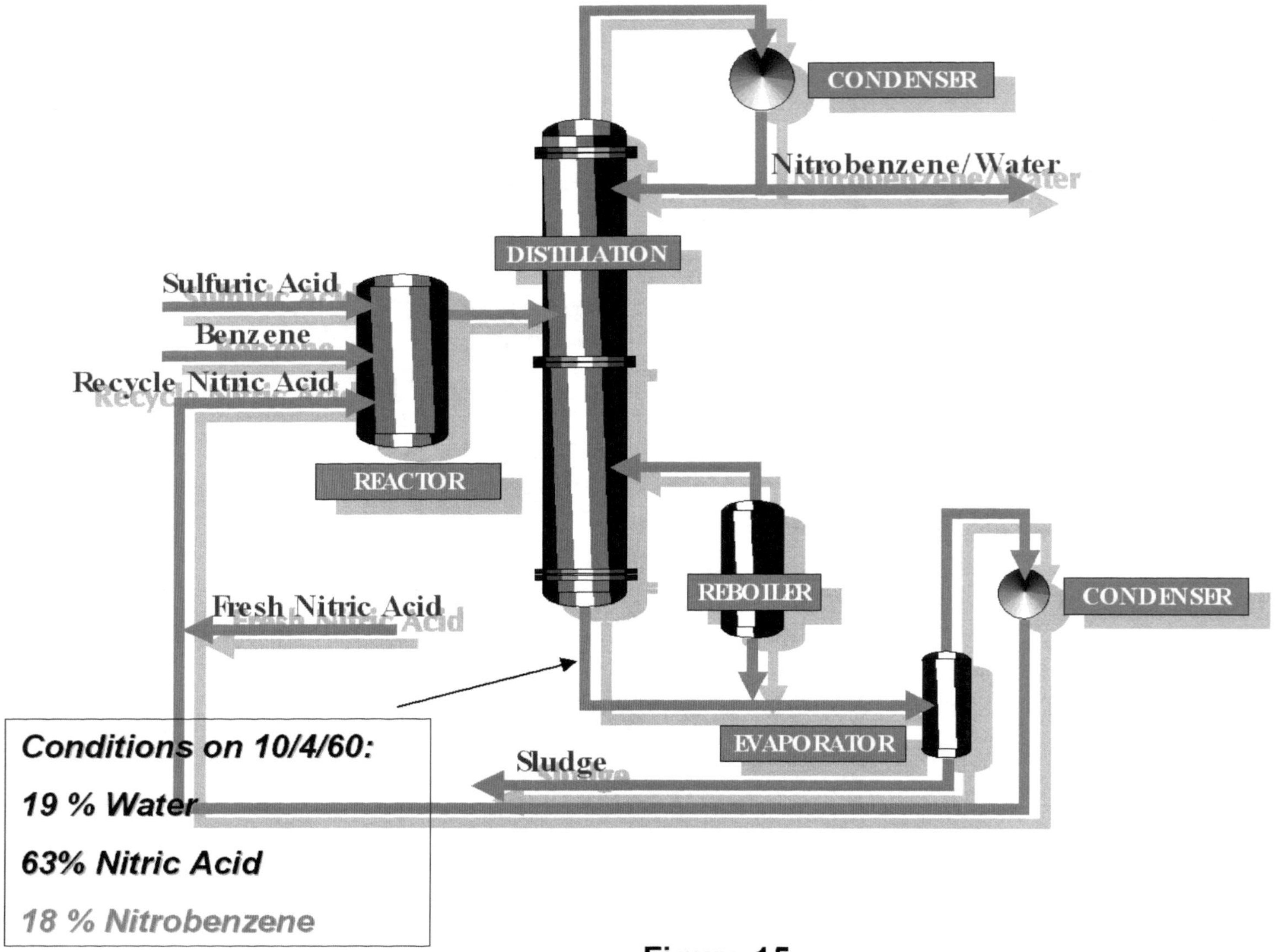
CONDENSER
Nitrobenzene/Water
DISTILLATION
Sulfuric Acid
Benzene
Recycle Nitric Acid
REACTOR
REBOILER
CONDENSER
Fresh Nitric Acid
EVAPORATOR
Sludge
Conditions on 10/4/60:
19 % Water
63% Nitric Acid
18 % Nitrobenzene

Figure 15

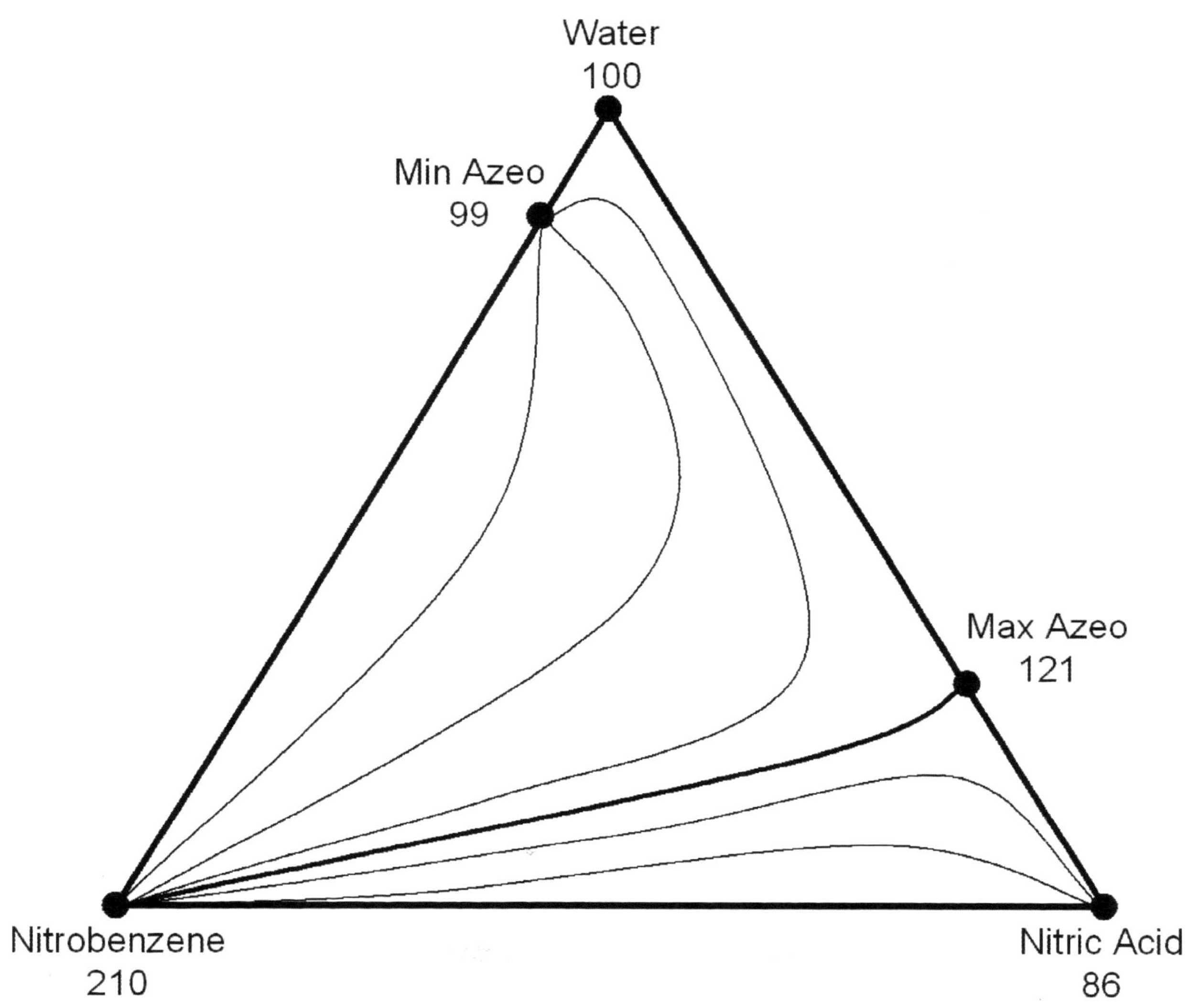
Water
100
Min Azeo
99
Max Azeo
121
Nitrobenzene
210
Nitric Acid
86

Figure 16

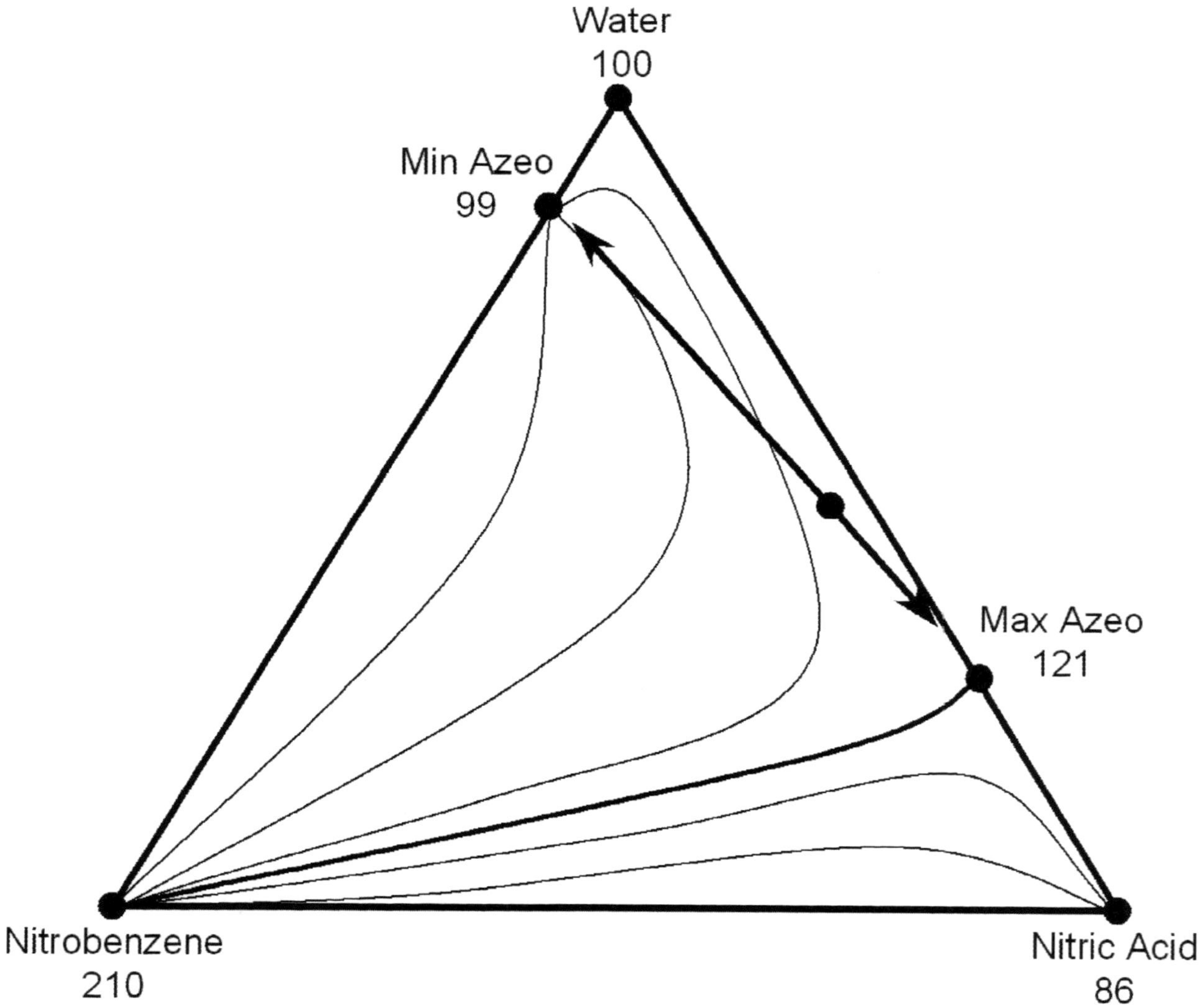
Water
100
Min Azeo
99
Max Azeo
121
Nitrobenzene
210
Nitric Acid
86

Figure 17

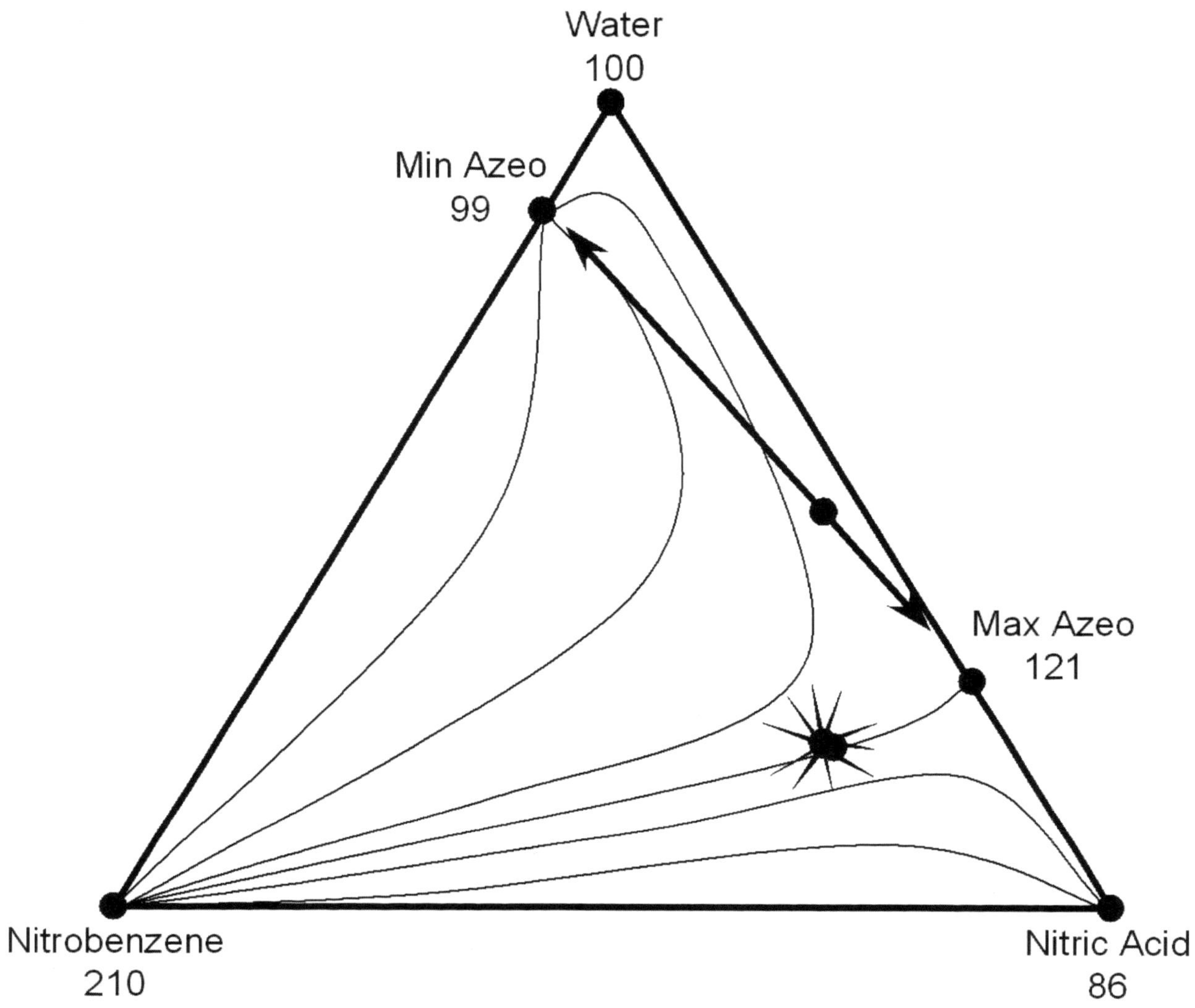
Water
100
Min Azeo
99
Max Azeo
121
Nitrobenzene
210
Nitric Acid
86

Figure 18

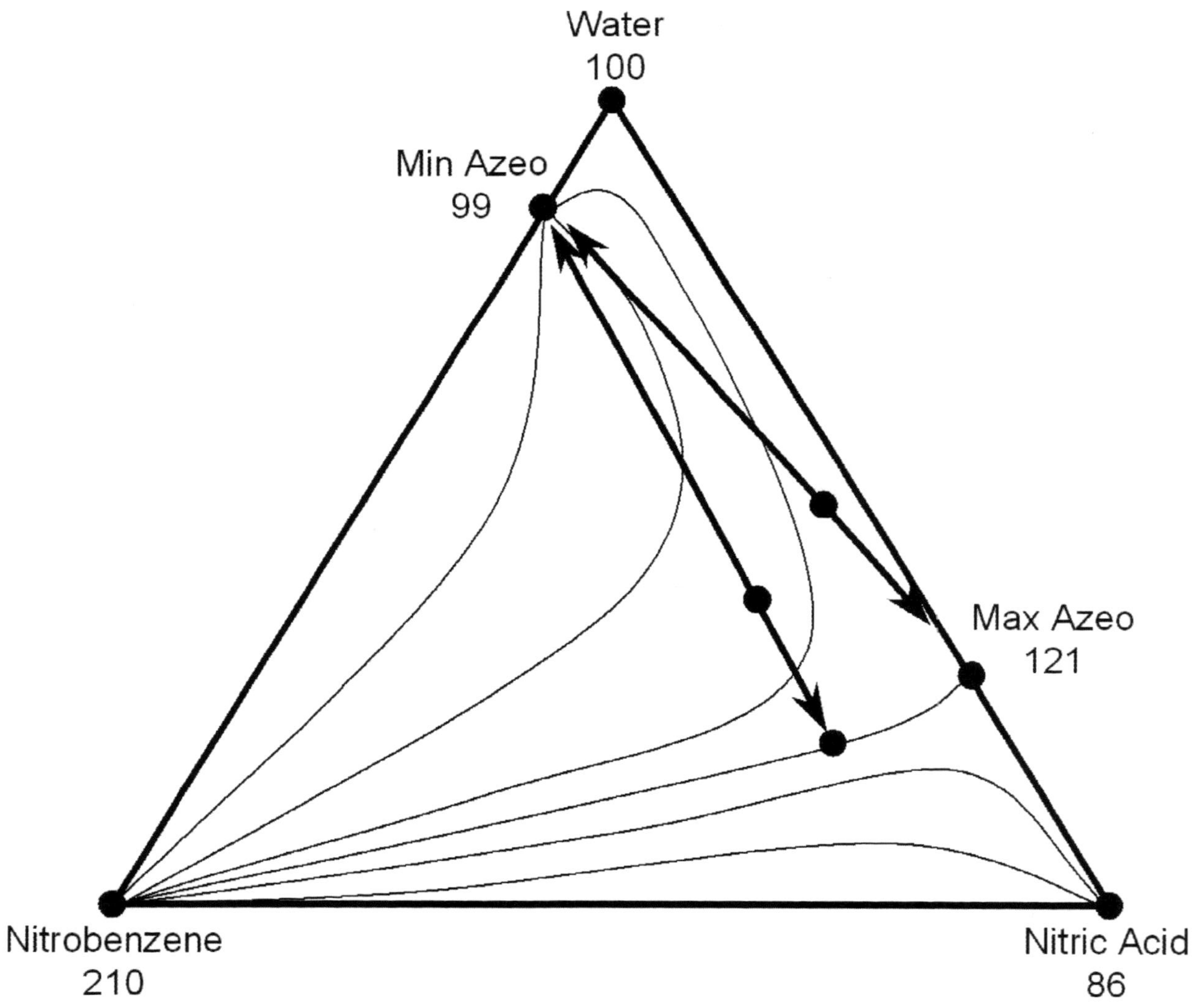
Water
100
Min Azeo
99
Max Azeo
121
Nitrobenzene
210
Nitric Acid
86

Figure 19

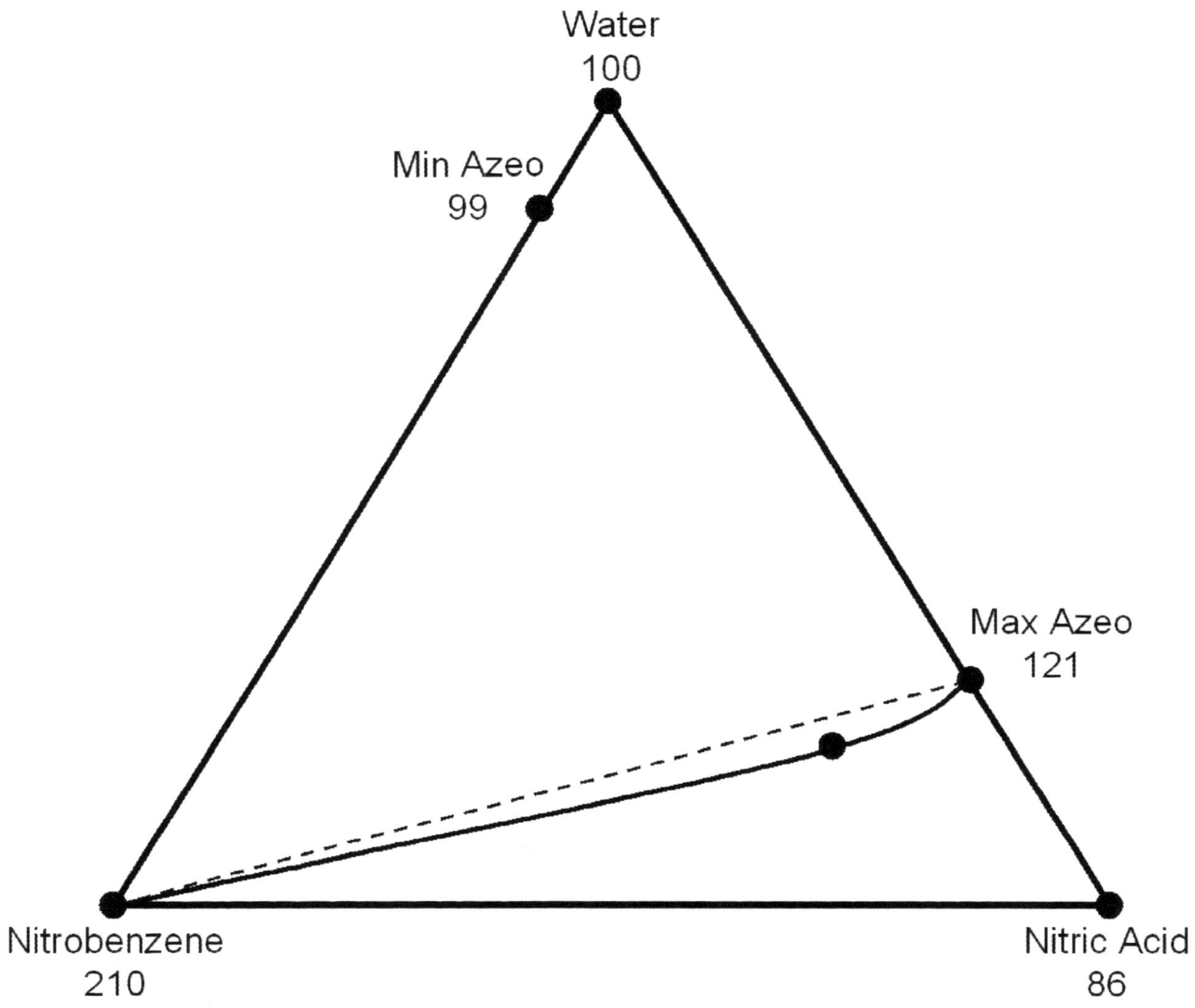
Water
100
Min Azeo
99
Max Azeo
121
Nitrobenzene
210
Nitric Acid
86

Figure 20

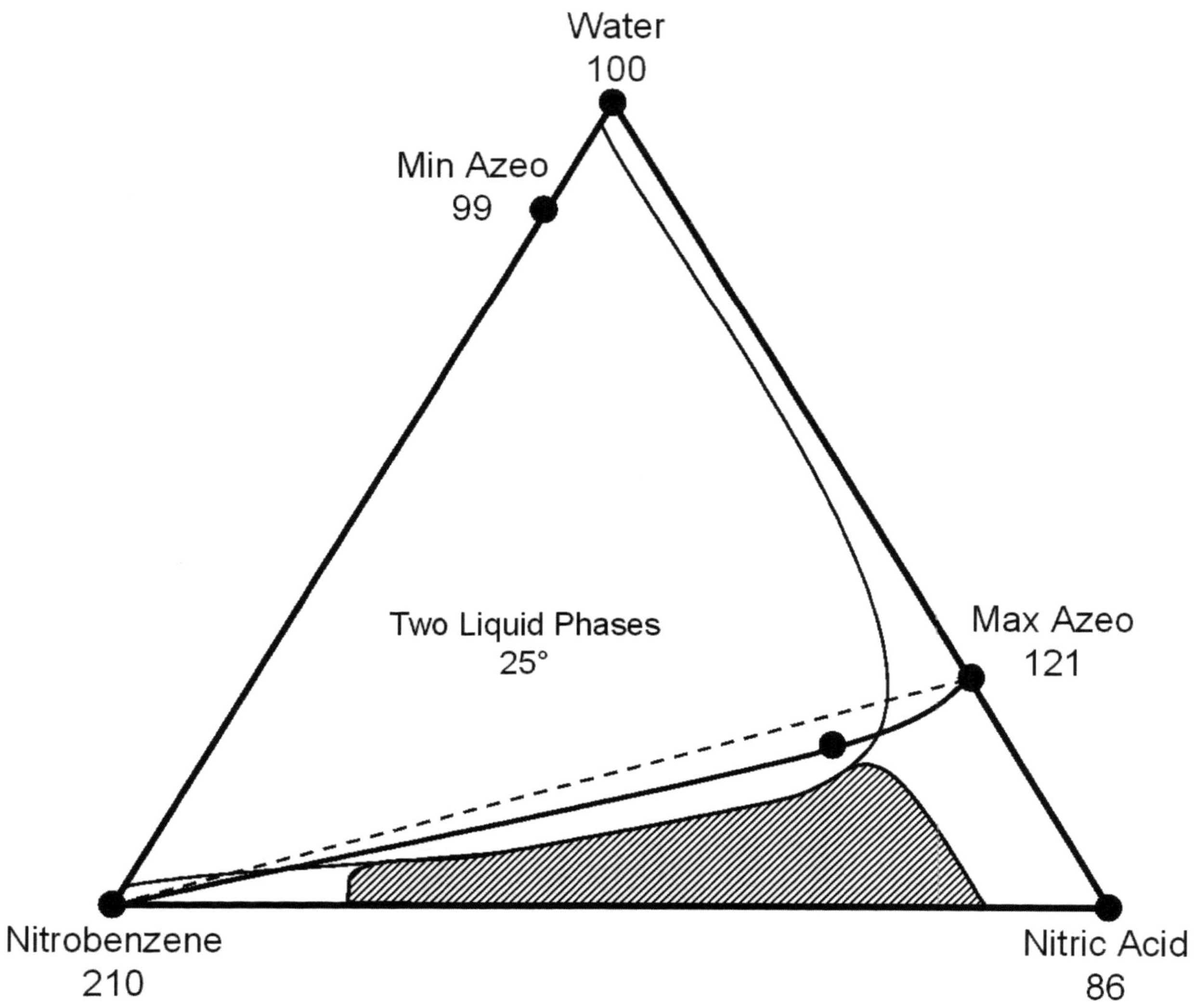
Water
100
Min Azeo
99
Two Liquid Phases
25°
Max Azeo
121
Nitrobenzene
210
Nitric Acid
86

Figure 21

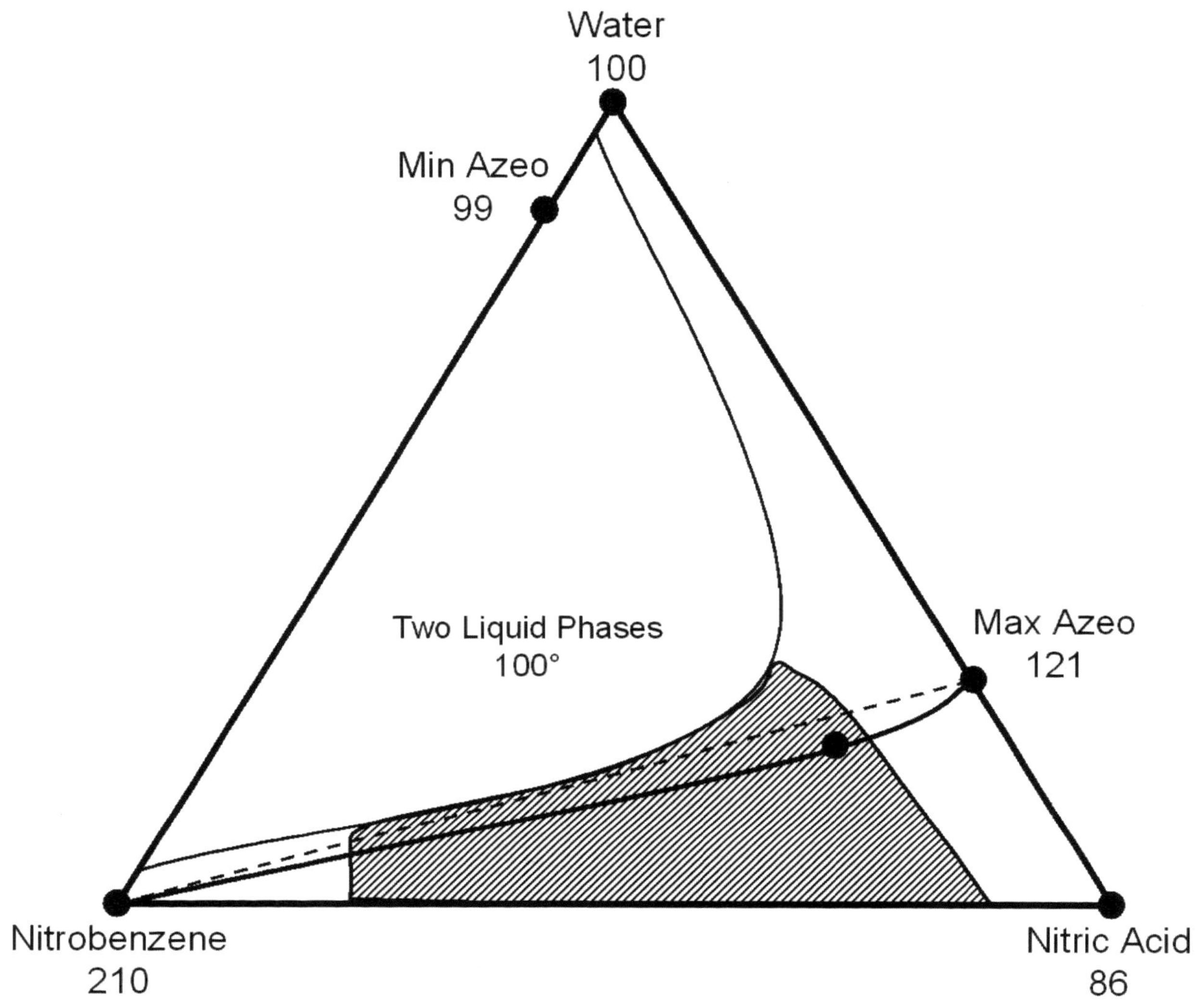
Water
100
Min Azeo
99
Two Liquid Phases
100°
Max Azeo
121
Nitrobenzene
210
Nitric Acid
86

Figure 22

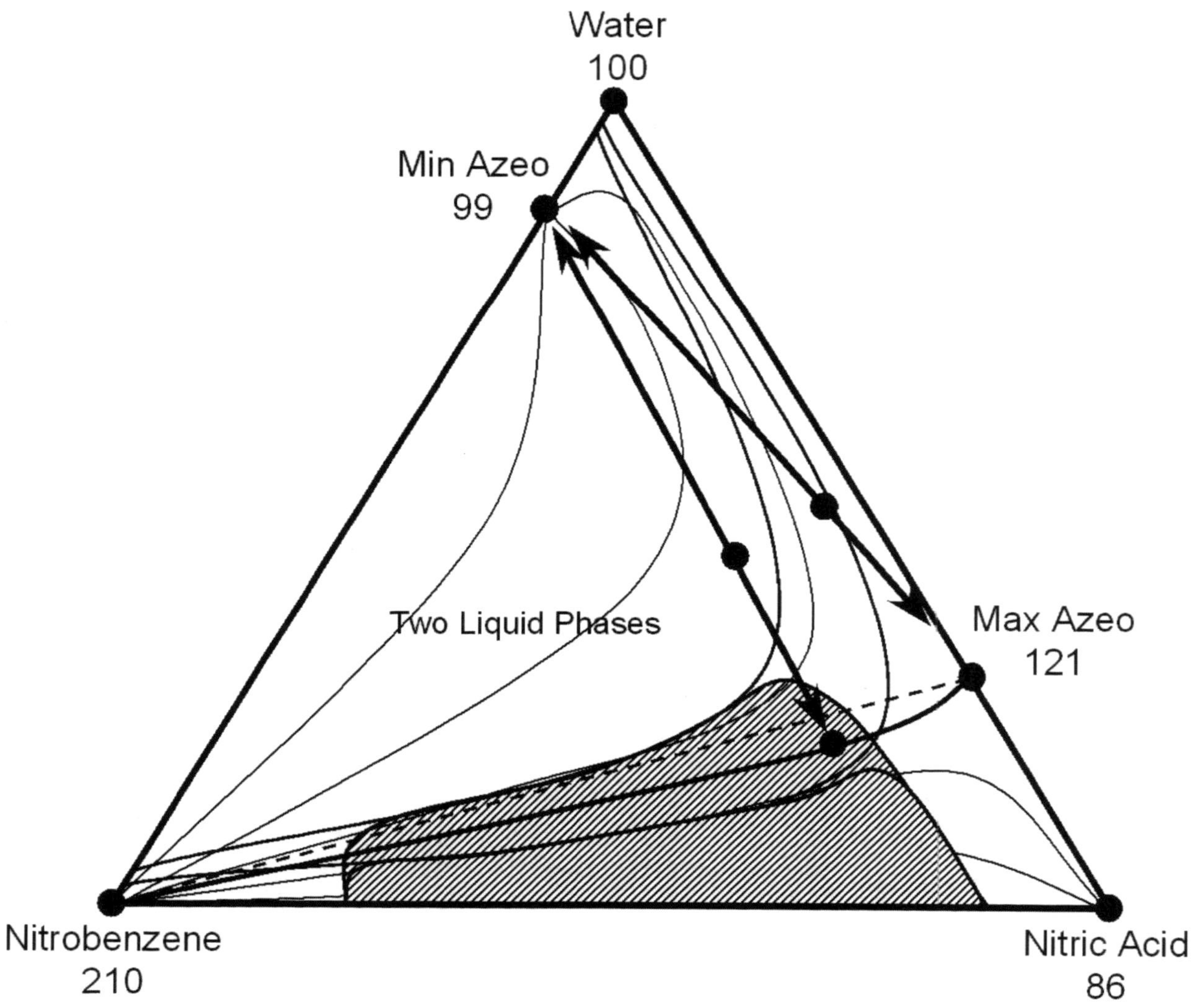
Water
100
Min Azeo
99
Two Liquid Phases
Max Azeo
121
Nitrobenzene
210
Nitric Acid
86

Figure 23

38th Annual Loss Prevention Symposium, American Institute of Chemical Engineers, New Orleans, Louisiana, April 2004.

LESSONS LEARNED FROM FIRES AND EXPLOSIONS INVOLVING AIR POLLUTION CONTROL SYSTEMS

Russell A. Ogle, Andrew R. Carpenter and Delmar "Trey" Morrison, III
Exponent
Chicago, Illinois

Six case studies of fires or explosions involving air pollution control (APC) systems are reviewed in this paper. These case studies have been generalized from actual accident investigations performed by the authors.

The case studies cover the APC technologies of thermal oxidation, catalytic oxidation, gas scrubbers, adsorption and condensation. Each of the technologies was used to control volatile organic compounds (VOCs) emissions. The accidents encompass a broad range of industries, ignition sources and circumstances. The causal factors for these accidents are compared with applicable safety guidelines and standards to show how safeguards could have prevented or mitigated these accidents.

The common theme that emerges from these accident investigations is that APC systems should not be specified and installed strictly by intuition or experience, but rather through engineering design. The key findings of this study are:

- Characterize the waste stream to be treated.
- Conduct a process hazard analysis for each APC system with particular emphasis on fire and explosion hazards.
- Design the APC system using good engineering practices.
- Operate the APC system within its design specifications.
- Periodically verify that the APC system performance satisfies its technical and regulatory objectives.
- Perform maintenance activities in accordance with manufacturer's recommendations.

Each of the accidents was the direct result of the omission of one or more of these basic tenets.

INTRODUCTION

The USEPA estimates that over 20,000 industrial facilities must comply with the requirements of the 1990 Clean Air Act Amendments (CAAA) [USEPA, 1998]. Most of these facilities install and operate air pollution control (APC) systems to achieve compliance with the environmental regulations. These APC systems have historically been and continue to be a source of fires and explosions. This correlation is due to the nature of the pollutants being controlled (i.e. volatile organic compounds or VOCs are often flammable). APC systems are often "plugged in" to an existing process with inadequate effort given to process hazard analysis. This paper presents the lessons learned from accident investigations involving APC systems performed by the authors. Specific details of the case studies have been omitted or altered to protect the identity of all concerned parties.

BACKGROUND

Although there is a wide range of chemicals regulated by the CAAA, the focus of this paper is on VOCs. The fire and explosion hazards associated with VOCs have been a concern of the chemical industry since the inception of the Clean Air Act (promulgated in 1970). Bodurtha [1976] reported that from 1951 to 1974, a time period of 24 years, only three fires or explosions involving APC equipment were reported in the Manufacturing Chemists Association database. Based on his own experiences at DuPont, and his informal discussions with representatives of Industrial Risk Insurers, he concluded that accidents involving APC equipment were being under-reported. He further stated, "…proper recognition of the hazard of the total system and not simply the pollution control device by itself is essential to safety" (emphasis in the original text) [Bodurtha, 1976].

In 2000 the American Petroleum Institute (API) published a survey of incidents involving air emission controls installed since 1990 [Ozog and Erny, 2000]. The survey was based on 20 API-member facilities (16 refineries, one gas plant, two chemical plants and one marketing terminal) and covered the period from 1990 to 1998. During that span, over 2,100 APC systems had been installed at these 20 facilities. Seventy incidents were reported in the survey: 16 accidents (i.e., an unplanned event resulting in either human injury or property damage) and 54 near-misses. Two treatment technologies stood out in this survey: carbon adsorption accounted for the largest number of incidents (19 out of 70) and thermal oxidizers accounted for 10 incidents.

Neither the Bodurtha accident statistics nor the API study gives a comprehensive picture of fire or explosion incidence rates with APC systems. Recognizing the limitations of this information, the data apparently indicate that the annual incidence rate for fires or explosions in APC systems has increased. This increase is likely due to the increased number of APC systems being installed and operated as a result of the CAAA. The statistical data are not available in the public domain to determine the accident incidence rate normalized against the number of APC installations. But the anecdotal data suggests that there is an urgent need to increase the awareness of fire and explosion hazards in APC systems in the technical community.

SYSTEM SAFETY ANALYSIS OF APC SYSTEMS

System safety is the systematic identification, evaluation and control of hazards in products and processes. The chemical process industries have had a certain degree of success in implementing system safety methodology into the design, operation and maintenance of their facilities. Much of this success has been spurred by process safety management regulations administered by the Occupational Safety and Health

Administration (OSHA) and the risk management program of the US Environmental Protection Agency (USEPA). There are numerous guidance documents available to the process engineer that illustrates the use of system safety techniques. For example, the Center for Chemical Process Safety (CCPS) of the American Institute of Chemical Engineers (AIChE) has published almost 40 books on applications of system safety to chemical processes. One book in particular [CCPS, 1993] discusses system safety considerations for the APC technology of carbon adsorption.

Unfortunately, the use of APC systems transcends the chemical process industries. A similar resource for system safety applications does not exist for the printing or food processing industries, two very large users of APC technology. Design books devoted to APC technology rarely discuss fire and explosion safety [e.g., see Cooper and Alley, 2002], and the design guides available from USEPA are largely silent on the topic [e.g., see *The EPA Air Pollution Control Cost Manual*, 2002]. The National Fire Protection Association (NFPA) has published a number of standards and guides that are applicable to fire safety in APC systems [see the review by Myers, Martin and Kytomaa, 2002]. The fire safety objectives typically advocated in the NFPA guidance documents can be expressed as a hierarchy from more protective to less protective of property:

1. Prevent ignition
2. If ignition occurs, prevent fire spread (or flame propagation)
3. If fire spread occurs, minimize damage

In the context of an APC system, this hierarchy might be modified as follows:

1. Prevent ignition
2. If ignition occurs, confine damage to equipment of origin
3. If fire spread occurs, confine damage to APC system

To achieve these objectives, one can employ several hazard control strategies. The system safety precedence is a hierarchy of hazard control strategies ranked from greater control to less control:

1. Eliminate the hazard
2. Use physical safeguards to control the hazard
    i. Passive physical safeguards
    ii. Active physical safeguards
3. Use warning devices (e.g., monitoring devices interlocked with alarms)
4. Use procedural safeguards
5. Accept the hazard

The first two strategies are often referred to as engineering controls and the second two as administrative controls.

The fire triangle is a useful concept for organizing hazard control actions. The fire triangle simply states that a fire (or explosion) requires fuel, air (or any other appropriate oxidizer) and an ignition source. Thus, the hazard control actions available to the process engineer are to control each of the three components of the fire triangle: the fuel, air or ignition source. These actions may be taken individually or in some combination. Table 1 illustrates a number of hazard control actions that can be implemented to protect against fire and explosion hazards in APC systems.

Table 1. APC hazard control strategies and actions

Hazard Control Strategy	Specific Actions
Eliminate hazard	Eliminate VOC vapor
Use passive physical safeguards	1. Bonding and grounding 2. Explosion-proof electrical service 3. Design for explosion containment 4. Construct APC system with noncombustible components 5. Flame arresters which quench flame propagation
Use active physical safeguards	1. Reduce the fuel concentration in gas stream below the lower explosive limit (LEL) using excess air 2. Reduce the oxygen concentration in the gas stream below the minimum oxygen level required for combustion using an inert gas 3. Use interlocks to shut APC system down in the event that a hazardous condition is detected 4. Install fire suppression systems (sprinklers, fire extinguishing systems) that actuate in the event a fire is detected
Use warning devices	1. Monitor fuel or oxygen concentration and signal alarm if hazardous condition is detected 2. Monitor and signal alarm if a fire is detected
Use procedural safeguards	Provide procedures for safe operation of the APC system: 1. Start-up 2. Normal operation 3. Recovery from a process upset 4. Normal shutdown 5. Emergency shutdown

SURVEY OF CASE STUDIES

The six case studies described below resulted in one fatality, 13 injuries and over $20 million in property damage. Two of these accidents occurred in the chemical process industries, the remaining four were in the general industrial sector. The case studies are organized according to the primary technology in use for the APC system.

Case Study 1: Thermal Oxidizer

A food processing plant had installed a new APC system approximately two years prior to the accident. The APC system treated one primary air pollution source, the exhaust from a continuous flow, fluidized bed dryer (oven). The feed stream to the APC system consisted of a hybrid mixture of flammable vapor (similar to vegetable oil) and a combustible dust in an air stream. The APC treatment technology involved a series of cyclones with a downstream thermal oxidizer. The feed to the oven unit was a bulk material that was transported through the oven on a fluidized conveyor system. The oven unit operated on manual control. Process measurements consisted solely of field-mounted equipment (temperature only). Measurement data was recorded manually on an infrequent basis. Typical operation required frequent operator intervention to resolve material handling problems with the bulk material.

The accident occurred during normal operation of the oven. The operator intervened to address a commonly observed conveyor drive failure. After fixing the problem, the operator restarted the feed stream and the oven's fluidized air blowers. A subsequent explosion and fire throughout the APC system destroyed the oven and killed the operator. A likely ignition source was the downstream thermal oxidizer.

The investigation revealed that over a period of several years, production had gradually increased to twice the design capacity of the APC system. No air monitoring was performed after the APC system was installed to verify its performance. There was no LEL monitoring on the feed stream to the thermal oxidizer and no flame arrester. The root causes of this accident were lack of an adequate design basis for the APC system. Specifically, the as-built design did not consider the flammability of the hybrid VOC/dust stream and did not incorporate flame suppression considerations.

Case Study 2: Catalytic Oxidizer

A printing facility operated an APC system for treating VOCs (mostly toluene, acetone, methyl ethyl ketone and methanol) in air. The APC system was fed by multiple VOC sources, some continuous and some episodic. The treatment technology was a catalytic oxidizer. A forced draft blower conveyed the effluent stream to the oxidizer. A continuous LEL detector monitored the flammable vapor concentration at the oxidizer inlet. The LEL detector was interlocked with a system of dampers. The purpose of the interlock was to prevent the flow of heavily solvent-laden air to the oxidizer if a process upset condition was detected, e.g., if the fuel concentration exceeded 50% of the LEL. In the event of a process upset, the interlock would divert the feed stream from the oxidizer and discharge it to the atmosphere.

On the day of the accident, a process upset condition (a flame-out in the oxidizer) was detected and the dampers automatically diverted all but one source of solvent-laden air from the oxidizer. The one VOC source not diverted was from a parts washing station. The parts washing station was an episodic source, i.e., it operated periodically for discrete periods of time. Unfortunately, it was being operated during the oxidizer upset. This allowed a slug of solvent-laden air to be conveyed to the oxidizer. Although the oxidizer was in a flame-out condition, the catalyst bed was at a temperature well above the autoignition temperatures of the VOCs being treated. Thus, the hot catalyst bed was the probable ignition source. The resulting explosion damaged the oxidizer and associated ductwork. There were no injuries.

An outside engineering firm designed the original APC system for an assortment of process and fugitive emission sources. The facility operator installed the parts washer approximately one year later without any engineering design review effort. The parts washer was installed without interlock to the LEL monitor. Therefore after the process upset was detected, the parts washer vapor effluent was not diverted to the atmosphere. The root causes were the failure to consider the safety of the parts washer integration to the APC and failure to install an interlock between the parts washer and the oxidizer.

Case Study 3: Regenerative Thermal Oxidizer/Gas Scrubber System

This facility recovered solvents from liquid hazardous wastes. The APC system treated multiple chlorinated VOC sources. The treatment technology was a regenerative thermal oxidizer followed by a gas scrubber to remove the hydrogen chloride vapor prior to atmospheric discharge. The APC system was installed as a design-build project. The facility conducted a process hazards analysis (specifically, a HAZOP) with the design-build contractor.

A power outage occurred forcing an emergency shutdown of the plant. Airflow to the thermal oxidizer, provided by an electrically operated blower, stopped immediately. The flow of scrubbing solution, provided by an electrically operated pump also stopped operating. The dampers for the thermal oxidizer were designed to fail either open or closed in the event of a power outage (depending on their location in the system). The fresh air damper failed in the open position and the blocking damper, which controlled the flow of the process stream, failed closed. After sitting idle for approximately eight hours, a fire was discovered in the scrubber. The scrubber, filled with plastic packing and constructed of glass reinforced plastic, was completely destroyed by the fire. There were no injuries.

The investigation concluded that fresh air flowing due to natural convection through the still hot oxidizer ignited the scrubber packing. Root causes included the improper design configuration of the oxidizer (fresh air intake located below the elevation of the oxidizer discharge), which allowed a natural draft to form (i.e., a "chimney effect"). The emergency quench in the supply duct was determined to be inadequate to fully cool the oxidizer during such a shutdown. Kletz [1998] has discussed other accidents caused by the chimney effect. Although the information is of an anecdotal nature, it appears that this type of accident resulting in scrubber damage or fire has happened to others [see Sorrell, 2002 and Banks, 2003].

Case Study 4: Adsorption Column/Recuperative Thermal Oxidizer

An electronic device manufacturing company installed an APC system to control their VOC emissions from a dilute air exhaust stream. The APC system consisted of a VOC concentrator unit and a recuperative thermal oxidizer. The VOC concentrator was necessary because the VOC concentration in the gas stream was too low to be effectively treated. The VOC concentrator was an adsorbent column that operated in alternating cycles of adsorption and desorption. The concentrated VOC stream generated by the desorption cycle was then treated in the thermal oxidizer. The adsorption system used a proprietary polymeric adsorbent.

After several months of operation the facility operator found that the APC system was not achieving the required VOC treatment level. The design-build contractor determined that the desorption cycle was not regenerating the adsorbent to the desired degree. Their recommendation was to increase the desorption cycle temperature. The facility operator followed this advice and three weeks later experienced a fire within the adsorption system. Although there were no injuries, the resulting fire damage destroyed the APC system and caused a significant interruption in the facility's operation.

Prior to suggesting a higher desorption temperature, the contractor had performed thermal stability testing on the adsorbent and determined that under stagnant conditions the adsorbent could undergo self-heating. The contractor failed to appreciate the significance of the test results. One of the root causes was the use of a desorption temperature, which was recommended by the contractor, that was higher than the thermal onset temperature for the adsorbent.

Case Study 5: Carbon Adsorption Drum

A polymer producer stored liquid monomer (a vinyl ester compound) in an atmospheric storage tank. A carbon adsorption drum system was installed on the tank's vent pipe to reduce VOC emissions. The adsorption drums were connected to the storage tank with pipes without integral pressure/vacuum relief devices or flame arresters. The atmospheric tank was equipped with a conservation vent open to the atmosphere to allow the tank to breath in and out.

Without warning the atmospheric storage tank ruptured and its liquid contents ignited. Facility personnel were able to quickly extinguish the fire. There were no injuries. The next day a maintenance mechanic noticed that one of the carbon drums was hot to the touch. The drum was removed from the vent piping and moved to an isolated location. The drum temperature was observed to increase from 48°C to 60°C in 24 hours. The drum was quenched with water.

The explosion occurred three days after spent carbon drums had been replaced with fresh ones. It was determined that the flashpoint of the liquid monomer corresponded roughly with the ambient temperature prior to the explosion. This implied that the fuel vapor concentration in the tank headspace and vapor piping would have been on the order of

>15,000 ppmv (parts per million by volume). Although the literature does not indicate that a vinyl ester compound would have a high potential for causing self-heating in an activated carbon bed, the high vapor concentration does indicate a potential hazard. The root causes included the failure to install flame arresters between the carbon drums and the tank and use of air as a headspace gas in the storage tank. These factors created a flammable condition and allowed flames to propagate back into the storage tank. This hazard scenario is predictable: the documentation of fire hazards with carbon adsorption systems is probably the most extensive for any of the APC treatment technologies reviewed in this paper [see e.g., Naujokas, 1979; CCPS, 1993; USEPA, 1997; and Zerbonia, et al, 2001].

Case Study 6: Refrigerated Condenser

A toll manufacturer operated a production unit that made a specialty nitrated organic compound. The process involved several batch reactors and separation units. The primary raw material, a chlorinated aromatic compound, was the source of VOC emissions. An inert gas stream transported the VOC vapor to a refrigerated condenser. Sometime after the initial installation, the manufacturer changed the APC system to recycle condensed VOC back into the front end of the process.

The process effluent vapor stream contained the primary VOC and a previously undocumented trace quantity of a strongly oxidizing compound. The trace compound accumulated within the process and subsequently affected process performance. During non-routine maintenance, the accumulated oxidizer was transferred out of a process vessel into a detention area containing a mixture of organics and other reactive compounds. The transfer led to a catastrophic explosion that destroyed that area of the facility but caused no injuries.

No hazard analysis was performed on the design changes prior to implementation, which caused a significant APC system change by capturing all condensable vapors. No sampling of in-process materials was performed to determine the cause for poor process yield performance. Analysis of these in-process samples may have identified the previously unknown presence of the oxidizing compound. The transfer that ultimately led to the explosion was a first time event that was contrary to the facility's typical stream handling procedures. The root causes of this accident included the failure to do a comprehensive hazard analysis to consider the consequences of incorporating the APC system into the process unit. On the contrary, the APC system was still treated like a downstream unit. Also, the unique stream transfer event that led to the accident was contrary to accepted practices.

LESSONS LEARNED

The six case studies presented above reveal several lessons that can be taken from these accidents. APC systems are engineered systems and therefore must be operated within the design envelope for safe and proper operation. It can be seen that deviation from the design or unanticipated upsets can result in damage to the unit and possibly the facility. Other more specific lessons learned are presented as follows:

Gas Scrubbers May Be Vulnerable to Heat Exposure

The thermal or catalytic oxidizers typically connected to gas scrubber towers typically operate at temperatures far greater than the ignition temperatures of the polymeric materials used in the construction of the gas scrubber. Care must be taken in the design the ensure that no flow conditions can develop which would allow very hot gasses to contact the combustible scrubber packing material or combustible structural material. Under normal operation, cooling water typically protects these materials, however in the event of a loss of cooling water or other unanticipated flow condition, this may result in heat-induced structural failure or ignition of the gas scrubber. Quench systems and flow dampers must be designed and sized to correctly eliminate this hazard.

Many Gas Adsorption Systems Are Susceptible to Self-Heating

The chemical processes of adsorption and desorption tend to liberate heat. Additionally, the material used as the adsorbent may also be capable of self-heating, either as virgin material or in it's solvent charged state. The airflows and operating conditions must be established to efficiently mitigate these heat sources. Further, the system must be designed to eliminate any sources of particle accumulation and stagnant regions where self-heating can lead to thermal runaway.

Characterize Your Waste Streams

The incoming waste stream and effluent streams of APC systems must be well characterized. Both the chemical composition and range of anticipated concentrations must be established to successfully establish the design envelope for the unit. A lack of understanding of the waste streams being fed into the unit creates a significant risk in the operation of the unit and further reduces the confidence that the unit is performing to the required efficiency. Failure to understand the chemical inputs into the system may effectively nullify the design basis and worse, may preclude the ability to safely operate the equipment.

CONCLUSIONS

This paper reviewed six case studies that covered APC fires and explosions in many industrial sectors. The common factor involved in the APC installations is the necessity of handling flammable vapors safely. The most stringent hazard control action of removing the hazard is typically not applicable to these systems. This places a greater emphasis on the need to adequately consider the chemical, thermal and physical properties of not only the APC feed streams, but also the APC equipment and its integration into the process. The effects of predictable process variations on the APC system's performance should be a prime factor in establishing the design envelope for safe operation.

Even in the more safety-aware chemical process industry, APC systems tend to be treated outside of the typical process unit safety systems. In fact, these case studies demonstrate that the typically high-temperature processes required for oxidation, adsorption and desorption cause these systems to be perhaps more susceptible to fires and explosions. This high hazard supports the need for engineering design of these systems. An integral component of this design process is to evaluate the hazards posed by varying feed streams and predictable upset conditions. Each initial design, design modification (e.g. addition of additional feed streams or process unit change), or APC operating parameter modification should be accompanied by chemical sampling to assess both the feeds and effluents from the APC unit. APC systems should be engineered systems based upon good engineering practice.

REFERENCES

Banks, D. Banks Engineering. "Scrubber Problems: Thermal Damage," at <http://www.banksengineering.com> (April 2003).

Bodurtha, F.T. "Explosion Hazards in Pollution Control," 81st AIChE National Meeting, Kansas City, Missouri (1976).

Center for Chemical Process Safety. *Guidelines for Engineering Design for Process Safety.* American Institute of Chemical Engineers, New York pp. 147-149 (1993).

Cooper, C.D. and F.C. Alley. *Air Pollution Control, Third Edition.* Waveland Press, Inc., Prospect Heights, Illinois (2002).

Kletz, T.. *What Went Wrong?, Fourth Edition.* Gulf Publishing Company, Houston, Texas, pp. 323-325 (1998).

Mussatti, D.C. *The EPA Air Pollution Control Cost Manual.* U.S. Environmental Protection Agency, EPA 453/B-02-001 (January 2002).

Myers, T.J., H.K. Kytomaa and R.J. Martin. "Fires and Explosions in Vapor Control Systems: a Lessons Learned Anthology," *Process Safety Progress*, 22: 195-199 (2003).

Naujokas, A.A.. "Preventing Carbon Bed Combustion Problems," Chemical Engineering Progress. Volume 12: Loss Prevention. American Institute of Chemical Engineers, New York, pp. 128-133 (1979).

Ozog, H. and W.J. Erny. "Safety Hazards Associated with Air-emission Controls," *Process Safety Progress*. 19: 25-31 (Spring 2000).

Sorell, G. Letter to the Editor of *Chemical Processing* (December, 2002) in response to the article by J. Mason and A. van Buren, "Handling the Heat," *Chemical Processing* (October 2002).

USEPA. Fire Hazard from Carbon Adsorption Deodorizing Systems. U.S. Environmental Protection Agency, EPA 550-F-97-002e (May 1997).

USEPA. Air Pollution Operating Permit Program Update. U.S. Environmental Protection Agency, EPA 451-K-98-002 (February 1998).

Zerbonia, R.A., et al. "Carbon Bed Fires and the Use of Carbon Canisters for Air Emissions Control on Fixed-Roof Tanks," *Journal of the Air & Waste Management Association*. 51:1617-1627 (2001).

Air Compressor Demister Fire and Explosion
A Case History

J. Kelly Thomas, Ph.D.
Baker Engineering and Risk Consultants, Inc.
3330 Oakwell Ct., Suite 100
San Antonio, TX 78218-3024
(210) 824-5960
kthomas@BakerRisk.com

Prepared for Presentation at

American Institute of Chemical Engineers
38th Annual Loss Prevention Symposium
AIChE 2004 Spring National Meeting

December 2003
UNPUBLISHED

ABSTRACT

This case history illustrates an explosion at a plant powerhouse in which a number of large air compressors were operating. A fire was initiated in the demister for a set of these compressors. The demister element fire most likely resulted from the ignition of compressor exhaust valve deposits due to the formation of a hot spot on an exhaust valve. The demister element fire heated the demister vessel to the point that it failed at system pressure. The demister vessel rupture resulted in the catastrophic failure of a valve in the line to an air receiver tank holding a considerable inventory of lubricating oil. Oil entrained in the flow of air from the receiver tank through the broken line produced a flammable mixture of oil mist and air, which subsequently exploded. The oil mist explosion toppled a nearby masonry wall and caused damage to other portions of the powerhouse. The oil mist explosion (i.e., the secondary explosion) produced blast loads that were more severe than the demister vessel failure (i.e., the initial explosion). Fortunately, due primarily to the distribution of personnel in the area at the time of the event, there were no significant injuries.

This event demonstrates that demister fires have the potential to initiate a serious accident. While it may be difficult to completely eliminate demister fires, there are a number of design options that can be employed to prevent a demister fire from triggering an explosion. One approach to prevent such an event is to install a safety interlock to shut down the compressors and isolate the air treatment system upon detection of a high temperature at the demister discharge.

INTRODUCTION

This case history illustrates an explosion involving air compressor components located within a plant powerhouse. Baker Engineering and Risk Consultants, Inc. (BakerRisk) conducted an investigation of this accident, and this paper discusses the results of the investigation. The investigation included an inspection of the accident site, interviews with eyewitnesses to the accident, analytical evaluation of potential scenarios, and a limited amount of testing involving the lubricants and filter material employed in the relevant processes. To familiarize the reader with the key components involved in the accident, a general overview of the plant powerhouse layout and equipment is provided first. The information gleaned from the observed damage and witness interviews is presented next, followed by the development of the accident scenario. The potential causes for this accident are then discussed, concluding with recommendations for the prevention of similar events at other air compressor installations.

POWERHOUSE LAYOUT AND EQUIPMENT DESCRIPTION

The powerhouse is approximately 190 feet long and 95 feet wide. A schematic of the powerhouse layout is given as Figure 1. This layout drawing is simplified and shows only the most relevant powerhouse equipment. There were three sets of compressors within the powerhouse, with a number of external temporary compressors located outside the west powerhouse wall. Not shown in this drawing are crossties between the compressor sets upstream of the air receiver tanks.

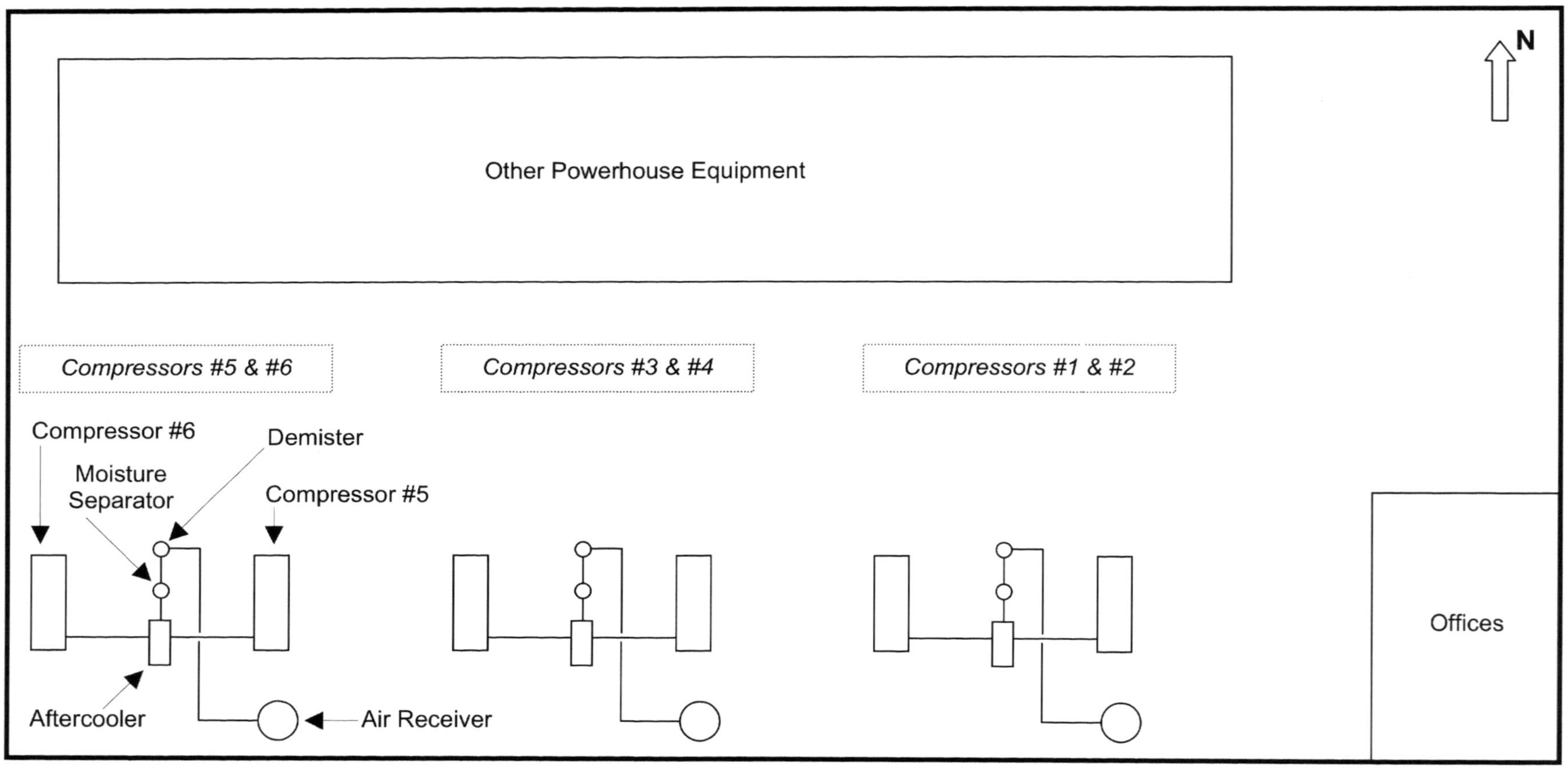

Figure 1. Schematic of Powerhouse Layout

The damage to the powerhouse was fairly limited, with the most obvious area of damage in the immediate vicinity of the #5 and #6 compressors and the air handling train for these compressors. Piping and valves connected to the #2 air receiver, which is part of the air handling train for the #3 and #4 compressors, were also damaged. The other compressor sets and systems in the powerhouse were essentially undamaged.

The #5 and #6 compressors are two-stage reciprocating units that deliver approximately 2200 cfm of air per compressor. An innercooler is utilized between the low- and high-pressure stages. The compressors utilized a synthetic (diester based) lubricant with an open cup flash point of 400°F on the airside and crankcase. The #5 compressor had just been rebuilt and was in the process of being restarted at the time of the explosion. Each of these compressors has a pressure relief valve and an automatic high temperature air shut down switch. The air for these compressors enters through a set of filtered intakes located on the powerhouse roof.

The #5 and #6 compressors share a common air handling train, as shown in Figure 1. Each compressor has its own vibration dampener tank ("surge bottle"). The exhaust pipes from the compressor vibration dampener tanks are joined into a common header. The header flow is directed into the air handling train, which consists of a two-stage aftercooler, a moisture separator, and a demister. The demister exhaust flow is directed to a large air receiver tank.

Auxiliary (rental) compressors located external to the powerhouse can also supply air to the plant. The air from the auxiliary compressors is fed into the common header to the internal compressor air handling trains. A single auxiliary compressor, running at approximately 1600 to 1800 cfm, was in use at the time of the event. The air for the auxiliary compressor is brought in through a local intake near the compressor. The auxiliary compressor in use at the time of the event utilized a solvent-dewaxed heavy paraffinic petroleum distillate (i.e., a traditional non-synthetic) lubricant with an open cup flash point of 400°F.

As noted above, the air handling train for a compressor set consists of a two-stage aftercooler, a moisture separator, and a demister, with the exhaust flow from the demister directed to a large air receiver. The aftercooler stages are a shell and tube design, with the air flowing through the tubes. A short length of u-bend piping connects the two stages. The moisture separator is a cyclone-type separator. A pressure relief valve (PRV) is situated between the moisture separator and demister. The PRV has a capacity of 3460 scfm with a set pressure of 125 psig, and relieved at 122 psig in a post-event test.

The demister is of particular relevance since, as discussed in the following sections, it suffered severe damage and played a key role in the event initiation. The demister vessel shell has a diameter and height of 2 feet and 10.5 feet, respectively, and is constructed of carbon steel plate approximately 0.19 inches thick (as measured in the field). The vessel has a flat top head and a dished bottom head. It has a nominal working pressure of 100 psig at 100°F, and is rated for 150 psig. The demister filter element is configured as a 2-inch thick annulus with a carbon steel wire cage that supports both the inner and outer surfaces of the element. The wire cage is closed at the bottom by a circular steel plate attached to the inner cage layer. The element and wire cage are supported from the top by a flanged connection that comprises the upper head of the vessel. At the bottom of the filter element, a number of tabs are connected to the outer wire mesh to keep the element centered within the vessel. Air flows into the vessel through the top, into the inner filter element annulus, through the filter, back up along the outer annulus, and exits through a pipe connected to the vessel side near the top. Liquid collected by the filter drains through the element and is collected in the bottom of the vessel.

The air receiver tank has a diameter of 5 feet and a height of 16 feet; the design pressure of the air receiver is specified to be 125 psig.

The demister and moisture separator are both equipped with automatic drain systems, which were reported to open every 5 minutes in order to preclude an accumulation of fluid in these vessels. The air receiver was reported to have a float-type fluid drain system, along with a manual drain system utilized on a daily basis.

The south wall of the powerhouse is concrete masonry unit (CMU) construction. The bulk of the south wall is unreinforced CMU, with the western edge of the south wall being reinforced CMU. The south wall is not a load-bearing wall, and there is a slight gap between the top of the wall and the roof. The west wall of the powerhouse is constructed of vertical concrete slabs that connect to the roof through a set of retention clips on the top (exterior) side of the roof.

OBSERVED DAMAGE

The most obvious area of damage was in the immediate vicinity of the #5 and #6 compressors and the air handling train for these compressors (i.e., in the southwest corner of the powerhouse). Figure 2 illustrates the damage to the powerhouse. The main damage observed is summarized below:

- Number 3 demister (demister for compressors #5 and #6) ruptured,
- Number 3 demister filter element showed localized melting of the wire mesh and filter medium, indicating that it was subjected to extremely high temperature,
- Fire damage in the immediate vicinity of the #5 and #6 compressors,
- Rupture of valves on pipes connected to the #2 air receiver (receiver for compressors #3 and #4),
- Rollup door to the west of the #3 demister blown out,
- South wall (CMU construction) of the powerhouse failed over portion of span,
- East wall (CMU construction) of the powerhouse bowed out slightly,
- Upper section of wall (CMU construction) on northeast side of powerhouse collapsed, and
- Retention clips for cast concrete panels on the west wall of the powerhouse deformed and/or torn loose from the panel,
- Windows on office in southeast corner of powerhouse blown in.

The demister vessel rupture appeared to initiate as a classic "fish mouth" failure (i.e., a split along the vessel long axis) near the upper section of the vessel. The split in the vessel turned and ran circumferentially approximately 1 foot below the top and 5 feet above the connection to the bottom head. The length of the axial split is roughly equal to the area of the demister vessel that was observed to be glowing red (see discussion of witness). The section of the vessel wall over which the axial split ran opened up completely along the circumference so that

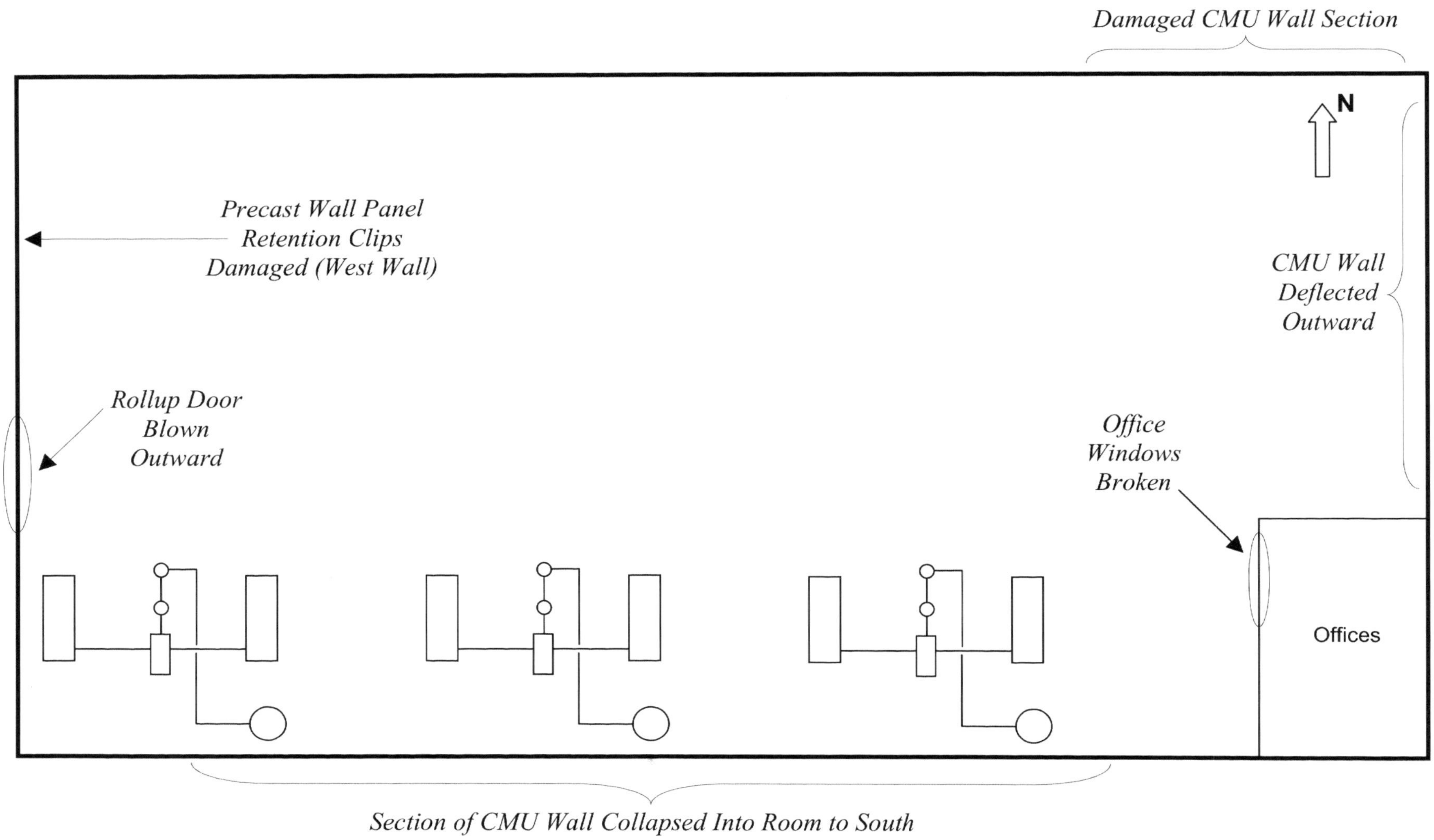

Figure 2. Schematic of Main Damage to Powerhouse

the wall "peeled" back, with the inner surface facing west. Limited portions of this section of the vessel wall were clearly oxidized, evidence of exposure to very high temperature. The failed demister vessel shell is shown in Figure 3. The failure at one side of the lower shell section is shown in Figure 4.

Figure 3. Demister Vessel Shell

Figure 4. Failure at Lower Portion of Demister Vessel Shell

The demister filter element was severely damaged. Large portions of the element were ejected from the supporting wire cage as a direct result of the demister vessel failure and the wire cage was severely deformed. A relatively small portion of the carbon steel cage near the upper middle section melted, and a collection of solidified pieces of metal and filter medium were present near the bottom cage support. Molten material from the filter element was sprayed onto the inner surface of the demister vessel and collected in the demister vessel bottom. The demister filter element was still connected to the vessel following the explosions, and was subsequently removed from the vessel. The filter element fire was difficult to put out, and continued to smolder periodically for a period of approximately three days. Figure 5 and Figure 6 show demister elements from the failed vessel and from one of the other demister vessels, respectively.

Figure 5. Demister Vessel Element After Removal from Vessel

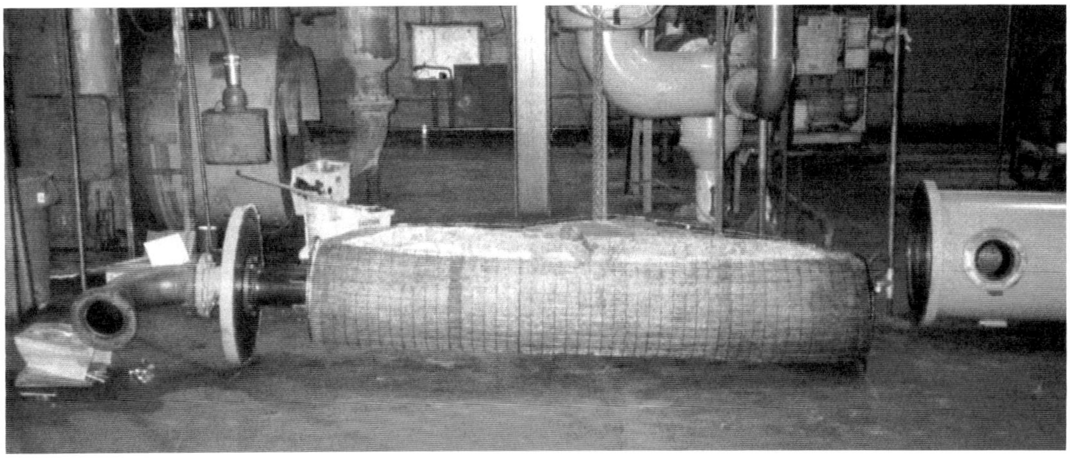

Figure 6. Undamaged Demister Vessel Element

The damage in the immediate vicinity of the demister vessel (i.e., near the #5 and #6 compressors and the air handling train for these compressors) was mainly due to fire and smoke. That is, very little blast damage was evident. The exception to this was that the drain system for the demister and moisture separator appeared to be significantly damaged.

Away from the demister, the only significant equipment damage was the failure of two cast iron valves on pipes connected to the #2 air receiver (receiver for compressors #3 and #4). It was apparent that a significant quantity of oil was sprayed out from one of the broken pipes onto the surrounding pipe sections and equipment. A liquid layer (organic and aqueous) 18 inches deep was still present in the #2 air receiver after the explosion.

The rollup door to the west of the #3 demister was blown out. This door is approximately 19' west of the demister and is not judged to have significant structural capacity. Hence, its failure is not surprising given the demister rupture and the orientation of the rupture (i.e. the vessel opened up in direction of the rollup door)

Several of the powerhouse walls failed or were damaged as a result of the explosions. The south wall of the powerhouse failed over a significant portion of its span. Most of this wall is unreinforced CMU, with the western edge of the wall being reinforced CMU. The east wall of the powerhouse (CMU construction) was bowed out slightly as a result of the explosions. The upper section of wall (CMU construction) on the northeast side of powerhouse collapsed as a result of the explosion; however, it was reported that at least the top few rows of blocks in this row were loose prior to the event. On several of the cast concrete panels that make up the powerhouse west wall, the retention clips connecting the panel to the roof were deformed and/or torn loose. The windows of the office on the east wall were also blown into the office as a result of the explosions.

Eyewitness Information

The key eyewitnesses interviewed included the powerhouse operator, the #5 compressor restart crew (three individuals), two employees in the room to the south of the powerhouse, a maintenance worker in a break room southeast of the powerhouse, and the foreman of the roofing crew present on the north end of the powerhouse roof at the time of the event. The most important observations provided by the eyewitnesses are as follows:

- The PRV situated between the moisture separator and #3 demister (demister for compressors #5 and #6) actuated just before or just after the demister fire initiated,
- Flames were observed coming directly off the #3 demister vessel,
- The #3 demister vessel began to glow red at a local spot, and the portion of the demister vessel glowing red spread across the circumference and extended from near the top of the vessel to somewhere near the middle of the vessel,
- A loud whistling noise was heard that persisted up to the first explosion,
- Two explosions separated by about 5 seconds occurred, with the second explosion significantly stronger than the first,
- Smoke and flames were observed coming over the top of the south wall, prior to its failure, at a location to the west of the #2 air receiver (receiver for compressors #3 and #4),

- The south wall initially deflected near the point at which flames and smoke were observed, and then collapsed,
- Flames were observed coming out of a roof ventilation opening just after the explosion.

The eyewitnesses also provided useful information with respect to the state of the equipment prior to and after the explosions.

The members of the compressor #5 restart crew and powerhouse operator were present in the powerhouse at the time of the event, and exited the powerhouse just before the first explosion. Their observations are particularly relevant. The first observation that a problem existed was actuation of the PRV situated between the moisture separator and #3 demister; PRV actuation was reported by the supervisor of the #5 compressor restart crew based on the location of the sound and prior experience. Shortly after, another member of the crew observed white and yellow smoke above the demister. He observed a glowing red spot on the demister vessel after moving in order to get a clear view of the demister; this spot was approximately 6 to 8 inches in diameter and 1/3 of the way down the demister from the top. The crew supervisor shut down the #5 compressor at this time and proceeded with the crew towards the exit along the east wall, near the powerhouse office. Several members of the restart crew looked back at the demister roughly halfway to the exit, and observed that the upper portion of the demister was glowing red. Several of the crew members looked back at the demister again just before exiting the powerhouse; one of the crew members reported seeing flames shooting from the top of the vessel at this point. The powerhouse operator was alerted to the situation by the #5 compressor restart crew supervisor. The operator also glanced at the demister prior to exiting with the restart crew, and observed flames near the top of the demister and that the demister vessel was glowing red from the top to approximately halfway down. All four of these eyewitnesses exited the powerhouse seconds before the first explosion.

ACCIDENT SCENARIO

Overview of Scenario

The first step in the development of the accident scenario is to understand the nature of the demister vessel failure. The normal pressure in the demister vessel is approximately 100 psig, and the vessel is rated for 150 psig at 300°F. Since the vessel failed, and there was no evidence of degradation, then the internal pressure rose well above the normal internal pressure and/or the strength of the carbon steel shell decreased significantly due to an increase in temperature. The observation that the vessel shell was glowing red indicates that it reached 1400°F to 1500°F. Based on thin-walled cylindrical stress theory, an internal pressure near 100 psig would be expected to cause vessel rupture in this temperature range. This is in agreement with the eyewitness observation that the PRV just upstream of the vessel actuated and reseated before the explosion; the PRV was set to relieve at 125 psig, and the post-event PRV test demonstrated that it was functional even after the explosion. It can therefore be concluded that the demister vessel failed because it overheated.

It was shown by analysis that the inventory of oil present on the demister filter element would be sufficient to raise the demister vessel wall temperature consistent with eyewitness observations. That is, the observed behavior does not require a process upset; specifically, the observed behavior does not require a malfunction of the demister drain system. Although it was observed that the drain system was severely damaged after the explosion, it may very well have been functioning properly prior to the event.

The demister rupture accounts for one of the explosions reported by the eyewitnesses. However, as discussed later, it would not account for the observed damage to the south wall of the powerhouse or the powerhouse office window failure. The second explosion was due to the formation of a relatively large cloud of oil mist resulting from the failure of a valve on a line connected to the #2 air receiver.

The compressed air inlet pipe to the #2 accumulator completely failed at a cast iron valve, such that the resulting pipe sections were completely separated. The pipe failure mechanism has not been established, but is most likely related to pipe motion induced by the demister rupture. A portion of the inventory of liquid residing in the bottom of the accumulator was sprayed out through the pipe section connected to the accumulator; 18 inches of liquid remained in the accumulator following the event. This loss of inventory is not surprising since the air pressure in the accumulator would have been near 100 psig. The large mass flow rate of air into the pipe would establish a significant air velocity field near the liquid surface that would effectively entrain liquid from the surface and carry it out of the vessel. Although the liquid inventory in the accumulator would be a mixture of liquid hydrocarbons and aqueous material, the liquid entrained would be predominantly hydrocarbon since it would float on the aqueous phase. The entrainment mechanism and sonic discharge would be effective at aerosolizing a significant fraction of the liquid entrained. The counter flowing discharge of air from the other end of the broken pipe would also be effective at aerosolizing the hydrocarbon discharged from the pipe section attached to the accumulator. Hence, it is quite likely that the failure of the pipe section attached to the #2 accumulator resulted in the formation of a significant cloud of flammable mist. It is expected that burning debris ejected from the demister would readily ignite the resulting vapor cloud. This would, of course, be consistent with the observations of several witnesses who reported hearing two explosions separated by a short period of time.

Explosion Characterization and Comparison with Observed Damage

The overpressure and impulse associated with the demister vessel rupture were characterized using the approach given in Reference 1 for a bursting pressure vessel. The blast overpressures are shown in Figure 7, along with those for the vapor cloud explosion discussed below. The positive phase duration associated with the bursting pressure vessel blast loads is in the range of 3 to 4 msec at the distances of interest.

The Baker-Strehlow-Tang vapor cloud explosion (VCE) blast load predictive methodology[2,3] was utilized to calculate the overpressure and impulse associated with the VCE. The VCE characterization is subject to more uncertainty than the demister vessel rupture since the flammable cloud size, composition, and location are not well known. To provide a reasonable estimate, it was assumed that the resulting mist cloud would occupy a volume equivalent to a 10-foot radius sphere. The volumetric combustion energy was assumed to be equivalent to half of that for a stoichiometric methane-air mixture. In addition to the total combustion energy involved in the explosion, the combustible material reactivity, degree of confinement, and degree

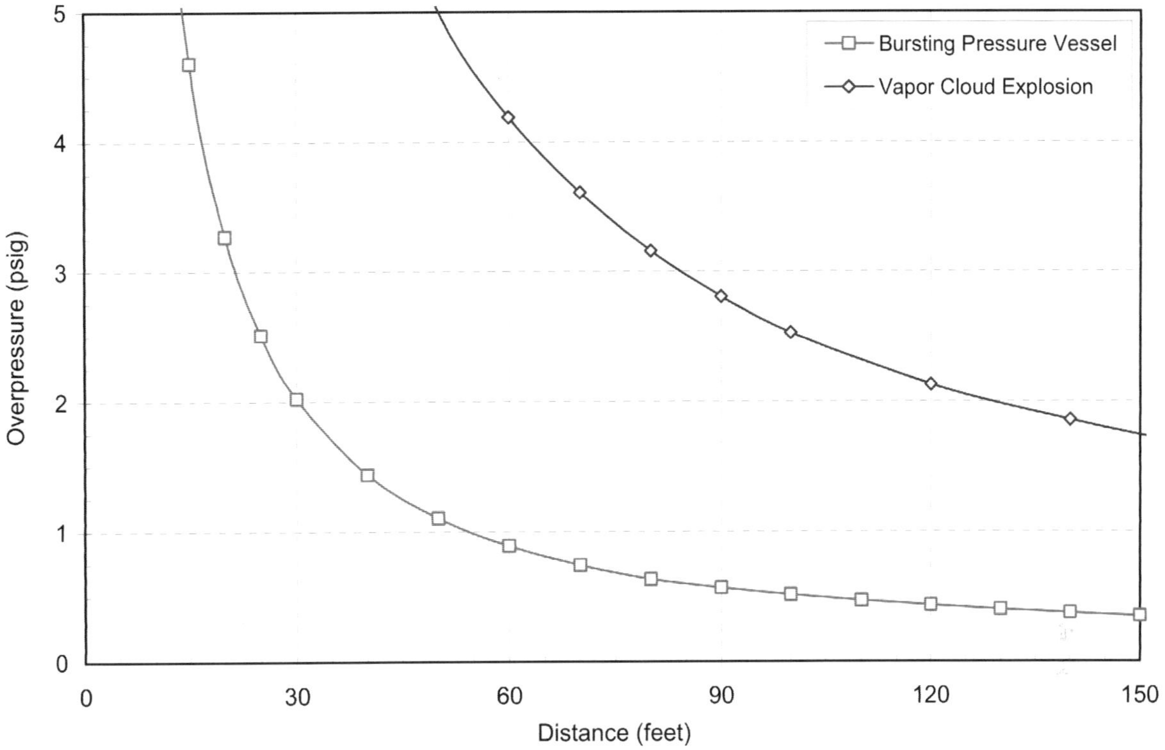

Figure 7. Calculated Blast Overpressures

of congestion in the vicinity of the cloud must be characterized in order to utilize the Baker-Strehlow-Tang methodology. It was assumed that the mist in the cloud would be reasonably fine due to the aerosolization mechanisms discussed above, so that a medium fuel reactivity to the cloud would be appropriate. The powerhouse roof and floor would offer significant confinement for the cloud sizes discussed above and hence two-dimensional confinement was assumed. This neglects the confinement offered by nearby walls, but this is offset by the presence of some room for expansion prior to the confining influence of the roof and floor being exerted. Most of the area above and around the #2 accumulator is only lightly congested. However, the turbulence associated with the receiver blowdown would increase the effective congestion rating, and there are some areas in the vicinity for which the degree of obstruction could be characterized as medium. A medium obstacle density rating was therefore employed. The VCE overpressures calculated as described above are significantly larger than those associated with the demister vessel failure, as shown in Figure 7, and the positive phase duration is 2 to 3 times that for the demister vessel failure.

The powerhouse south wall collapse and the failure of the powerhouse office windows were selected as blast indicators for comparison with the calculated blast loads. The overpressure required to fail the powerhouse office windows was estimated to be in the range of 1.0 to 1.5 psig for the positive phase durations of interest. The #3 demister is approximately 148 feet from the outer wall of the powerhouse office, while the center of the #2 air receiver is 82 feet from the office outer wall. Comparing the overpressures shown in Figure 7, it is clear that the demister vessel rupture could not have caused the failure of the office window while the VCE overpressure is in excess of the required value. Hence, it is concluded that the powerhouse office window failure was due to the VCE.

The overpressure required to fail the powerhouse south wall is strongly dependent on the blast wave positive phase duration. The estimated overpressures required for wall failure for a range of durations are provided below in Table 1.

Table 1. Blast Loads for Powerhouse South Wall Failure

Positive Phase Duration (msec)	Estimated Failure Pressure (psig)
10	24
50	4.7
100	2.4
150	1.6

The demister is approximately 27 feet from the south wall, while the center of the #2 air receiver is 5 feet from the south wall. Based on the distances and overpressure-duration combinations required for south wall failure as shown above, the demister vessel rupture could not have resulted in the failure of the powerhouse south wall while the VCE would be expected to fail the wall. Hence, it is concluded that the powerhouse south wall failure was also due to the VCE.

The overpressure and impulse resulting from the VCE would be significantly larger than those from the demister vessel rupture, which would be consistent with the observation that the second explosion was more severe than the first. The VCE would also account for the observed damage to the powerhouse south wall and the powerhouse office windows. Furthermore, the VCE would account for the observations of flames coming over the powerhouse south wall prior to its failure and flames coming out the roof ventilation opening; the demister vessel rupture would not account for these observations.

POTENTIAL CAUSES

There were two separate explosions, as discussed above. The first was due to the rupture of the demister vessel as a result of a fire inside the demister. The second was a VCE due to the rupture of a valve on a pipe connected to the #2 air receiver, with the valve rupture being a direct result of the demister vessel failure. The fuel for the demister fire was oil collected on the demister filter element, and the fuel for the VCE was generated by the aerosolization and dispersion of the organic phase in the air receiver due to the receiver tank blow down. The oxidant in each case was air. These conclusions follow directly from the available physical evidence and eyewitness interviews. However, it is not obvious what started the fire in the demister that led directly to these two explosions. That is, the available physical evidence and eyewitness interviews do not directly illuminate the root cause of the event. The potential causes for the initial demister fire are reviewed in this section. There are essentially two questions to be answered: (1) what was the fuel, and (2) what was the ignition source. Each of these is discussed separately below.

Potential Fuel Sources

The potential fuel sources considered in this investigation are listed below:

- Flammable materials utilized in roof maintenance activities,
- Utility fuels in use within or near the powerhouse (natural gas, propane, etc.),
- Flammable materials in use near the #5 or #6 compressors,
- Flammable material other than oil added to the system,
- Lubrication and crankcase oils for the #5 and #6 compressors,
- Lubrication and crankcase oils for the auxiliary external compressor,
- Other oils inadvertently introduced to the system, and
- Degraded oils resident within the system.

Roof Maintenance Activities

The potential for flammable materials utilized in roof maintenance activities was considered since the compressor inlets are on the powerhouse roof and these activities involved the use of tar and flammable gas torches. However, discussions with the roofing crew foreman indicated that no flammable materials were being utilized at the time of the event, and that the roofing crew was 75 feet from the #5 and #6 compressor air intakes.

Utility Fuels

There are multiple utility fuels used within the powerhouse (boilers, heaters, etc.). The boilers occasionally vent unburned gas through a pipe on the west side of the powerhouse. However, no utility fuel leaks or releases were reported near the time of the event, and it is unlikely that a release of gas through the vent would result in a significant concentration of flammable gas entering the compressor air intake.

Nearby Flammable Materials

There is no indication that flammable materials were in use near the compressors at the time of the event, so this potential fuel source was also eliminated from consideration. Similarly, there is no indication that flammable material other than oil was added to the compressor system.

Oils

The remaining potential fuel sources are the lubrication and crankcase oils intentionally added to the main and auxiliary compressors, other oils inadvertently introduced to the system, and degraded oils resident within the system. As noted previously, the main compressors utilize synthetic lube oil while the auxiliary compressor in use at the time of the event was using traditional non-synthetic lube oil. Both oils are reported to have a flash point of 400°F.

Tests were performed on various oil samples in order to examine their properties. The samples tested are listed below:

- Unused lubricant for the # 5 and #6 compressors,
- Lubricant taken from the lube oil reservoirs for the #5 and #6 compressors,
- Crankcase oil taken from the #5 and #6 compressors,
- Unused lubricant for the external auxiliary compressor,
- Lubricant taken from the lube oil reservoir for the external auxiliary compressor,

- Oil taken from the air side of the #3 aftercooler, and
- Oil/sludge sample from the pipe segment downstream of the #3 moisture separator.

Boiling curve tests were run on all samples to provide a comparison of different oil samples and a measure of relative flammability. The boiling curve tests did not indicate any of these oils had unusual properties that would impact their combustion behavior. The oil resident in the system for an extended period (i.e., the last two samples) did show a slightly lower initial boiling point, indicating that either some degradation may have occurred due to long term exposure within the system or that some foreign oil was present. Open cup flash point tests were also carried out with the unused lube oils, the crankcase oils, and the oil taken from the #3 aftercooler. The flash points for these samples ranged from 450°F to 493°F, which is well above the 400°F open cup flash point reported for these materials. The flash point of the sample taken from the aftercooler was approximately equal to that for new lube oil. It therefore does not appear that any other oils were inadvertently added to the system and the only likely fuel sources were the lubricating and crankcase oils intentionally added to the compressors and resident within the system. Furthermore, if other oils were added to the system at some point, their presence did not have a significant deleterious effect on the properties of the resident oil.

A test was also run with a sample of the #3 demister filter element soaked with the oil resident in the system (i.e., that from the #3 aftercooler) to examine its response to elevated temperatures (i.e., an auto-oxidation test). There was no evidence of exothermic reactions below 300°F. The data logs supplied by BFS show the #3 aftercooler inlet and exit temperatures were 240°F and 125°F, respectively, near the time of the event. Furthermore, the temperature downstream of the aftercooler would be well below this value.

Potential Ignition Sources

The main potential ignition sources considered in this investigation were:

- Within the compressor due to compression (i.e., dieseling),
- Within the compressor due to hot spots or sparks,
- Hot spot at compressor exhaust valve,
- Debris impact in vibration dampener ("surge bottle") or air handling train components,
- Static discharge within air handling train components, and
- Shock due to rapid valve actuation.

Dieseling Within Compressor

Since it does not appear that flammable gas could have been pulled into the compressor at a significant concentration, dieseling within the compressor is not a viable ignition source.

Hot Spots or Sparks Within Compressor

Ignition within the compressor due to hot spots or sparks was considered due to reports of small pieces of debris within the compressor. Intercooler fins had broken off in the past and passed through the high-pressure stage. Inadequate lubrication or equipment malfunction could also allow hot spots or sparks within the high-pressure stage. However, there was no evidence of combustion within either compressor #5 or #6, so it does not appear that this mechanism is a likely ignition source for this event.

Hot Spot on Compressor Exhaust Valve

Lubricating oil or oil deposits could be ignited directly on an exhaust valve if the valve developed a hot spot. Leakage past an exhaust valve, as would occur due to damaged or broken valve springs, channels, or gaskets, can lead to locally high exhaust valve temperatures. These compressors, particularly compressor #6, were reported to have a history of frequent maintenance due to exhaust valve issues. Compressor #6 had many of the discharge valves replaced and frequent valve maintenance activities in the year prior to the event. It was not uncommon to find exhaust valve component sections downstream of the compressor. Pieces of what appeared to be exhaust valve springs or channels were observed in the vibration dampener ("surge bottle") for the #6 compressor and in the aftercooler inlet. An examination of the #6 compressor discharge valves indicated that one of the valves had a broken channel, with the last 5 to 7 mm missing; the corresponding valve seat appears to have overheated at the location where this channel would rest. A piece of debris, possibly a section from the side of a valve channel was also found lodged in the same valve. These features can be seen in Figure 8 and Figure 9. A compressor #6 discharge cover seal was also severely degraded as a result of wear. Hence, it appears quite possible that at least one of the discharge valves on compressor #6 may have developed a hot spot capable of causing deposit ignition.

Valve cover deposits were reported by plant personnel and were observed during the investigation. Kletz[4] provides a recommendation that the exhaust air temperature should be kept below 284°F to avoid oil oxidation and the formation of carbonaceous deposits. Bodurtha[5] provides a similar, but slightly higher, maximum recommended safe outlet temperature. The high temperature shut down switches for the #5 and #6 compressors tripped at 405 and 355°F, respectively, in post-event testing, which is well above these recommended values. The temperature of the #6 compressor discharge valves, as measured with a laser pyrometer, ranged from 350 to 360°F just prior to the explosion. Excessive lubrication also promotes the formation of exhaust valve deposits[5], as well as increasing the oil inventory in downstream components, and the lubrication rate appeared to be above the manufacturer recommendations.

Hence, ignition of an exhaust valve deposit by a hot spot on the valve is judged to be a credible ignition source.

Debris Impact

Pieces of broken compressor exhaust valve springs, channels, and intercooler fins were present downstream of the compressors. Such debris could cause a friction spark within the vibration dampener or other air handling train components. There is no evidence to rule out this mechanism as a potential ignition source. However, it is not at all clear that debris carried downstream from the compressor would have sufficient velocity to cause a spark. Most of the debris would fall out of the air stream and be deposited at the bottom of the vibration dampener. Hence, while it is possible that debris impact was the ignition source for this event, it is judged to be less likely than a hot spot on compressor exhaust valve.

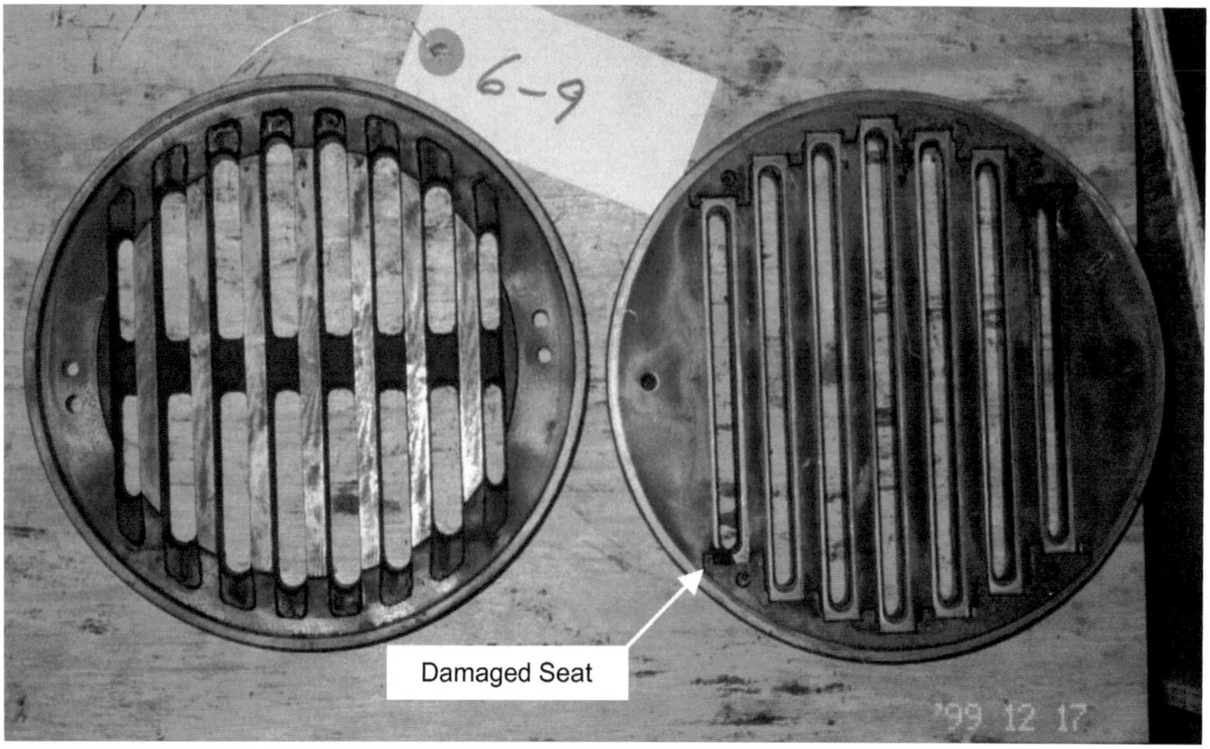

Figure 8. Damaged Valve Set

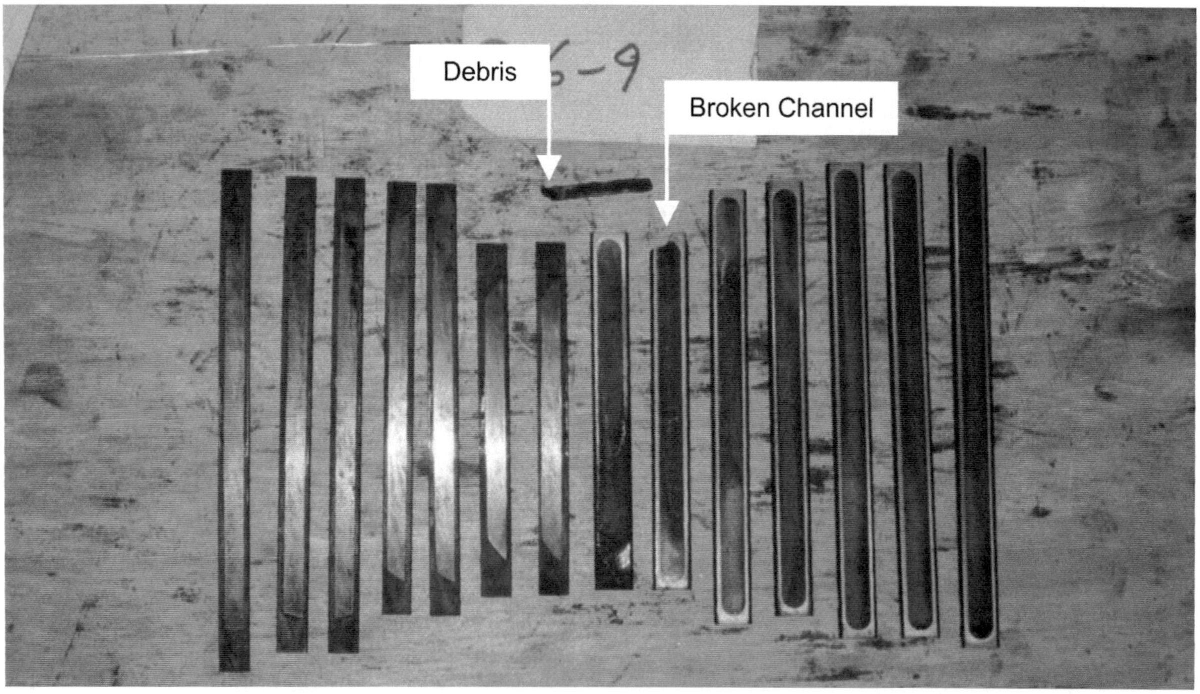

Figure 9. Leaf Spring Set for Damaged Valve Set (see Figure 8)

A similar mechanism would be impact between the tabs on the bottom of the demister filter element and the demister vessel. Pulsating flow within the demister could cause the tabs to strike the inner surface of the vessel. However, as discussed below, it does not appear that a weak spark within the demister vessel would directly ignite a demister fire.

Static Discharge

Ignition via static discharge was considered to be a potential ignition source since there are droplets of oil and water in the airflow downstream of the aftercooler. It is expected that the droplets can accumulate static charge, which could subsequently accumulate on the demister. A discharge of the accumulated static charge could then ignite a fire directly within the demister. A test was therefore conducted to examine the potential for direct ignition of the demister filter element from a weak ignition source. A sample of the #3 demister filter element soaked with the oil resident in the system (i.e., that from the #3 aftercooler) was tested at temperatures ranging from approximately 125 to 300°F using a glowing-wire filament ignition source placed a few millimeters from the element surface. No evidence of ignition over this temperature range was found. While it is recognized that the test is not completely representative of the conditions present in the demister, direct ignition of the demister fire via a static discharge appears unlikely based on these results.

Shock Due to Rapid Valve Actuation

Kletz[4] reports that it may be possible to ignite a thin oil film by the shock generated by rapid valve actuation. One of the eyewitnesses reported that the PRV actuated just before the event, and hence this mechanism was considered as a potential ignition source. However, it is judged unlikely that the rarefaction wave entering the pipe from PRV actuation would be sufficiently strong to serve as an ignition source. Furthermore, it appears much more likely that the PRV actuated due to the fire.

Recommendations To Prevent Reoccurrence

The most likely root cause (i.e., what initiated the demister fire) is judged to be ignition of an exhaust valve cover deposit due to the formation of a hot spot on the cover; ignition downstream due to debris impact is judged to be a credible, but much less likely, ignition source. However, the root cause of the explosions is not known with absolute certainty. Furthermore, it should be clearly recognized that even if the demister fire initiator for this particular case was known precisely, there are other mechanisms by which a fire in the demister or another system component could be initiated. While it is correct that fires in compressors and their air handling trains are not common events, they should not be viewed as rare events; this is particularly true for oil-lubricated reciprocating compressors[5]. BakerRisk is aware of other demister fire events, although none have led to the consequences observed in this investigation. In short, while vigorous effort should be applied to eliminating or reducing the frequency of potential ignition sources, it should be recognized that completely eliminating all possible ignition sources is not an achievable goal.

It is therefore recommended that the facility operation and/or equipment be modified such that the initiation of a demister fire would not lead to an explosion. That is, while a demister fire is not a desirable event, the system should be configured such that a demister fire could be tolerated. The most direct approach to accomplish this is to install a safety interlock that would shut down the compressors and isolate the air handling train upon detection of a high temperature at the discharge from the demister. A schematic of this type of system is given as Figure 10.

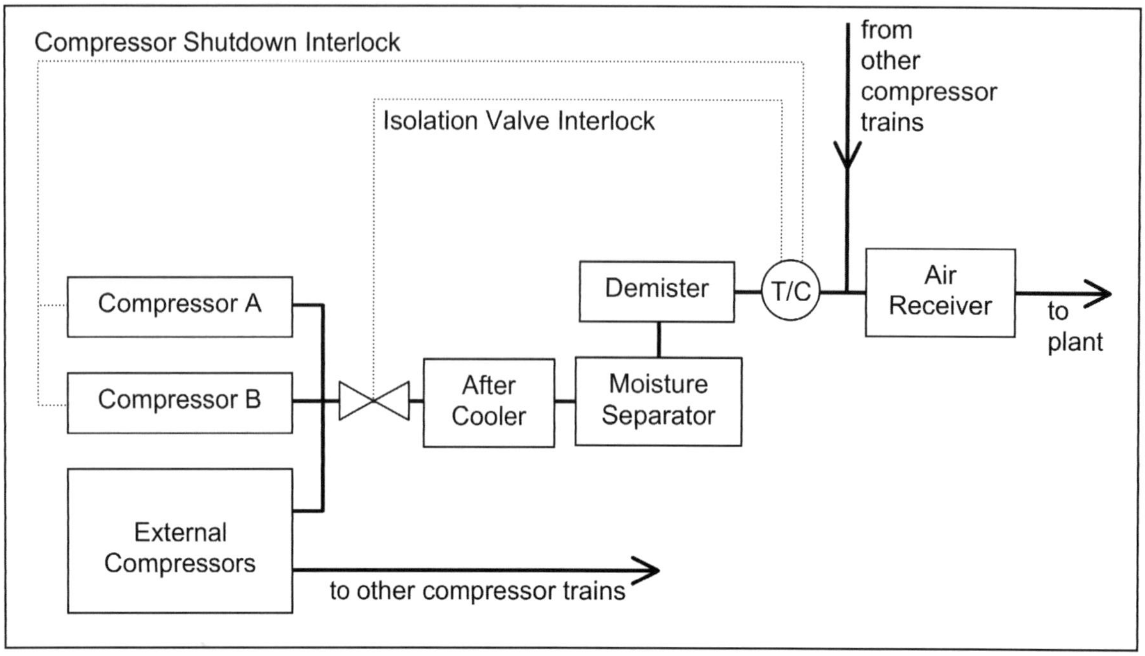

Figure 10. Schematic of Potential Safety Interlock Arrangement

The isolation valve interlock would isolate all components downstream of the compressor exhaust junction (e.g., aftercooler, moisture separator and demister) so that no additional air would be supplied to a fire in this portion of the system. Air from the other compressor trains could not backflow into these components since the flow path would dead end at the isolation valve. Hence, a fire in these systems would quickly be extinguished due to a lack of oxygen. Prudence would suggest that the isolation valve be configured to fail closed upon loss of signal and, if air operated, loss of air. The compressor shutdown interlock would shut off the main compressor units; this interlock may not be required if the compressors are already configured with a high-pressure shutoff switch since the exhaust side pressure would rise immediately in response to isolation valve closure.

It should be noted that this event would not have been precluded by the use of additional pressure relief valves (PRVs), either on or near the demister vessel. The demister vessel ruptured because the vessel wall overheated, not because the vessel was subjected to a pressure significantly in excess of the normal working pressure. Additional PRVs may indeed be a prudent design consideration to mitigate other potential process upsets, but they would not have affected the course of this accident.

There are a number of other steps that could be taken to reduce either the likelihood and/or the severity of a similar event occurring in the future, and the recommendations developed during the course of this investigation that fall into this category are summarized below. These recommendations are specific to the facility in question, and are not applicable to other facilities in all cases. It should also be recognized that the adoption of the safety interlock arrangement discussed above would eliminate the need for some of the recommendations provided below. In that case, most of these recommendations would only serve to reduce the frequency of fires within the compressors and their associated air handling trains.

- Control the lubricants utilized in external auxiliary compressors to the same standards as applied to the main compressors within the powerhouse,
- Decrease the compressor exhaust high temperature shutoff switch set point to at or below the recommended value,
- Reduce rate of lube oil consumption to within manufacturer recommendations,
- Periodically verify oil drain system operation,
- Explicitly evaluate acceptability of combustible materials in vicinity of compressor inlets,
- Reduce demister oil accumulation within the demister,
- Consider alternative piping layouts to reduce oil accumulation within entire system,
- Consider replacement of the cast iron valves,
- Improve the compressor maintenance program,
- Consider implementation of a debris control program,
- Consider using reinforcement when rebuilding the damaged CMU walls, and
- Consider improving the anchoring system used with west wall cast panels.

REFERENCES

1. Baker, W.E., P.A. Cox, P.S. Westine, J.J. Kulesz and R.A. Strehlow (1983) Explosion Hazards and Evaluation, Elsevier, Amsterdam, pp. 145-52.
2. Baker, Q.A., M.J. Tang, E. Scheier and G.J. Silva (1996) "Vapor Cloud Explosion Analysis," Process Safety Progress, 15(2): 106-9.
3. Baker, Q.A., C.M. Doolittle, G.A. Fitzgerald and M.J. Tang (1997) "Recent Developments in the Baker-Strehlow VCE Analysis Methodology," AIChE 31st Loss Prevention Symposium, 9-13 March 1997, Houston, TX.
4. Kletz, T. (1994) What Went Wrong? Case Histories of Process Plant Disasters, 3rd Edition, Gulf Publishing Co., Houston, TX, pp. 206-7.
5. Bodurtha, F.T. (1980) Industrial Explosion Prevention and Protection, McGraw-Hill Book Co., New York, pp. 130-1.

Hydrogen Peroxide Explosion

By

J. P. Lacoursiere, P.E.

University of Sherbrooke

Sherbrooke (Quebec)

Canada

Prepared for Presentation at the 37th Annual Loss Prevention Symposium

April 26-29, 2003

Case Histories and Lessons Learned Session

December, 2003

UNPUBLISHED

HYDROGEN PEROXIDE EXPLOSION

J. P. Lacoursiere, P.E.

University of Sherbrooke
Sherbrooke (Quebec) Canada

Two workers of a waste handling contractor lost their lives as a result of the explosion of a vacuum truck where they had inadvertently mixed sulfuric acid and hydrogen peroxide.

The consequences of the explosion are documented and were simulated.

Root causes and contributing factors were identified and in particular failure of the management systems is underlined.

1. INTRODUCTION

The Site is used to roast zinc ore. This operation produces sulfur dioxide (SO_2) which is reacted to produce sulfuric acid. SO_2 originating from the roasting operation goes through a series of washing steps to cool it and remove metal dusts. The gas is then sucked into electrostatic precipitators which remove acid mist and remaining dust. The wet gas is dried in a tower by direct contact with sulfuric acid 93% and a blower moves the gas through a series of heat exchangers into converter beds containing vanadium anhydride where SO_2 is oxidized in sulfur trioxide (SO_3). It then goes through a scrubber where it reacts with water to produce sulfuric acid. There are 3 adjacent gas treatment plants identified as Plant A1, A2 and A3. (Figure 1).

Tote tanks are used to drain acid at various points. There is a tote tank at each plant near the heat exchangers and near the stack at plant 1. They are 120 cm x 120 cm x 120 cm, they look similar and **they are not identified**. At each location there is a drain to the tote tank to collect sulfuric acid. (Figure 2).

At plant 1, there is another tote tank near a tank identified as "Tank for Scrubber". This tote tank is similar to the previous one. This tote tank contains hydrogen peroxide at 50%. On the steel frame, there are two labels: the WHMIS (Worker Hazardous Material Information System) indicating the content is hydrogen peroxide, a reactive, oxidizing and toxic substance and the TDG (Transport of Dangerous Goods) label indicating the content is hydrogen peroxide an oxidizing material. (Figure 3). A small metering pump is used to add hydrogen peroxide to a 150 ton sulfuric acid tank to destroy nitrous oxide. This is a temporary set up to run tests.

The tote tanks near the plant 1 stack, the one near the heat exchanger at plant 1 and the hydrogen peroxide tote tank are 30 m apart.

A sulfuric acid pumping station is located behind plant 2. It is a 5 m x 5 m building with a steel frame and a green steel siding. In this building there are 3 pumps and a collecting pit. The floor slopes towards the polyethylene lined collecting pit to recover spilled acid.

During the afternoon of August 26, 1998, maintenance work was done on one of the pumps in the pumping station. Sulfuric acid was drained to the collecting pit. This acid had to be removed.

The Site uses a Waste Handling Contractor with a service contract. When services are required, the Waste Handling Contractor dispatches a vacuum truck with a driver and a driver helper to perform the work. The Site has trained the Waste Handling Contractor staff on the work to be performed. On August 26, 1998, the Waste Handling Contractor is short of a trained driver and decides to use an untrained one from another site. He gives an ID card from another employee indicating he had followed the Company course. The ID cards do not have pictures.

Upon arrival at the site, the helper meets with the plant supervisor and gives him his ID card and the one provided to the truck driver. The plant supervisor knows him well and he asks him to empty the collecting pit in the green building (pumping station) and the usual tote tanks. The Waste Handling Contractor invoices 4 hours as minimum for each call and it is therefore the practice to do various pumping jobs while the unit is on site. No site representative has shown the crew operating the vacuum truck which tanks had to be pumped.

It is important to note that the Waste Handling Contractor driver helper that has shown up on the site that particular day, was known to be a hard worker that tried to empty as many tanks as he could for the time he was present on site.

The vacuum truck used for this particular job had a stainless steel tank with a capacity of 16,340 litres with a 345 kPa (50 psig) Maximum Working Pressure. This tank was cleaned by the Waste Handling Contractor but not checked by a site representative.

2. ACCIDENT DESCRIPTION

Around 18h30, there was an explosion behind plant 2 with a 45 m high white smoke plume.

The explosion had happened inside the vacuum truck. The cover on the man hole at the back of the truck was blown out through the pumping station. It hit the vacuum truck driver and his helper killing both of them on the spot. (Figure 4).

The man hole cover on top of the vacuum truck was also blown out and projected at 45 m. The safety valve piping on top of the truck was also broken. The pressure wave ripped the building siding but the steel frame stayed in place. Doors were bent.

The investigation has found that the Waste Handling Contractor employees thought that all tote tanks were containing sulfuric acid. They first pumped approximately 200 litres of hydrogen peroxide 50% from the tote tank and then went to the pumping station where they started pumping sulfuric acid 80% from the collecting pit. The explosion happened at this moment.

3. CHEMICAL REACTIONS

The following chemical reactions were identified:

- Mixing of concentrated sulfuric acid (80%) with water;
- Decomposition of hydrogen peroxide.

3.1. SULFURIC ACID AND WATER

When sulfuric acid is mixed with water there is production of heat. The enthalpy change could be estimated from the difference in enthalpy of the concentrated acid solution and the diluted solution using the following equation:

$$(\Delta H_f)_{SOLUTION} = (\Delta H_f)_{SOLUTE} + (\Delta H_f)_S(n)$$

where $(\Delta H_f)_{SOLUTE}$ is the solute enthalpy (here H_2SO_4);

and $(\Delta H_f)_S(n)$ is the mixture enthalpy; n is the number of moles of water per moles of sulfuric acid.

The temperature rise is about 20°C.

3.2. HYDROGEN PEROXIDE DECOMPOSITION

When sulfuric acid and H_2O_2 are mixed the most likely scenario is the following: The immediate dilution of the acid with H_2O_2 causes an increase of about 20°C. This small increase in temperature causes a decomposition of H_2O_2 which results in a significant increase in temperature (the reaction is strongly exothermic). After that, the system will be at elevated pressure and temperatures.

The final temperature and pressure could be estimated using the following methodology:

- The quantity of sulfuric acid that was pumped is unknown and could be estimated using the final dilution of sulfuric acid (28% residual measured in the vacuum truck): The dilution of acid was caused by the added water which comes from three sources: water originally in the acid, water form the H_2O_2 solution and water that comes from the decomposition of H_2O_2. The quantity of sulfuric acid is estimated at 100 kg.
- The heat added to the vacuum truck comes from decomposition of H_2O_2 and the dilution heat. This last one is about 10% of the heat generated which is estimated at 350,000 kJ. The tank was treated as an adiabatic process because the increase in temperature was very rapid;
- The pressure in the vacuum truck comes from oxygen (H_2O_2 decomposition) and evaporation of water. The partial pressure of O_2 is estimated from the decomposition of H_2O_2, while the water partial pressure comes from two balances:
  - Heat balance (unknown variable p_{H_2O} (or the evaporated quantity of water) and the final temperature)
  - Antoine equation: $\ln(p_{H_2O}) = A - B/(T + C)$. To relate the mass of evaporated water to its partial pressure, the law of perfect gas was used.
- The potential pressure in the vacuum truck was estimated at 16 atm (230 psi) and the final temperature at 188°C.

The vacuum truck could not withstand such a pressure and there was a physical explosion.

4. ACCIDENT CAUSES

The following causes were identified:

- Mixing of two incompatible products, hydrogen peroxide and sulfuric acid;
- The Waste Handling Contractor has used a untrained driver and has used the ID card of another employee to gain access to the site;
- The procedures of the Site and the Waste Handling Contractor are inadequate;
- The workers of the Waste Handling Contractor were left without supervision from the site personnel;
- The Waste Handling Contractor driver and driver helper have not taken into account the label on the hydrogen peroxide tote tank;
- The sulfuric acid tote tank was not identified;
- The Waste Handling Contractor employees had a very limited knowledge of chemical products and incompatibilities.

5. RECOMMENDATIONS

Site

Use of Vacuum Truck Procedure

The procedure "Use of vacuum truck for pumping liquids and sludge" was modified and explained to the site and Waste Handling Contractor personnel:

- A tank cleaning certificate is requested by the guard upon arrival on site;
- The site person responsible for the work shall verify the tank cleaning certificate and verify inside the tank to ensure it is clean;
- A pumping permit procedure was prepared: the identification of tanks, tote tanks, collection pits, etc. which need to be pumped are identified. The number of the tank, location and content are identified;
- When pumping is done with vacuum trucks, the Waste Handling Contractor personnel shall be supervised by trained site personnel. Cleaning of vacuum truck tank, hoses and cyclone followed with a visual inspection and a certificate of cleaning are required when another product is pumped.
- A certificate of cleaning done at the site must be issued each time a vacuum truck leaves the site or performs work in another department at the site.

Contractor Management

The existing contractor management procedure was updated. A section regarding the training of contractors was included. The ID card was modified to include the picture of the holder. The duration of the ID card was limited

Tank Identification:

Tanks were identified as required by the WHMIS and TDG Regulations.

Waste Handling Contractor

Pumping Procedure

The pumping procedure was updated. All workers have received training on the new site procedure.

Training :

During the winter 1998-1999, the following training was given to the Waste Handling Contractor workers:

- Worker Hazardous Material Information System (WHMIS);
- Incompatibility of chemical products;
- Transportation of Dangerous Goods (TDG) Regulation;
- Safety tours and checklist;
- Confined spaces;
- First Aid.

6. REFERENCE

Investigation report Philip Enterprise Inc. August 26, 1998, Valleyfield, Quebec, Canada, Quebec Safety and Health Board, Montreal, Qc.

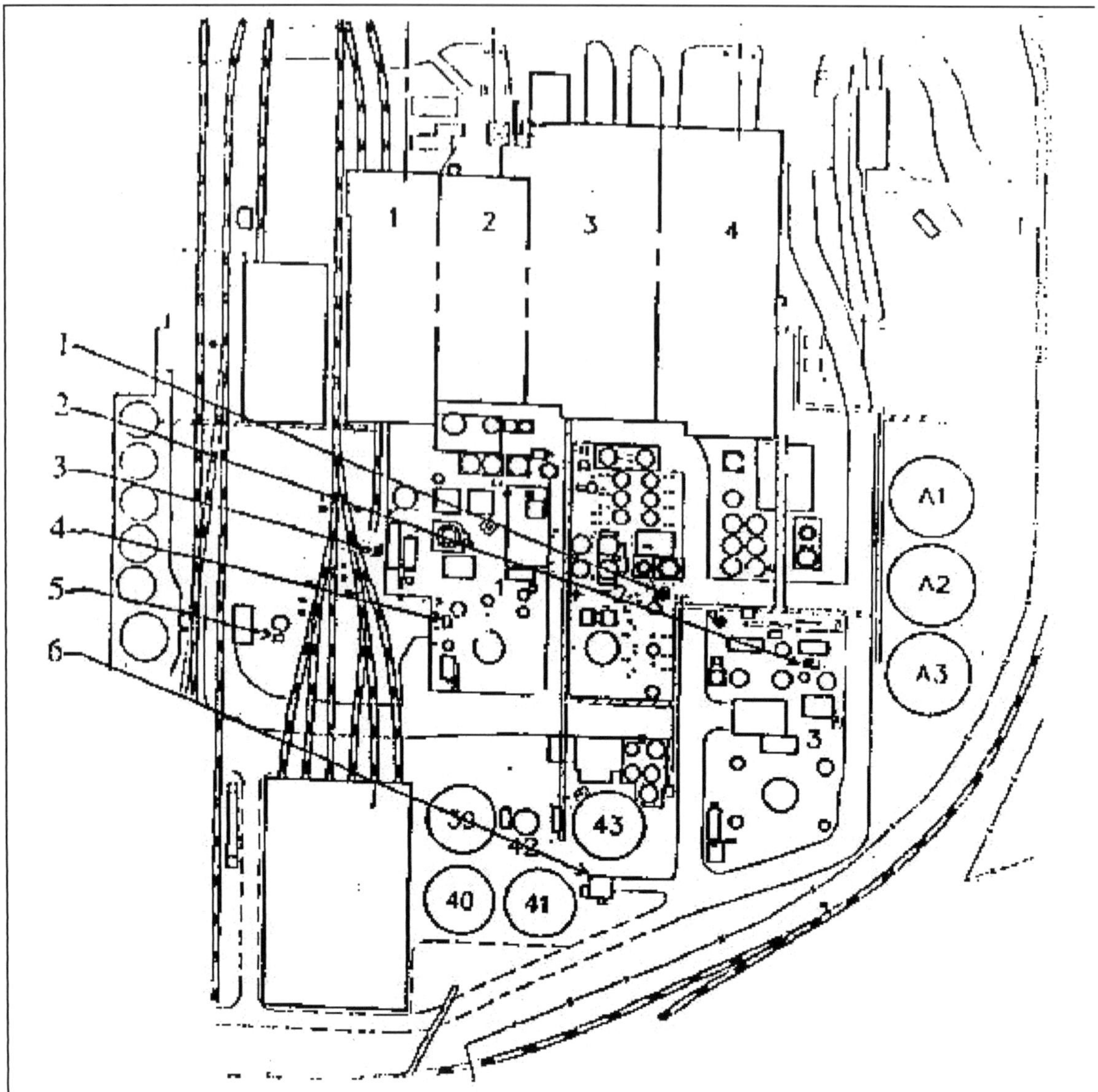

1- Tote tank at Plant 2
2- Tote tank at Plant 3
3- Hydrogen peroxide tote tank at plant 2
4- Tote tank at Plant 1
5- Tote tank near plant 1 stack
6- Sulfuric acid pumping station

Figure 1 Location of plants, tote tanks and pumping station

Figure 2 Tote tank near stack at plant 1

Figure 3 Hydrogen peroxide tote tank

Figure 4 Unloading opening

Figure 5 Vacuum truck and pumping station

AUTHOR INDEX

AUTHOR INDEX

American Institute of Chemical Engineers
3 Park Avenue
New York, New York 10016-5991

ISBN: 978-1-60423-457-2